MW01483462

BIOHYDROMETALLURGY AND THE ENVIRONMENT TOWARD THE MINING OF THE 21ST CENTURY

Part B
Molecular Biology, Biosorption, Bioremediation

Process Metallurgy
Advisory Editor: G.M. Ritcey

Process Metallurgy 9B

BIOHYDROMETALLURGY AND THE ENVIRONMENT TOWARD THE MINING OF THE 21ST CENTURY

Proceedings of the International Biohydrometallurgy Symposium IBS'99
held in San Lorenzo de El Escorial, Madrid, Spain, June 20-23, 1999

Part B
Molecular Biology, Biosorption, Bioremediation

Edited by

R. AMILS
Universidad Autónoma de Madrid

and

A. BALLESTER
Universidad Complutense de Madrid

1999

ELSEVIER
Amsterdam • Lausanne • New York • Oxford • Shannon • Singapore • Tokyo

ELSEVIER SCIENCE B.V.
Sara Burgerhartstraat 25
P.O. Box 211, 1000 AE Amsterdam, The Netherlands

First edition 1999

Library of Congress Cataloging in Publication Data
A catalog record from the Library of Congress has been applied for.

ISBN: 0 444 50193 2 (A&B)

⊗ The paper used in this publication meets the requirements of ANSI/NISO Z39.48-1992 (Permanence of Paper).
Printed in The Netherlands

FOREWORD

This two volume book presents the proceedings of the 13th International Symposium on Biohydrometallurgy held in San Lorenzo de El Escorial, Madrid, Spain, on June 20-23, 1999. The International Symposium on Biohydrometallurgy has become the forum where basic scientists and industrial experts join to interchange not only experimental results but also ideas about future developments in this interdisciplinary field.

This Symposium is quite different from the previous ones, held in Chile (IBS-95) and in Australia (IBS-97), where industrial applications had a prominent role. The active practice of Biohydrometallurgy in both countries justified their selection. In our case we decided to stress the importance of basic knowledge as the motor of future developments in the field.

Taking into consideration the importance that environmental issues already have in our society as well as their exponential increase in the coming century, the IBS-99 Organizing Committee decided to focus the meeting on the environmental aspects of biohydrometallurgy.

The theme of the Symposium is "Biohydrometallurgy and the Environment - toward the mining of the 21st century" because we feel that this is the area in which biotechnology will make its greatest contribution in the next century. The amount of papers from all over the world, 167 in all, covering most aspects of biohydrometallurgy, is the best way to prove that the field is ready to face the technological challenges of the next century.

All papers included in these books were previously reviewed by a group of experts selected from both the International Scientific Committee members and prestigious researchers in the Biohydrometallurgy field. Therefore, all the papers published have the appropriate requirements of scientific quality.

The Symposium was organized around four main topics:
- bioleaching
- biosorption
- bioremediation, and
- biology (micro and molecular)

with invited plenary lectures to review the state of the art and motivate discussion of future trends in the different areas of biohydrometallurgy, followed by a limited number of oral presentations selected by the Scientific Committee according to originality, scientific merit and topic. The rest of the communications were given in well-structured poster sessions organized according to the topics of the conference, followed by a general discussion of the posters' contents. Round tables were organized to discuss two important issues in biohydrometallurgy : "Direct versus indirect mechanism in bioleaching" and "Biohydrometallurgy and the Environment". Preliminary notes of the round table participants have been included in this proceedings to facilitate its discussion and as a reference for the specialists that could not attend the meeting.

All the papers selected for the Symposium have been included in these two volumes: Part A, contains papers dealing with basic and industrial scale bioleaching of base and precious metals, followed by contributions addressing the microbiological aspects of the process.

Part B, encompasses papers on the molecular biology of microorganisms responsible for the different biohydrometallurgical processes, followed by contributions dealing with environmental aspects of this biotechnology: biosorption and bioremediation.

The editors wish to thank the different Sponsors of the Symposium for their generous contribution which helped to put the meeting together, especially the Universidad Complutense de Madrid and the Universidad Autónoma de Madrid. The editors also wish to express their appreciation to their colleagues on the International Scientific Committee and other members of the biohydrometallurgy community for their prompt assistance with the peer review of the papers and for the selection of the oral presentations, which has not been an easy task. We also thank our colleagues on the Organizing Committee : Felisa González, M. Luisa Blázquez, Irma Marín and Jesús A. Muñoz for their valuable assistance and support. Finally, we want to thank all the participants to the Symposium and those who contributed with their work to this book, for without them it would not exist.

Antonio Ballester and Ricardo Amils

Biohydrometallurgy and the environment toward the mining of the 21st century

PART B

MOLECULAR BIOLOGY, BIOSORPTION, BIOREMEDIATION

TABLE OF CONTENTS

Chapter 4

BIOSORPTION

Chapter 5

BIOREMEDIATION

Biohydrometallurgy and the environment toward the mining of the 21st century

PART A

BIOLEACHING, MICROBIOLOGY

TABLE OF CONTENTS

Chapter 2

MICROBIOLOGY

xxii

Chapter 3

Molecular Biology

The molecular genetics of mesophilic, acidophilic, chemolithotrophic, iron- or sulfur-oxidizing microorganisms.

D. E. Rawlings

Department of Microbiology, University of Stellenbosch, Stellenbosch, 7600, South Africa.

The bacteria primarily responsible for decomposing metal sulfide ores and concentrates at temperatures of 40°C or below have been identified as *Thiobacillus ferrooxidans*, *Leptospirillum ferrooxidans* (or related *Leptospirillum* spp) *Thiobacillus thiooxidans* and recently, *Thiobacillus caldus*. These obligately acidophilic, autotrophic, usually aerobic, iron- or sulfur-oxidizing chemolithotrophic bacteria occupy an ecological niche that is largely inorganic and very different from that populated by the more commonly studied non-acidophilic heterotrophic bacteria. It has been of particular interest to discover how these 'biomining' bacteria are phylogenetically related to the rest of the microbial world. Based on 16S rRNA sequence data, the thiobacilli have been placed in the *Proteobacteria* division close to the junction between the β and γ sub-divisions. In contrast, the leptospirilli have been positioned within a relatively recently recognised division called the *Nitrospira* group. *T. ferrooxidans* is the only biomining bacterium whose molecular biology has been studied in some detail. Of the approximately 50 genes cloned or sequenced and published, by far the majority that can be tested are expressed and produce proteins which are functional in *Escherichia coli* (a member of the γ sub-division of *Proteobacteria*). These observations together with phylogenetic comparisons of most *T. ferrooxidans* protein sequences have confirmed the unexpectedly close relationship between *T. ferrooxidans* and *E. coli*. A special challenge has been the isolation of the various components of the iron-oxidation system and as a result of a global effort, this is almost complete. Several plasmids, transposons and insertion sequences have been isolated from *T. ferrooxidans*. These genetic elements are interesting because they may contain non-essential genes which are thought to improve the fitness of the bacterium and are frequently mobile. They have provided some fascinating insights into genetic exchanges that have occurred between *T. ferrooxidans* and other bacteria. There are clear indications that some of the other 'biomining' bacteria are even more important than *T. ferrooxidans* in many commercial biomining processes. The molecular biology of these bacteria is almost unstudied.

1. INTRODUCTION

A large variety of acidophilic, iron- or sulfur-oxidizing chemolithotrophic organisms capable of growth at temperatures from below 10°C to above 100°C have been identified. Some of these are facultatively chemolithtrophic microbes and under suitable conditions are able to grow non-autotrophically using organic matter (with more than a single carbon atom) as their carbon source. In order to limit the scope of this review, only the acidophilic, obligately chemolithotrophic iron- or sulfur-oxidizing bacteria which grow at or below 40°C will be considered. The bacteria which will form the main focus of this review are *Thiobacillus ferrooxidans, Leptospirillum ferrooxidans* (or related species of leptospirilli) *Thiobacillus thiooxidans* and *Thiobacillus caldus*. Although *T. caldus* is strictly-speaking moderately thermophilic (optimum growth temperature 45-48°C, 26), it is readily isolated from biooxidation plants that operate at 40°C (see elsewhere in this symposium) and has therefore been included. These microorganisms have made an impact on the field of microbiology in at least three areas. They have an interesting physiology, they are industrially useful and they have been implicated in several ecological processes including the mineralization and deposition of ores (6, 48). Iron- and sulfur-oxidizing chemolithotrophs usually occur as mixed bacterial consortia which are able to grow on iron and sulfur-containing minerals (such as iron pyrite) requiring in addition to the mineral only air, water and trace elements. The sulfuric acid produced during sulfur oxidation lowers the pH of their environment so that the bacteria are able to tolerate a pH of 1.5 or less even though their internal cytoplasmic pH is near neutral. From an industrial point of view, these bacteria are used in industrial processes such as the bioleaching copper from ores or the biooxidation of gold-bearing arsenopyrite ores (71). Ecologically, the most important contribution of these bacteria is that they have been implicated as the main culprits in the production of acid mine drainage water pollution.

2. PHYLOGENY OF THE CHEMOLITHOTROPHIC BACTERIA PRESENT IN BIOOXIDATION PROCESSES THAT OPERATE AT 40°C OR LESS

The inorganic, low pH environments in which mineral biooxidation takes place represents a highly specialised ecological niche. The obligately chemolithotrophic bacteria that thrive in such environments might be expected to have evolved in isolation from the more commonly studied bacteria. As a result of DNA, RNA and protein sequence information that has become available during the past 10 to 12 years it has been possible to determine how the obligately chemolithotrophic acidophiles are related to other bacteria in a way that was not possible prior to the advent of molecular taxonomy. The most widely used tool to separate organisms into groups at all levels is small subunit (SSU) ribosomal RNA sequence data (16S rRNA in the case of bacteria, 84). All the chemolithotrophic bacteria which grow at 40°C or less belong to the domain *Bacteria* whereas most of the organisms that grow at higher temperatures particularly those that grow above 65°C are placed within the domain *Archaea*. Since new bacteria are continually being discovered the number of divisions within the domain *Bacteria* is uncertain. As a result of the 16S rDNA sequence information obtained during the isolation of bacteria from a hot spring in the Yellowstone National Park (USA), it has recently been proposed that 36 divisions should be recognised (rather than 24 previously) within the domain *Bacteria* (28). However, little is known about many of the 12 new divisions and in most cases they are represented only by numbers allocated to uncultured bacteria. The positions of the iron- and sulfur-oxidising bacteria are indicated on a diagram showing the 24 bacterial divisions previously recognised. *T. ferrooxidans*, *T. thiooxidans* and *T. caldus* are closely related bacteria which fall into the *Proteobacteria* very close to the junction between the α and β subdivisions (26,46) (see Figure 1).

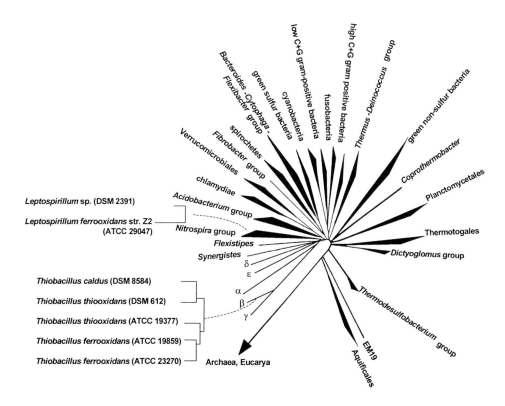

Figure 1. Phylogenetic relationship of *Leptospirillum* and relevant species of *Thiobacillus* to known bacteria based on 16S rRNA sequence data. Greek symbols α, β, γ, δ and ε indicate subgroups within the division *Proteobacteria*.

It came as a surprise to discover that the obligately chemolithotrophic thiobacilli were so closely related to *E. coli* (a member of the γ-*Proteobacteria*) and this is presumably the reason why so many *T. ferrooxidans* genes are expressed in *E. coli*. In contrast, based on 16S rDNA sequence information, the leptospirilli (for which sequence information is available) are placed into the division *Nitrospira* (28). As may be seen from 16S rRNA sequence, there is clearly more than one species within the genus *Leptospirillum* (27) (see Figure 1) and the possibility exists that there could be even more species.

The validity of a classification system based on only one class of macromolecules (e.g. rRNA sequence data) is questionable. It would be useful to obtain confirmation of the taxonomic position of the above bacteria using other criteria. Because no other suitable sequence information is available for the other bacteria, such confirmatory comparisons can be carried out for only *T. ferrooxidans*. The sequences of a number of *T. ferrooxidans* proteins which have also been sequenced from a sufficiently large number of other bacteria to enable phylogenetic comparisons to be made are available. These include the sequences for the RecA protein, glutamine synthetase and the α and β subunits of the F_1F_0 ATP synthase. Phylogenetic comparisons based on the RecA sequence has confirmed the position of *T. ferrooxidans* within the β-*Proteobacteria* (36) and taxonomy based on glutamine synthetase (unpublished) and the F_1F_0 ATP synthase α and β subunits (7) is consistent with this placement although the data sets are less complete than for 16S rRNA genes. There are two interesting exceptions to the placement of *T. ferrooxidans* within the β-*Proteobacteria.*

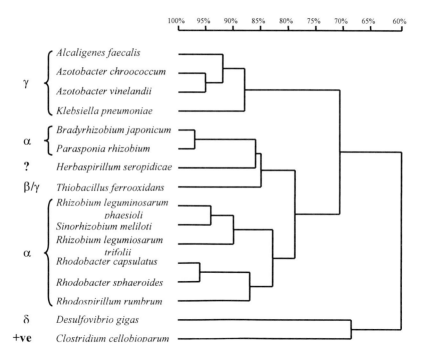

Figure. 2. The relationship of Fe proteins of the nitrogenase enzyme (γ-subunit) of members of the *Proteobacteria* indicating the unexpected finding that *T. ferrooxidans* is most closely related to a sub-group of α-*Proteobacteria*. Only bacteria for which complete *nifH* gene sequence data is available have been used with the Gram-positive *Clostridium cellobioparum* serving as an outgroup. The program DNAMAN was used for sequence alignment and dendrogram construction.

The amino acid sequence of the *T. ferrooxidans nifH* gene product (the nitrogenase iron protein) is most closely related to those of the α-*Proteobacteria* genera

Bradyrhizobium and *Parasponia* (see Figure 2) even though *T. ferrooxidans* is clearly not a member of this group if other criteria are used. This might be considered to be an example of lateral gene transfer were it not that phylogenetic placement based on the adjacent *nifDK* gene products (nitrogenase MoFe protein α and β subunits) was not as clear (65). A second exception is phylogeny based on subunits of the *T. ferrooxidans* pyruvate dehydrogenase complex. Although a more limited set of data is available (8 to 14 organisms), in this case the *T. ferrooxidans* enzyme appears unusual and falls within the region between the kingdoms *Bacteria* and *Eucarya* (58).

3. MOLECULAR GENETICS

The molecular genetics of the mesophilic iron- and sulfur-oxidizing chemolithtrophic bacteria other than *T. ferrooxidans* is almost unstudied. This review will therefore focus primarily on *T. ferrooxidans* but will include what little is known about the other physiologically similar bacteria. A previous review of the molecular genetics of *T. ferrooxidans* (70) covered the period up to mid-1993. Several significant advances have been made since then and this review is intended to focus mainly on developments over the past five years.

3.1 Genome size and genomics

The *T. ferrooxidans* ATCC 21834 genome has been estimated to be 2.9 mb in size (33) which is within the size range considered to be average for bacteria. This was determined digestion of the *T. ferrooxidans* genome with restriction enzymes which cut rarely within the genome followed by separation and of the fragments using pulse field gel electrophoresis and summation of fragment sizes. At the same time a physical map of the ATCC 21834 strain was generated. The size of 2.9 mb is similar to the 2.855 mb reported using *T. ferrooxidans* strain VKM-458 (39) and also to the 2.8 mb reported in the study by Holmes and co-workers using DNA reassociation kinetics (86). The study using strain VKM-458 was interesting in that the authors reported amplification of certain restriction enzyme fragments occurred when the cells became resistant to zinc or arsenic (39) and suggested that this was due to an increase in resistance gene copy number. Within the next few years *T. ferrooxidans* will enter into the field of bacterial genomics as the bacterium has joined the list of organisms whose entire genome is in the process of being sequenced. The sizes and mapping of the genomes of the other biomining bacteria covered in this review have not been reported.

3.2 Similarities between *T. ferrooxidans* and *E. coli*

E. coli is a heterotrophic bacterium that prefers to grow at a pH values close to neutral and is typically associated with the intestinal population of humans and animals. In contrast, *T. ferrooxidans* is an acidophilic, obligate chemoautolithotroph that is more commonly found in inorganic mineral environments. Indeed, *T. ferrooxdians* is so strictly autotrophic that it is not able to tolerate even the small amounts of free sugar present in most commercial agar preparations (83). In spite of this the two bacteria share much more in common from a molecular genetic point of view than might be imagined from their differences in physiology and the ecological niches which they occupy.

Table 1. Genes cloned or sequenced from the chromosome of *Thiobacillus ferrooxidans*

Gene(s)	Accession number	Function of gene product	Functional in *E. coli*	Expressed in *E. coli* from *T. ferrooxidans* promoter	Reference
alaS	gbX95571	alanyl-tRNA synthease	Yes	No	23
arsCBH	unpublished	arsenic resistance	Yes	Yes	unpublished
atpEF	gbM81087	F_0 subunits of ATP synthase	No	ND	8
atpHAGDC	gbM81087	F_1 subunits of ATP synthase	Yes	Yes	8
*cyc*1	embY07610	cytochrome *c*-552 (23 kDa)	NA	No	1
cysDNC	not sequenced	sulphate activation	Yes	Yes	21
glmS	gbL77909	glucosamine synthase	Yes	Yes	54
glnA	gbM16626	glutamine synthetase	Yes	Yes	2, 68
gltA	gbJ01619	citrate synthase	Yes	Yes	58
gltD	gbU36427	glutamate synthase small subunit	Yes	No	15
gshA	gbJ01619	γ-glutamylcysteine synthetase	Yes	Yes	59
iro	embX57324	high-redox-potential ferredoxin	NA	ND	45
leuB	gbD14585	3-isopropylmalate dehydrogenase	Yes	Yes	28, 36
merA	not deposited	mercury reductase	Yes	Yes	30, 32, 78
merC X 2	embX57326	mercuric ion transport (two copies)	Yes	Yes	32
merR X 2	embX57326	*merA* gene regulator (two copies)	Yes	Yes	32
nifHDK	gbM15238	nitrogen-fixing nitrogenase	ND	ND	61, 66
ntrA	gbM33831	RNA polymerase sigma factor, σ^{54}	Yes	Yes	4
ntrBC	gbL18975	two-component nitrogen gene regulators	Yes	Yes	38
pdhABC	gbU81808	pyruvate dehydrogenase complex	Yes	Yes	58
purA	embX57324	adenylosuccinate synthetase	Yes	Yes	43
rbcR	embD11141	Probable *rbcLS* gene regulator	NA	NR	40
rbpLS X 2	embD90113 embX70355	D-ribulose-1,5-biphosphate carboxylase (two copies present)	Yes	No	41, 44, 63
recA	gbM26933	homologous recombination and SOS response regulator	Yes	No	23, 64
recG	gbAF032884	ATP-dependent DNA helicase	ND	ND	54
recX	embX95571	putative recA regulator	ND	No	23
rnpB	embX16580	tRNA cleaving, RNaseP	ND	ND	82
*rrnT*₁ and *rrnT*₂	gbU18089	ribosomal RNA genes, partially sequenced	ND	ND	74
rus, rustA	embX95823 embX95324	rusticyanin	NA	No	3, 24
trxA	gbU20361	thioredoxin	Yes	Yes	60
tyrZ	embX79010	tyrosyl-tRNA synthetase	Yes	Yes (weak)	72
	gbAF005208	sulphur-regulated OMP	NR	NR	Buonfiglio, unpublished

ND = Not determined; NA = Not applicable as *E. coli* does not possess equivalent genes; NR = Not reported

Approximately 50 genes from about 20 regions of the chromosome have been sequenced to date (see Table 1). Cloned *T. ferrooxidans* genes may be tested in *E. coli* with a respect to product function˙ (if *E. coli* has similar genes) and gene expression. Almost all of the cloned gene products that can be tested by complementation of corresponding *E. coli* mutants have been found to be functional. An exception was the ATP synthase F_0 subunits c and b (*atpEF* gene products) which although shown to be expressed from a vector promoter using an *E. coli*-derived *in vitro* transcription-translation system, could not compliment *E. coli* mutants (8). However, this is a special case as the F_0 subunits are involved in the translocation of protons across the cytoplasmic membrane and the difference in pH across the membrane of *T. ferrooxidans* is far greater than *E. coli*. Likewise, most of the *T. ferrooxidans* genes that have been tested are most often expressed in *E. coli* from promoters that are located on the cloned *T. ferrooxidans* fragments. 'Housekeeping' genes which were not expressed are the linked *recA-recX-alaS* (23) and the *gltBD* (15) genes. The *rbcLS*1, *rbcLS*2 (41) and *rus* genes (3) were also not expressed, however as there is no counterpart of these genes in *E. coli*, suitable expression signals and regulators may be absent. In one study, the transcript start-site for the *T. ferrooxidans trxA* gene was determined using mRNA extracted from *T. ferrooxidans* and from *E. coli* which contained the cloned *T. ferrooxidans trxA* gene on a plasmid (60). The two strongest transcriptional start sites were identical irrespective of the source of mRNA. This study indicated that the transcription of heterologous genes in *T. ferrooxidans* should not present a major problem when attempting to genetically modify the bacterium.

3.3 Iron and sulfur oxidation

Thiobacillus ferrooxidans obtains its energy via the oxidation of ferrous iron to ferric or reduced sulfur compounds to sulfate. This chemolithotrophic metabolism is what makes the organism industrially important. It has therefore been of great interest to identify the essential components of the iron and sulfur oxidation systems, and to isolate the genes responsible and to study their regulation. The iron oxidation electron transport chain has been well-studied and it is thought that most of the components have been identified. *T. ferrooxidans* has been reported to have multiple cytochromes of the *c*-type, cytochromes of the a_1- and *b*- types, an iron-sulfur protein Fe(II)-cyt.c_{552} oxidoreductase, an a_1-type cytochrome oxidoreductase, high and low spin ferric hemes, a ferredoxin centre, ubiquinone 8 and several other possible components of electron transport chains (85). There also appears to be some variation in detail between strains, such as the exact sizes of the *c*-type cytochromes.

Components required for the oxidation of iron are thought to be; a 92 kD outer membrane porin (50), an Fe(II) oxidase (linked to cytochrome c_{552}), at least one c_4-type cytochrome c_{552} situated in the periplasm, a small 16 kDa blue copper protein called rusticyanin and a terminal cytochrome c oxidase (85). A model for the location of the components of the iron-oxidation system is shown in Figure 3. The exact role and position of each component in the electron transport chain remains to be resolved and there will almost certainly be some surprises. For example, the extensively studied small copper protein rusticyanin is considered to form part of the iron oxidation electron transport chain but recently it was reported that the aporusticyanin acts as specific receptor which stimulates the adhesion of the bacterium to pyrite (52). The first gene for a component involved in iron-oxidation to be cloned and sequenced was the *iro* gene (44). This gene was present in a transcriptional unit of its own and encoded a 90 aa protein (including a 37 aa signal sequence) which after processing is thought to become one of the 8 to 10 identical subunits that comprise the 63 kDa Fe(II) oxidase (reported and reviewed previously). After several attempts to isolate the gene for rusticyanin in a number of laboratories had been unsuccessful, a synthetic gene

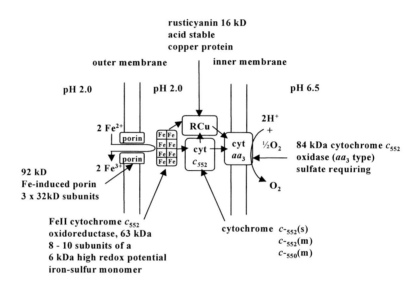

Figure 3. A model for the arrangement of the compontents of the iron-oxidation system of *T. ferooxidans*. Genes for all components with the exception of the outer-membrane porin have been cloned (modified from 67)

for rusticyanin was constructed which enabled the gene to be mutagenized (10). Subsequently, the cloning and sequencing of the gene for rusticyanin has been reported by two laboratories (3, 24). The amino acid sequence for rusticyanin from four *T. ferrooxidans* strains had been determined and as these varied by less than 10% it was possible to design primers to conserved regions. These were used in a polymerase chain reaction (PCR) to amplify a probe for use in isolating the *rus* gene by Southern hybridisation (3) and for identifying and cloning the amino acid sequence encoding portion of the gene using nested PCR (24). As rusticyanin represents up to 5% of the total proteins in iron-grown cells, the control *rus* expression is important. A single copy of the *rus* gene was present in *T. ferrooxidans* ATCC33020 and appears to be expressed either on its own or as the final gene in an operon. The *rus* gene has an upstream region which contains three putative σ^{70}-like promoter sequences and a strong ribosome binding site. As may be expected from a periplasmic protein, the predicted protein sequence included a 32 aa signal sequence. Interestingly, Northern hybridization experiments indicated that the *rus* gene was expressed even more strongly when cells were grown on thiosulfate than when grow on iron and the transcript size in sulfur-grown cells was slightly larger. This finding is unexpected as there have been several reports that the rusticyanin protein is present in considerably larger quantities in iron-grown cells (3).

More recently the *cyc1* gene which encodes a periplasmic cytochrome c_{552} has been cloned from *T. ferrooxidans* ATCC33020 (1). The gene which is present as a single copy was isolated by several rounds of inverse PCR with primers generated using limited amino acid sequence information obtained by sequencing the N-terminus of the purified protein. Although the exact start of the cytochrome c_{552} has not been established it appears to encode for a 23.2 kDa protein with a 29 aa signal peptide which when processed would result in a mature protein of 20 kDa. The cytochrome was found to be of the c_4 type and contained two heme-binding sites. Evidence for the size of the mature polypeptide and presence of the two heme binding sites has been supported by mass spectroscopy and EPR studies on a different *T. ferrooxidans* strain. When expressed in *E. coli* from a phage T7 vector promoter the cytochrome c_{552} polypeptide was processed by the *E. coli* secretion machinery even though the periplasmic pH of *E. coli* is 6.5 and that for *T. ferrooxidans* less than 3.0. However, it appeared that no hemes were incorporated into the apocytochrome produced in *E. coli*. Expression of the 650 bp cytochrome c_{552} mRNA transcript in *T. ferrooxidans* was approximately the same irrespective of the whether the cells were grown on ferrous iron, thiosulfate or elemental sulfur.

Recently a 8 kb sequence from the lab of Bonnefoy and co-workers has been deposited in the EMBL data base (accession numbers AJ006458, X95823, Y07610). Eight ORFs are presented which code for the structural genes (*cyc2* and *cyc1*) of a c_4 cytochrome (c_{552}), a hypothetical ORF, four subunits of an aa_3-type cytochrome oxidase and rusticyanin are linked on the chromosome of *T. ferrooxidans* ATCC33020. Details of this work are still to be published.

Very little is known in regard to the components of the iron-oxidation system of *Leptospirillum*. What is known is that the system differs considerably from *T. ferrooxidans*. Rusticyanin is absent and a soluble red cytochrome is present which is found in *Leptospirillum* but not in *T. ferrooxidans* (5). No molecular genetic studies of this iron oxidation system have been reported.

Compared with the iron-oxidation system of *T. ferrooxidans*, much less is known about the pathways and components that are required for sulfur oxidation. The biology of the oxidation of reduced inorganic sulfur compounds by the acidophilic thiobacilli has been reviewed (62). It is believed that thiosulfate is an important but unstable intermediate (75, 76). Sulfur appears to be stored in the polysaccharide layer of *T. ferrooxidans* when growing on pyrite where it serves as a reserve energy source when the iron has been oxidized or possibly when oxygen runs out and ferric iron serves as an alternate electron acceptor (72). Two genes associated with sulfur-oxidation in *T. ferrooxidans* have been identified. *T. ferrooxidans* ATCC 23270 produces a 40 kD sulfur-binding protein when grown on sulfur but not when grown on iron (53) which is possibly the same as the 40-kD protein identified by Mjoli and Kulpa (50). This protein formed thiol groups which strongly adhered to sulfur powder. Interestingly the sulfur-binding protein was located on the bacterial flagella. Recently the sequence of a gene for a sulfur-regulated outer membrane protein from *T. ferrooxidans* has been deposited in the Genbank database (accession number AF005208) but the primary publication on this work has still to appear. This sulfur-induced protein is larger than 40 kD and this protein and the sulfur-binding protein are probably different. In one of the few studies on *T. caldus* the ability of the bacterium to oxidize reduced inorganic sulfur compounds and the effect of uncouplers on the process was investigated by Hallberg and co-workers (25). As a result of this work the authors suggested that thiosulfate was oxidized in the periplasm while tetrathionate, sulfide, sulfur and sulphite were oxidized in the cytoplasm. With the exception of the sulfur-induced protein referred to above, nothing on the molecular genetics of sulfur oxidation is known.

The obligately chemolithotrophic bacteria also require a reverse electron transport mechanism. This is essential for the synthesis of NADH and/or NADPH considerable amounts of which are needed for the fixation of CO_2 and other processes. Little is known concerning the details of these processes in any of the acidophilic, chemolithotrophic bacteria.

3.4 Transposons on the chromosome of *T. ferrooxidans*

One of the most striking genetic similarities between *E. coli* and *T. ferrooxidans* was the discovery of a Tn7-like transposon (Tn*5468*) in the C-terminal region of the *T. ferrooxidans glmS* gene (54). There is a high frequency insertion site for Tn7 in the equivalent region of the *glmS* gene of *E. coli*. Tn*5468* is the closest relative to Tn7 characterised to date (see Figure 4). Southern hybridization experiments indicated that a

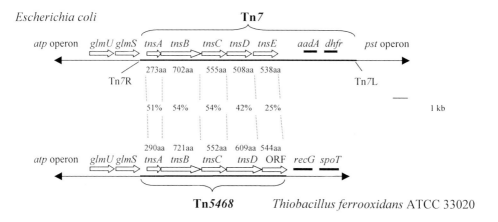

Figure 4. The structure and location of Tn7 at its specific insertion site immediately downstream of the *glmS* gene on the *E. coli* chromomosome compared with the structure and location of truncated transposon Tn*5468* on the chromosome of *T. ferrooxidans* ATCC 33020. Numbers between broken lines are the predicted sizes of the gene products and percentage amino acid similarity.

similar or identical transposon was present in all three *T. ferrooxidans* strains tested and which were isolated from different parts of the world. Insertion of Tn7 into specific sites has been detected in bacteria as divergent as *Agrobacterium tumefasciens*, *Caulobacter crescentus*, *Klebsiella pneumoniae*, *Pseudomonas aeruginosa*, *Rhizobium meliloti*, *Rhodospeudomonas capsulata* and *Xanthomonas campestris* but all these bacteria were artificially exposed to Tn7 in the laboratory. *E. coli* is the only bacterium in which Tn7 has been found naturally. Tn*5468* and Tn7 clearly share a common ancestor which appears to have been active within bacteria before they became as widely divergent physiolgically and ecologically as *T. ferrooxidans* and *E. coli* are today. Hopefully more Tn7-like transposons will be discovered in future and this will allow the tracking of the movement of this transposon family and provide some insight into how genetic elements have been able to move horizontally so that today they are found in such diverse bacteria.

Two repeated sequences present in 20 to 30 copies are located on the chromosome of *T. ferrooxidans* strains isolated from very regions and continents. These insertion sequences (IS*T1* and IS*T2*) are mobile within the chromosome (9). It has been suggested that movement of IS*T1* sequences may be associated with the phenotypic switching of colony morphology and the ability to oxidize iron (77). The evidence for a causal relationship between the movement of IS elements and phenotypic switching was discussed in a previous review as was the structure of IS*T2* (70). IS*T445* which appears to be identical or nearly identical to IS*T1* has since been sequenced (11). This 1.2 kb insertion sequence possesses 8 bp terminal inverted repeat sequences, three small open reading frames and was 54.4% identical at the nucleotide level to insertion sequence IS*AE1* from *Alcaligenes eutrophus.* As for Tn*5468*, once more ancestors of these insertion elements have been discovered it will be interesting to attempt to identify the evolutionary path by which these related IS elements have ended up in such different hosts.

3.5 Plasmids of *T. ferrooxidans* and other chemolithotrophic bacteria

Plasmids are extrachromosomal pieces of DNA that are non-essential for cell survival under certain conditions but which frequently contain genes which may improve host cell fitness under other conditions. It is of special interest to discover what types of genes are present on plasmids of bacteria from inorganic, low pH biomining environments. In addition it is of interest to discover whether features such as replication, mobilisation and stabilisation systems are unique to plasmids from low pH inorganic environments or similar to those found in other more commonly studied bacteria. Plasmids are frequently found in isolates of *T. ferrooxidans* and several of these plasmids have been studied in some detail.

Plasmid pTF-FC2 is a 12.2 kb broad-host-range mobilisable plasmid that has an origin of replication which is clearly related to those of the IncQ plasmids (see Figure 5) and a replication region which is distantly related to the TraI region of the IncP plasmids (70). This is the only *T. ferrooxidans* plasmid for which a functional accessory gene has been found. This gene is contained within a 3.5 kb transposon, Tn*5467* (14). This transposon is bordered by two 38bp inverted repeat sequences which are identical to those of the widely-distributed transposon Tn*21*. Tn*5467* has three accessory genes which code for a glutaredoxin, a MerR-like regulator protein and a 43 kD protein with similarity to the 12-loop trans-membrane spanning multidrug resistance transport proteins. Although it is unknown what properties pTF-FC2 confers on *T. ferrooxidans,* when cloned in *E. coli* the gene for the glutaredoxin complements *E. coli* thioredoxin mutants for several thioredoxin-dependent functions. It has recently been reported that pTF-FC2 has a proteic poison-antidote plasmid stability system called *pas* (plasmid addiction system) which is closely related to previously identified systems in regard to general mechanism of stability but not in the sequence similarity of its components (79). The poison-antidote system consists of three small proteins, an antidote-negative regulator (PasA), a toxin (PasB) and a protein

14

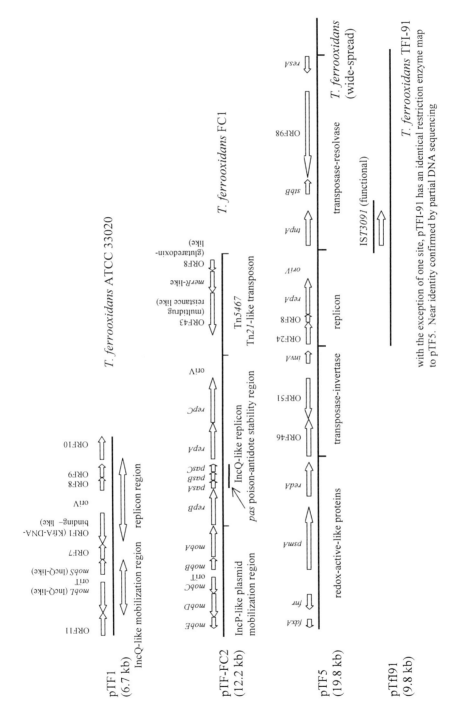

Figure 5. Features of the four *T. ferrooxidans* plasmids that have been studied in the greatest detail.

(PasC) which appears to enhance the binding of PasA to PasB (80, 81). The *pas* is located within the replicon on pTF-FC2 in such a way that deletion of *pas* does not result in a change in plasmid copy number.

The 6.7 kb mobilisable plasmid pTF1 is another *T. ferrooxidans* plasmid that has been extensively studied (see Figure 5). It appears to be a narrow-host-range plasmid which has a mobilisation region that is related to the IncQ plasmid RSF1010 and also to pSC101 (19). Plasmid pTF1 has a replicon with an ORF with similarity to the KrfA DNA-binding protein of the IncP plasmid RK2 and another ORF with similarity to a previously unidentified ORF in plasmid pSC101 (47). As far as is known pTf1 is cryptic.

Several authors have noted that members of a family of related *T. ferrooxidans* plasmids are widely distributed in strains that originated from different continents. Members of this pTFI91 family share a conserved replicon part of which is present on a 2.2 kb *Sac*I fragment. One member of this family has been isolated independently by workers in three different laboratories. Although the reported size varies from 19.6 to 20 kb the restriction enzyme maps of each plasmid were identical. One of these plasmids, pTF5, has been completely sequenced and found to contain at least fourteen ORFs (see Figure 5). Most ORFs had clear amino acid sequence similarity to proteins commonly found on plasmids such as replication-, partition- and plasmid stability functions. Several incomplete transposons and transposon 'scars' were also present with ORFs or partial ORFs with similarity to transposases, a resolvase and an invertase (17). Possibly the most interesting part of pTF5 is a 5.6 kb region which has four complete open reading frames which when translated have similarity to [3Fe-4S,4Fe-4S] ferredoxins, proteins of the FNR regulator family, prismane-like proteins and the NADH oxidoreductase subunit of a methane monooxygenase (18). Ferredoxins, prismane proteins and NADH oxidoreductase are redox-active proteins and FNR a regulator that responds to redox conditions. Proteins were expressed from three of the four ORFs using an *E. coli*-derived in vitro transcription-translation system. It is possible that these proteins represent an electron transport system of unknown function. An interesting finding was that DNA with homology to the 5.6 kb region containing the putative electron transport system but not the rest of the plasmid was also present on the chromosome of some *T. ferrooxidans* isolates. When the restriction enzyme maps of pTF5 and pTFI91 were compared, plasmid pTFI91 appeared to be contained entirely within pTF5 and it is the 5.6 kb region which is missing from pTFI91 (12). Studies using pTFI91 indicated that this plasmid contained an insertion sequence called IS*3091* which had highest similarity to IS*1086* of *Alcaligenes eutrophicus* and that insertion sequence IS*3091* was capable of transposition in *E. coli* (13). This is therefore the second insertion sequence (see IS*445* above) which has closest similarity to an insertion element from *A. eutrophicus*.

Much less research has been carried out on plasmids of *T. thiooxidans*, *T. caldus* or *L. ferrooxidans*. At least three plasmids have been found in a search for plasmids in three *T. caldus* isolates and one plasmid of 40-60 kb in a strain of *Leptospirillum* (by researchers in the author's laboratory) although nothing is known concerning the functions of these plasmids.

3.6 Gene transfer systems and the expression of foreign genes in *T. ferrooxidans*

After many years of unsuccessful attempts to transfer genes into *T. ferrooxidans*, success was achieved when a strain of the bacterium was transformed by electroporation using the *T. ferrooxidans merA* gene (mercury reductase) as a selectable marker and with natural plasmids from *T. ferrooxidans* serving as the vector (42). These workers were able to show that an IncQ-type of plasmid replicon was also able to reproduce in *T. ferrooxidans*. However, the transformation procedure was beset by several difficulties. Only one strain out of thirty tested was transformed, selection for mercury resistance was problematic because the cells take a long time to form colonies and mercury is naturally volatile. Approximately 50% of colonies which grew on the selection medium were not transformants. In addition the electroporation frequency was very low (120 to 200 colonies per μg of DNA).

The second report of gene transfer concerned the use of conjugation to transform *T. thiooxidans* and subsequently *T. ferrooxidans* was transformed using the same procedure (34, 56). Self-transmissible broad-host-range IncP plasmids were conjugated into *T. thiooxidans* by mating directly with *E. coli*. Mating was carried out by filter mating on plates using an inorganic thiosulfate-containing medium with a small quantity of yeast extract (0.05%) at pH 4.8. Either of the IncP plasmid kanamycin or tetracycline resistance genes but not the ampicillin resistance gene could serve as a selectable genetic marker and mating took place at a low but workable frequency (10^{-5} to 10^{-7} transconjugants per recipient). IncP plasmids could be transferred to isolates of *T. ferrooxidans* using a similar procedure and all seven strains tested were successful recipients. An IncQ plasmid was also conjugated into five *T. ferrooxidans* recipients using an IncP helper plasmid. Attempts were made to use kanamycin, ampicillin, tetracycline, streptomycin and mercury resistance genes as selectable markers, but only the kanamycin and streptomycin resistance genes were suitable. By using a narrow-host-range plasmid which was unable to replicate in *T. ferrooxidans* as a delivery vechile, it was shown that the kanamycin resistance encoding transposon Tn5 was able to transpose onto the chromosome of *T. ferrooxidans*. In a subsequent study, the arsenic resistance genes from plasmid pUC3 were cloned into an IncQ plasmid and transferred by conjugation from *E. coli* to *T. ferrooxidans* (55). As before, kanamycin resistance was used for the initial selection and transconjugants were then tested for expression of arsenic resistance. Only *T. ferrooxidans* transconjugants were able to grow in medium containing more than 10mM $NaAsO_2$.

No gene transfer systems for members of the genus *Leptospirillum* have been reported. Since only sulfur-based media have been successfully used in conjugation experiments and as these bacteria use only iron as an energy source the development of a direct *E.coli* to *Leptospirillum* conjugation system will present some additional challenges. Furthermore it is not clear how long the antibiotics which have been successfully used with *T. ferrooxidans* remain active on solid iron media of the type required by the leptospirilli.

3.7 Cloned genes of mesophilic, iron- or sulfur- oxidizing obligately chemolithotrophic bacteria besides *T. ferrooxidans*

The molecular genetics of acidophilic, chemolithotrophic bacteria besides *T. ferrooxidans* is almost unstudied. The isolation and sequencing of one protein-encoding gene from a *Leptospirillum* has been reported (16). This is the *lcrI* gene which codes for a chemotaxis receptor which is related to the family of methyl-accepting chemotaxis proteins (MCPs). The gene was expressed in *E. coli* and the product cross-reacted with antibodies to the *E. coli* Tar MCP but was not located in the membrane as it would be expected when expressed its natural *Leptospirillum* host. It was suggested that the difference in pH values across the membrane in *Leptospirillum* compared with *E. coli* might not have allowed for the correct folding of the protein. Two gene banks of two isolates of *Leptospirillum* have been constructed in the laboratory of the author. However, unlike what was found with *T. ferrooxidans* gene banks almost no clones from gene banks of *Leptospirillum* have been able to complement *E. coli* mutants (unpublished). The reason for this is unclear but unlike the thiobacilli, the leptospirilli are phylogenetically very different from *E. coli* and this is possibly the explanation for the lack of complementation.

4. CONCLUSIONS

It is clear from this review that most studies on molecular genetics of the mesophilic iron- and sulfur-oxidizing chemolithotrophic bacteria have focussed on *T. ferrooxidans*. Although *T. ferrooxidans* is readily isolated from natural environments its role in industrial heap and continuous-flow biooxidation plants has been reassessed. There a growing body of evidence to suggest that it plays much less important role than originally thought (20, 22, 57, 68). There is therefore a need to broaden the molecular genetic studies to include *Leptospirillum* and the sulfur-oxidizing bacteria. This review has been restricted to bacteria growing at 40°C or below since most of the biooxidation heap or stirred-tank processes take place at ambient temperatures (although internal heap temperatures may well exceed this) or at temperatures of 40°C (with the notable exception of the BacTech process, 49). There is little doubt that a new generation of biooxidation processes that operate at temperatures above 45°C, possibly even higher than 80°C will be developed. This will require that research is initiated into the molecular biology of moderate (35) and hyper-thermophilic (51) organisms (most hyperthermophiles are Archaea). Exactly which hyperthermophiles should become the main focus of attention will depend on which are most successful from an applied point of view. Exactly which organisms these will be is still unclear.

REFERENCES

1.   Appia-Ayme C., A. Bengrine, M. Chippaux and V. Bonnefoy, IBS97 Biomine97, Australian Mineral Foundation Glenside, Australia, (1997), P2.1.
2.   Barros M.E., D.E. Rawlings and D.R. Woods, J. Bacteriol., 164 (1985) 1386.
3.   Bengrine A., N. Guiliani, C. Appia, F. Borne, M. Chippaux and V. Bonnefoy, In, Biohydrometallurgical Processing (eds, C. Jerez, T. Vargas, H. Toledo & J. Wiertz) Vol. 2, Santiago, Chile, (1995) 75.
4.   Berger D.K., D.R. Woods and D.E. Rawlings, J. Bacteriol., 172 (1990) 4399.
5.   Blake R., E.A. Shute, J. Waskovsky and A.P. Harrison, J. Geomicrobiol., 10 (1992) 173.
6.   Brierley C.L., Sci. Am., 247 (1982) 42.
7.   Brown L.D. and D.E. Rawlings, Biohydrometallurgical Technologies, Vol 2, (eds A. Torma, M. Apel, & C. Brierley) TMS Press, Warrendale, Pa. (1993) 519.
8.   Brown L.D., M.E. Dennehy and D.E. Rawlings, FEMS Microbiol. Lett., 122 (1994) 19.
9.   Cádiz R., L. Gaete, E. Jedlicki, J. Yates, D.S. Holmes and O. Orellana, Mol. Microbiol., 12 (1994) 165.

10. Casimiro D.R., A. Toy-Palmer, R.C. Blake II, H.J. Dyson, Biochemistry, 34 (1995) 6640.

11. Chakraborty R., C. Deb, A. Lohia and P. Roy, Plasmid, 38 (1997) 129.

12. Chakravarty, L., T.J. Zupancic, B. Baker, J.D. Kittle, I.J. Fry and O.H. Tuovinen, Can. J. Microbiol. 41 (1995) 354.

13. Chakravarty L., J.D. Kittle, Jr. and O.H. Tuovinen, Can. J. Microbiol., 43 (1997) 503.

14. Clennel A-M., B. Johnston and D.E. Rawlings, Appl. Environ. Microbiol., 61 (1995) 4223.

15. Deane S.M. and D.E. Rawlings, Gene., 177 (1996) 261.

16. Delgado M., H. Toledo and C.A. Jerez, Appl. Environ. Microbiol., 64 (1998) 2380.

17. Dominy C.N., N.J.Coram and D.E. Rawlings, Plasmid., 40 (1998) 50.

18. Dominy C.N., S.M. Deane, D.E. Rawlings, Microbiol., 143 (1997) 3123.

19. Drolet M., P.Zanga And P.C.K. Lau, Mol. Microbiol., 4 (1990) 1381.

20. Espejo R.T. and J. Romero, Appl. Environ. Microbiol. 63 (1997) 1344.

21. Fry I.J. and E. Garcia, In Biohydrometallurgy-1989 (eds, J. Salley, R. McCready & P. Wichlacz), CANMET, Ottawa, Canada, 1989 171.

22. Goebel B.M. and E. Stackebrandt, Appl. Environ. Microbiol. 60 (1994) 1614.

23. Guiliani N., A. Bengrine, F. Borne, M. Chippaux and V. Bonnefoy, Microbiol., 143 (1997) 2179.

24. Hall J.F., S.S. Hasnain and W.J. Ingledew, FEMS Micro. Lett. 137 (1996) 85.

25. Hallberg K.B., M. Dopson and E. B. Lindström, J. Bacteriol., 178 (1996) 6.

26. Hallberg K.B., E.B. Lindström, Microbiol., 140 (1994) 3451.

27. Harrison A.P. and P.R. Norris, FEMS Micro. Lett. 30 (1985) 99.

28. Hugenholtz P., C. Pitulle, K.L. Hershberger and N.R. Pace, J. Bacteriol, 180 (1998) 366.

29. Inagaki K., H. Kawaguchi, Y. Kuwata, T. Sugio, H. Tanaka and T. Tano, J. Ferment. Bioeng., 70 (1990) 71.

30. Inoue C., K. Sugawara and T. Kusano, Gene., 96 (1990) 115.

31. Inoue C., K. Sugawara and T. Kusano, Mol. Microbiol., 5 (1991) 2707.

32. Inoue C., K. Sugawara, T. Shiratori, T. Kusano and Y. Kitagawa, Gene., 84 (1989) 47.

33. Irazabal N., I. Marin and R. Amils, J. Bacteriol, 179 (1997) 1946.

34. Jin S.M., W.M. Yanand Z.N. Wang, Appl. Environ. Microbiol., 58 (1992) 429.

35. Johnson, D.B. and F.F. Roberto, In, Biomining; Theory, Microbes and Industrial Processes (ed D. Rawlings), Springer-Verlag, Berlin, 259.

36. Karlin S., G.M. Weinstock and V. Brendel, J. Bacteriol., 177 (1995) 6881.

37. Kawaguchi H., K. Inagaki and H. Tanaka, Biosci. Biotechnol Biochem., 61 (1997) 2119.

38. Kilkenny C.A., D.K. Berger and D.E. Rawlings, Microbiol., 140 (1994) 2543.

39. Kondratyeva, T.F., L.N. Muntyan and G.I. Karavaiko, Microbiology 141 (1995) 1157.

40. Kusano T. and K. Sugawara, J. Bacteriol., 174 (1993) 1019.

41. Kusano T., K. Sugawara, C. Inoue and N. Suzuki, Curr. Microbiol., 22 (1991) 35.

42. Kusano T., K. Sugawara, C. Inoue, T. Takeshima, M. Numata and T. Shiratori, J. Bacteriol.,174 (1992) 6617.

43. Kusano T., T. Takeshima, C. Inoue and K. Sugawara, Curr. Microbiol., 26 (1993) 197.

44. Kusano T., T. Takeshima, C. Inoue and K. Sugawara, K. J. Bacteriol.,173 (1991) 7313.

45. Kusano T., T. Takeshima, K. Sugawara, C. Inoue, T. Shiratori, T. Yano Y. Fukumori and T. Yamanaka, J. Biol. Chem., 267 (1992) 11242.

46. Lane D.J., A.P. Harrison, Jr., D. Stahl, B. Pace, S.J. Giovannoni, G.J. Olsen and N.R. Pace, J. Bacteriol., 174 (1992) 269.

47. Lau P.C.K., M. Drolet, P. Zanga and D.S. Holmes, In, Biohydrometallurgical Technologies, Vol 2, (eds A. Torma, M. Apel, & C. Brierley) TMS Press, Warrendale, Pa. (1993) 635.

48. Lundgren D.G. and M. Silver, Annu. Rev. Microbiol., 34 (1980) 263.

49. Miller P.C., In, Biomining; Theory, Microbes and Industrial Processes (ed D. Rawlings), Springer-Verlag, Berlin, 81.

50. Mjoli N. and C.F. Kulpa, Biohydrometallurgy. Science and Tecnology Letters, Kew, United Kingdom, (1988) 89.

51. Norris P.R., In, Biomining; Theory, Microbes and Industrial Processes (ed D. Rawlings), Springer-Verlag, Berlin, 247.

52. Ohmura N. and R. Blake II, In, Biomine97, Australian Mineral Foundation Glenside, Australia, (1997), PB1.1.

53. Ohmura N., K. Tsugita, J. Koizumi and H. Saiki, J. Bacteriol., 178 (1996) 5776.

54. Oppon J.C., R.J. Sarnovsky, N.L. Craig and D.E. Rawlings, J. Bacteriol, 180 (1998) 3007.

55. Peng J., W. Yan and X Bao, Appl. Environ. Microbiol., 60 (1994) 2653.

56. Peng J., W. Yan and X. Bao, J. Bacteriol., 176 (1994) 2892.

57. Pizarro J., E. Jedlicki, O. Orellana, J. Romero and R.T. Espejo. Appl. Environ. Microbiol. 62 (1996) 1323.

58. Powles R., and D.E. Rawlings, Microbiol., 143 (1997) 2189.

59. Powles R., S. Deane and D.E. Rawlings, Microbiol., 142 (1996) 2543.

60. Powles R.E., S.M. Deane and D.E. Rawlings, Microbiol., 141 (1995) 2175.

61. Pretorius I., D.E. Rawlings and D.R. Woods, Gene., 45 (1986) 59.

62. Pronk J.T., R. Meulenberg, W. Hazeu, P. Bos and J.G. Kuenen, FEMS Microbiol. Rev., 75 (1990) 293.

63. Pulgar P., L. Gaete, J. Allende, O. Orellana, X. Jordana and E. Jedlicki, FEBS Lett., 292 (1991) 85.

64. Ramesar R.S., V. Abratt, D.R. Woods and D.E. Rawlings, Gene., 78 (1989) 1.

65. Rawlings D.E., (Unpublished observations)

66. Rawlings D.E., Gene., 69 (1988) 337.

67. Rawlings D.E., In, Biomining; Theory, Microbes and Industrial Processes (ed D. Rawlings), Springer-Verlag, Berlin, 229.

68. Rawlings D.E. In, Biohydrometallurgical Processing (eds, C. Jerez, T. Vargas, H. Toledo & J. Wiertz) Vol. 2, Santiago, Chile, (1995) 9.

69. Rawlings D.E., W.A. Jones, E.G. O'Neil and D.R. Woods, Gene., 53 (1987) 211.

70. Rawlings D.E. and T. Kusano, Microbiol, Rev., 58 (1994) 39.

71. Rawlings D.E. and S. Silver, Bio/Technology, 13 (1995) 773.

72. Rojas J., M. Giersig and H. Tributsch, Arch. Microbiol., 163 (1995) 352.

73. Salazar O., B. Sagredo, E. Jedlicki, D. Söll, I. Weygand-Durasevic and O. Orellana, J. Bacteriol., 176 (1994) 4409.
74. Salazar O., M. Takamiya and O. Orenella, FEBS Lett. 242 (1889) 439.
75. Sand W., T. Gerke, R. Hallmann and A. Schippers, Appl. Microbiol. Biotechnol., 43 (1995) 961.
76. Schippers A., P-G. Jozsa and W. Sand. Appl. Environ. Microbiol. 62 (1996) 3424.
77. Schrader J.A. and S. Holmes, J. Bacteriol., 170 (1988) 3915.
78. Shiratori T., C. Inoue, K. Sugawara, T. Kusano and Y. Kitagawa, J. Bacteriol., 171 (1989) 3458.
79. Smith A.S.G. and D.E. Rawlings, Mol. Microbiol., 26 (1997) 961.
80. Smith A.S.G. and D.E. Rawlings, J. Bacteriol., 180 (1998) 5458.
81. Smith A.S.G. and D.E. Rawlings, J. Bacteriol., 180 (1998) 5463.
82. Takamiya M., C. Inoue, Y. Kitagawa and T. Kusano, Nucleic Acids Res., 17 (1989) 9482.
83. Tuovinen O.H., S.I. Niemela and H.G. Gyllenberg, Biotech. Bioeng., 13 (1971) 517.
84. Woese C.R., Microbiol. Rev., 51 (1987) 221.
85. Yamanaka T. and Y. Fukumori, FEMS Microbiol. Rev., 17 (1995) 401.
86. Yates J.R. and D.S. Holmes, J. Bacteriol, 169 (1987) 1861.

Cloning of metal-resistance conferring genes from an *Acidocella* strain

S. Ghosh, N. R. Mahapatra and P. C. Banerjee*

Indian Institute of Chemical Biology, 4 Raja S. C. Mullick Road, Calcutta-700032, India
E-mail : iichbio@giascl01.vsnl.net.in

Metal resistance is regarded as the most suitable phenotypic trait for selection of genetically engineered bioleaching bacteria. Since the expression of metal resistance conferring genes is very limited in heterologous systems, search for these genes from homologous and related biomining bacteria is envisaged to be rewarding for genetic manipulation of this group of microorganisms. *Acidocella* strain GS19h, a bacterium having high resistance to cadmium and zinc, was chosen for cloning of metal resistance genes from its plasmids. Purified plasmid preparation from this bacterium was partially digested with *Sau*3AI, ligated at the *Bam*HI site of pBluescriptII KS+/-, and transferred into *E. coli* DH5α strain by transformation. The *E. coli* derivatives containing cloned segments were tested for metal resistance and a few cadmium resistant colonies were found to contain a recombinant plasmid having a 0.8 kb insert DNA. This DNA fragment was sequenced and analyzed for common sequences in other sources from gene data banks. Although partial homologies with a part of various transposons, *merR* genes and others were detected, it appears that the cloned gene is a novel one with no apparent similarity with the existing cadmium, copper, nickel, or zinc resistance-conferring genes.

1. INTRODUCTION

In the process of developing improved strains for leaching of ores employing acidophilic bacteria such as *Thiobacillus ferrooxidans*, construction of plasmid vectors is essential [1-5]. In the case of acidophiles, metal resistance is considered to be the most suitable phenotypic trait of such vectors [4-6]. But due to very low to non-existent levels of expression of this phenotypic characteristic in heterologous systems [7, 8], it is important to identify and isolate

* Corresponding author. This work was partly supported by the Department of Science and Technology, Govt. of India. Authors are grateful to Mr. Suvobrata Nandi and Mr. Chirajyoti Deb for their assistance in the study. S. G. and N. R. M. are thankful to Labonya Prova Bose Trust, Calcutta and Council of Scientific and Industrial Research, New Delhi for providing them with fellowships.

metal resistance encoding genes from *T. ferrooxidans* and other acidophiles of the same habitat, and preferably from their plasmids. It is presumed that such genes might be useful for genetic engineering of leaching microbes for the development of this biotechnology [6, 9, 10]. In the case of *T. ferrooxidans*, the evidence for plasmid-mediated metal resistance was only circumstantial. For example, among the *T. ferrooxidans* isolates four strains containing a 20 kb plasmid were highly resistant to UO_2^+ whereas in one strain absence of the plasmid coincided with a reduction in uranium resistance [11]. This well-studied leaching bacterium has so far been reported to contain its zinc, arsenic [12], mercury [13] and copper [14] - resistance determinants only in the chromosome. On the other hand, the resistance determinants for arsenic and mercury in an acidophilic heterotroph of acidic mine environment *Acidiphilium multivorum* AIU301 have been found to be encoded in its 56 kb plasmid [15]. The resistances to Cd^{2+}, Cu^{2+} and Zn^{2+} of another acidophilic heterotrophic bacterium *Acidocella* sp. strain GS19h have also been observed to be plasmid-mediated [16]. Thus, unlike *T. ferrooxidans* the acidophilic heterotrophs may harbour metal-resistance genes on their plasmids, which may be developed for construction of plasmid vectors for acidophiles. The present piece of work was undertaken to clone the metal resistant genes present in the plasmids of *Acidocella* strain GS19h via construction of a miniplasmid library.

2. MATERIALS AND METHODS

2.1. Bacterial strains, plasmids, and culture conditions
Acidocella species strain GS19h [17], *E. coli* DH5α [18] and pBluescript II KS+/- [18] were used in this study. The *Acidocella* strain was grown at 30°C with shaking in MGY medium [g l^{-1} composition: $(NH_4)_2SO_4$, 2.0; K_2HPO_4, 0.25; $MgSO_4.7H_2O$, 0.25; KCl, 0.1; glucose, 1.0; and yeast extract, 0.1] of pH 3. The *E. coli* cells were grown in LB broth [19]. Selection of transformants of *E. coli* DH5α was made on LB-agar containing ampicillin (100µg ml^{-1}) or $CdSO_4$ (8 mM and 10 mM) or $ZnSO_4$ (12 and 16 mM) or $CuSO_4$ (12 and 16 mM).

2.2. Isolation, purification and electrophoresis of plasmid DNA
Plasmid DNAs were isolated, in general, by the alkaline lysis method as described previously [16, 20], and purified by CsCl-ethidium bromide gradient centrifugation [19] or by polyethylene glycol treatment [21] whenever required. For nucleotide sequencing, isolation of plasmid DNA was carried out by strictly following the Perkin Elmer protocol [21] excepting that *E. coli* cells were grown in LB broth instead of 'terrific broth'. Plasmid DNAs were electrophoresed in 0.5-1.0% (w/v) agarose gels and detected after ethidium bromide (0.5 µg ml^{-1}) staining as usual [19].

2.3. Construction of miniplasmid library
About 1 µg of total plasmid DNA from *Acidocella* strain GS19h was digested with *Sau*3AI in order to get the most DNA fragments of size ranging between 0.5–6.0 kb. After dephosphorylation by calf intestinal phosphatase DNA fragments were ligated with *Bam*HI digested pBluescript II KS +/- by T4 DNA ligase as described previously [19]. The ligated DNA samples were used to transform *E. coli* DH5α cells made competent by $CaCl_2$ treatment

[19] and the transformants containing recombinant DNA molecules were selected on LB-agar with ampilicillin and requisite amounts of X-gal (5-bromo-4-chloro-3-indolyl-D-galacto-pyranoside) and IPTG (isopropyl β-D-thiogalactopyranoside). The transformed colonies were then checked for their metal resistance characteristics.

2.4. DNA sequencing

Nucleotide sequence of the DNA fragment inserted at the *Bam*HI site of the vector plasmid pBluescript II KS +/- was determined by ABI Prism Model 377 DNA sequencer (Perkin Elmer). Universal M13 primer was used for sequencing reaction.

3. RESULTS

3.1. Metal resistance characteristics of the transformants

The transformants were observed to be sensitive to 12 mM $CuSO_4$. But 13% of the total white colonies were resistant to $CdSO_4$ and about 5% of the same to $ZnSO_4$ - the MIC values being 10 mM and 16 mM respectively while those for *E. coli* DH5α (pBluescript II KS+/-) were 2 mM and 10 mM respectively.

3.2. Plasmid profile of the clones

Metal resistant *E. coli* transformants were of different types based on the sizes of the recombinant DNAs. From some Cd^{2+}-resistant transformants a recombinant plasmid designated as pSGX was isolated, and subjected to further studies. The size of pSGX was determined to be 3.7 kb as it gives 3.7 kb, 2.9 kb and 0.8 kb DNA bands on partial digestion with *Pst*I and *Not*I (Figure 1, lane 6).

3.3. Analysis of the nucleotide sequence of the cloned DNA fragment

The recombinant plasmid pSGX contains 814 bp DNA insert. The nucleotide sequence is shown in Figure 2. The forward sequence contains 149 A, 162 T, 242 G and 261 C bases; the overall (G+C) mol% being 61.9 which is almost the same as that of the chromosomal DNA [17]. About 80 restriction enzymes including *Alu*I, *Ava*I, *Dpn*I, *Dra*I, *Eco*RI, *Eco*RII, *Hae*II, *Hae*III, *Sau*3AI, *Sma*I and *Taq*I have cutting site(s) in the cloned region. On the other hand, about 150 enzymes do not have any restriction site in this piece of DNA. These include some enzymes commonly used in cloning experiments such as *Bam*HI, *Bgl*II, *Bgl*III, *Cla*I, *Dra*III, *Eco*RV, *Hae*I, *Hpa*I, *Kpn*I, *Nde*I, *Not*I, *Pst*I, *Pvu*I, *Pvu*II, *Sac*II, *Sal*I, *Sfi*I, *Spe*I, *Swa*I, *Xba*I and *Xho*I.

3.4. Sequence similarities of the cloned DNA with other genes

DNA sequence similarities between the cloned DNA sequence in pSGX and other sequences in the EMBL database were examined. It was found that the cloned plasmid region has insignificant sequence homology with other gene/DNA sequences in comparison with their large sizes vis-à-vis the small fraction of nucleotides with which the cloned DNA shows similarity. The maximum number of 347 bases (from nucleotide 464-814) showed 61.4 % homology with a *T. ferrooxidans* plasmid of 19.8 kb size. Majority of the other genes and DNA sequences (viz *Pseudomonus aeruginosa* multiresistance β-lactamase transposon

Tn*1412*, phage P1 *darA* operon, plasmid pMER05, mercury resistance transposons Tn*5053* and Tn*552*, *Rhodospirillum rubrum* plasmid pKY1, glucosidase gene of *Salmonella typhimurium*, *Enterobacter aerogens* R plasmid etc.) also showed 57-63 % similarity with the cloned DNA within the base 470 and 814. *P. fluorescens merR* gene, *A. faecalis merR* gene, *Enterobactor aerogens mer* gene for regulation, *P. testosteroni* merR gene, and *Rhizobium* sp. pNG have 55-58 % similarity with the first portion (from base 20-325) of the cloned DNA. The middle 135 bp portion (from base 315-450) of the cloned DNA segment did not show similarity with any of the reported genes to a significant extent. Thus, the cloned DNA although showing some homologies with a variety of different genes including some mercury resistance transposons and plasmids, it may be regarded as a novel one expressing Cd^{2+} resistance in *E. coli* DH5α.

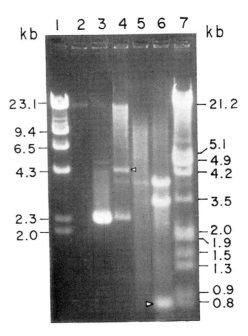

Figure 1. Agarose (1%, w/v) gel electrophoretogram showing generation of a ca. 0.8 kb insert DNA from recombinant clone pSGX through double digestion with restriction enzymes. Lane 1, λ *Hind*III digest; lane 2, chromosomal DNA of *E. coli* DH5α; lane 3, pSGX DNA; lane 4, *Pst*I digested pSGX (◁ indicates open circular form); lane 5, *Not*I digested pSGX; lane 6, *Pst*I and *Not*I digested pSGX(▷ indicates position of the cloned DNA frgament); lane 7, λ *Hind*III plus *Eco*RI digest. Numbers on both sides indicate molecular sizes of linear DNA markers.

```
1    GATCTTAGTCGTCTTGTTTTTCGCGTTACGCGACTGGCGG    40
                                             SD
41   ATTTCGCTGATCTATGGCGCCGCCGAAATCGCGCGCCTCA   80
       ORF1→       M
81   CCTTCATCTGCCGTGCCCGGGAACTCGGCTTCTCTCTCGA   120

121  CGAGGTACGCGGCCTTCTCAGCCTGGCCGAAAGAGATGAA   160

161  CGCCACTGTGAGGACGTGAAACAAGCTGCTATCCGCCATC   200
                       *
201  GTCAGGACGTGCGCCGCAAGATCGCCGACCTGCGGGCGGT   240

241  CGAGGTCACTCTGGGAACCCTCATTCGGCAATGCGAAGCA   280
                 SD          ORF2→    M
281  CGCGGGCCGGCGGAATGCCCCTTGATCGAAGCGCTATCTC   320

321  AACCGAAAGCGGCAGCGCCCGCGCCTTGAAGCGTCGCCGT   360

361  TTAAAACACCACCGGATGATCCGTCCATAATCTCAAACTA   400
                                 *
401  GCCAGATCAGAGCGATGCTGAGCAACGATGCGCGACACCG   440
         SD ORF3→M
441  TCGGTGCATTGATATTGTAAAGCCGAGCATTTCGGCCCCG   480

481  GACTTGCGGCCAGTGATGACGCTCTCGGCGATCTCGCGGC   520

521  GTTTGGCGGCATCGAGTTCTTGCGACGCCCACCGATGCGA   560

561  CTTTCGGCGCGGGCGGCTGCAAGGCCGGCGGAGGTCCGTT   600
                                     SD
601  CCCGGATCATGGCACGCTCGAATTCGGCGAAGCTGCCGAC   640
         ORF→4   M
641  CATCTGCATCATCATTCGGCCAGCCGGCGTCGTGGTGTCG   680

681  ATGTTCTCGGTTAGCGACCGGAAGCCTGCCCCGGCTTCCG   720

721  CGATGCGCTCCATGATGTGCAGCACGTCCTTCAGTGAGCG   760
                 *
761  TGACAGCCGGTCGAGCTTCCAGACAACGACGGTATCCCCC   800

801  TCCCGCAGATGATC                            814
               *
```

Figure 2. Nucleotide sequence of the 814 bp cloned DNA fragment from *Acidocella* strain GS19h plasmid. 'M' and '*' represent the underlined start and stop codons respectively; 'SD' represents the putative ribosome binding site; '↱' indicates direction of transcription.

3.5. Analysis of the deduced amino acid sequences

Deduced amino acid sequences (3 each from the forward and reverse directions of the nucleotide sequence) revealed that there are four complete open reading frames (ORFs) which are indicated in Figure 2, and are presented in Table 1. The putative proteins encoded by open reading frames (ORFs) were searched for similarity to other proteins from the EMBL database with program FASTA.

Table 1. Deduced amino acid sequences of the predicted ORFs

ORF	Nucleotides		Amino acid sequence
	from	to	
1	54	176	MAPPKSRASPSSAVPGNSASLSTRYAAFSAWPKEMNATVRT
2	271	387	MRSTRAGGMPLDRSAISTESGSARALKRRRLKHHRMIRP
3	415	732	MLSNDARHRRCIDIVKPSISAPDLRPVMTLSAISRRLAASSSCD AHRCDFRRGRLQGRRRSVPGSWHARIRRSCRPSASSFGQPAS WCRCSRLATGSLPRLPRCAP
4	609	809	MARSNSAKLPTICIIIRPAGVVVSMFSVSDRKPAPASAMRSMM CSTSFSERDSRSSFQTTTVSPSRR

ORF1 (Nucleotide sequence : 54-176)
The ORF1 codes for a small peptide of 41 amino acids. This ORF has a Shine-Dalgarno (SD)-like sequence (GGA) 12 base upstream of the ATG start codon. The putative polypeptide contains 8 serine(19.5%), 8 alanine (19.5%), 5 proline (12.2%), 3 arginine (7.3%). Apart from the N-terminal, it contains only another methionine residue.

ORF2 (Nucleotide sequence : 271- 387)
This ORF encodes a even smaller peptide with 39 amino acid residues, and is oriented in the same direction as ORF1. It has a SD like region (GGGAA) 13 base upstream of the start codon. This polypeptide also contains 9 arginine (23.1%), 5 serine (12.8%), 4 alanine (10.3%), 2 methionine (excepting N terminal), and 2 consecutive histidine residues.

ORF3 (Nucleotide sequence : 415-732)
This ORF encodes 106 amino acid residues,has a SD like sequence (AGAG), 2 bases upstream of the start codon and it is oriented in the same direction as ORF1. Notably, there are 21 arginine (19.8%), 17 serine (16%), 11 alanine (10.4%), 9 proline (8.5%), 7 cysteine (6.6%), and 3 histidine residues.

ORF4 (Nucleotide sequence : 609-809)

This ORF encodes only 67 amino acid residues and is oriented in the same direction as ORF1. It has a SD like sequence (GGAGG), 14 base upstream of the start codon. The putative polypeptide contains 15 serine (22.4%), 8 arginine (11.9%), 6 alanine (8.9%), 5 threonine (7.4%), and 2 cysteine residues.

4. DISCUSSION

The nucleotide sequence analyses of the 814 bp cloned DNA fragment from a plasmid of *Acidocella* strain GS19h show that the sequence does not have any homology with the conventional Cd^{2+}, Zn^{2+} or Cu^{2+} resistance genes. The deduced amino acid sequences also do not show any such similarity. It is however of interest to note that all the ORFs, especially ORFs 2 and 3 contain substantial amounts of the basic amino acid arginine. Additionally ORF3 contains 7 cysteine residues, the amino acid responsible for metal binding capacity of metallothioneins. As mentioned earlier, the nucleotide sequence corresponding to the ORF3 has homology with various genes including regulatory protein of mercury resistance operon. Therefore, it may be suggested that (i) in this acidophilic bacterium, bound arginine along with cysteine may play the same role as free histidine and metallothioneins do in other systems for relieving the cells from the heavy metal toxicity [22-25], and (ii) the resistance in this acidophilic heterotroph may arise via small proteins which have regulatory roles in gene expression.

REFERENCES

1. D. S. Holmes, J. R. Yates, J. H. Lobos and M. V. Doyle, Biotechnol. Appl. Biochem., 8 (1986) 258.
2. P. C. K. Lau, M. Drolet, P. Zanga and D. S. Holmes. In: A. E. Torma, M. L. Apel, and C.L. Brierley (eds.), Biohydrometallurgical technologies, TMS press, Warrendale, U.S.A, vol.2, 1993, p. 635.
3. D. E. Rawlings and S. Silver, Nature Biotechnol., 13 (1995) 73.
4. D. E. Rawlings and D. R. Woods. In: J. A. Thomson (ed.), Recombinant DNA and Bacterial Fermentation, CRC press Inc., Boca Raton, USA, 1988, p. 277.
5. D. E. Rawlings and D. R. Woods. In: C. Gaylarde, and H. Videla (eds.), Bioextraction and biodeterioriation of metals, Cambridge University Press, Cambridge, UK, 1995, p. 63.
6. T. Shiratori, C. Inoue, M. Numata and T. Kusano, Curr. Microbiol., 23 (1991) 321.
7. M. Mergeay, Trends Biotechnol., 9 (1991) 17.
8. E. Top, M. Mergeay, D. Springael and W. Verstraete, Appl. Environ. Microbiol., 56 (1990) 2471.
9. T. Kusano, G. Ji, C. Inoue and S. Silver, J. Bacteriol., 172 (1990) 2688.
10. D. E. Rawlings, R. A. Dorrington, J. Rohrer and A.-M. Clennel, FEMS Microbiol. Rev., 11(1993) 3.

11. P. A. W. Martin, P. R. Dugan and O. H. Tuovinen, Eur. J. Appl. Microbiol. Biotechnol., 18 (1983) 392.
12. T. F. Kondratyeva, L. N. Muntyan and G. I. Karavaiko, Microbiol., 141 (1995) 1157.
13. T. Shiratori, C. Inoue, K. Sugawara, T. Kusano and Y. Kitagawa, J. Bacteriol., 171 (1989) 3458.
14. T. Pramila, G. Ramananda Rao, K. A. Natarajan and C. Durga Rao, Curr. Microbiol., 32 (1996) 57.
15. K. Suzuki, N. Wakao, Y. Sakurai, T. Kimura, K. Sakka and K. Ohmiya, Appl. Environ. Microbiol., 63 (1997) 2089.
16. S. Ghosh, N. R. Mahapatra and P. C. Banerjee. Appl. Environ. Microbiol., 63 (1997) 4523.
17. P. C. Banerjee, M. K. Ray, C. Koch, S. Bhattacharyya, S. Shivaji and E. Stackebrandt, System. Appl. Microbiol., 19 (1996) 78.
18. T. A. Brown, T. Ikemura, M. McClelland and R. J. Roberts, Molecular biology labfax, BIOS Scientific Publishers Limited, Oxford, UK, 1991.
19. J. Sambrook, E. F. Fritsch and T. Maniatis, Molecular cloning : a laboratory manual, 2nd ed. Cold Spring Harbor Laboratory Press, Cold Spring Harbor, N.Y., 1989.
20. H. C. Birnboim and J. Doly. Nucleic Acids Res., 7 (1979) 1513.
21. ABI PRISM^{TM} 377 DNA Sequencing Analysis - Chemistry and Safety Guide, Perkin Elmer Corporation (1995) 2/9
22. L. Birch and R. Bachofen, Experientia, 46 (1990) 827.
23. D. H. Hamer, Annu. Rev. Biochem., 55 (1986) 913.
24. M. Joho, M. Inouhe, H. Tohoyama and T. Murayama, FEMS Microbiol. Lett., 66 (1990) 333.
25. U. Krämer, J. D. Cotter-Howells, J. M. Charnock, A. J. M. Baker and J. A. C. Smith, Nature, 379 (1996) 635.

Characterization of the genes encoding a cytochrome oxidase from *Thiobacillus ferrooxidans* ATCC33020 strain

C. Appia-Ayme, N. Guiliani and V. Bonnefoy

Laboratoire de Chimie Bactérienne, Institut de Biologie Structurale et de Microbiologie, C.N.R.S., Marseilles, France

Despite the importance of *Thiobacillus ferrooxidans* in bioremediation and bioleaching, we have only a rudimentary knowledge on its energetic metabolism. To characterize the *T. ferrooxidans* respiratory chains, we are studying the genes encoding electron transfer proteins.

In this paper, we report the sequence of the *cox* genes encoding the four sub-units of a cytochrome oxidase. Theses genes are located immediately upstream from the *rus* gene encoding rusticyanin and downstream from the *cyc2*, *cyc1* genes and ORF1, encoding respectively a high molecular weight cytochrome *c*, a c_4-type cytochrome (c_{552}) and a putative periplasmic protein of unknown function. Analysis of the polypeptides encoded by the four *cox* genes reveals different interesting points: (i) the higher homology with the sub-units I and II of *T. ferrooxidans* cytochrome oxidase is found with *Synechocystis* sp., *Anabaena* sp. and *Synechococcus vulcanus* enzyme, which is a likely candidate for the primordial oxidase; (ii) according to the topology of sub-unit I, the active site, where oxygen binds and is reduced to water, is located as in the other known cytochrome oxidases; (iii) nearly all the residues involved in the proton pathways are present in sub-unit I suggesting that this cytochrome is able to generate a proton motive force; (iv) this is a cytochrome and not a quinol oxidase since the residues involved in the copper binding site and in cytochrome *c* interaction are conserved (v) the molecular weight deduced from the sequence of *cox1* does not fit the molecular weight of the *T. ferrooxidans* aa_3 cytochrome oxidases sub-unit I characterized, suggesting the presence of two different cytochrome oxidases.

1. INTRODUCTION

Thiobacillus ferrooxidans is an extreme acidophilic Gram negative eubacterium able to derive all the energy required for growth from the oxidation of ferrous iron or reduced sulfur compounds present in ore. Its bioleaching and bioremediation abilities are due to its energetic metabolism.

Several electron transfer proteins have been identified in different *T. ferrooxidans* strains (see review 1). Even though most of these proteins have been very well characterized, there is no convincing argument to tell that they all belong to the same electron transfer chain and we do not know in which respiratory chain they are involved. Indeed, different pathways have been proposed in which the electron transporters characterized are not all present (2; 3; 4; 5; 6; 7). Furthermore, depending on the model, oxygen reduction occurs either in the periplasmic or the cytoplasmic side of the membrane (1; 2; 3; 4; 5; 6; 7). One could wonder whether this diversity reflects strain differences or if some of these redox components are involved in different respiratory chains. Therefore, it is important to obtain and to study the physiology of mutants in which these proteins are not anymore synthesized. As a preliminary approach, genes encoding these different redox proteins have been looked for and studied. The genes encoding rusticyanin (*rus*) (8), a high molecular weight cytochrome (*cyc2*) (7), a c_4-type cytochrome (*cyc1*) (7) and the HiPIP iron sulfur protein (*iro*) (9) from ATCC33020 strain have been already cloned by our group.

This paper reports the sequence of the genes encoding the four sub-units of a cytochrome *c* oxidase of *T. ferrooxidans* ATCC33020 strain.

2. MATERIAL AND METHODS

2.1. Strains, plasmids and growth conditions

The *T. ferrooxidans* ATCC33020 strain was obtained from the American Type Culture Collection. *T. ferrooxidans* was cultivated as described in (10).

2.2. DNA manipulations

General techniques were performed according to (11; 12). *T. ferrooxidans* genomic DNA preparation has been described in (8).

Inverse PCR was described in (13).

The DNA sequences were compiled, analysed and compared with the data banks through the Nestcape facilities.

3. RESULTS AND DISCUSSION

3.1. Sequencing of the region located downstream from the *cyc2*, *cyc1* and ORF1 genes

Recently, we have cloned and sequenced the *cyc1* and *cyc2* genes encoding respectively a c_4-type and a high molecular weight cytochrome. We have shown that these genes are cotranscribed with an ORF1 encoding a periplasmic protein of unknown function (7). Because the beginning of an open reading frame was found downstream from ORF1, we then wondered whether this operon could comprise other genes encoding more particularly other redox compounds or cytochrome *c* biogenesis proteins since cytochrome *c* biogenesis genes are often adjacent to cytochrome *c* structural genes. Chromosome walking by PCR and inverse PCR approaches downstream from ORF1 allowed us to find five open reading frames, the last one corresponding to the *rus* gene (8). The three open reading frames located downstream from ORF1 encode proteins related to the sub-units II, I and III of a cytochrome *c* oxidase and will be referred to now as *cox2*, *cox1* and *cox3* genes respectively. In eubacteria, the quinol and the cytochrome *c* oxidase complexes contain a small hydrophobic peptide, sub-unit IV, the sequence of which is not always conserved (14). When *cox2*, *cox1* and *cox3* genes are clustered, *cox4* gene is usually found downstream from *cox3* (15). By analogy with this genetic organization, we have supposed that the small open reading frame located downstream from *T. ferrooxidans cox3* gene encodes the fourth sub-unit of the cytochrome oxidase and will be tentatively referred to as *cox4*. The genes order is then *cyc2-cyc1*-ORF1-*cox2-cox1-cox3-cox4-rus* (Accession numberAJ006456). We will now focus on the four *cox* genes, the others having been described in (7; 8).

The *cox2* gene is located between position 3296 and 4060 and encodes a 254 amino-acid polypeptide of 28 240 Da. The *cox1* gene starts at position 4117 and ends at position 6000. The 627 amino-acid product it encodes has a molecular weight of 69 090 Da. The *cox3* extends from position 6019 to 6570 and the corresponding protein is 20202 Da and 183 amino-acids long. Finally, the 64 amino-acid putative fourth sub-unit of 7211 Da encoded by *cox4* gene is located between positions 6611 and 6805. These four genes are preceded by a correctly positioned ribosome binding site.

3.2. Analysis of the *cox1* encoded polypeptide

The primary structure of sub-unit I is the most conserved among the sub-units of the heme-copper cytochrome oxidase family (15; 16). Indeed, the polypeptide encoded by *T. ferrooxidans cox1* gene is related to quinol as well as cytochrome *c* oxidase sub-unit I from archaea, eucarya and bacteria. The higher score is obtained with cytochrome *c* oxidase sub-unit I from *Synechocystis* sp. strain PCC6803, *Anabaena* sp. strain PCC7120 and *Synechococcus vulcanus* (Figure 1).

T. ferrooxidans Cox1 consists of 14 transmembrane helices, that is the hydrophobic segment core common to all sub-unit I, and an extended N-terminus with 2 additional hydrophobic regions, presenting no significant homology in the protein data banks. The core region contains the active site, where oxygen binds and is reduced to water, and the proton conducting channels. The active site is binuclear with a pentacoordinated high-spin haem iron (haem a_3) and a copper (Cu_B). Electron transfer to this active site is mediated by a hexacoordinated low-spin haem (haem a). The histidines involved in binding these three metal atoms are conserved in *T. ferrooxidans* Cox1 sub-unit (H159, H333, H382, H383, H467 and H469), as well as the residues stabilizing Cu_B (W329 and Y337) and the residues involved in hydrogen bond with haem a or a_3 (W145, R529 and R530) (17) (Figure 1). It has to be pointed out that the localization of the 6 histidine residues involved in haems and copper binding is in agreement with the known Cox1 sub-units. This strongly suggests that the active site is like in other cytochrome oxidases in disagreement with the Kai *et al.* proposition (18; 19). Among the 2 residues involved in electron transfer between the two haems, only the invariant phenylalanine (F468) is present in *T. ferrooxidans* Cox1

(Figure 1). On the 2 residues coordinating the magnesium ion, which is supposed to play a role of communicator between sub-unit I and II, only the histidine (H459) is maintained (20). From the crystal structure of the *Paracoccus denitrificans* cytochrome *c* oxidase, Iwata *et al.* (21), Ostermeier *et al.* (22) and Hofacker and Schulten (23) have identified two proton transfer pathways, a "K" and a "D" channels. It has to be pointed out that none of the residues involved in these pathways is strictly invariant. Of the 7 residues supposedly involved in the "K" channel, 5 are conserved: Y337, T344, T407, S413 and T417 (Figure 1). On the other hand, out of the 9 residues likely forming the "D" channel, only 3 have been substituted, which are located at the entrance of the channel (Figure 1). This suggests that this *T. ferrooxidans* cytochrome oxidase is able to translocate protons from the cytoplasmic to the periplasmic side of the membrane.

```
(1)    1 MTIAAENLTANHPRRK--------WTDYFTFCVDHKVIGIQYLVTSFLFF    42
(2)    1 MTQAQLQETANIPARIEE-PGERHWRDYFGFNTDHKVIGLQYLVTSFIFY    49
(3)    1 --MAQAQLPLDTPLSLPEHPKAWKWYDYFTFNVDHKVIGIQYLVTAFIFY    48
(4)   74 MGSGVWEGWIRRAFGGKEAPTYTGIERYFRFGPDSKSAAVRYVILNILTF   123
         .   :    .            ** *   * *  .::*::   :: :

(1)   43 FIGGSFAEAMRTELATPSPD--FVQPEMYNQLMTLHGTIMIFLWIVPAGA    90
(2)   50 CIGGVMADLVRTELRTPEVD--FVSPEVYNSLFTLHATIMIFLWIVPAGA    97
(3)   49 LIGGLMAVAMRTELATADSD--FLDPNLYNAFLTNHGTIMIFLWVVPAAI    96
(4)  124 CFAGMAAMAIRIELLTPDSTSWWLSEIQYNQTFGIHGLMMLLGVVASAIV   173
         :.*    *  :* ** *..    ::.    **  :  *. :*::  :..*

(1)   91 A-FANYLIPLMVGTEDMAFPRLNAVAFWLTPPGGILLISSFFVG-APQAG   138
(2)   98 G-FANYLIPLMIGARDMAFPRLNAVAFWMIPPAGLLLIASLVVGDAPDAG   146
(3)   97 GGFGNYLVPLMIGARDMAFPRLNALAFWLNPPAGALLLASFLFG-GAQAG   145
(4)  174 GGVGYYLIPLMLGTRNVVFPKLLGLSWWLLPPATFAVFMSPTTG-GFQTG   222
         . .. **:***:*:.::.**:* .:::*: **.    :: *   * . ::*

(1)  139 WTSYPPLSLLSGKWGEELWILSLLLVGTSSILGAINFVTTILKMRIKDMD   188
(2)  147 WTSYPPLSLVTGQVGEGIWIISVLLLGTSSILGAINFLVTLLKMRIPSMG   196
(3)  146 WTSYPPLSTITATTAQSMWILAIILVGTSSILGSVNFIVTIWKMKVPSMR   195
(4)  223 WWGYPPLAQNSGS-GIVWYVLGAATILVASLLGAINIAGTMVYMRAKGMS   271
         * .****:  :.  .   :::.   :  .:*:*:**::*:  *:  *:  .*

(1)  189 LHSMPLFCWAMLATSSLILLSTPVLASALILLSFDLIAGTSFFN-PVGGG   237
(2)  197 FHQMPLFCWAMFATSALVLLSTPVLAAGLILLAFDLIAGTTFFN-PTGGG   245
(3)  196 WNQLPLFCWAMLATSLLALVSTPVLAAGLILLLFDINFGTSFYK-PDAGG   244
(4)  272 LGRVPIFVWGLFAAATTLVVESPATYTGALMDLSDMIAGSHFYTGPT--G   319
         :*:* *.::*::   ::.:*.  :. ::    *: *: *:. *   *

(1)  238 DPVVYQHLFWFYSHPAVYIMILPFFGVISEVIPVHARKPIFGYRAIAYSS   287
(2)  246 DPVVYQHMFWFYSHPAVYIMILPFFGAISEIIPIHSRKPIFGYKAIAYSS   295
(3)  245 NVVIYQHLFWFYSHPAVYLMILPIFGIMSEVIPVHARKPIFGYQAIAYSS   294
(4)  320 HPLAYLDQFWFLFHPEVYVFILPAFAIWLEILPAAAKRPLFA-RGWAIAG   368
         . : * . ***  ** **::*** *.    *::*   :::*:*. :. * :.
```

```
(1) 288 LA-ISFLGLIVWAHHMFTSGTPGWLRMFFMATTMLIAVPTGIKIFSWCGT 336
(2) 296 LA-ISFLGLIVWAHHMFTSGIPGWLRMFFMITTMIIAVPTGIKIFSWLAT 344
(3) 295 LA-ICCVGLFVWVHHMFTSGTPPWMRMFFTISTLIVAVPTGVKIFSWVAT 343
(4) 369 LVGVSMSGAMSGVHHYFTAVSDARMPIFMTI-TETVSIPTGFIYLSAIGT 417
         *. :.   *  :   .** **:       : :*:  :    :::***.  :*   .*

(1) 337 LWGGKIQLNSAMLFAFGFLSSFMIGGLTGVMVASVPFDIHVHDTYFVVGH 386
(2) 345 MWGGKIQFNSAMLFAAGFVGTFVIGGVSGVMLAAVPFDIHVHDTYFVVAH 394
(3) 344 LWGGKIRLNSAMLFAIGLMSMFVLGGLSGVTLGTAPVDIHVHDTYYVVAH 393
(4) 418 IWGGRLRINAAVLLVLMAMMNFLIGGLTGIFNADVPADLQLHNTYWVIAH 467
         :***:::*:*:*:.   :   *::**::*:   . .* *:::*:**:*:.*

(1) 387 FHYVLFGGSAFALFSGVYHWFPKMTGRMVNEPLGRLHFILTFIGMNLTFM 436
(2) 395 LHYVLFGGSVLGIFAAIYHWFPKMTGRMINEFWGKVHFALTIVGLNMTFL 444
(3) 394 FHYVLFGGSVFGLYAGIYHWFPKMTGRLLDERLGILHFVLTLIGTNWTFL 443
(4) 468 FHIRCFGGVIFTWIAALYWWFPKVTGRKINEFWGKFHAWWSFVFFNCTFF 517
         :*   ***    :  :.:* ****:*** ::*  *  .*    ::: * **:

(1) 437 PMHELGLMGMNRRIALYDVEFQPLNVLSTIGAYVLAASTIPFVINVFWSL 486
(2) 445 PMHKLGLMGMNRRIAQYDPKFTLLNEICTYGSYILAVSTFPFIFNAIWSW 494
(3) 444 PMHELGLKGMPRRVAMYDPQFEPVNLICTIGAFVLAFSIIPFLINIIWSW 493
(4) 518 PMFIAGLDGMNRRIAIYLPYLHDINLFMSISSFFLGAGFLIPLANLLYSW 567
         **.   ** ** **:* *    :   :* :  : .::.*. .  :    : * ::*

(1) 487 FKGEKAARNPWRALTLEWQTASP-PIIENFEEEPVLWCGPYD-FGIDTEL 534
(2) 495 LYGEKAGNNPWRALTLEWMTTSP-PAIENFDKLPVLATGPYD-YGLEKAS 542
(3) 494 NKGKIAGDNPWGGLTLEWTTSSP-PLIENWEVLPVVTKGPYD-YGIERRQ 541
(4) 568 RYGPKAEANPWGSNGLEWQIKSPTPYVPYPAGTEPEVVGPNDNYAAEAKD 617
          *  *  *** .  ***   ** * :       ** * :.  :

(1) 535 MDDEETVQTLIADAAGS---------------- 551
(2) 543 EGVPLSDPNPVLSAGPNSVLRAEPDEPYPTIES 575
(3) 542 ESTDEDHDE--QE-------------------- 552
(4) 618 PFIWVSTPSK---------------------- 627
```

Figure 1. Alignment of cytochrome c oxidase sub-unit I from *Synechocystis* sp. (1), *Anabaena* sp. (2), *Synechococcus vulcanus* (3) and *Thiobacillus ferrooxidans* (4). Residues involved in heme and copper binding are shown in bold face. Residues involved in proton channels are underlined." * ": identical residues. " : "and " . ": similar residues.

3.3. Analysis of the the *cox2* encoded polypeptide

The protein encoded by the *T. ferrooxidans cox2* gene presents a high homology to the sub-unit II of cytochrome c oxidases, and to a lower extend of quinol oxidases, from the three kingdoms of life. As with Cox1, the higher scores have been obtained with the cytochrome c oxidase sub-unit II from cyanobacteria (*S. vulcanus*, *Anabaena* sp. strain PCC7120 and *Synechocystis* sp. PCC6803) (Figure 2).

A putative signal sequence of 51 amino-acids could be predicted. The mature Cox2 has a molecular weight of 22 823 Da and contains 2 transmembrane helices and a large carboxy terminal hydrophilic domain, as in most sub-units II. The similarity is greater for the carboxy-terminal hydrophilic domain than for the amino-terminal transmembrane regions as expected.

In the periplasmic C-terminal domain, nearly all the ligands that have been implicated in the two copper atoms binding, are present (C222, C226, H181, M233, H230) (Figure 2) (24). Only the glutamate, whose carbonyl oxygen is a weak ligand of the second copper atom, has been substituted by a leucine (L224). This glutamate residue has been shown not to be a major ligand of copper (24). The residues which have been implicated from the structure of *P. denitrificans* cytochrome *c* oxidase, in shielding the CuA center from the solvent, are conserved in *T. ferrooxidans* Cox2 (W145 and D178) (Figure 2). This strongly suggest that Cox2 contains a dinuclear copper center and consequently is a cytochrome *c* and not a quinol oxidase. In agreement with this hypothesis, 3 of the 4 residues shown to be involved in cytochrome *c* interactions and which are highly conserved (Q144, D178, D193) are present in Cox2 (Figure 2) (25). It has to be noted that the fourth residue corresponds to the same glutamate residue described above as being involved in copper binding and which has been substituted by a leucine in Cox2$_{Tf}$.Furthermore, this glutamic acid residue coordinates the magnesium ion at the interface between Cox2 and Cox1 in *P. denitrificans* (20). The region of aromatic amino-acids at the beginning of the periplasmic C-terminal domain (145-WKWTFSY-151), which has been postulated to be involved in electron transfer between Cu$_A$ and Cox1 haem *a*, is also conserved (Figure 2) (26).

All together, these data are consistent with Cox2$_{Tf}$ being a cytochrome *c* oxidase and not a quinol oxidase.

```
(1)    1 -MKIPSSIWTLLIGIGLTLAS------LWYGQNHGLMPVAASDEADLVDG  43
(2)    1 -MKIPGSVITLLIGVVITVVS------LWYGQNHGLMPVAASADAEKVDG  43
(3)    1 MEQIPASIWTLTAGVVVTLIS------FWVGHHHGLLPEQASEQAPLVDN  43
(4)    1 MNAAKENLWKAFRGLVVVWIIGLAIFETLMAWGIGNWPILGSIQAHITAD  50
         .: .   *: :.           .   *  *  .* :*   . .

(1)   44 LFNTMMTVSAGIFLIVEGVLVYCVVKYRRRAGDHEDGPPVEGN--VPLEI  91
(2)   44 IFNYMMTIATGLFLLVEGVLVYCLIRFRRRKDDQTDGPPIEGN--VPLEI  91
(3)   44 FFDIMLTIGTALFLVVQGAIILFVIRYRRRAGEEGDGLPVEGN--LPLEA  91
(4)   51 ATTYLLWQAVFIYVLVGGAIVYSAFRFRASSMSDTAAPAYQKRTWAPFVV 100
         ::  .. ::::* *.::   .:*     .. . . :  . *:

(1)   92 LWTAIPAIIVIGISVYSFEVQIPQKQDAPGLGIVAPGIGSSPEKAGKPPE 141
(2)   92 LWTAIPTVIVFTLAVYSFEVSMGNMVAMAGDGDVALGIGLDSEEQGVNP- 140
(3)   92 FWTAIPALIVIFLGIYSVDIDATALLAAAQPP--EIGIGASPDVQGKAPD 139
(4)  101 TWLVLAIGINLANTIY----PG--MVGLEQLW----GIQLDT----KNP- 135
         *  .:.  *  :    :*              **  ..        *

(1)  142 LVVNVTGLQYAWIFTYPESGITTG-ELHVPIGREVQINMTANDVIHAFWV 190
(2)  141 LMVDVKGIQYAWIFTYPETGIISG-ELHAPIDRPVQLNMEAGDVIHAFWI 189
(3)  140 LVVDVAGMQYAWIFTYPDSGIVSG-ELHIPVGKDVQLNLSARDVIHSFWV 188
(4)  136 LVIDVTAQQWKWTFSYPKQGVTDVSQLVVPEGRTIYFVLRTKDVMHDFWV 185
         *:::*  . *:  *  *:**. *:    :*  *  .: : : :  : **:* **:
```

```
(1) 191 PEFRLKQDAIPGRQTEMRFTPKTAGD------YTLICAELCGPYHGAMRT 234
(2) 190 PQLRLKQDVIPGRGSTLVFNASTPGQ------YPVICAELCGAYHGGMKS 233
(3) 189 PQFRLKQDAIPG-VPTTRFKATKVGT------YPVVCAELCGGYHGAMRT 232
(4) 186 PAWGEKKDVIPNEVRHLFITPTMLGTTATNPMLRVQCSLICGNGHPLMRA 235
          *      *:*.**.       :...  *       : *: :**  *  *::

(1) 235 QVVVEPEEAFKKWTQEQLA...
(2) 234 VFYAHTPEEYDDWVAANAP...
(3) 233 QVIVHTPEDFETWRRQNQ-...
(4) 236 PVKVVTPADFKAWVANNSF 254
          .  .   :.  *    :
```

Figure 2. Alignment of cytochrome *c* oxidase sub-unit II from *Anabaena* sp. (1), *Synechocystis* sp. (2), *Synechococcus vulcanus* (3) and *Thiobacillus ferrooxidans* (4). Residues involved in copper binding are shown in bold face. Those involved in electron transfer are shadowed and those involved in cytochrome *c* interaction are underlined." * ": identical residues. " : " and " . ": similar residues.

3.4. Analysis of the *cox3* and the *cox4* encoded polypeptides

Low but significative scores were obtained with Cox3 from different organisms, such as *Mycobacterium leprae*, *Mycobacterium. tuberculosis*, *Anabaena* sp. strain PCC7120, *S. vulcanus*, *Synechocystis* sp. strain PCC6803, the mitochondrion of *Mytilus edulis*, and of *Mytilus trossulus* (Figure 3). The typical secondary structure of the cyanobacteria and mycobacteria Cox3 cited above, is maintained. Indeed, they have only 5 transmembrane regions instead of 7 and the 2 missing amino-terminal transmembrane helices of these Cox3 are not fused to Cox1, as it was observed in some bacteria (16). On the 30 residues which are generally conserved in this region of the cytochrome oxidase sub-units III (27), 8 are conserved and 11 are similar in $Cox3_{Tf}$ (Figure 3). The sub-unit III of the aa_3-type cytochrome oxidase from *T. ferrooxidans* is thus poorly conserved.

```
(1)  1 -----------------------------------------MTS-----TVGT  7
(2)  1 -----------------------------------------MTS-----AVGT  7
(3)  1 ----------------------------MQSQTIDPAKT-----ELNH 15
(4)  1 ---------MTSPMGPTARDFWQNSRNSATIQRLIVSPMQTTALSSDLNS 41
(5)  1 -------------------------------------MQG-----TVES  7
(6)    ..PSPWPFFVAISANGMAVGLILWLHRTPSFLLMGMSLGCMLLSTFSWWRDL
(7)    ..PSPWPFFVGISANGLAVGLILWLHRSPSYLLMGMSISCMMLSTFSWWRDL
(8)    -------------------------------------------------

(1)  8 LGTAITSRVHSLNRPNMVSVGTVVWLSSELMFFAGLFAMYFTARAQ---- 53
(2)  8 SGTAITSRVHSLNRPNMVSVGTIVWLSSELMFFAGLFAFYFSARAQ---- 53
(3) 16 HHT-AEAVGHHEEHPDHRLFGLFVFLVAEGMIFLGLFGAYLAFRS----- 59
(4) 42 TYTPGEAHGHHG-HPDLRMFGVVLFLVAESAIFLGLFTAYLIYRS----- 85
(5)  8 QGT-AIAVDHAHEHPDFRVLGLLVFLISESLMFGGLFAAYLLLRG----- 51
(6)    IRE-GDIGFHTRFVIKSFRDGVALFILSEVMFFSFFWTFFHNALSPSCE
(7)    IRE-GDMGFYTRFVIKSFRDGVALFILSEVMFFSFFWAFFHNALSPSCE
(8)  1 ---------MTDNSYAKLMDPASERAKRGAFFFLMLFAAIIFAMWD---L 38
                                       .*  :*
```

```
(1)  54 AGGKWPP---STELNLYQAVPVTLVLIASSFTCQMGVFSAERGDV-FGLR  99
(2)  54 AGGNWPPP--PTELNLYQAVPVTLVLIASSFTCQMGVFAAERGDI-FGLR 100
(3)  60 TLPVWPPA-GTPELELLLPGVNTVNLIASSFVMHNADTAIKKNDT-KGMR 107
(4)  86 VMPAWPPE-GTPELELLLPGVNSIILISSSFVMHKGQAAIRNNDN-AGLQ 133
(5)  52 MHEQWPPE-GT-EVELFVPTINTLILISSSFVIHYGDVAIKKDDV-RGMR  98
(6)     LGMRWPPPGIRTPNPSSTSLFETGLLISSGYSVTQAHKSMRLKDYDVGPF
(7)     LGMRWPPPGIRTPNPSSTSLFETGLLISSGLFVTQAHKSMRLKDYDVGPF
(8)  39 ARFLWGHS-VPATLSMGVGVALTVLMLVSLVPVMTARKKLDQGDD-AGIV  86
           *                   :  :: *        .      :

(1) 100 RWYVITLLMGLFFVLGQGYEYYHLITHGTTIPSSAYGSVFYLATGFHGLH 149
(2) 101 RWYVITFLMGLFFVLGQAYEYRNLMSHGTSIPSSAYGSVFYLATGFHGLH 150
(3) 108 TWLAITAAMGAIFLVGQVYEYTHLEFG---LTTNLFASAFYVLTGFHGLH 154
(4) 134 KWFGITAAMGIIFLAGQMYEYFHLEMG---LTTNLFASCFYVLTGFHGLH 180
(5)  99 KWYWITAAMGAVFLGGQVYEYLTLGYG---LRTNVFANCFYVMTGFHGLH 145
(6)     IGLVVTILCGTVFFLVQLREYYWNSYT---IADSVYGSVFYLLTGFHGMH
(7)     IGLLMTILCGAVFFLVQVREYCWNSYS---IADSVYGSVFYLLTGFHGAH
(8)  87 SSLATLMVVSLVMAGGIVYNWTTLTIG----SG--YGGIYDITSLWFLVH 130
            . .:         ::             ..  : : :  :.   *

(1) 150 VTGGLIAFIFLLARTTMS-KFTPAQATASIVVSYYWHFVDIVWIALFTVI 198
(2) 151 VTGGLIAFIFLLVRTGMS-KFTPAQATASIVVSYYWHFVDIVWIALFTVI 199
(3) 155 VTIGVLAIVAVLWRSRTQGHYSSEKHFGIEAAEIYWHFVDVIWIILFGLL 204
(4) 181 VTFGLLLILSVLWRSRQPGHYSRTSHFGVEAAELYWHFVDVVWIVLFILV 230
(5) 146 VFIGILLILGVIWRSRRPGHYNAQKHTGVAMAEIYWHFVDVIWIILFTLL 195
(6)     VVVGTLWLMVSLVRLWRG-EFSSQRHFGFEACIWYWHFVDVVWVALWCLV
(7)     VVVGTIWLMVSLVRLWRG-EFSSQRHFGFEACIWYWHFVDVVWVSLWFVV
(8) 131 FVAAILALLASIMKITRTPERAKRERWVSYNVLTFWGGVIVLWVAFFIVF 180
            .  .  ::     :      .           :*   *  ::*: :: :.

(1) 199 YFIR--------------------- 202
(2) 200 YFIR--------------------- 203
(3) 205 YLL--------------------- 207
(4) 231 YLL--------------------- 233
(5) 196 YILTRF------------------- 201
(6)     YVWFGGWLYMWWFKMWDGDVYTFKYP
(7)     YVWFGGWLYMWWFKMWDGDIYSFKYP
(8) 181 YIA--------------------- 183
         *.
```

Figure 3. Alignment of cytochrome *c* oxidase sub-unit III from *Mycobacterium leprae* (1), *Mycobacterium tuberculosis* (2), *Anabaena* sp. (3), *Synechocystis* sp. (4), *Synechococcus vulcanus* (5), *Mytilus trossulus* (mitochondrion) (6), *Mytilus edulis* (mitochondrion) (7) and *Thiobacillus ferrooxidans* (8). Residues usually conserved in sub-unit III are shown in bold face." * ": identical residues. " : " and " . ": similar residues.

In the case of the ORF found downstream from *cox3*, no significative homology was obtained in the protein data banks. Now, the bacterial cytochrome *c* oxidase complexes contain usually four sub-units whose genes constitute an operon with the order *cox2-cox1-cox3-cox4*. Then, this ORF has been proposed to be *cox4*. It has to be pointed out that the sub-unit IV of the *Paracoccus denitrificans* cytochrome *c* oxidase, which has been shown to copurify and to cocrystallise with the oxidase, presents only one transmembrane region. Furthermore, as for $Cox4_{Tf}$, no similarity was detected with the known sub-units of quinol and cytochrome c oxidases (14).

4. CONCLUSION

The polypeptides encoded by the *cox* genes described in this paper belong to the heme-copper oxidase superfamily since amino-acid similarity is found not only with sub-unit I, but also with sub-units II and III. Two different cytochrome oxidases (an aa_3-type and an a_1-containing oxidases) have been detected in several *T. ferrooxidans* (28; 29; 30). Furthermore, a aa_3 cytochrome oxidase has been purified from Fe-1, AP19-3 and OK1-50 strains (19; 32). However, the sub-unit I molecular weight of these cytochrome oxidases (53, 53 and 55 kDa, respectively) does not fit with the molecular weight deduced from the *cox1* gene sequence reported here (69 kDa). This suggests that the cytochrome oxidase described in this paper does not correspond to the ones already characterized.

The sub-unit I topology shows that the catalytic site is located as in the other known oxidases in disagreement with the electron transfer chain model of Yamanaka *et al.* (4). Since nearly all the residues involved in the proton pathways channels "D" and "K" are conserved, this cytochrome oxidase is likely a proton pump and thus is able to generate a proton motive force coupled to energy conservation.

In sub-unit II, not only the ligands of the two copper atoms are present but also some of the residues involved in the interaction with cytochrome c indicating that the physiological electron donor of this oxidase is a cytochrome c and not a quinol. Interestingly, the genes encoding this cytochrome oxidase are clustered with *cyc1* and *cyc2* encoding two c-type cytochromes in ATCC33020 *T. ferrooxidans* strain. This suggests that one of these cytochromes is the electron donor of the cytochrome oxidase. Furthermore, the *rus* gene is present at the same locus. We propose that the rusticyanin, the two c-type cytochromes and the cytochrome oxidase belong to the same respiratory chain. From the data presented in this paper and in (7) the following model can be postulated:

Fe^{2+} -----> ? ------> rusticyanin ------> $Cytc_4$ ------> Cytochrome oxidase.

REFERENCES

1. T. Yamanaka and Y. Fukumori, FEMS Microbiol. Rev., 17 (1995) 401.
2. W.J. Ingledew, J.C. Cox and P.J. Halling, FEMS Microbiol. Lett., 2 (1977) 193.
3. W.J. Ingledew and J.G. Cobley, Biochim. Biophys. Acta, 590 (1980) 141.
4. T. Yamanaka, T. Yano, M. Kai, H. Tamegai, A. Sato and Y. Fukumori, (Y. Mukohata, ed.), New Era of Bioenergetics, Academic Press, Tokyo, pp: 223 (1991).
5. R.C. Blake II and E.A. Shute, Biochemistry 33 (1994) 9220.
6. M.-T. Giudici-Orticoni, W. Nitschke, C. Cavazza, and M. Bruschi, (The Australian Mineral Foundation, ed.), Biomine pp. PB4.1 (1997).
7. C. Appia-Ayme, A. Bengrine, C. Cavazza, M.-T. Giudici-Orticoni, M. Bruschi,

M. Chippaux and V. Bonnefoy, FEMS Microbiol. Lett. 167 (1998) 171.

8. A. Bengrine, N. Guiliani, C. Appia-Ayme, E. Jedlicki, D.S. Holmes, M. Chippaux and V. Bonnefoy, Biochim. Biophys. Acta 1443 (1998) 99.

9. L. Cassagnaud, J. Ratouchniak, M. Chippaux and V. Bonnefoy, Unpublished results.

10. N. Guiliani, A. Bengrine, F. Borne, M. Chippaux, and V. Bonnefoy, Microbiology, 143 (1997) 2179.

11. F.M Ausubel, R. Brent, R.E. Kingston, D.D Moore, J.G. Seidman, J.A. Smith and K. Struhl, Current protocols in molecular biology, New-York: Greene publishing, (1992).

12. J. Sambrook, E.F. Fritsch and T. Maniatis, Molecular cloning; a laboratory manual Cold Spring Harbor Laboratory, Cold Spring Harbor N.Y. (1989).

13. H. Ochman, M.M. Medhora, D. Garza and D.L. Hartl, (M.A. Innis, D.H. Gelfand, J.J Sninsky, T.J. White, eds.), PCR protocols, a guide to methods and applications, Academic Press, Inc. Harcourt Brace Jovanovich, Publishers, pp. 219 (1990).

14. H. Witt and B. Ludwig, J. Biol. Chem. 272 (1997) 5514.

15. J.A. Garcia-Horsman, B. Barquera, J. Rumbley, J. Ma and R.B. Gennis, J. Bacteriol. 176 (1994) 5587.

16. M. Saraste, J. Castresana, D. Higgins, M. Lubben and M. Wilmanns, H. Baltscheffski (ed.), Origin and evolution of biological energy conversion, V.C.H., New-York, 1994, 255.

17. J.P. Shapleigh, J.P. Hosler, M.M.J. Tecklenburg, Y. Kim, G.T. Babcock, R.B. Gennis and S. Ferguson-Miller, Proc. Natl. Acad. Sci. USA, 89 (1992) 4786.

18. M. Kai, T. Yano, Y. Fukumori and T. Yamanaka, Biochem., Biophys. Res. Comm. 2 (1989) 839.

19. M. Kai, T. Yano, H. Tamegai, Y. Fukumori and T. Yamanaka, J. Biochem. 112 (1992) 816.

20. B.E. Ramirez, B. G. Malmstrom, J.R. Winkler and H.B. Gray, Proc. Natl. Acad. Sci., USA 92 (1995) 11949.

21. S. Iwata, C. Ostermeier, B. Ludwig and H. Michel, Nature 376 (1995) 660.

22. C. Ostermeier, S. Iwata and H. Michel, Curr. Opin. Struct. Biol. 6 (1996) 460.

23. I. Hofacker and K. Schulten, Proteins 30 (1998) 100.

24. M. Kelly, P. Lappalainen, G. Talbo, T. Haltia, J. van der Oost and M. Saraste, J. Biol. Chem. 268 (1993) 16781.

25. P. Lappalainen, N.J. Watmough, C. Greenwood and M. Saraste, Biochemistry 34 (1995) 5824.

26. G.J. Steffens and G. Buse, Hoppe-Seyler's Z. physiol. Chem. 360 (1979) 613.

27. M. Saraste, Q. Rev. Biophys. 23 (1990) 331.

28. J.G. Cobley and B.A. Haddock, FEBS Lett. 60 (1975) 29.

29. W.J. Ingledew and J.G. Cobley, Biochim. Biophys. Acta 590 (1980) 141.

30. R. Mansch and W. Sand, FEMS Microbiol. Lett. 92 (1992) 83.

31. K. Iwakori, K. Kamimura and T. Sugio, Biosci. Biotechnol. Biochem. 62 (1998) 1081.

Genetic transfer of IncP, IncQ, IncW plasmids to four *Thiobacillus ferrooxidans* strains by conjugation

Z. Liu, F. Borne, J. Ratouchniak and V. Bonnefoy

Laboratoire de Chimie Bactérienne, Institut de Biologie Structurale et de Microbiologie, C.N.R.S., Marseilles, France

The physiological function of some proteins involved in *Thiobacillus ferrooxidans* energetic metabolism has been difficult to assess. Our long term goal is to construct and to analyse mutants in which these proteins are no longer synthesized in order to determine in which respiratory chains they are involved. However, up to now, there is no report on the construction of mutants in this microorganism because of the lack of genetic tools. Recently, the transfer by conjugation from *Escherichia coli* of IncP plasmids and by mobilization of IncQ plasmids into private *T. ferrooxidans* strains has been reported, but the transfer frequency was very low. We have extended this research by establishing a genetic transfer system by conjugation of IncQ (pJRD215), IncP (pJB3Km1) and IncW (pUFR034) group plasmids from *Escherichia coli* not only to two private *T. ferrooxidans* strains (BRGM and Tf-49) but, more interestingly, to the two collection strains ATCC33020 and ATCC19859 from which most of the *T. ferrooxidans* genes characterized to date originate. IncQ plasmid pJRD215 had the higher transfer frequency and was stable in these four *T. ferrooxidans* strains. The conjugation conditions were optimized for the ATCC33020 strain by analysing the influence on the transfer frequency of: (i) the the donor strain (ii) the plasmid to be transferred (iii) the ratio donor/recipient cells; (iv) the adaptation of the donor and the recipient cells to the mating medium; (v) the composition of the mating medium. Under the best conditions, the transfer frequency could be increased from 10^{-7} to 5.10^{-5}.

1. INTRODUCTION

Thiobacillus ferrooxidans is an acidophilic chemolithoautotrophic bacterium that obtains its energy by oxidation of ferrous iron (Fe^{2+}) to ferric iron (Fe^{3+}) or reduced sulfur compounds to sulfuric acid. Because of this Fe^{2+} oxidizing activity, *T. ferrooxidans* has gained interest in mineral leaching. Furthermore, *T. ferrooxidans* is an attractive microorganism to study for its energetic metabolism. Considerable progress in understanding the physiology of this microorganism has resulted in the study of the biochemistry and the molecular biology of the bioenergenetics of iron oxidation. Many proteins involved in electron transport have been already characterized, and some related genes have been cloned and sequenced. However, their exact function still needs confirmation. The understanding of the physiological role of these *T. ferrooxidans* proteins can be facilitated by the analysis of mutants in which these proteins are no longer synthesized. Since genetic tools are not available in *T. ferrooxidans*, a marker exchange mutagenesis program has been initiated in our group. For this purpose, a genetic transfer system between *Escherichia coli* and *T. ferrooxidans* needs to be established. The introduction of plasmids into *T. ferrooxidans* private strains has been reported by electrotransformation (1) and by conjugation (2) but not into *T. ferrooxidans* collection strains. In this study, we have established the conditions for conjugation from *E. coli* to two ATCC strains (ATCC33020 and ATCC19859) and two private strains (BRGM and Tf-49) with plasmids from three incompatibility groups.

2. MATERIALS AND METHODS

2.1. Bacterial strains and plasmids

E. coli TG1 [*traD36 lacIΔ(lacZ-proAB) Δ(hsdM-mcrB) thi*] was used as the host strain for routine plasmid manipulation. *E. coli* HB101 [*Δ(gpt-proA) leuB6 ara14 galK2 lacY1 rpsL20 xyl5 mtl1 recA13 thi1 lacY1 hsdS$_B$20 supE44 mcrB$_B$*], MOS Blue [*endA1, hsdR17(rk⁻, mk⁻) supE44, thi1, gyrA46, recAI, lac/F'(lacIqZΔM15, proAB Tn10)*], MC1061 [*araD139 Δ(ara-leu)$_{7697}$ galE15 galK16 Δ(lac)$_{X74}$ rpsL hsdR2 mcrA mcrB1*] and S17.1 [*recA pro hsdR* (RP4-2, Tc::Mu, Km::Tn7)] (3) strains carrying the different broad-host range conjugative or mobilisable plasmids were used as donor strains in conjugation experiments. The *T. ferrooxidans* ATCC19859 and ATCC33020 strains were obtained from the American Type Culture Collection. The BRGM strain used in this study was kindly provided by Dr. Morin from Bureau des Recherches Géologiques et Minières (Orléans, France). The Tf-49 strain was isolated from a coal mine in Yunqin (China) (4). The IncQ pKT240 (5) and pJRD215 (6), the IncW pUFR034 (7) and the IncP pJB3Km1 (8) mobilizable plasmids were used.

2.2. Media and growth conditions

E. coli strains were usually grown in L-broth according to (9). *T. ferrooxidans* was cultivated as described in (10). The composition of the 2:2 solid medium is reported in (4). This medium has been improved to give the "DOP" medium which is described in the Results section. Kanamycin concentration in 2:2 medium is 200µg/ml.

2.3. Conjugation

At first, conjugation experiments were performed according to Peng *et al.* (2). Some modifications were added which are presented in the Results section. The *E. coli* donor strain was grown in 2:2 medium supplemented with 0.5% (w/v) yeast extract until the late exponential growth phase in the presence of one selective antibiotic. *T. ferrooxidans* were usually cultivated in sulfur 9K liquid medium for 5 days to reach stationary phase. Cells were collected by centrifugation and washed three times with basal salt medium (2:2 medium without energy source). The donor and recipient cells were mixed at a ratio of 1:2, and 0.1 ml of cell suspension (approximately 2×10^9 cells per ml) was spotted on the mating medium consisting of the 2:2 solid medium (0.8% agar) supplemented with 0.05% (w/v) yeast extract, 5×10^{-2} mM diamino pimelic acid and 0.05% (w/v) $Na_2S_2O_3$. After 5 days incubation, cells were suspended in 1.5ml of basal salt medium, and plated on 2:2 or "DOP" solid medium with or without kanamycin. The apparent transfer frequency is the ratio between the number of transconjugants (kanamycin resistant clones) and the number of recipient cells obtained on the non selective medium.

2.4. DNA analysis

Plasmid DNA from *T. ferrooxidans* was obtained using the Wizard DNA purification system from Promega. Restriction enzyme digestions were performed according to the manufacturer recommendation.

2.5. Stability analysis

A single colony of *T. ferrooxidans* conjugant was inoculated into 20ml of 9K liquid medium in the absence of antibiotic, shaken at 30°C to late-exponential-growth phase, and diluted 1/1000 into 9K liquid medium without antibiotic. Growth was continued to late-exponential phase, and the dilution-growth procedure repeated for three cycles. One aliquot was taken at the beginning of each cycle, diluted and plated onto solid 2:2 medium with or without kanamycin. Plasmid stability was calculated as the number of colonies on medium with kanamycin divided by that on the medium without kanamycin.

3. RESULTS

3.1. Search for a solid medium

To establish a genetic transfer by conjugation into *T. ferrooxidans* ATCC33020 strain, a solid medium was required. Different solid media were tested using agar or agarose as solidifying agent. We were able to get isolated clones only with the 2:2 medium described in (4). However, even after two months incubation at 30°C, the colonies remained small. To decrease the incubation time, different compounds were added in the 2:2 medium. Their effect is summarized in Table 1.

Table 1. Effect of different compounds on the yield and on the growth of the *T. ferrooxidans* colonies.

Product:	Concentration:	Clone number obtained with the following dilution:			Size of the colonies:
		10^{-5}	10^{-6}	10^{-7}	
2:2		ND	14	0	small
2:2 basal salts	3x	100	15	0	big
	6x	152	19	0	big
	10x	156	14	1	big
glycerol	1%	0	0	0	
	3%	0	0	0	
	6%	0	0	0	
	10%	0	0	0	
Casamino-Acids	0.7%	0	0	0	
Tryptophan	1 mM	125	19	1	tiny
	2 mM	134	11	0	tiny
	4mM	148	14	0	tiny
	20 mM	190	17	0	tiny
Cysteine	0.33 mM	162	15	1	small
	0.66 mM	183	26	3	small
	1.32 mM	100	18	2	small
	6.6 mM	150	0	0	small
Methionine	2.7 mM	39	0	0	
	5.4 mM	0	0	0	
	10.8 mM	1	1	0	
	54 mM	0	0	0	
Leucine	3 mM	161	19	2	big
	6 mM	149	12	0	big
	12 mM	175	25	0	big
	60 mM	170	12	2	big

Diamino	2.5 mM	170	22	1	big
Pimelic Acid	5 mM	186	17	1	big
	10 mM	180	17	2	big
	50 mM	159	27	0	big

ND: not determined.

A non fermentable carbon source such as glycerol, aromatic amino-acid such as tryptophan or potential nitrogen source such as methionine and cysteine, inhibited ATCC33020 growth on 2:2 medium. However, an increase in the basal salt concentration, addition of leucine and diaminopimelic acid, which is a component of the cell wall, improved the growth and the cell yield. By increasing the basal salt concentration and by adding leucine and diaminopimelic acid to the 2:2 medium, ATCC33020 clones appeared after about ten days incubation at 30°C. This medium will now be referred as the "DOP" medium.

3.2. Conjugative transfer of IncP plasmids to *T. ferrooxidans* ATCC33020 strain

Although the conjugation between *E. coli* and *T. ferrooxidans* has been reported (2), until now there is no report on conjugation from *E. coli* to any collection strains of *T. ferrooxidans*. Conjugation experiments between HB101 or MOS blue *E. coli* strains carrying IncP plasmid (RP4 or pRK2013) and ATCC 33020 were performed according to the method of Peng *et al.* (2). Kanamycin resistance was chosen as the selection marker since ATCC33020 has been shown to be sensitive to this antibiotic. Few kanamycin resistant clones were obtained but the frequency was lower than 10^{-8}.

3.3. Mobilization of IncQ plasmids by RP4 plasmid to ATCC33020

The broad-host-range pKT240 and pJRD215 mobilizable plasmids were chosen because they have been shown to replicate in *T. ferrooxidans* (1; 2) and because they carry distinct kanamycin resistance gene from different origins: Tn903 in the case of pKT240 and Tn5 in the case of pJRD215. Mobilization of these IncQ plasmids to the *T. ferrooxidans* ATCC33020 strain was tested with the helper plasmid RP4 either in a three partners or a two partners conjugation. In the first case, two *E.coli* strains were used: one (HB101) carrying the helper plasmid, RP4, and the second (MC1061) carrying the mobilizable plasmid, pKT240. In the second case, the *E. coli* donor strain (S17.1) carried the mobilisable pKT240 plasmid and has the helper plasmid RP4 integrated in its chromosome (3). In all cases, the transfer frequency was about 10^{-7}. Several parameters were then evaluated on the mobilization of pJRD215 plasmid from *E. coli* S17.1 strain to *T. ferrooxidans* ATCC33020.

3.4. Effect of the ratio between donor to recipient cells

The effect of the ratio between donor and recipient cells is given in the Table 2.

Table 2. Apparent transfer frequency according to the ratio between donor and recipient cells.

ratio E. coli /T. ferrooxidans:	Apparent transfer frequency :
1/4	$4.1\ 10^{-6}$
1/3	$7\ 10^{-6}$
1/2	**$1.2\ 10^{-5}$**
1/1	$2\ 10^{-6}$
2/1	$1.4\ 10^{-6}$
4/1	$6\ 10^{-7}$

A donor to recipient ratio of 1/2 gave the best result.

3.5. Effect of the growth medium

By adapting *E. coli* S17.1/pJRD215 strain to the 2:2 liquid medium supplemented with 0.5% yeast extract, the transfer frequency was increased from $0.7\ 10^{-6}$ to $2.0\ 10^{-6}$ per recipient cell.

Since energy is required during the mating process, the effect of the energy source was analysed. Furthermore, we have tested the adaptation of *T. ferrooxidans* to the 2:2 medium before the mating process. The results are presented in Table 3.

Table 3. Effect of the *T. ferrooxidans* growth medium on the apparent transfer frequency.

T. ferrooxidans growth medium:		Apparent transfer frequency:
basal salt:	energy source:	
9K	$FeSO_4$	10^{-6}
9K	**S_0**	**$3.2\ 10^{-5}$**
2:2	$Na_2O_3S_2 + FeSO_4$	$2.6\ 10^{-6}$

A higher transfer frequency was observed when a sulfur compound (S_0, $Na_2O_3S_2$) was present as an energy source, which is not unexpected since sulfur is known to be more energetic than ferrous iron (11). However, adaptation of *T. ferrooxidans* to the mating medium did not improve the frequency.

3.6. Effect of the mating medium composition

The physiological state of the cells during the conjugation process is an important factor that could influence the transfer frequency. The composition of the mating medium was modified in order to improve *E. coli* and *T. ferrooxidans* growth.

pH. The effect of the pH of the mating medium on the transfer frequency was tested from 4.0 to 6.0 (Table 4).

Table 4. Effect of the pH of the mating medium on the apparent transfer frequency.

pH of the mating medium:	Apparent transfer frequency:
4.3	10^{-7}
4.6	**1.8 10^{-6}**
4.8	**2 10^{-6}**
5	**2.2 10^{-6}**
5.2	**2 10^{-6}**
5.6	10^{-6}
6	10^{-7}

Higher transfer frequencies were obtained when the pH of the mating medium ranged from 4.6 to 5.2.

Thiosulfate. Thiosulfate is an energy source for *T. ferrooxidans* but could be also an inhibitor for *E. coli*. Different concentrations of thiosulfate were tested (Table 5).

Table 5. Effect of the thiosulfate concentration of the mating medium on the apparent transfer frequency.

Thiosulfate concentration (w/v):	Apparent transfer frequency:
0%	5,3 10^{-6}
0.05%	**6.9 10^{-6}**
0.1%	2.8 10^{-6}
0.2%	2 10^{-6}

The transfer frequency decreased as the concentration of thiosulphate increases. The concentration of 0.05% thiosulfate gave the highest transfer frequency. In the absence of thiosulfate, which is the *T. ferrooxidans* energy source, conjugation could occur. This was already observed (2).

Diaminopimelic acid. Addition of diaminopimelic acid (DAP), a component of the cell wall, in the 2:2 solid medium has been shown not only to increase the cell number but also the growth rate ("DOP" medium, see above). The effect of DAP on the transfer frequency was analysed. The results are shown in Table 6.

Table 6. Effect of diaminopimelic acid (DAP) on the apparent transfer frequency.

DAP concentration:	Apparent transfer frequency:
0	0.6 10^{-6}
0.25 10^{-4} M	6 10^{-6}
0.5 10^{-4} M	**8 10^{-6}**
10^{-4} M	4 10^{-6}
2 10^{-4} M	2 10^{-6}

In presence of $0.5 \ 10^{-4}$ M diaminopimelic acid, the apparent transfer frequency was increased about 8 times.

3.7. Conjugation with IncQ, IncW and IncP mobilizable plasmids in ATCC33020, ATCC19859, BRGM and Tf-49 strains

Now that genetic transfer of pJRD215 from *E. coli* to *T. ferrooxidans* by conjugation had been demonstrated, genetic transfer of mobilizable plasmids from three different incompability groups, IncP (pJB3Km1 (8)), IncQ (pJRD215) (6) and IncW (pUFR034 (7)) from *E. coli* to four *T. ferrooxidans* strains was tested. The two collection strains ATCC33020 and ATCC19859, from which originate most of the *T. ferrooxidans* genes characterized to date, and the two private strains BRGM and Tf-49 were chosen as recipients. The results are presented in Table 7.

Table 7. Apparent transfer frequency of pJRD215, pJB3Km1 and pUFR034 to ATCC33020, ATCC19859, BRGM and Tf-49 *T. ferrooxidans* strains.

donor: recipient:	pJRD215 (IncQ)	pJB3Km1 (IncP)	pUFR034 (IncW)
ATCC33020	$4.4 \ 10^{-5}$	$1.2 \ 10^{-6}$	$5 \ 10^{-6}$
ATCC19859	$1.4 \ 10^{-3}$	$6.2 \ 10^{-7}$	10^{-6}
BRGM	$2.5 \ 10^{-3}$	$1.3 \ 10^{-5}$	$2.2 \ 10^{-6}$
Tf-49	$2 \ 10^{-4}$	$2.7 \ 10^{-6}$	$6.8 \ 10^{-6}$

The IncP plasmid pJB3Km1 and IncW plasmid pUFR034 were transferred from *E. coli* S17.1 to the four *T. ferrooxidans* strains tested at a frequency depending on the recipient strain. The transfer frequencies of IncQ plasmid (pJRD215) were higher than the frequencies obtained with the two other incompatibility groups.

3.8. Transconjugants analysis

In all the conjugation experiments, eight kanamycin resistant clones were analysed.

In all the cases, a 283bp fragment internal to the *rus* gene, which is specific to *T. ferrooxidans*, was amplified by PCR. On the other hand, no amplification was obtained with two primer pairs corresponding to the regulatory region of the major nitrate reductase operon of *E. coli*. These results indicate that the kanamycin resistant clones were indeed *T. ferrooxidans* cells and not contaminating *E. coli* cells.

The presence of the kanamycin resistance gene was checked by PCR, either with two oligonucleotides hybridizing to Tn5 in the case of pJRD215, or to Tn903 in the case of pKT240, pJB3Km1 and pUFR034. This shows that the kanamycin resistant clones were due to the transfer of the plasmid and not to spontaneous mutation.

The presence of pKT240, pJRD215, pUFR034 and pJB3Km1 in the transconjugants was confirmed by the restriction analysis of the plasmids isolated from *T. ferrooxidans* kanamycin resistant clones and by transforming these plasmids to an *E. coli* TG1 strain.

From all these results, we can conclude that the different plasmids have been transferred from *E. coli* to *T. ferrooxidans* and were not integrated into the chromosome.

3.9. Stability

The maintenance of pJRD215, pUFR034 and pJB3Km1 in the transconjugants was studied as explained in the Materials and Methods section. The results are presented in Figure 1. IncQ plasmid pJRD215 was more stable than pUFR034 and pJB3Km1. More than 70% retention of pJRD215 was observed after 40 generations without kanamycin in the ATCC33020 and ATCC19859 strains (Figure 1A). pUFR034 was less stable in these four *T. ferrooxidans* than pJRD215, with 30%-50% retention after 10 generations (Figure 1B). The plasmid pJB3Km1 was extremely unstable in all the four *T. ferrooxidans* strains: after 20 generations, nearly all the plasmids were lost (Figure 1C).

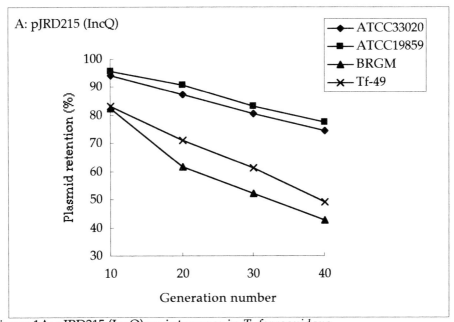

Figure 1A: pJRD215 (IncQ) maintenance in *T. ferrooxidans*

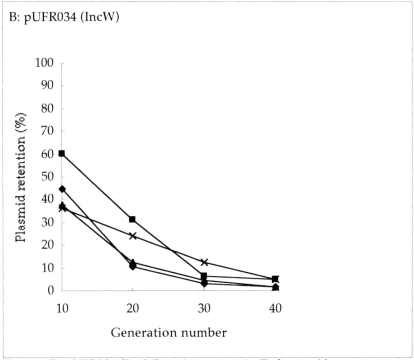

Figure 1B: pUFR034 (IncW) maintenance in *T. ferrooxidans*.

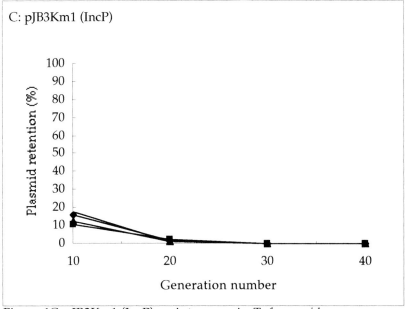

Figure 1C: pJB3Km1 (IncP) maintenance in *T. ferrooxidans*.

4. CONCLUSION

The apparent transfer of pJRD215 from *E. coli* S17.1 to ATCC33020 could be increased from 10^{-7} to 10^{-5}-10^{-4} per recipient cell (i) by growing *E. coli* in the 2:2 liquid medium containing 0.5% yeast extract; (ii) by growing ATCC33020 in the 9K liquid medium with sulfur as energy source; (iii) by conjugation on 2:2 solid medium which pH has been adjusted to 4.8 and which had been supplemented with 0.05% thiosulphate and $0.5 \ 10^{-4}$ M diaminopimelic acid.

Under these conditions, IncQ, IncP and IncW plasmids could be transferred by conjugation to the *T. ferrooxidans* private strains, BRGM and Tf-49, but also, to the two collection strains ATCC33020 and ATCC19859 from which originate most of the *T. ferrooxidans* genes characterized to date. Among these plasmids, the IncQ mobilizable plasmid pJRD215 gave the higher conjugation frequency and was the more stable. It can be used to introduce heterologous or homologous genes in *T. ferrooxidans*. On the other hand, IncP plasmids being unstable could be used as suicide vectors and enable marker exchange mutagenesis.

REFERENCES

1. T. Kusano, K. Sugawara, C. Inoue, T. Takeshima, M. Numata and T. Shiratori, J. Bacteriol. 174 (1992) 6617.
2. J.-B. Peng, W.-M. Yan and X.-Z. Bao, J. Bacteriol. 176 (1994) 2892.
3. R. Simon, U. Priefer and A. Pühler, Biotechnology 1(1983) 784.
4. J.-P. Peng, W.-M. Yan and X.-Z. Bao, J. Gen. Appl. Microbiol. 40 (1994) 243.
5. M.M. Bagdasarian, E. Amann, R. Lurz, B. Rückert and M. Bagdasarian, Gene 26 (1983) 273.
6. J. Davison, M. Heusterspreute, N. Chevalier, V. Ha-Thi and F. Brunel, Gene 51 (1987) 275.
7. R. DeFeyter, C.I. Kado and D.W. Gabriel, Gene 88 (1990) 65.
8. J.M. Blatny, T. Brautaset, H.C. Winther-Larsen, K. Haugan and S. Valla, Appl. Environ. Microbiol. 63 (1997) 370.
9. J.H. Miller, A short course in bacterial genetics: a laboratory manual and handbook for *Escherichia coli* and related bacteria. Cold Spring Harbor, New-York: Cold Spring Harbor Laboratory Press (1972).
10. A. Bengrine, N. Guiliani, C. Appia-Ayme, E. Jedlicki, D.S. Holmes, M. Chippaux and V. Bonnefoy, Biochem. Biophys. Acta 1443 (1998) 99.
11. T.D. Brock and M.T Madigan, Biology of microorganisms. Prentice-Hall International, Inc. (1991)

Characterization And Functional Role of Cytochromes Possibly Involved In The Iron Respiratory Electron Transport Chain of *Thiobacillus ferrooxidans*

M. T. Giudici-Orticoni, G. Leroy, R. Toci, W. Nitschke, and M. Bruschi

Laboratoire de Bioénergétique et Ingénierie des Protéines
C.N.R.S.- IFR1- Marseille- France - email: giudici@ibsm.cnrs-mrs.fr

The commercial extraction of metals from ores by microbial leaching is based on the iron oxidation capabilities of *Thiobacillus ferrooxidans* and requires an electron transport chain localized in the periplasmic space of the cell. Several redox proteins tentatively involved in this respiratory chain have been described. We have characterized various redox proteins located in the periplasmic space of *Thiobacillus ferrooxidans*. Among these, a new cytochrome c with a molecular mass around 30kDa has been purified. An electron transfer pathway between iron oxidation and cytochrome oxidase is proposed based on studies of the interaction between these periplasmic electron transfer proteins.

1. INTRODUCTION

Extremophilic microorganisms able to grow under extreme environmental conditions such as high temperature, high salt concentration or extreme pH values have attracted considerable interest in recent years due to their relevance regarding questions of the possible origin and limits of life as well as due to their biotechnological potential. Biomining techniques represent attractive alternatives to conventional methods of mining since in general they are less energy-intensive and less polluting than chemical ones. *Thiobacillus ferrooxidans* is one of the most important micro-organisms involved in bioleaching (1). This chemolithautotrophic bacterium acquires the energy necessary for its vital processes via the aerobic oxidation of ferrous iron.

Most of the metalloproteins involved in the respiratory chains are soluble acid-stable proteins that exhibit high redox potentials (2,3). Several redox proteins tentatively suggested to be involved in iron oxidation have been isolated from *T. ferrooxidans*: a blue copper protein, rusticyanin (4), a high potential iron - sulfur protein (HiPIP), (5-7) a membrane-bound cytochrome oxidase (8), a partially purified iron: rusticyanin oxidoreductase (9), a soluble and a membrane-bound cytochrome c_{552} (10,11) and more recently a diheme cytochrome of the c_4 type (12).

In this paper the interaction between cytochrome c_4 and rusticyanin is studied and the purification and characterization of a novel cytochrome which may take part in the respiratory electron transport chain of *T. ferrooxidans* are described.

2. MATERIALS AND METHODS

T. ferrooxidans was kindly supplied by Dr. D. Morin (Bureau des Recherches Geologiques et Minieres, Orleans, France). This bacterium has been isolated from drainage water at the Salsigne sulfur mine (France). It was grown at pH 1.6 in 9 K Silverman and Lundgreen medium (13) supplemented with $CuSO_4$ -5 H_2O at a concentration of 1.6 mM. Large scale cultivation of the organism was performed in 300 liter of the above medium with a home-made polyprene fermenter. Typical yields were in the range of 80 mg/l (w/v).

2.1. Purification

Cytochrome c_4 and rusticyanin were purified as previously described (12). Briefly, a sequence of purification steps using, carboxymethyl cellulose, Sepharose and mono S columns yielded 9 mg of pure cytochrome c_4 and 200 mg of pure rusticyanin from 100 g of wet cells.

2.2. Molecular mass determination

The molecular mass of the protein was determined by performing sodium dodecyl sulfate-polyacrylamide gel electrophoresis under reducing conditions (14) on a Pharmacia PhastSystem with PhastGel 8-25 % polyacrylamide and PhastGel SDS buffer strips. Proteins and hemoproteins were stained with Coomassie and tetramethyl benzidine, respectively (15). Mass Spectrometry analysis was carried out on a Perseptive Biosystem. The spectra were recorded as described previously (16).

2.3. Amino acid analysis and protein sequencing

For the aminoacid analysis, protein samples were hydrolysed in 200 µl of 6 M HCl at 110°C for 24 and 72h in sealed vacuum tubes and then analysed with a Beckman aminoacid analyser (System 6300). Heme was removed using Ambler's method (17) and the resulting apoprotein was isolated by gel filtration on Sephadex G 25 in 5 % (v/v) formic acid. Sequence determinations were carried out on the apoprotein with an Applied Biosystems gas-phase sequenator (Models 470 A and 473 A). Quantitative determinations were performed on the phenyl thiohydantoin derivatives by means of high-pressure liquid chromatography (Waters Associates, Inc.) monitored by a data and chromatography control station (Waters 840).

2.4. Optical absorption spectra

Visible and ultraviolet absorption spectra of the protein were determined with a Beckman DU 7500 spectrophotometer. Molar extinction coefficients at the absorption maximal were obtained from these spectra using protein concentrations based on amino acid analysis data.

2.5 Redox titration and EPR spectroscopy

Optical redox titration were performed at pH 3.0 according to Dutton (18) using a Kontron Uvikon 932 spectrophotometer. The following redox mediators were present at 5 µM: p-benzoquinone, TMPD and DAD for titration of cytochrome c; ferrocene dicarboxylic acid, ferrocene monocarboxylic acid and ferricyanite for experiments on rusticyanin.

EPR spectra were recorded on a Bruker ER300 spectrometer equipped with an Oxford Instruments liquid He cryostat. Complete oxidation of the sample was achieved by addition of iridium chloride followed by removal of the oxidant via passage through a PD-10 column. The sample contained 100 µM of rusticyanin in ammonium acetate (pH 4.8), MOPS (pH 7), Glycyl-glycyl (pH 8.2), Tricine-NaOH (pH 9) or 10 µM cytochrome c_4 and 10 µM rusticyanin in ammonium acetate (pH 4.8).

3. RESULTS AND DISCUSSION

Several hypotheses have been proposed for the organization of the iron oxidative electron transport chain in *T. ferrooxidans* (9). Even though the existence of multiple cytochromes had been reported, only a single species of cytochrome c has been included as a component of the respiratory chain in the various models with the exception of the model proposed by Ingledew (2).

We have studied the reduction of the dihemic cytochrome c_4 by rusticyanin in the presence of Fe^{2+} in order to show whether this cytochrome could be involved in iron respiration.

Kinetic experiments performed on cytochrome c_4 showed that whatever the Fe^{2+} concentration, cytochrome c_4 was not reduced by Fe^{2+}, but rusticyanin reduced the cytochrome c_4 in the presence of a large excess of Fe^{2+} (25 mM) (19, 20).

This finding appears paradoxical unless assuming significant changes in redox potentials of one or both of these electron transport proteins upon complex formation. If the E_m values (380mV and 480mV) of both cytochrome c_4 hemes remained constant, the E_m -value at pH 4.8 of the rusticyanin present in the complex (580mV) with cytochrome c_4 was indeed lower by about 100 mV with respect to that of free rusticyanin (680mV) at the same pH (fig1A). In addition to this redox potential modification, alterations of EPR spectral parameter of rusticyanin were observed in the complex (fig 1B).

These alterations of spectra are indicative of substantial rearrangements of unpaired electron density at the copper site upon complex formation. We have examined the pH-dependence of rusticyanin's EPR spectral parameters in the alkaline region (fig 1B). The change in the EPR spectra induced by pH was similar to the EPR spectra of rusticyanin in the complex.

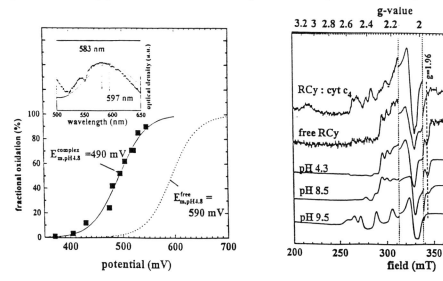

Figure 1. (A) Redox titration of rusticyanin - cytochrome c_4 complex (filled squares) at pH4.8. The dotted line indicates the titration curve of free rusticyanin at the same pH. (B) EPR spectra of oxidized rusticyanin prior (b) and after (a) complex formation with cytochrome c_4 at pH 4.8. EPR spectra of free oxidised rusticyanin at pH 4.8 (c), 8.5 (d) and 9.5 (e). For better comparison, EPR spectra of the complex and of free rusticyanin at various pH-values were normalised to compare signal amplitudes.

The ensemble of the above demonstrates that docking of rusticyanin to cytochrome c_4 induces changes at the copper site related to the redox-linked protonation state of amino acid residues. The modification of the redox properties of rusticyanin induced by the complex formation, make sense with regard to the thermodynamic constraints of electron transfer in Thiobacilli and therefore strongly facilitates electron transfer from rusticyanin to cytochrome c_4. We conclude that the rusticyanin catalysed transfer of an electron from Fe^{2+} to cytochrome c_4 is biologically relevant and that this reaction represents one individual step of the overall physiological electron transfer chain. Our present working hypothesis for the iron respiratory chain in *T. ferrooxidans* is presented in fig 2.

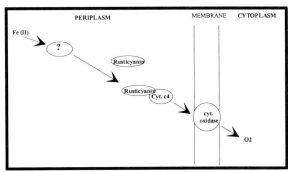

Figure 2: Scheme that describes the aerobic iron respiratory chain of *T. ferrooxidans*.

Electrons are passed on to rusticyanin, from there to cytochrome c_4 and ultimately to molecular oxygen via a membrane-bound oxidase. The direct interaction between cytochrome c_4 and oxidase has not been studied in *T. ferrooxidans* so far. Such an interaction, however, has been observed for cytochrome c_4 and cytochrome oxidase in *Azotobacter vinelandii* (21). We therefore propose an analogous reaction for the *T. ferrooxidans* system.

However, rusticyanin which accounts for up to 5% of the total soluble protein and which has been characterized in detail (22-27) does not seem to act as the initial electron acceptor (9).

With the aim to detect a possible first electron acceptor we have purified a new cytochrome c (cytochrome c_{30}) using carboxymethyl cellulose, S Sepharose and mono S columns. 6 mg of pure cytochrome c were obtained from 100 g of wet cells (figure 3).

This cytochrome is a basic soluble protein with a molecular mass deduced from SDS/PAGE analysis of about 30 kDa and . The mass spectra of the protein showed a major peak at 26900 Da. This result confirms the presence of only one polypeptide chain.

The UV-visible spectrum of the purified cytochrome is presented in fig 4. It exhibited a Soret peak at 411 nm in the oxidized state and peaks at 417, 523 and 552 nm in the ascorbate reduced form. The purity coefficient, expressed as the ratio between the absorbencies at given wavelengths $c = A^{red}_{552} - A^{red}_{570} / A^{ox}_{280}$ was found to be 1 in the case of the purified protein.

The amino acid composition of the protein was determined by performing amino acid analysis on the basis of the molecular mass of the protein (26900 Da). This analysis showed the presence of 17 Asp, 20 Thr, 19 Ser, 21 Glu, 23 Pro, 23 Gly, 38 Ala, 16 Val, 6 Met, 9 Ile, 17 Leu, 10 Tyr, 6 Phe, 6 His, 13 Lys and 11 Arg residues per molecule.

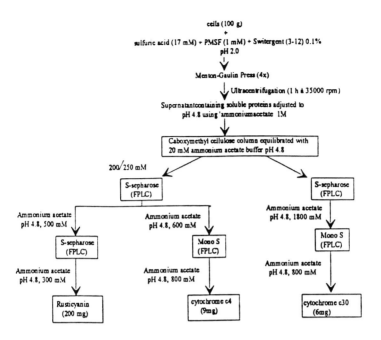

Figure 3. Survey of purification and yield of rusticyanin, cytochrome c_4 and 30kDa cytochrome c from *T. ferrooxidans*

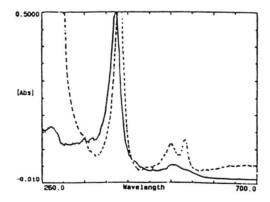

Figure 4. Absorbance spectra of oxidized (solid line) and reduced (dashed line) cytochrome c. Both spectra were determined in ammonium acetate 50 mM, pH 4.8, at 25°C.

56

Automated Edman degradation of 1 nmol of the apoprotein yielded the amino-terminal sequence: Ala-Ala-Ser-Ala-Pro-Asn-Pro-Lys-Ile-Val-Phe....
No homology with other cytochromes c has been evidenced.
Preminilary results show that this cytochrome is reduced in the presence of 5mM of Fe^{2+}. The 30 kDa cytochrome c therefore appears to be involved in iron respiration and might in fact represent the Fe: rusticyanin oxidoreductase described by Blake et al (9). In the near future we plan to study the kinetics of its reduction by iron and the interaction with the other redox proteins characterized in *Thiobacillus ferrooxidans*.

ACKNOWLEDGEMENTS

The authors gratefully acknowledge the Fermentation Plant Unit (LCB, Marseille, France) for growing the bacteria, and Nicole Zylber and Jacques Bonicel (Protein Sequencing Unit, Marseille, France) for performing amino acid analysis, N-terminal amino acid sequencing and mass spectrometry. This study was supported by a grant from l'Agence de l'Environnement et de la Maîtrise de l'Energie (ADEME) and le Bureau des Recherches Geologiques et Minieres (BRGM).

REFERENCES

1. Ewart, D.K. and Hugues, M.N., Advances in inorganic chemistry. 36 (1991) 103.
2. Ingledew, W.J., Biochim.Biophys.Acta. 683 (1982) 89.
3. Ingledew, W.J. and Houston, A., Biotech. and Appli. Biochem. 8 (1986) 242.
4. Cox, J.C. and Boxer, D.H., Biochem. J. 174 (1978) 497.
5. Fukumori, Y., Yano, T., Kai, M. and Sato, A., Dev. Geochem. 6 (1991) 267.
6. Kusano, T., Takeshima, T., Sugawara, K., Inoue, C., Shiratori, T., Yano, T., Fukumori, Y. and Yamanaka, T., J. of Biol. Chem. 267 (1992) 11242.
7. Cavazza, C., Guigliarelli, B., Bertrand, P. and Bruschi, M., FEMS Microbiol. Lett. 130 (1995) 193.
8. Kai, M., Yano, T., Fukumori, Y. and Yamanaka, T., Biochem. Biophys. Res. Commun. 160 (1989) 839.
9. Blake II, R*C. and Shute, E.A., Biochemistry. 33 (1994) 9220.
10. Sato, A., Fukumori, Y., Yano; T., Kai, M. and Yamanaka, T., Biochim. Biophys. Acta. 976 (1989) 129.
11. Tamegai, H., Kai, M., Fukumori, Y. and Yamanaka, T., FEMS Microbiol. Lett. 119 (1994) 147.
12. Cavazza C., Giudici-Orticoni M.T., Nitschke W., Appia C., Bonnefoy V. and Bruschi M., Eur. J. Biochem. 242 (1996) 308.
13. Silverman, M.P. and Lundgreen D.G., J. Bacteriol. 77 (1959) 642.
14. Haff, L.A., Fagerstam, L.A. and Barry, A.R., J. Chromatogr. 266 (1983) 409.
15. Thomas, P.E., Ryan, D. and Levin, W., Anal. Biochem. 75 (1976) 168.
16. Dolla, A., Florens, L., Bianco, P., Haladjian, J., Voordouw, G., Forest, E., Wall, L., Guerlesquin, F. and Bruschi, M., J. Biol. Chem. 269 (1994) 6340.
17. Peuke B., Ferenczi R. and Kovacks K., Anal. Biochem. 60 (1974) 45.
18. Dutton P.L., Methods Enzymol. 54 (1978) 411.

19. M. T. Giudici-Orticoni, W. Nitschke, C. Cavazza and M. Bruschi, Biomine 97 (1997) PB4.1.
20. C. Appia-Ayme, A. Bengrine, C. Cavazza, M.T. Giudici-Orticoni, M. Bruschi, M. Chippaux and V. Bonnefoy, FEMS Microbiol. Lett., 167 (1998) 171.
21. Ng, T.C.N., Laheri, A.N. and Maier, R.J., Biochim. Biophys. Acta. 1230 (1995) 119.
22. Nunzi, F., Haladjian, J., Bianco, P. and Bruschi, M., J. Electroanal. Chem., 352 (1993) 329.
23. Nunzi, F., Woudstra, M., Campese, D., Bonicel, J., Morin, D. and Bruschi, M. Biochim. Biophys. Acta. 1162 (1993) 28.
24. Hunt, A.H., Toy-Palmer, A., Assa-Munt, N., Cavanagh, J., Blake, RC. II and Dyson, J., J. Mol. Biol. 244 (1994) 370.
25. Walter R.L., Ealick S.E., Friedman A.M., Blake RC., Proctor P and Shoham M., J. Mol. Biol. 263 (1996) 730.
26. Botuyan M.V., Toy-Palmer A., Chung J., Blake R. Beroza P., Case D.A. and Dyson H.J., J. Mol. Biol. 263 (1996) 752.
27. Blake RC. and Shute E.A., J. Biol. Chem. 262 (1987) 14983.

The use of immunoelectron microscopy to analyze surface components of *Thiobacillus ferrooxidans* grown under different conditions

O. Coto[a], Y. Gómez[a], P. Varela[b], V. Falcon[a], J. Reyes[a] and C.A. Jerez[c]

[a]Departamento de Microbiología, Facultad de Biología, Universidad de la Habana y Centro de Ingeniería Genética y Biotecnología, La Habana, Cuba. ocoto@comuh.uh.cu

[b]Instituto de Ciencias Biomédicas, Facultad de Ciencias, Universidad de Chile, Santiago, Chile

[c]Departamento de Biología, Facultad de Ciencias, Universidad de Chile, Santiago, Chile. cjerez@abello.dic.uchile.cl

The surface components of bioleaching microorganisms such as *Thiobacillus ferrooxidans* apparently play an essential role in the adhesion of the bacteria to minerals. These components are synthesized in different amounts depending on several factors, such as the type of energy source and the lack of some nutrients. We have prepared antibodies against both whole cells and outer membrane preparations from *T. ferrooxidans* grown in ferrous-iron or elemental sulfur. The antibodies prepared were specific for *T. ferrooxidans* cells and allowed us to discriminate between iron- and sulfur-grown cells. The antigens were analyzed by Western immunoblotting and their localization on the surface of the cells was accomplished by immunoelectron microscopy using gold-labeled protein A. Interestingly, the sulfur-grown and not the iron-grown cells showed very long surface appendages reacting with the antibodies against outer membrane preparations of sulfur-grown *T. ferrooxidans*. These structures were similar to flagella and may be very important in adhesion of the microorganisms to solid sulfur and eventually to other minerals.

1. INTRODUCTION

The oxidation of minerals by *Thiobacillus ferrooxidans* requires that the microorganism interacts with the solid substrates (1-3). This adhesion could be mediated by flagella and pili (4, 5). Surface exopolysaccharides are also considered very important for the solid-bacteria interaction (6). Some of the molecular components in the bacterial surface are the lipopolysaccharides and the outer membrane proteins. A role for both of these kind of macromolecules has been described in bacterial adhesion (7, 8). *T. ferrooxidans* cells treated to remove part of their lipopolysaccharides showed altered adhesion properties (8). Also, the possible participation of the major outer membrane protein Omp40 in the attachment of *T. ferrooxidans* to substrates such as elemental sulfur has been suggested (8). We have observed that some of these surface components vary their expression during different environmental conditions (9-11), and the adherence of the microorganisms to sulfur also changed (8, 13).

When *T. ferrooxidans* is grown in ferrous iron or sulfur, changes in some outer membrane proteins have been observed (9, 12). Specifically, outer membrane proteins in the range of 50 kDa have been described as being synthesized in sulfur-grown cells and not in ferrous iron-grown microorganisms (12-14). Very recently, a sulfur-binding protein was described in *T. ferrooxidans* (5). This protein was synthesized in sulfur-grown but not in iron-grown cells. The thiol groups of this polypeptide formed a disulfide bond with elemental sulfur and mediated the strong adhesion between *T. ferrooxidans* cells and elemental sulfur.

In the present report, we have prepared polyclonal antibodies against whole cells and outer membrane preparations of sulfur-grown and iron-grown *T. ferrooxidans* cells. These antibodies were employed to localize some of the antigenic components in the surface of the cells by using immunoglobulin-gold conjugates and electron microscopy.

2. MATERIALS AND METHODS

2.1. Microorganisms and growth conditions

We employed *T. ferrooxidans* ATCC 19589 which was grown in modified 9K medium containing ferrous iron (11) or in the presence of sulfur prills as described before in our laboratory (8). The outer membrane preparations from both of these types of cells were done as we have described previously (10). As a control acidophilic microorganism we employed *Acidiphilium organovorum* M1, isolated from the Matahambre mining deposit in Pinar del Río, Cuba. This microorganism was grown in the medium for heterotrophic acidophiles described by Harrison (15). After growth, cells were collected by centrifugation (8,000 xg for 10 min) and were washed three times by resuspension and centrifugation in acidic water (pH 1.8).

2.2. Immunological methods

The antisera against whole cells of *T. ferrooxidans* were obtained as described before (16, 17). These antisera contain antibodies mainly against surface components such as outer membrane proteins and lipopolysaccharides (16). For the preparation of antisera against outer membrane preparations, a standard immunization protocol on New Zealand rabbits was employed. The sera obtained were against outer membrane preparations from iron-grown *T. ferrooxidans* and against outer membrane preparations from *T. ferrooxidans* grown in sulfur. Cross-reacting antibodies were removed by incubating the serum against the outer membrane of *T. ferrooxidans* grown in sulfur with ferrous ion-grown cells and viceversa, according to the protocol employed by Devasia et al. (18) for whole cells of the same microorganism.

Western blotting using total cells of *T. ferrooxidans* was done after electrophoretic separation of total proteins by SDS-PAGE followed by transfer to nitrocellulose membranes and detection by chemiluminiscence as described before (11, 19). For localization of the antigenic components on the cells we used immunoelectronmicroscopy with the different antibodies and protein A conjugated with colloidal gold particles (15 nm) in a ratio 1:20 in PBS. The antigen-antibody reactions were analyzed by using the transmission electron microscope JEOL JEM-2000 EX at an acceleration voltage of 89 kV.

2. RESULTS AND DISCUSSION

A major concern with immunolocalization techniques, particularly those using polyclonal antisera as the ones used here, is antibody specificity. Fig. 1 shows that when whole-mount cells of different microorganisms are immunolabeled with sera against

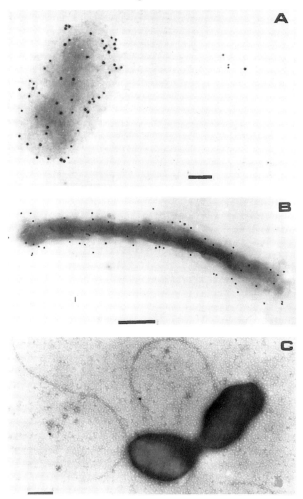

Figure 1. Surface localization of antigenic molecules in different acidophilic microorganisms. Representative electron micrographs showing profuse gold labeling on the surface of intact whole bacterial cells of *T. ferrooxidans* grown in ferrous iron (A) or sulfur (B) and no labeling of intact cells of *A. organovorum* (C). The bars represent 200 nm (A and C) or 500 nm (B).

62

whole cells of *T. ferrooxidans*, there was reaction only with *T. ferrooxidans* cells grown either in ferrous iron (Fig. 1A) or elemental sulfur (Fig. 1B).

On the other hand, when cells of *A. organovorum*, an acidophilic microorganism commonly found in bioleaching habitats was used, there was no reaction (Fig. 1C). These results indicate that the antibodies are specific for *T. ferrooxidans cells*. In this regard, we have previously prepared antibodies against whole cells of *T. ferrooxidans* which were highly specific, since they did not show crossreaction with *L. ferrooxidans* or *T. thiooxidans* and several other thiobacilli (16).

The binding of the antibodies was revealed with 15 nm gold-conjugated protein A. We can easily see that the antibodies covered the entire bacterial surface of *T. ferrooxidans*, whereas no antibodies were detected on the cell surface of *A. organovorum*. As we have shown previously, this kind of antibodies react preferentially with surface components of *T. ferrooxidans* such as outer membrane proteins and lipopolysaccharides (16). However, the reaction with other cellular components cannot

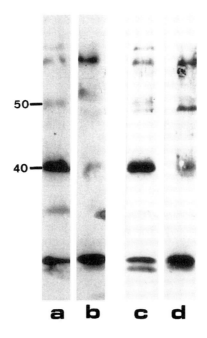

Figure 2. Western immunoblot of total cellular components from *T. ferrooxidans* with antisera against outer membrane preparations. Total cellular components from *T. ferrooxidans* grown in iron (lanes a and c) or grown in sulfur (lanes b and d) were separated by SDS-PAGE and immunoblotted and developed with antibodies against outer membrane preparations from *T. ferrooxidans* grown in iron (lanes a and b) or antibodies against outer membrane preparations from *T. ferrooxidans* grown in sulfur (lanes c and d).

be ruled out. For this reason, we prepared antibodies against outer membrane preparations of iron- or sulfur-grown *T. Ferrooxidans* cells. To analyze the outer membrane proteins that acted as antigens in each case, we performed a Western blotting experiment as seen in Fig. 2.

We can see that in general there were six or seven main proteins reacting with the antisera against outer membrane preparations from *T. ferrooxidans*. Total cellular components from *A. organovorum* gave no reaction under the same conditions (results not shown), indicating that these antibodies were also very specific for *T. ferrooxidans*. These antibodies did not show by Western blotting, the typical ladder-like migration of the lipopolysaccharides. This is an indication that the outer membrane preparations have little lipopolysaccharide, if any. Consequently, the antibodies would be recognizing mainly proteins.

Several of the proteins seen in Fig. 2 were present in both, cells grown in sulfur and those grown in iron, suggesting their constitutive nature. However, a protein of molecular mass around 48,000 kDa was present in much higher amounts in cells grown in sulfur. This protein was apparently recognized only by the antiserum against outer membrane proteins from cells grown in sulfur (compare lanes b and d in Fig. 2). A major outer membrane protein which corresponded to OMP40 (10), was apparently present in larger amounts in iron-grown cells (see lanes a and c). Also, proteins of sizes around 90 and 30 kDa were present in higher amounts in iron-grown cells.

These results are in general agreement with previous findings of Kulpa et al. (12) for proteins induced in iron-grown *T. ferrooxidans* cells and those of Buonfiglio et al. (14) for the induction of an outer membrane protein in the 48-50 kDa range in cells of *T. ferrooxidans* grown in sulfur.

To enrich the sera in the antibodies against the proteins characteristic of each growth condition, we preadsorbed the serum containing antibodies against outer membrane preparations of sulfur grown-cells with iron-grown whole cells and viceversa. The antisera treated in this way were employed for immunocytochemical localization of whole-mount cells as shown in Fig. 3. Panel C shows cells of *T. ferrooxidans* grown in iron labeled with antibodies against outer membrane preparations of iron-grown cells. We can see a strong reaction with surface antigens which probably corresponded to outer membrane proteins. When the same antibodies preadsorbed with whole cells grown in sulfur were employed (panel A), still there was a recognition of many surface antigenic components, indicating that some of these may be antigens specifically induced by growth of *T. ferrooxidans* in the presence of iron.

On the other hand, when iron-grown cells were reacted with antibodies against outer membrane preparations from *T. ferrooxidans* grown in sulfur and preadsorbed with iron-grown cells (panel B), no reaction was seen. This indicates that all the antibodies against the common antigens were removed by the preadsorption treatment. A similar situation of lack of reaction was observed when whole-mounted sulfur-grown *T. ferrooxidans* cells were treated with antibodies against outer membrane preparations from *T. ferrooxidans* grown in iron but preadsorbed with whole sulfur-grown cells (panel D).

These cells therefore not only removed the antibodies against the common antigens, but also those specific of the sulfur-grown cells (for example the 45-50 kDa sulfur-induced antigenic protein).

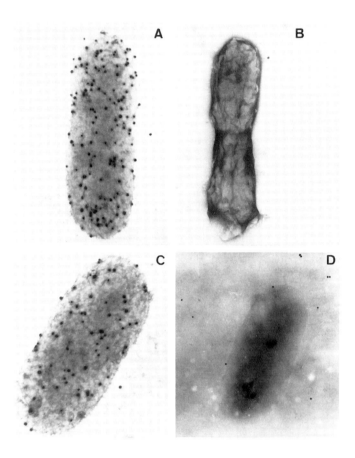

Figure 3. Comparative gold labeling of *T. ferrooxidans* cells grown under different conditions with antibodies against outer membrane preparations. Whole-mounted cells of *T. ferrooxidans* grown in ferrous iron (panels A, B and C) or in elemental sulfur (panel D) were reacted with antibodies against outer membrane preparations of *T. ferrooxidans* grown in iron (panel C) or against the same antibodies preadsorbed with whole cells grown in sulfur (panels A and D) or antibodies against outer membrane preparations of *T. ferrooxidans* grown in sulfur and preadsorbed with whole cells of *T. ferrooxidans* grown in iron (panel B).

When sulfur-grown *T. ferrooxidans* cells were labeled employing antibodies against outer membrane preparations from cells grown in sulfur and preadsorbed with *T. ferrooxidans* cells grown in iron, the very interesting results seen in Fig. 4 were obtained.

Figure 4. Gold immunolabeling appendages present in sulfur-grown *T. ferrooxidans* cells. Whole-mounted sulfur-grown *T. ferrooxidans* cells were reacted with antibodies against outer membrane preparations from sulfur-grown cells preadsorbed with whole iron-grown *T. ferrooxidans* cells. The arrowheads indicate the filamentous structures and the bar corresponds to 200 nm.

We can see gold particles in the body of the cells as detected in previous figures. In addition, we observed very long filaments (arrowheads) heavily decorated with gold particles. These results indicate that the outer membrane preparation from sulfur-grown *T. ferrooxidans* that we prepared contained in addition to outer membrane proteins, some proteins that may form part of structural filaments such as pili or flagella. Considering that the gold particles used have an average size of 15 nm, one can estimate a diameter of 15-20 nm for these structures. This size corresponds well with the size for flagellar structures that have been previously described for *T. ferrooxidans* (5). This antigen-antibody reaction was observed only in sulfur-grown cells and not in ferrous iron-grown cells. This is in agreement with the recent findings of Ohmura et al. (5), in which they described a sulfur-binding protein of 40 kDa which was synthesized in sulfur-grown but not in iron-grown cells. Furthermore, this protein was localized in the flagella of the bacteria (5). Since these surface structures and other components participate in the strong adhesion of *T. ferrooxidans* to solid substrates such as sulfur powder, it will be of great interest to define their participation during adhesion to ores subjected to bioleaching.

4. CONCLUSIONS

We have prepared antibodies specific against surface components of *T. ferrooxidans*. Our preliminary results indicate that these antibodies may discriminate between iron- and sulfur-grown cells by using immunoelectron microscopy and labeling with gold particles.

The antibodies prepared allowed us to detect surface appendages in sulfur-grown *T. ferrooxidans* cells, which may be important in cell adhesion to solid substrates. It is expected that the antibodies prepared will be useful in defining some of the main components involved in cellular adhesion.

ACKNOWLEDGEMENTS

This research was supported by an ICGEB grant N° 96/007.

REFERENCES

1. O. Tuovinen, In Microbial Mineral Recovery, H.L. Ehrlich and C.L. Brierley (eds.), pp. 55 (1990), McGraw-Hill Book Co., New York.
2. A.S. Myerson and P. Kline, Biotechnol. Bioeng. 28 (1986) 1669.
3. W. Sand, T. Gehrke, R. Hallmann and A. Schippers, Appl. Microbiol. Biotechnol. 43 (1995) 961.
4. A. Dispirito, M. Silver, L. Voss and O.H. Tuovinen, Appl. Environ. Microbiol., 43 (1982) 1196.
5. N. Ohmura, K. Tsugita, J-I. Koizumi and H. Saiki, J. Bacteriol. 178 (1996) 5776.
6. T. Gehrke, J. Telegdi, D. Thierry and W. Sand, Appl. Environ. Microbiol. 64 (1998) 2743.
7. A.A. Dispirito, P.R. Dugan and O.H. Tuovinen, Biotechnolo. Bioeng. 25 (1983) 1163.
8. Arredondo, R., A. García and C.A. Jerez, Appl. Environ. Microbiol. 60 (1994) 2846.
9. A.M. Amaro, D. Chamorro, M. Seeger, R. Arredondo, I. Peirano and C.A. Jerez, J. Bacteriol. 173 (1991) 910.
10. C.A. Jerez, M. Seeger and A.M. Amaro, FEMS Microbiol. Lett. 98 (1992) 29.
11. P. Varela, G. Levicán, F. Rivera and C.A. Jerez, Appl. Environ. Microbiol. 64 (1998) 4990.
12. N. Mjoli and C.F. Kulpa, In Biohydrometallurgy: proceedings of the International Symposium. Warwick (1988) pp. 89, Great Britain.
13. G. Osorio, P. Varela, R. Arredondo, M. Seeger, A.M. Amaro, and C.A. Jerez. In Biohydrometallurgical Technologies, A.E. Torma, M.L. Apel and C.L. Brierley, (eds.) (1993) pp. 565, The Minerals, Metals & Materials Society.

14. V. Buonfiglio, M. Polidoro, L. Flora, G. Citro, P. Valenti and N. Orsi, FEMS Microbiol. Rev., 11 (1993) 43.
15. A.P. Harrison, Jr., Annu. Rev. Microbiol. 38 (1984) 265.
16. R. Arredondo and C.A. Jerez, Appl. Environ. Microbiol. 55 (1989) 2025.
17. O. Coto, A.I. Fernández, T. León and D. Rodríguez, Microbiol. SEM 8 (1992) 76.
18. P. Devasia, K.A. Natarajan, D.N. Sathyanarayana and G. Ramananda Rao, Appl. Environ. Microbiol. 59 (1993) 4051.
19. M. Delgado, H. Toledo and C.A. Jerez, Appl. Environ. Microbiol. 64 (1998) 2380.

Molecular characterization of a chemotactic receptor from *Leptospirillum ferrooxidans*

M. Delgado, H. Toledo and C.A. Jerez

Departamento de Biología, Facultad de Ciencias, Universidad de Chile, Santiago, Chile. E-mail: cjerez@abello.dic.uchile.cl

Both indirect and direct forms of mineral attack require that the microorganisms adhere to specific sites on the surface of the minerals which they will oxidize to obtain their energy. This attachment will depend on the sensing by the microorganisms of a dissolved ion concentration gradient present in the immediate vicinity of the solid. On the other hand, it is known that some bacteria swim to sites that have a preferred redox potential. To find out more about the chemotactic behaviour of biomining microorganisms, we have cloned and sequenced a 2,262 bp chromosomal DNA fragment from the motile acidophilic bacterium *Leptospirillum ferrooxidans*. This DNA contained an open reading frame for a 577 amino acid protein (we named it LcrI) showing several characteristics of the bacterial chemoreceptors. This is the first sequence reported for a gene from *L. ferrooxidans* encoding for a protein. The *lcrI* gene showed both a σ^{28}-like and a σ^{70}-like putative promoters. The LcrI deduced protein contained two hydrophobic regions most likely corresponding to the two transmembrane regions present in all the chemotactic receptors. In conclusion, *L. ferrooxidans* possesses a putative chemotactic receptor which may be part of its system to sense and adapt to its bioleaching environment.

1. INTRODUCTION

Biomining microorganisms such as the chemolithoautotrophic acidophilic *Thiobacillus ferrooxidans*, *Leptospirillum ferrooxidans* and *T. thiooxidans* are motile by means of flagella (1, 2). They should therefore possess chemotactic responses to sense and adapt to their environment. This is specially important since the microorganisms have to adhere to specific sites on the surface of the minerals which they will oxidize to obtain their energy. The bacteria should be able to sense some of the chemical dissolution of the minerals. As suggested by Sand *et al.* (3), this dissolution would probably be controlled by electrochemical processes (such as generation of an anode and a cathode due to charge imbalances, faults, electron gaps, etc.)

We have previously demonstrated that in fact, *L. ferrooxidans* possesses a chemotactic response to aspartate and Ni^{2+} which is opposite to that observed in *E. coli*, since for the former aspartate acts as a repellent and Ni^{2+} as an attractant. In addition, Fe^{2+} is also an attractant for *L. ferrooxidans* (4, 5). On the other hand, a chemotactic response of *T. ferrooxidans* toward thiosulfate (6) and very recently towards elemental sulfur (7) have been reported.

We have cloned and sequenced a *L. ferrooxidans* 2,262 bp chromosomal DNA fragment which showed a region with high degree of identity with several chemotactic receptor genes from different microorganisms (8). In this report we describe some properties of this putative

L. ferrooxidans chemotactic receptor.

2. MATERIALS AND METHODS

2.1. Bacterial strains, plasmids and growth conditions

E. coli strains RP4372 (9), DH5α (10) and HCB721 (11) were cultivated aerobically in Luria-Bertani medium (LB) at 37°C. The strains harbouring plasmids were grown in the presence of ampicillin (100 μg/ml). Plasmid pNT201 in which the *tar* gene is under the control of P_{tac} (12) was kindly supplied by R. Bourret, California Institute of Technology, Pasadena, California, USA. The pNT201, pUC18, pGEM-3Z and the recombinant plasmids were all maintained in *E. coli* DH5α. *L. ferrooxidans* Z2, kindly supplied to us by Dr. A. Harrison, Jr., University of Missouri, Columbia, USA, was grown at 30°C in modified Mackintosh medium (13).

2.2. DNA sequencing

The *L. ferrooxidans* DNA fragment contained into the pLf13 recombinant plasmid was sequenced by the dideoxy chain termination method using the Sequenase version 2.0 kit (U. S. Biochemical Co.). The nucleotide sequence was determined for both strands. For DNA sequencing we employed the T7 and SP6 vector primers, as well as synthetic oligonucleotide primers constructed on the basis of the sequence being obtained. Computer analysis of the nucleotide sequence was performed using the PC-Gene programs. Homology searches were conducted against the GenBank, EMBL, DDBJ and PDB data bases using the BLAST (14) and FASTA (15) programs.

2.3. Nucleotide sequence accession number

The nucleotide sequence of the 2,262 bp DNA region containing the *lcrI* gene is available in the EMBL database under accession N° AJ002392.

3. RESULTS AND DISCUSSION

3.1. pLf13 plasmid construction.

Our previous results (4, 5) suggested that *L. ferrooxidans* may contain MCP proteins with conserved methylation domains. Therefore, we employed Southern blotting to analyze the chromosomal DNA from *L. ferrooxidans* using a 719 bp probe coding for part of the *tar* gene, including the methylated amino acid residues present in two regions of the Tar cytoplasmic domain (from amino acids 255 to 494). When the chromosomal DNA from *L. ferrooxidans* Z2 was digested with *Hin*dIII, a 3.5 kb DNA fragment hybridizing with the probe was obtained. This fragment was cloned into the vector pUC18 yielding the pLf3.5 recombinant plasmid (Fig. 1).

The pLf3.5 plasmid was digested with *Eco*RI enzyme, resulting in a 2.3 kb fragment still hibridizing with the probe. This fragment was subcloned into the expression vector pGEM-3Z to yield the plasmid pLf13. We could specify the orientation of the 2.3 kb fragment in pLf13 (Fig. 1) by digestion with *Hin*dIII.

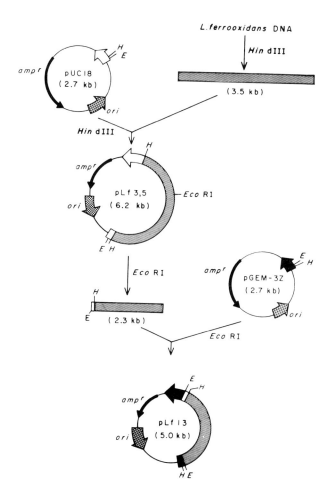

Figure 1. Construction of recombinant plasmids. pLf3.5 was obtained from a partial genomic library constructed in pUC18 with the *Hin*dIII digested chromosomal DNA from *L. ferrooxidans*. Digestion with *Eco*RI of the plasmid pLf3.5 yielded a 2.3 kb DNA fragment which was subcloned into pGEM-3Z to yield plasmid pLf13.

3.2. Sequence of the *lcrI* gene and properties of the LcrI protein.

The *L. ferrooxidans* DNA fragment contained in the recombinant plasmid pLf13 was sequenced in both strands. One complete open reading frame (LcrI) was found in the 2,262 bp *Eco*RI/*Hin*dIII insert of pLf13 by codon analysis, starting with an AUG codon in nucleotide 412 and stopping with an UGA codon in nucleotide 2,143 (Fig. 2). Identity search in databases with BLAST and FASTA programs indicated a strong similarity of this ORF with several chemotactic receptor genes. Therefore, it was called *lcr*I ("Leptospirillum chemotactic receptor I"). It was preceded by a plausible ribosome binding site with an AAAGAAAG core located 7 bases upstream from the initiating AUG codon (nucleotides 397 through 404) (Fig. 2).

Upstream of this ribosome binding site, a σ^{28}-like promoter sequence TAAA N_{15} CTCGAACT, similar to the consensus sequence for σ^{28}: TAAA N_{15} GCCGATAA (20) was present (Fig. 2, nucleotides 206 through 209 and 225 through 232 for the -35 and -10 regions, respectively). The presence of this σ^{28}-like promoter, which is characteristic of flagellar operons from *E. coli* and other microorganisms (23, 24, 25) strongly suggests that the protein from *L. ferrooxidans* coded in the sequenced gene participates in chemotaxis.

A plausible *E. coli* σ^{70}-like promoter sequence overlapping with the σ^{28}-like promoter sequence could also bee considered (Fig. 2). Whether one or both of these putative promoters function in the cell under different growth conditions, remains to be seen.

Downstream of the translational stop codon of LcrI, we could not find an inverted repeat that could function as a rho-independent transcription terminator. In addition, there was another open reading frame (ORF2) which could be cotranscribed with *lcr*I. This started with a GTG codon at nucleotide 2,169 and was interrupted on the 3' end by the restriction site used to clone this DNA. This ORF2 was also preceded by a plausible ribosome binding site (Fig. 2).

Upstream of the *lcr*I gene an incomplete open reading frame (ORF1) could also be seen in the same direction of *lcr*I. Its 5' end was interrupted by the restriction site flanking the cloned DNA fragment. This ORF1 codes for 108 amino acids, and its expression could respond to a rho-independent transcription termination site, consisting of an inverted repeat sequence between nucleotides 364 and 368 and 381 through 385 followed by a polyU tail (Fig. 2).

The deduced protein LcrI has 577 amino acids and a molecular mass of 63,957 Da. The LcrI amino acidic sequence showed a hydrophilicity profile indicating the presence of two highly hydrophobic regions which could correspond to transmembrane regions: TM1 from residues 8 through 26 and TM2 from residues 164 through 180.

A comparison of the hydrophilicity profile of LcrI with those from several MCPs is shown in Fig. 3. The putative TM regions from LcrI are present in positions similar to those found in other chemoreceptors.

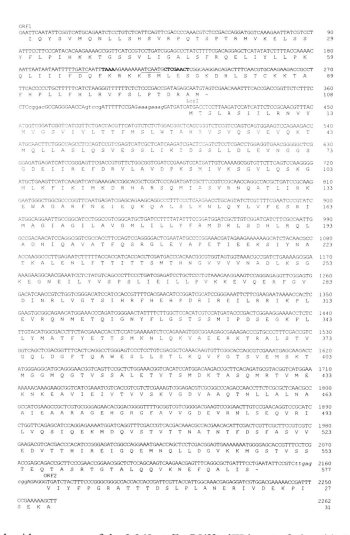

Figure 2. Nucleotide sequence of the 2,262 nt *Eco*RI/*Hin*dIII insert of plasmid pLfl3. ORF1 extends from nucleotides 1 through 326, ORF2 from nucleotides 2,169 through 2,262. The LcrI coding region extends from nucleotides 412 (first AUG) through 2,145 (stop codon). The following features are indicated in the nucleotide sequence: (underlined), a putative promoter region for σ^{70} (-35 and -10 region, at nucleotides 196 through 201 and 218 through 223 respectively); (in bold letters), a putative promoter region for σ^{28} (-35 and -10 region, at nucleotides 206 through 209 and 225 through 232 respectively); (in lower case letters in italics), two putative ribosome binding sites (nucleotides 397 through 404 for LcrI and 2,156 through 2,163 for ORF2). Under the nucleotide sequence, the deduced amino acid sequences for ORF1, LcrI and ORF2 are indicated. A potential transcription terminator is shown by lower case symbols (inverted repeat sequence cggac/gtccg from nucleotides 364 through 368 and 381 through 385).

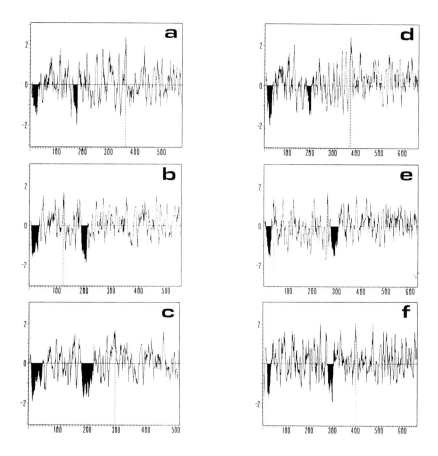

Figure 3. Hidrophilicity plots for LcrI and several chemoreceptors. The vertical axis shows the hydrophilicity units (-3 to +3) according to Hopp and Woods (16). Regions below zero are hydrophobic. The vertical broken line represents the maximum of hydrophilicity. The horizontal scale represents the amino acid residue number. Membrane spanning regions are indicated with black filling for Tar from *E. coli* (17) (b), Tas from *Enterobacter aerogenes* (18)(c), DcrA from *D. vulgaris* Hildenborough (19)(d), PctA from *Pseudomonas aeruginosa* (20)(e), McpB from *B. subtilis* (21)(f) and for the putative transmembrane regions present in LcrI from *L. ferrooxidans* (a).

The encoded amino acid sequence of LcrI was aligned with the corresponding sequence of the Tar protein from *E. coli* (Fig. 4).

Figure 4. Comparison of the amino acid sequences of Tar from *E. coli* and LcrI from *L. ferrooxidans*. The following features are indicated (in the 5' to 3' direction) in bold letters: two potential membrane-spanning regions, the signaling domain or HCD and putative methylation sites in both LcrI and Tar proteins. The asterisks indicate the amino acids identical in both proteins.

The putative periplasmic domain of LcrI (residues 27 through 163) shares an identity of 12% and a similarity of 55% with the corresponding domain of the Tar protein. The putative cytoplasmic domain from LcrI (residues 181 through 577) is 56 amino acids longer than the corresponding cytoplasmic domain of Tar (residues 213 through 553). They shared a 13% identity and 62 % of similarity. Within this possible cytoplasmic domain, LcrI possesses a region of 45 amino acids (residues 444 through 488) showing 67% identity and 96% of similarity with the HCD region of Tar (residues 361 through 405). HCD is the most highly conserved region within the chemotactic receptors (22). LcrI indeed contains an HCD region and that its sequence shows a very high degree of identity (ranging from 51% to 73%) and similarity (ranging from 85% to 96%) with the equivalent regions from twenty MCPs from different microorganisms, including some archaea (8). The proposed cytoplasmic domain of LcrI did not show regions similar to K1 and R1, the methylated regions present in the MCPs from enterobacteria. However, considering the 9 amino acid consensus sequence for the methylation sites present in MCPs from *E. coli*, *Bacillus subtilis* and possibly DcrH and DcrA

76

from *D. vulgaris* Hildenborough (19), we propose the glutamic acid residue 217 and the glutamine residue 556 as the possible methylation sites in LcrI (8).

The postulated cytoplasmic domain of LcrI has an isoelectric point similar to those from the cytoplasmic domains of MCPs from several microorganisms. This was expected, since all these bacteria, including *L. ferrooxidans*, would have similar intracellular pH values. On the other hand, the proposed periplasmic domain of LcrI, which would contain 14 less amino acids than the one corresponding to Tar, would be exposed to an acidic pH of 2-3 in the periplasm of an acidophilic microorganisms such as *L. ferrooxidans* (26). This putative periplasmic domain of LcrI has an isoelectric point of 10.43, which is very high when compared with the isoelectric point of most periplasmic domains in MCPs from several microorganisms. At neutral pH, if one assigns to each of the cationic amino acids arginine and lysine a charge of +1, the cationic amino acid histidine a charged of +0.5, the anionic amino acids glutamic acid and aspartic acids each a charge of -1, one can calculate the net charges of the periplasmic domains as the sum of the charges. This charge for the periplasmic domain of LcrI (Fig. 5) at pH 2.5 would be highly positive (+21).

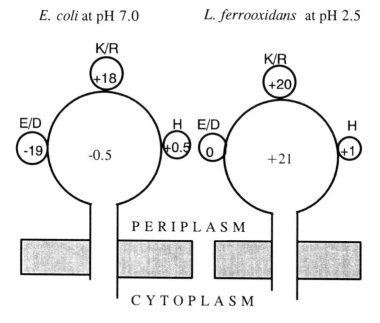

Figure 5. A scheme illustrating the differences in net charges of the periplasmic domains of the Tar receptor and the LcrI receptor at their corresponding periplasmic pH. The number of charges are indicated for the amino acid residues aspartic and glutamic acid (E/D), lysine and arginine (K/R) and histidine (H).

On the other hand, for a non-acidophilic bacterium such as *E. coli*, with a periplasmic pH of around 7, a net charge of -0.5 can be calculated for the periplasmic domain of a receptor such as Tar (Fig. 5). This charge difference may represent an special adaptation of acidophilic microorganisms such as *T. ferrooxidans* and *L. ferrooxidans* to sense effectors at the very low pH present in their periplasms.

It was not possible to show a chemotactic receptor function for LcrI expressed in *E. coli* (8). This was probably due to the fact that the periplasmic pH of *E. coli* does not allow the right conformation for the LcrI periplasmic domain, as already discussed. In addition, the lack of recognition by LcrI of the common *E. coli* chemotactic effectors is also possible. The lack of a genetic system in acidophilic chemolithoautotrophic bacteria and the appropriate *L. ferrooxidans* mutants, currently makes difficult to extend studies on the mechanisms of *L. ferrooxidans* sensing and adaptation.

4. CONCLUSIONS

The analysis and comparison of the *lcrI* sequence with those of *mcp* genes from different microorganisms indicates that the codified LcrI protein corresponds to an MCP protein. The MCPs from several bacterial species have been shown to contain conserved functionally significant regions. All of these features are present in LcrI in the expected regions: 1, two hydrophobic transmembrane segments; 2, an HCD domain; 3, two probable methylation sites. In addition, the protein not only possessed the expected molecular mass for a chemoreceptor, but it showed antigenic crossreaction with Tar from *E. coli* and it was localized in the cytoplasmic membrane of *E. coli* when expressed in this bacterium. Being HCD the region of the chemotactic receptor supposed to interact with CheA and CheW in *E. coli*, the results obtained for LcrI strongly suggest the existence of similar proteins in the signaling pathway of *L. ferrooxidans*.

AKNOWLEDGMENTS

This work was supported by FONDECYT grants 197/0417 (to CAJ), 4950008 (to MD), SAREC and ICGEB grant 96/007.

REFERENCES

1. A. Dispirito, M. Silver, L. Voss, and O. H. Tuovinen, Appl. Environ. Microbiol. 46 (1982) 1196.
2. N. Ohmura, K. Tsugita, J. I. Koizumi, and H. Saiki, J. Bacteriol. 178 (1996) 5776-5780.
3. W. Sand, T. Gerke, R. Hallmann, and A. Schippers, Appl. Microbiol. Biotechnol. 43 (1995) 961.
4. J. Acuña, I. Peirano, and C. A. Jerez, Biotechnol. Appl. Biochem. 8 (1986) 309.
5. J. Acuña, J. Rojas, A. Amaro, H. Toledo, and C. A. Jerez, FEMS Microbiol. Lett. 96 (1992) 37.
6. R. Chakraborty, and P. Roy, FEMS Microbiol. Lett. 98 (1992) 9.

7.   J.A. Rojas-Chapana, C.C. Bärtels, L. Pohlmann and H. Tributsch, Process Biochem. 33 (1998) 239.
8.   M. Delgado, H. Toledo and C.A. Jerez, Appl. Environ. Microbiol. 64 (1998) 2380.
9.   A. Krikos, M. Conley, A. Boyd, H. Berg, and M. Simon, Proc. Natl. Acad. Sci. USA 82 (1985)1326.
10. D. Titus, Promega protocols and applications guide. Second Edition. (1991) Promega Corporation, USA.
11. A. Wolfe, P. Conley, and H. Berg, Proc. Natl. Acad. Sci. USA 85 (1988) 6711.
12. K. Borkovich, N. Kaplan, J. Hess, and M. Simon, Proc. Natl. Acad. Sci. USA 86 (1989) 1208.
13. M. Mackintosh, J. Gen. Microbiol. 105 (1978) 215.
14. S. Altschul, W. Gish, W. Miller, E. Myer, and D. Lipman, J. Mol. Biol. 215 (1990) 403.
15. W. Pearson, and D. Lipman, Proc. Natl. Acad. Sci. USA. 85 (1988) 2444.
16. T. Hopp, and K. Woods, Proc. Natl. Acad. Sci USA. 78 (1981) 3824.
17. A. Krikos, N. Mutoh, A. Boyd, and M. Simon, Cell 33 (1983) 615.
18. M. Dahl, W. Boos, and M. Manson, J. Bacteriol. 171 (1989) 2361.
19. A. Dolla, R. Fu, M. Brumlik, and G. Voordouw, J. Bacteriol. 174 (1992) 1726.
20. A. Kuroda, T. Kumano, K. Taguchi, T. Nikata, J. Kato, and H. Ohtake, J. Bacteriol. 177 (1995) 7019.
21. D. Hanlon, and G. Ordal, J. Biol. Chem. 269 (1994) 14038.
22. J. Liu, and J. Parkinson, J. Bacteriol. 173 (1991) 4941.
23. D. Arnosti, and M. Chamberlin, Proc. Natl. Acad. Sci. USA. 86 (1989) 830.
24. J. Helmann, Mol. Microbiol. 5 (1991) 2875.
25. K. Kutsukake, Y. Ohya, and T. Ilno, J. Bacteriol. 172 (1990) 741.
26. W.J. Ingledew, and A. Houston, Biotechnol. Appl. Biochem. 8 (1986) 242.

Protein genes from *Thiobacillus ferrooxidans* that change their expression by growth under different energy sources

N. Guiliani and C. A. Jerez

Laboratorio de Microbiología Molecular y Biotecnología, Dpto. de Biología
Facultad de Ciencias, Universidad de Chile, Santiago, Chile
nguiliani@canela.med.uchile.cl - cjerez@machi.med.uchile.cl

We developed a protocol to obtain three different fractions from *Thiobacillus ferrooxidans* : soluble proteins (SP), inner membrane proteins (IMP) and outer membrane proteins (OMP). SDS-PAGE protein pattern comparisons of each of these fractions from *T. ferrooxidans* cells grown in iron or sulfur allowed us to connect the expression of some polypeptides with the energy source. We employed 2D-PAGE to isolate the individual polypeptides IMP30 and OMP40. To identify each protein, we determined the NH_2-terminal end sequences of the complete polypeptides and of some of the corresponding internal peptides. With these results, we obtained part of the coding region of the *omp40* gene by DOP-PCR and we began the identification of the *p30* gene by SSP-PCR. The Omp40 sequence obtained so far presented no similarities to other known outer membrane proteins. The partial sequence of the P30 protein showed a high level of similiraty to proteins CbbQ, NorQ and NirQ.

1. INTRODUCTION

The first step in ore oxidation by *T. ferrooxidans* requires a contact between the bacterium and the mineral (1). Adhesion of *T. ferrooxidans* to the solid results from interactions in which lipopolysaccharides and outer membrane proteins could be involved (2, 3). In the respiratory chain that *T. ferrooxidans* employes to oxidize ferrous iron aerobically, all the identified electron transport proteins have been localized in the periplasm or in the inner membrane (4, 5). However, it has been postulated that some of the first components of this chain are localized in the outer membrane (6). Moreover, they may participate in the bacterial adhesion mechanisms. We have previously compared the global protein expression of *T. ferrooxidans* cells grown in iron or sulfur media by employing two-dimensional PAGE (7). In the present report, we have employed this procedure to isolate and characterize the genes of some of these proteins.

2. MATERIALS AND METHODS

2.1. Bacterial strains, plasmids, and growth conditions

T. ferrooxidans ATCC 19589 strain was used in these studies. Growth on ferrous iron was done in modified 9K medium as before (8) and growth on elementary sulfur was done employing sulfur prills (2, 9). *Escherichia coli* JM109 strain was cultivated in Luria-Bertani medium at 37°C.

2.2. Preparation of *T. ferrooxidans* protein fractions

The cells were harvested in the late to mid-exponential growth phase by centrifugation (15,000 x g for 15 min at 4°C). The cell pellet was washed three times with acidic water (pH 2), three times with 10 mM sodium citrate pH 6.9 and one time with the sonication buffer (50 mM Tris-HCl, 10 mM EDTA, pH 8.15, 50 µg/ml RNase A). All the solutions contained 50 µg/ml PMSF. Finally, a 20 mg (wet/weight) cell pellet was resuspended in 2 ml of sonication buffer. Unless indicated otherwise all the following operations were done at 4°C. The cell suspension was sonicated (five times during 30 sec) and the lysate obtained was centrifuged at low speed (11,500 x g for 20 min) to eliminate the cellular debris. The supernatant was then centrifuged (100,000 x g for 2 h) to pellet the total membrane fraction. The supernatant was saved and employed as the soluble fraction (SP). The total membrane pellet was washed with the sonication buffer in the presence of 50 mM NaCl, resuspended in 600µl of 2% sodium laurylsarcosinate and incubated for one hour at 37°C. The suspension was centrifuged at low speed (11,500 x g for 20 min). The supernatant was centrifuged (100,000 x g for 2 h at 4°C) to pellet the outer membrane fraction. The supernatant was saved and employed as the inner membrane fraction (IMP). The pellet was then washed with the sonication buffer in the presence of 50 mM NaCl and solubilized in a 7.3% NP40, 0.18 M DTT and 9% β-Mercaptoethanol solution at 56°C for 30 min. The suspension was centrifuged at low speed (11,500 x g for 20 min) and the final supernatant was saved and employed as the outer membrane fraction (OMP).

2.3. Protein analysis

Standard 2-D PAGE (pH 5 to 7 in the first dimension) (10, 11) or 2-D non-equilibrium pH polyacrylamide gel electrophoresis (2-D NEPHGE) (pH 3 to 10 in the first dimension) (10, 11) was performed as described before for *T. ferrooxidans* (3, 7, 8). SDS-PAGE consisted of 7.5-15% polyacrylamide gradients (12).

2.4. DNA manipulations

Restriction enzyme digestions were performed according to the manufacturer's recommendations. Southern blotting was done with *T. ferrooxidans* total DNA digested with different restriction enzymes. The digested DNA fragments were seperated by electrophoresis in 0.9% agarose in TBE1X (overnight run at 35 volts). After electrophoresis, the DNA was denatured and transferred to a positively charged nylon membrane (Hybond-N$^+$, Amersham®) by the semi-dry capillary method (13). Prehybridization and hybridization reactions were performed at 42°C with the DIG Easy Buffer (Boehringer Mannhein®). Digoxigenin-labeled probes were obtained by PCR as described by Boerhinger Mannhein® with the non-degenerated primers Omp40NH2A-ND/P4023B-ND and P3023d/P3025d deduced from DOP-PCR *T. ferrooxidans* DNA fragment sequences. Detection of digoxigenin-labeled DNA fragments was accomplished by a chemiluminescent reaction with lumigen PPD as described by Boehringer Mannhein®.

The dideoxy chain termination method was employed to sequence DNA using $\gamma(^{33}P)$ dATP and the dsDNA Cycle Sequencing System from GIBCOBRL®. The DNA sequences were compiled and analysed with the UWGCG package (14).

2.5. Primers and PCR conditions

The oligonucleotide primers were purchased from "Fundación Para Estudios Biomédicos Avanzados" and Genset Corporation®. *Taq* polymerase and *Pwo* polymerase were from Promega® and Boehringer Mannhein®, respectively, and were used according to the manufacturer's recommendations. The fragments were recovered from 1% agarose gels, purified with Wizard PCR Prep (Promega®) and cloned in the pGEMt vector (Promega®). 20 mers degenerated oligonucleotides (DOPs) were designed on the basis of amino terminal end sequence determinations. 60 pmoles of each nucleotide and 25 ng of *T. ferrooxidans* total DNA were used in 50 µl reactions.
Amplification of flancking sequences was done by inverse PCR and SSP-PCR as was described before by Ochman et al. (15) and by Shyamala et al. (16), respectively.

2.5.1. P30 polypeptide

DOP-PCR. The oligonucleotide primers were (peptide sequences in parenthesis were employed to design the primers) : P30NH2A 5'-AAYATGCCNTAYTAYMGNACNGT-3' (NMPYYRTV), P3012A 5'-CAYCCNGAYTTYCARATHGT-3' (HPDFQIV), P3012B 5'-ACDATYTGRAARTCNGGRTG-3' (HPDFQIV) , P3015B 5'-TGNGCNGGRTARTTRAAR-3' (DFNYPAH).
DOP-PCR amplifications were as follows : 3 min at 95°C followed by 25 cycles at 95°C for 25 sec, 56°C for 30 sec and 72°C 30 sec, then 3 min at 72°C.
Southern Probe Labeling. It was done with the following primers : P3023d 5'-ACCGAAACGCTGTTTGGTAG-3', P3025d 5'-AGCACCGTAGCCGACGAAGT-3'.
SSP-PCR. We employed the P3023d and the pUC18 Forward Primers. SSP-PCR reactions with exact primers were performed on total *T. ferrooxidans* DNA digested by *Bam*HI and religated in pUC18, as follows : 3min at 95°C followed by 30 cycles at 95°C for 25 sec, 60°C for 30 sec and 72°C 1 min 30 sec, then 3 min at 72°C.

2.5.2. Omp40

DOP-PCR. The oligonucleotides primers were Omp40NH2A 5'-GTNTTYGGNTAYGCNCARAT-3' (VFGYAQI), P4017A 5'-TAYTAYATHCARGGNNCNTA-3' (YYIQGAY), P4017B 5'-TANGCNCCYTGDATRTARTA-3' (YYIQGAY), P4023A 5'-CAYGCNGAYGAYGTNATGGG-3' (HADDVMG) and P4023B 5'-CCCATNACRTCRTCNGCRTG-3' (HADDVMG).
DOP-PCR amplifications were as follows : 3 min at 95°C followed by 25 cycles at 95°C for 25 sec, 55°C for 30 sec and 72°C 45 sec, then 3 min at 72°C.
Southern Probe Labeling. It was done with the following primers : Omp40NH2A-ND 5'-ACCGGCGCCCAGCAGTTTGG-3', P4023B-ND 5'-TGGAGCATGGCGCCGGCGGA-3'
Inverse PCR. We employed the Omp40-1B 5'-GCACCAAAAATGAGGCCATT-3' and Omp40-2A 5'-GGCACCGCGGGTAATGAACT-3' primers. Inverse PCR reactions with exact primers were performed on total *T. ferrooxidans* DNA digested by *Ava*I and religated, as follows : 3min at 95°C followed by 30 cycles at 95°C for 25 sec, 67°C for 30 sec and 72°C 1 min, then 3 min at 72°C.

3. RESULTS AND DISCUSSION

3.1. Protein fraction isolation and analysis

We have analyzed by SDS-PAGE the three different protein fractions (SP, IMP and OMP) obtained from *T. ferrooxidans* grown in iron or sulfur (Fig. 1). In each protein fraction, it was possible to observe several proteins overexpressed in iron and other ones overexpressed in sulfur.

To search for the *p30* and *omp40* encoding genes, we purified the P30 and Omp40 polypeptides by 2-D PAGE. It was possible to determine the NH_2-terminal end sequences for P30 and Omp40 (table 1) and the NH_2-terminal end sequences of two internal peptides for each protein (table 2). We have used these sequences to obtain degenerate oligonucleotide primers to begin the corresponding gene analysis.

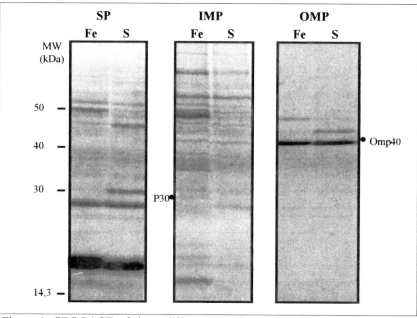

Figure 1. SDS-PAGE of three different protein fractions from *T. ferrooxidans* grown with different energy substrates. SP, Soluble Protein; IMP, Inner membrane Protein; OMP, Outer Membrane Protein. Fe, ferrous iron; S, Sulfur. The dot indicates the migrating position of Omp40 and P30.

Table 1. NH_2-terminal end sequences of proteins.

Protein	Sequence
P30	MSPEIDQYLVRNMPYYRTVA
Omp40	ADTSNADTGPVVFGYAQITGAQQFGT

Table 2. NH_2-terminal end sequences of internal peptides.

Protein	Internal peptide 1	Internal peptide 2
P30	FGALDFNYPAH(E/D)IIXEIVAH	(K)XELVQAHPDFQIVIXYNPG
OMP40	SAGAMLHADDVMGTG	GEAVPGVTYYIQGAY

3.2. Isolation of the *p30* gene

We have used the different P30 NH$_2$-terminal sequences (table 1 and table 2) to define four degenerate oligonucleotide primers. These pairs of primers were then employed in DOP-PCR experiments using purified genomic DNA from strain ATCC19589 as template. A 503 bp and a 114 bp DNA fragments were amplified with P30NH2A/P3015B and P3012A/P3015B DOPs pairs, respectively. These two DNA fragments were cloned in the pGEM-t vector and sequenced.

New primers, P3025d and P3023d were defined from the nucleotide sequence and employed to produce a digoxigenin probe which was used in Southern experiments against total DNA from *T. ferrooxidans*. After digestion with different restriction enzymes, only one DNA fragment hybridized with the probe, indicating that *T. ferrooxidans* ATCC19589 strain carries a single copy of the *p30* gene (data not shown).

```
           1                                                        50
CbbQ Ph    ~~~~~~~~MD LRNQYLVRSE PYYHAVGDEI ERFEAAYANR IPMMLKGPTG
P30 Tf     ~~~~~~~MSP EIDQYLVRNM PYYRTVADEV DLYEAAYSVR MPMMLKGPTG
NirQ Pa    ~~~~~~~~~~ ~~~~~MRDAT PFYEATGHEI EVFERAWRHG LPVLLKGPTG
NorQ Pd    ~~~~~MNAHV KTQGNGAVDA PLLPAAGDEV AVFEAAAAND LPVLLKGPTG
CONSEN     .................... .P.......E. .....A.... ....LKG TG

           51                                                       100
CbbQ Ph    CGKSRFVEYM AWKLGKPLIT VACNEDMTAA DLVGRFLLDK EGTRWQDGPL
P30 Tf     CGKTRFIEYM AWKLGKPLIT VACNEDMTAS DLVGRFLLDA SGTRWQDGPL
NirQ Pa    CGKTRFVQYM ARRLELPLYS VACHDDLGAA DLLGRHLIGA DGTWWQDGPL
NorQ Pd    CGKTRFVAHM AARLGRPLYT VACHDDLSAA DLIGRYLLKG GETVWTDGPL
CONSEN     CGK.R..E.M A......L.T VAC.....A. DL.GR.L... ..T.W.DGPL

           101                                                      150
CbbQ Ph    TTAARIGAIC YLDEVVEARQ DTTVVIHPLT DHRRILPLDK KGEVVEAHPD
P30 Tf     AIAARFGAIC YLDEVVEARQ DTTVVIHPLT DARRVLPLEK KGELVQAHPD
NirQ Pa    TRAVREGGIC YLDEVVEARQ DTTVAIHPLA DDRRELYLER TGETLQAPPS
NorQ Pd    TRAVREGAIC YLDEVVEARK DVTVVLHPLT DDRRILPIDR TGEEIEAAPG
CONSEN     ..A.R.G.IC YLDEVVEAR. D.T...HPL. D.RR.L.... .GE...A...

            151                                                     200
CbbQ Ph    FQIVISYNPG YQSAMKDLKT STKQRFAAMD FDYPAPEVES EIVAHESGVD
P30 Tf     FQIVISYNPG YQSLMKDLKQ STKQRFGALD FNYPA~~~~~ ~~~~~~~~~~
NirQ Pa    FMLVVSYNPG YQNLLKGLKP STRQRFVALR FDYPAAQQEA RILVGESGCA
NorQ Pd    FMLVASYNPG YQNILKTLKP STRQRFVAME FDFPEPAREV EIVARESGLD
CONSEN     F..V.SYNPG YQ...K..K. ST.QRF.... F..P...... ..........
```

Figure 2. Alignment of the partial P30 sequence from *T. ferrooxidans* with the CbbQ, NirQ and NorQ partial protein sequences (*Tf*). The underlined regions are presumed to be nucleotide-binding motifs. *Ph, Pseudomonas hydrotermophile* ; *Pa, Pseudomonas aeruginosa* ; *Pd, Pseudomonas denitrificans*. CONSEN, consensus sequence.

Moreover, the P3023d primer was used in SSP-PCR experiments as follows : the *T. ferrooxidans* DNA was purified and totally digested with *Bam*HI, which gave the smallest fragment as detected by Southern hybridization. The digested genomic DNA was ligated into the *Bam*HI linearized pUC18 plasmid.

The resulting population of ligated molecules was comprised of a complex mixture of chromosomal DNA and chromosomal-plasmid DNA hybrids. The plasmid derived region of the hybrid molecules provided the downstream priming site for the PCR amplification; the second site was given by the *p30* sequenced region. We have therefore amplified a 803bp DNA fragment that had been cloned in pGEM-t and sequenced.

This chromosome-walking approach led us to determine precisely the *p30* encoding region and 270 bp uspstream this region.The analysis of the 803 bp nucleotide sequence revealed that a possible Shine-Dalgarno sequence reading AGGAG located 6 bp upstream of the *p30* ATG. The partial P30 sequence presents a high level of similarity, near to 48%, with proteins CbbQ, NirQ, and NorQ (Fig. 2).

3.3. Isolation of the *omp40* gene

We have used the Omp40 NH_2-terminal end sequence (17) and the NH_2-terminal end sequences from Omp40 internal peptides (not shown) to define five degenerate oligonucleotide primers able to form four pairs of convergent primers.

These pairs were then employed in DOP-PCR experiments using purified genomic DNA from strain ATCC19589 as template. A 149 bp and a 444bp DNA fragments were amplified with Omp40NH2A/P4017B and Omp40NH2A/P4023B DOPs pairs, respectively. These two DNA fragments were cloned in the pGEM-t vector and sequenced (Fig. 3). Two new primers, Omp40NH2A-ND and P4023B-ND, were defined from the nucleotide sequence and used to produce a digoxigenin-labeled probe which was tested by Southern experiments against total DNA from *T. ferrooxidans*. In all the cases, only one DNA fragment hybridized with the probe indicating that the ATCC19589 strain carries a single copy of the *omp40* gene (data not shown).

Two others primers, Omp40-1B and Omp40-2A, were defined and employed in reverse PCR experiments as follows : the *T. ferrooxidans* DNA was purified and totally digested with the *Ava*I restriction enzyme that gave the smallest fragment by Southern analysis. We then amplified a 900bp DNA fragment that was cloned in pGEM-t. Its sequencing is currently in progress.

The analysis of the 444 bp nucleotide sequence revealed that only one open reading frame (Fig. 3) contained the correct sequences upstream and downstream of the initial peptide sequence used to define the degenerate oligonucleotide primers. The search in different data banks failed to give any positive similarity. However , to establish if Omp40 has a different structure due to the acidophilic character of *T. ferrooxidans*, the entire Omp40 sequence will be required.

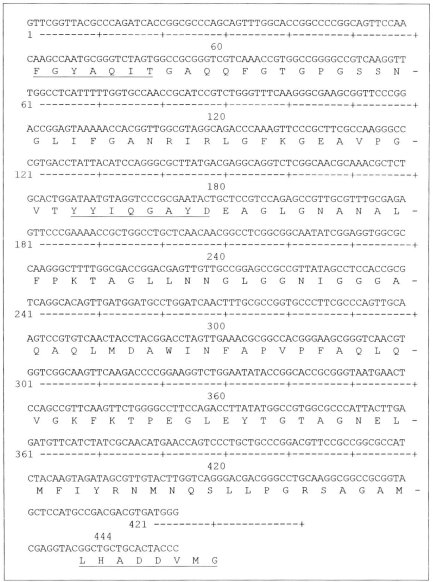

```
GTTCGGTTACGCCCAGATCACCGGCGCCCAGCAGTTTGGCACCGGCCCCGGCAGTTCCAA
1  ---------+---------+---------+---------+---------+---------+
                              60
CAAGCCAATGCGGGTCTAGTGGCCGCGGGTCGTCAAACCGTGGCCGGGGCCGTCAAGGTT
  F  G  Y  A  Q  I  T  G  A  Q  Q  F  G  T  G  P  G  S  S  N  -

TGGCCTCATTTTTGGTGCCAACCGCATCCGTCTGGGTTTCAAGGGCGAAGCGGTTCCCGG
61 ---------+---------+---------+---------+---------+---------+
                             120
ACCGGAGTAAAAACCACGGTTGGCGTAGGCAGACCCAAAGTTCCCGCTTCGCCAAGGGCC
   G  L  I  F  G  A  N  R  I  R  L  G  F  K  G  E  A  V  P  G  -

CGTGACCTATTACATCCAGGGCGCTTATGACGAGGCAGGTCTCGGCAACGCAAACGCTCT
121 ---------+---------+---------+---------+---------+---------+
                             180
GCACTGGATAATGTAGGTCCCGCGAATACTGCTCCGTCCAGAGCCGTTGCGTTTGCGAGA
   V  T  Y  Y  I  Q  G  A  Y  D  E  A  G  L  G  N  A  N  A  L  -

GTTCCCGAAAACCGCTGGCCTGCTCAACAACGGCCTCGGCGGCAATATCGGAGGTGGCGC
181 ---------+---------+---------+---------+---------+---------+
                             240
CAAGGGCTTTTGGCGACCGGACGAGTTGTTGCCGGAGCCGCCGTTATAGCCTCCACCGCG
   F  P  K  T  A  G  L  L  N  N  G  L  G  G  N  I  G  G  A  -

TCAGGCACAGTTGATGGATGCCTGGATCAACTTTGCGCCGGTGCCCTTCGCCCAGTTGCA
241 ---------+---------+---------+---------+---------+---------+
                             300
AGTCCGTGTCAACTACCTACGGACCTAGTTGAAACGCGGCCACGGGAAGCGGGTCAACGT
   Q  A  Q  L  M  D  A  W  I  N  F  A  P  V  P  F  A  Q  L  Q  -

GGTCGGCAAGTTCAAGACCCCGGAAGGTCTGGAATATACCGGCACCGCGGGTAATGAACT
301 ---------+---------+---------+---------+---------+---------+
                             360
CCAGCCGTTCAAGTTCTGGGGCCTTCCAGACCTTATATGGCCGTGGCGCCCATTACTTGA
   V  G  K  F  K  T  P  E  G  L  E  Y  T  G  T  A  G  N  E  L  -

GATGTTCATCTATCGCAACATGAACCAGTCCCTGCTGCCCGGACGTTCCGCCGGCGCCAT
361 ---------+---------+---------+---------+---------+---------+
                             420
CTACAAGTAGATAGCGTTGTACTTGGTCAGGGACGACGGGCCTGCAAGGCGGCCGCGGTA
   M  F  I  Y  R  N  M  N  Q  S  L  L  P  G  R  S  A  G  A  M  -

GCTCCATGCCGACGACGTGATGGG
                  421 ---------+-------------+
                              444
CGAGGTACGGCTGCTGCACTACCC
           L  H  A  D  D  V  M  G
```

Figure 3. Partial sequence of the *omp40* gene obtained by DOP-PCR. The amino acid sequences which allowed to design the degenerated oligonucleotides are underlined.

4. CONCLUSIONS

In this work, we have isolated two new genes from the *T. ferrooxidans* ATCC19589 strain.

The first one, *omp40*, encodes for the major outer membrane protein , Omp40. This protein presents some electrophysiological (18) and biochemical (Guiliani and Jerez, in preparation) aspects that suggest to us that it could be a porin-like protein. In *Neisseria gonorrhoeae*, the porin polypeptide contributes to the surface charge (19) and in *Rhanella aquatilis* the major outer membrane protein has a double functionality, as porin and as adhesin (20). In *T. ferrooxidans*, the surface charge depends on the energetic substrate in which the microorganism is grown (21). The cell adhesion of *T. ferrooxidans* depends on the outer membrane surface hydrophobicity, specially when the pH is less than 2.2 (2, 22). Because the Omp40 protein expression changes with the external pH (8), and it may contribute to cell adhesion (2), it will be of great interest to investigate the role of this protein in *T. ferrooxidans* cell attachment to ores.

The second gene we studied, *p30*, encodes a protein, P30, with two presumed ATP-binding motifs. The P30 sequence shows a high level of similarity to CbbQ, NirQ and NorQ. P30 does not contain a signal sequence, suggesting its cytoplasmic location. One possible hypothesis to explain its copurification with the PMI fraction, could be that, for its function, P30 is associated strongly to the internal side of the inner membrane. In *P. hydrogenothermophila*, the *cbbQ* gene is located downstream of the *cbbLS* operon that encodes for the two subunits of the ribulose 1,5-biphosphate carboxylase/oxygenase (RuBisCo) (23). In the *T. ferrooxidans* Fe1 strain, the presence of two sets of an *rbc* operon (equivalent to the *cbb* operon) has been demonstrated (24). No gene corresponding to the *cbbQ* gene was found in the two downstream operon sequences. The *nirQ* and *norQ* genes are located in the denitrification gene cluster from different bacterial species (25, 26, 27). However the CbbQ, NirQ and NorQ functions are unclear. They could affect the post-translational activation and/or assembly of oligomeric structures (23, 25). A recent study has revealed that the *nirQ* gene can activate the RuBisCo from *P. hydrogenothermophila* (28). Our hypothesis is that all of these proteins belong to the same protein family and could have some common functional aspects. For this reason, we think P30 could participate in the post-translational activation and/or assembly of an oligomeric structure located in the inner membrane and involved in *T. ferrooxidans* iron energetic metabolism.

AKNOWLEDGEMENT

This research was supported by FONDECYT P3960002, ICGEB 96/007 and Universidad de Chile.

REFERENCES

1. W. Sand, T. Gerke, R. Hallamann and A. Schippers, Appl. Microbiol., Biotechnol., 43 (1995) 961.
2. R. Arredondo, A. Garcia and C. A. Jerez, Appl. Environ. Microbiol., 60 (1994) 2846.

3.   A. M. Amaro, M. Seeger, R. Arredondo, M. Moreno and C. A. Jerez. (1993). *In* A. E. Torma, M. L. Apel and C. L. Brierley (ed.), The Minerals, Metals and Materials Society. Vol. II p. 577.

4.   R. C. Blake II, E. A. Shute, J. Waskovsky and A. P. Harrison Jr, J. of Geomicrobiol., 10 (1992) 173.

5.   T. Yamanaka, T. Yani, M. Kai, H. Tamegai and Y. Fukumori (1993). *In* A. E. Torma, M. L. Apel and C. L. Brierley (ed.), The Minerals, Metals and Materials Society. Vol II p. 453.

6.   C. F. Kulpa, N. Mjoli and M. T. Roskey, Biotechnol. Bioeng. Symp., 16 (1986) 289.

7.   G. Osorio, P. Varela, R. Arredondo, M. Seeger, A. M. Amaro, and C. A. Jerez (1993). *In* A. E. Torma, M. L. Apel and C. L. Brierley (ed.), The Minerals, Metals and Materials Society. Vol. II p. 565.

8.   A. M. Amaro, D. Chamorro, M. Seeger, R. Arredondo, I. Peirano and C. A. Jerez,. J. Bacteriol., 173 (1991) 910.

9.   D. Chamorro, R. Arredondo, I. Peirano and C. A. Jerez (1988). *In* P. R. Norris and D. P. Kelly (ed.), Biohydrometallurgy. Science and technology Letters, London, UK. p. 135.

10.   P. H. O' Farrell, J. Biol. Chem., 250 (1975) 4007.

11.   P. Z. O' Farrell, H. M. Goodman and P. H. O' Farrell, Cell 12 (1977) 1133.

12.   U. K. Laemmli, Nature, 227 (1970) 680.

13.   J. Sambrook, E. F. Fritsch and T. Maniatis (1989). Molecular cloning; a laboratory manuel. Cold Spring Harbour Laboratory, Cold Spring Harbor N. Y.

14.   Wisconsin Package Version 9.1, Genetics Computer Group (GCG), Madison, Wisc.

15.   H. Ochman, M. M. Medhora, D. Garza and D. L. Hartl (1990). *In* PCR protocols, a guide to methos and applications, M. A. Innis, D. H. Gelfand, J. J. Sninsky, and T. J. White (Eds) Academic Press, Inc. Harcourt Brace Jovanovich, Publishers. pp 219.

16.   V. Shyamala, E. Schneider and G.F.-L. Ames, EMBO J. 9 (1990) 939.

17.   C. A. Jerez, M. Seeger and A. M. Amaro, FEMS Microbiol. Lett., 98 (1992) 29.

18.   M. Silva, A. Ferreira, M. Rodríguez and D. Wolff, FEBS Lett., 296 (1992) 169.

19.   J. Swanson, D. Dorward, L. Lubke and D. Kao, J. Bacteriol., 179 (1997) 3541.

20.   W. Achouak, J. M. Pages, R. De Mot, G. Molle and T. Heulin, J. Bacteriol., 179 (1997), 3541.

21.   P. Devasia, K. A. Natrajan, D. N. Sathyanarayana and G. Ramananda-Rao, Appl. Environ. Microbiol., 59 (1993) 4051.

22.   J. A. Solari, G. Huerta, B. Escobar, T. Vargas, R. Badilla-Ohlbaum and J. Rubio, Colloid Surfaces, 69 (1992) 159.

23.   K. Yokoyama. N. R. Hayashi, H. Arai, S. Y. Chung, Y. Igarashi and T. Kodama, Gene, 153 (1995) 75.

24.   T. Kusano, T. Takeshima, C. Inoue and K Sugawara, J. Bacteriol., 173 (1991) 7313.

25.   A. Jüngst and W. G. Zumft, FEBS Lett., 314 (1992) 308.

26.   A.P. De Boer, J. Van Der Oost, W.N. Reijnders, H.V. Westerhoff, A.H. Stouthamer and R.J. Van Spanning, Eur J. Biochem., 242 (1996) 592.

27.   T.B. Bartnikas, I.E. Tosques, W.P. Laratta, J. Shi and J.P. Shapleigh, J. Bacteriol., 179 (1997) 3534.

28.   N.R. Hayashi, H. Arai, T. Kodama and Y. Igarashi, Biochim. Biophys. Acta, 1381 (1998) 347.

Strain diversity of *Thiobacillus ferrooxidans* and its significance in biohydrometallurgy

T. F. Kondratyeva, T. A. Pivovarova, L. N. Muntyan and G. I. Karavaiko

Institute of Microbiology, Russian Academy of Sciences, Pr. 60-letiya Oktyabrya 7, k. 2, Moscow, 117811 Russia[*]

Structural changes in chromosomal DNA of *Thiobacillus ferrooxidans* strains that occur under the influence of varied growth conditions were studied by pulsed-field gel electrophoresis. Strain diversity of *T. ferrooxidans* was manifested in different growth rates and oxidation rates of inorganic substrates under extreme conditions, in different resistance to metal ions and low pH values, and also in polymorphism of the chromosomal DNA fragments generated by the macrorestriction endonucleases. Adaptation of some strains to growth on media containing new substrates was accompanied by changes in the number and size of restriction fragments. Thus, new 177 and 164 kb DNA fragments were revealed after the substitution of the oxidation substrate from Fe^{2+} to FeS_2 or from Fe^{2+} to S^0, respectively, whereas 115 and 77 kb DNA fragments disappeared. The switching from Fe^{2+} to S^0 resulted in the change in the number of 27 kb DNA fragments. Another type of chromosomal DNA variability was found in the strains adapted to high concentrations of metal ions. A comparison of *XbaI*-restriction patterns in parent strains and in strains with acquired enhanced resistance to zinc (from 40 to 70 g/l) or arsenic (from 1.5 to 4.0 g/l) revealed amplification of 98 and 28 kb fragments, respectively. When both strains were subcultured on medium with Fe^{2+} without the inducing factors, amplification of DNA fragments was no longer detectable. However, the strain adapted tp 50 g/l of Fe^{2+}/Fe^{3+} had a mutation in the structure of chromosomal DNA. The data obtained on the natural and experimental genomic variability of *T. ferrooxidans* strains provide biotechnologists with practical recommendations for selection aimed at the intensification of bioleaching processes and testify about possibilities of strain monitoring in natural and technological conditions. Strains with the labile genome have an advantage in biohydrometallurgy.

1. INTRODUCTION

The role of bacterial strain variability in biohydrometallurgical processes has not yet received due attention. The efficiency of processes is usually evaluated by the activity of

*This work was supported by the Russian Foundation for Fundamental Research (grant no. 96-04-48287) and the State Program "Novel Methods in Bioengineering"

individual species of chemolithotrophic bacteria, e.g. *T. ferrooxidans*, *T. thiooxidans,* etc. It is however known that bacterial strains isolated from different ecological niches vary in such phenotypic properties as the resistance to metal ions, pH values, or the capacity for mineral substrate oxidation [1, 2]. This phenotypic variability is the result of the activity of genomic regulatory systems and results in the adaptation of the organism to new extremal environmental conditions. The relationship between this unique physiological variability of *thiobacilli* and the state and lability of their genome remains obscure. Genotypic variability is known to include changes in the DNA nucleotide sequences, chemical modification of nucleotides, duplications, insertions, inversions, excisions, translocations, alterations in the plasmid profiles, etc.

The aim of this work was to study phenotypic and genotypic strain diversity of *T. ferrooxidans* and analyze its significance in biohydrometallurgy.

2. MATERIALS AND METHODS

The strains of *T. ferrooxidans* used in this study were isolated from various environments (mine waters, ore deposits and ore concentrates, dense pulps obtained by processing complex sulfide concentrates), or adapted to growth in media with high concentrations of metal ions (Table 1, Fig.1,3).

Table 1
Highly active strains of *T. ferrooxidans* and their resistance to low pH values and metal ion concentration (g/l)

Strain	Resistance characteristics
TFL-1	dominates at pH 1.3
TFL-2	dominates at pH 1.5-1.8
TFL-3	dominates at pH 2.5-3.0
TFBk	resistant to pH 1.0 and Cu^{2+}, 20; Fe^{2+}, 40; Fe^{3+}, 40; As^{3+}, 3.0; As^{5+}, 1.5
TFV-1	resistant to Cu^{2+}, 17.5; As^{3+}, 2.0; As^{5+}, 2.0; Hg^{2+}, 0.4
TFI-Fe	resistant to pH 1.25 and As^{3+}, 3.0; Fe^{2+}, 50; Fe^{3+}, 50
458As2	resistant to pH 1.3 and As^{3+}, 4.0; As^{5+}, 2.0
T-2	resistant to As^{3+}, 2.0; As^{5+}, 3.0; Zn^{2+}, 60; Hg^{2+}, 0.2; Co^{2+}, 0.24; Cd^{2+}, 0.56; Ni^{2+}, 0.18
TFR2	resistant to Hg^{2+}, 0.0016; As^{3+}, 1.5; As^{5+}, 1.0

Batch cultivation of the strains was performed at 28^0C on a shaker (180 rpm) in 250-ml Erlenmeyer flasks containing 100 ml medium or in 2000-ml flasks containing 500 ml medium, or in aerated 5-l reactors containing 3 l medium. The initial pH of the medium was 1.8 - 2.0. Substrates used as energy sources were $FeSO_4$, S^0, FeS_2, FeAsS or gold-pyrite concentrates obtained from several deposits.

Preparation of intact chromosomal DNA of *T. ferrooxidans*, its digestion with the restriction endonuclease *XbaI* and separation of the DNA fragments by pulsed-field gel electrophoresis (PFGE) were described earlier [3]. The contents of zinc and copper in the medium were determined with Perkin Elmer 3100 atomic absorption spectrometer, the content of iron - by trilonometric titration.

3. RESULTS

3.1. Phenotypic and genotypic diversity of natural and industrial strains of *T. ferrooxidans*

Isolates used in this study differed in the kinetics of sulfide mineral oxidation (data not shown) and resistance to heavy metal ions and pH values (Table 1). Their phenotypic characteristics correlates with the conditions in the environments they were isolated from. The most resistant and efficient strains were isolated from dense pulps involved in technological processes: i.e. TFBk, TFN-d, TFG.

In addition to phenotypic variations found in *T. ferrooxidans* strains studied, pulsed-field gel electrophoresis of their DNA fragments generated by the restriction endonuclease *XbaI* allowed us to reveal strain polymorphism of the chromosomal DNA structure. Each strain was found to have its unique *XbaI*-restriction pattern of chromosomal DNA. Figure 1 shows *XbaI*-restriction patterns of several strains differing in the number and length of the DNA fragments generated by the endonuclase.

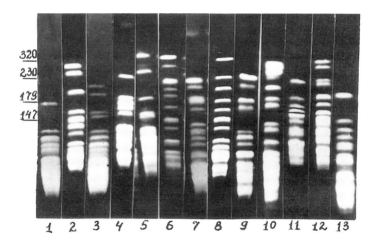

Figure 1. *XbaI*-restriction patterns of chromosomal DNA from different *T. ferrooxidans* strains: 1, TFD; 2, VKM B-1160; 3, TF1292; 4, TFG; 5, TFN; 6, TFBk; 7, TFM; 8, VKM B-458; 9, ATCC19859; 10, TFV-1; 11, TFW; 12, TF97; 13, TFR2. The PFGE conditions were 12 V/cm, 25-s pulse, 44 h run at 12-14^{0}C. The left panel is size markers, kb.

To reveal the reasons for strain polymorphism, in particular that one manifested in the chromosomal DNA structure, we studied the following three, in our opinion most essential, factors: the type of inorganic substrate used, the content of metal ions and the pH value of the medium. In these experiments we used *T.ferrooxidans* strains isolated from mine waters (VKM B-458) and from the pulp obtained by processing gold-arsenic concentrates (TFBk, TFN-d). Analysis of the alterations in the chromosomal DNA structure occurring during adaptation of strains to various oxidation substrates, other conditions (pH, Eh, the pulp density) kept unchanged, showed the following results. Ten passages of *T. ferrooxidans* VKM B-458 on media with FeS_2, FeAsS, S^0, and concentrates obtained from ores containing pyrrhotine, arsenopyrite, pyrite, or from pyrrhotine-free ores caused no changes in the *XbaI*-restriction patterns of chromosomal DNA (data not shown). Several passages of strain TFBk on media containing FeS_2, S^0 or ore concentrate containing pyrrhotine, pyrite and arsenopyrite caused changes in the *XbaI*-digested samples of the DNA (Fig. 2). Thus, after passages on the medium with FeS_2, a new band consisting of 177-kb fragments appeared in the restriction digests of chromosomal DNA (Fig. 2- line 1). After several subsequent passages on medium containing $FeSO_4$, strain TFBk restored the initial chromosomal DNA structure (Fig. 2- line 2). Five passages of strain TFBk on medium containing elemental sulfur as the only energy source resulted in the appearance of a new band consisting of 164-kb fragments in the *XbaI*-digested sample of the chromosomal DNA (Fig. 2- line 3). At the same time, a fragment of 77 kb disappeared and the number of 27-kb fragments decreased. The readaptation of strain TFBk to growth on $FeSO_4$-containing medium restored the initial chromosomal DNA restriction pattern characteristic of the culture grown on medium with

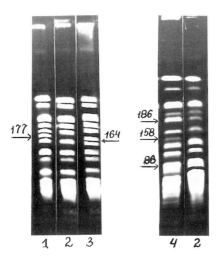

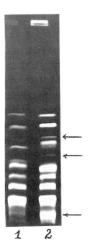

Figure 2. Changes in the structure of chromosomal DNA of *T. ferrooxidans* TFBk caused by its adaptation to various oxidation substrates. 1, FeS_2; 2, Fe^{2+}; 3, S^0; 4, gold-arsenic pirrhotine-containing concentrate. PFGE conditions the same as indicated in Fig. 1. The changed fragments are indicated by an arrow.

Figure 3. Changes in the structure of chromosomal DNA of *T. ferrooxidans* TFN-d caused by its adaptation to gold-arsenic pyrite-arsenopyrite concentrate. 1, Fe^{2+}; 2, concentrate. PFGE conditions the same as indicated in Fig. 1. The bands containing new fragments are indicated by an arrow.

ferrous iron. When strain TFBk was adapted to growth on pyrrhotine-containing gold-arsenic ore concentrate, a new band consisting of 158-kb fragments appeared in the *XbaI*-restriction pattern of DNA (Fig. 2- line 4). The changes were found in the fragments of smaller sizes, those of 88 and 77 kb and also in the increased fluorescence intensity of the band consisting of 186-kb DNA fragments. Passages of *T. ferrooxidans* TFN-d on medium containing gold-arsenic pyrite-arsenopyrrhitine concentrate also caused changes in the chromosomal DNA structure that were manifested in the size and number of fragments generated by the *XbaI*-restriction endonuclease (Fig. 3).

Metal ion concentration in the medium was another essential factor studied. We analyzed the influence of As, Zn, Fe and Cu on the genome structure of *T. ferrooxidans* strains adapted to each of the above elements. The strain TFY, resistant to 40 g Zn^{2+} per liter, was adapted to increasing zinc concentrations, which resulted in the selection of strain TFZ resistant to 70 g/l Zn^{2+}. In the *XbaI*-restriction pattern of chromosomal DNA from strain TFZ we observed amplification of the 98-kb fragment (Fig. 4- line 1, 2). Adaptation of strain VKM B-458 to increasing As^{3+} concentrations allowed obtaining strain 458As2 whose resistance to arsenic increased from 1.5 to 4.0 g/l. This caused amplification of the 28-kb fragment in the *XbaI*-restriction pattern (Fig. 4- line 3, 4). The amplification of restriction fragments revealed in the strains resistant to zinc and arsenic disappeared after several passages on iron-containing medium in the absence of the inducing factors. Therefore, these changes in the DNA structure were not caused by mutations as well by replication of plasmid DNA. T.ferrooxidans plasmids are cryptic and not responsible for metal resistance.

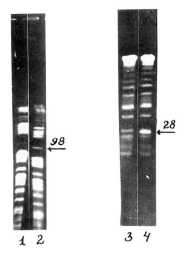

Figure 4. Amplification of the DNA fragments in *T. ferrooxidans* strains adapted to increased concentrations of Zn^{2+} and As^{3+}. 1, TFY; 2, TFZ; 3, VKM B-458; 4, 458As2. The PFGE conditions were 13 V/cm, 10-s pulse for the first two strains and 5-s pulse for the latter two strains, 68 run at 13^0C for the first two strains and at 20^0C for the latter two strains. Arrows indicate the amplified fragments.

A long-term continuous cultivation of *T. ferrooxidans* TFI (isolated from the flotation tails of zinc-bearing polymetallic sulfide ore processing) on the medium with increasing Fe^{2+}/Fe^{3+}

concentration resulted in the selection of strain TFI-Fe that was resistant to 50 g/l Fe^{2+}/Fe^{3+}. Pulsed-field gel electrophoretic separation of the $XbaI$-digested chromosomal DNAs revealed differences between the parent and adapted strain, i.e., the appearance of a new fragment (Fig. 5- line 2). Repeated passages of the adapted strain on medium containing 9 g/l Fe^{2+} did not result in its readaptation. The changes in the chromosomal structure appeared stable, which allowed the new strain to be regarded as a mutant.

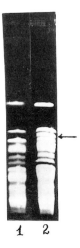

Figure 5. Changes in the structure of chromosomal DNA from *T. ferrooxidans* strain TFI caused by its adaptation to 45 g/l Fe^{2+}/Fe^{3+}. 1, the parent strain TFI; 2, the mutant strain TFI-Fe. PFGE conditions the same as indicated in Fig. 1. The band containing new fragments is indicated by an arrow.

Adaptation of *T. ferrooxidans* strains VKM B-458, TFV-1 and TFBk to increased concentrations of Cu^{2+} (20 passages) resulted in the increase in their resistance from 1.0 to 15, from 2.0 to 17.5 and from 2.5 to 20 g/l, respectively. However, no structural changes were revealed in the chromosomal DNA of either of three strains.

During the adaptation to growth at low pH values, strain VKM B-458 failed to grow below the pH 1.33, whereas strain TFBk, after 13 passages, could develop at pH 1.05. However, as well as in the case of adaptation to Cu^{2+}, this did not cause any changes in the $XbaI$-restriction pattern of its chromosomal DNA. It is not improbable that in the case of employing other restriction endonucleases some changes could be revealed in the nucleotide sequences located between the restriction sites of endonuclease $XbaI$.

3.2. Monitoring of *T. ferrooxidans* strains in technological processes

Pulsed-field gel electrophoresis was used for monitoring the dominating strains of *T. ferrooxidans* in several pilot-plant developments of biohydrometallurgical processing of gold-arsenic pyrite-arsenopyrite concentrates, including those containing pyrrhotine, and also zinc-containing industrial products. Several regularities were revealed that allowed the input of individual *T. ferrooxidans* strains into the processes realized in the reactors to be evaluated.

(1) In the absence of aboriginal *T. ferrooxidans* strains in the processed substrate, the highest competitive capacity is shown by the strains that possess the labile genome, i. e. by those with structural changes in the chromosomal DNA induced by alterations in the medium parameters. When a certain essential growth parameter, e. g., the pH value is changed in the technological process, another strain may come to dominate (Fig. 6).

(2) Aboriginal strains, which are most adapted to the complex of parameters characteristic of a given substrate, usually overgrow all other strains used for inoculation.

(3) As indicated above, when the process is long underway (for years) under the extreme conditions (50 g/l Fe^{2+}/Fe^{3+}), a mutant can appear with changes in the genome and new phenotypic characteristics.

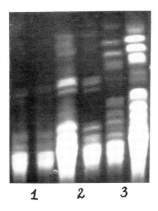

Figure 6. The change of the dominating *T. ferrooxidans* strains under the influence of experimental pH variations in the medium with zinc-containing industrial product. 1, strain TFL-1, pH 1.3; 2, strain TFL-2, pH 1.5-1.8; 3, strain TFL-3, pH 2.5-3.0. PFGE conditions the same as indicated in Fig. 1.

4. CONCLUSION

Peculiarities in the structure of chromosomal DNAs found in numerous strains of *T. ferrooxidans* isolated from various ore deposits and dense pulps obtained by processing complex sulfide concentrates, testify about their phenotypic and genotypic polymorphism. Each strain was found to have its unique *XbaI*-restriction pattern of chromosomal DNA. Studies on the influence of varied growth conditions (the oxidation substrate, metal ion concentration, pH value) on the chromosomal DNA structure of *T. ferrooxidans* allowed us in certain cases to reveal a genomic strain variability and ascertain the role of the oxidation substrate in this variability. It may be suggested that in the process of evolution, in various ecological niches certain structural genomic changes could become a stable genotypic characteristic, thus providing a strain polymorphism of the chromosomal DNA structure. Isolation of the *T. ferrooxidans* mutant resistant to 50 g/l Fe^{2+}/Fe^{3+} confirms the possibility of obtaining heritable genomic changes. However, the regulatory capacities of genome are so high in *T. ferrooxidans* that even the extreme factors of the medium (70 g/l Zn^{2+}, 4.0 g/l As^{3+}, 20 g/l Cu^{2+} or pH 1.05) did not cause the heritable changes in the genome.

The stable and unique structural characteristics of chromosomal DNA in *T. ferrooxidans* strains allowed us to use pulsed-field gel electrophoretic analysis for identification of strains and their monitoring in biohydrometallurgical processes.

REFERENCES

1. A.P.Jr. Harrison, Arch. Microbiol., 131 (1982) 68.
2. G. Karavaiko, J. Min. Met., 33 (1997) 51.
3. T.F. Kondratyeva, L.N. Muntyan and G.I. Karavaiko, Microbiology UK, 141 (1995) 1157.

Purification and characterization of 3-isopropylmalate dehydrogenase of acidophilic autotroph *Thiobacillus thiooxidans*

H. Kawaguchi[*], K. Inagaki, H. Matsunami, Y. Nakayama, T. Tano[*], and H. Tanaka

Department of Bioresources Chemistry, Faculty of Agriculture, Okayama University, Okayama, 700-8530, Japan

3-Isopropylmalate dehydrogenase was purified to homogeneity from the acidophilic autotroph *Thiobacillus thiooxidans*. The native enzyme molecule is a dimer of molecular weight 40,000. The K_m value for 3-isopropylmalate was estimated to be 0.13 mM and that for NAD$^+$ 8.7 mM. The optimum pH and temperature for the activity are 9.0 and 65, respectively. The properties of the enzyme are similar to those of the *Thiobacillus ferrooxidans* enzymes, except for substrate specificity. *T. ferrooxidans* 3-isopropylmalate dehydrogenase is able to utilize alkyl-malate as substrate in addition to 3-isopropylmalate. However, *T. thiooxidans* 3-isopropylmalate is not able to utilize malate as a substrate.

1. INTRODUCTION

3-Isopropylmalate dehydrogenase (EC 1.1.1.85), a key enzyme in leucine biosynthesis, catalyzes the oxidative decarboxylation of the substrate 3-isopropylmalate to 2-oxoisocaproate simultaneously with dehydrogenation. The enzyme has been found in a wide variety of bacteria and the amino acid sequence of the enzyme from various bacteria shows high similarity (1-10). *Thiobacillus thiooxidans* is a chemolithotrophic, acidophilic bacterium that obtains energy from the oxidation of reduced inorganic sulfur compounds. We previously cloned the *leuB* gene coding for 3-isopropylmalate dehydrogenase of an acidophilic chemolithotrophic bacterium, *Thiobacillus ferrooxidans*, in *Escherichia coli* (11) and purified the enzyme to homogeneity from *E. coli* cells harboring a recombinant plasmid containing the *leuB* gene (12). The *T. ferrooxidans* enzyme utilizes various alkyl-malate as the substrate as well as 3-isopropylmalate (13) similarly to an extreme thermophile, *Thermus thermophilus* (14) and thermoacidophilic archaeon, *Sulfolobus* sp. strain 7 (15). The crystal structure of *T. ferrooxidans* 3-isopropylmalate dehydrogenase complexed with 3-isopropylmalate at 2.0 resolution was also determined (16). The structure shows a fully closed conformation. The γ-isopropyl group of substrate 3-isopropylmalate is recognized by a unique hydrophobic pocket which includes Glu88, Leu91, Leu92 and Val193'. In this paper, we described the purification and characterization of the 3-isopropylmalate dehydrogenase from *Thiobacillus thiooxidans*.

---

* Present address: Kurashiki Sakuyo University, Kurashiki, Okayama, 710-0292, Japan

2. MATERIALS AND METHODS

2.1. Bacteria strains and medium

T. thiooxidans ON107 was grown on an iron-based medium with 0.25% $Na_2S_2O_3$ as a sole energy source at 30°C. The pH of bacterial broth and $Na_2S_2O_3$ concentration were kept 5.0 with K_2CO_3 and 0.25% (wt/vol) by feeding, respectively.

2.2. Enzyme assay

The standard assay mixture consisted of 100 mM Tris-HCl buffer (pH 9.0), 0.5 mM $MgCl_2$, 50 mM KCl, 6.7 mM NAD^+, 0.67 mM 3-isopropylmalate, and appropriately diluted enzyme preparation in a final volume of 1.5 ml (17). Enzyme activity was measured by monitoring the production of NADH at 340 nm on a Beckman DU-65 spectrophotometer. One unit of the enzyme is defined as the amount of enzyme that catalyzes the formation of 1 μmol of NADH per min at pH 9.0. Protein concentration was determined by the method of Bradford with bovine serum albumin as a standard (18).

2.3. Electrophoresis

Sodium dodecyl sulfate (SDS) disc gel electrophoresis was carried out according to Laemmli (19).

2.4. Sequencing of N-terminal amino acid

After SDS-polyacrylamide gel electrophoresis, the separated protein was transferred to PVDF membrane using Sartoblot II-S (Sartorius). The transferred protein was detected with 0.1% coomassie brilliant blue and sequenced by Edman degradation with an Applied Biosystems Model 477A gas liquid phase protein sequencer.

3. RESULTS

3.1. Enzyme purification

All procedures describes below were performed at 4°C.

Disruption of cells: Frozen cells (76 g) were suspended in 50 mM potassium phosphate buffer (pH 7.5) containing 0.01% 2-mercaptoethanol and 10% glycerol. The suspension was disrupted by ultrasonic oscillation (KUBOTA INSONATOR Model 201M) in an ice bath for 10 min. Cell debris and unbroken cells were removed by centrifugation at 105,000 x g for 60 min. The extract was dialyzed at 4°C for 12 hr against 10 mM potassium phosphate buffer (pH 7.5) containing 0.01% 2-mercaptoethanol and 10% glycerol (Buffer A).

DEAE-Toyopearl 650M column chromatography: The dialyzed solution was applied onto a column of DEAE-Toyopearl 650M (Tosoh) equilibrated with Buffer A containing 50 mM KCl. After washing with Buffer A, the enzyme was eluted with Buffer A containing 100 mM KCl.

Q-Sepharose Fast Flow column chromatography: The enzyme solution was applied to a Q-Sepharose Fast Flow column (Pharmacia) equilibrated with Buffer A. After washing with Buffer A, the active fractions were eluted with Buffer A containing 200 mM KCl. The fraction was combined, concentrated and dialyzed against Buffer A.

Butyl-Toyopearl 650M column chromatography: The enzyme solution was brought to

25% saturation with ammonium sulfate (Buffer A) and then applied to a Butyl-Toyopearl 650M column (Tosoh) equilibrated with Buffer A. After a thorough wash with the same buffer, the enzyme was eluted with a linear ammonium sulfate gradient of 25 to 0% in Buffer A. The active fractions were pooled.

Sephacryl S-200 column chromatography: The active fractions were pooled, concentrated and dialyzed against Buffer A containing 200 mM KCl and was applied to a Sephacryl S-200 column (Pharmacia) equilibrated with Buffer A containing 200 mM KCl. The enzyme was eluted with the same buffer.

Phenyl-Toyopearl 650M column chromatography: The concentrated enzyme solution was brought to 25% saturation with ammonium sulfate (Buffer A) and applied to a Phenyl-Toyopearl 650M column (Tosoh) equilibrated with Buffer A. After washing with the same buffer, the enzyme was eluted with Buffer A containing 17.5% saturation with ammonium sulfate.

Mono Q HR 5/5 anion exchange column chromatography: The concentrated active fraction was applied to a Mono Q column (Pharmacia) equilibrated with 20 mM Tris-HCl buffer (pH 7.5). The enzyme solution was eluted with a linear KCl gradient of 120 to 140 mM in Tris-HCl buffer (pH 7.5).

The purification procedure is summarized in Table 1.

Table 1
Purification of 3-isopropylmalate dehydrogenase from *Thiobacillus thiooxidans*

Step	Total Protein (mg)	Total Activity (units)	Specific activity[a] (units/mg)	Yield (%)
Crude extract	1810	77.3	0.4428	100
DEAE-Toyopearl 650M	58.7	21.2	0.361	27.4
Q-Sepharose Fast Flow	38.3	21.9	0.572	28.3
Butyl-Toyopearl 650M	5.07	14.4	2.84	18.6
Sephacryl S-200	1.59	4.28	2.69	5.5
Phenyl-Toyopearl 650M	0.13	0.88	6.69	1.1
Mono Q	0.09	0.67	7.44	0.9

[a] Specific activity is defined as μmol of NADH formed per mg of protein per min.

3.2. Molecular mass and subunit structure

The molecular mass of the native enzyme was determined to be 70,000 Da by gel filtration (Fig. 1). The subunit structure was examined by SDS-polyacrylamide gel electrophoresis. The molecular mass of the denatured enzyme with 0.1% SDS was estimated to be about 40,000 Da (Fig. 1). Thus, the enzyme appeared to be a dimer composed of two identical subunits, a structure in common with the *T. ferrooxidans* 3-isopropylmalate dehydrogenase.

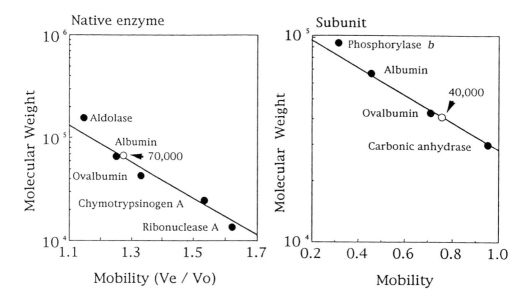

Fig. 1. The determination of the molecular weight of native (left) and subunit (right) 3-isopropylmalate dehydrogenase.

(Left) Purified enzyme was applied onto a Pharmacia fast-protein liquid chromatography system in Hiload 16/60 Superdex 200 column and eluted with 10 mM potassium phosphate buffer (pH 7.5) containing 0.2 M KCl at a flow rate of 0.5 ml/min.

(Right) Purified protein was treated with 2.0% SDS in 62.5 mM Tris-HCl buffer (pH 6.8) containing 5% 2-mercaptoethanol and 10% glycerol at 100 □C for 2 min, then the treated enzyme preparation was subjected to electrophoresis in the presence of 0.1% SDS with 7.5% polyacrylamide gels in Tris-glycine buffer.

3.3. Characterization of 3-isopropylmalate dehydrogenase

The enzyme exhibited maximal activity at pH 9.0 (Fig. 2). When the enzyme was assayed at various temperatures, maximal activity was found at 65°C (Fig. 3). The enzyme retained about 90% of the original activity after heating at 60°C for 30 min (Fig. 3). The presence of a divalent cation (e.g. Mg^{2+} or Co^{2+}, but not Mn^{2+}) was required for enzymatic activity. The K_m value for 3-isopropylmalate was estimated to be 0.13 mM and that for NAD^+ 8.7 mM. The ability of the enzyme to catalyze the dehydrogenation of various derivatives of 3-isopropylmalate was investigated, but none of them could be utilized as the substrate except for 3-isopropylmalate.

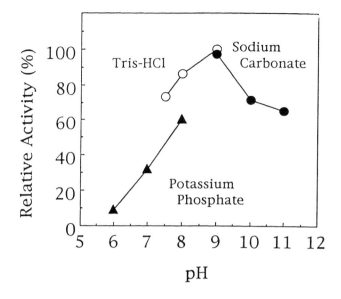

Fig. 2. Optimum pH for 3-isopropylmalate dehydrogenase from *T. thiooxidans*.
The reaction was carried out in each buffer at 37°C. The buffers used were: 0.1 M potassium
phosphate buffer (pH 6.0 - 8.0), 0.1 M Tris-HCl buffer (pH 7.5 - 9.0), 0.1 M Sodium
Carbonate buffer (pH 9.0 - 11.0).

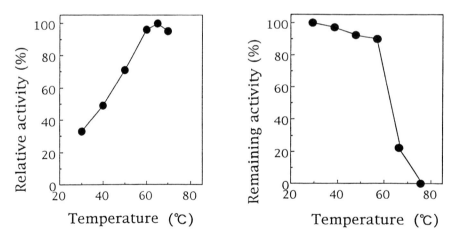

Fig. 3. Optimum temperature (left) and heat stability (right) of 3-isopropylmalate
dehydrogenase from *T. thiooxidans*
(Left) The enzyme reaction was done in 0.1 M Tris-HCl buffer (pH 9.0). The reaction
mixture without enzyme was preincubated for 10 min at each temperature.
(Right) The enzyme was subjected to heat treatment for 30 min in 0.1 M Tris-HCl buffer
(pH 9.0). The enzyme solution was then cooled and the activity remaining was assayed.

3.4. N-terminal amino acid sequence

The N-terminal amino acid sequence was determined using purified 3-isopropylmalate dehydrogenase from *T. thiooxidans* (Fig. 4) .

```
                              1                    10                    20
Thiobacillus thiooxidans    M K K • • • I A I F P G D G I G P E I V D A A

Thiobacillus ferrooxidans   M K K • • • I A I F A G D G I G P E I V A A A
Escherichia coli            M S K NY H I A V L P G D G I G P E VM T Q A
Salmomella typhimurium      M S K NY H I A V L P G D G I G P E VM A Q A
Bacillus subtilis           M K K • • R I A L L P G D G I G P E VL E S A
Bacillus coagulans          M K K • • K L A V L P G D G I G P E VM D A A
Bacillus caldotenax         M G N • Y R I A V L P G D G I G K E VT S G A
Thermus thermophilus        M K V • • • • A V L P G D G I G P E VT E A A
Thermus aquaticus           M R V • • • • A V L P G D G I G P E VL E A A
Lactococcus lactis          • • L KK K I VT L A G D G I G P E IM S A G
Candida utilis              M • P E K T I V V L P G D HV G T E I T A E A
Saccharomyces cerevisiae    M S A P K K I V V L P G D HV G Q E I T A E A
```

Fig. 4. Comparison of the N-terminal amino acid sequences of 3-isopropylmalate dehydrogenase from *T. thiooxidans* with those of enzymes from various microorganisms.

4. DISCUSSION

In this study we purified 3-isopropylmalate dehydrogenase from the acidophilic autotroph *Thiobacillus thiooxidans*. The purified enzyme was homogeneous as judged from disc gel electrophoresis.

The molecular weight of the native enzyme was calculated as 70,000 by gel filtration on FPLC. The molecular weight of the subunit enzyme was calculated as 40,000 by SDS-polyacrylamide gel electrophoresis. Thus, this enzyme is composed of two identical subunits. *Salmonella typhimurium* (17), *T. thermophilus* (20), *Sulfolobus* sp. (15), and *T. Ferrooxidans* (12) 3-isopropylmalate dehydrogenases have also been purified and their properties studied. Except for the archaeon *Sulfolobus* sp. enzyme, prokaryotic 3-isopropylmalate dehydrogenases are homodimeric. The properties of the enzyme from *T. thiooxidans* are similar to the reported data. However, the enzyme from *T. thiooxidans* is able to utilize 3-isopropylmalate only as the substrate. This indicates that the structure of the substrate binding site is different from *T. ferrooxidans* and *T. thermophilus* enzyme.

The N-terminal amino acid sequence of *T. thiooxidans* enzyme was determined. To date, the amino acid sequence of 3-isopropylmalate dehydrogenase from various microorganisms have been reported (Fig. 4). The *T. thiooxidans* sequence has approximately 60% homology with the enzyme from other microorganisms. In particular, it has 90% homology with the *T. ferrooxidans* enzyme, a species from the same genus. However, as mentioned previously, the *T. thiooxidans* enzyme is not able to utilize alkyl-malate as a substrate. Gene analysis of *T. thiooxidans leuB* should yield more information about the structure of *T. thiooxidans* 3-isopropylmalate dehydrogenase.

REFERENCES

1. H. Kirino, M. Aoki, M. Aoshima, Y. Hayashi, M Ohba, A. Yamagishi, T. Wakagi, and T Oshima, Eur. J. Biochem., 220 (1994) 275.
2. G. Kryger, G. Wallon, S. T. Lovett, D. Ringe, and G. A. Petsko, Gene, 164 (1995) 85.
3. R. Imai, T. Sekiguchi, Y. Nosoh, and K. Tsuda, Nucleic Acids Res., 15 (1987) 4988.
4. T. Sekiguchi, J. Ortega-Cesana, Y. Nosoh, S. Ohashi, K. Tsuda, and K. Shigenori, Biochim. Biophis. Acta, 867 (1986) 36.
5. T. Sekiguchi, M. Suda, T. Ishii, Y. Nosoh, and K. Tsuda, Nucleic Acids Res., 15 (1987) 853.
6. Y. Kagawa, H. Nojima, N. Nukiwa, M. Ishizuka,T.Nakajima, T. Yasuhara, T, Tanaka, and T. Oshima, J. Biol. Chem., 259 (1984) 2956.
7. H. Kirino, and T. Oshima, J. Biochem., 109 (1991) 852.
8. J. -J. Godon, M. -C. Chopin, and S. D. Ehrlich, J. Bacteriol., 174 (1992) 6580.
9. K. Hamasawa, Y. Kobayashi, S. Harada, K. Yoda, M. Yamasaki, and G. Tamura, J. Gen. Microbiol., 133 (1987) 1089.
10. A. Andreadis, Y. -P. Hsu, M. Hermodson, G. Kohlhaw, and P. Schimmel, J. Biol. Chem., 259 (1984) 8059.
11. K. Inagaki, H. Kawaguchi, Y. Kuwata, T. Sugio, H. Tanaka, and T. Tano, J. Ferment. Bioeng., 70 (1990) 71.
12. H. Kawaguchi, K. Inagaki, Y. Kuwata, H. Tanaka, and T. Tano, J. Biochem., 114 (1993) 370.
13. H. Matsunami, H. Kawaguchi, K. Inagaki, T. Eguchi, K. Kakinuma, and H. Tanaka, Biosci. Biotech. Biochem., 62 (1998) 372.
14. K. Miyazaki, K. Kakinuma, H. Terasawa, and T. Oshima, FEBS Lett., 332 (1993) 35.
15. T. Suzuki, Y. Inoki, A. Yamagishi, T. Iwasaki. T. Wakagi, and T. Oshima, J. Bacteriol., 179 (1997) 1174.
16. K. Imada, K. Inagaki, H. Matsunami, H. Tanaka, N. Tanaka, and K. Namba, Structure, 6 (1998) 971.
17. S. J. Parsons and R. O. Borns, J. Biol. Chem., 244 (1969) 996.
18. M. Bradford, Anal. Biochem., 72 (1976) 248.
19. U. K. Laemmli, Nature, 227 (1970) 680.
20. T. Yamada, N. Akutsu, K. Miyazaki, K. Kakinuma, M. Yoshida, and T. Oshima, J. Biochem., 108 (1990) 449.

Biochemical basis of chromate reduction by *Pseudomonas mendocina*

J. M. Rajwade, P. B. Salunkhe and K. M. Paknikar*

Division of Microbial Sciences, Agharkar Research Institute. G.G. Agarkar Road, Pune 411004, India.

A chromate reducing strain, *Pseudomonas mendocina* transformed hexavalent chromium to its trivalent form which precipitated as chromic hydroxide in the medium. Studies with respiratory inhibitors in conjunction with molar growth yield data indicated that chromate is not used as the terminal electron acceptor by this organism but probably serves as an auxiliary oxidant that helps in generation of oxidized cofactors necessary in cellular metabolism. Chromate reduction by *P. mendocina* is mediated by a periplasmic chromate reductase having molecular weight of approximately 183 KDa. The temperature and pH optima of the enzyme are 70°C and 8.5, respectively. Purified enzyme exhibited a K_m of 289 µM chromate and a V_{max} of 625 µM chromate reduced/min/mg protein.

1. INTRODUCTION

Hexavalent chromium [Cr(VI)] compounds are known to be toxic, mutagenic and carcinogenic (1). Microbial reduction of toxic hexavalent chromium to less soluble trivalent form represents a useful detoxification process that has been shown to be of practical importance for removal of chromium from industrial waste waters (2,3). A wide variety of bacteria are known to reduce Cr(VI) under aerobic or anaerobic conditions (4). The aerobic activity is generally associated with a soluble protein fraction utilizing NADH or NADPH as electron donors (5,6). Under anaerobic conditions, Cr(VI) may serve as a terminal electron acceptor through a membrane-bound reductase activity. For example, studies with *Enterobacter cloacae* HO1 (7,8) clearly indicated involvement of the respiratory chain components in the transfer of reducing equivalents to Cr(VI) through cytochrome *c*. However, no evidence was provided to show that such an electron transport to Cr(VI) could conserve enough energy to support growth. On the other hand, Shen and Wang (9) described a soluble chromium reductase and a minor activity associated with the respiratory chain in *Escherichia coli* ATCC 33456.

There are many reports of microorganisms enzymatically reducing a variety of metals in metabolic processes that are not related to metal assimilation. Microorganisms can conserve energy by coupling the oxidation of simple organic acids, alcohols, molecular hydrogen or

---

* Corresponding author. Fax: +91-20-351542, E-mail: paknikar@vsnl.com
JMR is thankful to Council of Scientific and Industrial Research, New Delhi for the award of a Post-doctoral Research Associateship.

aromatic compounds to the reduction of metals or may use them as terminal electron acceptors during anaerobic respiration (4).

A highly efficient microbial chromate detoxification process was developed in our laboratory using an indigenously isolated strain of *Pseudomonas mendocina* MCM B-180 (2,10). Chromate reduction in this strain could be mediated either by a soluble reductase or via the respiratory-chain-associated electron transport activity. Therefore, it is plausible to postulate that chromate reduction could either be a fortuitous activity or a mechanism for the generation of energy. The present paper attempts to confirm these possibilities.

2. MATERIALS AND METHODS

2.1. Microorganism and growth conditions

A chromate reducing bacterial strain was isolated previously in our laboratory from municipal sewage by two-stage enrichment method (10). The strain was identified as *Pseudomonas mendocina* and was deposited in MACS Collection of Micro-organisms (World Data Center Code No. 561, Accession No. MCM B-180). It was maintained on EG medium (composition in g/l : NH_4Cl, 0.03; $MgSO_4$. $7H_2O$, 0.01; NaCl, 0.01; yeast extract, 0.15; peptone 0.5; K_2HPO_4, 0.5; KH_2PO_4, 0.3; sodium acetate 2.0, pH = 7.5) and was stored at 5°C.

2.2. Chromate reduction in the presence of metabolic inhibitors

These experiments were carried out in 65 ml serum vials containing 50 ml EG medium supplemented with either rotenone (1 mM), carbonyl cyanide m-chloro phenylhydrazone (CCCP) (0.05 mM), sodium azide (1 mM), potassium cyanide (0.2 mM), 2,4 dinitrophenol (DNP) (0.5 mM), and cefotaxime (5 µg/ml). The vials were inoculated with washed cells of *P. mendocina* (0.5 ml cell suspension with a cell density of 2×10^9 cells/ml) and incubated at 30°C on a rotary shaker (Gallenkamp, UK) at 100 rpm for 30 min. Appropriate controls i.e. EG medium with and without cells were also run simultaneously. Potassium chromate (0.5 mM) was then added to the vials and incubated for a period of 4 h. Samples (2 ml) were withdrawn at 15, 30, 45, 60, 180, 240 min intervals from the vials for analysis of Cr(VI) in the supernatant. Total viable counts of *P. mendocina* prior to addition of chromium and after 4 h incubation were determined by plating suitable dilutions on EG agar.

2.3. Molar growth yield studies

A set of 40 ml serum bottles containing 20 ml EG medium supplemented with 1 mM chromate was used. The bottles were inoculated with 0.1 ml of 18 h grown culture of *P. mendocina*, sealed with sterile rubber septa and incubated on a rotary shaker (Gallenkamp, UK). A pair of bottles from the set were removed at an interval of 4 h, centrifuged at 10,000 x g for 10 min (at 4°C) and the cell pellet obtained was used for protein estimation. Supernatant after centrifugation was analyzed for Cr(VI) content. Values of (a) Y_M -protein (acetate) were determined from the plot of cell protein versus Cr(III) formed and (b) the protein : dry weight ratio was used to find Y_M (g dry weight/mol acetate). Standard value of $Y_{ATP} = 10$, for growth in inorganic medium containing acetate (11) was used to calculate estimated ATP formed (Y_M / Y_{ATP}) with chromate as electron acceptor.

2.4. Localization of chromate reducing activity

As a first step, chromate reducing activity in whole cells and cell-free medium was estimated as described in 2.7 below. Approximately 10^{10} cells were suspended in 50 ml EG medium (supplemented with 0.2 mM K_2CrO_4) in 65 ml vials that were incubated at 30°C on a shaker (100 rpm). Hourly samples (2 ml) were withdrawn from the vials, centrifuged (10,000 x g, 10 min) and chromate reducing activity was estimated. To test any periplasmic chromate reducing activity, *P. mendocina* was grown in the presence of antibiotics like cycloserine and polymyxin B sulfate (10 μg/ml) which inhibit cell wall synthesis and chromate reducing activity was checked in the cell-free medium. After antibiotics treatment, cell morphology in the centrifuged pellet was checked microscopically.

2.5. Purification of chromate reductase

P. mendocina was grown in 5 liter EG medium in a fermenter, cells were harvested by centrifugation and resuspended in different sets of 75 ml buffers. The buffers used comprised of hypertonic extraction buffer (20% sucrose in 20 mM Tris-HCl buffer, pH 7.5) supplemented with a detergent (Tween 20, Triton-X 100, sodium dodecyl sulfate and dodecyl trimethyl ammonium bromide, each at a concentration of 0.1%). Suspensions were incubated at 4°C for 12 h, centrifuged (at 12,000 x g for 20 min at 4°C) and cell-free supernatants (S1 fractions) were obtained. The S1 fractions were subjected to ammonium sulfate precipitation (80% saturation) to obtain partially purified enzyme preparations.

The S1 fraction showing maximum specific activity was heated at 70°C for 10 min, centrifuged (12,000 x g for 30 min, at 4°C) and the supernatant (67 ml) was loaded on DEAE-Sephacel column (10 mm x 130 mm) previously equilibrated with 20 mM Tris-HCl of pH 7.5. Protein fractions (5 ml each) were then eluted using 0 to 1 M KCl gradient in Tris-HCl buffer. The fractions were analyzed for protein content and chromate reductase activity. Fractions showing maximum enzyme activity were electrophoresced on 7.5% native polyacrylamide gels (Tris-glycine buffer, pH 8.3, 100 V) and those showing single bands upon silver staining (12) were further concentrated by freeze-drying (VirTis, Freezemobile lyophilizer, USA).

2.6. Characterization of chromate reductase

Effect of physiologically important anions on the expression of chromate reductase was checked by growing *P. mendocina* culture in EG medium supplemented with appropriate anions (1 mM nitrite or nitrate or sulfate), extraction of the enzyme and determination of its activity.

K_m and V_{max} values for the enzyme were calculated by determining the reaction velocity at substrate concentrations between 50-500 μM chromate. Optimum pH for the enzyme was determined by carrying out an enzyme assay in three buffer systems, viz. sodium citrate (pH 5.6-7.5), Tris-HCl (pH 7.5-8.6) and Glycine-NaOH (pH 9.6-10.6). The effect of temperature was evaluated in the range of 30-80°C.

To determine the molecular weight of the enzyme, purified enzyme preparation was subjected to 10% SDS-PAGE (13) after heat denaturation in the presence of β-mercaptoethanol. Protein molecular weight standards in the range of 14-97 KDa (Bangalore-Genei, India) were used as markers for SDS-PAGE. The purified enzyme was also applied to a gel-filtration column (Sephacryl S-200) along with gel filtration molecular weight markers (Sigma, USA).

2.7. Analyses

Cr(VI) reducing activity was assayed by measuring the decrease of Cr(VI). The assay mixture comprised of 0.2 mM chromate (as potassium chromate), 0.2 mM NADH and appropriate amount of the enzyme in 20 mM Tris-HCl buffer, pH 8.5. Appropriate controls without addition of enzyme were maintained. Incubation was carried out on a gyratory shaker-incubator at 30°C. Amount of residual Cr(VI) in the reaction mixture was estimated by 1,5-diphenylcarbazide colorimetric method (14). One unit of the Cr(VI) reducing activity was defined as the amount of enzyme which decreased 1 μmol of Cr(VI) per min at 30°C.

Protein content of the cells was estimated spectrophotometrically using Bradford's reagent (15) after digestion with 2 N NaOH followed by heating at 70°C for 60 min. Bovine serum albumin was used as a standard.

3. RESULTS

3.1. Chromate reduction in presence of metabolic inhibitors

It could be seen (Table 1) that chromate reduction by whole cells (cell density 2×10^8 cells/ml) in the absence of any metabolic poisons occurred in 4 h. Reduction of chromate was insensitive to sodium azide, potassium cyanide, rotenone and DNP. However, in the presence of both CCCP and cefotaxime, chromate reduction was completely inhibited. This observation was reflected also in the TVC counts obtained after 4 h incubation.

It was seen that *P. mendocina* cells remained viable in the presence of inhibitors such as sodium azide, potassium cyanide and DNP. However, with CCCP and cefotaxime there was a decrease in the total viable count.

3.2. Molar growth yield studies

Figure 1 depicts the disappearance of chromate, formation of Cr(III) and increase in protein when *P. mendocina* is grown in chromate containing medium. The slope of graph (Figure 2) represents a yield of 50.957 mg cell protein per mM Cr(III) formed which is the average Y_M -

Table 1
Effect of metabolic inhibitors on chromate reduction

Inhibitor	% Chromate reduced in 4 h	Total Viable Count (cfu/ml) after 4 h
Rotenone	>99.9	3.0×10^8
Sodium azide	>99.9	2.3×10^8
2,4-dinitrophenol	>99.9	2.5×10^8
Potassium cyanide	>99.9	1.5×10^8
CCCP	nil	1.0×10^7
Cefotaxime	nil	1.0×10^7
Control, uninoculated	nil	nil
Control, inoculated	>99.9	3.0×10^8

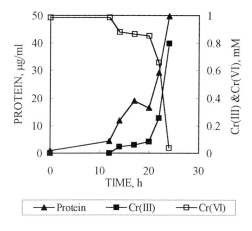
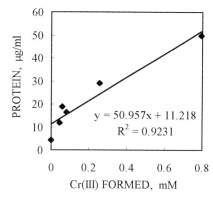

Figure 1. Chromate reduction and formation   Figure 2. Relationship between amount of
of Cr(III) during growth of *P.mendocina*.   cell protein made and Cr(III) formed.

protein (chromate). According to the equation given below, one chromate must be reduced to
Cr(III) in order for one acetate to be oxidized to two molecules of CO_2.

$$CH_3COO^- + CrO_4^{2-} + 7H^+ + 3e^- \rightarrow 2\,HCO_3^- + Cr^{3+} + 4H_2 \qquad \Delta G^{0'} = -409.34\ kcal/mol$$

Therefore, Y_M -protein (acetate) can also be expressed as 50.957 mg cell protein per mM
acetate.

3.3. Localization of chromate reductase activity

The cell-free broth obtained upon centrifugation did not reduce chromate and hence no
extracellular chromate reducing activity was detected. Specific activity of the enzyme
associated with whole cells was found to be 6.51 units. Further experiments incorporating
antibiotics inhibiting cell-wall synthesis showed an increase in the total activity in cell-free
medium confirming that the enzyme was periplasmic (Table 2). Microscopic observation of the
centrifuged pellet revealed that cells were broken in the presence of antibiotics. In control
flasks without antibiotics, no chromate reduction could be detected in cell-free medium. When
the cells were treated with hypertonic extraction buffer, a specific activity of 6.23 was
recovered in the cell free extract (S1). This indicated that the enzyme activity could be
periplasmic.

It is reported that presence of non-ionic detergents in extraction buffers facilitate release of
membrane bound enzymes. In experiments with *P. mendocina*, there was a decrease in
chromium reductase activity (Table 3) in cells treated with extraction buffer supplemented by
detergents (140-180 Units/ml) as compared to the detergent-free extraction buffer control(203
Units/ml). These results indicated that chromium reductase activity was not associated with
the membrane fraction.

Table 2
Effect of antibiotics on chromate reductase activity[a]

Antibiotic	Specific activity[b]	Total activity[c] (U)
Cycloserine	62.10	24.84
Polymyxin B sulfate	45.00	17.55
Control	0.00	0.00

a: partially purified (80% ammonium sulfate) preparation.
b: specific activity is expressed in Units/mg protein.
c: Total activity is the product of specific activity and total protein.

Table 3
Effect of detergents on chromate reductase activity

Detergent	Specific activity [a]
Triton X100	178.81
Tween 20	159.06
SDS	143.34
DTAB	188.79
Control	203.85

a: partially purified preparation (Units/mg protein).

3.4. Purification of chromate reductase

When the S1 fraction was subjected to heat treatment, anion-exchange chromatography and eluted with KCl, a sharp peak was obtained at about 0.4 M KCl, which showed highest activity. This sample gave a single band on polyacrylamide gel electrophoresis. A 29.8 fold purification relative to that of the crude extract was realized, as shown in Table 4. The apparent enzyme yield (2.41%) seemed very low. It was also found that maximum enzyme activity was at 70°C and pH 8.5 (Figure 3 and 4). However, the enzyme activity was retained over a wide range of temperatures (30-70°C) and pHs (5.6-10.6). An apparent Michaelis-Menten constant (K_m) of 289 μM chromate and a maximum velocity (V_{max}) of 625 μM/min/mg protein were computed from the Lineweaver-Burk plot (Figure 5).

Estimated molecular mass of the purified enzyme as determined by gel filtration chromatography was approximately 183 KDa and revealed 2 subunits (each with molecular weight ca. 63 KDa and 29 KDa) on 10% SDS-polyacrylamide gel electrophoresis (Figure 6). The enzyme required NADH or NADPH as an electron donor for reduction of chromate.

Table 4
Purification of chromate reductase

Step	Volume (ml)	Total protein (mg)	Total activity (U)	Specific activity (U/mg)	Yield (%)	Purification (fold)
Crude extract (S1)	75	185.01	1302	7	100	1
Heat treatment	67	149.28	1736	11	133	1.8
DEAE-Sephacel	4	0.15	31.46	207	2.41	29.8

Table 5
Chromate reductase activity in presence of oxyanions

Anions present (1mM)	Specific activity[a]
NO_2^-	93.46
NO_3^-	66.43
SO_4^{2-}	59.33

a: partially purified preparation (Units/mg protein).

3.5 Characterization of chromate reductase

Effect of physiologically important anions on the expression of chromate reductase was checked by growing *P. mendocina* culture in EG medium supplemented with appropriate anions (1 mM nitrite or nitrate or sulfate), extraction of the enzyme and determination of its activity. It was seen that chromate reductase activity was significantly higher when cells were grown in presence of nitrite as compared to nitrate and sulfate (Table 5). Interestingly, in our further experiments, when nitrite concentrations were decreased to 0.25 mM chromate reductase was found to increase to 193 Units/mg protein (data not shown).

4. DISCUSSION

Chromate reduction was found to be insensitive to azide and cyanide implying that cytochrome oxidase was not involved in the electron transfer to chromate. Complete inhibition of chromate reduction by CCCP which strongly inhibits transport processes in microorganisms suggested that a transport mechanism is essential for the entry of chromate into the cell and its subsequent reduction by intracellular mechanisms. Inhibition of chromate reduction in presence of cefotaxime points to the necessity of viable cells. Low concentrations of DNP are

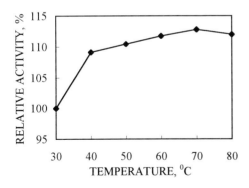

Figure 3. Effect of temperature on chromate reductase activity.

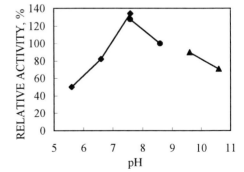

Figure 4. Effect of pH on chromate reductase activity.

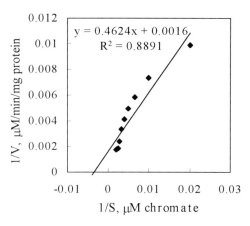

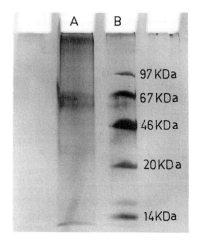

Figure 5. Lineweaver-Burk plot for chromate reductase.

Figure 6. Silver stained 10% SDS-polyacrylamide gel. Lane A: Purified chromate reductase showing two sub-units (63 & 29 KDa), Lane B: protein molecular weight standards.

known to stimulate oxygen consumption and inhibit phosphorylation. DNP increases the permeability of the inner membrane to H^+ ions by carrying protons through the cell membrane. Chromate reduction by *P. mendocina* in presence of DNP was probably due to these H^+ ions.

During the studies carried out on molar growth yields Y_M -protein (acetate) value obtained was 50.957 mg cell protein per mmol acetate. In *P. mendocina* the protein content of the cells was 59.24% dry weight. Using this data, the cell yield Y_M for *P. mendocina* grown on acetate plus chromate would be at least Y_M (chromate) = 86 g (dry weight/mol acetate). With this Y_M value, the estimated number of ATP made during growth can be calculated using Y_{ATP}. For growth on acetate in inorganic salts medium theoretical Y_{ATP} was estimated to be 10.0 g cells/mol ATP (11). In the present study, the calculated number of ATP made/mol acetate oxidized i.e. Y_M / Y_{ATP} was 8.6 ATP/mol acetate with chromate as electron acceptor.

It should be pointed out that since number of ATP made is only an estimate, and because maintenance corrections could not be taken into account, it is possible that actual number of ATP made per mol acetate oxidized is less than the values calculated above. If it is assumed that 50 kJ of standard free energy are required to synthesize 1 ATP (16), and energy conservation is 50% efficient, then the number of ATP that can be formed during growth on acetate plus chromate can be calculated using the free energies for growth reactions when chromate is the electron acceptor ($\Delta G^{0'}$ = -409.34 kJ/mol acetate). The theoretical number of ATP made during growth with chromate will be 409.34/100 = 4.09 ATP/mol acetate. It could be seen that the number of ATP/mole acetate oxidized as obtained in the experiments, i.e. 8.6 does not compare with the theoretical value of 4.09.

In a study on cell yield (Y_M) of *Thauera selenatis* (17) the estimated number of ATP gained during growth with selenate or nitrate were found to be similar to the theoretical number expected. Therefore, the authors argued that these compounds were respired and energy was conserved. In the present study the estimated number of ATP formed during chromate reduction did not match with the theoretical value. Therefore, it is probable that chromate does not serve as terminal electron acceptor during growth of *P. mendocina*. The hydrogen forming reaction of the metabolism (acetate conversion to CO_2 and H_2) is probably favored in *P. mendocina*, in which case, electrons produced during this reaction could be accepted by chromate ions resulting in their reduction to trivalent form.

In *Escherichia coli* ATCC 33456, Shen and Wang (9) reported that chromate reduction was largely due to a soluble reductase and was insensitive to azide, cyanide and rotenone but was stimulated by DNP. On the other hand, membrane bound chromate reductase of *Enterobacter cloacae* was inhibited by azide and cyanide (7). In most of the studies on chromate reduction, fermentable organic compounds were incorporated in growth media. In such cases it was postulated that chromate was cometabolized. The reactions involving fermentation of carbon source with chromate as electron sink were more energetically favorable than other types of fermentations not involving chromate reduction. Chromate reduction by *P. mendocina* appears to be similar to *Escherichia coli* ATCC 33456 and the data obtained suggests that the reducing equivalents from acetate metabolism can pass to auxiliary oxidants like chromate effecting reduction of latter. Such a reduction reaction could be used to dispose off excess electrons by the reoxidation of NADH, $FADH_2$, or quinones thus maintaining an optimal redox poise *in vivo*.

For the chromate reducing bacteria reported in literature two kinds of enzymatic mechanisms for chromate reduction have been proposed. The aerobic activity as described in case of *P. putida and P. ambigua* G-1 is generally associated with a soluble protein fraction utilizing NADH or NADPH as electron donors (18,6). Under anaerobic conditions, Cr(VI) may serve as a terminal electron acceptor through a membrane-bound reductase activity. For example, studies with *Enterobacter cloacae* HO1 (7,8) clearly indicated involvement of cytochrome C_{548} in the transfer of reducing equivalents to Cr(VI). In *Escherichia coli* ATCC 33456 the involvement of a soluble reductase and respiratory-chain-linked electron transport to Cr(VI) has been reported (9). Our studies with localization of chromate reductase in *P. mendocina* clearly showed that the activity was periplasmic. Such periplasmic activity has not been described for chromate reducers but since the Cr(III) formed upon reduction of chromate is extracellular, it is quite possible that chromate is reduced by a periplasmic enzyme and then effluxed out of the cell. A periplasmic selenite reductase, catalyzing the reduction of selenite to elemental selenium in *T. selenatis* (19) is an example similar to the one reported in this study.

Chromate pollution of the environment is mainly due to industrial activity. Therefore it seems reasonable to speculate that 'chromate reductase' is an additional activity of another enzyme with a different physiological substrate. Nitrite is a physiologically important anion and the data obtained during the present studies hint at the possibility that nitrite reducing enzyme could probably reduce chromium in *P. mendocina*. While working on *Enterobacter aerogenes*, Clark (20) did suggest that chromate reduction may be due to a formate linked nitrite reductase.

Enzymatic chromate reduction has been studied in different bacterial genera but there are only few reports on the detailed isolation and characterization of chromate reducing enzymes. In studies by Ishibashi and co-workers (18), the soluble chromate reductase of *P. putida* was not purified. However, the studies on its characterization revealed that it was heat labile and showed a K_m of 40 μM chromate and a V_{max} of 6 nmol/min/mg protein. The enzyme was not

114

inhibited by sulfate or nitrate. In *P. ambigua* G-1 (6), two enzymes reducing chromate have been described but only one was purified. The purified enzyme had a molecular mass 65 KDa on gel filtration and data suggested that it had a dimeric or trimeric structure. The apparent K_m of this enzyme was 13 μM chromate and V_{max} was 27 nmol chromate reduced/min/mg protein. The enzyme was found to be active within a wide range of temperatures (40-70°C) and pH (6-9). The other enzyme had a high molecular mass 130 KDa on gel filtration. The estimated molecular weight of chromate reductase from *P. mendocina* in the present study was found to be 183 KDa on gel filtration and that of the subunits was 63 and 29 KDa on SDS-PAGE. The ratio of these observed molecular weights was 1.9:1. This enzyme could thus have a dimeric structure. The enzyme isolated from *P. mendocina* has a much higher affinity of 289 μM chromate than that reported for the two enzymes from *P. putida* and *P. ambigua* G-1. It is also more thermostable than the enzyme from *P. ambigua* G-1. This indicates that the enzyme from *P. mendocina* is probably similar to the high molecular weight chromate reductase of *P. ambigua* G-1, which has not been purified and characterized so far.

REFERENCES

1. J.O. Nriagu and E. Niebor (eds.), Chromium in Natural and Human Environment, Wiley, New York, 1988.
2. J.V. Bhide, P.K. Dhakephalkar and K.M. Paknikar, Biotechnol. Lett., 18 (1996) 667.
3. H. Ohtake and J.K. Hardoyo, Wat. Sci. Tech., 25 (1992) 395.
4. D.R. Lovely, Annu. Rev. Microbiol., 47 (1993) 263.
5. H. Horitsu, S. Futo, Y. Miyazawa, S. Ogai and K. Kawai, Agric. Biol. Chem., 51 (1987) 2417.
6. T. Suzuki, N. Miyata, H. Horitsu, K. Kawai, K. Takamizawa, Y. Tai and M. Okazaki, J. Bacteriol., 174 (1992) 5340.
7. P. Wang, T. Mori, K. Toda and H. Ohtake, J. Bacteriol., 172 (1990) 1670.
8. P. Wang, T. Mori, H. Ohtake, I. Kusaka and I. Yabe, FEMS Microbiol. Lett., 78 (1991) 11.
9. H. Shen and Y-T. Wang, Appl. Environ. Microbiol., 59 (1993) 3771.
10. J.M. Rajwade and K.M. Paknikar, Proceedings of International Biohydrometallurgy Symposium, IBS BIOMINE 97, Sydney, Australia, (1997) p E-ROM4.1- 4.10 .
11. A.H. Stouthamer and C. Bettenhausen, Biochim. Biophys. Acta., 301 (1973) 53.
12. T. Rabilloud, G. Carpentier and P. Tarroux, Electrophoresis, 9 (1988) 288.
13. U.K. Laemmli, Nature (London), 227 (1970) 680.
14. APHA, AWWA, WPCF. Standard Methods for the Analysis of Water and Wastewater, American Public Health Association, Washington DC, 1992.
15. M.M. Bradford, Anal. Biochem., 72 (1976) 248.
16. K. Thauer and J.G. Morris, The microbe, Part II: Prokaryotes and Eukaryotes, Cambridge University Press, Cambridge, 1984.
17. J.M. Macy and S. Lawson, Arch. Microbiol., 160 (1993) 295.
18. Y. Ishibashi, C. Cervantes and S. Silver, Appl. Environ. Microbiol., 56 (1990) 2268.
19. H. DeMoll-Decker and J.M. Macy, Arch. Microbiol., 160 (1993) 241.
20. D.P. Clark, FEMS Microbiol. Lett., 122 (1994) 233.

Study of the proteins involved in the resistance of cadmium of *Thiobacillus ferrooxidans*

M. C. Taira, C. Reche, S. Porro and S. Alonso-Romanowski

Department of Science and Technology. National University of Quilmes. Roque Saenz Peña 180. (1876) Bernal. (Buenos Aires). ARGENTINA.

Microbial leaching is one of the most advantageous methods of removing heavy metals from sulfide ores and sewage sludges. Several species of bacteria were found to resist high levels of heavy metals, like copper, zinc and others.

In the present work total proteins from *Thiobacillus ferrooxidans*, have been fractionated by SDS-PAGE.

Cells grown on media with or without Cd^{2+} showed different patterns of proteins. The influence of the cadmium ion was tested at different concentrations. The maximum concentrations were 10,000 ppm. While some proteins remained unchanged in cells grown under different conditions (without and with several concentrations of metallic ion), others were present only under one of them and/or in different relative amounts.

We suggest that the enhanced resistance to cadmium in *T. ferrooxidans* is gained through enhanced synthesis of proteins involved in resistance.

1. INTRODUCTION

The capacity of *Thiobacillus ferrooxidans* for autotrophic CO_2 fixation at expense of the energy derived from the oxidation of sulfide minerals, elemental sulfur, and ferrous iron under low pH values and high concentrations of metal ions has determined its intense application in combined chemical and bacterial leaching of metals from ores, concentrates, and rocks. Studies of physiology and mechanisms of the energy substrate oxidation in *Thiobacillus ferrooxidans* allowed to optimize the conditions of its growth and substrate utilization. Intensification of the bacterial metal-leaching process was achieved through the development of the technology. Selection of new highly active and competitive strains of *Thiobacillus ferrooxidans* also offers great promise for intensification of the process. Industrially utilized strains should possess a high capacity for CO_2 fixation, biomass accumulation, and oxidation of the energy substrate; they should also be resistant to ions of heavy metals and acids and have some other practically valuable properties. Strains of *Thiobacillus ferrooxidans* isolated from natural habitat and those adapted to specific conditions of industrial practice show significant variations in the above-mentioned properties. The conditions of a metal-leaching operation favor the selection of efficient strains with enhanced oxidizing activity and resistance to high metal concentrations (1). In particular,

Thiobacillus ferrooxidans can be experimentally adapted to high concentrations of toxic metals. However, considering that adaptation represents an environmentally induced modificational variation of an organism and that the range of this variation, being restricted genetically, is unlikely to be extended, new approaches to improve industrial strain properties become a necessity.

Therefore, more promising, is the method of chemostat competitive selection of an efficient *Thiobacillus ferrooxidans* strain under conditions of continuous growth at low pH values and a high content of toxic metal ions. The selection could be conducted from the original combination of promising strains for the maximal growth rate and resistance to aggresive components of the medium. Selection of a spontaneous mutant possessing certain advantages over the other clones is also feasible.

Resistance to high concentrations of cadmium ions is a practically important characteristic of *Thiobacillus ferrooxidans* cells. According to current literature, the resistance of *Thiobacillus ferrooxidans* strains to Cd^{2+} reaches 56 g/liter (2). However, under such conditions the rate of oxidation is rather low.

The aims of the present work were: 1) to select a *Thiobacillus ferrooxidans* strain possessing high rates of growth and increased resistance to cadmium ions and 2) to identify proteins specifically related to cadmium resistance.

2. MATERIALS AND METHODS

2.1. Bacteria and growth conditions

The bacteria used in the present study where *Thiobacillus ferrooxidans* strain Tf18 (from Santa Rosa de Arequipa, Perú). The bacteria were grown in 9K medium (3) with and without 10000 ppm Cd^{2+}, in Erlenmeyer flasks at 30°C, pH 1.5, with shaking at 150 rpm. Cells were filtered through Whatman 41 paper to remove precipitates, and collected by centrifugation at 10000 g for 10 minutes. Pellets were washed and resuspended in 9 K medium with and without 10000 ppm Cd^{2+}

2.2. Measurement

Microbial growth was estimated by considering the bacterial metabolism corresponding to the iron oxidation of the substrate. Protein concentrations were deduced by a modified Coomasie blue G dye-binding assay described by Read and Northcote (4).

2.3. Cadmium tolerance

Cadmium tolerance was determined by culturing bacteria with modified media resulting from the 9K medium supplemented with several concentrations of cadmium sulphate ($CdSO_4.7H_2O$) corresponding to 10^{-6}, 10^{-5}, 10^{-4}, 10^{-3} and 5. 10^{-3} M of cadmium ions. Experiments were carried out in 250 ml erlenmeyer flasks containing 100 ml of medium under rotating agitation (150 rpm) at 30°C. Inoculations of these flasks were performed by 10 ml of 9K culture (1.0×10^8 cel/ml) during its exponential growing phase.

2.4. Bacterial adaptation to high cadmium concentration

This adaptation was obtained during serial subcultures in 9K medium supplemented with 10^{-2}, 5.10^{-2} and 9.10^{-2} M of cadmium ions. Cultures were performed in 250 ml flasks containing 100 ml of medium and incubated in a rotatory shaker (150 rpm) at 30°C. The first inoculation of these media was done by using a culture of *Thiobacillus ferrooxidans* resistant to 5.10^{-3} M of cadmium ions. Bacterial growth was regularly monitored by estimating the iron oxidation. When reaching the stage of 90% ferrous ions oxidized by the bacteria for their growth, 10 ml of the respective culture was used to inoculate the next one.

2.5. Preparation of membranes and soluble cell components

Cells were washed in cold 50 mM potassium phosphate, pH 7.2, and suspended in the same buffer (2.5 ml buffer/g wet weight). A few crystals of Dnase II and $MgCl_2$ were added. Cells were broken by sonic oscillation. Unbroken cells and cell debris were removed by centrifugation at 4,800 x g for 40 min. The resulting cell extract was centrifuged at 46,000 x g for 1.5 h to yield the soluble fraction (supernatant) and the membrane fraction (pellet). The soluble fraction contained periplasmic and cytoplasmic proteins.

2.6. Protein electrophoresis

SDS-PAGE was performed at room temperature on vertical 12 % polyacrylamide slab gels according to the method of Laemmli (5) using Mini Protean equipment (Biorad). Gels were stained for protein with silver stain. A low molecular mass calibration kit (Biorad) was used to derive the molecular masses.

3. RESULTS AND DISCUSSION

Bacteria such as *Escherichia coli* have a number of global regulatory networks that enable them to adapt rapidly to survive periods of adverse environmental conditions (6). In the case of bacteria such as *Thiobacillus ferrooxidans*, changes in environmental conditions have been shown to produce differential expression of some proteins (7, 8). However, the proteins involved in the molecular events responsible for cadmium resistance still remain mostly unknown.

In this paper we report an analysis of total proteins extracted from *Thiobacillus ferrooxidans* cells grown on medium with and without cadmium ions. This analysis was performed using SDS-PAGE electrophoresis. When cells grown on medium without cadmium were shifted to the medium with cadmium, several changes were observed in the cytoplasmic fraction (Fig.1) and the membrane fraction (Fig.2).

While most proteins remained unchanged, some seemed to be expressed in different relative amounts under the two conditions. Of particular interest are those, expressed exclusively in one or the other growth medium.

Protein 1, 2, and 3 in figure 1 were present only in cytosolic fraction of cells grown with cadmium, while proteins 4 and 5 were exclusively present in the cytosolic fraction of cells grown without cadmium.

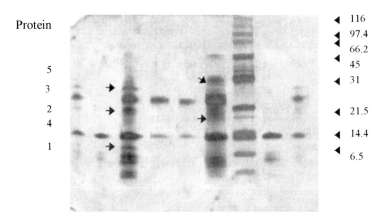

Figure 1. Silver stained SDS PAGE of cytosolic proteins. Lanes 1-3 and 8-9: proteins from *Thiobacillus ferrooxidans* growth with 10,000 ppm Cadmiun ions. Lanes 4-6: proteins from bacteria growth without Cadmiun ions. Lane 7: Molecular weigth markers. Arrow heads show differential protein of bacteria growth with and without Cadmiun.

The following standard proteins were used for molecular weight calibration:

1. β-galactosidase (116 kD), 2. Phosphorilase b (97.4 kD), 3. Bovine serum albumin (66.2 kD), 4. Ovalbumin (45 kD), 5. Carbonic anhydrase (31 kD), 6. Soybean trypsin inhibitor (21.5 kD), 7. Lysozyme (14.4 kD), 8. Aprotinin (6,5 kD).

Proteins 1, 2, and 3 have a molecular weight of about 7.3 kD, 20.8 kD, and 26.9 kD respectively, proteins 4 and 5 have a molecular weight of about 15.7 kD and 31 kD, respectively.

Proteins 6 and 7 in figure 2 were present only in the membrane fraction of cells grown with cadmium ions. Protein 6 has a molecular weight of about 9.9 kD , while protein 7 has a molecular weight of about 27.7 kD.

The preliminary results presented here indicate the existence of important differences in the expression of several proteins, depending on whether *Thiobacillus ferrooxidans* is grown on medium with or without cadmium. In future experiments we will determine the N-terminal amino acids of these proteins.

By combining the genomic analysis and the protein analysis of *Thiobacillus ferrooxidans*, it will be possible to obtain information about the regulation of the metabolic changes that take place, when the growth conditions are switched from medium without cadmium to medium with cadmium. This knowledge would be of great importance for obtaining genetically engineered microorganisms for improved biotechnological applications.

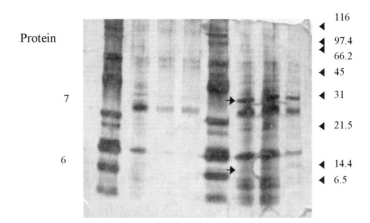

Figure 2: Silver-stained SDS-PAGE of membrane proteins. Lanes 1 and 5: Molecular weigth markers. Lanes 2-4: proteins from *Thiobacillus ferrooxidans* growth without Cadmiun ions. Lanes 6-8: proteins from bacteria growth with 10,000 ppm Cadmiun ions. Arrow heads show differential protein from bacteria cultured with Cadmiun.
The protein markers were the same of figure 1.

REFERENCES

1. G.I. Karavaiko, T.F. Kondrat'eva, V. P. Piskunov, V. G. Saakyan, L. N. Muntyan and O. E. Konovalova. Microbiology. 63 (1994) 132.
2. F. Baillet, J. P. Magnin, A. Cheruy and P. Ozil. Environmental Technology. 18 (1997) 631.
3. M. P. Silverman and D.C. Lundgren. J. Bacteriol. 77 (1959) 642.
4. S. M. Read and D. H. Northcote. Analytical Biochemistry. 116 (1981) 53.
5. U. K. Laemmli. Nature. 227 (1970) 680.
6. S. Gottesman. Annual Review of Genetics, 18 (1984) 467.
7. A. M. Amaro, D. Chamorro, M. Seeger, R. Arredondo, I. Peirano and C. A. Jerez. Journal of Bacteriology. 173 (1991) 910.
8. D. Chamorro, R. Arredondo, I. Peirano and C. A. Jerez . Proceedings of the 1987 International Symposium on Biohydrometallurgy, Warwick, UK, Jul 12-16, ed. P. R. Norris and D. P. Kelly, (1988) 135.

Site-directed mutagenesis of rusticyanin

Kazuhiro Sasaki*, Naoya Ohmura, and Hiroshi Saiki

Bio-Science Department, Abiko Research Laboratory, CRIEPI
1646 Abiko, Abiko-city, Chiba-pref. 270-1194 JAPAN

The site-directed mutagenesis of a blue copper protein, rusticyanin of *Thiobacillus ferrooxidans*, and the subsequent binding experiments to pyrite were carried out to clarify what amino acid residues were responsible for the binding to pyrite. Four different mutants of rusticyanin were prepared. Each of the four copper ligands (His85, Cys138, His143 and Met148) of the wild-type rusticyanin was individually replaced with alanine. All of the prepared mutants were capable of coordinating to a copper atom, therefore, the site-directed mutagenesis did not cause any structural imperfection for coordinating ability. When each mutant in the apo form was subjected to the pyrite particles, only His143Ala mutant could bind to pyrite while the other mutants did not. The conformational changes in the rusticyanin structure before and after copper incorporation was investigated by digestion pattern analysis using proteinase exposure. Rusticyanin in the holo form was much more resistant to proteinase digestion than was rusticyanin in the apo form. These results suggested ligand binding residues His 85, Cys138 and Met 148 were essential for the binding of apo rusticyanin to pyrite. Apo rusticyanin had a conformation different in structure from the holo rusticyanin.

1. INTRODUCTION

Bacterial leaching is one of the more important applications of bacteria in industry. There has been much interest as to how bacteria could leach metals from low grade ores. Many leaching profiles of various metals by several organisms have been noted by many investigators. However, very little fundamental information is available concerning the events involving bacterial

---

* To whom correspondence should be addressed

adhesion at the mineral surface during oxidation and leaching. Generally, bacterial adhesion is controlled by physical interactions such as electrostatic and hydrophobic interactions (1). The mineral pyrite has negatively charged and hydrophobic surface properties (2), while the bacterial cell has negatively charged and hydrophilic surface properties (2). These properties should provide insufficient attractive forces for bacterial adhesion. However, *T. ferrooxidans* cells were capable of adhering to pyrite and the adhesion was selective for pyrite even when pyrite was mixed with other minerals. This inconsistency between colloid theory and real phenomenon was solved by discovery of the pyrite binding protein of the bacterium (3). A sole binding protein to pyrite, which is located on the cell surface when cells grow on soluble iron, mediated this specific ability of adhesion. The binding protein was identified as the apo form rusticyanin (3). Rusticyanin was considered to be a component of the electron transport system (4-6) but it was shown to have another function during adhesion. Here we report which ligand-binding residues on the apo form of rusticyanin are essential for binding to pyrite using mutants produced by site-directed mutagenesis.

2. MATERIALS AND METHODS

2.1. Site-directed mutagenesis

The plasmid for the expression of recombinant rusticyanin was prepared (7) and then used as a template in subsequent mutagenesis studies. The following single mutants of rusticyanin, His[85]→Ala, Cys[138]→Ala, His[143]→Ala, Met[148]→Ala, were produced using a mutation kit, Gene Editor (Promega, Madison, WI). Respective mutagenic oligonucleotides were prepared as follows: "GTCAAAACTAGCACCGAATCC", "CCCCGGTATCTGAGCTACGT-AGTA", "GGTGGCGGCAGCCCCCGGTAT", and "GCCGAAAGCACCGGTGG-CG", and the 5' end was phosphorylated with T4 polynucleotide kinase. The mutagenesis reaction involved annealing the Selection Oligonucleotide, which was provided by the kit and hybridized to the β-lactamase gene to alter the substrate specificity of the enzyme and give cells increased resistance to analogs of ampicillin. At the same time mutagenic oligonucleotides were hybridized to the template rusticyanin gene, followed by synthesis of the mutant strand with T4 DNA polymerase and T4 DNA ligase. The heteroduplex DNA was then transformed into the repair minus *E. coli* strain BMH 71-18 mutS and the cells were grown in selective media containing an analog of ampicillin which was provided by the kit to select clones containing the mutant plasmid. Plasmids resistant to the novel antibiotic were then isolated and transformed into the final host strain, BL21 DE3, using the same selection conditions.

2.2. Purification of recombinant apo-rusticyanin

Cell-free extracts were prepared from recombinant *E. coli* that had been induced for the expression of apo-rusticyanin. The extracts were mixed with the same volume of 50 mM β-Alanine (pH 3.0) to produce the precipitate. The supernatant derived by centrifugation at 12,000 g for 20 minutes was dialyzed for one day against frequent changes of 20 mM Tris-HCl buffer (pH 7.5), and subsequently fractionated by ammonium sulfate precipitation at 80 % saturation. The pellet was resuspended in 20 mM sodium phosphate buffer (pH 6.5) and subsequently applied to an Econo-Pac 10DG (Bio-rad, Hercules, CA) desalting column equilibrated with the same buffer. The final step of the purification was performed using ion-exchange column chromatography with a Poros HS column (PerSeptive Biosystems, Tokyo, Japan) equilibrated with the same buffer on an FPLC system (Amersham Pharmacia Biotech Ltd., Uppsala, Sweden). Apo-rusticyanin that was bound to the resin was eluted by the application of a continuous gradient of 0 to 0.3 M KCl.

2.3. Preparation of pyrite

The pyrite used for this study was produced by the Mina Cerro mine in Peru. It was crushed, fractionated, and characterized as previously described (2).

2.4. Adhesion of apo-rusticyanin to pyrite in open tube

The pyrite was crushed into fine particles with a mortar and the particle diameters were adjusted from 53 μm to 75 μm using sieves. These particles were repeatedly washed with sulfuric acid solution (pH 2.0) containing 0.1 % (vol./vol.) triton X-100 until all tiny particles were completely removed. Then, 200 μl of the solution containing 200 μg of the purified recombinant apo-rusticyanin per ml in sulfuric acid (pH 2.0) with 0.1 % Triton X-100 was mixed with 0.2 g of the pyrite particles in a glass tube. The supernatant that contained unbound protein was removed, and its protein concentration analyzed as follows: the supernatant was put on SDS-PAGE and stained using a silver stain kit (Bio-rad). Its protein concentration was determined from the SDS spot using an image analyzing system (Scanner; Imagemaster DTS, Pharmacia, and Software; Quantity One, PDI Inc.).

2. 5. Drawing the tertiary structure of rusticyanin

The three-dimensional images of rusticyanin used in this paper were drawn by rasmol (RasWin Molecular Graphics Windows Version 2.6) obtained from the following site, URL; http://www.umass.edu/microbio/rasmol/. The three-dimensional data file of rusticyanin (PDBid: 1RCY) (8) was downloaded from the web site of the Protein Data Bank, URL; http://pdb.pdb.bnl.gov/pdb-bin/pdbmain.

2. 6. Partial digestion of rusticyanin

Apo-rusticyanin (15 µg) was digested with or without 4 mM $CuSO_4$ by 0.025 to 2.5 µg of chymotrypsin in 20 mM Tris-HCl buffer (pH 7.5) for 10 minutes at room temperature. Before the protein was digested in the presence of $CuSO_4$, pre-incubations were performed for 20 minutes at room temperature without chymotrypsin. The digested peptide fragments were quantified by SDS-PAGE and a density-meter as described in section 2.4..

3. RESULTS

3.1. Incorporation of a copper atom into mutant rusticyanins

Not only the intact rusticyanin but also the recombinant rusticyanin in the apo form could bind to pyrite, but rusticyanin in the holo form lost its binding ability (3,7). Therefore, it was expected that ligand sites for copper coordination might be required for the binding of the apoprotein to pyrite. To investigate this hypothesis, four different mutant rusticyanins were prepared with site-directed mutagenesis. Each copper ligand of rusticyanin (His85, Cys138, His143 and Met148) was replaced with alanine, which was inactive toward the metal coordination. The rusticyanin mutants were expressed in *E. coli* in a manners similar to the wild-type rusticyanin. However, it has been known that site-directed mutagenesis may affect protein folding. In order to clarify whether or not the mutation affected protein folding, the copper coordinating ability of each

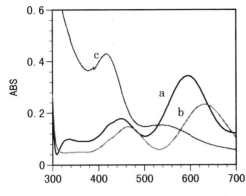

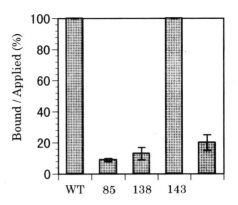

Fig. 1. Absorbance spectrum of Cu(II)-rusticyanin. (a) wild type. (b) His85Ala mutant. (c) Met148Ala mutant.

Fig. 2. Pyrite binding assay of rusticyanin. (WT) wild type. (85) His85Ala mutant. (138) Cys138Ala mutant. (143) His143Ala mutant. (148) Met148Ala mutant.

mutant was tested. Mutants in the apo form were reconstituted to the holo form by mixing copper ions and a stabilizing anion (9) to support incorporation of a copper atom. The absorption spectrum of each mutant in the holo form was measured (Fig. 1). It was known that wild-type rusticyanin in the holo form shows two absorption peaks (9). The wild-type holo protein in this study showed the same absorption maxima as a previous report (9) at 460 nm and 595 nm. The absorption spectrum of the His85Ala mutant showed two maxima at 449 nm and 630 nm. These maxima were also in good agreement with a previous report (9). Two highly absorptive maxima were observed on the Met148Ala mutant at 417 nm and 520 nm. The spectrum suggested that two apoproteins could coordinate a copper atom. However, the Cys138Ala and His143Ala mutants did not show a clear absorption in the visible wavelength range. Therefore, a copper atom in the protein was directly measured by inductive coupled plasma analysis. As a result of the copper quantification, it was found that all mutants including Cys138Ala and His143Ala contained a copper atom and that the ratio of copper atom per protein molecule was about 1.0 for every mutant (data not shown). Therefore, it was suggested that all mutants could coordinate a copper atom without alteration to the protein structure.

126

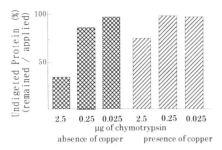 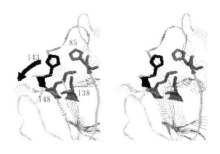

Fig. 3. Partial digestion of rusticyanin with proteinase. 15 μg of rusticyanin was digested with the indicated amount of chymotrypsin for the same time. The rates were calculated from the amount of undigested protein of the original molecular weight and the protein applied.

Fig. 4. Stereo graphics of pyrite binding site of rusticyanin. For our hypothesis, the loop containing His143 is opened in the direction of the arrow when rusticyanin is in the unfolded state.

3.2. Amino acids which were necessary for the adhesion to pyrite

It was clear that an individual mutant could coordinate a copper atom, therefore, the subsequent binding experiments were carried out with pyrite particles. Each purified mutant was mixed with pyrite particles in the binding solution to induce binding. The unbound mutant protein in the binding solution was quantified on SDS-PAGE gels. The amount of bound protein was calculated by subtracting the unbound protein from the total added protein. Amounts of bound protein are shown in Fig. 2. In the case of the His143Ala mutant, all of the applied proteins were bound to pyrite as well as the wild-type protein. On the other hand, His85Ala, Cys138Ala and Met148Ala mutants showed a much lower binding ability than His143Ala and the wildtype. These results suggested that if rusticyanin lost one of the His^{85}, Cys^{138} and Met^{148} ligands, it could not bind to pyrite. Therefore, His^{85}, Cys^{138} and Met^{148} are essential ligands for the binding to pyrite but not His^{143}.

3.3. Structural change between apo- and holo-rusticyanins

The copper ligands in the hydrophilic core are well protected by the coordination sphere in rusticyanin (8). If the copper ligands are capable of reacting with pyrite, the ligands should be located on the surface of the rusticyanin molecule. This is because binding should be complete only when bare ligands on the protein molecule are close to the solid surface of pyrite. It is expected that possible conformational changes in the protein structure may occur when rusticyanin coordinates to a copper atom. Therefore, the degradation ratio should be different between the apo and holo forms when each rusticyanin is exposed to proteinase, if the tertiary structure of the rusticyanin molecule changes before and after the copper coordination. Figure 3 shows the degradation ratios of both forms of rusticyanin when exposed to different amount of chymotrypsin. Chymotrypsin was selected for the degradation of aromatic amino acids that would be hidden inside of the protein molecule based on the reported molecular model. The holo form of rusticyanin showed strong resistance to chymotrypsin. Seventy five percent of the holoprotein remained without degradation even though 2.5 μg of chymotrypsin was used. In the apo form, rusticyanin was rapidly digested by chymotrypsin. Only 29% of apoprotein remained and 71% was digested when 2.5 μg of chymotrypsin was used. Copper ions did not influence the chymotrypsin activity in the control experiments using bovine serum albumin (data not shown). Therefore, it was expected that rusticyanin in the apo form would have a different structure when compared to that in the holo form.

4. DISCUSSION

We proposed that rusticyanin from *T. ferrooxidans* acts in both the respiration chain and cell adhesion onto pyrite (3). Bifunctional proteins are widely spread from bacteria to mammals (10-12). For example, in plants, the bc1 complex is bifunctional, being involved both in respiration and in protein processing (12). Therefore we wished to examine the functional regions are distributed in the bifunctional protein.

In this study, it was clarified that the copper coordination sites of rusticyanin could be pyrite binding sites. It appears that the same sites are active for both adhesion and electron transfer functions. Three out of the four residues, His[85], Cys[138] and Met[148], involved in the copper coordination sites, are needed for the binding to pyrite. However, it must be investigated as to how these ligands of rusticyanin in the apo form can interact with the pyrite surfaces. The crystal structure of rusticyanin in the holo form has been suggested in several reports. Coordination ligands including His[85], Cys[138] and Met[148] formed cavities and are buried in the inside of the protein molecule (8). Therefore, it is not possible for these ligands to with interact the pyrite surface in rusticyanin in the holo form. From chymotrypsin experiments, it was suggested that apo rusticyanin might have a different type of protein folding (Fig. 3). If the ligands were present on the surface of the apoprotein, it should be able to interact with the pyrite surfaces. If this is the case, the affinity of the apoprotein to pyrite depends on the three dimensional arrangement of the ligands. The geometry of the primary coordination shell around the binding site is supposed as illustrated in Fig. 4. His[85], Cys[138] and Met[148] ligands are needed for the binding (Fig. 2). The ligands would be closely arranged to form a primary coordination shell because together they coordinate a single metal atom. Only His[143] on the outside of the coordination shell is opened in the unfolding state in the apo form because its mutation did not effect the binding ability to pyrite (Fig . 2).

REFERENCE

1. N. Ohmura, K. Tsugita, J. Koizumi and H. Saiki: J. Bacteriol., 178 (1996) 5776.
2. N. Ohmura, K. Kitamura and H. Saiki: Appl. Environ. Microbiol., 59 (1993) 4044.
3. K. Sasaki, N. Ohmura, H. Kishida and H. Saiki, (1998) unpublished
4. J. G. Cobley and B. A. Haddock: FEBS Lett., 60 (1975) 29.
5. R. C., Blake II and E. A. Shute: J. Biol. Chem., 262 (1987) 14983.
6. R. C., Blake II and E. A. Shute: Biochemistry, 33 (1994) 9220.
7. N. Ohmura, K. Sasaki and R. C., Blake II, (1998) unpublished
8. R. L. Walter, S. E. Ealick, A. M. Friedman, R. C., Blake II, P. Proctor and M. Shoham: J. Mol. Biol., 263 (1996) 730.
9. D. R. Danlio, A. Toy-Palmer, R. C., Blake II and H. J. Dyson: Biochemistry, 34 (1995) 6640.
10. J. Nakagawa, S. Tamaki, S. Tomioka and M. Matsuhashi: J. Biol. Chem., 259 (1984) 13937.
11. W. Tirasophon, A. A. Welihinda and R. J. Kaufman: Genes Dev., 12 (1998) 1812.
12. E. Glaser, A. Eriksson and S. Sjoling: FEBS Lett., 346 (1994) 83.

Classification of rusticyanin structre genes from five different strains of *Thiobacillus ferrooxidans* into two groups

Chigusa Ida[a], Kazuhiro Sasaki[b]*, Naoya Ohmura[b], Akikazu Ando[a]
and Hiroshi Saiki[b]

[a]Department of Biotechnology, Graduate School of Science and Technology,
Chiba University, 648 Matsudo, Matsudo-city, 271-0092 Japan

[b]Bio-Science Department, Abiko Research Laboratory, Central Research
Institute of Electric Power Industry, 1646 Abiko, Abiko-city, 270-1194 Japan

The blue copper protein, rusticyanin, is an important component of the iron-oxidizing electron transport chain of *Thiobacillus ferrooxidans*. However, the rusticyanin proteins have not been thoroughly investigated in genetics yet. Here, we report the classification of rusticyanin proteins of five different strains. In all of the rusticyanins, the sequences corresponding to four ligands for a copper atom were conserved completely. However, there were clear differences in the sequences when compared with the proteins in the entire range. The sequences of ATCC 23270, JCM 3863 and JCM 7811 agreed well with the reported sequence of ATCC 33020, but those of JCM 3865 and IFO 14246 were different from the others. The sequence identities within each group ranged from 98 % to 100 % and those between the different groups ranged from 79 % to 80 %. The differences between the two groups were localized to certain regions that included signal peptides for protein processing and amino acids existing on the surface of the rusticyanin molecules. Those results suggested that *T. ferrooxidans* had two different types of rusticyanin.

---

*To whom correspondence should be addressed: Bio-Science, CRIEPI, 1646 Abiko, Abiko-city, Chiba, 270-1194 Japan. Fax: 81-471-83-3347.
E-mail: k-sasaki@criepi.denken.or.jp

1. INTRODUCTION

Rusticyanin is considered to be one of the electron transfer proteins in the acidophilic and chemolithotrophic bacterium, *Thiobacillus ferrooxidans* (1). It had been known that this copper protein was highly expressed in the bacterium. Five % of the total amount of protein was thought to be rusticyanin when the cells were grown on ferrous ions (2). This blue protein plays an important role in iron respiration of the bacterium to conserve energy for growth. Based on this physical importance of the protein, many researchers have been interested in studying the rusticyanin gene. The first report to describe the cloning of the rusticyanin gene was by Bengrine et al. (3). They reported that *T. ferrooxidans* ATCC 33020 had one copy of the rusticyanin gene and that a signal peptide consisting of 32 amino acid residues was removed from the mature protein. Another report showed that the rusticyanin gene was located immediately downstream of the cytochrome c4 gene (Cox3) (4). However, the complete sequence involved in the structural gene of rusticyanin was reported only for the ATCC 33020 strain. This report describes the characterization of rusticyanin genes that were obtained from five different strains of *T. ferrooxidans*. The rusticyanin genes were classified into two groups depending on the regions that encoded the signal peptide for protein processing and several amino acids on the surface of rusticyanin molecules.

2. MATERIALS AND METHODS

2.1. Bacterial strains, plasmids, and culture conditions

All of the *Thiobacillus ferrooxidans* strains were obtained from some culture collections. ATCC 23270 was purchased from the American Type Culture Collection. JCM 3863, JCM 3865 and JCM 7811 were purchased from the Japan Collections of Microorganisms. IFO 14246 and IFO 14262 were from the Institute Fermentation Osaka. DNA to be sequenced was cloned in competent *E. coli* JM109 cells (Invitrogen). Plasmids pCR2.1™ (Invitrogen) were used for gene cloning. *E. coli* was cultured on a standard LB medium at 37 °C. *T. ferrooxidans* was cultured on 9K medium at 30 °C (5).

2.2. Preparation of genomic DNA

T. ferrooxidans cells in 100ml of culture were harvested by centrifugation, and the cells collected were washed several times with sulfuric acid solution (pH 2.0) to remove iron precipitates. The washed cells were then used for preparation of chromosomal DNA using the method described by Murray et al. (6).

2.3. Plasmid constructions

The primers for amplifying the genomic region including the structural gene of rusticyanin were designed on the basis of the sequence of the rusticyanin gene of *T. ferrooxidans* ATCC 33020, (3,4). The rp-5-ATG and rp-3-UAA were used to amplify the rusticyanin structural gene (Fig. 1). Their

UAA were used to amplify the rusticyanin structural gene (Fig. 1). Their sequences were GGATAACAT**ATG**TATACAAACACGATG (rp-5-ATG) and AATTATGGATCC**TTA**CTTAACAACGATCTT (rp-3-UAA). The rp-5-ATG and rp-3-UAA have an NdeI site and a BamHI site, respectively. Those primers were designed to hybridize to both sides of the entire rusticyanin open reading frame as shown in Fig. 1 and were used for the following PCR amplification with genomic DNA obtained from each strain. All of the PCR reactions in this study were carried out using a Model 9600 thermal cycle system (Perkin-Elmer, Norwalk, CT) and reagents prepared with a PCR reaction kit (Sawaday Technology, Tokyo, Japan). The PCR reactions involved a hot start at 95 °C for 5 min followed by 25 cycles each of 30 sec at 95 °C, 30 sec at 51 °C, and 30 sec at 72 °C. The reaction was completed with 5 min incubation at 72 °C by the addition of 3' adenosine-overhang post-amplification. The PCR products were fractionated by agarose gel electrophoresis with a DNA fragment purification system (SUPELCO, Bellefonte, PA) and cloned into the PCR 2.1™ plasmid vector (Invitrogen, Carlsbad, CA). The recombinant plasmids were named pR-1 (JCM 3863), pR-2 (JCM 3865), pR-3 (JCM 7811), pR-4 (IFO 14246) and pR-6 (ATCC 23270). These plasmids were transferred into *E. coli* JM109. Every clone was collected from three individual colonies and sequenced. A part of the sequence was confirmed by the following method. Additional primers were designed (Fig. 1). The sequences of the 5' and 3' primers were TGGAATCGTCTACAACTGGA (prp-5) and AGTCTTGCCGCTGTAGGTAA (prp-3), respectively. The prp-5 primer hybridized to Cox3 located about 1 kbp upstream from the rusticyanin start codon. The prp-3 primer hybridized to the region 207 bp downstream from the start codon. Then, the products should contain the 207 bp of the structural rusticyanin gene. The products were cloned into the plasmids in the same manner mentioned above. Finally, the cloned fragments were sequenced for comparison with the sequence obtained from another primer set (rp-5-ATG and rp-3-UAA)

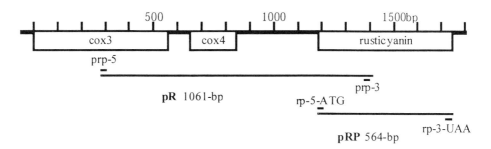

Fig. 1 Physical map of T.ferrooxidans genomic DNA around the rusticyanin structure gene.

The pR and pRP were produced by PCR reaction using the PCR primer, prp-5, prp-3, rp-5-ATG and rp-3-UAA, and subsequently cloned.

2.4. DNA sequencing

Plasmids for sequencing the template were purified using a Flexi Prep™ Kit (Amersham Pharmacia Biotech Ltd., Uppsala, Sweden). The sequencing reaction was carried out with a Gene Amp PCR System 9600 (Perkin-Elmer) and an ABI PRISM™ Dye Primer Cycle Sequencing Core Kit (Perkin-Elmer). The sequencing data were analyzed with Genetyx-Mac 10.0 (Software Development Co., Ltd., Tokyo, Japan).

2.5. Western Blotting

T. ferrooxidans cells harvested from 2 liters of culture were washed with sulfuric acid solution (pH 2.0) and resuspended in 1 ml of 20 mM Tris-HCl buffer (pH 7.5). The cell extracts were prepared by sonication. Twenty μg of the cell free extract (1mg/ml) was applied on SDS-PAGE. The proteins on the gels were transferred onto Immobilon™-P, a PVDF membrane (Millipore). The bands corresponding to the rusticyanin protein were detected by anti-rusticyanin polyclonal antibody, which was purified by affinity chromatography with rusticyanin protein, using a vectastain abc-ap kit and an alkaline phosphatase substrate kit (Vector Laboratories). Protein markers were stained with CBB.

2.6. Protein structure

The three-dimensional image of the rusticyanin molecule used in this paper was drawn by rasmol (RasWin Molecular Graphics Windows Version 2.6) on the basis of the three-dimensional data file of rusticyanin (PDBid: 1RCY) (7), which was downloaded from the web site of the Protein Data Bank, URL; http://pdb.pdb.bnl.gov/pdb-bin/pdbmain.

3. RESULTS

The existence of rusticyanin in the stains used in this study has not been reported. Therefore, a series of experiments to detect the rusticyanin protein was carried out first. Each strain was cultured on soluble ferrous ions (Fe^{2+}) to induce expression of rusticyanin. The cell-free extracts, which were prepared from the cells, were subjected to SDS-PAGE (Fig. 2). There were no significant differences between the band patterns for every strain. A large clear band at the molecular weight of 18 kDa was observed on each lane where the extracts were loaded. This molecular weight size was considered to correspond to the rusticyanin protein because the purified rusticyanin from ATCC 23270 showed a band of the same size. Therefore, to confirm that the protein corresponded to the uniform band on every strain, western blotting was carried out with immuno staining of rusticyanin (Fig. 3). The antibody used for staining was produced by immunization of rabbits with the purified rusticyanin protein from ATCC 23270. An affinity column immobilized with the purified rusticyanin was used to purify anti-rusticyanin polyclonal antibody. As a result of immuno staining,

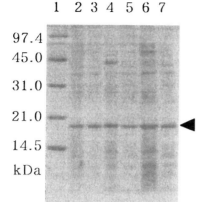

Fig. 2 Comparison of the protein expression of 6 different strains.

Cell extract (20mg) was applied to SDS-PAGE respectively. lane 2 ; ATCC 23270. lane 3 ; JCM 3863. lane 4 ; JCM 3865. lane 5 ; JCM 7811. lane 6 ; IFO 14246. lane 7 ; IFO 14262. lane 1 ; protein size marker. The arrow indicates the position of rusticyanin.

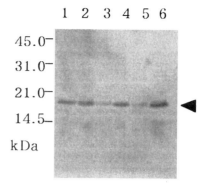

Fig. 3 Comparison of the rusticyanin expression of 6 different strains.

The SDS-PAGE was performed as Fig. 2. anti-ATCC 23270 rusticyanin IgG was performed as primary antibody. lane 1 ; ATCC 23270. lane 2 ; JCM 3863. lane 3 ; JCM 3865. lane 4 ; JCM 7811. lane 5 ; IFO 14246. lane 6 ; IFO 14262. The arrow indicates the position of rusticyanin.

antigen-antibody reactions were observed with all extracts. However, the band of 18 kDa, which was expected to be rusticyanin by CBB,
was stained on every lane loaded with each extract. Based on those results, it was clear that rusticyanin was highly expressed in every strain when grown on ferrous ions.
The rusticyanin structural gene has been reported only for ATCC 33020. We had succeed in cloning the rusticyanin open reading frame from the following five strains, ATCC 23270, JCM 3865, JCM 3863, JCM 7811 and IFO 14246. All of the fragments were cloned as the corresponding pR series plasmids and sequenced entirely. The sequences of the genes were translated to the corresponding amino acid sequences as shown in Fig. 4 with the addition of the sequence of ATCC 33020 (3) and of ATCC 23270 (8). The summarized similarities of amino acids sequences and identities of DNA sequences are also shown in Table 1. The amino acid sequences of rusticyanin of ATCC 23270, JCM 3863 and JCM 7811 were quite similar to that of ATCC 33020. The similarities and identities between them were more than 98.9 % and 93.6 % respectively. Those strains could be classified into the same group. The sequences of JCM 3865 and IFO 14246 were significantly different from that of

134

```
ATCC 33020   1 MYTQNTMKKNWYVTVGAAAALAATVGMGTAMAGTLDSTWKEATLPQVKAMLEKDTGKVSGDTVTYSGKT
ATCC 23270   1 ..............................................T.....................
JCM 3863     1 -------.......................................T.....................
JCM 7811     1 ..............................................T.....................
IFO 14246    1 -------XQGSACRRRLSTI...AL.IS....AP..TS..M.........L.A..S.....K.......
JCM 3865     1 -------XQGSACRRRLSTI...AL.IS....AP..TS..M.........L.A..S.....K.......
               .............***..*..****..**..**..*******.*.**.**.**.********

ATCC 33020  70 VHVVAAAVLPGFPFPSFEVHDKKNPTLEIPAGATVDVTFINTNKGFGHSFDITKKGPPYAVMPVIDPIV
ATCC 23270  70 ...................................................................
JCM 3863    70 ...............................R...................................
JCM 7811    70 ...................................................................
IFO 14246   70 ................GI..V.....D.....I............L..................I
JCM 3865    70 ................I..V.....D.....N..I..........L..................I
               ******************..**.*****.**.**.**.***********.***************.

ATCC 33020 139 AGTGFSPVPKDGKFGYTDFTWHPTAGTYYYVCQIPGHAATGMFGKIIVK       187
ATCC 23270 139 ..............................................V..       187
JCM 3863   139 ...........................................------       187
JCM 7811   139 ...........................................------       187
IFO 14246  139 ..........S.G........R.A...................------       187
JCM 3865   139 ..........S.G........R.A...................------       187
               **********.*.*********.*.********************
```

Fig. 4 Comparison of rusticyanin amino acid sequences.

The amino acid sequences of rusticyanin from the each strains were aligned. Asterisks indicate the residues conserved in all strains. Bars show primer regions. The ▼ indicates processing site. The sequence of ATCC 33020 is cited from reference 4. The part of sequence (64-187) of ATCC 23270 is cited from reference 8.

Table 1 Comparison of the identities of the DNA sequences and the similarities of the amino acids sequences of the rusticyanin structure genes.

nucleotide(%) amino acid(%)	ATCC 33020	ATCC 23270	JCM 3863	JCM 3865	JCM 7811	IFO 14246
ATCC 33020		93.6	93.8	73.9	94.3	74.8
ATCC 23270	98.9		98.5	73.9	99.1	74.8
JCM 3863	98.9	99.4		73.1	99.0	56.7
JCM 3865	79.0	79.5	79.5		73.1	98.8
JCM 7811	99.4	100	99.4	79.4		73.0
IFO 14246	79.7	80.2	80.0	98.9	79.4	

Table entries have been placed in two groups. The upper section indicates the identities of the DNA sequences. The lower section indicates the similarities of amino acid sequences.

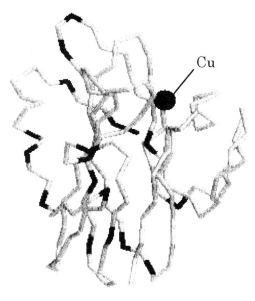

Fig. 5 Differences in amino acid sequences of rusticyanin from ATCC 33020 and JCM 3865.

Difference in rusticyanin sequence from ATCC 33020 and JCM 3865 are colored black and gray. Differences surface amino acids are indicated in black and others are in gray.

ATCC 33020. There were most 80.2 % similarity and 74.8 % identity between groups. However, the between JCM 3865 and IFO 14246 had 98.9 % similarity and 98.8 % identity. These results clearly showed that the rusticyanin genes of the five strains should be classified into two different groups.

The sequences of rusticyanin suggested another important finding. The amino terminal sequence of rusticyanin had been reported (9,10), therefore, the rusticyanin precursor should be processed to the mature protein by removing 32 amino acids from the amino terminal. Many differences between the two groups in the amino acid sequences were localized on the region including 32 amino acids for the protein processing as shown in Fig. 4. In ATCC 33020, ATCC 23270, JCM 3863 and JCM 7811, there were no differences in the sequences of the region. However, 17 out of 26 amino acids in the regions of JCM 3865 and IFO 14246 were different when compared with that of ATCC 33020. These results suggested that the rusticyanin protein of each group had different sequences for protein processing.

4. DISCUSSION

It has been thought that rusticyanin is a component of the electron transfer chain in the bacterium (1,11,12). However, we had shown recently that rusticyanin in the cells was located in three different places, the periplasmic space, the cell envelope and the inside of the inner membrane (13). Therefore, rusticyanin should be transferred to each of these places in the cells. From this point of view, it could be expected that there might be several signal sequences in the N-terminal region of the structural gene for rusticyanin. Many differences in sequence between the two groups were observed in the region of the signal sequence that is required for protein processing (Fig. 4). This fact might reasonably explain why rusticyanin could be transferred to different places. This hypothesis raises another question. If the different sequences in the region for protein processing are related to protein delivery, each strain should have three different genes for rusticyanin because the protein was located in three places in the cell. Bengrine et al. had reported that gene for rusticyanin was present as only one copy in ATCC 33020 (3). We are now trying to detect other rusticyanin genes in *T. ferrooxidans* using several primers in the same manner mentioned in this study. As preliminary result, we found that more than two rusticyanin genes were observed on both IFO 14246 and JCM 3865 strains by southern blot experiments (Data not shown). Those genes might have different sequences in the region for protein processing to enable the delivery of rusticyanin to three places.

In recent years, the three-dimensional protein structure of rusticyanin has been described in several reports (7,9). The protein consists of a single polypeptide chain containing 155 amino acid residues and is capable of coordinating a copper atom to a cluster of four amino acid residues (His85, Cys138, His143 and Met148). The sequences corresponding to the four ligands were completely conserved in the two groups of rusticyanin (Table 1). However, different sequences between the two groups were observed in the region coding the mature rusticyanin protein. There were 19 differences in amino acid sequences as shown in the following, 1GlyAla, 2ThrPro, 6ThrSer, 9GluMet 18MetLeu, 20GluAla, 23ThrSer, 29AspLys, 56ValIle, 59LysVal, 65GluAsp, 71ThrAsn, 74ValIle, 87PheLeu, 106ValIle, 117AspSer, 119LysGly, 128HisArg and 130ThrAla. In mature rusticyanin, differences were found in 12.8 % of all of the amino acids. Therefore, the differences in the mature protein were analyzed using a tertiary structural model to determine what parts of the protein were replaced. The model of the rusticyanin molecule was constructed on the basis of the amino acid sequences of JCM 3865 and is shown in Fig. 5. The model constructed was compared with the rusticyanin of ATCC 23270 (7). The differences in amino acids of both models are shown as colored parts of the molecule. As a result, 12 out of 19 differences were observed between the mature rusticyanins of ATCC 23270 and JCM 3865 in amino acids located on the surface of the rusticyanin molecule. However, the physiological relevance of the differences in amino acids on the protein surface is still under investigation.

REFERENCES

1. J. G. Cobley, and B. A. Haddock, FEBS Lett., 60 (1975) 29.
2. J. C. Cox et al., Biochem. J., 174 (1978) 497.
3. A. Bengrine, N. Guiliani, C. Appia, F. Borne, M. Chippauxm, and V. Bonnefoy, in Biohydrometallurgical Processing (C. A. Jerez, T. Vargas, H. Toledo, and J. V. Wiertz eds.), Universidad de Chile, (1995) pp 75.

4. A. Bengrine, et.al, GenBanc, LOCUS [TFAJ6456], ACCESSION [AJ006456 X95823 Y07610], National Center for Biotechnology Iinformation. (1998). URL: http://www2.ncbi.nlm.nih.gov/cgi-bin/genbank.

5. N. Ohmura et al., Appl. Environ. Microbiol., 59 (1993) 4044.

6. M. G. Murray, and W. F. Thompson, Nucleic. Acids. Res., 8 (1980) 4321.

7. R. L. Walter et al., J. Mol. Biol., 263 (1996) 730.

8. Hall, J. F. et al., FEMS Microbiol., Lett., 137 (1996) 85.

9. M. Ronk, et al., Biochemistry, 30 (1991) 9435.

10. T. Yano et al., FEBS Lett., 288 (1991) 159.

11. R. C. Blake II, and E. A. Shute, J. Biol. Chem., 262 (1987) 14983.

12. R. C. Blake II, and E. A. Shute, Biochemistry, 33 (1994) 9220.

13. K. Sasaki et al., (1998) unpublished

The use of insertion sequences to analyse gene function in *Thiobacillus ferrooxidans*: A case study involving cytochrome c-type biogenesis proteins in iron oxidation

D. S. Holmes[a], E. Jedlicki[b], M. E. Cabrejos[b], S. Bueno[b], M. Guacucano[a], C. Inostroza[a], G. Levican[b], P. Varela[b] and E. Garcia[c].

[a] Dept. of Biological Sciences, University of Santiago (USACH), Santiago, Chile.

[b] Institute of Biomedical Sciences, University of Chile, Santiago, Chile.

[c] Lawrence Livermore National Laboratory, Ca., USA.

With DNA sequence information available for *Thiobacillus ferrooxidans* it will become much easier to understand its physiology and genetics. However, based on DNA sequence information alone a number of its genes, especially those involved in the more unusual or specific functions of the microorganism such as iron oxidation, could remain without a clear assignment of function. We show how naturally occurring insertion sequences can be used to generate mutants by knocking out gene function and how these knock-out mutants can be analyzed to reveal the gene involved in the mutation.

An iron oxidizing mutant of *T. ferrooxidans* ATCC19859 is shown to contain a copy of the insertion sequence IST1 in a gene, which we term resB, that is probably involved in the biogenesis of a c-type cytochrome. Downstream of resB, and perhaps part of the same transcriptional unit, is another gene, resC, also probably involved in the biogenesis of a c-type cytochrome. Some of the predicted properties of these biogenesis proteins and their evolutionary relations are described.

The target cytochrome(s) of the putative cytochrome-c type biogenesis proteins remains unknown, although just upstream of resB is a gene potentially encoding a cytochrome c1, making it a likely candidate as the target cytochrome. The target cytochrome is essential for growth on iron but not for growth on thiosulfate and may prove specific for the pathway of energy transduction and/or electron flow during iron oxidation.

1. INTRODUCTION

Recently, the Institute for Genome Research (TIGR) and the USA Dept. of Energy initiated a program to sequence the entire genome of *Thiobacillus ferrooxidans*. It may have been completed and released to the public by the time this paper is published. Clearly, the full sequence of all the genes of *T. ferrooxidans* will aid enormously in the understanding of the physiology and genetics of this microorganism. However, if *T. ferrooxidans* is similar to other microorganisms that have been sequenced, then functions will be able to be assigned to only about 65% of its genes based on their similarity to other known genes. This will leave approximately 35% of all its genes without a defined function. Amongst this 35% are likely to be genes that are more specific to the physiology of the microorganism such those involved in the oxidation of iron and sulfur and in its ability to survive in extremely acid conditions. Because genes such as these will provide us with a clearer picture of the unusual physiology of this microorganism it will repay our efforts to try to analyze them and deduce their function. But this presents a problem.

The genetic manipulation of *T. ferrooxidans* is still at a rudimentary stage of development. In other microorganisms, techniques such as transformation and conjugation permit researches to introduce wild type DNA back into mutant strains selecting for the capacity of the wild type DNA to complement the missing function in the mutant. In this way the specific gene that complements the mutant function can be isolated, sequenced, analyzed and assigned a function. However, reproducible and general techniques for the introduction of DNA back into cells of *T. ferrooxidans* are not available even though there exist some specific and isolated cases of the use of such techniques (Kusano et al., 1992, Peng et al., 1994).

In the absence of such genetic tools investigators have focused their efforts on the use of other techniques to isolate and characterize genes from *T. ferrooxidans*. A valuable approach has been the use of complementation of defined mutants of other microorganisms, such as *E. coli*, with cloned DNA from *T. ferrooxidans*. For example, the several genes that encode the F1 subunits of the *T. ferrooxidans* ATPase pump were identified by complementing appropriate *E. coli* mutants with cloned genes of *T. ferrooxidans* (Brown et al., 1994). The F1 subunits of *T. ferrooxidans* probably reside in the cytoplasm which has a similar pH to that of *E. coli* and so there was a reasonable chance that the complementation experiments would succeed. However, this approach is limited to the identification of genes that can complement functions in other well characterized microorganisms and genes encoding specialized proteins of *T. ferrooxidans* such as those involved in iron oxidation or genes whose products function only under specialized conditions, such as those that operate in the low pH environment of the periplasm, could be missed using this approach. For example, the gene encoding the Fo subunit of the ATPase pump of *T. ferrooxidans* was unable to complement the appropriate mutant of *E. coli*, most likely because the Fo subunit of *T. ferrooxidans* has a surface exposed to the low pH environment of the periplasmic space and probably exhibits significant structural differences from its *E. coli* counterpart (Brown et al., 1994).

Another approach that has yielded a substantial amount of genetic information involves the use of reverse genetics. A protein of interest is first identified by biochemical criteria, and is then isolated and sequenced. From the sequence, or partial sequence, DNA probes can be designed that, by hybridization or PCR, have the capacity to identify the corresponding gene from a genebank of cloned *T. ferrooxidans* DNA. For example, a gene that encodes a Fe(II) oxidase was identified and characterized by this technique (Kusano et al., 1992).

In this paper we present an overview and preliminary results utilizing another technique that may be particularly useful for the identification of the function of genes of *T. ferrooxidans*. The approach could also verify the role of genes that have received an assignment of function through biochemical studies or by DNA sequence homology to other genes of known function.

The strategy relies on the ability of naturally occurring insertions sequences (ISes) to transpose and inactivate genes causing mutations. If the IS causes a conditional mutation with an observable phenotype then DNA can be prepared from the mutant and any novel repositioning of the ISes can be identified. Subsequently, using DNA   probes or PCR amplification the mutant gene containing the IS element can be isolated, sequenced and its protein product deduced. This provides evidence that the gene could be involved in the pathway that leads to the observed phenotype but does not prove the contention. It is possible that the repositioning of an IS element coincided with the appearance of the mutant phenotype but was not the cause of it. However, the contention that the IS element caused the mutant phenotype is strengthened if revertants of the mutant back to the wild type are associated with the loss of the IS element from the gene in question.

In this paper we report results of the use of the insertion sequence IST1 (Yates and Holmes, 1987; Holmes et al., 1987; Zhao et al., 1998) to knock out gene function in *T. ferrooxidans*.

2. MATERIALS AND METHODS

T. ferrooxidans   ATCC19859 was obtained from the American Type Culture Collection. *T. ferrooxidans* was grown on modified 9K-ferrous iron medium (Yates and Holmes, 1986). *E. coli* C600 were grown in LB broth. An iron oxidizing mutant of *T. ferrooxidans* was isolated as described previously (Schrader and Holmes, 1988).

DNA cloning, Southern blot   hybridization, DNA sequencing and other molecular biology techniques were carried out by standard procedures (Maniatis et al., 1988). DNA and protein data banks were searched using the Blast Beauty program (Altschul et al., 1990; Worley et al., 1995).

Multiple protein sequence alignments were made using the ClustalW program: (http://www.ibc.wustl.edu/msa/clustal.html) .

Predictions of membrane protein structure were made using the TMPRED program: (http://www.isrec.isbsib.ch/software/ TMPRED_form.html).

Gene organization was analysed using the Kegg encyclopedia: (http://www.genome.ad.jp/kegg/kegg2.html)   and   the   Puma   workstation: (http://wit.mcs.anl.gov/WIT2/wit.html).

3. RESULTS

T. ferrooxidans was grown for many generations on thiosulfate and the pattern of integration of IST1 in the genome was investigated by Southern blot analyses using DNA digested by the restriction enzyme BamH1 which does not cut inside the IST1 element (Schrader and Holmes, 1988). The cells were then plated on a solid medium containing both ferrous iron and thiosulfate. Wild type cells form small reddish colonies not unlike those typically found when *T. ferrooxidans* is grown on solid medium containing ferrous iron. In contrast, mutant cells unable to oxidize ferrous iron utilize the thiosulfate as an energy and electron source and grow as large whitish colonies (Schrader and Holmes, 1988). One of the mutant colonies was selected and DNA was prepared from it. The Southern blot pattern of the DNA from the mutant after digestion with BamH1 was compared to that from wild type DNA and a novel band was observed in the mutant DNA, indicating that an IST1 element had transposed to a new location (data not shown).

142

The novel band was excised from the agarose gel and was cloned and sequenced. Since the sequence of IST1 was previously known (Zhao, 1995; Zhao et al., 1998) its position in the newly sequenced DNA could be deduced. It was found to reside in an open reading frame, termed resB whose 3' terminus extended beyond the 3' end of the cloned DNA. We subsequently isolated and cloned a second fragment of DNA that extended the DNA sequence of resB to its 3' end and which also included two additional ORFs (resC and hyp1). In addition, a partial ORF (cyt1) was found upstream of resB. We argue below, based on homology studies and expression of RNA, that these ORFs are functional genes. The relative positions of these genes and that of the inserted IST1 are diagrammatically represented in figure 1.

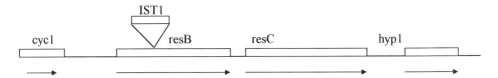

Figure 1. Diagrammatic representation of a region of the *T. ferrooxidans* genome associated with the capacity to oxidize Fe+2. In the mutant an IST1 element has inserted in resB in the position indicated. The arrows indicate the direction of transcription of the genes: cyt1 encoding a hypothetical apocytochrome c1, resB encoding a hypothetical cytochrome c-type biogenesis protein (GenBank Acc. No. AF089765), resC encoding a hypothetical cytochrome c-type biogenesis protein (GenBank Acc. No. AF089766) and hyp1 encoding a hypothetical protein of unknown function (GenBank Acc. No. AF089767). Total lenght of fragment illustrated is approximately 5kb.

Searches of the DNA and protein data banks reveal that the putative product of cyt1 has significant homology to the apocytochrome c1 gene of *Chromatium vinosum*, and that the products of the genes resB and resC are homologous to putative cytochrome c-type biogenesis proteins ResB and ResC respectively of *Aquifex aeolicus*. The gene hyp1 has homology to a hypothetical protein also of *A. aeolicus*. These, and additional homologies of the putative proteins, are listed in table 1.

Preliminary RT-PCR experiments have demonstrated that resB, resC and hyp1 encode a policistronic RNA (data not shown). No clear function can be assigned to the product of ORF 4 and it will not be discussed further.

It can be seen from table 1 that the putative products of resB and resC are probably cytochrome c-type biogenesis proteins with similarity to the ResB and ResC family of proteins respectively, based upon the *B. subtilis* convention of naming. Alternative names for the resB gene are ccs1 and ycf44 and for resC are ccsA and ycf5, depending on the organism where they were first discovered.

The predicted molecular weights and isoelectric points of the putative protein products of resB and resC are shown in Table 2. The DNA sequence of cyt1 remains incomplete and so no determination of its molecular weight or pI can be made.

Table 1
Protein homologies of the putative products of cyc1, resB and resC

proposed gene name	protein homology to: (organism)	P value
cyc1	apocytochrome c1 (Chromatium vinosum) (Rhodopseudomonas viridis)	7×10^{-7} 1.6×10^{-4}
resB	cytochrome c-type biogenesis protein ResB (Aquifex aeolicus) (Synechoccus PC 7002) (Bacillus subtilis)	7.6×10^{-19} 5.2×10^{-11} 9.7×10^{-5}
resC	cytochrome c-type biogenesis protein ResC (Aquifex aeolicus) (Odontella sinensis) (Synochocystis sp.)	7.4×10^{-49} 4.6×10^{-45} 2.5×10^{-44}

Results of a Blast-Beauty search of GenBank and Swiss-Prot sequence data banks for the products of genes resB, resC and hyp1. P is a measure of the probability with which the similarity of the proteins could occur randomly. A P value of less than 10-5 is considered to show, with a high degree of confidence, that the respective proteins are biologically related.

Table 2.
Predicted molecular weights and isoelectric points (pI) of the putative protein products of resB and resC.

Gene	Putative protein:	Predicted molecular weight (kilodaltons)	pI
resB	cytochrome c-type biogenesis (similar to resB, ccs1, ycf44)	65.6	9.48
resC	cytochrome c-type biogenesis (similar to resC, ccsA, ycf5)	42.7	9.36

A computer analysis of the hypothetical proteins cytochrome c1 and the two cytochrome c biogenesis proteins using the program TMED reveals that it is highly likely that they are all integral inner membrane proteins. The cytochrome c1 has a membrane anchor domain and the two cytochrome c biogenesis proteins have 5 and 9 predicted transmembrane domains respectively and are probably oriented in the membrane as shown in figure 2. These proposed computer generated structures are consistent with existing models for homologous cytochrome c-type biogenesis proteins found in other organisms.

The pIs of the domains of the putative cyc1, resB and resC products that are predicted to face the periplasmic are in the range of 9.2 - 9.5 and thus would be highly protonated at the acid pH of 2 of the periplasmic space *T. ferrooxidans*. In addition, the putative product of resC has a predicted C terminal rich in tryptophan, which is a characteristic of the resC family of proteins.

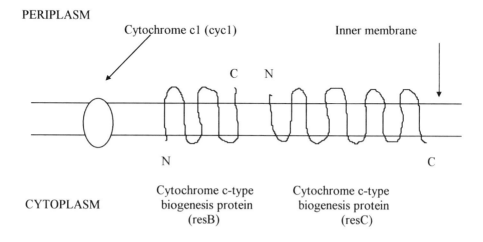

Figure 2. Computer predicted transmembrane domains and orientation in the inner membrane of the hypothetical cytochrome c-type biogenesis proteins ResB and ResC and the protein cytochrome c1.

4. DISCUSSION

4.1. Identification of a novel protein involved in iron oxidation

The insertion of IST1 into the gene resB is strongly correlated with the generation of mutants of *T. ferrooxidans* that lack the ability to oxidize iron (Fe- phenotype). Since IST1 contains strong translational stops in all reading frames, its insertion into resB would prevent its proper expression. Therefore, we propose that the Fe- phenotype is caused by the IST1 element interrupting the expression of resB. The coding region of resC is only 35 nucleotides downstream from the translation termination of resB and this intergenic region lacks both an obvious rho independant type transcriptional stop and an obvious promoter. Furthermore, preliminary reverse PCR experiments indicate that resB and resC are cotranscribed raising the possiblity that resB and resC are part of an operon. If this is the case then the insertion of IST1 into resB could also prevent the expression of resC. It has been demonstrated (Zhao, 1995) that wild type revertants of the Fe- mutant lack the IST1 sequence in gene resB supporting the view that the presence of IST1 in resB is the cause of the mutant phenotype and is not just coincidentally associated with it.

In all systems studied, heme addition to c-type cytochromes does not take place in the cytoplasm, rather the apocytochrome c is first passed through the membrane by the Sec transport system and correct folding and heme addition takes place on the periplasmic side aided by various cytochrome biogenesis proteins. It follows that heme must also be passed through the membrane and this is achieved using additional cytochrome biogenesis proteins (see review by Thony-Meyer, 1997). However, the number and type of cytochrome biogenesis proteins involved differs between organisms. Kranz et al. (1998) have proposed that there are basically three different systems involved in cytochrome c-type biogenesis. System 1 organisms, which includes alpha and gamma proteobacteria, plant mitochondria and Archaea, use the Hel family of proteins (and other proteins), whereas System 2 organisms, which include Aquifex, Gram positive bacteria, cyanobacteria and plastids, use the Res family of proteins. System 3 organisms, which include fungal, invertebrate and vertebrate mitochondria, have a much reduced method of cytochrome c-type biogenesis that involves neither the Hel nor the Res proteins. It is curious that *T. ferrooxidans*, which is a member of the gamma proteobacteria, does not use the System 1 Hel proteins but rather appears to use the Res proteins pertaining to System 2 organisms. It remains to be seen if *T. ferrooxidans* has copies of the hel genes in addition to res family.

It is also curious that the maximum homology of the *T. ferrooxidans* resB and resC genes is with homologs from *Aquifex aeolicus* which is considered to be close to the root of divergence between the Archaea-Eukaryote line and the Eubacterial line of descent. Aquifex is also an extreme thermophile. There is substantial homology between the domains of resB and resC that face into the periplasmic space with the equivalent domains on the Aquifex homologs. But in the case of *T. ferrooxidans* these domains would be exposed to an acid pH whereas, in the case of Aquifex they would be exposed to a more neutral but high temperature periplasmic space. Thus the extent of homology between the periplasmic domains is perhaps surprising. A detailed analysis of these domains could reveal important information regarding the stabilization of the proteins in an acidic or thermophilic environment respectively.

The product of hyp1, downstream from resC, remains unknown but preliminary evidence also indicates that hyp1 is co-transcribed with resB and resC. Between cyc1 and resB is a sequence with a very strong similarity to an *E. coli* type rho independant stop suggesting that cyc1 may be transcribed independently of resB and resC.

The homologies and proposed transmembrane structures of the products of resB and resC make it highly likely that the two hypothetical proteins are cytochrome c-type biogenesis proteins. However, we do not know what cytochrome(s) they assist. In some organisms the genes encoding cytochrome biogenesis proteins reside in proximity to the gene for the specific cytochrome that they assist. If this is true for *T. ferrooxidans* it could mean that the products of resB and resC assist in the maturation of the hypothetical apocytochrome c1 encoded by cyc1, which resides just upstream of resB.

What we demonstrate in this paper is the correlation of the Fe- phenotype of *T. ferrooxidans* with the insertion of IST1 in resB that potentially encodes a cytochrome c-type biogenesis protein. What we propose is that the interruption of resB, and possibly also resC, is the cause of the Fe- phenotype. If this is true, it would be the first genetic evidence for the involvement of a protein(s) in iron oxidation in this microorganism. It seems quite plausible that the lack of an adequately processed cytochrome c1 (or other cytochrome c) could result in a Fe- phenotype. With the identification of the cytochrome involved, experiments could be devised to test its role in iron oxidation. At the moment we can only say that the cytochrome, whichever one it may be, is required for iron oxidation but is dispensable for thiosulfate oxidation.

Clearly much work remains to be done to elucidate and expand upon the several aspects of the proposed model. The model makes several testable predictions. Amongst these are: 1) the insertion of IST1 in resB results in an aberrant mRNA and protein from resB and possible also from resC, 2) the hypothetical protein products of resB and resC have predicted molecular weights, pKs and inner membrane locations and transmembrane domains, 3) The Fe-mutant lacks a properly processed, or lacks entirely, a cytochrome c-type protein and 4) the cytochrome and the cytochrome c-type biogenesis proteins are not required for thiosulfate oxidation and they may not be expressed during growth in thiosulfate medium.

4.2. The use of insertion sequences to identify gene function in *T. ferrooxidans*

In this paper we propose that the naturally occurring insertion sequence IST1 of *T. ferrooxidans* ATCC19859 can be used to generate distinctive phenotype whose mutant gene can be subsequently recovered and analyzed by virtue of the presence within it of an IST1 element. The IST1 element then acts as a tag to pull out the mutant gene for further analysis. This is a powerful technique for identifying gene function and will probably be especially useful when the full DNA sequence of *T. ferrooxidans* becomes available. IST1, and other IS elements, are present in several strains of *T. ferrooxidans* (Haq and Holmes, 1989) and so the technique could be widely applicable.

The above approach for gene identification requires that the phenotypic consequence of the mutation caused by the IST1 is observable. In addition, any lethal or detrimental mutations must also be conditional, because it is necessary to have a condition that permits the growth of mutant cells for subsequent DNA isolation and analysis. An additional point to be considered is that mutations caused by IST1, at least in the case under study, are reversible and thus are unstable. Therefore, mutant cells must be checked periodically for the appearance of wild type revertants and, if necessary, repurified.

In addition to the mutation described in this paper, one can envisage the possibility of generating a wide range of other mutants using IST1 transposition. For example, additional mutants that involve other genes in the iron oxidation pathway could be isolated, or mutants of sulfur or reduced sulfur oxidation that can grow on iron. The technique would be also applicable to the generation of pH dependent mutants, temperature sensitive mutants and so on.

An important challenge will be to identify the function of those genes that do not fit the requirements described above. Here again, perhaps insertion elements could be useful in this context. Smith et al. (1996) utilized an inducible transposon, Ty1, to identify the function of yeast genes whose sequence was known but not their function. A large number of yeast cells were grown in one type of condition (condition A) and the transposition of Ty1 was transiently induced. Theoretically, the Ty1 could transpose to any position in the genome. Cells were then permitted to resume growth in condition A and were simultaneously transferred to other growth conditions B, C, and D etc. and samples of DNA were extracted at different times thereafter. The samples were then analyzed for the position of the Ty1 element. If Ty1 had entered a gene causing lethality in, for example, condition B it would not be found in this position in DNA from cells grown in condition B but would be found in DNA from cells grown in condition C if the Ty1 was not lethal in that gene in that condition. Also, if the Ty1 entered a gene that reduced survivability it might be present in that gene at first, but when the culture was continued the cells with the Ty1 in the above mentioned gene would be at a selective disadvantage and the proportion of cells with the Ty1 in that location would be diminished with time and perhaps ultimately be eliminated altogether. Thus, subtle contributions of genes to growth could be detected. In this way the investigators were able to confirm previously known mutant phenotypes and to identify new phenotypes for about 30% of the genes.

Theoretically it should be possible to use IST1 to generate information regarding gene function in a manner similar to the yeast study if it could be induced to transpose at a sufficient rate and if its transposition was random. Therefore, it could repay efforts to see whether such conditions can be achieved or to identify other possible IS elements or transposons with the necessary properties.

ACKNOWLEDGEMENTS

This work was supported by FONDECYT grant 1980665. The part of the work carried out at Lawrence Livermore National Laboratory was performed under the auspices of the U.S. Dept. of Energy, contract no. W-7405-Eng-48.

REFERENCES

- S. F. Altschul, W. Gish, W. Miller, E. W. Myers and D. J. Lipman, J. Mol Biol., 215 (1990) 403.
- L. D. Brown, M. E. Dennehy and D. E. Rawlings, FEMS Microbiol Lett., 122 (1994) 19.
- D. S. Holmes, J. R. Schrader and J. R. Yates, Biohydrometallurgy, eds P. R. Norris and D. P. Kelly, Science and Technol. Letters, London, (1987) 153.
- D. S. Holmes and U. Haq, Biohydrometallurgy, eds J. Salley, R. McCready and P. Wichlacz, Canmet, Ottawa, Canada, (1989) 115.
- R. Kranz, R. Lill, B. Goldman, G. Bonnard and S. Merchant, Mol. Microbiol., 29 (1998) 383.
- T. Kusano, T Takeshima, K. Sugawara, C. Inoue, T. Shiratori, T. Yano, Y. Fukumori and T. Yamanaka, J. of Biol. Chem., 267 (1992) 112442.
- T. Kusano, K. Sugawara, C. Inoue, T. Takeshima, M. Numata and T. Shiratori, J. Bacteriol., 174 (1992) 6617.
- T. Maniatis, E. F. Fritsch and J. Sambrook, Molecular Cloning: A Laboratory Manual. Cold Spring Harbor Laboratory, NY, (1988).
- J. Peng, W-M. Yan and X-Z Bao, J. Bacteriol., 176 (1994) 2892.
- J. A. Schrader and D. S. Holmes, J. Bacteriol., 170 (1988) 3915.
- V. Smith, K. N. Chou, D. Lashkari, D. Botstein and P. 0. Brown, Science, 274 (1996) 2069.
- L. 'Ihony-Meye, Microbiol. Mol Biol. Rev., 61 (1997) 337.
- K. C. Worley, B. Wiese and R. F. Smith, Genome Research, 5 (1995) 173.
- J. R. Yates and D. S. Holmes, J. Bacteriol., 169 (1987) 1861.
- J. R. Yates, R. P. Cunningham and D. S. Holmes, Proc. Nat. Acad. Sci. USA., 85 (1988) 7284.
- H. Zhao, Ph.D Thesis, Clarkson University, NY, USA, (1995).
- H. Zhao, E. Jedlicki, M-E Cabrejos, A. Bengrine, V. Bonnefoy and David S. Holmes. (1999, submitted).

Comparative genomic characterization of iron oxidizing bacteria isolated from the Tinto River

E. González-Toril, F. Gómez, N. Irazabal, R. Amils and I. Marín

Centro de Biología Molecular Severo Ochoa, Universidad Autónoma de Madrid, Cantoblanco, Madrid 28049, Spain

The genomic organization of different iron oxidizing bacteria belonging to the *Thiobacillus* and *Leptospirillum* genus isolated from the Tinto River was studied using pulsed field gel electrophoresis (PFGE). The electrophoretic analysis of intact DNA prepared from different isolates showed that all have a circular chromosome with a variable size, ranging from 2.15 to 2.70 Mb for thiobacilli and 1.64 to 2.53 for leptospirilli. Extrachromosomal elements of important size were only detected in one leptospirilli isolate. The rest of the strains might have supercoiled plasmids of small size, like many of the thiobacilli reference systems or none at all. Low-frequency restriction fragment analysis (LFRFA) was carried out to determine macrorestriction patterns for rare cutters (*Spe*I, *Xba*I, *Swa*I and *Pme*I), which were used for taxonomic characterization (karyotyping), genome size determination and physical map generation. The results obtained strongly suggest the existence of at least three different clusters in the group of isolates identified phenotypically as *T. ferrooxidans* strains, while the isolates identified as *L. ferrooxidans* strains exhibit distinct restriction profiles for different rare cutters, suggesting that they might correspond to different species. A revision of the taxonomic status of these important acidophilic chemolithotrophs is suggested.

1. INTRODUCTION

The discovery that some microorganisms could not only survive but also thrive at extreme conditions has changed our conservative concept of the limits of life. Of all the extreme conditions, acidity is considered one of the most critical factors affecting organism growth. Natural acidic environments can be of biological and geological origin. Attention was focused on the former after it was established that Acid Mine Drainage (AMD), a problem of environmental concern, was a consequence of the metabolic activity of chemolithotrophic microorganisms (1) and, more recently, since the biotechnological potential of these microorganisms was established (2,3,4). In mining areas, where most of the studies of acidophilic microorganisms have been performed, iron and sulfur are readily oxidized by chemolithotrophic microorganisms, which are, at the same time, responsible for both the production and maintenance of the extreme conditions found in these habitats (1,5). The products of these metabolic reactions: ferric ion and protons are also powerful oxidizing agents able to attack other sulfide minerals, thus facilitating the oxidation of other metallic cations. The microbial ecology of these acidic environments showed that they are mainly dominated by chemolithotrophic bacteria belonging to the genus *Thiobacillus* and *Leptospirillum*, together with obligate chemoorganotrophs of the *Acidiphilium* genus. Some moderate thermophliles and archaeal thermophiles have also been isolated from acidic habitats (6). The lowest limit of pH is shared by *Thiobacillus thiooxidans*, capable of obtaining energy only from reduced forms of sulfur, and *Leptospirillum ferrooxidans*, limited to oxidizing ferrous iron (7).

The genus *Thiobacillus* has been defined by its main feature: its ability to obtain energy from oxidation of inorganic sulfur compounds. Although, most of the species are autotrophic, many can grow under organotrophic conditions, exhibiting variable patterns of utilization of organic compounds as a source of carbon and/or energy. A large number of strains belonging to this genus have been described (8). The first isolates were strict chemolithoautotrophs (*T. thioparus, T. denitrificans, T .thiooxidans* and *T. ferrooxidans*), all of which exhibit highly specialized metabolisms. Of those only *T. ferrooxidans* is able to grow oxidizing ferrous ion to generate energy. An extensive characterization of different thiobacilli has allowed different physiological groups to be defined (9). Preliminary genotypic data based on G + C content and interspecific DNA-DNA hybridization values, revealed a great range of genetic heterogeneity. Later, phylogenetic analysis confirmed the need for a taxonomic revision of the genus, which is under way (10,11).

Although the *Leptospirillum* genus was defined a long time ago, little attention was paid to it due to its strict dependence on ferrous iron as its sole energy source, which was considered a rather inefficient system when compared to the thiobacilli capable of oxidizing reduced sulfur compounds. The recent demonstration that *L ferrooxidans* is uniquely associated with the first steps of AMD production (20) and an important element in the industrial bioleaching processes (21,22,23) contrary to *T. ferrooxidans*, which seems to be absent or in rather low numbers in these habitats, is a nice example of an ecological artifact produced by the use of a extremely selective enrichment medium. Obviously this word of caution does not intent to mean that *T. ferroxidans* is not an interesting chemolithotrophic model, but that iron oxidation is proving to be a more important energy source than we had expected from simple thermodynamic calculations.

In spite of the fundamental interest as well as the biotechnological potential of acidophilic chemolithotrophic microorganisms, very little progress has been made in developing the tools needed to generate genetic information on their unique mode of energy transduction. The use of pulsed field gel electrophoresis (PFGE) techniques can produce genomic information about microorganisms for which conventional techniques are difficult to implement. A comparative study of the genomic organization of different acidophilic iron oxidizers isolated from the Tinto River, characterized phenotypically as members of the genus *Thiobacillus* and *Leptospirillum*, in terms of size, number, topology of their genomic elements, and their macrorestriction profiles obtained by PFGE are presented and its taxonomic implications discussed.

2. MATERIALS AND METHODS

2.1 Bacterial strains

The following strains were obtained from culture collections : *T. thiooxidans* ATCC 19377, *T. ferrooxidans* ATCC 21834, *T. ferrooxidans* ATCC 23270, *T. ferrooxidans* ATCC 19859 and *L. ferrooxidans* DSM 2705. Enrichment cultures from water and sediment samples from different stations along the Tinto River were used to isolate different strains identified phenotypically as *T. ferrooxidans* (RT3, RT4, T4, O2 and Z) and *L. ferrooxidans* (L3.2, Lnda, Lndana and L1) (12,13).

2.2 Total intact DNA and pulsed field gel electrophoresis

The basic method described by Smith and Cantor (15) with the modifications introduced by Marín et al. (12) have been used for the preparation of intact DNA from the different microorganisms analyzed in this work. Contour-clamped homogeneous electric field electrophoresis (CHEF) (14) was performed in a Pharmacia-LKB apparatus. Ortogonal-field alternating gel electrophoresis (OFAGE) was carried out in a Pharmacia-LKB Pulsaphor system. Gels were made up of 1% agarose (Seakem LE Agarose, FMC) in modified 0.5 X TBE buffer (100 mM Tris, 100 mM boric acid and 0.2 mM EDTA, final pH 8) and run at 15° C in the same buffer at different resolution windows : running time, voltage and pulse time.

2.3 DNA restriction digestions

Restriction enzymes were purchased from New England Biolabs or Boehringer. Prior to restriction, plugs were washed extensively in buffer containing 10 mM Tris-HCl, 0.1 mM EDTA, pH 8. For digestion, slices of plugs containing genomic DNA (1.5 µg) were placed in the appropriate restriction buffer containing 0.1 mg of acetylated bovine serum albumin per ml and the selected enzyme (20U) in a final volume of 100 µl. Incubations were carried out for 3 to 6 h at the appropriated temperature. After incubation the solution was pipetted off and 250µl of ESP solution (0.5 M EDTA pH 9.5, 1% lauroyl sarcosine and 1mg/ml of proteinase K) added and incubated for 2 h at 50° C before samples were loaded on a PFGE gel.

3. RESULTS AND DISCUSSION

As mentioned in the introduction an important problem concerning the genetic studies of strict chemolithoautotrophic acidophilic microorganisms is the shortage of reliable techniques that can be used to characterize the genes involved in their unique metabolism. Due to the difficulties of conventional genetic techniques, we decided to study the possibilities offered by PFGE to gain genomic information on the iron oxidizing bacteria.

PFGE allows the resolution of DNA at the megabase range, providing information about the different genomic elements. By obtaining macrorestriction patterns for different rare cutters comparisons can be made for taxonomic purposes and for the generation of the physical maps, which can then be used to construct low resolution genetic maps by location of genes using homologous and heterologous probes. One of the limitations to the application of PFGE techniques is the absolute requirement for intact DNA. The unavoidable presence of heavy metals in most of the strict acidophilic chemolithotrophs cell preparations is an additional

152

challenge, due to their capacity to produce nonspecific breaks into DNA. Several protocols have been developed which allow the preparation of intact DNA for strict acidophilic chemolithotrophs.(12).

3.1 Number of genomic elements

Most prokaryotes contain a single circular chromosome, although several exceptions have been documented for both chromosomal number and topology (16). PFGE is a simple method to determine both variables. Figure 1 shows the electrophoretic mobilities at long pulse times of intact DNA from different iron oxidizing bacteria. As can be seen, most of the DNA remains in the wells, strongly suggesting that they correspond to circular chromosomes. Some samples showed a band below the compression zone which corresponds to the linearized form of the chromosome, generated by mechanical stress during the electrophoretic run. In these conditions some of the extrachromosomal elements can be resolved, although short pulse time runs are required for a better characterization.

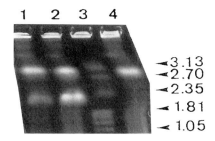

Figure 1. PFGE analysis of intact DNA from different thiobacilli. Fragments were separated using a 5 V cm^{-1} , 2000 s pulse time for 5 days. Lane 1: isolate O2, lane 2: ATCC 19859, lane 3: *H. wingei* chromosomes, lane 4: ATCC 21834.

3.2 Extrachromosomal elements

PFGE allows the number, topology and size of extrachromosomal elements to be determined. Linear elements run faster than supercoiled forms. Also, linear forms do not show changes in their relative mobility when compared with linear DNA at different pulse times. Preliminary results showed the absence of megaplasmids in all the Tinto isolates characterized as *T.ferrooxidans*. In this respect these isolates resemble the genomic organization obtained for reference systems from culture collections : *T.ferrooxidans* ATCC 21834, *T.ferrooxidans* ATCC 23270, *T.ferrooxidans* ATCC 19859 and *T.ferrooxidans* ATCC 19377. The type of analysis performed up to now does not allow the presence of supercoiled plasmids of small size to be ruled out as reported for *T.ferrooxidans* ATCC 21834 (8.6 kb) (16), *T.ferrooxidans* ATCC 33020 (6.7 kb) (17) and *T.ferrooxidans* FC (12.2 kb) (18). *T. ferrooxidans* ATCC 33020 (Marsella clone) showed an additional megaplasmid of around 0.3 Mb, so far this is the only megaplasmid characterized by PFGE in strains of *T. ferrooxidans*. On the other hand, the leptospirilli isolates showed a wider range of extrachromosomal elements: from none in isolate

LM2, to one small supercoiled plasmid in isolate L3.2 and up to three possible linear megaplasmids of 0.54, 0.34 and 0.19 Mb in isolate L1.

3.4 Macrorestriction patterns

A number of different restriction enzymes have been tested to select those that produce a reasonable amount of fragments of different size which can be then resolved by PFGE. Figure 2 shows the *Spe*I restriction profiles obtained for different thiobacilli isolates together with

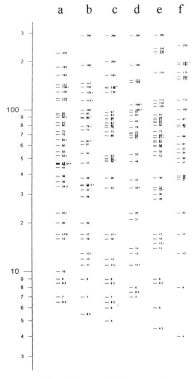

Figure 2. *Spe*I restriction patterns for different *T.ferrooxidans* strains. Lane a: ATCC 21834, lane b: isolate RT4, lane c: isolate RT3, lane d: isolate T4, lane e: isolate O2, lane f: isolate Z

several reference strains. Figure 3 shows the statistical analysis of the restriction profiles obtained for the different *T. ferrooxidans* isolates when compared with the type collection reference systems.

The similarity dendogram showed three distinct groups of *T. ferrooxidans* strains : A first group comprises the three reference systems included in this study : *T. ferrooxidans* ATCC 19859, *T. ferrooxidans* ATCC 21834 and *T.ferrooxidans* ATCC 23270, and two *T. ferrooxidans* isolates from the Tinto River, O2 and RT3. A second group is formed by a *T. ferrooxidans* isolate from the Tinto River, RT4, which clusters together with the reference

strain of *T. thiooxidans* ATCC 19377. And a third group is made up of two *T. ferrooxidans*
Tinto isolates, Z and T4, which differ with the other *T. ferrooxidans* strains not only in their

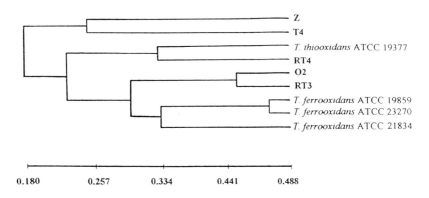

Figure 3. Similarity dendogram generated with the *Spe*I fragments of different *T.ferrooxidans*
strains

restriction patterns but also in their genome size. This group of acidophilic thiobacilli are well
separated from the rest of the thiobacilli and related bacteria analyzed so far (24). Similar
results were obtained by Lane et al. (10) using partial 16S rRNA sequences. In this case most
of the *T. ferrooxidans* strains analyzed formed a cluster in which two *T. thiooxidans strains*,
ATCC 19377 and DSM 612, were included. These data suggest that the ill defined *T.
ferrooxidans* species based only on a rather small number of phenotypic properties (shape,
energy sources and optimal pH) requires extensive revision. Further genomic analysis
complemented with rDNA sequence will be required to clarify their taxonomic and
phylogenetic status. In this context it is important to emphasize that although LFRFA analysis
can not be used to compare microorganisms belonging to different genus, it is a very powerful
technique with which to adscribe new isolates and to differentiate strains that exhibit a high
level of rDNA homology.

The *Spe*I macrorestriction patterns for the *Leptospirillum* Tinto isolates showed quite
distinct patterns among all the strains analyzed (Figure 4). Isolate Lndan is the only strain that
exhibited sufficient homology with the reference strain DSM 2705, although its chromosome is
much smaller (see below). Similar results were obtained using other restriction enzymes (data
not shown). rDNA sequences showed that *Leptospirillum ferrooxidans* have very little
phylogenetic relationship with most of the acidophilic chemolithotrophic bacteria, including *T.
ferroxidans* (10). Recently, using the same type of analysis it has been shown that *Nitrospira*, a
new group of anaerobic nitrite oxidizing bacteria, is closely related with the leptospirilli (19).
To further clarify the relationship of the Tinto *L. ferrooxidans* isolates the partial 16S rRNA
sequence of isolate L1 was used to generate the phylogenetic tree shown in Figure 5, in which

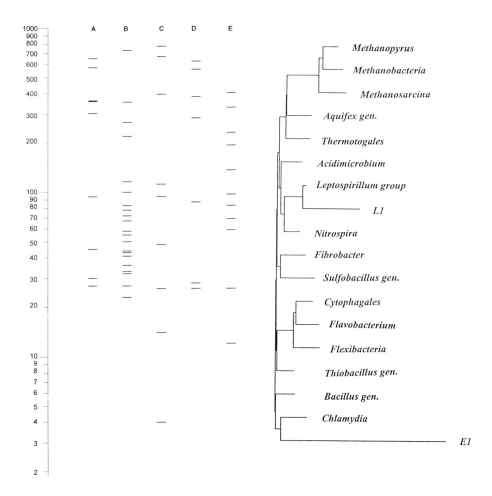

Figure 4. Comparative *Spe*I macrorestriction patterns for different *L.ferrooxidans* strains

Figure 5. 16S rRNA phylogenetic tree of different bacterial groups including *L. ferrooxidans* isolate L1

it can be seen that the Tinto isolate clustered together with the rest of the leptospirilli whose sequences are accessible in the public data bank. The leptospirilli cluster maintains its close relationship with the *Nitrospira* group. Further analysis of the 16S rRNA sequences of all Tinto leptospirilli isolates should complement the LFFRA analysis to further characterize the taxonomic status of this important group of acidophilic iron oxidizing bacteria.

3.5 Genome size

As mentioned above, PFGE analysis of intact DNA using long pulse times or γ-irradiation can produce partial linearization of circular genomic elements, allowing a preliminary characterization of prokaryotic genome size. Macrorestriction profiles with different enzymes are used for an accurate estimation of the genome size. As can be seen in Table 1 the genomes of the acidophilic iron oxidizing bacteria isolated from the Tinto River are quite variable, ranging from 2.15 to 2.87 Mb for the thiobacilli and from 1.64 to 2.53 Mb for the leptospirilli. The thiobacilli isolates that showed a separate cluster when their macrorestriction profiles were taken into consideration showed a consistant smaller genome than the rest of the isolates analyzed, which stresses their possible taxonomic segregation. The reference type collection strains showed a rather similar genome size which also agrees with the similarity exhibited by their LFRFA. The genome size of the leptospirilli isolates studied is rather different. One of them, isolate L1 has an extremely small genome, although it must be considered that this strain showed the presence of linear megaplasmids which have not been considered in its genome size determination.

Table 1. Genome size for different acidophilic iron oxidizing bacteria in Mb

T. ferrooxidans strain	size	L. ferrooxidans strain	size
ATCC 23270	3.33	DSM 2705	2.45
ATCC 19377	3.14	L3.2	2.53
ATCC 21834	2.93	Lnd	2.14
ATCC 19859	2.87	Lndan	2.0
O2	2.70	L1	1.64
RT3	2.69		
RT4	2.57		
Z	2.29		
T4	2.15		

ACKNOWLEDGEMENTS

This work was supported by the following grants : AMB97-0547-C02-02 (CICYT) and 07M/0180/1997 (CAM) and by an institutional grant to the CBMSO from the Fundación Areces. E.G.T. is a predoctoral fellow from the Comunidad Autónoma de Madrid,, F.G. is a posdoctoral fellow from the Centro de Astrobiología and N.I. was a predoctoral fellow from the Gobierno Vasco.

REFERENCES

1- Colmer, A.R., Hinkle, M.E. Science, 106 (1947) 253.

2- Torma, A.E. Bitech. Bioeng. Symp., 16 (1986) 49.

3- Maloney, S., Moses, V. Biotechnology, the science and the business, Moses, V., Cape, R.E. (eds), Harwood Academic Publs., London (1991) 581.

4- Hutchins, S.R., Davidson, M.S., Brierley, J.A., Brierley, C.L. Ann. Rev. Microbiol., 40 (1986) 311.

5- Ehrlich, H.L. J. Bacteriol., 86 (1963) 350.

6- Harrison, A.P., Jr. Ann. Rev. Microbiol., 38 (1984) 256.

7- Hallman, R., Friedrich, A., Koops, H.P., Pommerening-Röser, A., Rohde, K., Zenneck, C., Sand, W. Geomicrobiol. J., 10 (1992) 206.

8- Kelly, P.D., Harrison, A.P. Bergey's Mannual of Systematic Bacteriology, Staley, J.T., Bryant, M.P., Pfening, N., Holt, J.G. (eds), Williams & Wilkins, Baltimore, vol 3 (1989) 1842

9- Harrison, A.P. Jr. Int. J. Syst. Bacteriol., 33 (1983) 211.

10- Jane, D.J., Harrison, A.P. Jr, Stahl, D.A., Pace, B., Giovannoni, S.J., Olsen, G.J., Pace, N.R. J. Bacteriol., 174 (1992) 269.

11- Moreira, D., Amils, R. Int. J. Syst. Bacteriol., 47 (1997) 522.

12- Marín, I., Amils, R., Abad, J.P., Gene, 187 (1997) 99

13- Irazabal, N., Marín, I., Amils, R. J. Bacteriol., 179 (1997) 1946.

14- Chu, G., Vollrath, D., Davies, R.W. Science, 234 (1984) 1582.

15- Smith, C.L., Klco, S.R., Cantor, C.R. Genome analysis : a practical course, Davies, K. (ed), IRL Press, Oxford (1988) 41.

16- Drlica, K., Riley, M. The bacterial chromosome, Drlica,K., Riley, M. (eds), ASM, Washington, (1990) 3.

17- Holmes, D., Lobos, J.H., Bopp, L.H., Welch, G.C. J. Bacteriol., 157 (1984) 324.

18- vanAswegen, P.C., Godfrey, M.W., Miller, D.M., Haines, A.K. Miner. Metallurg. Processing, 8 (1991) 188.

19- Ehrich, S., Behrens, D., Lebedeva, E., Ludwig, W., Bock, E. Arch. Microbiol., 164 (1995) 16.

20- Schrenk, M.O., Edwards, K.J., Goodman, R.M., Hamers, R.J., Banfield, J.F. Science, 279 (1998) 1519.

21- Espejo, R.T., Pizarro, J., Jedliki, E., Orellana, O., Romero, J. Biohydrometallurgical Processing, Jerez, C.A., Vargas, T., Toledo, H., Wiertz, J.V. (eds), U. de Chile, Santiago (1995), 1.

22- Rawlings, D.E., Biohydrometallurgical Processing, Jerez, C.A., Vargas, T., Toledo, H., Wiertz, J.V. (eds), U. de Chile, Santiago (1995), 9.

23- García, A., Jerez, C.A. Biohydrometallurgical Processing, Jerez, C.A., Vargas, T., Toledo, H., Wiertz, J.V. (eds), U. de Chile, Santiago (1995), 19.

24- Amils, R., Irazabal, N., Moreira, D., Abad, J.P., Marín, I. Biochimie, (1999), in press.

Chapter 4

Biosorption

Biosorption for the Next Century

Boya Volesky

http://www.mcgill.ca/biosorption/biosorption.htm

Chemical Engineering Department, McGill University,   e-mail: boya@chemeng.Lan.mcgill.ca
3610 University St., MONTREAL, Canada H3A 2B2

The potential of metal concentration by certain types of dead biomass has been well established over the last two decades. This phenomenon can probably make the most significant impact in using it for removing toxic heavy metals from industrial effluents. An interdisciplinary approach seems essential for bringing the phenomenon to a successful process application stage. Challenges in the novel biosorption process development are briefly summarized here for scientists and entrepreneurs alike.

METALS:
ENVIRONMENTAL THREAT

By far the greatest demand for metal sequestration comes from the need of immobilizing the metals 'mobilized' by and partially lost through human technological activities. It has been established beyond any doubt that dissolved particularly heavy metals escaping into the *environment* pose a serious health hazard. They accumulate in living tissues throughout the food chain which has humans at its top. The danger multiplies. There is a need for controlling the heavy metal emissions into the environment.

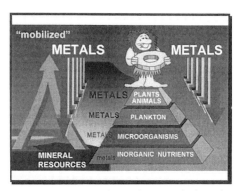

The food-chain pyramid receives metals through man's activities. On top of the pyramid, man receives pre-concentrated metal toxicity.

Environmental Pressures

• Stricter regulations with regard to the metal discharges are being enforced particularly for industrialized countries.

• Toxicology of heavy metals confirms their dangerous impacts.

• The currently practiced technologies for removal of heavy metals from industrial effluents appear to be inadequate and expensive. They often create secondary problems with metal-bearing sludges.

Biosorption is competitive and cheap

Advantages of biosorption would tend to outweigh the very few shortcomings of biosorbents in applications.

Heavy metals need to best be removed at the source in a specially designed 'pre-treatment' step which has to feature low costs to be feasible. The search is on for efficient and particularly cost-effective remedies.
Biosorption promises to fulfill the requirements.

Biosorption uses biomass raw materials which are either abundant (seaweeds) or wastes from other industrial operations (fermentation wastes). The metal-sorbing performance of certain types of biomass can be more or less selective for heavy metals. That depends on:
- the type of biomass,
- the mixture in the solution,
- the type of biomass preparation,
- the chemico-physical environment.
It is important to note that concentration of a specific metal could be achieved either during the sorption uptake by manipulating the properties of a biosorbent, or upon desorption during the regeneration cycle of the biosorbent.

Biosorption process of metal removal is capable of a performance comparable to its closest commercially used competitors, namely the ion exchange treatment. Effluent qualities

in the order of only ppb (μg/L) of residual metal(s) can be achieved. While commercial ion exchange resins are rather costly, the price tag of biosorbents can be an order of magnitude cheaper (1/10 the ion exchange resin cost).
This is the main attraction of biosorption - cost effectiveness.
While ion exchange can be considered a 'mature' technology, biosorption is in its early developmental stages and further improvements in both performance and costs can be expected.

Yes, biosorption can become a good weapon in the fight against toxic metals threatening our environment. While the biosorption process could be used even with a low degree of understanding of its metal-binding mechanisms, better understanding will make for its more effective and optimized applications. That poses a scientific challenge and continued R&D efforts. In addition, even the same type of industrial activity can produce effluents which differ from each other a great deal. Close collaboration with each 'client' industrial operation is absolutely essential: a consulting-engineering type of approach. Engineering skills become quite important because it is a process operation one is aiming at and dealing with.

"Treatability studies" which are usually carried out in close cooperation with the client provide the backbone for assessing the optimum treatment sequence.
Biosorption does offer a competitive wastewater treatment alternative, the basis of which needs to be well understood in order to prevent application failures.

The potential pitfalls in introducing the new biosorption alternative are quite similar to those encountered with any other novel technology close to the application stage. However, there is little doubt that steadily mounting environmental pressures provide a powerful driving force for new business opportunities.

When it comes to a new "biosorption" enterprise, there are two aspects to such:
1) *products*: new family of biosorbents;
2) *services* involved in:
- assessing the effluent problem;
- assessing biosorption applicability;
- developing customized treatment;
- designing and building the plant;
- eventually even operating the effluent treatment process, and even
- recovering metal(s) for resale/re-use.

Metal Removal/Recovery "Priorities"

An example of the priorization for recovery of ten metals is in TABLE 1 which may be simplistic but provides a useful direction by ranking into 3 general priority categories:
(1) environmental risk (ER);
(2) reserve depletion rate (RDR);
(3) a combination of the two factors.

Environmental risk assessment could be based on a number of different factors which could even be weighed.

The RDR category is used as an indication of probable future increase in the market price of the metal. When coupled with the ER in this example there is an indication that Cd, Pb, Hg, Zn are a high priority. However, the technological uses of Hg and Pb may be considered declining, while the Cd use is on the increase. These projections and the degree of risk assessment sophistication could change the priorities among the metals considered.

Biosorption and entrepreneurial activities

Growth industries and point-source effluents are of primary concern.

TABLE 1: Ranking of metal interest priorities

Relative Priority	Environ. Risk [a]	Reserve Depletion	Combined Factors
HIGH:	Cd	Cd	Cd
	Pb	Pb	Pb
	Hg	Hg	Hg
	--	Zn	Zn[c]
MEDIUM:	--	Al	--
	Cr	--	--
	Co	Co	Co
	Cu	Cu	Cu
	Ni	Ni	Ni
	Zn	--	(see High)
LOW:	Al	--	Al
	--	Cr	Cr
	Fe	Fe	Fe

164

STRUCTURE OF
A BIOSORPTION PROJECT

With new discoveries of highly metal-sorbing biomass types there is a real potential for the introduction of a whole family of new biosorbent products which are likely to be very competitive and cost-efficient in metal sorption. As a potential competition for synthetic ion exchange resins, capable of doing the same 'job', the costs of biosorbents must be maintained very low. That could be guaranteed by low-cost raw material and minimum of processing.

Some types of industrial fermentation waste biomass are excellent metal sorbers. It is necessary to realize that some "waste" biomass is actually a commodity, not a waste: this applies particularly for ubiquitous brewer's yeasts sold on the open market for a price, usually as animal fodder. Activated sludge from wastewater treatment plants has not demonstrated high enough metal-sorbing capacities.

Some types of seaweed biomass offer excellent metal-sorbing properties. Local economies can benefit from turning seaweeds into a resource.

Visit: **http://www..mcgill.ca/biosorption/ biosorption.htm**

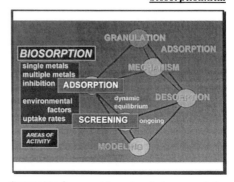

Screening for new biosorbents is essential

As a fall-back, high metal-sorbing biomass could even be specifically propagated relatively cheaply in fermentors using low-cost or even waste carbohydrate-containing growth media based on e.g. molasses or cheese whey.

Screening for Adsorption:
Batch ***equilibrium*** sorption experiments are used for screening for suitable biomass types. Unfortunately, there are to many errors in the literature betraying little understanding of equilibrium sorption concepts.

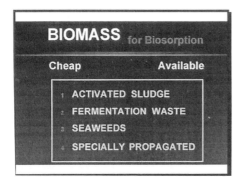

Using waste biomass for preparing new biosorbents is particularly advantageous.

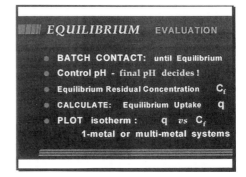

Standard procedure for evaluating simple sorption systems - see in the home-page.

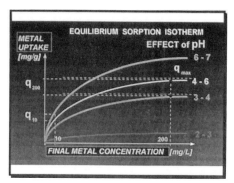

Example of (bio)sorption isotherms affected by the pH of solution

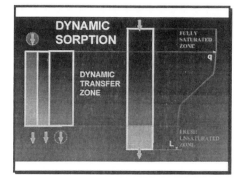

Fixed-bed column is the most powerful sorption process arrangement

Solution chemistry of the metals to be examined for biosorption should be well understood for explanation of experimental results. For this purpose, a widely available computer data-base program MINIQL+ is extremely useful.

The most appropriate method of assessing the biosorbent capacity is the derivation of a whole sorption isotherm. Anything else represents a potentially misleading shortcut which may lead to outright erroneous conclusions. While experimental volume increases almost exponentially with the number of metallic species present in the solution, evaluation of multimetal sorption systems offers a special challenge.

'Enough' time is allowed for equilibrium contact sorption experiments. Kinetics tests show the time-concentration profile for sorption. The sorption reaction itself is inherently an extremely fast one. It is mainly the particle mass transfer which controls the overall sorption kinetics (sorbent particles size, porosity and mixing in the sorption system).

Environmental factors such as the solution pH, ionic strength, to a lesser degree temperature, etc. are likely to affect the sorption performance. The range of conditions for biosorbent screening should be carefully selected.

Dynamic sorption studies are invariably more demanding. The most optimal configuration for continuous-flow sorption is the packed-bed column which gets gradually saturated from the feed to the solution exit end. Correct and non-trivial interpretation of experimental results is important and becomes scientifically rather involved. However, it is expected.

In the sorption column contactor the saturated zone is moving along the column length pushing the transitional dynamic sorption zone ahead of itself. With multimetal sorption systems featuring different affinities of ions toward the sorbent the whole system becomes even more complex as chromatographic effects and simultaneous displacement of deposited ions take place. It is obvious that simplistic observations of the experimental "break-through" curve resulting from the conventional operation of a flow-through sorption column will not suffice. They are usually narrowly specific and cannot be used elsewhere.

Desorption:

The possibility of regeneration of loaded biosorbent is crucially important to keeping the process costs down and to opening the possibility of recovering the metal(s) extracted from the liquid phase. The deposited metals are washed out (desorbed) and biosorbent regenerated for another cycle of application. The desorption process should result in:
- high-concentration metal effluent;
- undiminished metal uptake upon re-use;
- no biosorbent physico-chemical damage.

The desorption and sorbent regeneration studies might require somewhat different methodologies. Screening for the most effective regenerating solution is the beginning.

Different affinities of metal ions for the biosorbent result in certain degree of metal selectivity on the uptake. Similarly, another selectivity may be achieved upon the elution-desorption operation which may serve as another means of eventually separating metals from one another if desirable.

The Concentration Ratio *(CR)* is used to evaluate the overall concentration effectiveness of the whole sorption-desorption process:

$$CR = \frac{\text{Eluate metal concentration}}{\text{Feed metal concentration}}$$

Obviously, the higher the CR is the better is the overall performance of the sorption process making the eventual recovery of the metal more feasible with higher eluate concentrations.

Recovery of the metal from these concentrated desorption solutions is carried out in a different plant by electrowinning.

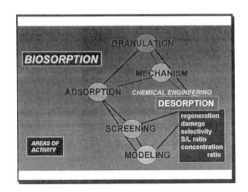

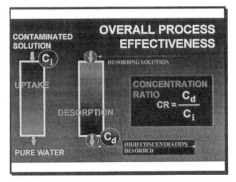

Following desorption of the metal(s), the column may still be pre-treated (e.g. pre-saturated with protons, Ca, K, etc.) for optimum operation in the subsequent metal uptake cycle. The types of this pre-treatment may vary and could be used to optimize the column performance.

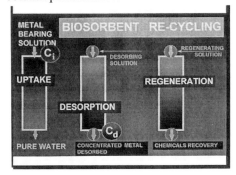

Complete biosorbent regeneration may take two or more operations.

Mechanism of metal biosorption:

Adsorption and desorption studies invariably yield information on the mechanism of metal biosorption: how is the metal bound within the biosorbent. This knowledge is essential for understanding of the biosorption process and it serves as a basis for quantitative stoichiometric considerations which constitute the foundation for mathematical modeling of the process.

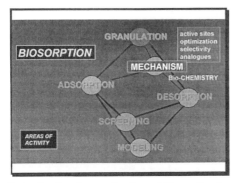

Understanding the mechanism of biosorption is important even for very practical reasons

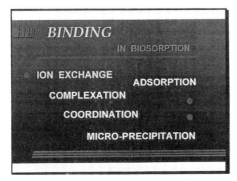

While other mechanisms might also contribute, ion exchange prevails

A number of different metal-binding mechanisms has been postulated to be active in biosorption such as:
- chemisorption: by ion exchange, complexation, coordination, chelation;
- physical adsorption, microprecipitation.

There are also possible oxidation/reduction reactions taking place in the biosorbent. Due to the complexity of the biomaterials used it is quite possible that at least some of these mechanisms are acting simultaneously to varying degrees depending on the biosorbent and the solution environment. More recent studies with fungal biomass and seaweed in particular have indicated a dominant role of ion exchange metal binding. Indeed, the biomass materials offer numerous molecular groups which are known to offer ion exchange sites: carboxyl, sulfate, phosphate, amine, could be the main ones.

When the metal - biomass interaction mechanism(s) are reasonably understood, it opens the possibilities of:
- optimizing the biosorption process on the molecular level;
- manipulating the biosorption properties of biomass when it is growing;
- developing economically attractive analogous sorbent materials;
- simplifying and effectively guiding the screening process;
- 'activating' biomaterials low-level biosorbent behavior.

Simple and economically feasible pretreatment procedures for suitable biomaterials may be devised based on better understanding of the metal biosorbent mechanism(s).

168

Modeling:

Mathematical modeling and computer simulation of biosorption offers an extremely powerful tool for a number of tasks on different levels. It is essential for process design and optimization where the equilibrium and dynamic test information comes together representing a multivariable system which cannot be effectively handled without appropriate modeling and computer-based techniques. The dynamic nature of sorption process applications (columns, flow-through contactors) makes this approach mandatory. When reaction kinetics is combined with mass transfer which is, in turn, dependent on the particle and fluid flow properties only a rather sophisticated apparatus can make sense out of the web of variables.

The mission of biosorption process modeling must be *predicting* the process performance under different conditions. Computer simulations can then replace numerous tedious and costly experiments.

Advanced sophistication in this area and availability of very powerful computer hardware and software makes contribution of the process modeling/simulation activity very realistic and indispensable indeed.

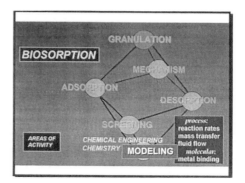

Advanced scientific approach aids in understanding the phenomenon and in developing biosorption for applications

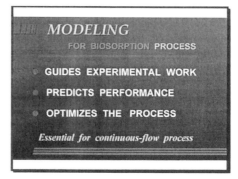

Process modeling is sophisticated and should be done very pragmatically

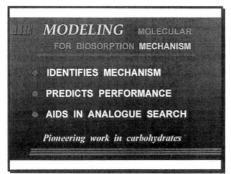

Contemporary molecular modeling software is extremely powerful can be very useful

A whole new area is opening up in modeling of molecules, their parts and interactions. "Seeing" how the biosorbent works on a molecular level would aim at purposefully preparing, 'engineering', a 'better biosorbent'. While significant inroads have been made in revealing protein and nucleic acid structures and their behavior, carbohydrate chemistry which seems to be at the basis of the biosorption behavior still has not significantly benefited from these advanced computer modeling techniques.

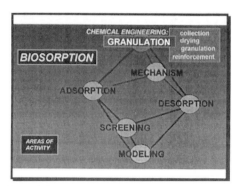

Essential process 'development' type of work for
flow-through sorption applications

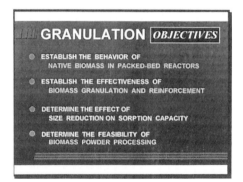

Different biomass types require different 'pre-
processing' after which the sorption
performance has to be always tested

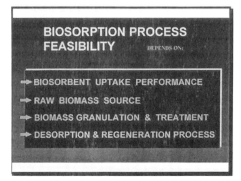

Establish the overall process feasibility.

Granulation:

The last but not the least area to be developed in the field of biosorption is the granulation of biosorbent materials. It is rather empirically based but without it reliably delivering *granulated* biosorbents there may not be any scaled-up biosorption applications. The most effective mode of a sorption process is undoubtedly based on a fixed-bed reactor/contactor configuration. The sorption bed has to be porous to allow the liquid to flow through it with minimum resistance but allowing the maximum mass transfer into the particles as small as practical (0.7-1.5 mm) for a reasonable pressure drop across the bed. Biosorbents have to be hard enough to withstand the application pressures, porous and/or 'transparent' to metal ion sorbate species, featuring high and fast sorption uptake even after repeated regeneration cycles. Considering the vast variety of and differences in the raw biomass materials, this is a tall order.

Conventional granulation technologies are rather advanced and their adaptation(s) will likely yield desirable biosorbent granules. At the same time, the broad variety of biomass types will undoubtedly require extensive experimentation for the purpose. There may be also some 'logistical' problems because of transportation of raw biomass. Microbial biomass comes with a high water content and is prone to decay. Its drying may be required if it cannot be processed and/or granulated directly on location in the wet state.

Processing or 'granulation' of biomass materials into suitable cost-effective biosorbents is a crucial step for the success of biosorption processes.

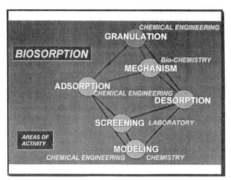

Different areas of the project can benefit most from specific scientific disciplines

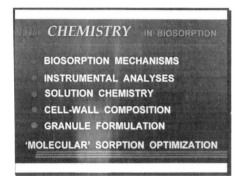

Challenges for chemistry and biochemistry

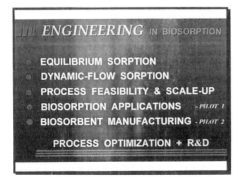

Process engineering will have to develop the process with its 2 pilots

Project disciplines:

It is obvious that many different and challenging contributions can be made on the path to developing biosorption from a scientific curiosity to useful applications. There is no doubt that there is a potential in this field. Apart from individual scientific challenges there is a special one in crossing the boundaries of conventional science disciplines to accomplish the goal. Individual projects undertaken best be effectively interdisciplinary.

The two types of backgrounds which might undoubtedly contribute most in developing the science basis of biosorption in the direction of its applications are chemistry, including biochemistry, and (chemical process) engineering. Applied microbiology needs to elucidate the composition of microbial and algal cell walls which are predominantly responsible for sequestering the metals.

Following equilibrium sorption and dynamic sorption studies, the quantitative basis for the sorption process is established, including process performance models. The biosorption process feasibility is assessed for well selected cases. It is necessary to realize that there are 2 types of pilot plants to eventually be run hand in hand:
- Biomass processing pilot plant;
- Biosorption pilot plant.

The biomass supplies need to be well secured. That, in turn, brings the 'whole world' into the picture whereby it may become attractive for developing countries with biomass resources to participate in further development of the new biosorption technology.

Biosorption of metals. The experience accumulated and the outlook for technology development

M. Tsezos

National Technical University of Athens. Department of Mining and Metallurgical
Engineering. Laboratory of Environmental Engineering
Heroon Polytechniou 9, 157 80 Athens, Greece

Over twenty years ago, when I first started working in this area, biosorption was more of a curiosity rather than a defined process. Several workers, mostly microbiologists, had already observed that microbial cells had the ability to concentrate, in their cellular mass, manyfold metals that existed in dilute concentrations in their aqueous environment. Researchers, like Polikarpov[1], attempted to quantify these observations by reporting "concentration factors" which indicated the number of times a specific metal concentration was increased into the cellular mass as compared to its concentration in the cells aquatic environment. In an indirect way biosorption of metals was also being used by microbiologists in the process of staining microbial cells in order to increase their electron density so they could be better studied under the electron microscope.

During the 1970's, the increasing awareness and concern about the environment motivated research for new efficient technologies that would be capable to treat,inexpensively, waste waters polluted by metals with emphasis to radionuclides. This search brought biosorption to the foreground of scientific interest as a potential basis for the design of novel waste water treatment processes. At that time, concepts of classical Chemical Engineering were also brought into this research effort. The equilibrium and the kinetics of biosorption started being investigated in a systematic way, utilizing tools from the field of adsorption such as activated carbon adsorption[2,3].

The work on biosorption continued to expand and in the early 1980's the first patents appeared claiming the application of specific microbial biomass types, as biosorbents for the treatment of contaminated waste waters. At first, the biomass was proposed to be used in its native form. Very quickly, techniques for immobilising microbial biomass were also developed, tested, patented and proposed[3,4,5,6]. These patents made use of immobilization approaches such as biomass encapsulation[5] or biomass chemical processing, hardening and then granulation[3]. The immobilization of the microbial biomass was shown to be an indispensable requirement for any potential technology development based on biosorption. At the same time, the immobilization of the microbial biomass offered to biosorption based technologies the use of well known and well developed traditional chemical engineering reactor configurations, such as upflow or downflow packed bed reactors, fluidized bed reactors etc.

Pilot installations and few commercial scale units were constructed in the USA[7] and in Canada[8]. The pilot plants confirmed the applicability of biosorption as the basis for a metals sequestering/recovery process, especially in the case of uranium where it was tested in order to be combined with in situ bioleaching giving rise to an integrated biotechnology based uranium production scheme[8]. These pilot plants, also helped us to realize the limitations surrounding the industrial application of biosorption. The first issue that emerged out of this experience was the requirement for a reliable supply of waste microbial biomass of the type that would be suitable for each one of the intended biosorption applications. Fermentation industry was reluctant or unable to secure a steady supply of waste microbial biomass as the inexpensive raw material that would be used for the production of the new biosorbents, taking advantage of the economy of scale. The cost for producing the required biomass for the sole purpose of transforming this biomass into biosorbents was shown to be too expensive. Logistical problems having to do with the immobilized biomass distribution, regeneration and reuse made the above issues even more complex. Furthermore, the negative effect of solution matrix co-ions on the immobilized microbial biomass targeted metal's uptake capacity upon recycling and reuse made matters even more difficult[9].

Three attempts to commercialize immobilized biomass biosorption in the fields of waste water treatment[4,5,8] and metal value recovery[6] finally did not manage to succeed. As of today, two more attempts to market two different types of immobilized microbial biomass, one by BV SORBEX[10] and the other by the US Bureau of Mines[10] are not known to have made a successful commercial application in the market.

By the beginning of 1990's, the research work in the field of biosorption had focused again on the better elucidation and understanding of biosorption fundamentals, such as the competing ions effect and selectivity, rather than on the biosorption process design.

We understand now a lot more about how microbial biomass sequesters metals and which are the requirements for a potential process design. However, we have not managed, as yet, to produce an equivocal market success story. Biosorption is not yet a proven technology. The question one asks, and I must admit I also ask to myself, is the following : What is the future of biosorption as a potential technology base? The answer, I think , is that there does exist a future. Biosorption is a process with some unique characteristics. It can effectively sequester dissolved metals out of dilute complex solutions with high efficiency and quickly (rapid intrinsic kinetics). These characteristics make biosorption an ideal candidate for the treatment of high volume low concentration complex waste waters. It is therefore a desirable component in the design of flow sheets that we can describe as hybrid technologies. These are flow sheets that make use of a combination of various processes, including biosorption, in order to reach their target.

Hybrid technologies can be intrabiotechnological, that is to say they make use of various biotechnology based processes in their flow sheet as for example biosorption, bioreduction and bioprecipitation. They could also be described as intertechnological, as they can integrate into their flow sheets biotechnology based processes along with other non biotechnology based processes as for example chemical precipitation, electrochemical processes etc. Either type of hybrid technologies can make successful use of biosorption as one of the implemented processes benefiting from the advantages of biosorption.

The combination of biosorption along with metabolically mediated processes as for example bioreduction (eg $Cr^{6+} \rightarrow Cr^{3+}$, $Se^{4+} \rightarrow Se^{0}$ etc). and bioprecipitation is also possible inside novel reactor designs. We can even make use of combinations of biological and chemical processes into effective hybrid processing schemes inside new single reactor designs. We should keep in mind that metabolically active organisms can be produced inside our new reactors thus overcoming the problem of the reliable, appropriately controlled biomass supply at a specific site. Such reactor systems have been proposed and are being tried successfully at pilot stage[11]. More such ideas are needed. We must also continue the fundamental research into the better understanding of the mechanism of biosorption, on what drives the selectivity of biosorptive and bioaccumulatory processes. In the process of these new studies, we must avoid duplicating or repeating work already performed. There also appears to be a need to follow the well tried and documented methodologies for the study of the biosorptive phenomena. It is best not to invest research energy and time without first profiting from the reported state of the art which is changing quite rapidly.

We can say that there is a need to develop and propose to the market reliable, robust, simple and effective process designs in order to arrive to a success story. Many of these designs are likely to be hybrid technologies.

References

1. Polikarpov, G. G., Radioecology of Aquatic Organisms, North Holland Publishing Company-Amsterdam, 1966.
2. Yang R., T., Gas separation by adsorption processes, Butterworths, 1987.
3. Tsezos M. and B. Volesky, Biotechn and Bioeng., 23 (1981) 583.
4. • Brierley, J. A., C. L. Brierley, R. F. Decker and G. M. Goyak. Treatment of mocroorganisms with alkaline solution to enhance metal uptake properties. U. S. Patent No 4, 690,894 (1987).
   • Brierley, J. A., C. L. Brierley, R. F. Decker and G. M. Goyak. Metal recovery. U. S. Patent No 4, 789, 481 (1988).
   • Brierley, J. A., C. L. Brierley, R. F. Decker and G., M., Goyak, Metal recovery.U.S. Patent No 4, 898,827 (1990).
   • Brierley, J. A., C. L. Brierley, R. F. Decker and G., M., Goyak. Metal recovery. U.S. Patent No 4, 992, 179 (1991).
5. Tsezos M., Noh S. H., Particle encapsulation technique. U. S. Patent 4828882 (1987).
6. Garnham G. W., Biosorbents for Metal Ions, Edited by John Wase and Christopher Forster, Taylor and Francis, (1997) 11.
7. Full scale AMT biosorption units: Eagle-Picher, Inc. (a lead-acid battery manufacturing facility), Socorro, New Mexico. Bestop, Inc (a zinc galvanizer), Broomfield, Colorado. AMT Pilot Plant : Black Hills Jewelry, Rapid City, South Dakota.
8. Tsezos M., The design and monitoring of a modified uranium biosorption pilot plant, Final Report, Head CANMET Biotechnology Section, Contract file SSC 015SQ23440-0-9055.
9. Tsezos M., Z. Georgousis and E. Remoudaki, Biotechn. and Bioeng., 55 (1997) 16.
10. Biosorbents for Metal Ions, Edited by John Wase and Christopher Forster, Taylor and Francis, 1997.
11. Removal and recovery of heavy metals from waste water by sandfilters inoculated with metal biosorbing or bioprecipitationg bacteria, Brite Euram Project BRPR CT 96 0172.

Competitive biosorption of copper, cadmium, nickel and zinc from metal ion mixtures using anaerobically digested sludge

A. Artola, M.D. Balaguer and M. Rigola

Laboratori d'Enginyeria Química i Ambiental, Facultat de Ciències, Universitat de Girona, Campus de Montilivi, 17071-Girona, Spain.

The biosorption of four metal ions, cadmium, copper, nickel and zinc from multi-component mixtures in aqueous solution by anaerobically digested sewage sludge is reported. The metal competition study was performed on binary metal systems where the two metals present had the same initial concentration. An increment of the total metal adsorption was obtained in all tests when compared to the single metal systems. The same behaviour was observed when a mixture of all the four metal ions was studied. The metal-sludge affinity order found was the same in single and multi-component (binary and quaternary) mixtures: copper, cadmium, zinc and nickel. Copper has been demonstrated to have the capacity to desorpt cadmium and zinc previously bound to the sludge.

1. INTRODUCTION

Heavy metals can be found in a number of industrial waste water effluents. High concentrations of these pollutants are present in effluents from electroplating and metal-finishing processes as well as from metal extractive operations. Public awareness of the long term toxic effects of water containing dissolved heavy metal ions has been growing in recent years. According to more restrictive regulatory standards the concentration of these pollutants has to be reduced. Consequently, improved and innovative methods of waste water treatment are being developed (1). New treatment methods focus on the improvement of the metal removal rate, the reduction of the final volume of waste generated as a consequence of the treatment applied and the possibility of metal recovery and reuse. When economically feasible, recovery is a preferred option. Economical viability of the treatment is a primary consideration.

The biosorption process is one of the proposed techniques for the removal of heavy metals from waste water. Biosorption can be considered as a collective term for a number of passive accumulation processes such as physical and chemical adsorption, ion exchange, complexation, chelation and micro-precipitation taking place essentially in micro-organisms cell wall (2). Both living and dead biomass have biosorptive properties (3).

The cost of pure adsorbents can be considered a limitation in the application of sorption process to detoxification of waste water loaded with heavy metal ions (4). The possibility of using waste biomass makes biosorption a cost-effective process. The biological sludge in excess from a conventional sewage treatment plant has demonstrated to be a good biosorbent (5,6,7).

In spite of many papers published on biosorption systems using waste biomass and a single heavy metal ion, scarce information is available on biosorption of multi-metal mixtures (8,9). However, multi-metal systems are usually present in effluents from industrial metallurgical processes (10). The presence of other metal ions can derive in the modification of the biosorption equilibrium parameters determined from a single-metal system. The possible competition between heavy metals for the binding sites on the micro-organisms cell wall requires a detailed study of the effects derived from the presence of other metal ions on the adsorption of the desired one.

The objective of this work is the determination of the adsorption capacity of anaerobically digested sludge in presence of copper, cadmium, nickel and zinc from different mixtures of these metals. Values obtained on these tests will be compared with the results obtained from single-metal systems.

2. MATERIALS AND METHODS

2.1 Sludge characteristics

Anaerobically digested sludge was obtained from a conventional sewage treatment plant in the city of Girona (Spain). This plant treats a wastewater flow of 42,000 m^3/d approximately. The characteristics of the sludge used in the tests were the following: pH, 7.2-7.6; total solids, 17.5-26.2 kg/m^3; volatile solids, 9.4-14.8 kg/m^3; initial content of selected metals was: [Cu], 0.30-0.38 g Cu/kg TS; [Zn], 0.37-0.40 g Zn/kg TS; [Cd] and [Ni] were under detection limits.

2.2 Analytical methods

Total and volatile solids (TS, VS) were measured according to Standard Methods (11). A specific electrode was used to measure pH. Metal concentration in the samples was determined by flame atomic absorption (AA) using a VARIAN SPECTRAA-300 spectrophotometer. Metal content in the original sludge was also measured by flame atomic absorption with prior acidic digestion of the samples as described bellow.

2.3 Digestion of sludge samples

Acid digestion was used for the determination of the sludge metal content. 30 ml of concentrated HNO_3 plus 10 ml of concentrated HCl were added to a 50 ml sludge sample. Three sludge samples were used in each test. The sludge-acid mixture was digested under reflux during 2 hours at a temperature of 100°C. After this period, each sample was filtered through a micro-fiber filter (WHATMAN GF/C) and diluted for its AA analysis. A blank test was also undertaken following the same procedure but substituting the sludge sample by 50 ml of distilled water.

2.4 Determination of the heavy metal binding capacity

Heavy metal binding capacity was determined in batch experiments. A 30-ml sample of original sludge was added to 200 ml of metal chloride solution in a 250-ml open Erlenmeyer flask. The metal solution was prepared dissolving the chloride salts of the different divalent metals in a fixed volume of distilled water. All samples were prepared with the same molar concentration of all metals present. A magnetic stirrer was used to maintain well-mixed conditions and to minimise the mass transfer resistance. After one hour of contact, a 5 ml sample was taken from the flask and filtered through a micro-fiber filter (WHATMAN GF/C). The volume of filtrate taken in each experiment was different in order to obtain the convenient concentration of the different metals for the spectrophotometric analysis after dilution with distilled water.

The adsorption tests were repeated at different initial concentrations of total metal with a constant sludge concentration in the flask of 3.31 g TS/l. In each test, the initial metal concentrations were also determined by adding 30 ml of distilled water to 200 ml of metal solution and following the same procedure used with the samples.

The pH of the sludge-metal system was measured but not modified. Two measures of this parameter were performed. First, at the beginning of the tests just after the addition of the sludge to the metal solution, and again after one hour of contact before the extraction of the sample. The pH of the blank test (metal chloride solution plus 30 ml of distilled water) was also measured initially and checked after one hour contact time remaining approximately constant.

2.5 Desorption tests

Three identical portions of anaerobically digested sludge which had previously adsorbed cadmium and zinc from a mixture of the two metals at an initial concentration of 8 mmols/l were taken. One of the portions was used for the determination of the total solids content. A second one was digested to determine its metal content as described in point 2.3. The third portion was resuspended in 200 ml of 4 mmols/l copper solution. The contact between sludge and metal solution was maintained during 19 hours. Samples were taken after 1 and 2 hours of contact and at the end of the contact time.

3. RESULTS AND DISCUSSION

3.1 Binary systems

The results from competition experiments of equimolar binary solutions of Cu, Cd, Zn and Ni at two different initial metal concentration, 2.5 and 8 mmols/l, are presented in Tables 1 and 2 respectively. In order to reflect the capacity of the sludge for each metal, results without any added competing metal are also presented. These tables summarise the percentage of removal and adsorption capacity values obtained in each test. This last value is expressed on the basis of mass of metal (mg) uptake per gram of biomass but also on molar units (mmols of metal per gram of biomass) to compare metal uptake as the total number of metal ions adsorbed (last column of each table). Total solid concentration in the reactor was 3.3 g/l.

For an initial metal concentration of approximately 2.5 mmols/l (Table 1), the removal percentage obtained for each metal in single metal systems was higher than 90% for copper, cadmium and zinc. Nickel was the metal presenting lower values (74.4% of removal). The percentage of metal removal is maintained for copper in the binary mixtures. The affinity of copper for the sludge is maintained when another metal is present in the solution.

In the other cases, cadmium, nickel and zinc, when a competing metal was present, adsorption capacity values decrease respect of the values obtained in single metal tests. This reduction is specially relevant for nickel in front of cadmium in which case the adsorption capacity for nickel is reduced by a half. The adsorption of cadmium and zinc decreases in a 45 and a 50% respectively when both metals are present in the adsorption test.

Table 1
Competition results in binary systems at metal initial concentration around 2.5 mmols/l.

Metal	Metal adsorption (single-metal systems)			Metal	Metal adsorption (binary systems)			Total metal adsorbed
	(mg/g)	*(mmol/g)*	*% rem.*		*(mg/g)*	*(mmol/g)*	*% rem.*	*(mmol/g)*
Cu	48.34	0.76	98.70	Cu	48.28	0.76	98.56	1.40
([Cu]$_{in}$= 2.6mM)				Cd	71.82	0.64	71.32	
				Cu	48.06	0.75	99.95	1.11
				Zn	23.94	0.36	56.61	
				Cu	47.25	0.74	99.09	1.25
				Ni	29.94	0.51	56.36	
Cd	80.72	0.72	91.37	Cd	71.82	0.64	71.32	1.40
([Cd]$_{in}$= 2.5mM)				Cu	48.28	0.76	98.56	
				Cd	45.41	0.40	53.49	0.79
				Zn	25.23	0.39	58.07	
				Cd	68.27	0.61	79.27	0.95
				Ni	20.15	0.34	38.07	
Zn	38.98	0.60	94.62	Zn	23.94	0.36	56.61	1.11
([Zn]$_{in}$= 2.2mM)				Cu	48.06	0.75	99.95	
				Zn	25.23	0.39	58.07	0.79
				Cd	45.41	0.40	53.49	
				Zn	36.41	0.56	83.70	0.99
				Ni	25.63	0.43	48.05	
Ni	39.70	0.67	74.43	Ni	29.94	0.51	56.36	1.25
([Ni]$_{in}$= 3.1mM)				Cu	47.25	0.74	99.09	
				Ni	20.15	0.34	38.07	0.95
				Cd	68.27	0.61	79.27	
				Ni	25.63	0.43	48.05	0.99
				Zn	36.41	0.56	83.70	

The total metal adsorbed (last column of Table 1) increases in all cases compared to the values obtained for each metal in single metal tests. In spite of that, the total capacity of adsorption is always lower than the sum of the individual adsorption capacities of each metal taking part in the test.

The decrease of adsorption capacity compared to the single metal systems observed for all metals with exception of copper, reflects the existence of a competition between the four metals studied for the binding sites present in bacterial cell wall.

However, when total initial metal concentration in solution was of 2.5 mmols/l, values obtained for the adsorption capacity of copper were lower than the maximum adsorption capacity value obtained in previous work from single metal tests (1.4 mmols Cu/ g TS) (12). It can be deduces that all the copper present in the binary mixtures is adsorbed, but free sites are still left to bind part of the second metal present in the solution. To verify if the same conclusion can be extended to higher concentrations of metal, a new set of tests was undertaken at higher initial metal concentration in solution (around 8 mmols/l). At this high concentration the metals may reach the maximum adsorption capacity obtained in individual tests. The results for the single metal and competitive experiments are presented in Table 2 which has the same structure as Table 1.

As observed in the tests undertaken at initial metal concentration of 2.5 mmols/l, the adsorption capacity and the percentage of removal obtained for copper in single metal systems are maintained in binary-metal systems. At initial metal concentration of 8 mmols/l none of the other metals is significatively adsorbed when copper is present in solution.

Table 2
Competition results in binary systems at metal initial concentration around 8 mmols/l.

Metal	Metal adsorption (single-metal systems)			Metal	Metal adsorption (binary systems)			Total metal adsorbed
	(mg/g)	*(mmol/g)*	*% rem.*		*(mg/g)*	*(mmol/g)*	*% rem.*	*(mmol/g)*
Cu	80.15	1.26	52.73	Cu	76.62	1.21	51.81	1.21
([Cu]$_{in}$=7.91mM)				Cd	0	0	0	
				Cu	79.27	1.25	53.93	1.25
				Zn	0	0	0	
				Cu	79.73	1.25	53.81	1.25
				Ni	0	0	0	
Cd	117.37	1.04	43.48	Cd	0	0	0	1.21
([Cd]$_{in}$=7.94mM)				Cu	76.62	1.21	51.81	
				Cd	52.27	0.47	18.39	0.90
				Zn	28.31	0.43	16.64	
				Cd	53.78	0.48	21.20	0.63
				Ni	8.76	0.15	6.13	
Zn	50.15	0.77	29.46	Zn	0	0	0	1.25
([Zn]$_{in}$=8.61mM)				Cu	79.27	1.25	53.93	
				Zn	28.31	0.43	16.64	0.90
				Cd	52.27	0.47	18.39	
				Zn	48.50	0.74	23.16	0.79
				Ni	2.72	0.05	1.94	
Ni	37.76	0.64	24.5	Ni	0	0	0	1.25
([Ni]$_{in}$=8.68mM)				Cu	79.73	1.25	53.81	
				Ni	8.76	0.15	6.13	0.63
				Cd	53.78	0.48	21.20	
				Ni	2.72	0.05	1.94	0.79
				Zn	48.50	0.74	23.16	

The adsorption capacity for cadmium, nickel and zinc in the binary systems decreases when compared to the single metal tests as it happen at lower metal initial concentration. Values of this parameter are very low for nickel in binary mixtures with cadmium and zinc, approaching zero in the case of zinc. It is also noteworthy that in the binary mixtures with cadmium the total adsorption capacity of both metal ions looks subordinated to the capacity of the partner metal.

Total adsorption capacity obtained in the binary systems does not suggest a significant increment in the number of active ligands although there is a clear apparent competition between metals for the binding sites.

The affinity order of anaerobically digested sludge for the four metals under study has been established as:

$$copper > cadmium > zinc > nickel$$

in agreement with the affinity order obtained from single metal tests.

3.2 Four metal system

To confirm the competition between metals for the binding sites of anaerobically digested sludge established in binary metal systems, adsorption tests were undertaken from different equimolar mixtures of all four metals. Tests were performed for different total initial concentrations of the metals (ranging from 0.5 to 8 mmols/l) while the concentration of total sludge solids in the reactor was kept at 3.4 g/l. The results obtained in these tests are summarised in Figure 1. This figure presents the adsorption capacity measured for the four metals in each test as well as the sum of these values given as total adsorbed metal.

All four metals are fully adsorbed at the lowest metal initial concentration tested (0.5 mmols/l). The tests from 3 to 8 mmols/l show the clear predominance of copper as the main adsorbed metal. The behaviour of copper is very similar to that obtained from single and binary metal systems reaching a maximum adsorption capacity of 1.35 mmols Cu/g TS, close to the value found in those tests (1.25 mmols Cu/g TS).

The adsorption increases for all metals till the initial metal concentration is 3 mmols/l. Nickel and zinc show decreasing values for total initial metal concentrations higher than 3 mmols/l. The adsorption of cadmium reaches the maximum at an initial cadmium concentration around 5 mmols/l decreasing for the highest metal initial concentration tested, 8 mmols/l.

The increment in the adsorption of copper at high metal concentrations derives in a reduction of the adsorption of cadmium, nickel and zinc. On the other hand, even if the adsorption of the other metals is low at initial metal concentration of 8 mmols/l it is not zero as was in binary mixtures when copper was present. Equilibrium pH obtained for the quaternary metal mixture at 8 mmols/l metal initial concentration was 5.8. If this value is compared to that obtained in binary copper mixtures (5.9) the difference between them does not justify the different adsorption behaviour of the rest of metals in front of copper.

As can be seen in Figure 1, total metal adsorption reaches a value of 2 mmols metal/g TS. This value is significatively higher than the total metal adsorption found in binary systems where maximum values of 1.2-1.3 mmols/l were observed. It seems that the total metal adsorption capacity of the sludge increases when increasing the number of metals present even if in the value of the total adsorption capacity is bellow the sum of the capacities of each individual metal. This fact supports the asumed competition between metals for the sludge binding sites.

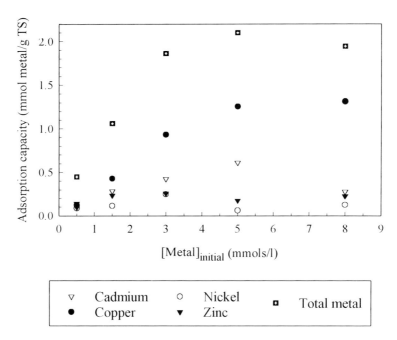

Figure 1. Anaerobically digested sludge adsorption capacity for cadmium, copper, nickel and zinc from equimolar solutions of the four metals.

Measurements shown in Figure 1 confirm the affinity order of the studied metals for the sludge established from binary metal systems.

3.3 Competitive metal desorption

The selectivity of anaerobically digested sludge for copper in the presence of other metal ions suggested the possibility of the substitution of previously bound metal ions by copper.

Desorption of 0.38 mmols of zinc and 0.57 mmols of cadmium previously adsorbed on the anaerobically digested sludge was performed with 200 ml of a 4 mmols/l copper solution. The contact was maintained during 19 hours. The results obtained are shown in Figure 2 where the amount of free metal (in mmols) is represented in front of the contact time.

As can be seen in Figure 2, both cadmium and zinc are progressively desorbed in presence of copper during the contact time. At the same time copper is adsorbed at the free sites that the other two metals leave at sludge surface.

The desorption of cadmium is slower that the desorption of zinc. At the end of the contact time all the cadmium previously bound to the sludge is desorbed (0.57 mmols). A small quantity of zinc remain bound to the sludge after the same contact time (0.38 mmols of zinc were adsorbed and 0.3 mmols of zinc are desorbed after 19 hours of contact). There is not a direct relationship between the mmols of zinc and cadmium desorbed and the mmols of copper adsorbed.

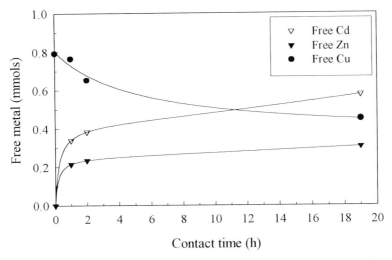

Figure 2. Cadmium and zinc desorption in contact with a 4 mmols/l copper solution.

The desorption of metals previously bound to the sludge is possible by means of the adsorption of another metal that shows a highest affinity for the sludge binding sites. The selective recovery of this metal could be possible after the use of the same sludge for successive contact steps with an effluent containing this metal.

4. CONCLUSIONS

The four metals studied, copper, cadmium, nickel and zinc, present a competitive adsorption when they are in contact with a fixed amount of anaerobically digested sludge. This competition has been demonstrated in adsorption tests performed from binary and quaternary mixtures of the four metals. The metal affinity order observed was copper>camium>zinc>nickel.

The adsorption capacity of sludge for copper was maintained in multi-metal systems if compared with the adsorption capacity values obtained in single metal systems. The total metal adsorbed in binary mixtures was higher than the metal adsorbed in individual tests. The maximum adsorption value (1.4 mmols metal/g TS) was obtained in the cadmium-copper system at initial metal concentration of 2.5 mmols/l. This quantity increased when the adsorption was undertaken from a mixture of all the four metals reaching a maximum value of 2 mmols of metal/g TS when initial metal concentration in solution was 5 mmols/l for all the metals present. The total capacity of adsorption is always lower than the sum of the individual adsorption capacities of each metal taking part in the test.

The desorption of zinc and cadmium previously bound to the sludge using a copper solution has been demonstrated to be effective. In the case of cadmium a complete desorption was obtained after 19 hours of contact. A small quantity of zinc remained bound to the sludge after the same contact time. This fact could be used for the selective recovery of a metal showing a higher affinity for the sludge binding sites than the other metals in solution. In this case, copper.

REFERENCES

1. S.J. Allen and P.A. Brown, J. Chem. Tech. Biotechnol., 62 (1995) 17.
2. Y. Sag and T. Kutsal, Process Biochem., 31 (1996) 573.
3. G.M. Gadd, Chem. and Ind. (Biotechnology), July (1990) 421.
4. R. Apak, E. Tütem, M. Hügül and J. Hizal, Wat. Res., 32 (1998) 430.
5. B.R. Fristoe and P.O. Nelson, Wat. Res., 19 (1983) 771
6. A. Artola and M. Rigola, Biotechnol. Lett., 14 (1992) 1199.
7. P. Battistoni, G. Fava and M.L. Ruello, Wat. Res, 27 (1993) 821.
8. Y. Sag and T. Kutsal, Process Biochem., 31 (1996) 561.
9. Y.P.Ting, F. Lawson and I.G. Pince, Biotechnol. and Bioeng, 37 (1991), 445.
10. K.H. Chong and B. Volesky, Biotechnol. and Bioeng., 49 (1996) 629.
11. APHA, Standard Methods for the Examination of Water and Wastewater, Washington, 1989.
12. A. Artola, M.D. Balaguer and M. Rigola, Wat. Res., 31 (1997) 997.

Activated sludge as biosorbent of heavy metals

A. Hammaini, A. Ballester, F. González, M. L. Blázquez and J.A. Muñoz

Departamento de Ciencia de Materiales e Ingeniería Metalúrgica,
Facultad de Ciencias Químicas, Universidad Complutense,
Av. Complutense s/n-28040 Madrid, Spain

Attempts were made to recover cadmium, copper, nickel, lead and zinc in aqueous solution by three different types of biomasses: Activated sludge (A.S.), brewer's yeast and barley's root. The former one was the most effective to uptake metal ions. pH solution was found to be critical to metal ions uptake, the optimum pH range was 4-5. Effect of biomass loading was also investigated, the optimum biomass concentration was 1 g L^{-1}. Temperature (20-60 °C) had no effect on the level of accumulation of all metal ions tested. Adsorption behaviour of the five metals fitted Langmuir isotherm more than Freundlich model.

1. INTRODUCTION

Rapid industrialisation has led to increased disposal of heavy metals and radionuclides into the environment. Removal of heavy metals and radionuclides from metal-bearing wastewater is usually achieved by physico-chemical processes before discharging the effluents into natural water-body systems. Physico-chemical processes in use for heavy metal removal from wastewater include precipitation, coagulation, reduction processes, ion exchange, membrane processes (such as ultrafiltration, electrodialysis and reverse osmosis) and adsorption. Conventional treatment technologies like precipitation and coagulation become less effective and more expensive when metal concentrations are in the range of 1-100 mg L^{-1}. High costs, process complexity and low removal efficiency of membrane processes have limited their use in heavy metal removal. Adsorption on activated carbon is a recognised method for the removal of heavy metals from wastewater. The high cost of activated carbon limits its use in adsorption. A search for a low-cost and easily available adsorbent has led to the investigation of materials of agricultural and biological origin, along with industrial by-products, as potential metal sorbent.

Microorganisms such as bacteria, fungi, yeast and algae [1] can remove heavy metals and radionuclides from aqueous solutions in substantial quantities. The uptake of heavy metals by biomass can take place by an active mode (dependent on the metabolic activity) known as bioaccumulation or by a passive mode (sorption and or complexation) termed as biosorption. Biosorption may be defined as "a non-direct physico-chemical interaction that may occur between metal/radionuclide species and the cellular compounds of biological species". Because metal uptake by nonliving biomass involves different types of adsorption processes, biosorption is affected by various physical and chemical factors such as pH, temperature, oxidation-reduction potential, ionic strength, metal concentration in solution, presence of complexing agents, etc.

In this paper, the results concerning the use of some waste biomaterials as dry biomasses for biosorption of cadmium, copper, nickel, lead and zinc are presented. The most important parameters affecting the phenomenon, such as pH, concentration of biomass and temperature, have been studied.

2. MATERIALS AND METHODS

10 L of activated sludge supplied by a sewage treatment plant in Arroyo de la Vega (Madrid, Spain) were centrifuged. The pellets obtained were dried at 50 ºC for 6 days and then, ground under 0,1 mm particle size. Two more types of biomass have been used in this study: a brewer's yeast and a barley's root, both supplied by Mahou S.A. The former one was treated the same as the A.S., whereas the second was used as supplied.

1 litre metal stock solutions of 1000 mg L^{-1} of Cd^{2+}, Cu^{2+}, Pb^{2+}, Zn^{2+} and Ni^{2+} were prepared from $CdSO_4 \cdot 8/3\ H_2O$, $CuSO_4 \cdot 5\ H_2O$, $Pb(NO_3)_2$ (Panreac Pa), $ZnSO_4$ and $NiSO_4 \cdot 7\ H_2O$ (PRO-BUS), respectively, and stored at room temperature. Samples of stock solutions were diluted to prepare working metal concentrations. Initial pH was adjusted with dilute NaOH and additions of dilute H_2SO_4 for Cd, Cu, Ni and Zn or dilute HNO_3 for Pb.

A known quantity of biomass was suspended in 250 ml of a monometalic solution, then the mixture was stirred for 2 hours giving ample time for adsorption equilibrium to be reached. pH was periodically controlled and, at the same time, 2 ml samples were taken. Samples were centrifuged at 5000 rpm for 20 min and the supernatant was separated and analysed by atomic absorption spectroscopy. The metal concentration bound to the biomass was calculated as a difference between metal concentration before and after the sorption process.

3. RESULTS AND DISCUSSION

3.1. A comparative study of the adsorption capacity of three different biosorbents

The purpose was to find a way of decreasing the above mentioned ions concentration using the three biomasses already described. Biosorption test were carried out at initial pH values of 4 and at initial metal concentration of 30 mg L^{-1} for Zn, 50 mg L^{-1} for Cd, Cu and Pb and 90 mg L^{-1} for Ni. The results concerning the test of the three biomasses are summarised in Table 1. As is resulting from this data, A.S. have much efficiency in cation metal removal than barley's root. Except for Pb, brewer's yeast was unable to uptake any other metal.

Figure 1 shows that the rate of metal uptake by A.S. was very fast. Within the first 5 min of contact, 90 % of metal uptake was completed. Adsorption equilibrium was reached 30 min after biomass addition. From these results, A.S. was chosen to carry out the study of the influence of pH, biomass concentration and temperature on the biosorption phenomenon.

Table 1
Metal ions removal by the three biomasses

Metal	% of metal ion removal		
	activated sludge	Barley's root	Brewer's yeast
Cd	68,1	14,7	0
Cu	48,7	16,1	0
Ni	18,2	13,8	3,0
Pb	88,2	68,8	62,5
Zn	52,0	16,7	0

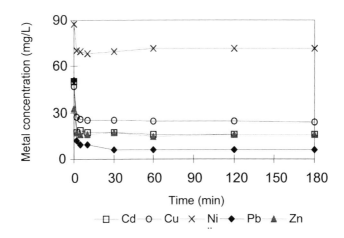

Figure 1. Time profile of metal ions sorption by A.S.

3.2. Effect of pH

Initially, pH was adjusted at values below the metal precipitation to assure complete dissolution of each metal ion. All tests were carried out with 1 g L^{-1} of biomass and 50 mg L^{-1} of metal. The influence of pH on the biosorption capacity of the A.S. for the different metals is shown in Figure 2. Similar trend was observed for all metals. At very low pH values (pH 1 to 2) metal uptake was negligible. The metal uptake increased as the pH was increased up to 4, beyond this pH value no improvement in the adsorption capacity was observed.

At very low pH value (i.e., with a high proton concentration)cell-wall ligands would be closely associated with H_3O^+, and access to ligands by metal ions as a result of repulsive forces would be restricted. As pH levels are increased, more ligands with negative charge would be exposed with a subsequent increase in attraction for positively charged metal ions [2].

Figure 3 shows the pH evolution versus time for all experiments carried out with Zn. Similar trend was observed for all the rest of metals. At initial pH values lower than 3, pH remained constant during the time. For initial pH values $\geq$ 3, the trend was to reach, within the first 5 minutes, a pH value ranged between 5 and 6, remaining constant for the rest of biosorption process. This behaviour should be conditioned by the functional groups (carboxyl, hydroxyl, amine, etc) placed on the cell wall, each of them has an acidic constant (pK_a) which control its equilibrium state with the corresponding conjugate base. So that, when the biomass is in contact with a metallic solution of pH lower than pK_a, the equilibrium is displaced towards the formation of the acidic specie with a protons consumption and therefore the medium pH increases. Otherwise, when the metallic solution pH value is higher than pK_a, the equilibrium shifts toward the formation of the basic specie, which is accompanied by a supply of protons from the biomass and therefore the medium pH decrease.

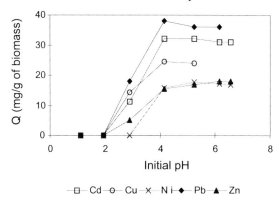

Figure 2. Effect of pH on the biosorption of the five metals
by A.S.

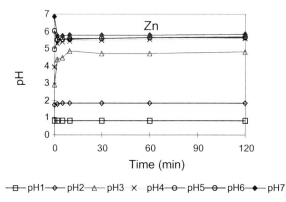

Figure 3. Evolution of pH versus time

3.3. Effect of biomass loading

0,25; 0,5; 1,0; 2,0 and 3,0 g L^{-1} of biomass were tested with 50 mg L^{-1} at room temperature, in order to obtain the optimum biomass concentration corresponding to a maximum binding capacity. Initial pH values were 4 for Cd, Cu and Pb and 5 for Ni and Zn. Figure 4 shows the evolution of adsorption capacity as a function of biomass concentration. For all metals tested, the metal uptake increased as the biosorption concentration increased up to 1 g L⁻¹, beyond this cell density value the specific metal uptake decreased as the biomass concentration increased. From these results 1 g L⁻¹ was fixed as the optimum biomass concentration. From Figure 5 it may be observed that the percentage of metal removed from solution increased as biomass concentration increased as long as metal in solution was not totally sequestered.

The biomass concentration is an important variable that can affect the metal uptake from solution. At a given equilibrium concentration the biomass adsorbs more metal ions at low cell densities than at high cell densities. High biomass concentration could make a "screen" effect, unless true equilibrium is reached, of the dense outer layer of cells, protecting the binding sites from metal. As a result, the specific metal uptake, that is amount of metal removed per unit biomass, is found to be lower at high biomass densities. High biomass densities imply the presence of large amount of biomass, and therefore, the total metal removed from the solution is found to be higher at high biomass densities. Therefore, high biomass densities are required in order to maximised the percentage of metal removal from solution [2].

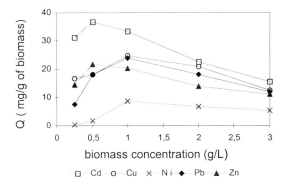

Figure 4. Effect of biomass loading on biosorption capacity of A.S.

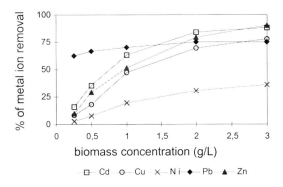

Figure 5. Effect of biomass concentration on % of metal removal by A.S.

3.4. Effect of temperature

The effect of temperature on the five metal ions biosorption is represented in Figure 6. Over the range investigated (20° - 60 °C), temperature-related effects were not significant.

The metabolism of growing cells is strongly affected by temperature. However, biosorption by nonliving biomass is metabolism-independent, and therefore, temperature is not expected to have a significant effect on the metal uptake [2].

For economic considerations, room temperature was chosen as the optimum for the adsorption of the five metal ions by A.S.

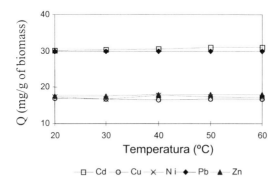

Figure 6. Effect of temperature on the adsorption of the
five metals by A.S.

3.5. Isotherms

As a basis for the line of investigation considered, conventional one-metal sorption isotherms were experimentally determined for A.S. These isotherms were derived at the optimum pH value corresponding to each metal ion tested, room temperature and with 1 g/L of biomass concentration. Figure 7 shows the bioadsorption isotherm corresponding to Zn. The same trend was observed for the rest of metal tested.

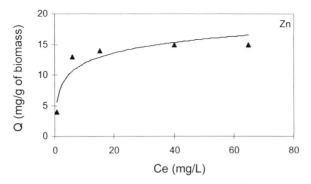

Figure 7. Biosorption isotherm for Zn

The data pertaining to the sorption dependence upon metal ion concentration were subjected to the Langmuir and Freundlich adsorption isotherms. [3] The Langmuir adsorption isotherm predicts about the physical monolayer adsorption of adsorbate over adsorbent in gaseous state and is also applicable for liquid state. The Langmuir isotherm equation is as follows:

$$C_e/Q = 1/Q_{sat} + C_e/B \cdot Q_{sat} \tag{1}$$

Where C_e is the equilibrium concentration of metal in solution (mg L^{-1} or mole L^{-1}) and Q is the equilibrium concentration of metal adsorbed onto the adsorbent (mg g^{-1} or moles g^{-1}). Q_{sat} is the limiting concentration and B is the intensity of the adsorption. In the case of Zn, when a graph was plotted between C_e/Q versus C_e a straight line was obtained indicating the data fitted the isotherm under test. In our case the Langmuir adsorption isotherm is followed over a very range of metal ion concentration (10-150 mg L^{-1}) as shown in Figure 8. The data were also subjected to Freundlich adsorption isotherm as given below:

$$Log\ Q = Log\ A + 1/n\ Log\ C_e \tag{2}$$

Where Q is the adsorbed metal concentration (mg g^{-1} or moles g^{-1}), A is the maximum sorption capacity, 1/n is the steepness of adsorption while C_e is the metal ion concentration in solution al equilibrium in mg L^{-1} or moles L^{-1}. The Freundlich adsorption isotherm describes a multilayer adsorption. It also includes a large energetically homogenous sorption site over the adsorbent surface. The values of A and 1/n are characteristic for a particular sorption system. Figure 9 shows the Freundlich isotherm for Zn.

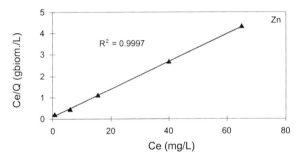

Figure 8. Langmuir adsorption isotherm for Zn at A.S.

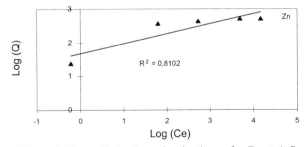

Figure 9. Freundlich adsorption isotherm for Zn at A.S.

Experimental C_e and Q data were used to evaluate the Langmuir constants: Q_{sat} and B for all metal tested. Results are summarised in Table 2, from which it may be seen that, the order of the sorption was: Pb> Cu > Cd $\cong$ Zn > Ni. A reason for different sorption efficiencies might be the electro-chemical properties of the metal ions. The order of ion according to their electrode potential is similar to the order of affinity of A.S [4].

Table 2.
Langmuir biosorption isotherm parameters for single-metal uptake.

Metal	Q_{sat} (mmol g^{-1})	B (l mmol^{-1})	R^2
Cd	0.247	35.97	0.99
Cu	0.296	7.63	0.94
Ni	0.153	8.80	0.94
Pb	0.690	20.72	0.95
Zn	0.236	39.88	0.99

4. CONCLUSIONS

The use of activated sludge as biosorbent may offer an effective way to decrease Cd, Cu, Ni, Pb and Zn concentration in wastewater.

pH was an important variable that affect markedly the phenomenon of biosorption. The optimum pH value was 4 for the uptake of Cd, Cu and Pb and 5 for Ni and Zn.

The biosorption equilibrium was conditioned by the acidic constant pK_a of micro-organism cell wall functional groups.

The optimum biomass concentration was 1g L^{-1}

Temperature-related effects on the five metal ions biosorption were not significant.

The data pertaining to the sorption dependence upon metal ion concentration fitted Langmuir isotherm better than Freundlich model.

The order of the sorption was the same as ion electrode potentials: Pb> Cu> Cd $\cong$ Zn> Ni.

ACKNOWLEDGEMENTS

The authors wish to expres their sincere gratitude to Comision Interministerial de Ciencia y Tecnología (CICYT) of Spain for funding this research.

REFERENCES

1. B. Volesky, Biosorption of heavy metals, CRC press, Tnc U.S.A. ISBN 0-8493-4917-6,1990.
2. J.M. Modak and K.A. Natarajan, Biosorption of metals using nonliving biomass-A review, Minerals and metallurgical processing, (1995) 189.
3. A.M. Khalid, A.M. Shemsi, K. Akhtar and M.A. Anwar, Uranium biosorption by *Trichoderma harzianum* entrapped in polyester foam beads, Biohydrometallurgical technologies, A.E. Torma, M.L. Apel and C.L. Brierley, (1993) 309.
4. B. Mattuschka & G. Straube, Biosorption of metals by a waste biomass, J. Chem. Tech. Biotechnol., 58 (1993) 57.

Biochemical characteristics of heavy metal uptake by *Escherichia coli* NCP immobilized in kappa-carrageenan beads*

S.S. Bang[a] and M. Pazirandeh[b]

[a]Department of Chemistry and Chemical Engineering, South Dakota School of Mines and Technology, 501 E. St. Joseph Street, Rapid City, SD 57701, USA

[b]Naval Research Laboratory, Code 6900, 4555 Overlook Ave. SW, Washington, DC 20375, USA

A biosorbent was developed with *Escherichia* coli NCP immobilized in 2.25% κ-carrageenan to remove heavy metals from solutions. The NCP is a recombinant *E. coli*, expressing the *Neurospora crassa* metallothionein gene that codes for a high-affinity, metal-binding peptide. The NCP has previously been shown to sequester low levels of cadmium (Cd) and other metals from solutions. κ-Carrageenan gel encapsulated with the NCP forms an ellipsoidal bead with average diameters of 5.82±0.31mm along the long axis and 2.70±0.27mm along the short axis. This study reports physiological and biochemical characteristics of Cd uptake by the biosorbent.

Encapsulated cells grew up to 1.82×10^9 cells within individual beads. However, a small amount of the immobilized cells continuously leaked out of the bead during the biosorption experiments when cell numbers reached 9×10^8 cells/bead. Scanning electron micrographs evidenced that encapsulated cells migrated toward the surface area, leaving small clusters of cells embedded in the gel matrix. The maximum uptake of Cd by the biosorbent was achieved within three hours, while cells were still in the exponential growth phase. Individual cells sequestered the maximum amount of Cd at 2-hour incubation, after which the specific uptake per cell declined significantly as cells appeared to escape from the bead. Results of the specific metal uptake by immobilized cells and free cells were compared to identify the transport limitation and availability of metals in polymer matrices. There was no apparent microbial limitation present in ellipsoidal-shaped, 2.25% κ-carrageenan beads. Rather, the immobilized individual cells sequestered more Cd when they were grown in the LB medium than free cells dissolved in saline solutions. Results indicate that the immobilized recombinant cells function as more effective biosorbents than the free cells.

1. INTRODUCTION

The use of biosorbents in remediation of heavy metal wastes has become increasingly popular in recent years (1,2). It is a cost-effective biological treatment in the removal of a wide spectrum of toxic wastes. Biosorbents are constructed with microorganisms and polymer, the networking matrix of which provides the space to house a high concentration of

---

*The research was supported by funding from the US Naval Research Laboratory.

microorganisms. Several types of natural polysaccharides and synthetic polymers have been used to encapsulate microorganisms. Properties of the polymer are greatly important not only in retaining microorganisms but also for the uptake of metals (3).

A large group of negatively charged macromolecules on the microbial cell wall have shown selective affinities to various types of metal cations (4,5,6,7). However, the use of free cells as a biosorbent has a limited potential because biomass has low physical strength and difficult separation potential from the surroundings. Immobilization would provide several advantageous strategies; increasing mechanical strength of the biosorbent, protecting the cells from hostile environments, and maintaining higher cellular metabolic activities. There are also potential disadvantages of the immobilized cells. Substrates including oxygen cannot penetrate into the polymer matrix and products may not be able to diffuse out. Therefore, chemical properties of each type of polymer play an important role in its application as a biosorbent. The structure and ionic strength of the matrix attribute to the efficacy of adsorption of metals by the biosorbent.

Previously, the authors reported the expression of the *Neurospora crassa* metallothionein gene in *E. coli* NCP and the use of recombinant microorganisms in removal of heavy metals such as Cd^{2+} (8). The genetically engineered NCP strain exhibited not only a high, selective affinity to the cadmium uptake but also a release of metals under different ionic strengths. Findings suggested the positive potential of a metallothinein-based biosorbent in removal of toxic metals. It was also observed that biomass of the NCP immobilized in polymer also performed better than other sorbents in removal of Cd, Hg, and Pb from the factory wastewater (9).

Our recent study compared the physical and biochemical characteristics of immobilized microorganisms in beads made of κ-carrageenan, alginate, and alginate coated with chitosan (10). Besides its strong physical integrity, the κ-carrageenan bead encapsulated with microorganisms worked as a superior biosorbent in Cd sequestration to the beads of other polymers. κ-Carrageenan originally extracted from the red seaweed consists of alternating copolymers of β–*D*-galactose-4-sulfate and 3,6-anhydro-*D*- or *L*-galactose (11). The objective of this study was to characterize the biochemical behavior of Cd uptake by the NCP immobilized in κ-carrageenan. Values of the specific uptake by free cells and immobilized cells have been compared to better understand the diffusion hindrance of the gel matrix.

2. MATERIALS AND METHODS

2.1. Microorganisms and growth conditions

The recombinant *E. coli* strain NCP and the control *E. coli* strain TB1 without the plasmid were maintained as described previously (8). Both cultures were grown in Luria-Bertani (LB) medium (Bacto) with ampicillin (10 μg ml^{-1}) and induced with isopropyl-β-D-thiogalactopyranoside (IPTG), following the procedure of Pazirandeh *et al.* (8). Cells of the late exponential growth (0.8-0.9 OD_{600nm}) were harvested, washed, and suspended in 50 mM Tris-HCl, pH 7.5 prior to cell immobilization.

2.2. Immobilization and dissolution of κ-carrageenan beads

With the modified procedure of Mattiasson (12), 0.9 g κ-carrageenan (Type III, Sigma) was dissolved in 30 ml of physiological saline and autoclaved. The gel solution was allowed to cool down to 45°C, and then was mixed with 10 mL cells to obtain the final concentration of 2.25% κ-carrageenan. Beads were made as the κ-carrageenan-cell suspension was dropped via a sterile syringe into a cold 300 mM KCl solution. Beads were cured in the KCl solution for an hour at 4°C and then stored in a buffer solution containing 50 mM Hepes and 1 mM KCl, pH 7.5. κ-Carrageenan beads were completely dissolved within 5 minutes in a solution of 0.85% NaCl at 37°C while shaking.

2.3. Scanning electron microscopy (SEM)

Details of the sample preparation for electron micrography were described previously (10). All prepared samples were examined by a scanning electron microscope (JEOL JSM-840A).

2.4. Cell growth and metal sorption experiments

To characterize the growth curve of encapsulated NCP cells in the absence of metals, approximately 40 beads were added to a 500-mL flask containing 100 mL LB broth with ampicillin. Two beads were withdrawn from the flask every hour and dissolved in 10 mL of saline. Dissolved cells were determined directly by reading optical density at 600 nm with a spectrophotometer (Spec 20) or after serial dilutions counting viable cells in LB plates containing ampicillin.

Metal uptake experiments were carried out with both NCP and TB1 strains. Metal sorption was initiated by adding 4 beads encapsulated with cells into a 50-mL Erlenmeyer flasks containing 10 mL of Luria broth dissolved with 100 μM $CdCl_2$ and ampicillin. Each set was prepared in triplicate. At 30-minute intervals for 3 hours, the reaction of each set was terminated; two flasks were used for the metal uptake analysis and one for cell counting. Incubation of bacterial culture and other metal sorption experiments were carried out in LB broth containing Cd and ampicillin at 37°C in a water bath shaker (Lab-line, Model 3540).

2.5. Metal analysis

Upon removal, beads were placed in 10 ml of saline. After the complete dissolution of beads, the solution was treated with an equal volume of 30% HCl. With a standard solution of Cd (Aldrich Chemical Co.), contents of the metal were determined by atomic absorption analysis (Perkin-Elmer, AA/5000). Sum of the metal recovery from beads and broth was 98.72 ± 0.36 %.

3. RESULTS AND DISCUSION

3.1. Microbial growth and Cd uptake in LB broth

Figure 1 shows a typical growth pattern of the immobilized NCP in LB broth without metals. In a 2.25% carrageenan bead, cells grew to the maximum concentration of approximately 1.82×10^9 cells/bead. Figure 2 graphs viable cell counts in relation to the optical density (OD) at 600 nm. As expected, the OD reading is roughly proportional to the cell count at low densities, while the linearity becomes less evident at higher densities. Comparing to the OD readings of free *E. coli* cells (13), the encapsulation yields a lower optical density reading per cell, indicating that immobilized cells are relatively smaller in size than free cells. This suggests a possibility that growth of the encapsulated cells is restricted within the network of the polymer matrix.

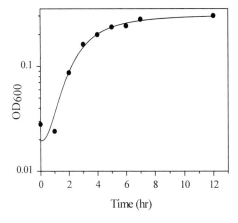

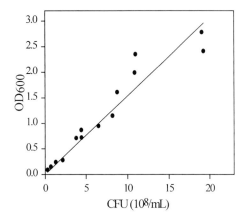

Figure 1. Growth of the NCP encapsulated in a κ-carrageenan bead.

Figure 2. Correlation of OD readings to colony forming units (CFUs): $y = 64.74x - 0.35$, where y represents OD and x represents 10^8 CFU/mL, ($R^2 = 0.917$).

κ-Carrageenan was formed as an ellipsoidal bead with average diameters of 5.82±0.31mm along the long axis and 2.70±0.27mm along the short axis. An average size of rod-shaped, encapsulated *E. coli* is estimated to be approximately 1.8 μm long and 0.46 μm in diameter. Based on a calculation of the volumes of both, one bead can hold approximately up to 1.3×10^{12} cells within it. However, during the growth, it was noticed that cells started to escape after 2 hours at which the cell concentration attained above the concentration of approximately 9×10^8 cells/bead. Approximately 0.1% of encapsulated cell populations were estimated to leak out into the medium.

3.2. Electron micrography of immobilized cells

SEM pictures of cells encapsulated in κ-carrageenan gel beads are presented in Figure 3, which shows that cells were evenly distributed in freshly-made beads (Fig. 3a). However, after 12 hours of incubation in LB broth with cadmium, it was found that most cells migrated near the surface area where cells form dense distribution layers of distinct spherical microcolonies as they multiplied (Fig. 3b). A higher magnification shows that a thin layer of gel surrounds microcolonies with a high cell density on the surface of the beads (Fig. 3c), where some cells escaped from the sac are found loosely attached to the surface of the microcolonies, suggesting a possible leakage. Clusters of a low cell density embedded in the matrices of the gel bead are also distinct in a higher magnification (Fig. 3d).

Migration of the microorganisms seems inevitable due to the diffusion limitation of the nutrient and oxygen through the barrier of the gel matrix. Highly populated surface areas were also increased, forming unique sac-like structures to accommodate growing cells. Even though the individual beads had enough space to hold growing cells (1.3×10^{12} cells), the continuous exposure of cells to the environment, while still multiplying on the surface, might result in a constant cell leakage.

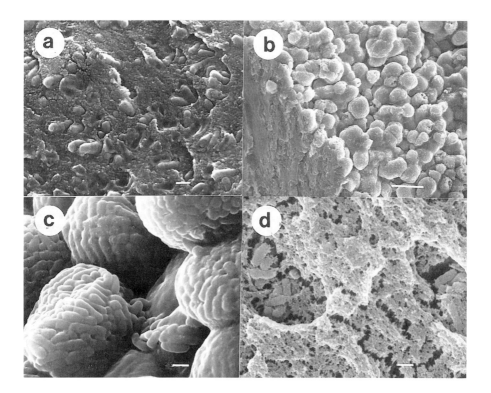

Figure 3. Scanning electron micrographs of the encapsulated NCP. ***a***. cross section of the bead after immobilization, bar, 1 μm; ***b***. cross section of the bead after 12 hour incubation in LB broth, bar, 10 μm; ***c***. microcolonies on the surface of the bead shown in ***a***, bar, 1 μm; ***d***. clusters of cells embedded in the matrix of the gel, bar, 1 μm.

There are several physiological factors that can also influence the cell leakage; e.g., polarity changes during the course of the cell growth and ionic strength of the medium. The mechanical strength of the gel, however, plays an extremely important role in retaining the cell population. The higher the concentration of gel was prepared, the stronger the mechanical strength of the biosorbent was expected. However, the cell growth rate was significantly impaired mainly due to lack of the transport of nutrients via the matrix barrier (14). To reduce the cell leakage, it may be necessary to develop an alternative biosorbent with a higher concentration of gel with adequate mechanical strength if cell growth inside the matrix is not needed. Coating glass beads with the gel suspension mixed with genetically engineered cells would be a possible solution to this.

3.3. Cd uptake in LB medium containing Cd

Previously the authors reported that the maximum metal uptake took place within 3 hours (10). In the present study, Cd uptake experiments were carried out for 3 hours and metal sorption was analyzed at 30-minute intervals. Results of the metal uptake by immobilized cells are plotted in Figure 4. A rapid uptake took place during the first-hour incubation, where nearly 70% of the total metal adsorption took place. When the specific uptake of Cd per cell was plotted against time, it became obvious that the cell capacity to sequester the metals peaked at 2 hour-incubation where cells were growing exponentially in the beads without showing any leakage yet (Figure 5). However, a significant decrease in the specific uptake was observed after 2 hours.

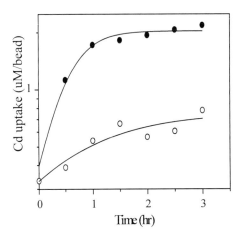

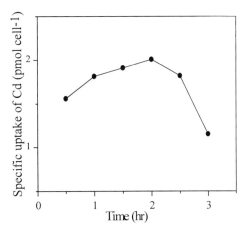

Figure 4. Cd uptake by the encapsulated NCP (●) and TB1 (O) (values are means of duplicates).

Figure 5. Specific Cd uptake by the encapsulated NCP.

Encapsulated cells grown in LB broth captured approximately 1.8 pmole Cd per cell at 1-hour incubation, at which one free cell sequestered approximately 1pmole of Cd dissolved in saline (8). This noticeable increase may be, in part, due to the presence of the supporting matrix of κ-carrageenan, which protects cells from the toxic metals and further secures the selective binding of metals on the cell surface. It is also possible that the immobilization of recombinant cells has improved plasmid stability, resulting in an elevated level of gene expression to accommodate the influx of heavy metals (15).

The development of biosorbents from microbial cells as an approach for heavy metal removal is a promising area of research. The ultimate utility of such a system depends on binding capacity, affinity, stability, and ease of production. Kinetic studies on Cd uptake are currently in progress to identify kinetic constants and specific uptake rate involved in metal removal. Although the current work utilizes recombinant *E. coli*, the expression of metal binding peptides in other environmental organisms needs to be studied for further development. Efforts are also being made to scale up the metal sorption experiments using a continuous bioreactor system (BioFloIII, New Brunswick).

REFERENCES

1. G. M. Gadd and C. White, Trends in Biotechnol., 11 (1993) 353.
2. P.L. Mattison, in P.L. Mattison (ed.), Bioremediation of Metals, Cognis, Santa Rosa, CO, 1992.
3. J.A. Brierley and D.B. Vance, in D.P. Kelly and P.R. Norris (eds.), Biohydrometallurgy 87, Sci. Technol. Lett., Kew Surrey, UK, 1987.
4. T.J. Beveridge, Ann. Rev. Microbiol., 43 (1989) 147.
5. S.A. Churchill, J.V. Walters, and P.F. Churchill, J. Environ. Eng., 121 (1995) 706.
6. R.M. Sterritt and J.N. Lester, in Eccles, H. and Hunt, S. (eds.), Immobilization of Ions by Bio-Sorption, Ellis Harwood, UK, 1986.
7. W.A. Corpe, Dev. Ind. Microbiol., 16 (1975) 249.
8. M. Pazirandeh, L.A. Chrisey, J.M. Mauro, J.R. Campbell, and B.P. Gaber, Appl. Microbiol. Biotechnol., 43 (1995) 1112.
9. J. Brower, R.C. Ryan, and M. Pazirandeh, Environ. Sci. Technol., 31 (1997) 2910.
10. S.S. Bang, and M. Pazirandeh, J. Microencap., in press.
11. J. Anderson, W. Campbell, M.M. Harding, D.A. Rees, and J.W.B. Samuel, J. Mol. Biol., 45 (1969) 85.
12. B. Mattiasson, in B. Mattiasson (ed.), Immobilized Cells and Organelles, CRC Press, Boca Raton, FL, 1983.
13. J.L. Ingraham, O. Maaloe, and F.C. Neidhardt (eds.), Growth of the Bacterial Cell, Sinauer Associates, Inc. Sunderland, MA, 1983.
14. S-H. Moon and S.J. Parulekar, Biotechnol. Prog, 7 (1991) 516.
15. J.C. Castet, M. Craynest, J-N. Barbotin, and N. Truffaut, FEMS Microbiol. Rev., 12 (1994) 63.

Potentiometric Titration: a Dynamic Method to study the Metal Binding-Mechanism of Microbial Biomass

G. Naja[a], S. Deneux-Mustin[b], C. Mustin[a], J. Rouiller[a], C. Munier-Lamy[a] and J. Berthelin[a]

[a]Centre de Pédologie Biologique. UPR 6831 du CNRS associée à l'Université de Nancy Henri Poincaré, 17 rue Notre Dame des Pauvres, BP n°5, 54501 Vandœuvre-lès-Nancy Cedex, France.

[b]Laboratoire Environnement et Minéralurgie. URA 235 du CNRS. Ecole Nationale Supérieure de Géologie-INPL, rue du doyen Marcel Roubault, BP n°40, 54501 Vandœuvre-lès-Nancy Cedex, France.

Sorption of metals is not only dependent on solution conditions but also on the particle size of the biomass. Acidities of different size classes of mycelium of *Rhizopus arrhizus* were determined by potentiometric titration. It appears that the [50-200]µm size class has the largest capacity of fixation, 2 or 4 times greater than the other size classes. Despite these differences, the pKa of the different acidities, obtained using Henderson-Hasselbach's equation, are quite similar for all of them.
Metal binding experiments are then performed, before the equilibrium, using an original method of potentiometric titrations with an ion selective electrode. This technique provides informations about the differences between the binding sites of the size classes during lead sorption.
Moreover binding sites and mainly carboxylic groups are identified by Infrared Spectroscopy (FTIR).

1. INTRODUCTION

Sorption is one of the important chemical processes in soil. It affects greatly the fate and mobility of contaminants in soils and waters. Biosorption occurs mainly through interactions of the metals ions with functional groups contained in or on the cell-wall biopolymers of either living or dead organisms. Fungal mycelia were generally recognised as efficient biosorbents of heavy metals [1-13]. The use of fungal biomass to extract heavy metals from effluents is an area of intensive research and development activity [14]. However, chemical mechanisms responsible for the biosorption of metals ions by non-living biomass are open to debate. It has been suggested [15-16] that specific functional groups were involved in the binding processes of heavy metals. Fungal cell-walls contain polysaccharides, proteins having amino, carboxyl, phosphate and sulphate groups. During the reaction, cations compete with protons for the fungal exchange and complexing cell-wall sites.

The physico-chemical properties of these organic constituents (diversity of the functional groups, heterogeneity of their repartition, polyelectrolytic nature) have suggested that the potentiometric titration method, previously used to study the formation of soluble organo-metallic complexes [17-18], can help to estimate the metal binding properties of dead microorganisms and in particular of cell-wall constituents.

This paper aims to quantify and characterise by chemical and physico-chemical means, the functional groups of the biomass in lead biosorption. Lead (II) ions found in waste water are typical of heavy metal environment pollutants derived from industrial processes... In this paper, *Rhizopus arrhizus* have been examined by reacting them with Pb(II).

2. MATERIALS AND METHODS

The biomass used in this work was the fungus *Rhizopus arrhizus* collected from pharmaceutic industries or cultivated in the laboratory. The industrial biomass contains an important quantity of diatomite (a filtration adjuvant). The cultivated biomass (BC) (DSM 905) was collected after 5-6 days of growth, washed twice with distilled water and dried at 40°C. Before use for titration or biosorption studies, these biomasses were ground, washed, lyophilised and stored under vacuum conditions.

2.1. Particle size distribution and carbon analysis

Granulometric fractions were isolated by successive and wet sifting: 2g of the global fraction (FG) of the biomass was added to 200ml distilled water. Different size classes were obtained: the [2000-200]µm (F1), [200-50]µm (F2) and [50-0]µm fractions.

The [50-0]µm fraction was dispersed, during 12 hours, with 200 ml distilled water and 100ml of Na-Amberlite IR resin. After sieving at 200 µm (to remove the resin) and at 50 µm, the suspension was transferred into 1l beaker and made up to a 1l volume with distilled water. Particle size distribution was achieved using the pipette method according to the Stockes law. The [50-20]µm (F3), [20-2]µm (F4) and [2-0]µm (F5) fractions were then obtained.

After lyophilisation, hydrogen, total carbon and nitrogen were determined using a CHN analyser (1106-Carlo Erba). The organic carbon was quantified by combustion in a Carhmograph (12-Wöstoff) Analyser.

2.2. Potentiometric titrations

To identify the different acid functional groups, an acid-base titration have been carried out at 25°C in an enclosed, thermostated, stirred glass cell under nitrogen atmosphere. 100mg of sample were suspended in 50 ml of $NaClO_4$ (μ =0.1M). Titration was performed, using an automatic titrator, from an initial pH=2.3 to a final pH=10.5 using NaOH (0.025 N).

The titration curves were achieved using a computer connected to the titrator. Calculation of the different acidities and evaluation of the functional groups was based on the Gran's method [19].

In complement, to investigate the binding process of metallic cations and to point out the competition between protons and cations, an original method of potentiometric titrations with a Pb^{2+} selective electrode was used.

The system involves:
- the previous stirred glass reactor containing 100 mg of biomass in 50 ml of $NaClO_4$ (0.1M),
- a 2-way ionometer (pH and selective electrode potential),
- a pH electrode combined with a reference silver electrode,
- an ion-selective electrode (Pb^{++}) with a solid-state membrane,
- an automatic burette for the addition of micro-volume (down to 1μl) of the metallic solution ($Pb(NO_3)_2$, $10^{-4}M$),
- a computer to control the frequency and the rate of metallic solution injections, to adjust the value of pH and to acquire automatically data.

2.3. Determination of pK_a and pK_m

The different acidities of the biomass were obtained by the combination of the derived function (dpH/dv) of titration curves and the Gran's function G:

$$G=(v_i+v)10^{-pH} \text{ (for pH<7) and } G=(v_i+v)10^{-pOH} \text{ (for pH>7)} \tag{1}$$

where v_i is the initial volume and v is the added volume of NaOH.

VaN and Ve, corresponding to the equivalence volume of, respectively, the control and the sample experiments, were determined from the first derived function. Va and Vb were determined by the acidic and basic slopes of the Gran's function (Figure 1).

The application of the Gran's function is very useful for the validation of the electrode (linearity in the pH range) and the partition of the total acidity of samples in three types: (i) the strong acidities calculated from the difference between the NaOH added volume [VaN-Va], (ii) the weak acidities evaluated from the volume [Ve-Va] and (iii) the very weak acidities calculated from the volume [Vb-Ve].

The value of the dissociation coefficient (α_a) and the acidity constant K_a depending on pH are obtained using the following equation:

$$K_a=[H^+] \cdot \alpha_a/(1-\alpha_a) \tag{2}$$

The pK_m value, defined as the apparent constant corresponding to the dissociation of 50% of the acid groups for each type of acidity, are derived from the value of pK_a when $\alpha_a=0.5$. The Henderson-Hasselbach's equation (3) [20] allows to relate the parameters of the dissociation sites and the pKm value to the pH and to determine a pK_m constant (a mean of the acidity constant):

$$pH=pK_m + nLog (\alpha_a/(1-\alpha_a)) \tag{3}$$

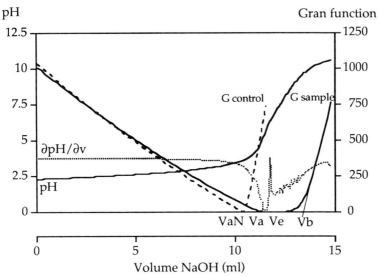

Figure 1. Gran's function, first derived and titration curves of the global fraction (FG).

2.4. Infrared Spectroscopy (FTIR)

To complete the characterisation of the functional groups, an IR investigation was performed with a Fourier Transform Infrared (FTIR) spectrometer (Brüker Vector 22). Each 1mg of sample was mixed up with 200mg of KBr. Spectra are reported in absorbance units and the influence of atmospheric water and CO_2 were always subtracted.

3. RESULTS

3.1. Particle size distribution and carbon analysis

The biomass (FG) and its different fractions F1, F2, F3, F4 and F5 were characterised. The nitrogen, hydrogen, organic and mineral carbon contents were compared in Table 1.

The granulometric "breakdown" have shown that the F3, F4 and F5 fractions represent 79%w/w of the total biomass and that diatomites are mostly involved in the F3 and F4 fractions.

The C/N ratio decreased with the particles size and indicated an enrichment in peptide function of the small particles. For the F5 size class, which contained the majority broken cell walls, this ratio is equal to 8.12.

The F1 and F4 fractions showed exactly opposite characteristics: F1 have a high %Corg but a weak % in mass whereas it is the contrary for F4.

However, F2 is the most organic fraction (46% of the total organic carbon of the biomass).

Table 1
Particle size distribution. Organic carbon and nitrogen analysis.

Granulometric classes in μm	% of biomass	Corg %	N %	C/N	% of total N	% of total C org
Total Biomasse (FG)	100	15.34	1.89	8.11	100	100
2000-200 (F1)	6.8	43.76	4.14	10.57	15.77	20.52
200-50 (F2)	20.9	31.86	3.91	8.14	45.78	45.92
50-20 (F3)	1.2	4.31	0.74	5.82	0.50	0.36
20-2 (F4)	52.5	3.16	0.55	5.74	16.18	11.44
< 2 (F5)	18.6	16.98	2.09	8.12	21.78	21.78

3.2. Potentiometric titrations

The potentiometric titration permits the quantification of the acid functions (carboxylic COOH, amine NH_3^+...). Titration curves of the different fractions of the biomass are presented in figure 2.

Two types of curves can be distinguished: the curves of the F2 and BC fractions and the others. Every change of slope implicates changes in the nature of sites.

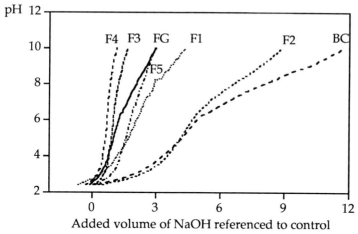

Figure 2. Basic potentiometric titration of the industrial biomass (FG) and its fractions. Titration of the cultivated biomass (BC).

Moreover, Gran's plots and derived curves revealed the existence of 3 types of acidity in biomasses (Table 2):

- strong acidities can be attributed to the presence of mineral (phosphoric) or organic (carboxylic) acidic groups (pH< 4.5),
- weak acidities corresponding to carboxylic groups (4.5<pH<7),
- very weak acidities corresponding to phenolic and amid groups of fungi cell walls (pH >8).

The determination of the different acidities, their pK_a and pK_m (Table 2) are carried out using the derived curves $\partial pH/\partial v$, the extrapolation of the Gran's function and the Henderson-Hasselbach representation.
It appears that the [200-50]µm size class has the largest capacity of fixation, 2 or 4 times greater than the other fractions. The quantification of acidities showed that the F2 fraction is close to the pure cultivated biomass. Moreover, the presence of different sites corresponding to the weak and very weak acidities is observed.
Despite these differences, the pKa of the different acidities obtained using Henderson-Hasselbach's equation, are quite similar for all the size classes.

Table 2
Calculation of the different type of acidity of biomass fractions.

Fractions in µm	Total organic acidity		Strong acidity	Weak acidity		Very weak acidity	
	me/g	pKm	me/g	me/g	pKa	me/g	pKa
Total biomass (FG)	0.823	6.89	0.153	0.13	4.28	0.54	8.2
2000-200 (F1)	1.25	6.99	0.23	0.49	5.21	0.53	9.1
200-50 (F2)	2.41	5.99	0.76	0.61	4.49	1.04	8.15
50-20 (F3)	0.49	5.50	0.2	0.12	5.28	0.18	9.81
20-2 (F4)	0.32	4.80	0.12	0.05	4.71	0.14	9.19
2-0 (F5)	0.78	4.78	0.31	0.12	4.53	0.35	8.22
Pure biomass (BC)	3.14	6.8	0.67	0.48	3.91	1.99	7.86

3.3. Sorption experiences and discussion

As the totality of the metallic cation are sorbed during the first hour of the experiment, reactions of biosorption of a metallic cation on biomass have been performed before the equilibrium.
We note that:
- the titration of the global fraction is characterised by a progressive decrease of pH and the apparition of several steps (Figure 3a). The change of the slopes may indicate the presence of several sites of sorption. The fixation of lead at pH(4, 5, 6) on this fraction is done with a small quantity of released protons: only 54% of the reactive sites of the total area are involved in sorption.
- the sorption on F1 is characterised by a rapid decrease of pH (Figure 3b) with one change of the slope. This size class is identified by a great amount of organic carbon but a small ratio of sorption (16% of the reactive sites are involved).
- for the other size classes (< 200 µm), an increase of pH (+0.2) during the sorption of lead is observed (Figure 3c). This phenomenon could be interpreted by the existence of a multitude of free sites in the matrix of the biomass, which permit the sorption of protons and metallic cations. Consequently, the competition between the two cations can be showed: the proton small and charged and lead heavy and doubly charged (Figure 4).

3.3.1. pH influence

The common phenomenon for all the size classes is the dependence of the sorbed quantity of lead with the pH of solution. In fact, the quantity of released protons during biosorption is proportional to the quantity of sorbed metal (Figure 4).

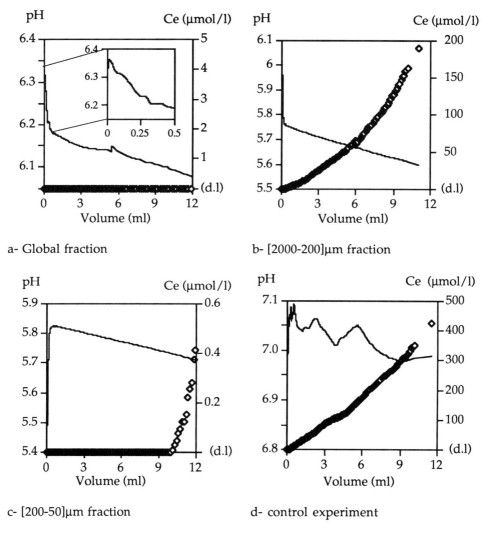

a- Global fraction

b- [2000-200]μm fraction

c- [200-50]μm fraction

d- control experiment

Figure 3. pH (line) and concentrations of lead (diamond) during the process of biosorption versus the added volume of the metallic solution (at pH=5).

When the concentration of metal increases, the reaction of sorption is favoured and the concentration of protons in solution increases (4). Moreover, the dissociation of the carboxylic groups release their protons which improved the uptake of cations.

$$R\text{-}H + M^+ <=> R\text{-}M + H^+ \tag{4}$$

where M^+ represents a cation of the solution and R-H a functional group of the biomass.

208

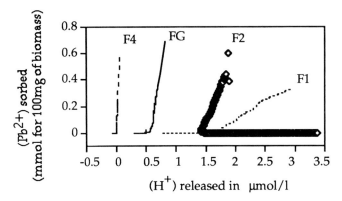

Figure 4. Variation of the sorbed quantity of lead versus the quantity of released protons at pH=5.

3.3.2. Calculation of the affinity and identification of the nature of the sites

Values of the affinity b (Table 3) at two different pH have been estimated from the curves of figure 5.

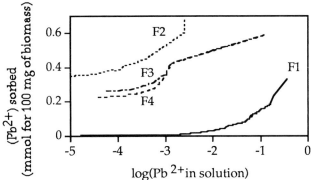

Figure 5. Variation of sorbed quantity versus the logarithm of the concentration of lead in solution.

The values indicate that F2 and F5 classes have an important affinity for lead. In fact the affinity of the < 2μm size class is due to the high specific area.

At pH=5, the competition H^+/Pb^{2+} is too weak. Thus, the sorbed quantity is important and the affinity value is high. Whereas, at pH=4, the interaction of the organic matter and H^+ is important which reduced both affinity and sorbed quantity.

By using the potentiometric titration, the nature of the sites have been identified: (i) insertion sites in the biomass matrix or at the mineral surface [SiO]. These sites take part during the sorption of lead and are characterised by a slight variation of pH and (ii) exchange sites (carboxylic, phosphates..) where the sorption produces a release of cations (H^+, Na^+...) and a decrease in pH.

Table 3
Calculation of the value of b and of the sorbed quantity of lead. Influence of pH.

	pH=4		pH=5	
	b	Q_{sorbed} mg/g	b	Q_{sorbed} mg/g
Global fraction	1.6	3.186	26.5	3.345
[200-50μm] (F2)	5.9	2.535	269.5	3.369
< 2μm (F5)	27.1	2.798	∞	3.385

$b=K_a/K_d$ is the affinity constant, where K_a and K_d are the adsorption and desorption constants.

The precipitation of lead onto the surface of biomass must be taken into account. It is probable that, after the formation of the first lay of metal sorbed, a bond Pb-Pb is formed. The sudden change of the slope at the end of titration illustrates this hypothesis (Figure 5).

3.3.3. Determination of lead binding sites

The IR spectra of *Rhizopus arrhizus* before and after biosorption are presented in figure 6. The IR spectrum shows the characteristic absorption bands of OH, NH at 3600 and 3000 cm^{-1}, of alkyl chains (3050 to 2800 cm^{-1}), of C=O at 1658 cm^{-1} and of C-N and NH (1658-1950 cm^{-1}) for amid bonds. After sorption, a new band appeared at 1516 cm^{-1}, frequency of C=O associated, corresponding to the carboxylic groups bonded to lead. Other bands disappeared (1710 cm^{-1}) corresponding to the C=O free acidic groups and others are slightly shifted (1073 cm^{-1}).

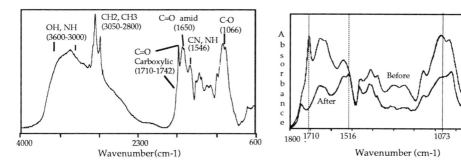

Figure 6. Infrared spectra of the biomass *Rhizopus arrhizus* before and after lead biosorption.

4. CONCLUSION

With assumption that proton binding sites on biomasses are available to metal ions, the metal binding ability and the maximum adsorption capacity of fungal mycelia and other biosorbents can be measured efficiently by the potentiometric titration. Moreover, this method allows to determine the different types of acidity of functional groups involved as metals binding sites. The potentiometric titration can be used to compare and to select biosorbents or to determine the best conditions for the bioremediation of metal contaminated effluents or to study the fixation and accumulation of metals in disturbed or natural environments such as soils and sediments.

Finally, infrared spectroscopy analysis of the loaded biomass shows that carboxylic sites are majority involved in the binding of lead.

REFERENCES

1. J. Berthelin, Microbial Weathering Processes. In W.E. (ed) Microbial Geochemistry. Blackwell Scientific Publ., Oxford, U.K (1988) 223.
2. M. Tsezos, Biotechnol. Bioeng., 25 (1983) 2025.
3. S.M. Siegel, B. Margalith Galun and B.Z. Siegel, Water, Air and soil Pollution, 53 (1990) 335.
4. N. Kuyucak, B. Volesky, Biotechnol. Lett., 10 (1988) 137.
5. E. Luef, P. Theodor, P. Christian and Kubicek, Appl. Microbiol. Biotechnol., 34 (1991) 688.
6. L.E. Macaskie and A.C.R. Dean, Env. Technol. Lett., 3 (1982) 49.
7. C. Munier-Lamy, P. Adrian, J. Berthelin and J. Rouiller, Org. Geochem., 9 (1986) 285.
8. E. Guibal and C. Roulph, Journal Francais d'Hydrologie, 21 (1990) 229.
9. J. Berthelin and C. Munier-Lamy, Environ. Biogeochem., R. Hallberg (ed), Ecol Bull., Stockolm., 35 (1983) 395.
10. C. Munier-Lamy, P. Adrian and J. Berthelin, Toxicol. Environ. Chem., 31-32 (1991) 527.
11. Y. Bizri, M. Cromer, J.P. Scharff, B. Guillet and J. Rouiller, Geochim. Cosmochim. Acta., 48 (1984) 227.
12. M. Aplincourt, A. Bee-Debras and J.C. Prudhomme, Science du Sol, 26, 3 (1988) 157.
13. S. Bartnicki-Garcia, In Phytochemical Phylogeny, J. B. Harborn (ed)., Academic Press, New York, 1970.
14. B.N. Noller, R.A. Watters and P.H. Woods, J. Geochem. Explor., 58 (1997) 37.
15. J.M. Tobin, D.G. Copper and R.J. Neufeld, Appl. Env. Microbiol., 47 (1984) 821.
16. B. Volesky, Biosorption of Heavy Metals, Mc Gill University (ed), Montreal Canada, 1989.
17. G. Brunelot, P. Adrian, J. Rouiller, B. Guillet and F. Andreux, Chemosphere. 19 (1989) 1413.
18. D. Fremstad, 4th Symposium on Ion Selective Electrodes, (1984) 383.
19. G. Gran, Part II, Analyst., 77 (1952) 661.
20. H. Rossotti, In the study of ionic equilibria. An introduction, Longman (ed), London, 1978.

A bioelectrochemical process for copper ion removal using *Thiobacillus ferrooxidans*

A. Boyer, J.-P. Magnin, P. Ozil

Laboratoire d'Electrochimie et de Physico-chimie des Matériaux et des Interfaces, UMR 5631-CNRS-INPG-UJF, BP 75, 38 402 Saint Martin d'Hères cedex, France

A new process for copper ion removal was studied, coupling electrochemistry and biology. An electrochemical reactor using a packed bed cathode of $10 \times 2.5 \times 30$ cm of activated carbon was selected and built. The electrochemical deposition of copper was studied in this reactor in the absence and in the presence of biomass adsorbed on the cathode. A pre-concentration phase of copper ion on the surface of the activated carbon was shown to increase of 30 % the deposition efficiency. In the presence of biomass this efficiency was increased of 20 % more. An elimination of 54% was achieved in this latter case.

1. INTRODUCTION

The treatment of effluents containing heavy metal ions and the recovery of metal ions from mine leakage are similar processes only differing by their aim : metal elimination in the first case and recuperation in the second. Nowadays, treating effluents loaded with heavy metal ions offers the double interest of preserving environment and recycling metals, getting closer to the aim of mine leakage treatment.

Precipitation is the most classical and cheapest method to remove metal ions from wastes. However at low concentrations, such a metal recovering is still inefficient and disallows a further reuse. On the other hand, ion exchange or membrane processes are efficient for metal concentration from low concentration solutions and allows metal reuse. Using both of these processes leads to a ionic form of the metal. Then, obtaining a metallic form requires a further transformation which can be achieved by an electrochemical process. Electrolysis is an efficient way for high metal ion concentrations and thus can be directly applied to mine waters.

In the case of effluents or poor ores, a pre-concentration step is necessary because no efficient recovery process exists at low concentrations. Such processes either lead to final metal ion concentrations higher than the required standards (case of wastes) or imply a great energy consumption.

The aim of the present work was to develop a new energy-saving process able to treat copper ion solutions at low concentrations. This process couples electrochemistry and biology and its design exploits two basic assumptions to be verified : the electrochemical deposition of a metal is enhanced by a previous adsorption of metal ions on the electrode surface, the presence of a bacterial biofilm on surface increases the metal ion adsorption on this surface.

The process involves two steps : firstly a preliminary adsorption of copper ions on the electrode surface modified by the biofilm, secondly an electrochemical deposition.
The first step implies to find a biomass able to improve the fixation of copper ions on the activated carbon. Such an enhancement has been already observed for cadmium with a biomass of *Enterobacter aerogenes* [1]. In our case, *Thiobacillus ferrooxidans* was preferred since some previous results showed a great capacity of this bacterium to fix metal ions [2, 3, 4] : specific uptake capacities can reach optimised values as high as 0.75 g Cu^{2+} / g dry weight, 1.5 g Cd^{2+} / g dry weight and 1.8 g Cr^{6+} / g dry weight. Other studies report that the same bacterium from another strain is able to form a biofilm on activated carbon within 2 hours [5]. Consequently, *T. ferrooxidans* was chosen and tested for the bioelectrochemical process under study.

On the other hand, the electrode material must firstly have surface properties allowing biofilm formation (no toxicity, suitable roughness and surface tension…). Moreover it has to be appropriate to electrochemical deposition at low concentrations (high porosity and conductivity, large contact surface…). These considerations lead to select activated carbon as the material for the cathode. A packed bed electrode was preferred to a plane electrode because of its higher efficiency to treat dilute solutions due to a large contact and reactive surface area [6].

The next sections are dedicated to verify the bases of the bioelectrochemical process and to detail it.

2. MATERIALS AND METHODS

2.1. *Thiobacillus ferrooxidans*

The culture of the bacterium *T. ferrooxidans*, strain DSM 583, in 9K medium [7] was conducted in batch reactors. Pre-cultures were realised in 100 mL Erlenmeyer, under 30°C and rotating agitation. 20 mL were inoculated with 7 % v/v and cultures were performed either in a 1 L Erlenmeyer for small productions (500 mL) or in a 50 L reactor, at 30°C under air flow stirring, for larger productions (40L) [8]. The bacterial growth on ferrous sulphate was monitored by measuring the Fe^{2+} consumption using the ortho-phenantroline colorimetric method [9]. The protein concentration, which is directly correlated to the substrate consumption, was measured by the Lowry colorimetric method [10].

2.2. Activated carbon

Two types of activated carbon were used. The first one, used for the rotating plate electrode, was a vegetal carbon supplied under pellet form by Prolabo (diameter 3mm , length 5 mm, specific surface 1733 m^2 m^{-3}, porosity 0.37). The second one, used for biofilm formation and in the reactor, was the Picachem 150 activated carbon supplied by Pica industries under platelet form (average size 1.5 mm, specific surface area 8380 m^2 / m^3, porosity 0.9). The resulting packed bed offers a 0.7 porosity and a 2540 m^2 / m^3 specific area.

2.3. Preliminary copper electrochemical deposition

Experiments were performed on a rotating plate electrode built with 7 pellets of activated carbon and offering a $49.5 \ 10^{-6} \ m^2$ surface area. 5 mL of a copper solution were used. Copper ion concentration was chosen at 10^{-4} M, $CuSO_4$ dissolved in sodium sulphate (pH 2). The electrochemical deposition potential was determined from cyclic voltammetry measurements as - 600 mV /$HgSO_4$. The current intensity influence was studied : the optimum current was found to be –2 mA [8].

The reduction runs were conducted under solution stirring, with and without a preliminary 30 minute adsorption of copper ions on the electrode surface. Each run was replicated at least 3 times to check the experimental reproducibility. The copper ion concentration was measured by using polarography as a function of time in the solution from 5µL samples diluted to 5mL.

2.4. Copper adsorption on activated carbon

The adsorption of copper on activated carbon was studied for both modified and unmodified activated carbon. In the first case, the biofilm formation is done before the adsorption of copper itself.

Biofilm formation

For biofilm formation, biomass was harvested at the end of the exponential growth phase and re-suspended in sulphuric acid at pH 1.4. Seven grams of desaerated activated carbon were mixed with 50 ml of the bacterial solution (50 mg / L). The stirring speed was fixed at 100 rpm and temperature at 30 °C. The protein concentration in the solution was measured as a function of time by regular sampling to determine the fixed protein amount.

Filtration through a nylon grid separated the activated carbon from the bacterial solution. The carbon is then rinsed with sulphuric acid (pH 1.4).

Copper ion adsorption

Seven grams of activated carbon (modified or not) were placed in a 250 ml-flask. 100 ml of copper sulphate solution were added (10^{-2} M, pH 1.4, 100 rpm). The concentration of soluble copper was determined on samples by using atomic adsorption spectrophotometry and polarography. These measurements allowed to determine both the kinetics and the efficiency of adsorption.

2.5. Bioelectrochemical process

The reactor (height 0.3 m, width 0.1 m and thickness 0.025 m) was built in Plexiglas with two stainless steel (316L) electrodes (Figure 1). Filling the reactor with the carbon was achieved by separating the anodic and cathodic compartments, putting 280 g of activated carbon, and then building again the reactor, like a press-filter.

The copper ion solution (flow rate 3.5 L h^{-1}, copper concentration 10^{-2} M, pH 2) was re-circulated through the bed using a peristaltic pump, from the bottom to the top (Figure 2).
The packed bed cathode worked in a radial configuration (orthogonal flow and electric field) to enhance the efficiency at low concentrations. The anodic compartment was filled with small glass Raschig cylinders.

214

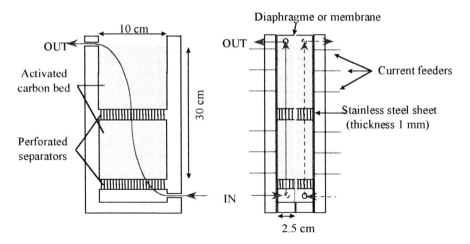

Figure 1. Electrochemical reactor : face and side view.

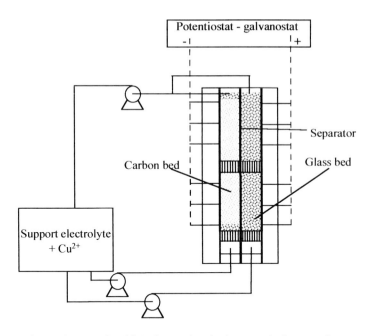

Figure 2. Experimental set up for (bio)-electrochemical removal of copper ions.

The process involved two successive steps :
- a 50-minute adsorption of copper ions on the electrode surface, modified or not by the biofilm, under recirculation and without current application,
- the electrochemical deposition under a current intensity fixed to - 0.7 A.

During both the periods, 0.5 mL samples were regularly taken at the inlet and outlet of the reactor so that the kinetic and the efficiency of the process could be determined. The copper ion concentration was determined by absorption atomic spectrophotometry after dilution to 4.5 mL in ultrapure water.

3. RESULTS AND DISCUSSION

3.1. Enhanced copper deposition by adsorption

Figure 3 shows the evolution of the copper ion concentration in solution during the electrolysis. The kinetics without any pre-concentration phase were determined from the average results of 6 runs (with the confidence interval at 95 %). For the other case, only three experiments were performed (error bars).

In the absence of the adsorption phase, ninety percents of the initial copper were eliminated within about 100 minutes. This duration was decreased down to 70 minutes after a 30-minute pre-concentration step. Thus the same elimination ratio can be obtained with a saving of 30 minutes of current supply or a copper ion concentration lower than $4 \ 10^{-5}$ M can be reached with the same duration of current application.

This improvement of copper deposition kinetics is due to the impact of adsorbed ions on the germination step of metal formation. Each adsorbed ion is instantly reduced and leads to a germ for the other ions in solution.

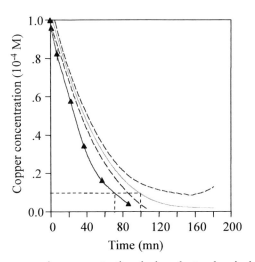

Figure 3. Evolution of the copper ion concentration during electrochemical reduction with (▲ and error bars) or without(—— and ---- confidence interval at 95 %) a pre-concentration phase (30 minutes).

3.2. Enhanced copper adsorption thanks to the biofilm

The formation of the *T. ferrooxidans* biofilm on activated carbon is the first step of copper adsorption. It appeared to be rather fast: 90 % of bacteria were fixed within 5 minutes, while

30 minutes were necessary to reach equilibrium [11]. For an initial protein concentration of 50 mg / L, a biofilm of 0.286 mg protein per gram of activated carbon was obtained, and for 350 mg / L of proteins a biofilm of 1.4 mg / g of carbon. The second step consists in copper adsorption itself. Experiments showed a two-phase kinetics of copper ion fixation on activated carbon modified or not by the bacteria (Figure 4). 60 % of the copper were fixed during a first fast phase within 15 minutes while the equilibrium was only reached after 2 hours of contact. These two phases have be explained by the double scale of porosity of the activated carbon, and by the mechanism of biosorption by *T. ferrooxidans* [12].

Moreover, these experiments highlighted that copper ion adsorption on activated carbon is effectively enhanced by the presence of the bacterial biofilm as initially supposed as a basic assumption for the process: 3.81 mg of copper were adsorbed on pure carbon and 4.77 mg on modified carbon by 0.286 mg protein / g C. This 25 % increase of the adsorption capacity lead to an elimination of 52 % of the initial copper in solution. However a thin biofilm appeared to be more efficient to adsorb copper than a thick one (4.77 instead of 4.12 mg Cu).

The improvement of copper adsorption could result from the biofilm formation which implies the creation of adhesive exopolymers by the bacterium, acting as additional adsorption sites for copper ions. Moreover, bacteria have been shown to be able to bioprecipitate copper ions from aqueous solutions [8].

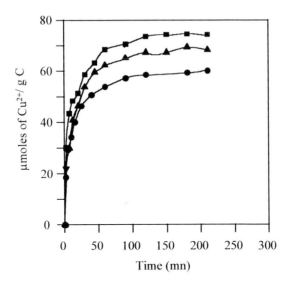

Figure 4. Kinetic experiments of copper adsorption on activated carbon : 0 (●), 0.286 (■), or 1.4 (▲) mg protein / g carbon.

3.3. Bioelectrochemical process

The first experiment was conducted with pure activated carbon and without a pre-concentration step for determining the kinetics of reduction. A rapid decrease in copper ion concentration was observed, corresponding to the first liquid circulation in the packed bed, and then a slower reduction phase occurred. The concentration decreased from its initial value 10^{-2} M to $0.68\ 10^{-2}$ M after 50 minutes and $0.6\ 10^{-2}$ M after 180 minutes, with a current efficiency of 5 % in the reactor.

The second experiment was performed by starting with a 50 minute pre-concentration phase and using an electrode without biofilm coating (Figure 5). The copper ion concentration reached $0.74\ 10^{-2}$ M after 50 minutes of adsorption. A reduction period of 180 minutes lead to a copper ion concentration of $0.55\ 10^{-2}$ M and a current efficiency of 6 %. Only 125 minutes were necessary to obtain a final concentration equivalent to the first case i.e. $0.60\ 10^{-2}$ M.

The third experiment was done with a biofilm and a pre-concentration phase (Figure 5). The end of the adsorption phase (50 minutes) was characterised by a copper ion concentration of $0.68\ 10^{-2}$ M. After a 180 minute electrolysis, the concentration decreased to $0.46\ 10^{-2}$ M. Only 75 minutes were necessary for a 40 % decrease in copper ion concentration.

Comparing these two last experiments shows that the presence of the biofilm increased in 33 % the adsorption capacity for copper ion in a packed bed configuration. The result obtained in a stirred tank reactor was a little lower with only a 25 % increase. Such a difference is due firstly to the biofilm thickness which was not exactly the same in the two reactors. Secondly, the shear-stress is less important in the packed bed reactor, so preserving the biofilm from being stressed and less efficient.

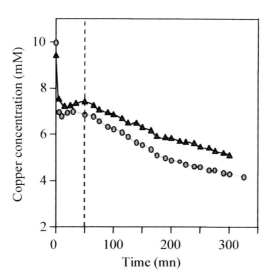

Figure 5. Evolution of copper ion concentration in solution as a function of time during the pre-concentration phase and the reduction phase, in presence of a biofilm (●) or not(▲).

Besides, the current efficiency in the first minutes of the reduction was close to 8 % without the biofilm, and about 11 % with the biofilm: the presence of bacteria on the electrode surface not only enhanced the copper adsorption but also the electrochemical reduction. Thus the bacterium fixed on the surface of the electrode did not behave as a barrier for the electric current as it could be expected, so reducing the efficiency. In fact, the process was more efficient in presence of the biofilm, after a pre-concentration step.

4. CONCLUSION

A new bioelectrochemical process was studied for copper ion removal. This process involves three steps : biofilm formation on the cathode material, copper ion adsorption on the electrode surface, electrochemical reduction of copper ions into its metallic form.

This process was based on two basic considerations which were checked and validated. Firstly, the presence of a biological biofilm of *T. ferrooxidans* enhances the adsorption capacity of copper ions on activated carbon. A 25 % increase was observed in a stirred-tank reactor and a 33 % increase in a packed bed reactor.

Secondly a pre-concentration phase of copper ions by adsorption on the electrode surface allowed either to reduce the time required for electrolysis (30 minutes with the rotating plate electrode) or to increase the elimination of ions.

The process tested in an electrochemical packed bed reactor confirms the advantages of using both a bacterial biofilm and a preliminary adsorption phase for copper ion recovery. In that latter case, the treated solution only contained $0.46 \ 10^{-2}$ M of copper after a 5 hour electrolysis, instead of $0.6 \ 10^{-2}$ M without biofilm and adsorption. Only 75 minutes were necessary for obtaining the same concentration ($0.6 \ 10^{-2}$ M) after a preliminary adsorption on the biofilm, so saving 105 minutes of current application.

REFERENCES

1. J.A. Scott and A.M. Karanjakar, Biotechnol. Lett., 14(8) (1992) 737.
2. A. Boyer, J.-P. Magnin, P. Ozil, Biotechnol. Lett., 22 (1998) 187.
3. F. Baillet , J.-P. Magnin, P. Ozil, Biotechnol. Lett., 20 (1997) 95.
4. F. Baillet , J.-P. Magnin, P. Ozil, Environ. Technol., 18(1997) 631.
5. A.S. Myerson and P. Kline, Biotechnol. Bioeng., 25 (1983) 1669.
6. A. Stork and F. Coeuret, Lavoisier (eds), Eléments de Génie électrochimique, Collection technique de documentation, Paris, 1984.
7. O.H. Tuovinen and D.P. Kelly, Arch. Microbiol., 88 (1973) 285.
8. A. Boyer, Etude d'une biomasse de *Thiobacillus ferrooxidans* et de sa contribution à la récupération électrochimique de cuivre en solution, thesis (1997).
9. M.K. Muir and T.N. Andersen, Metall. Trans. B., 88 (1977) 517.
10. O.H. Lowry, N.J. Rosebrough, A.L. Farr, R.J. Randall, Biol. Chem., 193 (1951) 267.
11. A. Boyer, J.-P. Magnin, P. Ozil, 4th International Symposium on Waste Management Problems in Agro-industries, Istanbul, Turkey, September 23-25 (1998).
12. A. Boyer, J.-P. Magnin, P. Ozil, International Seminar "Biosorption and Bioremediation II", Prague, Czech Republic, July 12-17 (1998).

Sorption sites in dried leaves

R.P. de Carvalho, K. J. Guedes and K. Krambrock

Departamento de Física, ICEx, Universidade Federal de Minas Gerais,
CP 702, 30123-970, Belo Horizonte, MG, Brazil

We report on the biosorption of copper ions in dried leaves from the Brazilian flora. By Electron Paramagnetic Resonance (EPR) it is shown that the copper ions are incorporated in the leaves in a strongly axial site. Washing with an acidic solution does not remove the copper from the leaves, but does modify the axial site symmetry. Infrared absorption seems to be compatible with a site which is located near to a simple or a double carbon - oxygen bond. Dried leaves are a good and commercially interesting alternative for sorption of metal ions compared to other biomasses.

1. INTRODUCTION

The biosorption (sorption of metallic ions from a solution) occurs in live or dried biomasses and offers an alternative technique for the depollution of industrial effluents as well as the recovering of the metals contained therein [1]. Although the phenomenon has been studied in various types of biomasses as algae, yeast or bacteria [2], which all showed good sorptive behaviour, very few studies were done in order to identify the sorption sites and the mechanisms involved in the process.

In this work we present the results of sorption of copper ions by dried leaves of the Brazilian flora studied by Atomic Absorption Spectroscopy (AAS), Electron Paramagnetic Resonance (EPR) and Fourier Transform Infrared Absorption (FTIR) analysis. The results show that dried leaves have sorption capacities comparable to other biomasses. The copper ions are incorporated in the dried leaves in a strongly axial site.

2. MATERIAL PREPARATION AND EXPERIMENTAL METHODS

Sorption and desorption of copper ions were studied using the following biomasses: tea leaves (*Thea sinensis*), dried leaves of the Brazilian flora (*Aspidosderma tomentosum, Qualea parviflora, Maitena truncata*) as well as fibre residuals of *Maitena truncata*. The leaves were washed with distilled water, dried at 30°C for 7 days and at 70°C for 24h, and ground.

Sorption samples were prepared with equilibrium experiments as described elsewhere [3] and the sorptive capacities were determined using the initial and final concentrations in the solutions [4]. Concentration determinations were done by AAS spectroscopy.

Cu-charged biomasses were analysed by EPR using a custom-build spectrometer for X band frequencies (9-11 GHz). The dried leaves were prepared as a very fine powder. The EPR spectra were analysed using simulation as described elsewhere [5]. The same biomasses were also investigated by FTIR absorption spectroscopy.

For the desorption studies, Cu-charged *M. truncata* biomass was washed in distilled water, with the pH adjusted to values varying between 3 and 9 with HCl and NH$_4$OH solutions. After 1h with agitation at room temperature, the biomass was recovered by filtration, dried and taken for EPR analysis. The Cu concentration of the final solution was determined by AAS in order to calculate the metal desorption.

3. RESULTS

Figure 1 shows the sorption isotherms for various biomasses compared with values from the literature [3, 6] (activated carbon has a reported sorption efficiency of 3% to 5%, corresponding to q = 30 to 50 mg metal / g biomass). All of our samples showed the same sorption capacity as previously studied biomasses and are comparable to activated carbon. Even if the studied biomasses have a slightly smaller efficiency, they can be considered as commercially interesting due to the low cost of preparation.

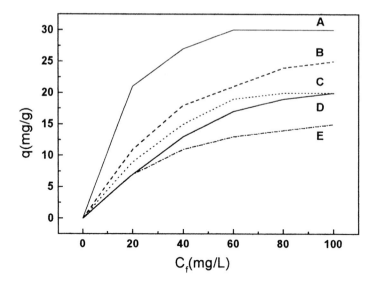

Figure 1: Sorption isotherms for different biomasses as determined from AAS spectroscopy from the initial and final concentrations of metal ions in the solutions: A - *A. nodosum* (from ref. [3]), B - *A. tomentosum*, C - *Q. parviflora*, D - *M. truncata* and E - *T. sinensis*.

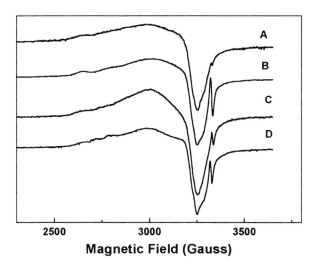

Figure 2: EPR spectra of different biomasses measured at room temperature with a microwave frequency of 9.12 GHz: A - *A. tomentosum*, B - *Q. parviflora*, C - *M. truncata* and D - *T. sinensis*.

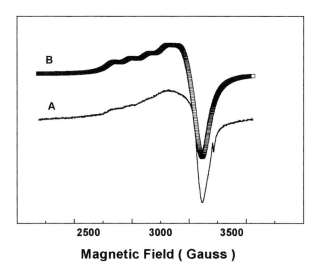

Figure 3: EPR spectrum for *M. truncata*: A - meaured and B - simulated with the parameters described in the text.

Figure 2 shows the EPR spectra for various Cu-charged biomasses measured at room temperature. The spectra were normalized in order to compare the shapes of the curves. All of our Cu-charged samples showed similar EPR spectra: the copper ion is seen in a site with axial symmetry.

Figure 3 represent a simulated EPR spectrum assuming that the Cu ion is located in a site of nearly axial symmetry. For the simulation we used the following Hamiltonian:

$$H = \beta S \underline{\underline{g}} B + S \underline{\underline{A}} I - \beta_n g_n S I \tag{1}$$

with S = 1/2, an axial g tensor with values $g_\parallel$= 2.00, $g_\perp$ = 2.28, an axial hyperfine tensor for the Cu with I = 3/2 (100 % abundance) and $A_\parallel$ = 140 G, $A_\perp$ = 10 G and a linewidth of 75 Gauss. Unfortunately, the linewidths of the EPR curves are large, however, the quartet hyperfine structure of the Cu - ions is resolved (I = 3/2).

FTIR absorption spectra before and after Cu-sorption are shown in figure 4 for *M. truncata*. The spectra were normalized to take account for the thickness differences in the prepared samples and for comparison and determination of changes in the absorption lines in the presence of the metal ions. It can be seen that the presence of copper modifies the absorption in the regions of 1300 cm^{-1} and 1600 cm^{-1}. Those absorption peaks can be attributed to vibrations of carbon-oxygen single bonds (C-O) and carbon-oxygen double bonds (C=O), respectively [7]. In the measured range from 500 cm^{-1} to 3000 cm^{-1} no other changes in the absorption lines have been observed.

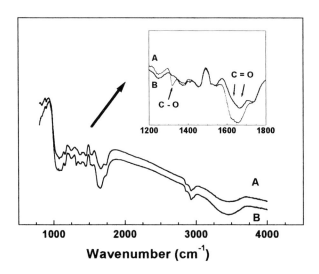

Figure 4: FTIR absorption spectra of *M. truncata* measured at room temperature: A - natural and B - Cu-charged.

Table 1: Recovering of Cu from M. truncata with different pH treatments. Data were obtained from AAS and EPR.

pH	AAS (%)	EPR (%)
2	70	55
3	35	20
4	13	15
5	9	10
9	<2	3
10	<2	<2
11	<2	<2

Sorption experiments analysed by EPR and FTIR were also done in fibre residuals of *M. truncata* and showed similar results with those found for the complete leaves.

Table 1 shows the results of Cu desorption experiments in *M. truncata* with pH varying from 3 to 9. Desorption rates were deduced from the Cu concentrations in the final solution. These values are compared with the size of the main EPR peak intensity for the spectra of samples treated with the various pH values. The correlation of the results demonstrates that the desorbed copper ions are the same that cause the EPR signal, that is, they are located in an axial symmetry. It is noted that the sorbed metal is partially recovered with an acidic treatment of the biomass. However, for an effective desorption, other treatments must be used [8].

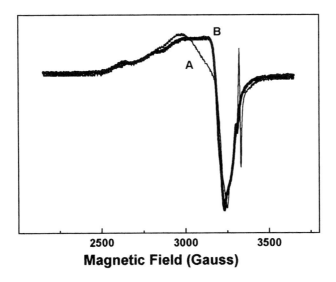

Figure 5: EPR spectra of Cu-charged *M. truncata* after desorption treatment with different pH: A - pH 3 and B - pH 9.

Figure 5 shows the EPR spectra done on the biomasses after desorption treatments. EPR spectra were normalized in order to compare the shapes of the curves. For basic treatment, although the desorption of the metal is not obtained, it is noted a change in the spectral shape. This indicates a change in the neighbourhood of copper ions inside the biomass.

4. DISCUSSION

Based on the results presented above we recommend the use of dried leaves as sorbents that can be commercially interesting due to the low aquisition and preparation cost of the biomasses, as compared, for example, with activated carbon.

The EPR results show that the Cu ions are incorporated in the dried leaves in a strongly axial site. The similarity between the spectra of the different biomasses and those obtained from fibre residuals from *M. truncata* indicate that the sorption sites and mechanisms must be related with components of fibres of the biomass (dextrines, mainly formed with C,O and H combinations). FTIR absorption experiments confirm this hipothesis, since they indicate a change in the C-O and C=O bonds due to the presence of the Cu ions.

Washing with acid solution results in a modification of the Cu neighbourhood as seen in the modified EPR spectra. However, the linewidths of the EPR spectra is large. Therefore, no superhyperfine interaction with the neighbors could be resolved. Electron Nuclear Double Resonance (ENDOR) experiments are in progress to investigate more directly the site and the neighbourhood of the Cu ions.

5. CONCLUSIONS

We conclude that dried leaves offer a good alternative to the other known biomasses, that the sites responsible for the sorption must be the same in all types of leaves, have an axial configuration and are located near to a single or double carbon-oxygen bond in the fibres of the biomass.

ACKNOWLEDGEMENTS

We wish to thank Drs. Roberto Moreira and Julio Cesar D. Lopes for the FTIR measurements, the Laboratório de Absorção Atômica of our Institute for the AAS determinations, the Laboratório de Fitoquímica and the Departamento de Botânica of our University for the provided samples, Manoel G. Lutkenhaus, Pollyana S. Moerira and Juliana R. Freitas for the sample preparation and Dr. Maurício V.B. Pinheiro for the discussion of the EPR results. This work was supported by the Brazilian financing agencies CNPq, CAPES, FINEP and FAPEMIG.

REFERENCES

[1]   B. Volesky, Biosorption and Biosorbents, in *Biosorption of Heavy Metals*, ed. B. Volesky, p. 3, Boca Raton, FL: CRC Press, Inc. (1990).
[2]   C.L. Brierley, Metal Immobilization Using Bacteria, in *Microbial Mineral Recovery*, eds. H.L. Ehrlich and C.L. Brierley, p. 303, New York: Mc Graw-Hill (1990).
[3]   R.P. de Carvalho, K.-H. Chong and B. Volesky, Biotech. Lett., 16, (1994) 875.
[4]   R.P. de Carvalho, K.-H. Chong and B. Volesky, Biotech. Prog., 11 (1995) 39.

[5]   see for example: J.E. Wertz and J.R. Bolton, Electron Spin Resonance, chap. 7, p. 154, Mc Graw-Hill (1972).

[6]   S.D. Faust and O.M. Aly, Adsorption Process for Water Treatment, p.1, Boston: Butterworths (1987).

[7]   R.M. Silverstein, Spectrometric Identification of Organic Compounds, New York: John Wiley and Sons, 5th ed. (1991).

[8]   B. Volesky, Removal and Recovery of Heavy Metals by Biosorption, in *Biosorption of Heavy Metals*, ed. B. Volesky, p. 8, Boca Raton, FL: CRC Press, Inc. (1990).

Evaluation of potential use of immobilized *Penicillium griseofulvum* in bioremoval of copper

M. P.Shah[a], S. B.Vora[b] and S. R. Dave[a]

[a]Department of Microbiology, School of Sciences,
Gujarat University, Ahmedabad 380 009, India.

[b]Gujarat Mineral Development Corporation, Ahmedabad 380 009, India.

Biomass of spent *P. griseofulvum* was immobilized with various immobilizing agents. Among the different immobilizing materials screened, entrapment of cells in a naturally occurring protein with glutaraldehyde showed most favourable results for copper recovery. Entrapped cells of *P. griseofulvum* were successfully used even after 8 cycles of copper sorption. Glutaraldehyde concentration of 2% in the system and contact time of 60 minutes functioned ideally for optimum copper sorption. Presence of 10 ml of natural protein as compared to 5 ml in the immobilized system increased the copper removal by 1.3 fold. Both, blank beads and immobilized biomass proved efficient to recover the copper from solutions having low copper concentration. When blank beads, immobilized biomass and free biomass were repeatedly charged with copper bearing solution, after 8 cycles, immobilized biomass showed nearly 71 fold higher copper removal as compared to blank beads and 1.5 fold higher than free biomass. Under the experimental conditions immobilized system showed as high as 20.47 mg.g^{-1} copper loading as opposed to 15.51 mg.g^{-1} by free biomass. Challenging the system with high copper concentration of 63.5 μg.ml^{-1} (total 50.80 mg) showed 19, 56 and 44 % copper removal with blank beads, immobilized biomass and free biomass respectively. Free biomass reached saturation point with 38.10 mg of copper charging. Blank beads showed diphasic mode after 4th cycle. In the initial 2 to 3 cycles, free biomass sorbed more copper as compared to immobilized biomass. After 3 cycles, immobilized biomass showed significant positive difference in copper sorption. In the 6th cycle immobilized biomass gave 13 times more copper removal as compared to free biomass. Depending on the cycles of copper charging, the percent copper removal ranged from 2.3 to 94.0. The immobilized *P.griseofulvum* cells performed quantitative copper removal both from pure copper solution as well as from industrial effluent.

Corresponding author : S. R. Dave

1. INTRODUCTION

Intensive research has been carried out worldwide for employing biosorption as a polishing technology to remove trace metals from industrial effluents. Both living and dead biomass are competent for removing metal ions and radionuclides which have direct implications for biotechnological exploitation. (1,2,3). But the practice of bioremediation process using viable cells is limited by the requirements to maintain stringent conditions for cell growth. Consequently the process becomes costly for application at large scale. Use of dead biomass offers greater or equivalent sorption of metals and resists shock loads and toxicity. Availability of various dead biomass as waste products from industries make them attractive option to conventional ion exchange process. (4,5,6,7). But the main restraint posed by freely suspended biomass for column operation includes small particle size and mechanical fragility. Moreover, equivalent density of biomass to the effluent may affect the efficient separation of bioadsorbent from wastewater. Thus it is essential to fashion the biomass into immobilized form. Immobilized biomass particles have advantages like control of particle size, mechanical durability, easy recovery and regeneration, minimum clogging under continuous flow and better potential to apply in packed bed or fluidized bed reactors. (8,9,10,11). A variety of immobilization approaches have been studied to achieve superior biosorbent preparations (12,13,14,15).

Penicillium spp. has been reported for remediation of lead and chromium ions (16,17). *Penicillium griseofulvum* biomass can be acquired in large quantities from pharmaceutical industries in and around Gujarat, India. In this context the biomass was selected and was grafted into various immobilizing materials to select the best for copper bioremediation. Biosorption ability of the free and immobilized biomass was studied using sorption cycles with wide range of copper concentrations.

2. MATERIALS AND METHODS

2.1. Biomass
Penicillium griseofulvum was supplied by Cadila Healthcare Pvt. Ltd., Gujarat, INDIA. Biomass was thoroughly washed with deionized water and dried in oven at 50°C. Particle size of the biomass ranged between -30 +120 # (B.S.S).

2.2. Copper solution
Analytical grade sulfate salt of copper was dissolved in distilled water having pH 4.5 adjusted with 0.1 N HNO_3.

2.3. Immobilization techniques
P.griseofulvum pellets were immobilized by standard procedure with calcium alginate, polyacrylamide (18), urea-formaldehyde (19), 20% natural protein crosslinked with glutaraldehyde (20) and polyvinyl alcohol (21). Blank beads refer to control with no biomass.

2.4. Biosorption trials
Copper biosorption was carried out in 250 ml Erlenmeyer flasks having 0.5 gm free and immobilized biomass in 100 ml solution containing copper concentration of 63.5 mg/l.

Flasks were agitated on Environmental shaker at 30±2° C at 150 rpm for one hour (or otherwise mentioned). Biomass was seperated by centrifugation (10,000 rpm) or filtration and equilibrium copper concentration was estimated by diethyldithiocarbamate indicator (22) and AAS (Varian 175). To determine the maximum copper sorption capacity of the biomass above mentioned process was repeated upto 8 cycles. Copper sorption ability was checked at the end of each cycle. Gross copper removal and metal loading (mg.g^{-1}) was calculated at the end of 8 cycles.

The developed *P. griseofulvum* biomass was tested for copper removal from industrial waste containing copper, ammonia and ferrous at concentration of 2000, 4000 and 40 ppm respectively. The pH of the sample was 1.8±2. For biosorption trials, waste was 25 times diluted and pH was adjusted to 4.5 which was the optimum pH for copper sorption.

3. RESULTS AND DISCUSSION

3.1. Immobilization of *P.griseofulvum*

As can be seen from Figure 1 copper sorption ability of *P.griseofulvum* was least obscured by protein glutaraldehyde entrapment. On the other hand urea-formaldehyde immobilization drastically reduced the copper sorption ability of biomass resulting in only 12.37% removal. Cells immobilized in calcium alginate showed highest total copper removal of 84% but plain beads itself sorbed 66% copper under the experimental conditions giving less than 20% activity due to immobilized cells. Polyacrylamide entrapped cells demonstrated 67% copper removal against 40% removal by blank beads. Exploitation of polyacrylamide as immobilizing material is well supported by the reports of Macaskie et.al and Nakajima et.al. for cadmium and uranium removal respectively (14,23). But the inefficiency of polyacrylamide to use it at large scale warrants its application for commercial purpose (24). Subsequently, *P. griseofulvum* was immobilized in protein glutaraldehyde matrix. This preparation resulted in more than 80% copper removal by immobilized cells as compared to 40% removal by blank beads resulting in 2 fold increase in copper uptake by immobilized *P. griseofulvum* preparation. Polyvinyl alcohol film was dissolved during the biosorption trials thus it was not studied further.

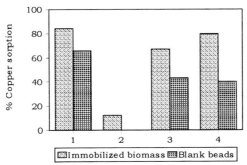

Figure 1. Influence of immobilizing material on copper sorption by *P.griseofulvum*
1. Calcium alginate ; 2. Urea formaldehyde ; 3. Polyacrylamide gel ; 4. Protein glutaraldehyde

Among the immobilizing materials evaluated, protein glutaraldehyde matrix displayed a superior metal uptake capacity; thus it was used to immobilize *P. griseofulvum* in all further experiments.

3.2. Effect of glutaraldehye concentration

As glutaraldehyde could be responsible for improved rigidity, strength and resistance to microbial attack; influence of various concentrations of glutaraldehyde in the range of 0.5 to 2.5 ml on copper removal is shown in Figure 2. Beads with 0.5 ml glutaraldehyde concentration were found to be unsuitable for the biosorption study. Protein biomass crosslinking using 1.0 ml glutaraldehyde exhibited best results for 3 cycles of copper removal giving 6 to 12% more copper sorption as compared to higher glutaraldehyde concentration in the system. As can be observed from Figure 2, there was less than 5% difference in copper sorption at 1.5, 2.0 and 2.5 ml glutaraldehyde concentration. But *P. griseofulvum* cells immobilized with 1.0 ml and 1.5 ml crosslinker were biodegraded during the process. Thus, 2.0 ml glutaraldehyde concentration was used in the immobilization system.

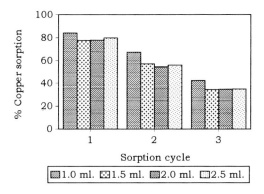

Figure 2. Effect of glutaraldehyde concentration on copper sorption

3.3. Influence of natural protein

In an attempt to improve copper removal by immobilized *P. griseofulvum*, two concentrations of protein and glutaraldehyde were tested (Figure 3). Copper sorption was found to be directly proportional to protein concentration used in immobilization system irrespective of the glutaraldehyde concentration used for crosslinking. When amount of protein was increased by two fold, copper sorption was enhanced by 1.33±2 fold under the experimental conditions.

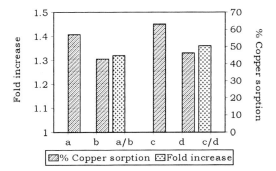

Figure 3. Effect of amount of natural protein in
immobilization system on copper sorption

a : High protein high glutaraldehyde;   b : Low protein high glutaraldehyde;
c : High protein low glutaraldehyde;   d : Low protein low glutaraldehyde

3.4. Influence of contact time

Table 1 presents the copper remediation as a function of contact time with free and
immobilized biomass. Results showed that 60 to 90 minutes of incubation time was
optimum for copper sorption by free and immobilized cells. More than 80% copper was
removed within 30 minutes of contact time supporting biosorption as main mechanism for
copper remediation. Immobilized biomass showed 66±1% copper removal in comparision
to 80 to 85% by free cells.

Table 1
Influence of contact time on copper sorption

Time (minutes)	% Copper sorption	
	Free biomass	Immobilized biomass
30	83.56	ND
60	85.45	65.57
90	85.45	65.00
120	79.12	ND

ND : Not determined

3.5. Sorption cycles

Results of repeated copper charging at low and high copper concentrations on copper sorption ability of free and immobilized *P. griseofulvum* are depicted in Figure 4a and b. Compared to immobilized biomass, free biomass proved to be more efficient for copper removal for first four and two cycles with low (6.35 µg ml⁻¹) and high (63.5 µg ml⁻¹) copper concentration respectively. On the other hand immobilized biomass retained its capacity to remove more than 50 % of copper even after 8 cycles of copper charging at low copper concentration. In presence of higher copper concentration this ability was seen upto 5th cycle. Free biomass reached its saturation within 6 cycle of sorption resulting in just 2-3 % removal as opposed to 41% removal by immobilized biomass. Blank beads showed diphasic mode of copper removal at higher copper concentration. At low copper concentration empty beads gave less than 10% copper removal and showed declination in copper removal from 7 to 2 % from 1st to 3rd cycle.

As can be seen from the data copper sorption by immobilizing material was found to be significantly dependent on the initial copper concentration in the solution. This point requires further investigation, as no data is available in literature regarding the metal sorption by immobilizing material used.

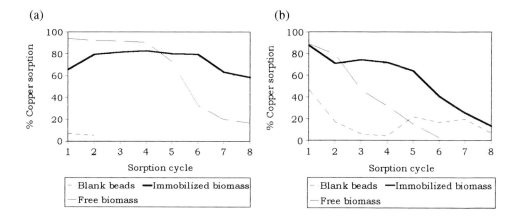

Figure 4. Copper sorption cycles at (a) low and (b) high copper concentration

When data of total percent removal and metal loading were compared (Table 2), it was observed that immobilized biomass showed better metal loading and percent copper removal over the range of copper concentration tested. At higher copper charging, free biomass was exhausted for its metal removal ability at 38100 µg copper addition as opposed to immobilized biomass retaining its metal removal ability even after 50800 µg copper charging. When percent copper removal was considered in terms of total addition of 50800 μg copper, it resulted in 33.17% and 56.11% by free biomass and immobilized biomass respectively.

Table 2
Copper removal profile at various initial copper concentrations

System	Copper concentration (μg)	Metal loading (mg/g)	% Copper sorption
Immobilized Biomass	6985	10.91	79.23
	26670	15.69	38.38
	50800	37.88	56.11
Free Biomass	6985	7.28	52.11
	26670	15.51	29.08
	38100	33.71	44.23 (33.17)

3.6. Copper removal from industrial effluent

The copper removal efficiency of free and immobilized biomass to remove copper from industrial effluent was also studied. Comparative data of percent removal from waste and aqueous copper solution are presented in Figure 5. Both free and immobilized biomass were found to be efficient copper sorbents for the waste water which also contained high ferrous impurity. More than 60% and 70% of copper was removed from the waste by free and immobilized biomass respectively. The observed difference could be due to the other cations present as pollutant in waste sample.

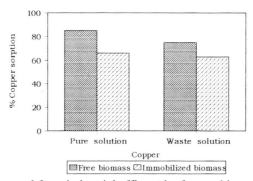

Figure 5. Copper removal from industrial effluent by free and immobilized biomass

234

4. CONCLUSION

(a) Among the immobilizing materials used, protein glutaraldehyde was found to be the best.
(b) Free biomass got saturated with lower copper charging.
(c) Judicial combination of glutaraldehyde and natural protein was found to be optimum for copper removal by immobilized biomass.
(d) The developed system was found to be efficient for waste with high ferrous content.

ACKNOWLEDGEMENT

We are thankful to Gujarat Mineral Development Corporation for research grant and fellowship to one of the authors (M.P.Shah).

REFERENCES

1. C.L. Brierly, Geomicrobiology Jr., 8 (1991) 201.
2. T.J. Beveridge in M.J. Klug and C.A. Reddy (eds.), Current perspectives in microbial ecology, ASM press, (1984) 601.
3. G.M. Gadd in H. Eccles and S. Hunt (eds.), Immobilization of ions by biosorption, Ellis Horwood, Chichester (1986) 135.
4. T. Sakaguchi, T. Tsuji, A. Nakajima, T. Horikoshi, Eur. J. Appl. Microbiol., 8 (1979) 207.
5. A.B. Norberg and H. Persson, Biotech. Bioeng., 24 (1984) 239.
6. K.A. Matis, A.I. Zouboulis, A.A. Grigoriadon, N.K. Lazaridis, L.V. Ekateriniadour, Appl. Microbiol. Biotechnol., 45 (1996) 569.
7. B. Mattuschka, K. Junghans, G. Straube in A.E. Torma, M.L. Apel, C.L. Brierly (eds.), Biohydrometallurgical Technologies Vol II , TMS Pennsylvania, USA (1993) 125.
8. M. Tsezos, R.G.L. Mcready and J.P. Bell, Biotech. Bioeng., 34 (1989) 10.
9. Z.R. Holan, B. Volesky and I. Prasetyo, Biotech. Bioeng., 41 (1993) 820.
10. G.M. Gadd in J.C. Fry, G.M. Gadd, R.A. Herbert, C.W. Jones, I.A. Watson-Craik, Cambridge University press, 1992.
11. M. Tsezos in H.L. Ehrlich and C.L. Brierly (eds.) Microbial Mineral recovery New york : McGraw-Hill Publishing Company,(1990) 323.
12. Z.R. Holan and B. Volesky, Biotech. Bioeng., 43, (1994) 1001.
13. J.Z. Xie, H.L. Chong and J.J. Kilbane, Bioresource technology 57 (1996) 127.
14. L.E. Macaskie, J.M. Wates, A.C.R. Dean, Biotech. Bioeng., 30 (1987) 66.
15. E. Rus, S. Sofer, F. Lakhwala, Bioprocess Eng. 13 (1995) 13.
16. N. Hui, X.S. Xue, J.H. Wang, B. Volesky, Biotech. Bioeng.,42 (1993) 785.
17. S. Siegel, P. Keller, M. Galam, H. Zehr, B. Siegel, E. Galum, Water Air Soil Pollut., 27 (1986) 69.
18. I. Chibata, T. Tosa, T. Sato, in K. Mosbach (ed.) Methods in enzymology,44, Academic Press, NewYork, SanFrancisco, London (1976).

19. K.M. Paknikar and A.D. Agate (eds.) Laboratory manual of international workshop on metal microbe interactions and their appliations, Pune Mircen, 1995.
20. P. Brodelius and E.J. Vandamme in Biotechnology vol 7a J.F. Kennedy (eds) VCH verlagsgesellschaft mbH, D-6940 Weinheim, 1987.
21. R. Bagai and D. Madamwar, Appl. Biochem. Biotechnol. 62 (1997) 213.
22. A.I. Vogel: A text book of quantitative inorganic analysis, (3rd edition), Longmans, Great Britain.
23. A. Nakajima, T. Horikoshi and T. Sakaguchi, Eur. J. Appl. Microbiol. Biotechnol. 16 (1982) 88.
24. G.W. Bedell and D.W. Darnall in B. Volesky (ed.) Biosorption of heavy metals Boca Raton, CRC press, 1990.

Environmental Protection from Cadmium Ions in Liquid Effluents by Biosorption

Aïda Espinola[a], Rupen Adamian[a] and Lola Maria Braga Gomes[b]

[a]COPPE/UFRJ, C.P. 68505, Rio de Janeiro, Brazil, ZIP 21945-970 -- FAX 55.21.290-6626

[b]Instituto de Química/UFRJ, Rio de Janeiro, Brazil -- FAX 55.21.290-4746

Several types of materials of biological origin have been used in the removal of metal ions from aqueous solutions, including peat, vegetal coal, algae. The authors verified that coconut fiber (*Cocos nucifera*), in Brazil, popularly called "coco da Bahia", has high efficiency in the removal of Cd(II) from aqueous effluents. The use of this material *in natura* is an innovation; alternatives of length of the fragmented fiber and its pre-washing have been studied. The capacity of the coconut fiber for ion removal depends on the complex chemical composition of its surface, where active chemical groups provide the mechanism of the chemical adsorption. We have already confirmed that there are also interfering compounds on the surface, which generate soluble coordination compounds stabilizing the cadmium ion in solution. The optimum conditions of the ratio mass of fiber/solution; of contact time fiber/solution and pH are established, leading to 99% removal efficiency of Cd(II) from mineral and metallurgical industries effluents.

1. INTRODUCTION

Biomass materials - wood, animal, forest and agricultural wastes - have been used for the purpose of generating energy since the beginning of the civilization. They have multiple uses, some classical, in beverages or food; some as raw materials for chemical and pharmaceutical industries; and some more recently in biosorption processes because of the world's increasing preoccupation with environmental pollution from a variety of industries. In this latest application, biomass presents relative advantage of low cost, in addition to the utilization of a material that, by itself, represents a waste. Another advantage over the most usual process for waste treatment, the chemical precipitation as sludges, is that chemical reagents added may be deleterious to the environment and the accumulation of sludges represents a detrimental effect, by geochemical redistribution of metallic ions. In addition, chemical precipitation by alkalis may not be effective for some metallic ions, such as cadmium, because of the high K_{sp} of cadmium hydroxide. Much alkali is then required and thus contaminates the effluent for disposal in waterways. A modern clean technology is the electrochemical treatment of liquid effluents, but its use has been limited by the cost of the electrical current *vs* current and energy efficiency of the electrolytic process.

Several agricultural wastes have received interest because of their low cost associated with the interest in their elimination; they have been proposed as metal ion scavengers in biosorption processes for the purpose of treatment of industrial wastes contaminated by toxic metals. Some industrialized products, such as ion exchange resins, artificial zeolites, chitosan (a synthetic polymer product of the deacetylation of the natural product chitin, with sodium hydroxide) and a commercial p-amino styrene polymer have been reported. Several barks - of black oak, of several pine trees like the Douglas fir and Brazil wood (*Caesalpinia echinata*) some leaves, seeds and orange peel - were essayed to absorb mercury [1].

Pinus silvestris bark has been shown by Gaballah [2] to adsorb cadmium ion with the bark *in natura*, either dried and comminuted, or previously processed to remove compounds such as tannin, which darkens the effluent. The efficiency of the metal removal increased proportionally to pH; efficiencies on the order of 98% were obtained in pH>5, using starting solutions with 10 and 100 ppm Cd(II); the residual Cd(II) concentrations were 0.2 and 2 ppm. It is to be noted that, by alkaline precipitation, such low residual concentrations would require pH of 10.5 and 9.8, respectively. In a saturation test under conditions of pH = 6.8, the retention capacity was 47 mg/g of dried substrate. Using industrial effluent with around 0.5 ppm Cd content, which had been treated by the process of neutralization/precipitation, the complete elimination of cadmium and chromium was obtained, as well as 90% of the copper, nickel, lead and zinc content. These results show that the process is efficient to lower the cadmium ion content to <20 ppb, which is much lower than that established by the European Union, 200 ppb, for discarding in waterways. Additionally, several tree barks and wood have been described by Gaballah, as metal scavengers, both *in natura* and pre-treated with compounds such as formaldehyde, alkalis (NaOH, KOH or NH_3); sulfides (Na_2S or $(NH_4)_2S$, or acids such as HCl, HNO_3, $HClSO_3$, or H_3PO_4, all of them with the purpose of preventing the lixiviation of coloured compounds, such as tannins.

Another biomass tested for the removal of cadmium ions from a synthetic waste was coal produced from peanut shell [3]. Results were favorable to the peanut shell coal in comparison with commercial activated carbon, as the minimum mass of coal required to remove Cd(II) from a solution containing 20 ppm Cd(II) was only 0.7 g/L of peanut shell coal compared to 12 g/L of commercial carbon. Cadmium adsorption increases with the pH. Maximum removal of the ion is obtained in the interval of pH 3.0-9.5.

Mattuschka and Straube [4] have studied the biomass from *S. nuoursei*, for adsorbing silver ions. They obtained 90% removal of Ag(I) from pure solutions, as well as from solutions containing other ions. It was observed that Cr(III), Pb(II) and Cu(II) were also strongly adsorbed, much lesser Zn(II), Co(II), Cd(II) and Ni(II). The order of adsorption of these metallic ions followed that of the electrochemical scale. The biomass could be reused after lixiviation with HCl, but, after four or five cycles, the adsorption capacity for these metals decreased to a minimum.

Biomass of *Cladophora crispata* has been studied by Özer [5], for the removal of Cr(VI) and Pb(II) from aqueous solutions. It was found that the sorption was consistent with the Langmuir and Freundlich isotherms, and was much higher for lead that for chromium.

Peat shows excellent adsorption capacity for heavy and toxic metals and may be used as a scavenger for them. It may be used *in natura*, simply dried and screened, otherwise thermally and chemically pre-treated. Couillard [6] lists several of these treatments that increase the peat's adsorption capacity. D`Avila *et al.* [7], using a process similar to ion exchange, increased by five times the efficiency of peat as a metal scavenger for copper, zinc, cadmium, chromium, lead and nickel, each metal in pure solutions or in mixtures of the metallic ions. In pure ions solutions, the best efficiencies obtained were around 70, 60, 40, 28, 94 and 80% for copper, zinc, cadmium, nickel, chromium and lead, respectively.

Coconut shell, processed into active coal has been used by Arulananthan *et al.*[8] as a scavenger for cadmium and lead ions. The sliced coconut shell was treated with 1:1.5 w/w H_2SO_4, at 140-160°C for 24 hours and washed, ground and vapor treated for 30 minutes. The excess of acid was neutralized by immersion in an $NaHCO_3$ solution followed by water washing. Excellent removal of metal ions, close to 100%, was obtained with starting contents of 20 ppm each, at pH 4.5-9.5 for cadmium and 3-6.5 for lead.

The authors of this paper used the fibers from the mesocarp of the coconut *in natura*, which is a much simpler, much less expensive method, for effluent purification. The coconut shell used is the agricultural waste of the so called "coco-da-Bahia", which is extremely abundant all over the Brazilian coast. The green coconut is consumed for its milk (which in Brazil is called "coconut water") leaving a bulky shell waste which has not, as yet, found a practical industrial application, except for a very small production of a flour which is locally consumed as a natural product of pharmaceutical use. Exact data on the volume of the coconut produced, as a whole, in Brazil, is not yet available. For instance, the coconut consumed along the beaches of Rio reaches the order of hundreds of thousand pieces/year. Additionally, in the region of Fortaleza, Ceará, in the northeast, industrialization of bottled coconut water leaves a considerable volume of coconut shells waste, plus that mentioned for production of flour.

In Table I, a number of the reported materials are listed, with information of their efficiency.

Table I. Some vegetable wastes investigated as cadmium ions scavengers

ADSORBENT	INITIAL CONC.	EFFICIENCY	ADSORPTION CAPACITY	
	ppm	%	mg /g	mmol / g
Coconut fiber [9]	10	99.9	1.0	0.01
Coconut fiber [9]	200	50.7	10.1	0.09
Black oak bark [2]	22400	___	25.9	0.23
Quercus spp bark [2]	N. A. (*)	___	25.8	0.23
Sequoia sempervirens bark [2]	N. A. (*)	___	32.0	0.28
Pinus sylvestris bark [2]	1000	___	47.0	0.42
Peanut hulls carbon [3]	20	___	89.3	0.79
Coconut shell carbon [8]	N. A. (*)	___	119	1.06
Tsuga heterophylla bark [2]	56	11	___	___
Sequoia sempervirens bark [2]	56	85	___	___
Cryptomeria japonica [2]	85	33	___	___
Pinus sylvestris bark [2]	10	98	___	___

(*) N. A. : non available

2. MATERIALS AND METHODS

2.1. Materials

Coconut fiber - The fiber used was homogeneous in composition. In order to ensure this homogeneity, the coconut shells used in this research were always collected in a selected area.

The process to isolate the fiber consisted of removing the green husk to cut the mesocarp into pieces. Two alternative processes were employed: (1) Pieces of the mesocarp were wet ground by rotating blades. The resulting material consisted of fibers with a variable length, up to 3 cm. It was spread on trays and dried in an oven at 50°C for 48 hours, whereupon it became light brown. (2) The pieces of mesocarp around 2 cm long were dried in an oven at 50°C for 48 hours, followed by disintegration in a hammer mill. The product was -16 mesh (Tyler), or 1.0 mm.

Cadmium solution - A stock solution of 1000 ppm was prepared from 3 $CdSO_4 \cdot 8H_2O$ (Carlo Erba).

2.2. Methods

Cadmium removal test - Bench scale tests consisted of adding 25 ml of the cadmium ion solution to the weighed coconut fiber, in 50 ml test tubes with stoppers, which were mounted on a rotating disc, and shaken for a selected period of time (generally 2 hours), at 4 rpm. The pulp was vacuum filtered, and cadmium ion concentration in both the initial and final solutions was determined by atomic absorption spectrophotometry.

Chemical analyses – Quantitative determination of Cd(II) ions, has been made by atomic absorption spectrophotometry (Perkin Elmer 3300). Infrared spectra were obtained with Nicolet Magna-IR 760 spectrometer.

3. RESULTS AND DISCUSSION

All measurements were made under condition of equilibrium between the cadmium retained onto the fiber and in the solution. Figure 1 shows the influence of contact time on the residual cadmium ion concentration. When using the ground fiber, equilibrium was attained in 60 minutes. For the comminuted fiber, with or without previous washing, the contact time was reduced to around 20 minutes.

The efficiency of cadmium ion removal from solution depends on the following variables:

i) *pH*. According to the authors in a previous publication [9] the efficiency of cadmium ion removal is directly proportional to the pH value, up to pH=4, then, levels out, up to pH=10. Between pH=10-12, the efficiency of removal decreases, indicating that the formation of $Cd(OH)_2$ (K_{sp} = 2.2 x 10^{14} at 25°C) is deleterious to the adsorption of cadmium ion. The interval of pH=4-10 represents, then, the optimum interval; pH=7 was selected for the process, to comply with conditions established by FEEMA (Environmental Engineering Foundation, the environmental control agency of Rio de Janeiro) for the discard of liquid wastes in waterways, in the State of Rio de Janeiro [11].

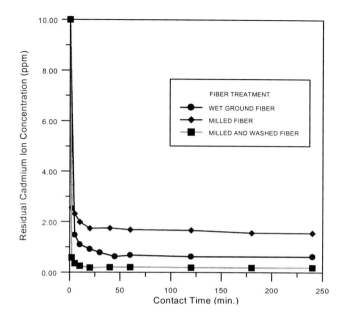

Figure 1 - Dependence of cadmium ion residual concentration/fiber contact time; initial concentration 10 ppm.

ii) *Influence of the mass of fiber*. Figure 2 shows the influence of this variable, for several types of fiber processing.

In curves 1 and 2 of Fig. 2, it may be seen that the efficiency of cadmium ion removal increases up to 0.5 g/100 ml, due to the increase in sites for adsorption; from then on, it decreases.

The authors attribute this decrease in efficiency to the presence of hydrosoluble compounds on the surface of the fiber, which combine with the cadmium ions, possibly as coordination compounds, to stabilize them in solution. This proposition was confirmed by studying fibers that had been pre-washed with distilled water. As seen in curves 3, 4 and 5, of Fig. 2, which represent these tests, after hydrosoluble compounds removal by pre-washing, the decrease in efficiency no longer appeared in curve 5, for the first washing of ground fibers in rotating blade mill; also in curve 4, for the twice washing of the hammer mill ground fiber.

The hammer mill ground fibers present a larger surface area, consequently, a larger amount of hydrosoluble compounds, which contribute to a higher stabilization of Cd(II) in solution, in opposition to the ground fibers. Additionally, the wet grinding represents a pre-washing process itself.

The use of pre-washed fibers increased the Cd(II) ion removal efficiency considerably, up to around 100%, as seen in curves 4 and 5 of Figure 2.

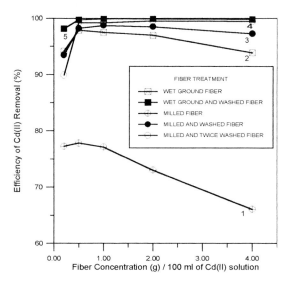

Figure 2 - Comparison of Cd(II) ion removal efficiency with the mass of fiber, for raw and washed fiber at initial concentration of 10 ppm; pH=7.

The identification of the hydrosoluble compounds from the fibers surface was done by infrared analysis. A preliminary spectrum is shown in Figure 3.

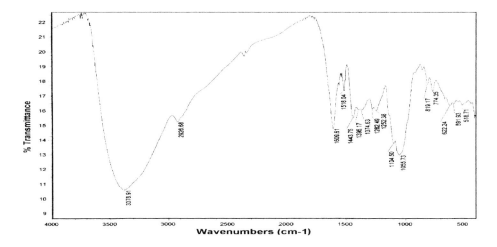

Figure 3. Infrared spectrum of the aqueous extract from coconut fiber

The 3378 cm^{-1} band is characteristic of the axial vibration of O-H, from alcohol or phenol; those bands of 1518 and 1609 cm^{-1} are attributed to a C-C vibration of an aromatic ring, and that of 1055 cm^{-1} is due to the axial vibration of C-O [12]. It is important to note that no bands of C=O, which would appear around 1700 cm^{-1} exist; this indicates that aldehydes, ketones, esters and carboxylic acids are either absent or below the sensitivity limit of the used instrument.

The nature of the hydrosoluble compounds is clearly shown in Figure 4. Figures 4a and 4b exhibit infrared partial spectra of the 2000 - 500 cm^{-1} interval, for fibers in original condition (no pre-wash) and pre-washed with water, respectively. They demonstrate our assertion that the pre-washing with distilled water removes hydrosoluble compounds hereby identified as aromatic compounds, indicated by bands in 700 to 850 cm^{-1} region. It is relevant to compare the 1736.13 cm^{-1} band in Figures 4a and 4b; as it turned more intense and detailed in Fig. 4b, we assume that compounds with C=O radical have been apparently concentrated, since hydrosoluble compounds were removed by the pre-washing.

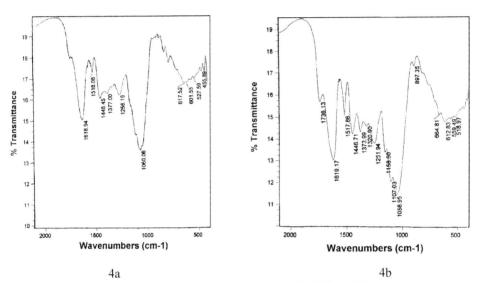

4a   4b

Figure 4. Infrared spectra of original fiber (4a) and pre-washed fiber (4b)

4. CONCLUSIONS

1. Coconut fiber is an efficient adsorber for the removal of cadmium ions from aqueous solution, with efficiencies as high as 99%, producing purified solutions with less than 0.1 ppm of remnant Cd(II) ion. This is adequate for disposal in waterways in the State of Rio de Janeiro, according to the FEEMA [10] established limits.
2. Conditioning the proportion of mass fiber to volume of solution, as well as of pH is required, in order to obtain a high efficiency.
3. The grinding process is another important variable, especially when using no pre-washed fibers. Wet grinding in blade mills produces larger fiber fragments, leading to the best efficiency.

4. Pre-washed fibers with distilled water produce the best purification of aqueous effluents by sorption of Cd(II) ions, since the pre-washing removes the hydrosoluble compounds which, otherwise, would interfere by combining with Cd(II) ions, possibly in the form of coordination compounds, thus, stabilized in solution, instead of concentrated onto the fiber surface.

ACKNOWLEDGEMENTS

Thanks are due to the CNPq, by one of the authors, Researcher 1A, A.Espínola; and to Carlos Riehl for the infrared analyses.

REFERENCES

1. M. S. Masri, F. W. Reuter and M. Friedman, J. Applied Polymer Science, 18 (1974) 675.
2. I. Gaballah et al., EPD Congress 1994, G. Warren, Ed. (1994) 43.
3. K. Periasamy and C. Namasivayam, Ind. Eng. Chem. Res., 33 (1994) 317.
4. B. Mattuschka and G. Straube, J. Chem. Tech. Biotechnol., 58 (1993) 57.
5. D. Özer et al., Environmental Technology, 15 (1994) 439.
6. D. Couillard, Water Res., 28 (1994) 1261.
7. J. S. D'Ávila, C. M. Matos and M. R. Cavalcanti, Wat. Sci. Technol., 26 (1992) 2309.
8. A. Arulanantham, N. Balasubramanian and T. V. Ramakrishna, Metal Finishing, (1989) 51.
9. A. Espinola, R. Adamian and L.M.B. Gomes, XVII Encontro Nacional de Tratamento de Minérios e Metalurgia Extrativa, ANAIS (1998) 115.
10. A. Espinola, R. Adamian and L. M. B. Gomes, EPD Congress 1998, B. Mishra, Ed. (1998) 39.
11. FEEMA, Manual do Meio Ambiente, Fundação Estadual de Engenharia do Meio Ambiente, Rio de Janeiro (1983).
12. R. M. Silverstein, G. C. Bassler and T. C. Morrill, Identificação Espectrométrica de Compostos Orgânicos, Guanabara Dois (Rio de Janeiro), 3ª ed., 1979.

Fungal biomass grown on media containing clay as a sorbent of radionuclides

M.A. Fomina[a], Kadoshnikov V.M.[b] and Zlobenko B.P.[b]

[a]Institute of Microbiology and Virology of National Academy of Sciences of Ukraine, Zabolotnogo 154, Kiev 252143, Ukraine

[b]National Academy of Sciences of Ukraine Scientific Centre of Environmental Radiogeochemistry, Palladin av. 34, Kiev 252142, Ukraine

The use of clay and biomass mixtures for the removal of toxic metals or radionuclides attracts attention because of the impact of sorption capacity of both sorbents. It was recently shown that biomass of melanin-producing microfungi *Cladosporium cladosporioides* grown on media containing bentonite could be successfully used as a biomineral sorbent of radiocaesium. The aim of present investigation was to demonstrate the possibility of using of fungal biomass of *Cladosporium cladosporioides* grown on media containing different clays from Ukrainian deposits to remove radiostrontium (^{90}Sr) and radiocaesium (^{137}Cs) from aqueous solutions. Adsorption isotherms were examined by varying the concentration of Sr and Cs from 1-20 Bq/ml. The two generic types of kaolinites from the Prosyanovskoe and Glukhovskoe deposits and palygorskite and natural mixture of palygorskite and montmorillonite from Cerkassy deposit were used.

The clay minerals altered the growth, size and sorption properties fungal pellets. The addition of 0.5% of clays into media led to reduction of the specific area of obtained biomineral sorbents and their sorption capacity. The increase of the concentration of all tested clays up to 5% caused a two-fold decrease in fungal pellet size. Fungal biomass grown on media containing 5% of Glukhovsky kaolinite demonstrated enhanced ability to sorb ^{90}Sr (2670 Bq/g) compared to pure clay (1951 Bq/g) and pure biomass (1551 Bq/g). Less significant increase of sorption capacity to caesium-137 and strontium-90 was found in the case of 5% of Prosyanovsky kaolinite. The positive correlation between an increase of sorption capacity and a decrease of fungal pellet size while increasing concentration of clays up to 5% was observed. The difference between predicted and experimentally obtained values of radionuclides uptake in the mixtures of clays and biomass was demonstrated.

1. INTRODUCTION

In the environment microorganisms are in close contact with solid adsorbents in soils, waters and sediments. Microbial growth and metabolic activity can be strongly influenced by adsorbents. Numerous investigations have clearly demonstrated that inorganic colloids such as clays may stimulate microbial growth and metabolism [1-7].

Both clays and microbial biomass are investigated due to their properties to sorb toxic metals. But most studies concerned with metal ion removal by soil components have concentrated on the uptake of metals by either clay minerals or microbial biomass as individual and separate entities [8-18]. However, few studies have been made on sorption properties of mixtures of clay mineral (montmorillonite, kaolinite) and microbial biomass (algae, fungi) [19-21]. The detailed investigation of sorption of toxic metals by clay minerals and soil fungi *Rhizopus arrhizus* and *Trichoderma viride* [21] showed that on a dry weight basis the clay minerals took up greater amounts of metal (up to 435 μmol g^{-1}) than the fungi (up to 78 μmol g^{-1}). However, when data were expressed in terms of metal bound per unit surface area, the fungi showed much greater sorption capacities (up to 36.6 μmol m^{-2}) than clays (up to 3.1 μmol m^{-2}). Mixtures of montmorillonite and fungal biomass showed reduced uptake of metals, depressed below calculated values by up to 37% at pH 4 , possibly because of masking of exchange sites [21].

Microfungi *Cladosporium cladosporioides* produce a lot of melanin pigment (up to 35% in mycelium) and their biomass was recommended as effective sorbent of toxic metals (Patent of Ukraine 1523,15.09.93) [22]. It is well known that fungal melanins have a high biosorptive capacity for a variety of metal ions and the presence of melanin in pigmented cell walls is responsible for the higher levels of biosorption in comparison with albino cells [16, 22-24]. Thus investigation of biosorption of metal ions to fungal pigmented biomass grown on clay is especially prospective. It was recently shown that biomass of *Cladosporium cladosporioides* grown on media containing bentonite could be successfully used as a mixed biomineral sorbent of radionuclides [25].

In the present work we investigated the possibility of using fungal biomass of *Cladosporium cladosporioides* grown on media containing different clays (kaolinite, palygorskite and natural mixture of palygorskite and montmorillonite) from Ukrainian deposits to remove radiostrontium (^{90}Sr) and radiocaesium (^{137}Cs) from aqueous solutions.

2. MATERIALS AND METHODS

2.1. Microorganism and cultivation
Cladosporium cladosporioides (Fresen) de Vries 396 was obtained from the collection of The Department of Taxonomy and Physiology of Micromycetes of Institute of Microbiology and Virology of National Academy of Sciences of Ukraine in Kiev (Patent of Ukraine 1523,15.09.93) [22]. The fungi were grown in Erlenmeyer flasks at 25° C and 160 rpm on liquid modified media, comprising per litre distilled water: NH$_4$NO$_3$, 2 g; KH$_2$PO$_4$, 1 g; MgSO$_4$.7H$_2$O, 0.5 g; sucrose, 15 g. The spore' inocula (10^6 conidia ml^{-1}) were used. Conidia for inoculum were obtained from the fungal surface culture grown on Malt agar medium.

2.2. Clays
Four different clays from Ukrainian deposits were used. There were two generic types of kaolinites from the Prosyanovsky deposit (kaolinite Prosyanovsky - **KP**) and Glukhovsky deposit (kaolinite Glukhovsky - **KG**), palygorskite from the third stratum of Cherkassy deposit (palygorskite Cherkassky - **PCh**) and natural mixture of palygorskite and montmorillonite (1/1) from the fourth stratum of Cherkassy deposit (palygorskite+montmorillonite Cherkassky - **PMCh**). All studied clays were of natural form. The mean values of clay particles were 1-2 μ for kaolinite KP, 0.2-0.8 μ for kaolinite KG, 0.05-0.1 μ for palygorskite PCh and 0.01-0.05 μ for PMCh. The index of water adsorption was 160-180 % for PCh and 180-220 % for PMCh. The cation exchange capacity (CEC) in

meq/100g was 1.5-2 for KP, 8-10 for KG, 16-20 for PCh and 35 for PMCh. As exchangeable cations all tested clays possessed Ca^{+2} (80 %), Mg^{+2}, Na^{+1} and K^{+1}.

The clays were added to liquid medium to given concentration of 0.5 % and 5 %. The control medium was deprived of clay.

2.3. Analysis

Cellular material (or mixture of fungal mycelium and clays) was obtained by filtration in the stationary phase of fungal growth, approximately 140 hours after inoculation. To estimate biomass production the material from each flask was dried to a constant weight. To quantify fungal biomass in the mixture of mycelium and clays the amount of organic carbon in the samples was estimated using Carbon Analyzer AN 2960 (Belorussia). The morphology and size of fungal pellets were observed using light microscopy. As a result of fungal growth on media containing clays biomineral sorbents were obtained. They were separated, washed with water, dried and ground to provide a mixture of mycelium and clay.

To investigate sorption of radionuclides onto biomineral sorbents the standard method was used. To 1 g of sorbent in Teflon flasks nitrate of radiostrontium (^{90}Sr) or radiocaesium (^{137}Cs) was added to achieve certain concentrations of 1-20 Bq/ml of radionuclides so that the ratio solid to liquid phase was 1/10. The time of reaction was 24 hours to reach equilibrium. Quantification of radionuclides in the dried samples was performed using β-radiometry (radiostrontium) and γ-spectroscopy (radiocaesium).

The specific area of the biomineral sorbents was evaluated using Brunauer-Emmett-Teller (BET) method with argon adsorption at low temperatures.

Statistical treatment of the results was carried out by estimating the standard deviation of population. Standard deviation of population in the experiments estimating specific area and sorption of radionuclides did not exceed 10-12 % of the mean values.

3. RESULTS

3.1. Influence of clays on fungal growth

It was shown that fungal growth was altered by clays (Fig. 1). The effect of different clays on the processes of production of fungal biomass depended on clay species and clay concentration in the media. The presence of all clays: kaolinites (KP and KG), palygorskite (PCh) and mixture of palygorskite and montmorillonite (PMCh) in the concentration 0.5 % in the media did not change considerably the production of fungal biomass compared to the control conditions.

The increase of clays concentration up to 5 % in the media enhanced significantly biomass synthesis. The addition of both types of kaolinites (KP and KG) in concentration 5 % led to a 2.4-fold increase of fungal biomass, whereas added in the same concentration both palygorskite and mixture of palygorskite and montmorillonite (PCh and PMCh) caused only a 1.7-fold increase of biomass production.

3.2. Influence of clays on size of fungal pellets

Clay species and concentrations altered not only the processes of biomass synthesis but fungal morphology. It was found that size of fungal pellets changed in the presence of clays in the media (Fig. 2). The addition of 0.5 % of kaolinite KP, palygorskite PCh and palygorskite

248

and monmorillonite mixture PMCh did not change pellet size compared to control conditions with medium deprived of clay.

The increase of concentration of the clay species (excluding palygorskite and montmorillonite mixture PMCh) up to 5 % considerably reduced the values of pellet size. Approximately two-fold reduction of pellet size under these conditions was observed. Kaolinite Glukhovsky KG was found to be the most active in the reduction of pellet size. The addition of KG into the medium even in the concentration 0.5 % led to the decrease of pellet' size similar to that of 5 %.

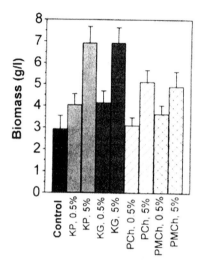 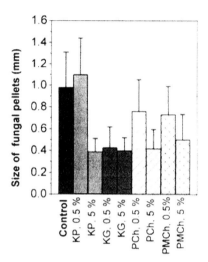

Figure 1. Effect of clay species and concentrations (0.5 % and 5 %) on biomass production by *Cladosporium cladosporioides.*

Figure 2. Effect of clay species and concentrations (0.5 % and 5 %) on size of fungal pellets of *Cladosporium cladosporioides.*

Clay species: **KP** - kaolinite Prosyanovsky, **KG** - kaolinite Glukhovsky, **PCh** - palygorskite Cherkassky, **PMCh** - palygorskite and montmorillonite mixture from Cherkassky deposit.

3.3. Specific area of obtained biomineral sorbents

Our results demonstrated that the values of specific area of of mixtures of fungal mycelium and clays were altered by both clay species and clay concentrations (Table 1).

The values of specific area of biomineral sorbents were reduced compared to the specific area of corresponding clay species. The addition of 0.5% of clays into media led to the considerable reduction of the specific area of obtained biomineral sorbents, especially for palygorskite PCh and palygorskite and montmorillonite mixture PMCh.

The significant increase of specific areas of biomineral sorbents was observed if compared to the values of specific area of pure mycelium of *C. cladosporioides.*

3.4. Sorption of radionuclides to biomineral sorbents

Sorption capacity of mycelium and clays mixtures depended on nature of radionuclides, species of clays and their concentration in the media and ,as a result, content of fungal biomass in the obtained biomineral sorbents.

Table 1
Specific area of biomineral sorbents

Sample	Specific area, m^2g^{-1}
Kaolinite KP, control of pure clay	18.0
Kaolinite KP, 0.5% in media	8.1
Kaolinite KP, 5% in media	14.3
Kaolinite KG, control of pure clay	70.0
Kaolinite KG, 0.5% in media	14.0
Kaolinite KG, 5% in media	44.0
Palygorskite PCh, control of pure clay	265.0
Palygorskite PCh, 0.5% in media	50.0
Palygorskite PCh, 5% in media	133.0
Palygorskite+Montmorillonite PMCh, control of clay	289.5
Palygorskite+Montmorillonite PMCh, 0.5% in media	36.0
Palygorskite+Montmorillonite PMCh, 5% in media	148.0
Biomass of *C.cladosporioides*, control	1.4

Radiocaesium sorption onto the biomineral sorbents differed considerably from radiostrontium sorption (Table 2 and 3). It was found that radiocaesium sorption capacity of the mixtures of biomass and clay did not change considerably compared to control of corresponding pure clay (Table 2). This tendency was observed for both 0.5 % and 5% concentrations of clays in media. The growth of fungi on the media containing clays significantly enhanced radiocaesium sorption ability of biomass/clay mixture compared to pure fungal mycelium possessing much lower radiocaesium sorption.

Experimentally obtained data of radiocaesium sorption were compared to the predicted data calculated with values of radiocaesium sorption to pure clay and to pure mycelium and concentration of biomass in the fungal-clay mixtures (Table 2). It was found that the presence of all tested clay species in the concentration 0.5% in the medium led to the significant increase of radiocaesium sorption capacity of biomineral sorbents compared to the corresponding predicted data. Kaolinites KP and KG in the concentration 5% in the medium caused less increase of radiocaesium sorption, whereas palygorskite PCh and palygorskite and montmorillonite mixture PMCh did not change radiocaesium sorption compared to predicted data.

Radiostrontium sorption data showed that the presence of all clays at 0.5 % in the medium did not change (KP) or even significantly reduced (KG, PCh and PMCh) sorption capacity of biosorbents compared to control of corresponding pure clay (Table 3). For palygorskite PCh and palygorskite and montmorillonite mixture PMCh the increase of clay concentration in the medium up to 5 % did not change considerably sorption properties of biomass and clay mixtures compared to pure clay control. In contrast, the addition of kaolinite KG into medium

250

led to the significant increase of radiostrontium sorption (2670 Bq/g) compared to pure kaolinite KG (1951 Bq/g) and pure mycelium (1551 Bq/g). Less significant increase of radiostrontium sorption was manifested for 5 % of kaolinite KP.

It was found that radiostrontium sorption data for all clay species (excluding kaolinite KP) added in the concentration 0.5% into medium were considerably less (especially for palygorskite and montmorillonite mixture PMCh) than corresponding predicted values. The presence of clays: kaolinites KP and KG and palygorskite PCh in the media in the concentration 5% led to values of radiostrontium sorption increased compared to the predicted values. The most significant increase was observed for kaolinite KG. In the case of palygorskite and montmorillonite mixture PMCh in the concentration 5% the experimentally obtained values of radiostrontium coincided with predicted ones.

Table 2
Sorption of radiocaesium to biomineral sorbents

SAMPLE	Mycelium amount in sorbent, %	Sorption of ^{137}Cs, Bq/g	
		predicted	experimental
Kaolinite KP, control of pure clay	-	-	3722
Kaolinite KP, 0.5% in media	41.4	2264	3535
Kaolinite KP, 5% in media	12.8	3272	3950
Kaolinite KG, control of pure clay	-	-	4221
Kaolinite KG, 0.5% in media	45.1	2407	4225
Kaolinite KG, 5% in media	12.9	3702	4296
Palygorskite PCh, control of pure clay	-	-	4362
Palygorskite PCh, 0.5% in media	35.3	2893	4248
Palygorskite PCh, 5% in media	10.8	3913	4324
Palygorskite+Montmorillonite PMCh, control of clay	-	-	4388
Palygorskite+Montmorillonite PMCh, 0.5% in media	35.4	2906	4257
Palygorskite+Montmorillonite PMCh, 5% in media	10.4	3953	4305
Biomass of C.cladosporioides, control	100	-	200

4. DISCUSSION

It is well known that adsorption between micoorganisms and mineral solid particles is very important ecological phenomenon of microbial existence which helps soil microorganisms to hold themselves in higher layers of soil and to prevent stress and lethal environmental effects [26, 27]. This phenomenon is more studied for the bacterial cells than for mycelial microorganisms. The main nutritional substances are usually concentrated on the border between solid and liquid phases, so microorganisms immobilized to solid surface have physiological advantages in continuous liquid cultures [28]. The process of adsorption depends on properties of both microorganisms and adsorbents as well as conditions of contact between cells and adsorbent's surface [5]. Interaction between cells and solid particles is subjected to complex of strengths of both physicochemical (electrostatic, ionic, hydrophobic and other) and biological (chemotaxis, production of specific enzymes, polysaccharides, lectines and other adhesines, etc.) nature [4, 5]. Under conditions of physiological values of pH the cells of microorganisms have negative charge and mostly amino- and carboxylgroups on their surface [5, 29]. It was shown that positively charged edges of slices of montmorillonite were electrostatically attracted by negative carboxylgroups of bacterial surface, whereas positive sites of cell surface can attach negatively charged lateral flatness of montmorillonite particles [30].

Table 3
Sorption of radiostrontium to biomineral sorbents

SAMPLE	Mycelium amount in sorbent, %	Sorption of ^{90}Sr, Bq/g	
		predicted	experimental
Kaolinite KP, control of pure clay	-	-	1471
Kaolinite KP, 0.5% in media	41.4	1504	1343
Kaolinite KP, 5% in media	12.8	1482	1837
Kaolinite KG, control of pure clay	-	-	1951
Kaolinite KG, 0.5% in media	45.1	1713	1443
Kaolinite KP, 5% in media	12.9	1899	2670
Palygorskite PCh, control of pure clay	-	-	2288
Palygorskite PCh, 0.5% in media	35.3	2027	1671
Palygorskite PCh, 5% in media	10.8	2208	2567
Palygorskite+Montmorillonite PMCh, control of clay	-	-	2851
Palygorskite+Montmorillonite PMCh, 0.5% in media	35.4	2391	1555
Palygorskite+Montmorillonite PMCh, 5% in media	10.4	2715	2771
Biomass of C.cladosporioides, control	100	-	1551

Fungal pellets are one of the main mycelial forms of submerged fungal growth. Pellet formation and structure depends to a large extent on physicochemical properties of spores and hyphae as well as on their physiological properties [31-33]. The changes in fungal morphology (pellet' size and density) reflect the differences in their physiological activity. The presence of clay within pellets may influence positively the diffusion of nutrients and oxygen or remove toxic metabolites.

There exists a lot of studies of microbial physiology showing that interaction of microorganisms with solid surfaces led to an increase of biomass yielding and growth rate, synchronization of microbial development, changes in the duration of lag period, effectiveness of substrate utilization, respiration activity and enzymes and metabolites production [1-7].

Most of them demonstrated stimulation of physiological activity of microorganisms by clay particles. Our results showed that all studied clay species added in concentration 5 % into medium led to the appearance of fungal pellets of two-fold less size and to the considerable stimulation of biomass production.

The possible mechanisms of stimulatory effects of clay on microbial activity are discussed in the literature. First, the maintenance of a more favorable pH is possible during the incubation or the exchange of Ca^{2+} or other cations from the clay for cell surface H^+-ions [1] and , second, the specific adsorption of one or more metabolic inhibitors (as well as nutrients and growth stimulators) may occur [3, 5, 34]. The possible action of these two mechanisms depends onto mutual distribution of microbial cells and solids. If the cells and solid particles exist separately in liquid media, clay can act indirectly removing from media certain substances (toxins, nutrients and growth stimulators). If the solids are of size much less than the microbial cells, clay particles can surround the surface of the microorganisms, create microzones with specific conditions and lead to changes in cell wall and membrane [5, 35]. These processes, probably, occurred during the growth of mycelial fungi like *C.cladosporioides*. Fungi not only adsorbed clay particles but used them for building of their pellets bodies. All these processes are very complicated and their study would contribute to both biotechnological and ecological considerations.

As was shown in the present study the sorption properties of both clay species and fungal mycelium could be changed in the biomineral sorbents obtained after growth of fungi in the media containing clays. Sorption ability depended on clay species and concentration and nature of radionuclide. The decrease of metal retention was usually explained as being the result of cation-enhanced aggregation between cells and clays, blocking of active sites and reducing the cation binding capacity of the biomass/clay aggregate [21]. These processes could occur during reduction of radiostrontium sorption by biomineral sorbents in the case of clays concentration in the media 0.5%. But it was found that the increase of clays concentration to 5% enhanced sorption capacity of biomineral sorbents in most cases. Probably, such changes in sorption properties compared to predicted data and control may be explained by the appearance of new regions and sites for cation binding on the surface of fungal hyphae and clay particles. As a rule, the increase of sorption capacity of biomineral sorbents coincided with the reduction of pellet' sizes, but this correlation seems not to be strictly obligate for all studied clays. Kaolinite KG was the most active in both the reduction of fungal pellet' size (in concentration 0.5% and 5%) and the increase of sorption capacity of fungal/clay mixture (5%). The values of the specific area of obtained biomineral sorbents reduced in the presence of 0.5% of clays in media, but the significant decrease of sorption was observed only for radiostrontium and 0.5% of palygorskite PCh and mixture of palygorskite and montmorillonite PMCh and much more less for kaolinite KG.

The phenomenon of fungal growth on medium containing clay with its changes in physiological activity, pellets' structure and fungal-clay mutual influence is of a great importance for creation of biomineral sorbents for toxic metals and bioremediation. By varying clay species and concentration in the medium and species of microorganisms we can enhance sorption ability of a biomineral sorbent.

REFERENCES

1. G. Stotzky, Can. J. Microbiol., No.12 (1966), 831.
2. K. Haider, Z. Filip and J.P. Martin, Arch. Mikrobiol., 73 (1970), 201.
3. J.P. Martin, Z. Filip and K. Haider, Soil Biology and Biochemistry, 8 (1976), 409.
4. M. Fletcher, Microbiological Sciences, 4, No. 5 (1987), 133.
5. D.G. Zvyagintcev, Soil and Microorganisms, Ed. Moscow University, Moscow, Russia, 1987.
6. K.C. Marshall Can. J. Microbiol., 34, No. 4 (1988), 593.
7. H. Clause and Z. Filip, Soil Biology and Biochemistry, 22, No.4 (1990), 483.
8. H. Farrah and W.F. Pickering. Australian Journal of Chemistry, 29 (1976), 1649.
9. R. Fujiyoshi, A.S. Eugene and M. Katayama, Applied Radiation and Isotopes, 43 (1992), 1223.
10. R. van Bladel, H. Halen and P. Cloose, Clay Minerals, 28 (1993), 33.
11. Y. Andres, H.J. MacCordick and J.-C. Hubert, Applied Microbiology and Biotechnology, 39 (1993), 413.
12. Huang Chin-pin, Huang Chin-pao and A.L. Morehart, Water Research, 24 (1990), 433.
13. M.D. Mullen, D.C. Wolf, F.G. Ferris, F.G. Beveridge, C.A. Flemming and G.W. Bailey, Applied and Environmental Microbiology, 55 (1989), 3143.
14. I.S. Ross, Stress Tolerance of Fungi, Ed. by D.H.Jennings, Marcel Dekker: New York, 1993, 97.
15. N.N. Zhdanova, A.I. Vasilevskaya, V.I. Gavrilyuk, E.L. Sholokh and L.A. Koval, Mycologiya I Fytopatologiya, 24, No.2 (1990), 106.
16. G.M. Gadd, Microbial Mineral Recovery, Ed. by H.L.Ehrlich & C.L.Brierley, McGraw-Hill: New-York, 1990, 249.
17. G.M. Gadd, New Phytologist, 124 (1993), 25.
18. G.W. Garnham, G.A. Codd and G.M. Gadd, Microbial Ecology, 25 (1991), 71.
19. V.M. Kadoshnikov, B.P. Zlobenko, N.N. Zhdanova and T.I. Redchitz, Proc. HLM, LLW, Mixed Wastes and Environmental Restoration - Working Towards A Cleaner Environment, WM'95, 26.02-02.03, 1995, Tucson, Arizona, C.D., 61.
20. G.F. Morley and G.M. Gadd, Mycol. Res., 99, No.12 (1994), 1429.
21. N.N. Zhdanova and A.I. Vasilevskaya, Extreme ecology of fungi in nature and experiment, Kiev: Naukova Dumka, 1982.
22. G.M. Gadd and L. de Rome, Applied Microbiology and Biotechnology, 29 (1988), 610.
23. R.V. Fogarty and J.M. Tobin, Enzyme and microbial Technology, 19 (1996), 311.
24. M.A. Fomina, V.M. Kadoshnikov and B.P. Zlobenko, Proceedings of International Regional Seminar "Environmental Protection: Modern Studies in Ecology and Microbiology", Uzhgorod, 13.05-16.05, 2 (1997), 86.
25. H. Babich and G. Stotzky, Applied and Environmental Microbiology, 33 (1977), 696.
26. S. McEldowney and M. Fletcher, Journal of General Microbiology, 132 (1986), 513.
27. J.W. Costerton, M.J. Marrie and K.J. Cheng, Bacterial adhesion: Mechanisms and physiological significance Ed. by D.C.Savage & M.Fletcher, Plenum Press: New York, London (1985), 3.
28. D.T. Plummer and A.M. James, Biochem. et biophys. acta, 53, No. 3 (1961), 453.

29.K.C. Marshall, Bacterial adhesion: Mechanisms and physiologycal sygnificance (ed. D.C.Savage & M.Fletcher), Plenum Press: New York, London (1985), 131.

30.T. Yanagita and F. Kogane, Journal of Gen. and Applied Microbiology, 9 (1963), 171.

31.R. Wittler, H. Baumgartl, D.W. Lubbers and K. Schugerl, Biotechnology and Bioengineering, 28 (1986), 1024.

32.J.I. Prosser, The growing fungus. Ed. by Gow N.A.R. and Gadd G.M., , Chapman & Hall, London (1994), 301.

33.J. Macura and L. Pavel, Folia Microbiol., 4 (1959), 82.

34.M. Fletcher and J.H. Pringle, Journal of Colloid and Interface Science, 104, No. 1 (1985), 5.

Study of some biosorption supports for treating the waste water from uranium ore processing

P.D. Georgescu[a], Nicoleta Udrea[a], F. Aurelian[a] and I. Lazar[b]

[a]Uranium National Company-R&D Institute for Rare and Radioactive Metals, Bucharest, Bd.Carol I, no.78, Romania

[b]Institute of Biology to the Romanian Academy, Bucharest, Romania

The aim of this study was to research the possibility of decontamination of uranium alkaline or acid waste waters using immobilized biomass on biosorbent supports (sand shell, bentonite, volcanic tuff, molecular screen, synthetical zeolits, activated carbon and dry waste yeast biomass). We executed the biosorption tests using percolation columns for testing the sorption capacity of the mentioned supports and for studying different immobilized culture of bacteria on the biosorbent supports. The results showed that the most efficiently biosorbent support is activated carbon. The bacterial system studied included bacterial pure cultures of *Bacillus* and *Pseudomonas*, cultures producing exopolysaccharides like *Xantomonas campestris pv.* and *Pseudomonas sp. 577*, bacterial strains from the indigenous microflora, denitrification bacterial strains, yeast cells and *Spirulina platensis*. In this paper are presented the experimental results obtained by using this nonconventional process as an application for treating waste water from uranium ore processing.

1. INTRODUCTION

Waste water, which results from different industrial sectors, especially from mining, are usually discharged to the drainage system. This process causes environmental possible affects on the quality of life.

Because the effluents from mining processes have a small content of toxic heavy metals, it is not economic to apply a treatment for recovery at these metals and their discharge can thus lead to toxic effects in the environment.

International laws establish the accepted pollution limits for waters, including the potable water. For potable water the international standards provide the following maximum metal ion concentrations: 0.01 mg/l Se, 50.0 mg/l Mg, 0.1 mg/l Fe, 0.3 mg/l Mn.

The bioremediation methods for waste water with high concentrations of metal ions involve many micro-organisms or biological compounds. Micro-organisms may be used living or as dead biomass resulting from different pre-treatments.

Metal accumulation from industrial effluents by microbial methods may take place at the cell walls by association between metallic ions and biological compounds like exopolysaccharides.

Few classes of micro-organisms may be identified.

The first class is represented by: *Zoogloea ramigera, Alteromonas atlantica, Arthrobacter viscosum, Pseudomonas putida, Klebsiella aerogenes* etc. These bacteria produce biopolymers (polysaccharides) which have the capacity to sorb metal ions from industrial effluents (1 - 6). The polysaccharides produced are able to accumulate Cu^{2+}, Cd^{2+}, Ni^{2+}. For UO_2^{2+} and Cr^{2+} removal by solution, remarkable results are obtained using a natural polymer named chitosan (7).

A second class of micro-organisms used in waste water decontamination are heterotrophic aerobic bacteria. This class of micro-organisms is represented by bacterial strain of *Pseudomonas, Alcaligenes eutrophus, Arthrobacter, Acinetobacter, Citrobacter* etc. *Pseudomonas stutzeri RS34*, was found to adsorb a variety of different metal like: Zn^{2+}, Ni^{2+}, Cu^{2+}, Mn^{2+}, CrO_4^{2-}, Al^{3+}, Pb^{2+}, *Pseudomonas mendocina AS 302* was used to remove Hg, Sb, As, Th, *Pseudomonas syringae* for remove Cu, *Alcaligenes eutrophus ER 121 for* Zn^{2+}, Co^{2+}, Cd^{2+} (8), *Arthrobacter sp.* for Mn^{2+}, Cu^{2+}, Cd^{2+}, Ni^{2+}, Pb^{2+}, CrO_4^{2-}, UO_2^{2+} (9) removal and *Citrobacter* for UO_2^{2+} etc.

A third class of micro-organisms are fungal biomasses. Bosecker, K. (10), studied 80 types of fungal biomasses (*Penicillium, Cladosporium, Trichoderma harzianum*). This class of micro-organisms could be used to remove metal ions of Cd^{2+}, CrO_4^{2-}, Cu^{2+}, Ni^{2+}, Zn^{2-}.

Uranium adsorption from solutions, the influence of other cations (K^+, Na^+, Ca^{2+}, Fe^{2+}) and the pH effect was studied for different types of culture (immobilized or free). Dry biomass from waste industrial waters was used for heavy metals removal (Zn^{2+}, Al^{3+}, Cu^{2+}, Fe^{2+}, Ni^{2+}, Co^{2+}, Mg^{2+}, CrO_4^{2-}).

Marine algae form the fourth class of micro-organisms. These are able to accumulate metal ions. Efficiently results were obtained using treated or untreated algal biomasses (*Sergasum, Ascophillum and Euglena*) to adsorb metal ions of Fe^{2+}, Ni^{2+}, Cu^{2-}, Mo^{2-}, Th^{2+}, Cd^{2-}, Zn^{2+}.

The fifth class of micro-organisms is represented by aquatic plants. Metal biosorption by dry biomass of *Potamogeton lucens, Salvinia herzogii and Eihhornia crassipes* takes place by a ion exchange reaction.

Good results were obtained for Ni^{2+}, Cu^{2+}, Zn^{2+}, Pb^{2+} biosorption. In the last few years was observed a growing interest in using micro-organisms or other biomasses as sorbents for metal ions removal from waters.

The paper presents the results of uranium and other metal ions biosorption from waste waters using different biosorbent supports.

2. MATERIALS AND METHODS

2.1. Industrial effluents

For the biosorption tests we used an industrial effluent provided from an uranium preparation plant. The physical-chemical characterization is presented in Table 1 and the microbiological one, in Table 2.

The industrial effluent is an alkaline solution which contains a high density of aerobic heterotrophic bacteria and N_2-fixing bacteria.

Table 1
Physical-chemical characterisation of the effluent

Characteristic	Sample I	Sample II
Colour	greenish	greenish
Temperature (°C)	18.0	20.0
pH at 20°C	10.0	10.0
Conductivity (mS/cm)	18.5	17.8
$Na^+ + K^+$ (mg/l)	5748.0	5070.0
Mo (mg/l)	15.0	20.0
NO_3^- (mg/l)	200.0	222.0
NO_2^- (mg/l)	5.1	4.7
Cl^- (mg/l)	1838.0	1767.0
HCO_3^- (mg/l)	2648.0	2281.0
CO_3^{2-} (mg/l)	3180.0	3120.0
SO_4^{2-} (mg/l)	1350.0	1240.0
PO_4^{3-} (mg/l)	12.0	10.2
Oxidability-$KMnO_4$ (mg/l)	150.0	141.0
Silica	50.0	55.0
Solid suspensions (mg/l)	520.0	670.0

Table 2
Microbiological characterisation of the effluent

Type of micro-organism	Cells number/ml
Heterotrophic aerobic bacteria	
Nutrient medium	1.5×10^6
Heterotrophic anaerobic bacteria	4.5×10^2
Sulphate reducing bacteria	1.5×10^2
Iron-oxidizing bacteria	2.5×10^2
Denitrifying bacteria	2.0×10^2
N_2-fixing bacteria	9.5×10^5

2.2. Biosorbent supports

In the present study we used the following biosorbent supports: activated carbon, molecular screen, volcanic tuff, bentonite, seashell sand.

The types and the characteristics of the biosorbent materials are represented in Table 3.

Table 3
Adsorbant supports

Type of adsorbant support	Weight of adsorbant support,g	Characteristics
Seashell sand	10	Obtained by breaking and screening the seashell from Black Sea Coast (Romania)
Bentonite	10	Argilliceous sedimentary rock (montmorillonite)
Volcanic tuff	10	Natural zeolite
Molecular screen	5	Synthetic zeolite, diameter 2 mm
Activated carbon	5	From coke, diameter 2 mm
Waste yeast dry biomass	5	By fermentation process in pharmaceutical industry

Before using, the support was washed with HCl solution, 0.1 N, then washed with distilled water (few times) and dried at 250°C.

2.3. Biosorption equipment

The biosorption tests were carried out in percolation equipment with six glass columns using different types of biosorbent materials. In each column was introduced two layers of biosorbent material and between the layers was intercalated a layer of sand. The sand was screened to provide a uniform granulation, washed with HCl solution 1N and distilled water and finally dried at 105°C. The recirculation pump assured the solution flow and the aeration of the microbial cultures, which is necessary for the metabolic processes.

2.4. Microbial cultures on the adsorbant support

In biosorption tests the most usually method is immobilization by biosorption on a porous material. The microbial cells are able to form a continuous cell layer at the surface of the solid particles, called a biofilm.

The most efficient biosorbent material was activated carbon. On this support was immobilized types of bacterial materials. Table 4 shows the microbial systems used in these tests.

The cultivation of these strains was realized by mixing (200 rpm) at 28°C for 24-72 hours. After cultivation was obtained a quantity of living biomass with $9.5 \times 10^{10} - 7.3 \times 10^{11}$ cell/ml density.

The immobilization technique was as follows: the bacterial inocul was passed through the activated carbon column, till saturation. The column was closed for 24 hours at 28°C.

Biosorption tests were realized in two ways:
a) Waste water sample was passed through biosorbent support. For each test the conditions were as follows: temperature 28°C, time 24 hours, sample volume 500 ml, recirculation number 10;

b) Biosorption tests using seven microbial systems immobilized on the activated carbon support. After the removing of the nutrient medium, 500 ml waste water was passed through each column at 28°C, for 10 times.

Table 4
Microbial systems

Microbial system used	Type of bacterial strain
Bacterial strain *Bacillus* type	- *Bacillus subtitles* 1; W_2; 17; 25; 3214
	- *Bacillus cereus*
	- *Bacillus licheniformis* W_7
Bacterial strain *Pseudomonas* type	- *Pseudomonas fluorescens*
	- *Pseudomonas aeruginosa*
	- *Pseudomonas putida*
	- *Pseudomonas sp.*
Bacterial strain- exopolysaccharides produced	- *Xanthomonas campestris pv. 153K*
	- *Pseudomonas sp. 577*
Bacterial strain from effluent microflora	- *Bacillus sp.*
	- *Pseudomonas sp.*
	- *Micrococcus sp.*
	- *Arthrobacter sp.*
Denitrification bacterial strain type	- *Pseudomonas sp. W1*
	- *Micrococcus sp. W0*
Yeast strain	- *Saccharomyces cerevisiae*
Bacterial strain *Spirulina* type	- *Spirulina platensis*

3. RESULTS AND DISCUSSION

The tests for uranium and other ions removal from industrial effluents involved the following objectives:
- optimal adsorbent support identification;
- behaviour of different immobilized microbial systems in the biosorption process, when the pH effluent was varied

The biosorption tests took place at 28°C, the waste water was recirculated on the biosorbent support by 10 cycles using for each column a quantity of 500 ml waste water.

Summaries of observation made about the process are given in Table 5.

Table 5
Observation about biosorption process on different supports

Biosorbent support	Initial pH	Finally pH	Flow rate	Clarity ; Colour
Seashell sand	10.0	8.97	very rapid	+++; colourless
Bentonite	10.0	8.57	edium	+++; colourless
Volcanic tuff	10.0	8.73	very rapid	++; colourless
Molecular screen	10.0	9.00	very rapid	+++; colourless
Activated carbon	10.0	7.96	medium	++++; colourless
Waste yeast dry biomass	10.0	7.12	slowly	++++; colourless

++ - ++++ = intensity of the phenomena

The results of the adsorption tests on the biosorbent supports are represented in Table 6.

Table 6
Results of the adsorption tests

Adsorbent Support	Elements content (g/l)					
	U	Mo	Cl⁻	CO_3^{2-}	HCO_3^-	SO_4^{2-}
Effluent sample	0.00332	0.00252	0.39	0.53	1.76	0.412
Seashell sand	0.00294	0.00260	0.39	0.32	1.93	0.461
Bentonite	0.00278	0.00245	0.39	0.11	1.93	0.424
Volcanic tuff	0.00294	0.00260	0.39	0.21	1.93	0.432
Molecular screen	0.00282	0.00258	0.39	0.42	1.85	0.412
Activated carbon	0.00176	0.00172	0.39	0.01	1.93	0.457
Waste yeast dry biomass	0.00130	0.00235	0.39	0.01	1.26	0.393

In the adsorption process we were interested about the following anions: uranil-tri-carbonate $[UO_2(CO_3)_3]^{4-}$, molybdate $[MoO_4]^{2-}$, chloride $[Cl]^-$, carbonate $[CO_3]^{2-}$, bicarbonate $[HCO_3]^-$ and sulphate $[SO_4]^{2-}$.

Experimental data were reorganized and represented in Table 7. The best results were obtained for activated carbon support and for waste dry yeast biomass. Activated carbon removed from the effluent uranyl-tri-carbonate, molybdate and carbonate. The uranium content decreased by 40% from initial value, molybden by 31% and carbonate by 98.2%. Dry yeast biomass was found to be able to accumulate 60.9% of the uranium content, 98% of carbonate and 28.4% of bicarbonate.

Table 7
Experimental data obtained from adsorption on the supports

Adsorbant Support	Percentage decrease of elements concentration					
	U	Mo	Cl⁻	CO_3^{2-}	HCO_3^-	SO_4^{2-}
Seashell sand	11.5	0.0	0	39.6	0	0
Bentonite	16.3	2.8	0	79.3	0	0
Volcanic tuff	11.5	0.0	0	60.4	0	0
Molecular screen	15.1	0.0	0	20.8	0	0
Activated carbon	47.0	31.8	0	98.2	0	0
Waste yeast dry biomass	60.9	0.0	0	98.2	28.4	0

We were tested the microbial systems immobilized on activated carbon with two effluent types: acid effluent (pH = 1.7) and alkaline effluent (pH = 10).

In Table 8, is represented the behaviour of different microbial systems immobilized on activated carbon in contact with the acid effluent. The effect of the bacterial activity in the acid medium is pH modification.

Table 8
Observation about biosorption process using immobilized biomass on activated carbon

Microbial system	Initial pH	Final pH	Flow rate	Clarity ; Colour
Bacterial strain *Bacillus*	1.7	3.66	very rapid	++++; colourless
Bacterial strain *Pseudomonas*	1.7	4.04	very rapid	+++; colourless
Bacterial strain-exopolysacharides produced	1.7	3.32	medium	+++; colourless
Bacterial strain from effluent microflora	1.7	4.14	very rapid	+++; colourless
Denitrification bacterial strain	1.7	3.54	very rapid	++++; colourless
Yeast strain	1.7	6.13	slow	+ ; pink colour
Bacterial strain *Spirulina*	1.7	7.11	slow	++++;blue colour

+ - ++++ = intensity of the phenomena

Denitrification bacterial strain, *Bacillus* bacterial strain, yeast and *Spirulina platensis* provided the best results, so, we will continue the research in this field.

Living biomass from *Bacillus*, *Pseudomonas* and yeast strain immobilized on activated carbon, were tested with the industrial alkaline effluent (pH=10). The results of these tests are shown in Tables 9, 10. The best results were obtained for *Bacillus* strains (uranium content decrease with 40%) and *Pseudomonas* (uranium content decrease with 20.5%). Furthermore, the content of the sulphate ions decreased by 28,7% in the case of *Bacillus* and 27% for *Pseudomonas*.

Table 9
Results of the biosorption tests

Microbial	Elements content (g/l)					
System	U	Mo	Cl⁻	CO_3^{2-}	HCO_3^-	SO_4^{2-}
Industrial effluent	0.00332	0.018	0.866	1.95	2.92	0.412
Bacterial strain *Bacillus*	0.00204	0.017	0.849	1.91	2.86	0.294
Bacterial strain *Pseudomonas*	0.00264	0.016	0.849	1.91	2.86	0.301
Yeast strain	0.00289	0.018	0.708	1.17	3.53	0.354

Table 10
Biosorbtion process performances

Microbial	Percentage decrease of elements concentration					
System	U	Mo	Cl⁻	CO_3^{2-}	HCO_3^-	SO_4^{2-}
Bacterial strain *Bacillus*	40.0	0	0	0	0	28.7
Bacterial strain *Pseudomonas*	20.5	0	0	0	0	27.0
Yeast strain	13.0	0	0	0	0	14.1

Alkaline industrial effluent (pH=10) may be treated using yeast biomass to reduce the uranium content, or using Bacillus strain to accumulate sulphate ions. If these strains are immobilized on activated carbon support, uranium adsorption occurs as well.

REFERENCES

1. M.J. Brown and J.M.Lester, Water research, 13 (1979) 117.
2. M.J. Brown and J.M.Lester, Water research, 16 (1982) 1539.
3. L.K.Jang and G.G.Geesey, Procedings of Internatiomal Biohydrometallurgy Symposium IBS-93), Jackson Hall, Wyoming, (1993) 75.
4. F.Roe and Z.Lewandowski Procedings of Internatiomal Biohydrometallurgy Symposium (IBS-93), Jackson Hall, Wyoming, (1993) 145.
5. J.A.Scott, G.K.Sage and S.J.Palmer, Biorecovery 1 (1988) 51.
6. J.M.Chartier and E.Guibal, Procedings of Internatiomal Biohydrometallurgy Symposium (IBS-95),Viña del Mar, (1995) 267.

7. N.C.M. Gomes and E.R.S. Camargos, Procedings of Internatiomal Biohydrometallurgy Symposium (IBS-95), Viña del Mar, (1995) 401.
8. C.D. Boswell and R.E.Dyck, Procedings of Internatiomal Biohydrometallurgy Symposium (IBS-95), Viña del Mar, (1995) 299.
9. K. Bosecker, Procedings of Internatiomal Biohydrometallurgy Symposium (IBS-93), Jackson Hall, Wyoming, (1993) 55.

Platinum Recovery on Chitosan-Based Sorbents

E. Guibal[a], A. Larkin[b], T. Vincent[a] and J. M. Tobin[b]

[a]Ecole des Mines d'Alès - Laboratoire Génie de l'Environnement Industriel
6, avenue de Clavières - F-30319 Alès cedex - France

[b]Dublin City University - School of Biological Sciences
Dublin 9 - Ireland

Chitosan proved efficient at removing platinum in dilute effluents. The maximum uptake capacity exceeds 280 mg g^{-1} (almost 1.5 mmol g^{-1}) under optimum conditions: obtained at pH 2. A glutaraldehyde crosslinking pre-treatment is necessary to stabilize the biopolymer in acidic solutions. Sorption isotherms were studied in function of the pH, the particle size of the sorbent and the crosslinking ratio. The extent of the crosslinking has no significant influence on uptake capacity. Competitor anions such as chloride or nitrate induce a large decrease in the sorption efficiency. Sorption kinetics also show that uptake rate is not significantly changed by increasing the crosslinking ratio, and the particle size of the sorbent. Mass transfer rates are more significantly affected by the initial platinum concentration and by the conditioning of the biopolymer. While for molybdate and vanadate ions, mass transfer was governed by intraparticle mass transfer, for platinum both external and intraparticle diffusion control uptake rate. Platinum, in contrast with the latter, is not able to form polynuclear hydrolysed species, which are responsible for steric hindrance to diffusion into the polymer network.

1. INTRODUCTION

Though the value of this precious metal is changing with stock exchange variations, platinum is one of the most commonly used among noble metals, due to its uses in catalytic processes. In the last decades, several processes have been used and carried out to recover and separate platinum from synthetic or industrial effluents, such as sorption [1-2], liquid-liquid extraction or membrane processes [3]. Impregnated resins and liquid membranes are powerful processes [4], however they can involve some environmental drawbacks such as a release of solvent or extractant traces or limitations in technical and economic efficiencies for dilute effluents.

Since the beginning of the 80's, many studies focussed on metal ion biosorption [5] using several biosorbents such as fungi [6-7] or algae [8]. However, the literature is not very abundant on noble metal recovery using such biosorbents [9-10]. A wide range of industrial synthetic sorbents or ion-exchange systems is available for platinum recovery, however our research focussed on natural-based products such as chitosan sorbents due to their availability and their environmentally friendly behaviour especially for the final treatment of the exhausted sorbents. Platinum can be recovered by elution using several extractants such as thiourea, however taking into account the high value of the metal and that of the sorbent, it recovery can be also performed by the thermic degradation of the organic sorbent, resulting combustion products are less hazardous for the environment than those of common ion-exchange resins.

Chitosan is derived from the alkaline deacetylation of chitin, the most abundant biopolymer in nature after cellulose. It is soluble in dilute mineral and organic acid solutions (except sulphuric acid). Most industrial effluents containing platinum are strongly acid and chitosan can be used for platinum recovery only after a crosslinking step to increase its chemical stability. Several chemical agents have been proposed such as epichlorohydrin or glutaraldehyde [11-13]. In the present work, glutaraldehyde was selected for chemical stabilization.

Chitosan and derived products have been studied for the recovery of a wide range of metals such as uranium, lead, cadmium, chromium, molybdenum [14-17], but also noble metals such as gold and platinum [18-19]. However in the latter case, the studies were carried out using medium to high concentrations or using very costly sorbents proceeding from complex modifications of the chitosan backbone. Few studies have been dedicated to the recovery of platinum from low concentration solutions encountered in wastewaters from mining and metallurgical industries. To address these research and development needs, this study presents the influence and optimization of the pH for platinum sorption and studies the sorption isotherms, taking into account several parameters, such as the presence of competitor anions and the influence of particle size. The sorption isotherms are compared with curves obtained with chitosan gel beads produced by an alkaline cast procedure. Sorption kinetics are also presented with a special emphasis on the influence of the particle size, metal concentration, sorbent dosage, and crosslinking yield.

2. MATERIAL AND METHODS

2.1 Materials

Chitosan was supplied by ABER-Technologie (France) as a flaked material, with a deacetylation percentage ca. 87 % (defined by IR spectrometry [20]). The mean molecular weight was measured at 110 000 (using a size exclusion chromatography method coupled with a differential refractometer and a multi-angle laser light scattering photometer). The moisture content of sorbent particles, for both crosslinked and uncrosslinked sorbents, was determined at around 10 %; unless specified, sorbent masses will be expressed on a wet basis. The chitosan bead fabrication has already been described [13].

Chemical crosslinking of chitosan chains with the bifunctional reagent glutaraldehyde occurs by a Schiff's reaction of aldehyde groups on glutaraldehyde with amine groups on the chitosan biopolymer chain [21]. The glutaraldehyde concentration of the crosslinking bath contained was 0.78 M. The ratio of glutaraldehyde to chitosan was set at 2.22 on a molar basis. Crosslinking lasted for 16 hours. The crosslinked chitosan particles were extensively rinsed with demineralized water. The general procedure was applied to manufacture crosslinked chitosan flakes at four particle sizes: G1<125, G2<250 μm, G3<500 μm, G4<710 μm. A similar procedure was adopted to prepare crosslinked chitosan beads with different contact conditions to take into account the expanded size of the beads in comparison with flakes but the glutaraldehyde/amino ratio was maintained at constant values 1:1 and 4:1.

2.2 Experimental procedure for platinum sorption

Platinum solutions were prepared from dihydrogen hexachloroplatinate (IV) ($H_2PtCl_6.6H_2O$, ChemPur) in demineralized water (except for the influence of chloride and nitrate ions, added in the form of NaCl and KNO_3). The pH of the solutions was controlled using sulphuric acid and sodium hydroxide concentrated solutions (5 M). Except for the experiments carrying out the optimization of the pH, this parameter was kept constant throughout the sorption step.

For sorption isotherms, performed at room temperature (20 °C ± 1 °C), a known volume of platinum solutions (200 or 1000 mL) at fixed concentration (6, 25 or 50 mg L^{-1}) was mixed with several sorbent amounts (10 to 80 mg wet mass). After 5 days of agitation, in a reciprocating shaker, solutions were filtered on 1.2 μm filtration membrane. Solutions were analysed by the $SnCl_2$/HCl spectrophotometric method (Shimadzu UV-160A) [22] or alternatively using ICP analysis (Jobin-Yvon). The platinum concentration in the sorbent, was obtained using a mass balance equation and was expressed as mg Pt g^{-1} sorbent, without reference to actual chitosan content in the sorbent, which depends on the crosslinking ratio. Unless specified, the sorption capacity has been expressed in function of actual sorbent mass.

For sorption kinetics, a standard procedure was applied [13]. 1 L of platinum solution at fixed pH was mixed with 50 mg of sorbent, in a jar-test mixed system (240 rotations per minute). 5 mL samples were withdrawn at specified times and filtrated on a 1.2 filtration membrane and analysed as previously specified.

3. RESULTS AND DISCUSSION

3.1 Influence of the pH on platinum removal.

Representative data for the pH effect on platinum recovery using crosslinked chitosan flakes are shown in Figure 1. The initial concentration is around 18 mg L^{-1} and the crosslinked chitosan flakes were obtained from an initial GA/mol of $-NH_2$ ratio equal to 2.22.

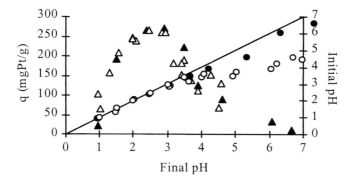

Figure 1. Influence of the pH on sorption capacity (including pH variation during sorption) (open symbols: free variation of pH, filled symbols: pH control)

The figure shows that the removal efficiency sharply increases from pH 1 to pH 2-2.5, in this pH range the efficiency reaches a maximum followed by a slight decrease. Two experimental conditions were used regarding the pH control during the sorption step, however the sorption efficiency does not seem to be significantly modified when the pH is controlled for the beginning of the mixing. The figure shows that platinum sorption and/or mixing of the biopolymer with the solution involves a significant pH decrease. For the more acidic solutions the variation in pH is difficult to detect, while for the near neutral zone the change in pH can exceed 1 unit. The larger the pH change, the lower the uptake efficiency. Potentiometric titrations allow the overall charge of the sorbent to be obtained as a function of the pH. From the results of Domard [23], it appears that the pK_a of the chitosan strongly varies with the dissociation degree for chitosan of low acetylation fraction (lower than FA=0.15) while for higher FA, the pK_a hardly varies with the dissociation degree around 6.4-6.5. For a pH lower than the pK_a value by 3 units, the protonation yield reaches 99.9 %.

So, the shape of the curve (Figure 1) can be explained by a defavourable sorbent charge for the near neutral pH and by a competitor effect of counter ions at the lowest pH. At pH 1, the polymer charge is strongly positive (high protonation level of nitrogen sites) and attracts counter anions such as chloride and sulphate or hydrogeno-sulphate ions and involves a strong competition between these counter ions and hexachloroplatinate.

3.2 Influence of competitor anions (Cl⁻, NO₃⁻) on sorption capacity

Figure 2 illustrates the effect of chloride and nitrate salts on platinum recovery at pH 2, with a metal concentration of 70 mg L^{-1} and a sorbent concentration of 222 mg L^{-1}. It clearly demonstrates that both anions strongly interfere with metal uptake and decrease sorption capacity to a few dozens of mg g^{-1}, while the distribution coefficient does not exceed 500 with molar solutions of competitor anions. Nitrates have a slightly more marked effect on the decrease of sorption performances by comparison with chloride ions, perhaps due to steric effect. The strong influence of ionic strength confirms that platinum uptake proceeds by ion-exchange rather than by pure adsorption process. The sorption levels reached with molar solutions of competitor anions are consistent with the levels cited by Inoue et al. with hydrochloric acid solutions [19].

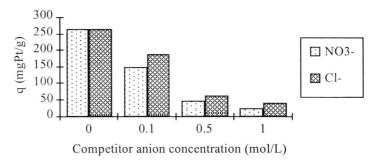

Figure 2. Influence of competitor anions on platinum sorption at pH 2.

The influence of HCl concentration seems to result from both pH and chloride concentration. These results are conflicting with results cited by Mitani et al. [24] who have shown that cobalt and nickel sorption on chitosan beads is increased by an increase in ionic strength, especially when using sulphate ions. Their interpretation is related to a change in the crystalline structure of the polymler due to sulphate action. However an increase in crystallinity has been correlated to a decrease in both equilibrium and kinetic sorption performances [13,25]. So, the difference in the behaviour of chitosan sorbents for several metals can be attributed to a difference in uptake mechanism regarding complexation and/or ion-exchange steps. The influence of competitor anions is an important parameter for the prediction of sorption performances for further application on industrial effluents. The decrease in sorption capacities due to the matrix has to be taken into account for the definition of the metallurgical process when the liquid effluents proceed from solid extraction.

3.3 Particle size effect on sorption isotherm

Figure 3 shows the sorption isotherms obtained with G1, G2, G3 and G4 particle sizes and with beads (diameter 2.3 mm). The steepness of the curves seems to depend on the particle size, however the trend is not linear between the four size fractions and it is difficult to interpret such small differences. The maximum uptake capacity seems to be independent of the particle size of crosslinked flakes. This result indicates that hexachloroplatinate ions are able to diffuse through the biopolymer and to be sorbed on each of the internal sites of the sorbent. Table I shows the coefficients of the Langmuir model for experimental data.

Sorption isotherms are comparable for crosslinked chitosan whatever the conditioning of the biopolymer. This behaviour is significantly different to previous results obtained with other metals using chitosan or crosslinked chitosan flakes. While uranium sorption on chitosan depends on sorbent particles, the modification of the biopolymer by substitution of organic acids (ascorbic acid for example) involves a modification of the polymer structure and allows the diffusion of the metal ions into the whole mass of the sorbent [17]. Similar trends were obtained with crosslinked chitosan flakes and molybdate. The maximum molybdenum uptake capacity depends on the sorbent diameter, while uncrosslinked chitosan and crosslinked chitosan beads have sorption levels which are not influenced by the particle size. The expansion of the polymer network by gel bead fabrication enhances the diffusion properties and the accessibility to internal sites [13]. The differences in the behaviour of the sorbents in function of the metal ions can be attributed to hydrolysis mechanisms and to the ability of several metals, such as molybdate or uranium dioxide, to form hydrolysed polynuclear species, whose ionic radius is considerably larger than those of free species [13]. On the other hand, hexachloroplatinate ions are mononuclear species, though the chloride ions bound to a platinum center increase the ionic size of the molecule, it is still smaller than those of polynuclear metal ions. Consequently, the diffusion of hexachloroplatinate ions is less controlled by polymer structure and particle size. The particle size can influence the sorption kinetics and the time required to reach equilibrium but does not influence the exhaustion of the sorption or ion-exchange sites.

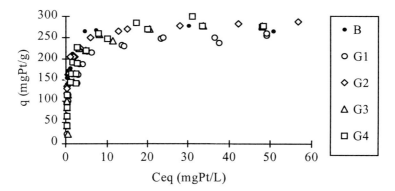

Figure 3. Influence of particle size on sorption isotherms at pH 2 (Gi: flakes, B: beads)

Table I:
Effect of particle size on sorption isotherm parameters (Langmuir model)

Particle size	q_m (mgPt g^{-1})	b (l mg^{-1})	R^2
Beads	269.2	3.38	0.94
G1	246.5	1.23	0.81
G2	266.6	2.78	0.94
G3	272.7	1.17	0.88
G4	279.5	1.21	0.88

3.4 Sorption kinetic modeling

Mass transfer involves several steps including (i) bulk diffusion, (ii) film diffusion, (iii) intraparticle diffusion and (iv) sorption and/or ion-exchange process, to which heat transfer may be added. However due to the heat transfer properties of water, kinetic limitations due to heat transfer can be neglected. Providing a sufficient agitation to avoid particle and solute gradients in the batch reactor allows the bulk diffusion to be neglected also. Moreover it is usually accepted that sorption processes can be considered as instantaneous processes and the kinetic control is mainly attributed to film and intraparticle diffusion. Simplified models were carried out such as the homogeneous diffusion model (HDM), and the shrinking core model (SCM) with special definitions for film diffusion, intraparticle diffusion and chemical reaction control respectively [26]. However these simplified models do not fit experimental data over the total time range. Yiacoumi and Tien [27-28] describe and compare sorption kinetics controlled by either reaction or diffusion mechanisms. Several models exist which include both external, intraparticle diffusion and sorption isotherm characteristics [29-30]. Except in the case of linear isotherms, such models cannot be solved analytically and numerical analysis is required. The first approach in this work will consist in separating external and intraparticle diffusion steps to obtain the order of magnitude of the external and intraparticle diffusion coefficients.

To obtain the external mass transfer coefficient, a simplified procedure previously described by McKay and Allen [31] was used. Coupled mass transfer equations and boundary and intial conditions can be solved algebraically to give the following equation:

$$\frac{C_t}{C_o} = \frac{1}{1 + m K} + \frac{m K}{1 + m K} \exp\left(-\frac{1 + m K}{m K} k_f S t\right) \qquad (1)$$

where m represents the sorbent dosage (g L^{-1}), K is the adsorption equilibrium constant, $K = q_m$ b (in g L^{-1}), k_f is the external mass transfer coefficient, S the outer surface of sorbent particles per unit of volume of particle-free solution (m^{-1}), which is obtained using the following equation:

$$S = \frac{6 m}{d_p \rho (1 - \varepsilon)} \qquad (2)$$

where d_p is the mean particle diameter (m), r the density of sorbent particles (kg m^{-3}) and e the porosity of sorbent particles. The external mass transfer coefficient, k_f is obtained from the slope of the plot (eq. 3):

$$\ln\left(\frac{C_t}{C_o} - \frac{1}{1 + m K}\right) = f(t) \qquad (3)$$

Crank [32] proposed a model whereby diffusion is controlled only by intraparticle mass transfer for well-stirred solution of limited volume (V), assuming the solute concentration being always uniform (initially C_o), and the sorbent sphere to be free from solute. Under these conditions, the total amount of solute M_t (mg g^{-1}) in a spherical particle after time t, expressed as a fraction of the corresponding quantity after infinite time (M_g, mg g^{-1}) is given by:

$$\frac{M_t}{M_\infty} = 1 - \sum_{n=1}^{\infty} \frac{6\alpha\,(\alpha+1)\exp\!\left(-Dq_n^2 t / d_p^2\right)}{9 + 9\alpha + q_n^2\alpha^2} \tag{4}$$

where D is the intraparticle diffusion coefficient (m^2 min^{-1}). The fractional approach to equilibrium, FATE, may be used to estimate the intraparticle diffusion coefficient D, when the external diffusion coefficient being neglected. a is the effective volume ratio, expressed as a function of the equilibrium partition coefficient (solid/liquid concentrations ratio) and is obtained by the ratio $C_{eq}/(C_o-C_{eq})$. q_n represent the non-zero solutions of equations (5)

$$\tan q_n = \frac{3\,q_n}{3 + \alpha\,q_n^2} \qquad \text{and} \qquad \frac{M_t}{VC_o} = \frac{1}{1+\alpha} \tag{5}$$

The infinite sum terms are summed until the summation does not vary. In this study, equation (3) was used to determine the overall intraparticle diffusivity which best fitted experimental data (minimizing the sum of the square of the differences between experimental results and calculated data).

3.5 Influence of platinum concentration on sorption kinetics

Figure 4 presents sorption kinetics for G2 particles with a crosslinking ratio CR equal to 2.22. Sorption of platinum using chitosan-derived sorbents is a fast mecanism by comparison with other metals such as molybdate or uranium. The crosslinking treatment in the case of molybdate showed a significant decrease in sorption velocity due to restriction in diffusion for flake particles, while it does not influence kinetic performances for beads (due to the expansion of the polymer network during gel formation). A long contact time is required to reach the actual equilibrium (usually 3 to 4 days), however more than 90 % of the total sorption occurs in a time depending of the concentration but ranging between 2 and 5 hours.

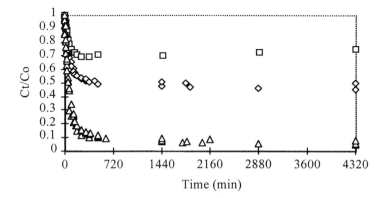

Figure 4. Influence of the initial concentration on sorption kinetics at pH 2 ($\underline{d}$: 6.5, ?: 23, $\underline{a}$: 47 mg L^{-1}).

Table II summarizes the kinetic parameters, external and intraparticle diffusion coefficients for chitosan flakes. It shows that for increasing concentrations, the external diffusion coefficient decreases, while the intraparticle diffusivity increases. However the variations are not very important and in any case the diffusivities change by less than one order of magnitude.

Table II:
Influence of initial concentration and particle size on diffusion coefficients and Biot number

Particle size	C_0 (mg L^{-1})	$k_f 10^4$ (m s^{-1})	$D 10^{13}$(m^2 s^{-1})	Biot Number
G1	6.5	0.23	0.023	6.8
G2	6.5	1.95	1.82	1.9
G3	6.5	1.27	2.85	1.7
G4	6.5	0.26	1.82	0.8
G1	23	0.55	1.01	1.2
G2	23	1.10	11.7	0.6
G3	23	0.58	14.2	0.5
G4	23	1.80	43.0	0.8
G2	47	0.59	20.3	0.4
Beads	6.5	0.25	100	0.06

The surface diffusion modified Biot number, Bi_S, has been defined to give a criterion for the predominance of surface diffusion against external diffusion:

$$Bi_S = \frac{k_f\, d\, C_0}{2\, \rho\, D_s\, q_0} \tag{6}$$

where C_0 is the initial concentration and q_0 the solid phase concentration in equilibrium with a liquid concentration C_0. External diffusion predominates when $Bi_S \ll 1$, while surface diffusion controls diffusion when $Bi_S \gg 1$.

The Biot number decreases, indicating that for high initial concentrations film diffusion is the controlling step. This trend is surprising: Hellferich [33] has stated that mass transfer is mainly controlled by film diffusion for systems with ion exchangers of high concentration of fixed ionic groups, low degree of crosslinking or small size, with dilute solutions and with inefficient agitation. Rorrer et al. [17] propose a pore blockage mechanism to explain the increase in intraparticle diffusion with increasing sorbate concentration. At low external concentration there is a low flux exist between the solution and the inner part of the sorbent, the solute is adsorbed on the external sites of the polymer and chelate complexes clog the pores. On the other hand, at high concentration, the high concentration gradient between the solution and the sorbent internal layers involves a high metal anion flux through the porous network. Nestle and Kimmich [34] demonstrate that for ion-exchange systems, concentration-dependent diffusion coefficients can be related to sorption isotherm properties. The intraparticle diffusivity can be deduced from free-diffusivity and model equations, especially for systems described by the Langmuir equation. While the corresponding equation cannot fit experimental data correctly for intraparticle diffusivities, this model states that the intraparticle diffusivity increases with the equilibrium concentration and consequently initial metal level.

The increase in the diffusivities enhances the intraparticle diffusion step, and its contribution in the overall kinetic control is reduced, which can explain the decrease in the Biot number. However, the Biot number is frequently equal or near 1, indicating that sorption occurs with a simultaneous kinetic control by both film and intraparticle mass transfer. These conclusions contrast with previous results on molybdate sorption on chitosan beads. The difference in the behaviour of the sorbent between platinum and molybdenum may be attributed to a difference in the chemical properties of the metals and especially to the ability of molybdenum to form polynuclear complexes whose diffusion into the polymer network is controlled by steric hindrance. On the other hand, hexachloroplatinate is a mononuclear species whose diffusion is less hindered by steric properties and thus the intraparticle diffusion does not restrict the access to internal sites to the same extent.

3.6 Influence of particle size on sorption kinetics

Figure 5 shows sorption kinetics for 4 particle sizes at 2 initial concentrations. It appears that the particle size has a weak effect on the shape of the experimental curves, for flakes, while the kinetic rate is significantly lower for chitosan beads.

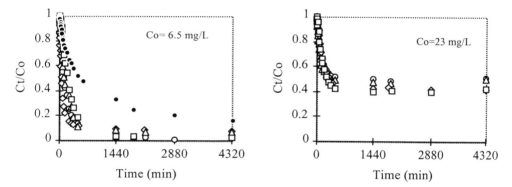

Figure 5. Influence of particle size on sorption kinetics at 2 initial concentrations (G1:o, G2:?, G3:d, G4:a, Beads:C).

Unexpected results are obtained in the comparison of sorption kinetics for flakes and beads sorbents: while gel bead conditioning has enhanced sorption properties for molybdenum and vanadium recovery, with platinum the gel conditioning reveals a restricting treatment, mainly on uptake kinetics. Such conditioning treatment is justified for column systems in limiting head losses and clogging mechanism.

This result confirms the preliminary conclusion proposed in the preceding section. The intraparticle diffusion which was so important in the diffusion of molybdate species has a limited effect on the platinum sorption rate [13], the differences in the shape of the curves are unsignificant by comparison with those of molybdate, for crosslinked chitosan flakes. Table II gives the corresponding coefficients for both external and intraparticle diffusion. Surprisingly, the Biot number is higher at low concentrations than at medium concentration, this is consistent with the results of the preceding section. Though the diffusion coefficients slightly change with experimental conditions (remaining in the same order of magnitude), it is not possible to show a uniform trend. These random variations can be attributed to a non-uniform heterogeneous crosslinking and the wide range of particle size in each fraction: particles are separated in broad size scales in which the size profiles are not determined. It is interesting to observe that the diffusion coefficients for both film and surface diffusion are significantly

274

higher than those obtained for the sorption of molybdate ions by chitosan flakes, with at least 1 or 2 orders of magnitude [13]. The orders of magnitude for the intraparticle coefficient seem surprisingly nearer to those obtained with crosslinked chitosan beads and molybdate. For these sorbents, the expansion of the polymer network is supposed to increase accessibility to internal sites. The mononuclear species of the hexachloplatinate salts are less sensitive to steric hindrance and to the opening of the three-dimension structure of the polymer after crosslinking treatment. On the other hand, for chitosan beads the intraparticle mass transfer coefficient for hexachloroplatinate is significantly higher than those obtained with flakes, while the external diffusion coefficient is lower than the corresponding data for flakes. The Biot number for chitosan beads indicates that the kinetics is mainly controlled by external diffusion in this case.

4. CONCLUSION

Chitosan modified by glutaraldehyde-crosslinking is an efficient sorbent for platinum recovery in dilute solutions. The reinforcement of chitosan stability by di-aldehyde crosslinking allows the sorbent to be used in very acid solutions such as encountered in industrial effluents. Indeed, the study of the pH effect has shown that the sorption capacity increases with the pH till an optimum in the range 2-2.5, while above pH 3, the sorption efficiency progressively decreases. The first part of the curve is explained by the competition between anioninc competitor ions brought by the acidification of the solution. The second section of the curve can be interpreted as the consequence of an unfavorable charge of the sorbent which interferes on both binding and/or ion-exchange mechanisms. The sorption capacity reaches levels as high as 300 mgPt g^{-1} (approximately 1.4 mmolPt g^{-1}), and the sorption isotherms, which could be modelled using the Langmuir equation, are characterized by steep curves in the initial concentration range: sorption isotherms are highly favourable. The solid concentration rises 100 mg Pt g^{-1} and higher levels even at so low concentrations as 0.1-0.2 mg Pt L^{-1}. Thus, the process is very efficient for the economic recovery of this noble metal. However the sorption properties are strongly influenced by the presence of competitor anions (nitrate, chloride anions) in the matrix of the effluent. Another interesting characteristics of these systems is the fast sorption rate obtained in platinum recovery. While several days are required to reach equilibrium, almost 90 % of the total metal removal is achieved in the first 2 to 5 hours of contact. For many metal ions, the limiting step in sorption is intraparticle diffusion due to the low porosity of chitosan. The crosslinking treatment is able to decrease crystallinity , it favours accessibility to internal sites. Here, the calculation of Biot numbers has shown that both intraparticle and film diffusions are involved in the kinetic control: the enhancement of intraparticle diffusion coefficients allows the external mass-transfer resistance to become significant. The intraparticle diffusivities range between 10^{-14} and 10^{-12} m^2 s^{-1}, 2 to 3 orders of magnitude lower than the molecular diffusivities in water, but higher than those obtained with other metal ions such as molybdenum, or vanadium. Both sorption capacities and intraparticle diffusivities are little varied by increasing sorbent size.The main effect in mass-transfer coefficients is obtained with increasing metal concentration: it involves a faster diffusion of metal ions. Surprisingly, gel bead conditioning has little effect on sorption capacity and strongly depresses the mass transfer rate. The effect of the crosslinking ratio is currently under investigation. Dynamic sorption is also being carried out as well as the study of platinum desorption in order to recycle the sorbent and to scale-up the process for industrial purposes.

REFERENCES

1. R.A. Beauvais and S.D. Alexandratos, React. Funct. Polym., 36 (1998) 113.
2. J.L. Cortina, E. Meinhardt, O. Roijals and V. Marti, React. Funct. Polym., 36 (1998) 149.
3. J. Fu, S. Nakamura and K. Akiba, Sep. Sci. Technol., 32 (1997) 1433.
4. J.L. Cortina, N. Miralles, M. Aguilar and A.M. Sastre, Hydrometallurgy, 40 (1996) 195.
5. B. Volesky and Z.R. Holan, Biotechnol. Prog., 11 (1995) 235.
6. M. Tsezos and B. Volesky, Biotechnol. Bioeng., 26 (1982) 385.
7. E. Guibal, C. Roulph and P. Le Cloirec, Wat. Res., 26 (1992) 1139.
8. J.R. Duncan, D. Brady, A. Stoll and B. Wilhelmi, B. in: Biohydrometallurgical processing, C.A. Jerez, T. Vargas, H. Toledo and J.V. Wiertz (Eds.) University of Chile, Santiago, vol. II, p. 237-246, 1995.
9. N. Kuyucak and B. Volesky, Biorecover, 1 (1989) 189.
10. J.J. Byerley, J.M. Schareri and S. Rioux, in: J.Salley, RG.L. McCready and P.L. Wichlaz (Eds.) Biohydrometallurgy, CANMET SP89-10, p. 301-316, 1989.
11. Y. Kawamura, M. Mitsuhashi, H. Tanibe and H. Yoshida, Ind. Eng. Chem. Res., 32 (1993) 386.
12. T.Y. Hsien and G.L. Rorrer, Ind. Eng. Chem. Res., 36 (1997) 3631.
13. E. Guibal, C. Milot and J.M. Tobin, Ind. Eng. Chem. Res., 37 (1998) 1454.
14. C.A. Eiden, C.A., Jewell, and J.P. Wightman, J. Appl. Polym. Sci., 25 (1980) 1587.
15. M. Gonzalez-Davila, M. Santana-Casiano and F.J. Millero, J. Colloid Interface Sci., 137 (1990) 102.
16. G.L. Rorrer, T.Y. Hsien and J.D. Way, Ind. Eng. Chem. Res., 32 (1993) 2170.
17. E. Guibal, M. Jansson-Charrier, I. Saucedo and P. Le Cloirec, Langmuir, 11 (1995) 591.
18. Y. Baba and H. Hirakawa, Chem. Lett., (1992) 1905.
19. K. Inoue, T. Yamaguchi, M. Iwasaki, K. Ohto and K. Yoshizuka, Sep. Sci. Technol., 30 (1995) 2477.
20. A. Baxter, M. Dillon, K.D.A. Taylor and G.A.F. Roberts, Int. J. Biol. Macromol., 14 (1992) 122.
21. G.A.F. Roberts, Chitin chemistry. MacMillan, London, 1992.
22. G. Charlot, Dosages absorptiométriques des éléments minéraux. Masson ed., Paris, 1978.
23. A. Domard, Int. J. Biol. Macromol., 9 (1987) 98.
24. T. Mitani, C. Nakajima, I.E. Sungkono and H. Ishii, J. Environ. Sci. Health, A30 (1995) 669.
25. E. Piron, M. Accominoti and A. Domard, Langmuir, 13 1997) 1653.
26. R.-S. Juang and C.-Y. Ju, Ind. Eng. Chem. Res., 37 (1998) 3463.
27. S. Yiacoumi and C. Tien, J. Colloid Interface Sci., 175 (1995) 333 and 347.
28. S. Yiacoumi and C. Tien, Kinetics of metal ion adsorption from aqueous solutions: models, algorithms and applications. Kluwer Academic Publishers, Norwell, MA, 1995.
29. C. Tien, Adsorption calculations and modeling. Butterworth-Heinemann, Boston, 1994.
30. W.J. Weber jr. and F.A. DiGiano, Process dynamics in environmental systems, John Wiley & Sons, Inc., New York, 1995.
31. G. McKay and S.J. Allen, Can. J. Chem. Eng., 58 (1980) 521.
32. J. Crank, The mathematics of diffusion. 2nd ed., Clarendon Press, Oxford, 1975.
33. F. Hellferich, Ion exchange. Dover Publications, Inc., Mineola, NY, 1995.
34. N.F.E.I. Nestle and R. Kimmich, J. Phys. Chem., 100 (1996) 12569.

As(V) removal from dilute solutions using MICB (molybdate-impregnated chitosan beads)

L. Dambies, A. Roze, J. Roussy and E. Guibal*

Ecole des Mines d'Alès - Laboratoire Génie de l'Environnement Industriel, 6, avenue de Clavières - F-30319 Alès cedex - France

A new adsorption process for As(V) ion removal from an aqueous solution was studied using molybdate-based chitosan gel beads. Arsenate ions were strongly adsorbed in the pH range from 2.5 to 3.5 with a minimum release of molybdate ions. The sorption mechanism assumed here is a complexation between arsenate ions and molybdate ions. Even at low equilibrium concentration, the sorption capacity is high, and allows the process to be used as a polishing treatment. Phosphates ions significantly depress arsenate collection because of a competing reaction for the active sites. Simultaneously with the arsenate sorption, molybdenum is released to a significant extent (about 15-20 %). However this molybdenum release can be reduced using an orthophosphoric pre-treatment of MICB, which allows the weakly bound molybdenum to be removed from the sorbent and therefore its release does not exceed 2 %. A binding constant of 1.2 L mg^{-1} and a capacity constant of 197.6 mgAs g^{-1}Mo were obtained, using the Langmuir model. A selective and total elution can be carried out using a 0.1 mol L^{-1} orthophosphoric acid solution. The spent sorbent can be re-used for further sorption of arsenate ions with the same performance. The process has been successfully carried out with real industrial effluents from Mining and Microelectronics.

1. INTRODUCTION

Arsenic is widely distributed: in natural waters and wastewaters. In industry: it is currently used in microelectronics for the manufacturing of GaAs supports [1], as well as in pesticides and wood preservatives. Increased activity in gold mining results in arsenic contamination of streams and groundwaters. Other mining activities also generate large amounts of contaminated wastewaters since arsenic is present as an admixture in molybdenum, lead and copper ores. The wastewater generated during the hydrometallurgical treatment of these minerals usually contains high arsenic levels. Usual treatments for arsenic removal consist of flotation [2], co-precipitation with ferric chloride, sulfide precipitation [3] or lime softening, involving the production of highly toxic sludges, which must be further treated before being environmentally safe for disposal.

As an alternative to these treatments, sorption techniques have been experimented with, using activated carbon [4], fly ash [5], coral limestone [6] or biological materials such as

---

* Author to whom all correspondence should be addressed

living or non-living biomass. However taking into account the low sorption capacities of these materials and/or the difficulties of using them in column systems (fungal or bacterial biomass frequently causes clogging and head losses in such continuous systems), several derived products of biological origin, such as alginate [7], chitin, chitosan [8-9] have been studied as alternatives for metal recovery.

While chitosan has been shown to be very efficient for the removal of metals such as uranium,vanadium and molydenum,with uptake capacities ranging between 2 and 8 mmol g^{-1} (dry weight, d.w.)[10-12] for arsenic, uptake levels do not exceed 0.1 mmol g^{-1} (unpublished results) comparable to sorption levels reached using amine modified coconut coir [13].

Recently Min and Hering [14] have used an impregnation technique for the improvement of arsenate sorption on alginate beads, using mainly Fe(III)-doped materials. The mechanism involved is often an ion exchange/precipitation between the impregnated metal and arsenate ions. However, in most cases, the reuse of the sorbent is difficult to achieve, so the development of new processes allowing recycling of the sorbent is thus needed. To address this objective, in this study we use a metal which has the ability to complex arsenate ions. Thus, arsenate can be eluted from the sorbent selectively by using another complexing agent. In this study, molybdate ions were used for this impregnation. Moreover, metal ion sorption has been shown to be restricted in many cases by diffusion limitations in raw chitosan due to the low porosity of the native material. Diffusion can be enhanced by a conditioning of the biopolymer, including a gel formation treatment [12,15]. Arsenic sorption on chitosan gel beads impregnated with molybdate is herein investigated.

2. MATERIAL AND METHODS

2.1 Materials

Chitosan was provided by ABER-Technologies (Brest-France). Chitosan characteristics were : $pK_a = 6.2$, molecular weight $= 1.3\ 10^5$, and deacetylation percentage $= 87\%$. Chitosan gel beads were manufactured using the alkaline coagulation/precipitation procedure of an acetic acid solution of chitosan [12,15]. Water content of chitosan gel beads was about 95 %. Reagents were purchased from Merck: $(NH_4)_6Mo_7O_{24}.4\ H_2O$ and $Na_2HAsO_4.7\ H_2O$.

2.2 Molybdate-impregnation procedure

The optimum pH for molybdenum sorption is around pH 3 [12]. Known amounts of wet chitosan beads were put in contact with known amounts (volume and concentration) of ammonium heptamolybdate at pH 3. Dry mass was obtained by weight loss at 105 °C. The molybdate concentration in the sorbent was obtained by mass balance between liquid and solid phases (ICP-AES was used for molybdate analysis in solution). The MoO_3 content was also obtained by weight loss at 480 °C, until weight remained constant, and the molybdate content deduced from the difference with dry mass at 105 °C. Both methods are consistent with an experimental error lower than 8%. Total Mo and P concentrations in the gel phase were determined (when necessary) by disrupting the gel phase with hydrogen peroxide, since this reagent is known to depolymerize chitosan. At least 20 impregnated beads were transferred to a 25 ml volumetric flask, 2 ml of 50 % v/v hydrogen peroxide was added and the flask was heated at 90°C in a thermostated bath for 20 minutes. After cooling, the flask was filled with demineralized water. The solution was then filtered through a Whatman filter membrane (pore size 1.2 µm) and Mo, P concentrations were determinated by ICP-AES. Experiments were run in duplicate. MICB sorbents treated with orthophosphoric acid (MICB-

PO4) were prepared by mixing MICB in a 0.1 molar solution of orthophosphoric acid for 24 hours followed by an extensive rinsing with demineralized water. Total Mo and P concentrations in the gel phase were determined by the analytical procedure described above.

2.3 Experimental procedure for arsenic sorption and desorption

Except for experiments dedicated to the study of pH effects, the pH was adjusted to a fixed optimum pH and was not controlled during metal ion sorption; only the final pH was measured. Sorption isotherms were obtained through contact of a varying number of beads with a known volume (50 mL) of arsenate solution (C_0: 5, 10, 20 mg L^{-1}) at room temperature, for at least 120 hours. Arsenic concentration was measured by ICP-AES, after filtration. A control check was carried out to make sure that no unwanted arsenate sorption took place on experimental apparatus. Concentration in the solid was obtained by mass balance and hydrolysis procedure using mineralisation of chitosan sorbent with H_2O_2 and further ICP-AES analysis. For kinetic studies, 200 beads were mixed in one liter of As (V) solution. At pre-determined times, samples were withdrawn and filtered through a 1.2 μm pore size filter before analysis.

Several elutants were investigated, such as tartaric acid, citric acid and phosphate, by putting a fixed number of chitosan beads (ca. 20 beads) of known arsenate concentration in contact with a fixed volume of elutant (ca. 10 mL) at a controlled pH. Arsenic content was determined by the mineralization procedure described above.

Dynamic arsenic removal was carried out using a column (diameter: 0.05 m, depth: 0.6 m). Chitosan beads (960 g wet weight, 76.9 g dry weight) were impregnated by mixing at pH 3 with 80 L of a molybdate solution (initial concentration: 500 mg L^{-1}). The molybdate content in beads was obtained by mass balance and weight loss and reached 866 mgMo g^{-1} dry weight. The total volume of the column, BV (bed volume) was 1.178 L, the sorbent volume was 0.841 L and the voidage was 0.29. The column was pre-conditioned at pH 3 with sulphuric acid solution. The concentration of the feed solution was 23.2 mgAs L^{-1}. The solution was pumped up through the column with a flow rate of 1.165 L h^{-1}. The superficial velocity was 0.6 m h^{-1} and the volumetric velocity was almost 1 BV h^{-1}. Samples were regularly withdrawn until the outlet concentration of the column reached 90 % of the initial concentration. The desorption was performed with 5 L of 0.1 M solution of orthophosphoric acid pumped down (first step) and up (second step), and samples were collected over the desorption step for As, Mo, and P analysis. The column, between desorption and adsorption steps, was conditioned with sulphuric acid solution to pH 3.

3. RESULTS AND DISCUSSION

3.1 pH optimization

Figure 1-a shows the change in As(V) sorption capacity with pH using molybdate impregnated-chitosan beads: pH 3 appears to be the optimum pH for arsenic removal. In these acid conditions, the overall charge of the raw biopolymer is positive. This property was used to sorb molybdate and saturate the nitrogen sorption sites (with uptakes capacities as high as 1000 mgMo g^{-1} dry chitosan). Apart from this electrostatic effect, the effective mechanism of sequestration is not yet demonstrated: identification of the stoichiometry between Mo and amino functions, as well as the molybdate species adsorbed on the biopolymer (mono- or hydrolysed polynuclear forms) is still under discussion, due to the various polynuclear species involved in the hydrolysis phenomena [16]. The overall charge of the sorbent saturated with molybdate ions is thus not clearly identified. Arsenate ions at pH ranging between 3 and 4 mainly occur as H_3AsO_4 and $H_2AsO_4^-$ [17-18].

The mechanisms involved in arsenate removal by molybdate-impregnated chitosan beads may consist of electrostatic attraction, precipitation on molybdate and/or ion exchange [13,17]. Whatever the mechanisms, for this anion a pH lower than 4 is required to reach a significant sorption level. It corresponds to the predominance of neutral arsenate species (H_3AsO_4). This optimum pH range is consistent with those suggested by Huang and Vane [18] but it is different to the optimum pH range for IRA-900 (pH 5.5-6.5) [13]. Min and Hering [14] obtain similar results for arsenate sorption using Fe(III)-doped alginate beads, with a sharp optimum around pH 3. Ohki et al. [6] point out that the calcium carbonate of the coral limestone, which dissolves slightly in aqueous solution, works as a buffer to keep the pH of the solution constant and slightly alkaline.

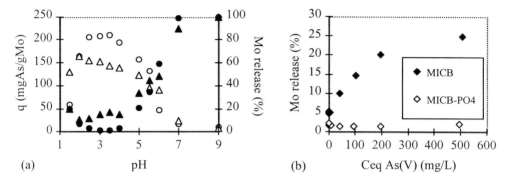

Figure 1. (a) pH influence on As(V) sorption (open symbols) and Mo(VI) release (filled symbols) on MICB ($\square$/$\blacksquare$) and MICB-PO4 ($\bullet$/$\circ$) (C_0: 100 mgAs L^{-1}, sorbent dosage: 180 mg L^{-1} (d.w.)); (b) Mo(VI) release as function of As(V) equilibrium concentration: comparison of MICB (1033 mgMo g^{-1} (d.w.)) and MICB-PO4 (972 mgMo g^{-1} (d.w.)). (20 beads in 50 mL).

Figure 1-b shows the difference in the molybdenum release for both MICB and MICB-PO4 sorbents. It clearly demonstrates that the molybdenum release does not exceed 2 %, even at high residual arsenic concentration, for the latter, while the native sorbent (MICB) releases almost 25 % in large excess of arsenic. The assumed competitive binding occuring in solution between molybdenum and arsenic which involves molybdate release from the sorbent, is not observed to the same extent with the modified sorbent. The systematic leaching of large amounts of molybdenum observed on the raw material suggests that molybdate ions present in the sorbent may be separated in two fractions: molybdenum strongly bound to the polymer network, and molybdenum simply absorbed in the gel phase of the sorbent. The latter is the weakly bound molybdenum easily displaced by arsenic binding in solution. In the case of MICB-PO4, orthophosphoric acid treatment displaces 20 % of the initial amount of molybdate, it allows the weakly bound molybdate to be removed and to overcome the problem of molybdate release when significant amounts of arsenate are present in solution.

3.2 As(V) sorption isotherms

Isotherms were carried out with MICB-PO4 (Figure 2). MICB-PO4 is characterized by a higher capacity for arsenate ions than MICB, maximum uptake capacity is increased by the orthophosphoric acid treatment. This enhancement is directly connected to the removal of weakly bound molybdate. Molybdate release in the solution is decreased and arsenate ions are not complexed by molybdate species. The change in sorption capacity is thus attributed to an increase in the relative proportion of active molybdenum in the sorbent (displacement of the weakly bound molybdate due to the orthophosphoric acid pre-treatment). It induces also a reduction in the competition of free-molybdate in solution with sorbent-fixed molybdate for arsenic binding. The uptake levels are comparable for IRA-900 [13] and the molybdate-impregnated sorbent around 0.7 mmol g^{-1} in the concentration range investigated, while a comparison with Fe(III)-doped alginate beads (Min and Hering, 1998) shows that the latter is significantly higher than the levels exhibited by MICB. Obviously, the arsenate sorption is followed by molybdate release. The molybdate desorption reaches a minimum in a broad range of pH around pH 3. This pH can be seen as the best pH for both As(V) sorption and Mo(VI) release and was selected for subsequent experiments. At pH 9 molybdate was almost entirely released and arsenate ions were not collected at all. Hence, arsenate sorption can be correlated to the release of molybdate ions from the beads. The high relative sorption capacities obtained with MICB is balanced by the serious drawback represented by the large Mo(VI) release occuring by As(V) sorption.

The Langmuir model best fits the isotherm data. Non-linear least square regression was used for the determination of the isotherm constants and q_m and b were found to be respectively q_m: 197.6 mgAs g^{-1}Mo and b = 1.2 L mg^{-1}. These results are consistent with the study of Min and Hering [14] in which constants were found to be 352 mgAs g^{-1}Fe and 1.68 L mg^{-1} for arsenate sorption by Fe (III)-impregnated alginate beads.

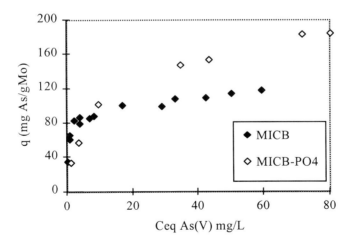

Figure 2. Sorption isotherm of As(V) on MICB-PO4 at pH 3 (experimental points and Langmuir modeling curve) (Mo-Charge: 972 mgMo g^{-1} (d.w.)).

3.3 Desorption of As(V)

To be economically attractive, the treatment of arsenic wastes using MICB, requires the regeneration of the spent sorbent. This regeneration was studied by determining the ability of the arsenic to be stripped selectively from the functionalized sorbent, using several elutants such as citric acid, tartaric acid and phosphate. Several elutants are able to desorb arsenate efficiently, however, in many cases molybdate is also released, preventing an effective separation of arsenate and molybdate and restricting the potential re-use of the sorbent (Table 1). The best results were obtained with phosphate (Figure 3-a).

Whatever the fixed pH, desorption efficiency increases with phosphate concentration, while for a fixed concentration in phosphate, the influence of pH is non-significant. Whatever the pH, an increase in phosphate concentration involves an increase in molybdate desorption. At a fixed phosphate concentration, increasing the pH involves a large increase in molybdate release. The optimization of the regeneration of the sorbent has to take into account both pH and phosphate concentration. To address this objective, desorption was carried out at three orthophosphoric acid concentrations (Figure 3-b). The higher the acid concentration, the higher the arsenic desorption. Total arsenic recovery and a 15 % release of molybdate were obtained with a 0.1 molar solution.

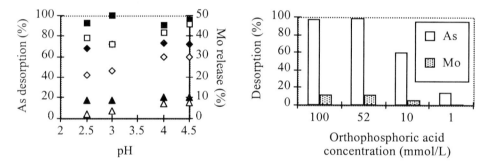

Figure 3. As (V) desorption (filled symbols) and Mo(VI) release from MICB with phosphates ions (a) and orthophosphoric acid (b) at different concentrations (A/a: 0.1 M; B/?: 0.01 M; D/d: 0.001 M)(20 beads in 10 mL, Mo-Charge: 972 mgMo g^{-1} (d.w.)).

3.4 Dynamic sorption and desorption of As(V) on MICB

Figure 4 shows the breakthrough curves for As(V) sorption on MICB. It clearly demonstrates that the sorption capacity is maintained even after the desorption of the column and the treatment with orthophosphoric acid. This result was expected as the treatment with phosphoric acid contributes to the formation of MICB-PO4, which has proved to be more efficient at removing arsenic than the raw MICB material. The breakthrough occurs at about 80 L, corresponding to almost 70 BV. However, the second adsorption step shows a small increase in the breakthrough volume. It can be explained by the phosphoric acid treatment preventing further release of molybdenum. The molybdenum concentration outside of the sorbent being lower, the competition between free- and sorbed-molybdenum for arsenic binding is less active and the efficiency of the column is maintained for a longer time and volume.

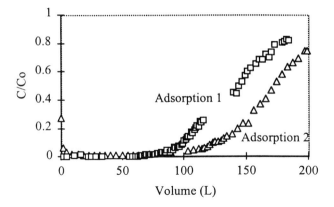

Figure 4. Breakthrough curves for As(V) sorption on MICB/MICB-PO4 sorbents at pH 3 (C_0: 20 mg L^{-1})

Figure 5 shows the desorption of the exhausted columns with phosphoric acid. It appears that the desorption curves are not symmetrical. This could be explained by several reasons: (i) hydrodynamic constraints such as dead volume, (ii) non-linearity of the sorption isotherm. The Langmuir model fits experimental data best, and a symmetrical curve is obtained with a linear system [19]. It appears that the desorption curves follow the same trend with a slight shift in the beginning of the breakthrough: for the second desorption, the breakthrough occurs earlier, as well as the end of the desorption process, than for the first desorption. This shift can be attributed to the better conditioning resulting from the orthophosphoric acid treatment and the removal of weakly bound molybdate. At the second desorption, arsenic is concentrated in a low volume by comparison with the first step. The pre-treatment involves an increase in the concentration factor.

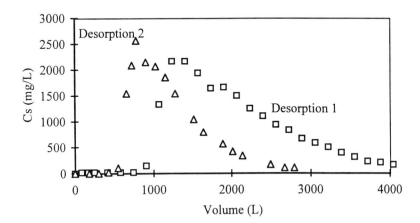

Figure 5. Desorption curves for exhausted MICB/MICB-PO4 sorbents using orthophosphoric acid (0.1 M).

It is also interesting to compare the breakthrough curves corresponding to arsenic desorption, molybdenum release and phosphorus release over the desorption step (Figure 6). The breakthrough for each of these compounds occurs at the same volume at 0.9 L: phosphate ions are sorbed on the sorbent, molybdate ions are stable on the sorbent and arsenic is not desorbed. In a second stage of the process, arsenic appears in the effluent simultaneously with molybdate and phosphate ions. Phosphate ions substitute molybdate and arsenate ions involving their displacement to the liquid phase. Each of the three breakthrough curves has the same slope, while in the third stage, molybdenum and arsenic concentrations outlet the column strongly decrease, phosphate concentration slightly increases with a change in the slope of the curve. A correlation exists between the displacement of the three compounds, however, it seems difficult to give a clear and unequivocal interpretation of these relationships.

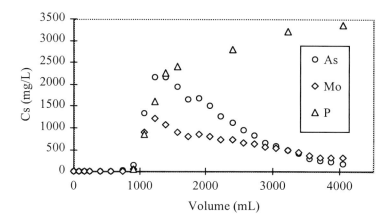

Figure 6. Correlation between As(V) desorption, Mo(VI) and P(V) release during the first step.

Table I reports the sorption and desorption performance observed during the two sorption/desorption cycles. It confirms that the adsorption capacity is higher for the second cycle than for the first, the desorption efficiency on arsenic exceeds 99 %, while the sorbent looses about 5 % of the initial molybdenum content at each sorption and desorption step, when the column has been pre-treated with phosphoric acid.

Table I
Mass balance in adsorption and desorption steps (initial Mo amount: 28.6 g)

Step	As removal / desorption (g)	Mo release (g)	Residual Mo amount (g)	Mo relative loss (%)	q (mgAs gMo⁻¹)
Adsorption 1	3.32	2.90	25.7	10.1	116.0
Desorption 1	3.27	2.09	23.7	7.8	-
Adsorption 2	3.95	1.08	22.6	4.6	166.8
Desorption 2	3.94	1.08	21.5	4.9	-

3.5 Efficiency of MICB and MICB-PO4 for the treatment of industrial effluents

The process is currently carried out for the treatment of 2 industrial sites: an abandoned mining site which produces acid mining drainage (AMD) and a microelectronic company involved in the manufacturing and recycling of GaAs supports.

The AMD effluents are characterized by the presence of several metal cations at pH 3, among them, iron is the predominant element with concentration reaching 500 mg L^{-1}, and the arsenic concentration exceeds 50 mg L^{-1}. On the other hand, the microelectronic effluent is strongly alkaline (pH 10.2), contains some traces of Ga and a large concentration of colloidal silicates (exceeding 50 g L^{-1}), while the arsenic concentration reaches 20 mg L^{-1}. Effluents were previously controlled at the optimum pH for arsenic removal (ca. pH 3).

Preliminary results are presented in Figure 7 and show that the presence of competitor cations for the AMD effluent significantly affects the efficiency of the process. For equivalent residual concentration the sorption capacity decreases by a factor higher than 4. Lead is also removed as well as iron: the concentration decreases by 10 %. However in the case of lead, surprisingly, increasing the number of beads increases the residual concentration.

For the effluents of the microelectronic unit, the presence of silicate, In such a large concentration, involves a strong decrease in sorption capacities. The influence of silicate, at low concentration (the Si/As molar ratio varies between 1 and 10), has been examined and it appears that these elements do not interfere in the As(V) sorption (Figure 8-a), molybdate release does not exceed 1.5 % and the sorption capacity is of the same order as in absence of silicates. In the industrial effluent, the concentrations are significantly higher, under these conditions, silicates are present in part as colloidal species which could involve a competition with MICB-PO4 sorbent: the silicate colloids are suspected to adsorb or bind to molybdate ions. The availability of molybdate sorption sites is thus reduced, which explains the loss of sorption capacities. The maximum uptake capacity is 5 fold lower than the standard level. For increasing arsenic concentrations, the sorption capacity decreases significantly (Figure 8-b), in this concentration range, experiments were performed with a small number of beads and molybdate is known to be easily displaced by the arsenic in the medium, while a high sorbent concentration involves simultaneous reduction of arsenic concentration and molybdate release.

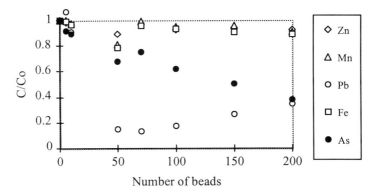

Figure 7. Influence of the number of beads (MICB) on the sorption of As(V) and removal of Pb(II) in AMD.

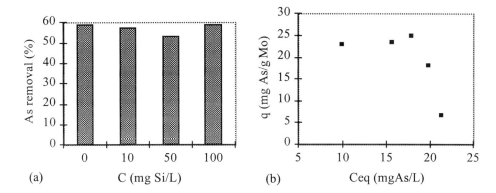

Figure 8. Influence of silicate ions on (a) As(V) sorption on MICB-PO4 (AsV = 20 mg L^{-1}, pH=3.2) and (b) As(V) sorption isotherm in the treatment of the microelectronic unit effluent.

4. CONCLUSION

Molybdate-impregnated chitosan beads (MICB) were shown to be effective means of removing arsenate, reaching high relative sorption capacities even at low residual arsenic concentrations. At pH 3, arsenate sorption is maximum and the release of molybdate is minimum. The yield of Mo(VI) release is mainly a function of the arsenic concentration in solution, and is almost independent of the Mo(VI) concentration in the sorbent. This may be due to a competition effect induced by arsenate in solution, displacing molybdate from the solid to the liquid phase.

The study of the effect of pH demonstrated that arsenate sorption can be correlated to the release of molybdate ions from the sorbent, while arsenate ions are bound to molybdate on impregnated sorbent. According to the speciation diagram of arsenate, arsenate species are more negatively charged at pH 4 than at pH 2.5 (80 and 10 % of $H_2AsO_4^-$ form at pH 4 and 2.5 respectively). However, the extent of As (V) sorption remained unchanged within this pH range, leading to the conclusion that the adsorption does not occur by ion exchange.

The mechanism is likely to occur through complexation between arsenate and molybdate ions, since the molybdoarsenate complex is well-known and currently used for the colorimetric determination of arsenate ions in solution [20]. However, the chemistry of the molybdo-arsenic anions is very complex and only a predominance of several species has been established depending on pH and concentration of both anions. Thus, in solution when both As(V) and Mo(VI) are present at pH lower than 3, the predominant species is the uncharged complex $H_3AsMo_{12}O_{40}$ [2,18]. Adsorption of cations or anions such as molybdate on chitosan brings about a considerable rise in pH. The change in pH during metal sorption can be explained by both metal chemistry in solution and protonation of the amino groups of chitosan. However during arsenate sorption, pH varies little even with a high concentration of beads. Over the sorption process, neither OH$^-$ nor H$^-$ is released in the solution. It confirms that the mechanism is different to the usual mechanism involved in metal ion accumulation on chitosan-based sorbents. This is confirmed by the lack of effect of chloride and nitrate ions on As(V) removal, since a reduction in sorption performances would be expected for ion-exchange process.

Arsenate desorption can be performed in an acid medium in the presence of phosphate: desorption efficiency can exceed 95 %. The optimum conditions for sorbent regeneration are controlled by both pH and phosphate concentration: a 0.1 molar solution of orthophosphoric acid appears to be the optimum for high arsenic desorption and low molybdenum release.

The process has been carried out with two wastewaters from mining and microelectronic industries, the presence of high concentrations of competitor cations as well as very high silicate levels decreases sorption capacity by a factor 4 or 5.

REFERENCES

1. United States Environmental Protection Agency (1995). Profile of the electronics and computer industry. September 1995.
2. Y. Zhao, A.I. Zouboulis and K.A. Matis, Hydrometallurgy, 43, 143.
3. D. Bhattacharyya, A.B. Jumawan and R.B. Grieves, Sep. Sci. Technol., 14 (1979) 441.
4. C.P. Huang and P.L. Fu, J. Water Pollut. Control. Fed., 56 (1984) 233.
5. E. Diamadopoulos, S. Ioannidis, and G.P. Sakellaropoulos, Wat. Res., 27 (1993) 1773.
6. A. Ohki, K. Nakayachigo, K. Naka, and S. Maeda, Appl. Organometal. Chem., 10 (1996) 747.
7. J. Chen and S. Yiacoumi, Sep. Sci. Technol., 32 (1997) 51.
8. C.M. Elson, D.H. Davies and E.R. Hayes, Wat. Res., 14 (1980) 1307.
9. R.A.A. Muzzarelli, F. Tanfani and M. Emanuelli, Carbohyd. Polym., 4 (1984) 137.
10. E. Guibal, M. Jansson-Charrier, I. Saucedo and P. Le Cloirec, Langmuir, 11 (1995) 591.
11. E. Guibal, I. Saucedo, M. Jansson-Charrier, B. Delanghe and P. Le Cloirec, Wat. Sci. Technol., 30 (1994) 183.
12. E. Guibal, C. Milot and J.M. Tobin, Ind. Eng. Chem. Res., 37 (1998) 1454.
13. A.E. Baes, T. Okuda, W. Nishijima, E. Soto and M. Okada, Wat. Sci. Technol., 35 (1997) 89.
14. Min, J.H. and J.G. Hering, Wat. Res., 32 (1998) 1544.
15. G.L. Rorrer, T.Y. Hsien and J.D. Way, Ind. Eng. Chem. Res., 32 (1993) 2170.
16. C.F. Baes and R.E. Mesmer, Hydrolysis of cations, Wiley & Sons, New York, 1976.
17. L. Lorenzen, J.S.J. van Deventer and W.M. Landi, Miner. Eng., 8 (1995) 557.
18. Y. Zhao, A.I. Zouboulis and K.A. Matis, Sep. Sci. Technol., 31 (1996) 769.
19. C.P. Huang and L.M. Vane, J. Water Pollut. Control. Fed., 61 (1989) 1596.
20. W. Kast and W. Otten, Ind. Chem. Eng., 29 (1989) 197.
21. G. Charlot, Dosages absorptiométriques des éléments minéraux, Masson, Paris, 1978.

On the melanin and humic acids interaction with clay minerals

V. Kadoshnikov[a], N. Golovko[a], M. Fomina[b], B. Zlobenko[a], J. Pisanskaya[a]

[a]National Academy of Sciences of Ukraine Scientific Center of Environmental Radiogeochemistry, Palladin av. 34, Kiev 252142, Ukraine

[b]Institute of Microbiology and Virology of National Academy of Sciences of Ukraine, Zabolotnogo 154, Kiev 252143, Ukraine

When developing technologies of ionselective biomineral sorbents produced by dark-colored micromycetes cultivation in media containing clay minerals, it is of a great importance to define processes taking place on the boundary of mineral phase and organic phase. Particularly, the interaction between sorption-active components of dark-coloured microfungi melanins and surface of clay minerals (montmorillonite, kaolinite) was studied by infrared spectroscopy. Keeping in view that the melanin structure closely resembles that of humic acids, we examined humic acids and clay minerals associates obtained by heterocoagulation of peat humic acid hydrosol with clay mineral particles. It was established that the organic phase fixation on the clay minerals surface occurred as a result of (I) the interaction between -COOH groups of the organic phase and exchange groups of mineral after the ligand exchange and (II) hydrogen bonds formation between corresponding groups of organic and mineral phase. The direct interaction between the organic phase and mineral surface with coordination bonds is also possible.

1. INTRODUCTION

One of the most important problem now is to protect environment against radioactive and heavy metals pollution. Natural high-dispersity alumino-silicates are used in the development of geochemical barriers. Microorganisms are ubiquitous in soil environment and present in the natural geochemical barriers. Microorganisms, especially dark-coloured microfungi (micromycetes), could play the significant role in geochemical barriers affecting the changes of their filtration and sorptive properties. High-dispersity materials alter the growth and the activity of soil microorganisms [1]. Some micromycetes are able to dissolve hot particles and at the same time to sorb mobile forms of radionuclides and heavy metals [2]. It is well known that fungal melanin have a high biosorptive capacity for a variety of metal ions [3]. The feasibility of using microbial biomass mixed with clay in the sorption of toxic metals attracted a special attention [4, 5]. As we have recently shown, biomass of dark-coloured fungi grown on the media containing clays can form biomineral sorbent with enhanced sorption ability compared to controls of pure clay and pure biomass [6, 7]. We supposed that the changes of sorptive properties of biomineral sorbents occur thanks to certain structural organization of fungal melanin on the surface of clay minerals. The aim of present investigation is to study interaction between clay minerals (montmorillonite, kaolinite) and fungal melanin using method of infrared (IR) spectroscopy. Taking into account that the main structural units of melanin and humic acids are very similar [8], the comparative study of interaction of fungal melanin and peat humic acids with clay minerals was carried out.

2. MATERIALS AND METHODS

2.1. Melanin

Melanin of *Cladosporium cladosporioides* (Fresen) de Vries 396 obtained from the collection of The Department of Taxonomy and Physiology of Micromycetes of the Institute of Microbiology and Virology of the National Academy of Sciences of Ukraine in Kiev (Patent of Ukraine 1523,15.09.93). The fungi were grown in Erlenmeyer flasks at 25° C and 160 rpm on liquid modified media, comprising per liter distilled water: NH_4NO_3, 2g; KH_2PO_4, 1g; $MgSO_4.7H_2O$, 0.5g; sucrose, 15g. The spore' inocula (10^6 conidia ml^{-1}) were used. Conidia for inoculum were obtained from the fungal surface culture grown on Malt agar medium.

Melanin was isolated from the dried mycelia using a modification of the method of acidic separation of pigment from biomass [9]. The essence of this modification consists of a one-stage boiling of ground fungal biomass in mixture of concentrated acetic and sulfuric acids for 4 hours followed by removal of hydrolysis products with hot 80 % acetic acid and then with hot distilled water at a pH of 5.5-6.0. The suggested modification of the method gave stable and statistically reliable melanin yields from *C. cladosporioides* biomass and was recommended for melanin extraction from other melanin-producing fungi. All chemical and spectral characteristics testified to the melanin nature of the obtained pigment [9].

2.2. Humic acids

To obtain humic acids from peat (Belorussia) the method of M. Kononova [10] was used. Dried and sifted peat was treated with 0.1 N acid to destroy bonds of humic substances with mineral components of peat and then was treated with 0.1 N solution of NaOH during 12-14 hours. After that the coagulation of alkali extracts with sulfuric acid solution to reach pH 2-3 was carried out. Sediments of humic acids were washed several times with NaOH and purified using ion-exchange resins. Purified samples of humic acids were dried to a constant weight at 60°C.

2.3. Clays

Montmorillonite, isolated from bentonite of Cherkassky deposit, the second stratum (Ukraine), and high-dispersity disordered kaolinite (fire-clay), isolated from secondary kaolinite of Glukhovsky deposit (Ukraine), were used in this study. Some principal physic-chemical properties of these clay minerals are shown in the Table 1.

Table 1.
Physic-chemical properties of clay minerals

Properties	Kaolinite	Montmorillonite
Content of main mineral (%)	90-95	90-95
Mean size of particles (μ)	0.2-0.5	0.05-0.10
Specific area (m2/g)	60-70	110-120
Ion-exchange capacity (MEQ/100g)	10-12	70-80
Exchangeable cations (%)	Ca^{+2} - 80 Mg^{+2} - 20	Ca^{+2} - 80 $Mg^{+2}+, Na^+, K^+$ - 20

2.4. Infrared Spectroscopy

To examine the IR spectra of samples spectrophotometer UR-20 was used. The samples were ground to obtain size of particles 1μ, were mixed with powdered crystalline salt of potassium bromide and then pressed into a transparent disk. A disk of pure potassium bromide was used as a blank. Using this method samples do not disperse radiation at wavenumbers less than 400 cm^{-1} and more than 5000 cm^{-1} and their optical density can be quantified.

3. RESULTS AND DISCUSSION

IR spectra of melanin, montmorillonite and biomineral sorbents, obtained using melanin and montmorillonite were compared (Fig.1). For IR spectrum of biomineral sorbent compared to that of melanin it was found that:

- intensity of absorption at wavenumbers 1385 cm^{-1} and 1415 cm^{-1} (deformation oscillations in bonds CH-, OH-, CO=) decreased;
- the extension of the bandwidth from 1640 cm^{-1} to 1625-1640 cm^{-1} , but intensity of its absorption decreased too;
- intensity of absorption band 1720 cm^{-1} (valency oscillations of bond CO=) considerably reduced and its maximum of absorption shifted into the lower-frequency band ($\Delta\nu\approx10$ cm^{-1});
- absorption intensity at wavenumbers 2860-2960 cm^{-1} (valency oscillations of bond CH-) significantly decreased too, meanwhile absorption intensity in the region of valency oscillations of OH-groups (3250 cm^{-1}) considerably increased;
- absorption band 3285 cm^{-1} did not developed due to the considerable extension of the bandwidth 3300 cm^{-1} and its maximum shifted to the higher-frequency band ($\Delta\nu\approx10$ cm^{-1}).

Comparative study of IR spectra of melanin, kaolinite and biomineral sorbent, obtained using melanin and kaolinite, showed that the tendency in changes of IR spectrum of melanin-kaolinite biomineral sorbent compared to melanin IR spectrum was very similar to that found for melanin-montmorillonite biomineral sorbent.

The structures of melanin pigments and humic acids are very similar. IR spectra of clay mineral and humic acids complexes, obtained using adsorption of humic acids from hydrosol on the surface of high-dispersive clay minerals, were investigated. The humic acids adsorption by clay minerals (kaolinite) from hydrosol was found to occur quickly and complete within the first 15 minutes of interactions. Further adsorption of humic acids occurred slower. The quantity of adsorbed humic acids depended mainly on the specific area of clay mineral.

The main attention of IR spectroscopic study was paid to wavenumbers associated with functional groups responcible for the interactions between humic acids and clay mineral surface [11]. There were: the absorption band 1715 cm^{-1} associated with valency fluctuations of bond CO=, the absorption bands 2850 and 2920 cm^{-1} (methyl and methylene groups of aliphatic chains), the wide slightly structured absorption band 3000-3600 cm^{-1} with two maximums 3400 cm^{-1} (OH- fluctuations) and 3250 cm^{-1} (NH-fluctuations) (Fig. 2, 3).

It was found that IR spectra of both humic acid and melanin adsorbed to the clay surface exhibited almost full disappearance of absorption band 1715 cm^{-1} attributive to pure humic acids or melanin. The extension of the bandwidth of high frequency boundary of absorption maximum of the band 1650 cm^{-1} was observed. Thus, absorption band 1715 cm^{-1} shifted to the lower-frequency regions. A new absorption band 1400 cm^{-1} appeared.

The extension of absorption band with maximum 1650 cm^{-1} could result from the appearance of the additional absorption in this region due to: (I) the shift of the band 1715 cm^{-1} to the lower-frequency band after formation of stronger hydrogen bonds with OH-groups of

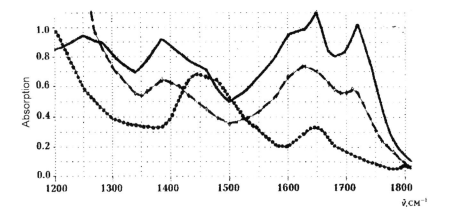

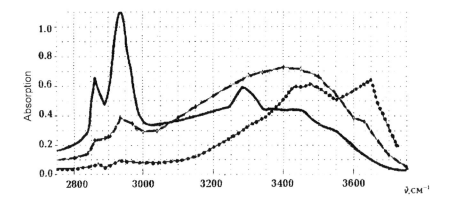

Fig. 1. Infrared spectra :

————— – melanin

———— – biomineral sorbent

•••••• – montmorillonite

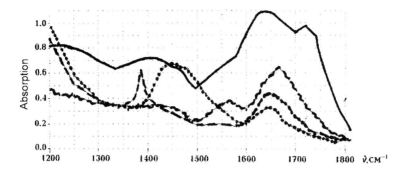

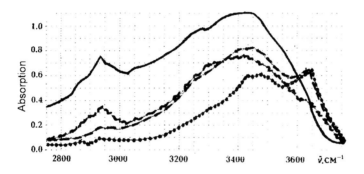

Fig. 2. Infrared spectra :
    ▬▬ — humic acid,
    •••• — montmorillonite,
    ▬ ▬ — organomineral sorbent (5% of humic acid),
    ∼∼ — organomineral sorbent (0.5% of humic acid).

294

clay minerals; (II) the disappearance of carbonyl groups absorption band and the following appearance of symmetric (band 1400 cm⁻¹) and antisymmetric (region of overlapped bands at 1650 cm⁻¹) vibrations of COOH-groups as a result of the interaction of an anion of humic acid-like polymer (melanin) with a cation of the clay surface.

Interaction may occur in two possible ways:

$$M^{+2\delta} \underset{O^{-\delta}}{\overset{O^{-\delta}}{\diagup}} C-R \qquad\qquad M-O \underset{O}{\overset{}{\diagup}} C-R$$

It was previously supposed [12] that at pH above 3 polyvalent exchange cations of clay minerals can interact with humic acid-like polymers by the mechanism of ligand exchange leading to the formation of surface substance:

$$\left[R-C \underset{O}{\overset{COO}{\diagup}} Fe \underset{OH}{\overset{OH}{\diagdown}} (H_2O)_2 \right]^-$$

In the region of valency fluctuations of OH-groups of humic acids adsorbed by the clay surface the increase of wavenumber of a maximum from 3400 to 3415 cm⁻¹ was observed. Meanwhile the relative intensity of this absorption band was the same. This can be expected in the case of superimposition of absorption band of water OH-bonds on initial band as a result of adsorption of water molecules on clay surface. Thus, the ratio of involving of OH-groups in hydrogen bonds in the micelles of free humic acids or melanin it can be assumed to decrease during adsorption by clay, probably, due to dissociation of colloid aggregates and certain orientation of humic acid-like polymers on the surface of clay mineral particles. As a result, the superstructure of colloid aggregates changed and the surface organo-mineral associates were formed. The significant desorption of water molecules from the surface of clay particles occurred because humic acid as more active adsorbate can substitute for adsorbed molecules of water. Treatment with acid solution led partly to re-adsorption of water.

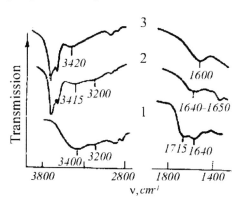

Fig. 3. Infrared spectra :
1. – kaolinite,
2. – humic acids,
3. – organomineral sorbent kaolinite – humic acids.

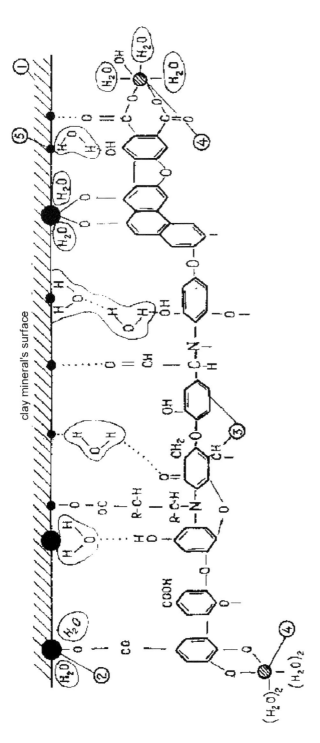

Fig. 4. Scheme of bonds of humic acid-like polymer with clay minerals

1. – clay mineral,
2. – exchange cation of clay,
3. – humic acid-like polymer,
4. – cations of humic acids,
5. – active site of clay.

Some authors supposed [11,12] that functional groups of humic acids can substitute for molecules of water from the first coordination sphere and interact directly with the exchange cations.

Taking into account our results it can be assumed that during humic acid-like polymers sorption to the surface of clay particles the adsorption layer of adsorbate with the open structure is formed as a result of partial dissociation of colloids of humic acids or melanins connected with breaking of intramolecular hydrogen bonds and appearance of additional adsorption sites [13] (Fig. 4). This supposition is testified by data of radionuclides sorption from aqueous solutions by biomineral sorbents made from clay and biomass of *Cladosporium cladosporiodes*, its melanin [7] and humic acids (Table 2, 3).

Table 2.
Sorption radiocaesium by biomineral sorbents [7]

Sorbent	Specific sorption, Bq/g	Distribution coefficient Kd[*], ml/g
Bentonite	8960	8220
Mycelium	4820	113
Melanin	8149	885
Mechanical mixture bentonite-mycelium	8885	4803
Biomineral sorbent 1 montmorillonite-mycelium	8970	9079
Biomineral sorbent 2 montmorillonite-melanin	9020	17860

Table 3.
Sorption of radiocaesium by organomineral sorbents (our unpublished data)

Sorbent	Distribution coefficient Kd[*], ml/g
Bentonite	2415
Humic acids	220
Mechanical mixture bentonite-humic acids	672
Organomineral sorbent bentonite-humic acids	4176

* Kd is used as a main parameter, which presents a ratio of equal concentrations of radionuclides in solid phase and solution. $Kd = A_S/A_l$, where A_s – concentration of radionuclides in solid phase, Bq/g; A_l – corresponding concentration in solution, Bq/ml.

4. CONCLUSIONS

The fixation of humic acids and melanins to clay minerals is assumed to be a result of physical and physic-chemical adsorption.

1. Physical adsorption means a fixation of the humic acids (melanins) to the surface of clay particles with Van der Waals forces. Such humic acids are slightly bonded with the mineral part and can be easily washed out with neutral water solution of salts.

2. Mechanism of the physic-chemical adsorption is more complicated and can be described as followed:
   a) physic-chemical adsorption due to the interaction between carboxyl and phenolic groups of the organic phase and the exchange cations of the external surface of clay particles;
   b) adsorption as a result of hydrogen bonds formation between corresponding groups of humic acids (melanins) and surface of the clay particles;
   c) adsorption thanks to the direct interaction of active sites of mineral particles with corresponding functional groups of the organic phase (humic acids and melanins).

REFERENCES

1. 1.D.G. Zvyagintcev, Soil and Microorganisms, Ed. Moscow University, Moscow, Russia, 1987.
2. N.N. Zhdanova, A.I. Vasilevskaya, V.I. Gavrilyuk, E.L. Sholokh and L.A. Koval, Mycologiya I Fytopatologiya, 24(2) (1990) 106.
3. R.V. Fogarty and J.M. Tobin, Enzyme and Microbial Technology, 19 (1996) 311.
4. G.W. Garnham, G.A. Codd and G.M. Gadd, Microbial Ecology, 25 (1991) 71.
5. G.F. Morley and G.M. Gadd, Mycol. Res., 99, No.12 (1994) 1429.
6. V.M. Kadoshnikov, B.P. Zlobenko, N.N. Zhdanova and T.I. Redchitz, Proc HLM, LLW, Mixed Wastes and Environmental Restoration - Working Towards A Cleaner Environment, WM'95, Tuscon, Arizona, C.D., (1995) 61.
7. M.A. Fomina, V.M. Kadoshnikov and B.P. Zlobenko, Proceedings of International Regional Seminar "Environmental Protection: Modern Studies in Ecology and Microbiology", Uzhgorod, 2 (1997) 86.
8. M.S. Boils, Photochemistry and Photobiology Rev., 3 (1978) 115.
9. N.N. Zhdanova, A.B. Melezhik, A.T. Shkolny and O.I. Sinyavskaya. Mikrobiologichesky Zhurnal (in Russian with English Summary), 55(1) (1993) 79.
10. M.M. Kononova, Organic substances of soil, Ed. AN USSR
11. F.J. Stevenson, Humus Chemistry, John Wiley & Sons Inc., N.-Y., 1994.
12. Yu.I. Tarasevich and F.L. Ovcharenko, Adsorption by clay minerals, Naukova Dumka, Kiev, Ukraine, 1975.
13. V.M. Kadoshnikov and V.J. Manichev, Proceedings of International Mineralogical Association, 16-th General Meeting, Italy, 1994.

Biosorption of rare earth elements

A. A. Korenevsky, V. V. Sorokin, and G. I. Karavaiko

Institute of Microbiology, Russian Academy of Sciences,
Prospekt 60-letiya Oktyabrya 7/2, 117811 Moscow, Russia

By using X-ray microanalysis, the mechanism of sorption of rare earth elements (REE) and their localization in cells of *Candida utilis* were found to depend on the metal ion speciation in solution, the permeability of the cytoplasmic membrane (CPM), and elemental composition of cells. Sorption capacity of the yeast cells increased with the increase in the pH of solution, which is connected with the extent of metal hydrolysis. Cells with native permeability of CPM did not sorb either scandium at pH values below 4.5 or lanthanum and samarium at pH values below 5.0. Such cells accumulate rare earth elements on surface structures. Only the cells with impaired CPM could sorb REE from the acid solutions. In this case, REE were accumulated inside the cells due to the interaction with phosphorus-containing compounds; the amount of sorbed REE depended on the content of phosphorus in the yeast cells. The yeast cells were shown to have extremely high affinity to scandium which thus can be selectively sorbed from solutions containing other REE, iron, and aluminum.

1. INTRODUCTION

The rare earths are more abundant in the Earth crust than many better known metals. The other metals are better known in part because nature has concentrated them in forms that make their recovery economical.

From the geochemical point of view rare earths are "dispersed" elements, i.e., spread around among many common minerals rather than concentrated into a select few. Rare earths occur mainly as trace constituents of the major, rock-forming minerals of common rocks and partly in accessory minerals in which the rare earths are either essential constituents (e.g., monazite) or are concentrated (e.g., apatite) [1].

Rare earth elements (REE) are present in titanium, zirconium, tin, tungsten, uranium ores, bauxites, and some others. These ores that now yield REE as by-products or co-products, may become principal sources for REE production. During complex treatment of these ores, REE are accumulated in the wastes (solutions, slags, muds, etc.) in concentrations exceeding manifold their initial content in the ores.

Leaching of the above-mentioned slags and muds yields solutions containing REE and associated elements (Fe, Al, Ti in high concentrations and some other elements) [1].

Conventional methods of REE recovery from solutions, i.e., precipitation, extraction, and ion exchange, often appear inefficient due to the low concentrations of REE (less than 20 mg/l) and high concentrations of the associated metals (iron, aluminum, titanium, calcium, etc.)

Investigations by different authors have shown high sorption affinity and selectivity of microbial biomass and biosorbents for some elements (U, Ag, Au, Mo, Cr, etc.) [2]. We found that the employment of biosorption for the recovery of REE from acid leach solution of red mud (the by-product of bauxite treatment by the Bayer method) allowed scandium and yttrium to be separated from each other and from Fe, Al, Ti, Si, and Ca [3].

The aim of this study was to investigate the mechanisms of REE biosorption and the reasons of the selectivity of this process.

Various spectroscopic methods are employed in studies of the mechanisms of metal sorption by microbial biomass and biopolymers: IR-spectroscopy, NMR, EPR, and also X-ray microanalysis [4-7]. X-ray microanalysis is one of the few methods that allow one to perform multielemental analysis of the cell, reveal the localization of sorbed metals, and determine the composition of a discrete metal deposit. Therefore, we employed X-ray microanalysis to study the influence of elemental composition of the yeast cells on their capacity to sorb REE.

2. MATERIALS AND METHODS

Batch culture of the yeast *C. utilis* VKM Y-1668 was used in the experiments. The yeast was grown in medium of the following composition (g/l): sucrose, 5.0; $(NH_4)_2SO_4$, 1.0; K_2HPO_4, 0.5; $MgSO_4 \cdot 7H_2O$, 0.1; the Pfennig trace element solution, 1 ml/l; the pH was 6.5. For inoculation, the yeast culture grown for 12 h in the above medium was used. The yeast cultures were grown in 250-ml Erlenmeyer flasks containing 100 ml of medium at 28°C on a shaker (180 rpm). After 16 h of growth, yeast biomass was separated from the culture liquid by centrifugation and washed three times with deionized water. The yeast suspension was introduced in 0.5 mM solutions of nitrates of rare earth elements in deionized water. The desired values of pH in REE solutions were adjusted with HNO_3 or NaOH. The yeast cells were incubated in REE solutions at 20°C for 20 min, then separated by centrifugation and washed three times with deionized water.

To obtain yeast cells with the impaired CPM, the cells were treated with 40% ethanol and then washed three times with deionized water.

For X-ray microanalysis, the cell suspensions were applied to collodion-coated copper grids sputtered with carbon and examined on a JEOL JEM-100CXII electron microscope equipped with an EM-ASID4D scanning unit and a Link 860 X-ray analyzer with an E5423 detector.

3. RESULTS AND DISCUSSION

Electron microscopy studies of yeast cell preparations revealed the occurrence of several morphologically distinct types of cells: single nondividing cells, cells with buds, lysed and damaged cells. Cells containing deposits of metal hydroxides on their cell surface were also encountered.

The influence of elemental composition of cells on their capacity to sorb metals was studied on morphologically uniform cells, namely, on cells without buds. The damaged cells and cells with hydroxide deposits on the surface were not examined.

The ability of *C. utilis* cells to sorb REE depended on pH of the solution. No sorption of scandium was observed at pH values below 4.5. Lanthanum, samarium, and yttrium were not sorbed at pH values below 5.0. At pH values above 7.0, the deposits of REE hydroxides appeared on the surface of cells.

It is known that trivalent cations of REE cannot enter the live cells with the help of energy-dependent or passive transport [8], hence, it may be concluded that these metals are sorbed by the surface structures of the yeast cells. In *C. utilis*, the microcapsule, cell wall, and CPM may be regarded as such structures.

It is known that in solutions, the REE ions are involved in the processes of hydration, hydrolysis, polymerization, and complexing with the anions of mineral and organic acids. When present at concentrations below 10^{-3} M, REE exist in the form of hydrated monomeric ions. The stable hydrolyzed forms of $Ln(OH)_n^{(3-n)+} \cdot aq$ are formed in a multistep process that depends on the pH in the medium. Scandium, the most liable to hydrolysis element, has the smallest ionic radius (0.083 nm) and the increased electrostatic field intensity as compared with other REE cations. Hydrolysis of scandium ions begins at pH values above 3.0, hydrolysis of the other REE ions — above 5.0 [9]. By the degree of hydrolysis and stability constants of hydroxycomlexes, scandium considerably differs from other REE whose stability constants are lower by 2-3 orders of magnitude [10].

REE sorption by the surface structures of cells begins upon the appearance of hydrolyzed metal ions in the solution. Therefore, it is the hydrolyzed forms of metals that are sorbed by the cell.

Literature provides evidence that the metals accumulated by microorganisms are localized either inside the cells or within their cell wall. Thus, uranium, sorbed by the cells of *Saccharomyces cerevisiae* and *Rhizopus arrhizus*, and lanthanum, sorbed by *Pseudomonas aeruginosa*, *Escherichia coli*, and *Bacillus subtilis*, were accumulated on the cell surface or within the cell wall [5, 6,11]. At the same time, the cells of *P. aeruginosa* were shown to accumulate uranium intracellularly [11]. Localization of uranium and gold in the cells of microscopic algae and yeast (*S. cerevisiae*) was shown to depend on the duration of contact and the concentration of metals in the solution. A short-term contact mainly resulted in the accumulation of metals on cell wall and in periplasm, whereas after a long-term incubation, gold and uranium were accumulated both on the cell wall and intracellularly [12].

Localization pattern of sorbed metals is evidently associated with the integrity of cytoplasmic membrane (CPM) of the microbial cell. The metals can accumulate inside the cell either with the help of the transport system or due to diffusion through the impaired CPM. Microbial CPM can be damaged by various substances: the efficient complexing agents, detergents, organic solvents, etc., including the heavy metal ions. The ions of Ag^+, Hg^{2+}, Ni^{2+}, Cu^{2+}, and Pb^{2+} were shown to damage the CPM of microorganisms [7, 13, 14]. As a consequence, the cells became partially or completely deprived of potassium and the intracellular contents of phosphorus and sulfur were also reduced [13]. The damaging effect of heavy metal ions on the CPM is attributed to their interaction with the sulfhydryl groups of the membrane.

In our experiments, the rare earth elements did not exhibit a damaging effect on CPM of the yeast cells. Therefore, to study the influence of the CPM integrity on the REE sorption, the cells of *C. utilis* were treated with 40% ethanol. This treatment resulted in a virtually complete disappearance of intracellular sodium and potassium. It is obvious that the ethanol treatment impaired the integrity of the CPM, thus providing for the free access of REE inside the cell. The presence of the potassium peak in the spectrum ($\overline{K}\alpha1$ — 3.312 keV) is an evidence of the CPM integrity and, consequently, of the cell viability. The treatment of *C. utilis* cells with ethanol at concentrations below 20%, did not produce any effect on their CPM and or elemental composition.

Cells treated with 40% ethanol were found to sorb REE even at pH 0.5. This could only occur in the case of intracellular sorption, since no REE sorption by the cell surface structures could proceed in the acid medium.

Correlation analysis showed a direct relationship between the amount of sorbed REE and the content of phosphorus in the cells with impaired permeability of the CPM (Table 1). The amount of REE sorbed from both the acid and neutral solutions increased with the increase of the phosphorus content in the yeast cells treated with ethanol. No relationship between the amount of sorbed REE and the content of phosphorus was found in the cells of *C. utilis* with the intact CPM. The relationships between the sorption of scandium and the content of phosphorus in *C. utilis* cells are presented in Fig. 1; this is a vivid example of the effect produced by ethanol pretreatments of the cells on the relationship under consideration. No stable relationship between the contents of Cl, K, Mg, and Ca and the amount of sorbed REE has been found.

Correlation coefficients between the sulfur content and the content of sorbed REE reach 0.86–0.88, although they are less than those between the contents of phosphorus and REE (Table 1). However, in cells treated with 40% ethanol a correlation between the contents of phosphorus and sulfur became evident (the correlation coefficients comprised 0.74–0.91). To reveal the influence of the correlation between the contents of sulfur and phosphorus on the relationship between these elements and the amount of sorbed REE, we calculated the partial correlation coefficients (Table 1). Partial correlation coefficients between the contents of sulfur and REE varied from 0.02 to 0.36, whereas their variation between the contents of phosphorus and REE was 0.74–0.99. Hence it may be assumed that in such cases the correlation between sulfur and REE is mainly determined by the correlation between the contents of sulfur and phosphorus.

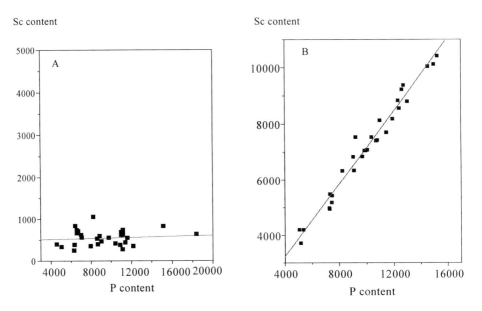

Figure 1. Correlation between the intracellular content of phosphorus and scandium sorbed by cells of *C. utilis* (pH=5.0; P=0.95); untreated (A) and treated (B) with 40% ethanol.

Table 1.
Correlation coefficients between the amount of metals sorbed and the amounts of
phosphorus and sulfur in the cells of *C. utilis.*

Sorption condition / Treatment	Standard correlation			Partial correlation	
	P : Me	S : Me	P : S	P : Me -S	S : Me -P
Control, pH=7.0	—	—	0.09	—	—
Control, pH=7.0; 40% ethanol	—	—	0.91	—	—
Control, pH=2.0; 40% ethanol	—	—	0.85	—	—
Sc 0.5 mM, pH=5.0	0.10	0.05	0.58	—	—
Sc 0.5 mM, pH=5.0; 40% ethanol	0.99	0.88	0.86	0.94	0.36
Sc 0.5 mM, pH=2.0; 40% ethanol	0.99	0.39	0.40	0.87	0.21
La 0.5 mM, pH=7.0	0.32	0.13	0.29	—	—
La 0.5 mM, pH=7.0; 40% ethanol	0.87	0.85	0.88	0.77	0.10
La 0.5 mM, pH=2.0; 40% ethanol	0.95	0.12	0.38	0.99	0.02
Sm 0.5 mM, pH=7.0	0.24	0.18	0.08	—	—
Sm 0.5 mM, pH=7.0; 40% ethanol	0.98	0.89	0.88	0.74	0.20
Y 0.5 mM, pH=7.0	0.08	0.15	0.22	—	—
Y 0.5 mM, pH=2.0; 40% ethanol	0.96	0.38	0.42	0.93	0.09
Fe 0.5 mM, pH=2.0	0.12	0.24	0.53	—	—
Fe 0.5 mM, pH=1.0; 40% ethanol	0.95	0.90	0.93	0.79	0.18
Al 5 mM, pH=2.0; 40% ethanol	0.88	0.40	0.38	0.86	0.17

Moreover, in cells preincubated in acid solutions as well as in cells sorbing REE at pH 2.0, the correlation degree between the contents of sulfur and phosphorus decreases, and the correlation coefficients decline from 0.91 to 0.85 in the control variants and from 0.88 and 0.86 to 0.38 and 0.40 in cells sorbing lanthanum and scandium, respectively. At the same time, the correlation between the contents of phosphorus and REE remains costant. It is apparent that REE interact with the phosphate-containing compounds of the yeast cells, i.e., phosphates, polyphosphates, nucleic acids, phospholipids, and phosphorylated polysaccharides.

Studies on the interactions of REE with phosphates and polyphosphates showed that the high affinity of REE for phosphate groups provides for the high stability of their complexes, which can be compared with that of the hydroxycomplexes. Thus, lgβ for $[Sc(OH)]^{2+}$ and $[Sc(HPO_4)]^+$ comprise 9.26 and 10.12, respectively [10].

High values of the correlation coefficients (approaching 1.0) indicate that the compounds with a definite ratio of REE:P are formed.

A similar regularity was observed in the studies of uranium sorption by the cells of *Streptomyces longwoodensis* [4]. The amount of the uranium sorbed increased with the growth of the culture and the increase of the phosphorus content in the cells. The stoichiometric ratio of the phosphorus content to the content of uranium sorbed, was 1:1. The interaction between heavy elements (uranium, lanthanum) and phosphate-containing

304

compounds of microbial cells was also reported by other authors [6,7]. It was also shown that sorption of lanthanum by the immobilized cells of *Citrobacter* sp., the producer of acid phosphatase, resulted in accumulation of lanthanum phosphates on the surface of cells [15].

We found that in the pH range from 0.5 to 5.0, *C. utilis* cells with both intact and impaired CPM, sorbed only scandium from the solutions containing equal amounts (0.5 mM) of Sc, Y, La and Sm. At the pH values above 7.0, the peaks of all four REE, present in the solution, were revealed in the X-ray spectra of the yeast cells (Fig. 2).

The mechanism of binding ferric iron and aluminum by the cells of *C.utilis*, was found to be the same as that of REE binding.

Sorption of ferric iron and aluminum by the cells with the intact CPM, proceeded at pH values above 1.6–1.7 and above 4.0, respectively. The yeast cells treated with ethanol, were found to sorb ferric iron even at pH 0.5, whereas aluminum was insignificantly sorbed at pH 2.0 and was not sorbed at pH 0.5. It is notable that a considerable accumulation of aluminum in the yeast cells occurred during the biosorption from the solutions of the red mud acid leaching (pH 0.6) [3].

The sorbed ferric ions significantly inhibited the REE sorption. At a Fe:Sc ratio of 5--10, the ethanol-treated cells sorbed no more than 10–15% of Sc as compared with the solution without Fe^{3+}. However, when the Fe:Sc ratio was 1, scandium completely inhibited the sorption of ferric iron.

As evident from Table 1 and Fig. 3, in cells with the impaired CPM that sorbed iron or aluminum from the acid solutions, there was a stable correlation between the intracellular content of phosphorus and the amount of sorbed metals.

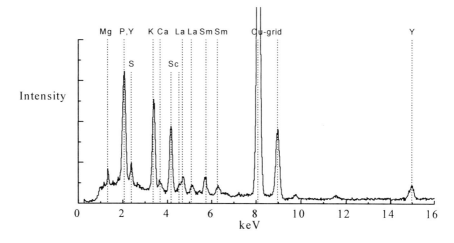

Figure 2. X-ray spectrum of the cells of *C. utilis* after sorption Sc, Y, La, and Sm at pH=7.5.

Fe content   Al content

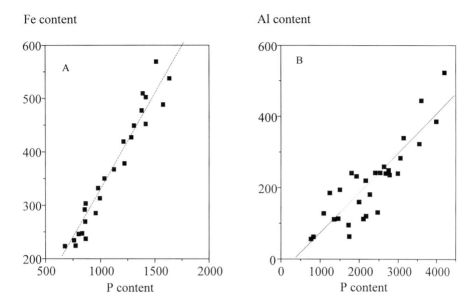

P content   P content

Figure 3. Correlation between the content of phosphorus in the cells of *C. utilis* treated with 40% ethanol and the amount of ferric iron sorbed (A); pH=1.0; and aluminum sorbed (B); pH=2.0; (P=0.95).

Filatova and co-authors showed that the selectivity of sorption of trivalent cations (by zirconium and titanium phosphates from acid solutions depends on the ability of a given cation to form a four-member cycle ($\equiv$P$<^O_O>$Me-). The stability constant for the phosphate complex of scandium was found to be higher by more than two orders of magnitude than the constants for the other REE, ferric iron, and aluminum phosphate complexes. The stability constant of phosphate complexes is mainly determined by the value of ionic radius (r) of the cation: r=0.083 nm is the maximum radius providing the formation of a four-membered cycle possessing the least tension [16].

It is evident that during intracellular sorption, the trivalent cations of REE, aluminum and iron compete for the phosphate groups. The increased selectivity of scandium sorption is determined by the fact that its radius closely matches the above-mentioned radius.

The analyses performed, allowed us to reveal two different ways of metal binding by the yeast cells, which depended on the integrity of CPM and the pH of the solution: the intracellular sorption and sorption with the help of the cell surface structures. REE sorption by the cell surface structures is strongly pH dependent. In contrast, intracellular sorption of REE proceeds in a broad range of pH and is mainly determined by the content of phosphate-containing compounds of the cell.

REFERENCES

1. L.A. Haskin and T.P. Paster, In: Handbook on the Physics and Chemistry of Rare Earths., V.3, K.A. Gschneider, Jr and L. Eyring (eds.), North-Holland Publ. Company, 1979.

2. B. Volesky and Z.R. Holan, Biotechnol. Prog., 11, (1995) 235.

3. G.I. Karavaiko, A.S. Kareva, Z.A. Avakian, V.I.Zakharova, and A.A. Korenevsky, Biotechnol. Lett., 18, (1996) 1291.

4. N. Friis and P. Myers-Keith, Biotechnol. Bioeng., 28, (1986) 21.

5. M. Tsezos and B. Volesky, Biotechnol. Bioeng., 24, (1982) 385.

6. M.E. Bayer, M.H. Bayer, J. Bacteriol., 173, (1991) 141.

7. M.D. Mullen, D.C. Wolf, F.G. Ferris, T.J. Beveridge, C.A. Fleming and G.W. Bailey, Appl. Environ. Microbiol., 55, (1989) 3143.

8. C.F. Evans, Trends Biochem. Sci., 83, (1983) 445.

9. V.A. Nazarenko, V.I. Antonovich and E.M. Nevskaya, Metal ion hydrolysis in dilute solutions, Atomizdat, Moscow,1979.

10. L.I. Komissarova, Zh. Neorg. Khim., 25, (1980) 143.

11. G.W. Strandberg, S.E. Shumate II, and J.R. Parrot, Appl. Environ. Microbiol., 41, (1981) 237.

12. N. Kuyucak and B. Volesky, Biotechnol. Lett., 10, (1988) 137.

13. A.A.Korenevsky, G.I. Sorokin, and G.I. Karavaiko, Microbiology, 62, (1993) 428.

14. V.S. Lebedev, E.Yu. Deinega, O.S. Savluk, and Yu.I. Fedorov, (1989) Biol. Membrany., 6, 1313.

15. L.F. Tolley, L.F. Strachan, and L.E. Macaskie, J.Indust.Microbiol., 14, (1995) 271.

16. L.N. Filatova., T.N. Shelyakina, and T.N. Kurdyumova, In: Reagents and ultrapure compounds. Trudy IREA 46, (1984) 72.

An overview of the studies about heavy metal adsorption process by microorganisms on the lab. scale in Turkey

Y. Sağ and T. Kutsal

Hacettepe University, Chemical Engineering Department, 06532, Beytepe, Ankara, Turkey

Heavy metal ions - namely Cr(VI), Pb(II), Cu(II), Fe(III), Ni(II) and Zn(II)- were removed from synthetic aqueous solutions by using *Z. ramigera* and *R. arrhizus*. The adsorption equilibrium data fit both the Langmuir and the Freundlich models. Heavy metal adsorption on free and immobilized cells was also investigated in various reactor types such as stirred tank reactors (single and multi-stage), packed bed and fluidized bed reactors. The multi-component biosorption of Cr(VI), Fe(III), Cu(II), Pb(II) and Ni(II) from binary and ternary mixtures was described, compared to single metal ion situations in solution. The multi-component sorption phenomena were expressed by the competitive adsorption models developed for optimum adsorption conditions.

1. INTRODUCTION

Bioremoval involves a combination of active and passive transport mechanisms. The first stage, usually referred to as passive uptake, is an initial rapid and reversible accumulation step [1]. This kind of uptake is also termed ' biosorption'. Biosorption can be considered as a collective term for a number of passive accumulation processes such as physical and chemical adsorption, ion exchange, coordination, complexation, chelation and microprecipitation [2]. There are two principal mechanisms involved in biosorption: 1) ion exchange, wherein ions such as Na, K, Mg and Ca become displaced by heavy metal ions; 2) complexation between metal ions and various functional groups, such as carboxyl, amino, thiol, hydroxy, phosphate and hydroxy-carboxyl, that can interact in a coordinated way with metal ions [1,3]. The second stage, usually referred to as active uptake, is a slower intracellular bioaccumulation, is often irreversible and is related to the metabolic activity [1,4]. This slow phase of metal uptake can be due to a number of mechanisms, including covalent bonding, surface precipitation, redox reactions, crystallization on the cell surface or, most often, membrane transport of the metal ions into cell cytoplasm and binding to proteins and other intracellular sites [1,5]. During the adsorption, a rapid equilibrium is established between the adsorbed metal ions on the cell (q_{eq}) and unadsorbed metal ions in solution (C_{eq}) and can be represented by either the Freundlich or the Langmuir adsorption isotherms, which are widely used to present data for water and waste water treatment applications[6-8].

2. MATERIALS AND METHODS

Zoogloea ramigera , an activated sludge bacterium, and *Rhizopus arrhizus,* a filamentous fungus, were obtained from the US Department of Agriculture Culture Collection. The growth media of *Z. ramigera* and *R. arrhizus* were given elsewhere[9-11]. In the stationary phase of growth (120 h), *Z.ramigera* cells were centrifuged at 5000 rev min^{-1} for 5 min and then dried at 60°C for 24 h. After the growth period (96 h), *R. arrhizus* was inactivated using 1% formaldehyde and dried at 60°C for 24 h. For the biosorption studies, 1.0 g of dried cells was suspended in 100 ml of distilled water and homogenized for 20 min in a homogenizer at 8000 rev min^{-1}. The microorganism suspension (10 ml) was mixed with 90 ml of the desired metal solutions in an Erlenmeyer flask. The flasks were agitated on a shaker at constant temperature for 24 h, which is more than ample time for adsorption equilibrium. The samples were centrifuged at 6030 g for 3 min and the supernatant liquid was analysed for metal ions. Preparation of Ca-alginate gel-immobilized *Z. ramigera* column was also reported in our previous paper[12]. Analysis of heavy metal ions were given elsewhere[9-16].

3. RESULTS AND DISCUSSION

The kinetic results are given as the initial adsorption rate, r (mg g^{-1} min^{-1}), equilibrium results as the adsorbed metal ion quantity, $q_{eq,}$ per unit weight of dried biomass (mg g^{-1}), and the unadsorbed metal ion concentration, C_{eq} (mg l^{-1}), in solution at equilibrium, the equilibrium adsorption yield, Y_{eq} and the total equilibrium adsorption yield, $Y_{t,eq}$. The initial adsorption rate was obtained by calculating the slope of a plot of the adsorbed metal ion quantity, q, per gram of dried microorganism (mg g^{-1}) versus time (min) at t=0. The equilibrium adsorption yield was defined as the ratio of the adsorbed metal ion concentration, C_{ads} (mg l^{-1}) at equilibrium to the initial metal ion concentration, C_i (mg l^{-1}). The total equilibrium adsorption yield was defined as the ratio of the adsorbed total metal ion concentration, $C_{ads,t}$, at equilibrium to the total initial metal ion concentration, $C_{i,t}$.

3.1. Single-metal biosorption on *Z. ramigera* and *R. arrhizus* in batch stirred reactors

Bioremoval of single species of Pb(II), Ni(II), Cu(II), Fe(III), Cr(VI) and Zn(II) ions on *Z. ramigera* and *R. arrhizus* was investigated as a function of initial pH, temperature, initial metal ion concentration and microorganism concentration in batch stirred reactors. The optimum biosorption conditions for each microorganism-metal system, the initial adsorption rates, the adsorbed metal ion quantity per gram of dried biomass at equilibrium and the equilibrium adsorption yields obtained at these conditions and at a constant initial metal ion concentration of 100 mg l^{-1} are given in Table 1 [10,11,15,17].

The initial adsorption rates of metal ions on *Z. ramigera* increased with increasing metal ion concentrations up to 75-200 mg l^{-1}. Maximum initial adsorption rates for Pb(II), Ni(II), Cu(II), Fe(III) and Cr(VI) ions were found to be 10.4; 7.5; 3.3; 3.8 and 3.4 mg g^{-1} min^{-1} at 150-200; 200; 125; 100 and 75 mg l^{-1} initial metal ion concentrations, respectively and at 25°C [15]. The initial adsorption rates of metal ions on *R. arrhizus* increased with increasing metal ion concentrations upto 50-300 mg l^{-1}. Maximum initial adsorption rates for Pb(II), Ni(II), Cu(II), Fe(III), Cr(VI) and Zn(II) ions were determined as 15.5; 10.3; 4.8; 3.9, 8.4, 0.95 mg $g^{-1}$$min^{-1}$ at 300; 150; 125; 125-150, 125-150 and 50 mg l^{-1} initial metal ion concentrations, respectively and at 25°C (Figure 1).

To investigate the effect of microorganism concentration on the biosorption of heavy metal ions, the quantity of dried cells was increased while the metal concentration and solution volumes were kept constant. A increase in the microorganism concentration increased the adsorbed metal concentrations because of increasing adsorption surface area, whereas the adsorbed metal ion quantity per gram of dried cell decreased by increasing the biomass quantity. This can be attributed to the fact that the aggregates formed during biosorption reduce the effective adsorption area and lead to interference between binding sites [14].

3.2. Use of non-competitive adsorption models and a comparison of biosorption characteristics

In the biosorption studies of metal ions by *Z. ramigera* and *R. arrhizus*, adsorption equilibrium was reached in 5-15 min. Relevant adsorption parameters have been calculated according to the non-competitive Langmuir and Freundlich adsorption equations. For the biosorption of heavy metal ions on *Z. ramigera* and *R. arrhizus*, the individual Langmuir constants calculated at the optimum pH values and at 25°C are listed in Table 2 [9,15]. q_s is the amount of adsorbate per unit weight of adsorbent to form a complete monolayer on the surface. b is the ratio of adsorption/desorption rates and the ratio b/a gives the theoretical monolayer saturation capacity, q_s.

a° and b° values have been obtained by evaluating the non-competitive Freundlich adsorption isotherms and are also presented in Table 2 [9,10,11,15]. The magnitude of a° and b° shows easy separation of metal ions from waste water and also indicates favourable adsorption. The intercept of the linearised Freundlich equation, a°, is an indication of the adsorption capacity of the adsorbent; the slope, b°, indicates the effect of concentration on the adsorption capacity and represents the adsorption intensity. Comparing initial adsorption rates, equilibrium adsorption yields and adsorption constants for single-metal situation shows that the relative capacities of *Z. ramigera* and *R. arrhizus* were in the order Pb(II)>Ni(II)>Cu(II)>Fe(III)>Cr(VI) and Cr(VI)>Pb(II)>Ni(II)>Cu(II)>Fe(III)>Zn(II),

Table 1
Initial biosorption rates and equilibrium metal removal at optimum biosorption conditions

	pH	T (°C)	C_i (mg l^{-1})	r (mg g^{-1} min^{-1})	q_{eq} (mg g^{-1})	Y_{eq} (%)
Z.ramigera						
Pb(II)	4.0-4.5	35	102.4	12.8	66.7	65
Cu(II)	4.0	25	100.1	3.3	27.0	27
Cr(VI)	2.0	25	100.0	3.4	34.9	35
Fe(III)	2.0	45	100.0	7.1	58.9	59
Ni(II)	4.5	25	101.2	5.6	35.0	35
R.arrhizus						
Pb(II)	5.0	35-45	100.7	7.6	48.2	47
Cu(II)	4.0	25	104.5	4.8	27.4	26
Cr(VI)	1.0-2.0	35-45	105.9	7.7	68.2	64
Fe(III)	2.0	35	91.9	3.9	44.4	46
Ni(II)	4.5	25	99.9	6.1	20.5	21
Zn(II)	5.0	25	100.0	0.9	8.8	9

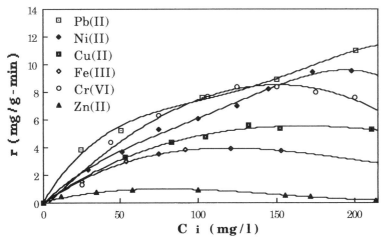

Figure 1. The effect of initial metal ion concentration on the biosorption rate of *R. arrhizus* at 25°C (the pH is constant at the optimum pH of each metal ion, microorganism conc.= 1.0 g l^{-1}; stirring rate= 150 rev min^{-1}).

respectively and *R. arrhizus* is a better metal accumulator than *Z. ramigera*. The adsorption capacities of *Z. ramigera* and *R. arrhizus* were also compared with other biosorbents and reported in our earlier papers[10,17,18]. The metal removal performance of only some species of algae known as good biosorbents can be compared to that of *R. arrhizus*.

3.3. Heavy metal adsorption by *R. arrhizus* in batch stirred reactors in series

The adsorption of Cu(II), Ni(II) and Pb(II) ions on *R. arrhizus* was investigated in three batch stirred reactors operated in series. The adsorption in multistage reactors can be considered as multistage equilibrium operation, which depends on two constraints: that of equilibrium and that of mass balance. The sorption phenomenon was expressed by the Freundlich adsorption isotherm developed for optimum adsorption conditions[14]. When aqueous solutions containing Cu(II) at a concentration of 99.9 mg l^{-1}, Ni(II) at a concentration of 97.9 mg l^{-1} and Pb(II) at a concentration of 101.7 mg l^{-1} were fed to the first reactor, 65.7 mg of copper per gram of dried biomass[14], 90.2 mg of nickel per gram of dried biomass[14] and 99.8 mg of lead per gram of dried biomass were removed in the solution, leaving a third reactor, at a chosen V_O/X_O ratio of 1.00.

3.4. Heavy Metal Adsorption by Ca-alginate and Immobilized *Z. ramigera*

The identification of the polymer-mediated metal uptake is an important advance in heavy metal uptake process. Ca-alginate itself is a good adsorbent. When blended with other metal-binding biomass, it can also serve as an excellent matrix. *Z. ramigera* was immobilized to Ca-alginate in order to obtain higher adsorption capacity and yields. The biosorption of Cu(II) [12], Pb(II) and Cr(VI) by *Z. ramigera* immobilized on Ca-alginate was investigated in packed columns operated in continuous mode. The equilibrium Cu(II) removal percentages with respect to the inlet metal ion concentration are shown in Table 3. Copper adsorption in the immobilized-cell column was initially limited by diffusion of the solute through the pores. It was interestingly found that the initial removal of Cu(II) by the immobilized microorganism system was therefore lower than initial removal of Cu(II) by the Ca-alginate system, but equilibrium Cu(II) removal percentages and total Cu(II) removal percentages were higher than those of Ca-alginate. Because of obtaining such diverse results, the dynamic behaviour of the system was investigated by stimulus-response technique and the kinetic results were re-evaluated by moment analysis, and the adsorption rate constant of Cu(II) ions on the immobilized *Z. ramigera* - Ca-alginate matrix was found to be 0.6955 cm^3 g^{-1} s^{-1}[19]. Copper recovery in 0.005 M H_2SO_4 solution, at pH 2.0, was achieved. The metal-unloaded biosorbents can be reused. For example, after regeneration, the metal-unloaded immobilized cells were loaded with fresh feed containing copper of 150 mg l^{-1} concentration, and maximum metal removal by immobilized *Z. ramigera* was obtained as 53.7% of metal presented.

Table 2
Individual Langmuir and Freundlich adsorption constants obtained at the optimum
pH values and at 25°C for the biosorption of heavy metal ions on Z. ramigera and
R. arrhizus

	Langmuir constants			Freundlich constants	
	q_s	a	b	a^o	b^o
	$(mg\ g^{-1})$	$(L\ mg^{-1})$	$(L\ g^{-1})$	$(mg^{b^o+1}\ g^{-1}\ L^{-b^o})$	
Z. ramigera					
Pb(II)	81.230	2.762	0.034	5.680	0.540
Cu(II)	34.050	1.396	0.041	3.780	0.450
Ni(II)	57.430	1.723	0.030	4.410	0.500
Fe(III)	36.360	1.054	0.029	2.890	0.500
Cr(VI)	27.530	0.661	0.024	2.240	0.480
R. arrhizus					
Pb(II)	45.780	41.339	0.903	22.939	0.161
Cu(II)	33.875	1.495	0.044	8.325	0.257
Ni(II)	30.125	0.798	0.027	2.026	0.524
Fe(III)	34.849	2.300	0.066	6.136	0.363
Cr(VI)	58.617	2.755	0.047	6.170	0.476
Zn(II)	8.720	1.267	0.145	1.886	0.336

The results obtained confirmed that Ca-alginate itself had a high adsorption
capacity to bind Pb(II) ions. However, metal uptake yields obtained by using
immobilized Z. ramigera were higher than those of Ca-alginate. The desorption of
lead was achieved in 0.05 N HNO_3 solution, at pH 1.3. Ca-alginate for the uptake of
Cr(VI) ions can be thought as an inert support matrix.The results obtained also
showed that immobilized Z. ramigera had low capacity and weak affinity to bind
Cr(VI) ions (Table 3). The Cu(II) biosorption on Ca-alginate and immobilized Z.
ramigera was also investigated in a fluidized bed but lower equilibrium adsorption
yields were obtained, compared with those obtained in the packed column.

3.5. Simultaneous biosorption of heavy metal ions on R. arrhizus
Although single toxic metallic species rarely exist in natural and waste waters,
the uptake of single species of heavy metal ions by microbial biomass has been
extensively studied, very little attention has been given to the study of multi-metal
ion systems. The dependence of the bioremoval process upon pH, temperature,
initial metal ion concentration and   microorganism concentration has been
optimised for only one metal at a time. Even though conditions of optimum
biosorption of heavy metal ions to various species of microorganisms are known,
the presence of a multiplicity of metals leads to interactive effects and affects the
physiological and biochemical processes of the various organisms.

Table 3
Equilibrium Cu(II) and Pb(II) removal percentages and maximum Cr(VI) removal percentages by Ca-alginate and immobilized *Z. ramigera.*

Biosorbent	Initial Metal Ion Concentration		
	50.0(mg/l)	l00.0(mg/l)	l50.0(mg/l)
	Packed column		
	Adsorbed copper amounts at equilibrium (%)		
Ca-alginate	74.0	76.2	35.0
Ca-alginate+*Z.ramigera*	94.3	89.4	60.0
	Adsorbed lead amounts at equilibrium (%)		
Ca-alginate	75.7	75.8	73.5
Ca-alginate+*Z.ramigera*	94.0	97.0	96.0
	Maximum adsorbed chromium amounts (%)		
Ca-alginate	24.1	25.3	25.7
Ca-alginate+*Z.ramigera*	77.0	48.8	48.9
	Fluidized bed		
	Adsorbed copper amounts at equilibrium (%)		
Ca-alginate	70.0	45.0	32.0
Ca-alginate+*Z.ramigera*	90.0	61.0	46.0

As seen from the single-metal biosorption, different metals have different pH optima, possibly due to the different solution chemistry of metal ions. Cu(II) and Cr(VI) ions have different pH optima. For that reason, to investigate the interactive effects of a mixture of metals on *R. arrhizus*, these metal ions were selected for experimental use as suitable representatives of metal ions adsorbed prefentially at a definite pH value. The concentrations of Cu(II) ions adsorbed were significantly low while Cr(VI) ions were preferentially adsorbed at pH 2.0. The effect of Cr(VI) ions on the uptake of Cu(II) ions was found to be antagonistic at pH 4.0, whereas the total interactive effects of Cu(II) and Cr(VI) ions on *R. arrhizus* can be thought to be synergistic at this pH value [20].

Cr(VI) and Fe(III) ions are more effectively adsorbed to the biomass at low values of pH and at high values of temperature. Therefore, pH and temperature will not provide selectivity for the required purification. However, this property was used for the simultaneous biosorption of Cr(VI) and Fe(III) ions from wastewaters and these metal ions were selected for experimental use as suitable representatives of fully competitive adsorption. Since the instantenous, equilibrium and maximum uptake of Cr(VI) and Fe(III) was reduced by the presence of increasing concentrations of the other metal, the combined action of Cr(VI) and Fe(III) on *R. arrhizus* was generally found to be antagonistic. The removal of Cr(VI) from the binary metal mixtures in the presence of Fe(III) was always greater than the uptake of Fe(III) under the same experimental conditions [21]. The equilibrium biosorption data of Cr(VI) and Fe(III) ions from binary mixtures by *R. arrhizus* were analysed using the competitive Langmuir model, the modified Langmuir model and the competitive

Freundlich model. The behaviour of adsorption isotherms can be well approximated to Langmuir type as compared to Freundlich type in case of Cr(VI) and Fe(III). To obtain an excellent fit at low concentrations of the competing metal ions, the modified Langmuir model equations were written for both the metal ions, solved simultaneously and the correction factors were determined from the competitive adsorption data. The correction factors for both the metal ions were close to 1, validating the proposed models for adsorption data [22]. The simultaneous biosorption of three metal ions-Cr(VI), Fe(III) and Cu(II)- has been studied using *R. arrhizus*. Although Cr(VI) ions were adsorbed selectively from the ternary metal mixtures, Fe(III) ions competed strongly with Cr(VI) ions to bind to active sites on the fungus [23].

The simultaneous biosorption of Ni(II) and Pb(II) on *R. arrhizus* from binary metal mixtures is shown to be a function of pH, the number of metals competing for binding sites in the fungal cells and metal concentrations. The effect of Pb(II) ions on the uptake of Ni(II) ions was found to be antagonistic, whereas the total interactive effects of Ni(II) and Pb(II) ions on the biosorption of Pb(II) ions by *R. arrhizus* can be considered to be synergistic at a defined concentration interval [24].

Table 4
Comparison of the equilibrium Cr(VI), Fe(III) and Cu(II) removal from binary and ternary mixtures by *R. arrhizus* at pH 2.0 and at 25°C.

$C_{Cr,i}$ (mg l^{-1})	$C_{Fe,i}$ (mg l^{-1})	$C_{Cu,i}$ (mg l^{-1})	$q_{Cr,eq}$ (mg g^{-1})	$q_{Fe,eq}$ (mg g^{-1})	$q_{Cu,eq}$ (mg g^{-1})	$Y_{Cr,eq}$ (%)	$Y_{Fe,eq}$ (%)	$Y_{Cu,eq}$ (%)
25.4	45.2	0	12.4	18.5	-	49	41	-
52.2	49.0	0	18.6	17.0	-	36	35	-
78.1	49.7	0	24.5	14.6	-	31	29	-
102.1	51.9	0	28.0	13.1	-	27	25	-
124.8	49.8	0	32.0	11.1	-	26	22	-
24.8	0	54.2	18.0	-	5.5	73	-	10
49.8	0	50.3	23.0	-	6.0	46	-	12
74.7	0	50.4	31.5	-	5.9	42	-	12
99.5	0	49.6	35.0	-	5.7	35	-	12
124.8	0	49.8	36.9	-	4.1	30	-	8
24.3	49.7	51.1	13.0	19.0	2.8	53	35	5
46.3	50.0	50.4	16.9	17.2	2.7	36	34	5
74.8	49.7	50.4	20.0	16.2	2.5	27	33	5
99.0	49.7	49.3	25.1	14.5	1.6	25	29	3
123.3	45.2	52.2	24.7	10.0	1.5	20	22	3

Table 5
Comparison of the equilibrium Cr(VI) and Cu(II) removal from binary mixtures by *R. arrhizus* at pH 4.0 and at 25°C.

$C_{Cu,i}$ (mg l^{-1})	$C_{Cr,i}$ (mg l^{-1})	$q_{Cu,eq}$ (mg g^{-1})	$q_{Cr,eq}$ (mg g^{-1})	$q_{t,eq}$ (mg g^{-1})	$Y_{Cu,eq}$ (%)	$Y_{Cr,eq}$ (%)	$Y_{t,eq}$ (%)
24.9	0	14.9	-	14.9	60	-	60
50.0	0	19.1	-	19.1	38	-	38
74.9	0	20.9	-	20.9	28	-	28
99.0	0	22.9	-	22.9	23	-	23
124.8	0	25.9	-	25.9	21	-	21
150.0	0	27.9	-	27.9	19	-	19
22.7	10.0	13.1	7.6	20.7	58	76	63
44.3	9.9	17.5	6.9	24.4	40	70	45
75.0	10.1	19.6	7.4	27.0	26	73	32
95.4	10.1	20.9	7.9	28.8	22	78	27
130.2	10.0	25.3	7.6	32.9	19	76	24
150.0	10.1	26.7	7.9	34.6	18	78	22
24.3	27.4	11.1	6.3	17.4	46	23	34
48.6	28.3	15.4	9.8	25.2	32	35	33
76.3	24.5	17.5	8.1	25.6	23	33	25
101.7	25.4	19.7	8.6	28.3	19	34	22
121.2	25.0	23.9	9.1	33.0	20	36	23
150.0	25.0	25.0	4.8	29.8	17	19	17

Table 6
Comparison of the equilibrium Pb(II) and Ni(II) removal from binary mixtures by *R. arrhizus* at pH 4.5 and at 25°C.

$C_{Pb,i}$ (mg l^{-1})	$C_{Ni,i}$ (mg l^{-1})	$q_{Pb,eq}$ (mg g^{-1})	$q_{Ni,eq}$ (mg g^{-1})	$q_{t,eq}$ (mg g^{-1})	$Y_{Pb,eq}$ (%)	$Y_{Ni,eq}$ (%)	$Y_{t,eq}$ (%)
22.8	0	18.4	-	18.4	81	-	81
24.3	24.4	22.4	5.0	27.4	92	20	56
24.8	50.1	22.8	12.7	35.5	92	25	47
22.9	75.4	20.4	15.3	35.7	89	20	36
24.6	100.0	21.4	18.7	40.1	87	19	32
24.0	124.8	19.9	21.2	41.1	83	17	28
23.3	148.7	18.6	24.4	43.0	80	16	25
50.0	0	34.5	-	34.5	69	-	69
49.9	25.3	30.1	5.5	35.6	60	22	47
50.1	50.6	29.1	9.8	38.9	58	19	39
49.1	75.2	28.5	13.1	41.6	56	17	33
49.8	96.8	27.4	15.6	43.0	55	16	29
49.0	121.0	26.5	17.4	43.9	54	14	26

4. CONCLUSION

Biosorption is being illustrated to be a potential alternative to convential systems for the removal of toxic metals from industrial effluents. However, there is a need to research and develop bioremoval to realise systems that are flexible and cost-effective in the treatment of waste waters. There are two main areas in this field of research: improving the efficiency of metal removal technology and modelling the mechanism of metal removal to reach a better understanding of the process. The development of the bioprocess requires further investigation in the direction of modelling of multi-component systems, of regeneration and re-use of the biosorbent, and of testing biomasses with real industrial waste waters.

REFERENCES

1.   E. W. Wilde and J.R Benemann, Biotech. Adv., 11 (1993) 781.
2.   D. Brady, A. Stoll and J.R. Duncan, Environ. Technol., 15 (1994) 429.
3.   H.-B. Xue, W. Stumm and L. Sigg, Wat. Res., 22 (1988) 917.
4.   Y. P. Ting, F. Lawson and I.G. Prince, Biotech. Bioeng., 34 (1989) 990.
5.   D. Khummongol, G.S. Canterford and C. Fryer, Biotech. Bioeng., 24 (1982) 2643.
6.   J.M. Smith, Chemical Engineering Kinetics, third edition, McGraw-Hill, New York, 1981.
7.   J.C. Bellot and J.S. Condoret, Process Biochem., 28 (1993) 365.
8.   W. Fritz and E.U. Schluender, Chem. Eng. Sci., 29 (1974) 1279.
9.   Y.Sağ and T.Kutsal, Biotechnol. Lett., 11 (1989) 141.
10. Z. Aksu, Y. Sağ and T. Kutsal, Environ. Technol., l3 (l992) 579.
11. Y. Sağ, D. Özer and T. Kutsal, Process Biochem., 30 (1995) 169.
12. Y. Sağ, M. Nourbakhsh, Z. Aksu and T. Kutsal, Process Biochem., 30 (1995)175.
13. F.D. Snell and C.T. Snell Colorimetric Methods of Analysis, third   edition, Van Nostrand, New York, 1959.
14. Y. Sağ and T. Kutsal, Chem. Eng. J., 58 (1995) 265.
15. Y. Sağ and T. Kutsal, Chem. Eng. J., 60 (1995) 181.
16. Y. Sağ and T. Kutsal, Process Biochem., 32 (1997) 591.
17. M. Nourbakhsh, Y. Sağ, D. Özer, Z. Aksu, T. Kutsal and A. Çağlar,   Process Biochem., 30 (1994) 175.
18. Z. Aksu, Y. Sağ and T. Kutsal, Environ. Technol., l1 (l992) 33.
19. M. Mutlu, Y. Sağ and T. Kutsal, Chem. Eng. J., 65 (1997) 81.
20. Y. Sağ and T. Kutsal, Process Biochem., 31 (1996) 561.
21. Y. Sağ and T. Kutsal, Process Biochem., 31 (1996) 573.
22. Y. Sağ, Ü. Açıkel, Z. Aksu and T. Kutsal, Process Biochem., 33 (1998) 273.
23. Y. Sağ and T. Kutsal, Process Biochem., 33 (1998) 571.
24. Y. Sağ and T. Kutsal, Process Biochem., 32 (1997) 591.

Interactions between marine bacteria and heavy metals

V.O. Ivanitsa[a], T.V. Vasilyeva[a], A.E. Buchtiyarov[a], E.B. Lindström[b] and S. McEldowney[c]

[a]Department of Microbiology and Virology, Odessa Mechnikov State University, Odessa, Ukraine*

[b]Department of Microbiology, University of Umeå, Sweden

[c]Department of Biosciences, University of Westminster, London, England**

The aim of this study was to determine the resistance of a range of marine heterotrophic bacteria to selected heavy metals. The bacteria were isolated from coastal and open regions of the Black Sea contaminated by different levels of heavy metals and other pollutants. The heterotrophic bacteria were isolated over a period of two years, 1991 and 1992. The isolates were identified as strains of *Vibrio, Pseudomonas, Cytophaga, Acinetobacter, Alcaligenes, Chromobacterium, Aeromonas, Bacillus, Arthrobacter, Micrococcus* and *Plancococcus*. The resistance of the isolates to selected heavy metals was examined and compared for 1991 and 1992 isolates. The genera *Cytophaga* and *Pseudomonas* were studied in more detail for resistance to metals and metal accumulation. Tolerance to metals was determined by point inocula onto peptone-yeast extract agar containing metal salts at a range of concentrations. The cultures were grown aerobically at 22°C for 48 hrs after which time the presence or absence of growth was determined. Metal uptake was assessed by incubating cell suspensions, containing 5 mg (wet weight) biomass in 10 ml seawater or seawater with 20 mg/l $FeSO_4$, $Mn(CO_3)_2$, $Ni(NO_3)_2$, $CdCl_2$, or $ZnSO_4$ for 7 days at 22°C. The amount of metal accumulated under each condition was determined by flame atomic adsorption spectrophotometer.

The accumulation of heavy metals by 15 *Pseudomonas* strains and 4 *Cytophaga* strains isolated from metal polluted sites varied with metal and with the strain of bacterium. The order of metal accumulation, however, was the same for both genera and was Zn>Ni>Fe>Mn>Co>Cd. The pattern of accumulation by individual strains varied between seawater and seawater with added metal ions. The resistance of the bacteria to heavy metals also varied considerably with metal and microbial type, with highest resistance shown by

* P.O. 2 Dvoryanskaya st., 270026, Odessa, Ukraine

Internet: ivanitsa@te.net.ua

** Researches supported by INTAS-UKRAINE Grant N 95-01116

Pseudomonas strains. There appeared to be no relationship between the levels of metal resistance shown by *Pseudomonas* strains and *Cytophaga* strains and the accumulation of metals by the isolates. There were marked differences in the resistance shown by 1991 isolates and 1992 marine isolates.

1. INTRODUCTION

Heavy metals are among the most common compounds polluting marine habitats, especially coastal regions of seas and oceans [1]. Micro-organisms may accumulate metals and are often the key in the transformation of heavy metals, contributing significantly to metal biogeochemical cycling [2]. Bacterial accumulation or transformation of metals may contribute to the subsequent removal of metals from the water column either through deposition in bottom sediments [2] or through attachment to solid surfaces [3]. The accumulation of potentially toxic metals by bacteria has a number of further implications. Bacteria with an accumulated metal load will serve as an entry port into marine food webs increasing the metal exposure load of filter feeders and thence organisms throughout the web. The ability of micro-organisms to recycle nutrients may be impaired through the toxicity of the accumulated metals. In addition, metal transformations may result in the formation of potentially more toxic metal compounds [2]. Bacterial accumulation of metals from seawater and their resistance to toxic metal ions are, therefore, of particular interest. The resistance of bacteria to the toxic effects of metal cations has been found previously to vary with the concentration of the metal ions in seawater [4,5].

The types and concentration of metal ions in seawater varies within a broad range. If the source of pollution is remote then metals are often present only in trace amounts [1,2]. The trace concentrations of metal ions are unlikely to have any toxic impact on many micro-organisms. However, the ability of bacterial cells to accumulate heavy metals internally has been found previously to result in the metals ultimately reaching toxic concentrations within the cell [1, 3].

The aims of this study were to determine the resistance of various marine heterotrophic bacteria to selected heavy metals, and assess metal accumulation by these bacteria.

2. MATERIALS AND METHODS

2.1. Organisms and storage conditions

Heterotrophic bacterial isolates originating from the Black Sea were used in this study. The dominant heterotrophic bacteria isolated included *Vibrio*, *Pseudomonas*, *Cytophaga*, *Acinetobacter*, *Alcaligenes*, *Chromobacterium*, *Aeromonas*, *Bacillus*, *Arthrobacter*, *Micrococcus* and *Planococcus* [6,7]. The cultures had been isolated from coastal and open regions of the Black Sea contaminated by different levels of heavy metals and other pollutants. The isolates were maintained on medium consisting of 0.1 % (w/v) peptone and 2.0 % (w/v) agar in filtered sterilized seawater at 4 °C.

2.2. Determination of metal ion accumulation by bacterial isolates

Marine *Pseudomonas* and *Cytophaga* strains were grown on peptone agar (see above) at 22°C before harvesting and resuspending 5 mg (wet weight) of biomass in 10 ml filter sterilized seawater only; and 10 ml filter sterilized seawater containing 20 mg/l $FeSO_4$, $MnCO_3$, $CoCl_2$, $Ni(NO_3)_2$, $ZnSO_4$, $CdCl_2$. The suspensions were incubated for 7 days at 22 °C.

Cells were then harvested by centrifugation and their dry weight determined. The dry biomass was then acid digested and the heavy metal content determined by atomic-absorption spectrophotometer AAS-1 [5]. Metal ion accumulation was expressed as μg metal/ mg cell dry weight.

2.3. Determination of the toxicity of selected metal ions to marine isolates

All isolates were grown on peptone agar (above) at 22°C, harvested and suspended in filtered seawater to give cell suspensions of 10^9 cells/ml. Heavy metal resistance was determined by point inoculation of each cell suspension onto nutrient agar containing $Hg_2(NO_3)_2$, $Pb(NO_3)_2$, $CdCl_2$, $NiSO_4$, $CoCl_2$ or $CdCl_2$ at a range of concentrations. The cultures were grown aerobically at 22 °C for 48 hr after which time the presence or absence of growth was determined. The minimum inhibitory concentration (MIC) of the metal was considered to be the minimum concentration at which no growth occurred.

3. RESULTS AND DISCUSSION

3.1. Metal accumulation by marine isolates

A total of 15 different *Pseudomonas* sp. strains and 4 strains of *Cytophaga* sp. isolated from the Black Sea were used to study the ability of marine bacteria to accumulate metal. Both the the *Pseudomonas* strains and the *Cytophaga* strains were found to vary greatly in their ability to accumulate metals and the response differed with metal (Table 1). This variation with organism has frequently been observed in the past [2]. It is interesting to note that little or no Co and Cd ions were accumulated from filtered seawater by the pseudomonads even though these ions were present in solution. In contrast Ni and Fe ions were accumulated in large quantities. Cadmium was not accumulated by *Cytophaga* strains either and in general neither was Co. Three out of the 4 *Cytophaga* strains studied did not accumulate any Co ions. Although broadly comparable, there were differences between uptake by the *Pseudomonas* strains and *Cytophaga* strains. *Cytophaga* strains accumulated relatively high levels of Mn in comparison with the pseudomonads (Table 1). Such variations in accumulation with microbial strain and metal ion may find their basis in several different factors including: differences in cell surface characteristics between strains affecting cell surface sorption of the metals [2,3]; variations in internal uptake through activity driven systems [8]; resistance mechanisms resulting in efflux of metal cations [9]; and metal chemistry [2].

The pattern and extent of accumulation was very different when 20 mg/l of the selected metal salts was added to the incubation medium. Some strains showed a 10-100 times increase

Table 1. The range of metal accumulation (μg metal /mg dry weight cells) in marine *Pseudomonas* and *Cytophaga* strains after incubation in filtered seawater

	Fe	Mn	Zn	Ni	Co	Cd
Pseudomonas sp.	0.3-2.5	0.1-0.7	0.1-0.7	0.6-2.7	0.0-0.4	0.0-0.2
Cytophaga sp.	0.6-2.1	0.3-3.8	0.3-0.4	0.8-3.4	0.0-1.1	0.0-0.0

Table 2. The range of metal accumulation (μg metal /mg dry weight cells) in marine *Pseudomonas* and *Cytophaga* strains after incubation in filtered seawater containing added metal salt solution (20 mg/l).

	Fe	Mn	Zn	Ni	Co	Cd
Pseudomonas sp.	0.5-50.0	0.5-61.3	0.1-23.7	0.6-40.0	1.8-149.4	23.0-189.9
Cytophaga sp.	1.3-40.6	0.7-88.2	0.5-23.7	1.8-43.6	2.5-154.7	7.2-182.3

in accumulation compared with that in filtered seawater, although for the majority of strains the increase was by a factor of 2-5 times (Table 2, 3 and 4). The maximum uptake varied with metal ion but the range of accumulation between strains was broadly similar for the two genera studied (Table 2). In general, the majority of strains was found in the lower part of the range of accumulation. The high levels of uptake were only shown by a few strains of *Pseudomonas* sp. and *Cytophaga* sp. (Table 3 and 4).

Table 3. Metal accumulation by *Pseudomonas* sp. and *Cytophaga* sp. strains from seawater with no added metal salts.

Bacterial strains	Metal accumulation (μg metal /mg dry weight)					
	Fe	Mn	Zn	Ni	Co	Cd
Pseudomonas sp.						
1	1.7	0.7	0.4	2.4	0.4	0.2
3	1.1	0.3	0.2	1.8	0.0	0.0
4	0.5	0.3	0.4	1.3	0.0	0.0
5	1.4	0.2	0.1	0.7	0.0	0.0
6	1.3	0.5	0.5	1.2	0.0	0.0
7	2.5	0.5	0.7	2.7	0.0	0.0
8	1.4	0.7	0.5	2.5	0.0	0.1
9	1.7	0.7	0.3	1.9	0.0	0.0
10	1.6	0.4	0.3	1.6	0.0	0.0
18	0.7	0.3	0.3	0.9	0.0	0.0
20	0.3	0.3	0.1	0.6	0.0	0.0
32	0.5	0.1	0.2	0.8	0.0	0.0
36	0.4	0.1	0.2	0.6	0.0	0.0
43	1.3	0.3	0.5	1.2	0.0	0.0
93	0.8	0.2	0.2	1.2	0.0	0.0
Cytophaga sp.						
12	0.6	0.3	0.3	0.8	0.0	0.0
13	2.1	1.4	0.3	3.4	0.0	0.0
15	1.5	2.7	0.4	1.8	1.1	0.0
16	2.0	3.8	0.4	2.0	0.0	0.0

Table 4. Metal accumulation by *Pseudomonas* sp. and *Cytophaga* sp. strains from seawater with added metal salts.

Bacterial strain	Metal accumulation (μg metal /mg dry weight)					
	Fe	Mn	Zn	Ni	Co	Cd
Pseudomonas sp.						
1	7.8	20.0	6.3	10.9	22.9	43.4
3	2.8	0.9	0.6	1.8	3.6	5.0
4	1.7	1.9	0.8	1.4	6.4	8.3
5	2.1	2.3	1.3	1.3	7.5	10.1
6	3.8	4.8	2.0	2.0	12.3	17.4
7	50.0	53.3	18.4	32.7	158.0	189.9
8	40.6	61.3	23.7	40.0	149.4	187.3
9	3.1	2.7	1.3	2.0	11.1	15.2
10	0.5	0.7	0.3	1.0	2.1	2.4
18	0.7	0.5	0.4	0.6	1.8	2.3
20	3.1	3.1	0.6	2.9	9.0	10.6
32	1.1	1.2	0.3	2.1	3.8	4.8
36	0.6	0.6	0.1	1.2	2.2	2.4
43	0.5	0.6	0.2	0.7	2.5	2.8
93	2.0	1.9	0.5	1.9	5.4	7.2
Cytophaga sp.						
12	1.3	2.1	0.6	1.8	4.2	7.2
13	3.3	0.7	0,5	3.1	9.1	103
15	7.5	10.1	3,0	4.0	2.5	36.1
16	40.6	88.2	23,7	43.6	154.7	182.3

Interestingly, the extent of accumulation of individual metals by strains in seawater did not necessarily reflect the level of accumulation in the presence of added cations. For example, *Pseudomonas* spp. strain 20 accumulated more Ni than other metals from seawater, but from the cation solutions this strain accumulated Fe, Mn and Ni to similar extents. The pattern of uptake by *Pseudomonas* spp. strain 7 was very different between seawater and seawater with added metals (Table 3 and 4). All *Pseudomonas* and *Cytophaga* strains showed a marked difference in the accumulation of Co and Cd between uptake from seawater only and seawater with the added metal salts. In general, no Co or Cd was accumulated from seawater, but after the addition of the metal salts then accumulation occurred in all strains often to high levels (Table 3 and Table 4).

The metal ion concentration is, inevitably, higher in seawater with added metal salts then in filtered seawater alone. This will increase metal accumulation until cell saturation is reached [2,3]. The difference in metal ion concentration between the two media, however, is insufficient to account for the very major increases in accumulation shown by many strains, particularly for Co and Cd. A number of factors may be the basis for the changes in uptake characteristics with incubation conditions. First, the cells themselves may be different in terms of activity and cell wall characteristics between the two media, which would affect cell surface binding of the metals [3]. Secondly, it is possible that the form of metal available for uptake was different between the two media. In the seawater with added metal salts, much of the metal may have been present as free soluble ions. In seawater alone the metal ions may not have been free but may have been bound to organic ligands and potentially not available for uptake. Alternatively, there may have been inhibition of uptake induced by the presence of other anions or cations in seawater [2]. These types of restrictions on uptake may have been relatively insignificant or overcome given the highly increased metal ion load in the seawater with added metal salts.

These major differences in metal uptake shown between seawater and seawater with added metal salts raises a real problem with regard to determining bacterial responses to metals in the natural environment. The amount of metals accumulated by bacteria in laboratory based experiments using soluble metal salts as the metal source may bear little resemblance to bacterial metal accumulation in natural environments.

3.2. Resistance of marine *Pseudomonas* strains and *Cytophaga* strains to toxic heavy metals

There were significant differences in the pattern of resistance shown by *Vibrio*, *Pseudomonas*, *Cytophaga*, *Acinetobacter*, *Alcaligenes*, *Chromobacterium*, *Aeromonas*, *Bacillus*, *Arthrobacter*, *Micrococcus* and *Planococcus* isolated in 1991 in comparison with those isolated in 1992 (Fig.1 and 2).

In general 1991 isolates from the Black Sea were most tolerant to Ni ions, showing a modal MIC value of 1024 µg metal salt /ml (Fig. 1). Bacteria isolated in 1992 were at least 4 times more sensitive to this metal salt (Fig. 2). Similarly, Pb resistance appeared greater for isolates from 1991. Although the range overlapped, the modal MIC value was 512 µg metal salt / ml in 1991 and almost 2 times lower in 1992 (Fig. 1 and 2).

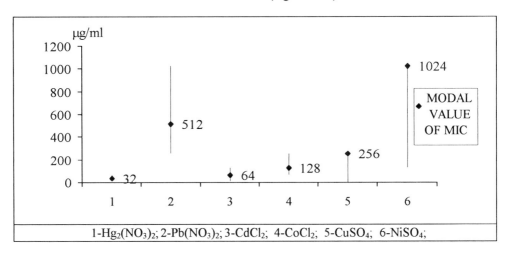

Figure 1. Minimum inhibitory concentration (mode and range) of selected metals shown by marine strains isolated from Black Sea in 1991

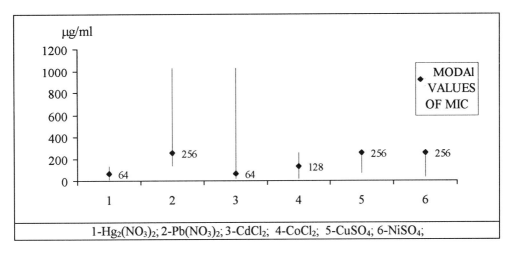

Figure 2. Minimum inhibitory concentration (mode and range) of selected metals shown by marine strains isolated from Black Sea in 1992

In the case of mercury, however, resistance was higher in 1992 compared to 1991. The modal MIC for Hg salts was 64 µg/ml with a range up to 128 µg/ml in 1992, but the mode for MIC was 32 µg/ml in 1991 with the range very little above this. The bacterial strains showed similar resistance levels to the other metal salts studies, $CdCl_2$, $CoCl_2$ and $CuSO_4$, in 1991 and 1992. The probable explanation for these results lies with variations in metal pollutant levels in the Black Sea between 1991 and 1992. Bacterial populations can evolve rapidly in response to selection pressures in the environment. Any rise or fall in bioavailable heavy metal concentration would undoubtedly elicit an adaptive response from individual populations, ultimately mirrored at community level. Changes in the pattern of metal ion resistance shown by the marine isolates in this study are probably a reflection of this and underline the potential for rapid community genetic response to heavy metal pollutants. It is interesting to speculate on using measures of bacterial resistance to heavy metals as a monitoring technique for changes in bioavailable heavy metal pollutants in natural ecosystems.

There appeared to be no relationship between the resistance of *Pseudomonas* and *Cytophaga* strains to particular metal salts and their capability to accumulate that metal either from seawater alone or seawater with added metal salts (Table 3, 4 and 5). Some of the highest accumulators, however, showed the highest resistance to a range of metal ions for example *Cytophaga* sp. strain 16 and *Pseudomonas* sp. strain 7. It is difficult to make direct comparisons in the availability of the metals for uptake between the seawater, the seawater plus metal salts and the agar medium used for replica plating. In all these conditions the metals may have been present in different forms and therefore their availability for uptake may have been different. Even so, given that *Cytophaga* sp. strain 16 and *Pseudomonas* sp. strain 7 took up among the largest amounts of metal from seawater and seawater with added metal salts it is probably that they continued to do so from the agar medium. It may be the case that their resistance mechanism was either through the production of metal binding proteins [10] or

Table 5. MIC for selected heavy metal salts shown by *Pseudomonas* sp. strains and *Cytophaga* sp. strains isolated from the Black Sea

Strain	MIC (μg metal salt/ml)					
	$Hg_2(NO_3)_2$	$Pb(NO_3)_2$	$CdCl_2$	$NiSO_4$	$CoCl_2$	$CdCl_2$
Pseudomonas sp.						
1	64	256	512	1024	512	256
3	1	16	0.5	8	32	1
4	2	16	128	512	64	128
5	1	32	256	512	128	128
6	1	16	256	512	256	256
7	64	256	512	1024	1024	512
8	64	512	256	1024	512	512
9	2	32	128	32	256	128
10	0.5	16	1	16	32	4
18	0.25	16	8	32	32	1
20	2	256	128	512	128	64
32	1	32	256	256	64	32
36	1	16	1	8	32	1
43	0.5	16	0.5	16	32	4
93	2	32	256	16	32	32
Cytophaga sp.						
12	1	16	256	16	32	32
13	1	16	128	16	64	64
15	2	32	128	16	32	64
16	4	128	256	512	512	256

perhaps through binding at the cell surface e.g. to exopolymer [11], making the metal unavailable for internal uptake and thereby negating any toxic effect.

6. DISCUSSION

Bacteria are undoubtedly capable of accumulating heavy metal ions from marine environments and concentrating them to levels orders of magnitude greater than those in the environment. Care must be taken, however, in extrapolating accumulation data obtained from laboratory experiments using free metal ions to events in individual habitats where the form of the metals and therefore their bioavailability may be different. Studies on the effect of bacterial metal accumulation on the environmental fate and impact of heavy metals must be undertaken with care. Given the rapid nature of bacterial evolution under selective pressures such as those exerted by heavy metal pollutants, it may be possible to use metal resistance shown be bacterial communities to monitor changes in pollution levels with time.

REFERENCES

1. Y.A. Izrael and A. V. Tsyban. Ocean anthropogenic ecology, Leningrad, Gidrometioizdat., 1989.
2. D. Hardman, S. McEldowney and S. Waite. *Pollution: Ecology and Biotreatment.* Addison, Wesley,Longman. Essex, England.
3. S. McEldowney, Appl.Environ.Microbiol., 60 (1994) 2759.
4. V.O. Ivanitsa. Microbiol J., 56 (1994) 61.
5. V.O. Ivanitsa. The state and variability of microbial community in marine ecosystems. Dr. of Science dissertation. - Inst. of Microbiolgy and Virology, Nat. Acad. of Sci. of Ukraine, Kiev, 1996.
6. N.N. Panchenko and M.A. Kiaynickay. Methodological bases of complex ocean monitoring, Moscow, Gidrometeoizdat, 1988.
7. V.O. Ivanitsa, G.V. Khudchenko, N.N. Panchenko A.E. Buchtiayrov and V.I. Medinets, Ecosystem study of The Black Sea, Odessa, 1994, 54.
8. I.S. Ross, Int. Ind. Biotechnol. 6 (1986) 184.
9. Z. Tynecka, Z. Gos and J. Zajac, J. Bacteriol. 147 (1981) 313.
10. D.P. Highman, P.J. Sadler and M.D.Scawen, Science 225 (1984) 1043.
11. J.A. Scott and A.M. Karanjkar. Biotechnol. Lett., 14 (1992) 737.

Biosorption of long-lived radionuclides

N.N. Lyalikova-Medvedeva[a], T.V.Khijniak[b]

[a] Laboratory of Ecology and Geochemical Activity of Microorganisms, Institute of Microbiology Russian Academy of Sciences, Prospect 60-letija Oktjabrja 7/2, Moscow 117811, Russia

[b] Laboratory of Radiological Researches, Institute of Physical Chemistry Russian Academy of Sciences, Leninski prospect 31, Moscow 117915, Russia

The aim of our work was to study the possibility to reduce the volume of radioactive wastes, using the sorption of radioactive elements by biomass or transfer in less soluble state. We dealt with technetium-99 (half-life is $2.12*10^5$ y, specific activity - 1.7 mCi/ mmol). The fungi *Aspergillus niger*, *Scopulariopsis brevicaulis*, yeasts, several species of *Pseudomonas*, thionic and sulfate-reducing bacteria were used in the experiments. The radioactivity of solution (Tc - 24 mg/l) was decreased on 58, 65 and 40% after contact with *A. niger*, *S. brevicaulis* and yeast *Pichia wicker*, respectively. Anaerobic consortium of chromate-reducing bacteria removed technetium-99 almost completely. The concentration factor was $4,1*10^3$. In experiments with sulfate-reducing bacteria the radioactivity was decreased on 25-48%. Sulfides of technetium Tc_2S_7 and TcS_2 were formed. Vanadate-reducing bacteria decreased the radioactivity on 25% (concentration of Tc - 40 mg/l). The sorption by living cells was 4-5 times higher than by dead biomass. Thionic bacteria - *Thiobacillus ferrooxidans* and *Thiobacillus thiooxidans* can oxidize ferrous iron and sulfur in anaerobic conditions, using pertechnetate as electron acceptor. The decrease of radioactivity was 50-70%. By paper chromatography we shown that heptavalent technetium reduced to 4 and 5 valent states by cultures *Pseudomonas vanadiumreductans* in neutral conditions and thionic bacteria in acid media. Dead biomass of *A. niger* - a waste of citric acid plant was used in experiments for removing U, Am, Ce, Cs, Eu, Pa, Sb at joint presence. All elements were removed in different degree, which depends on their chemical properties and concentration.

1. INTRODUCTION

Pollution of environment by toxic metals and radionuclides take place as a result of increasing industrial activity of mankind. The problem of radioactive wastes arise due to experimental explosions of nuclear weapons and wastes of reprocessing plants of nuclear fuel cycle and using isotopes for medical purposes. Sometimes significant pollution of environment due to the accidents like Chernobyl.

There are many papers about using of biotechnological techniques for purification from uranium, strontium, cesium and iodine [1,2,3].

In our study we paid attention to radionuclides which have concentration less than above mentioned but having long half-life. Due to that fact impact of this radionuclides in whole radioactivity growing with time. The most radionuclides exist as cations and have good sorption ability, when natural and chemical sorbents are used. Among long-lived radionuclides technetium stand out because it is the most stable in anionic form as pertechnetate ion (TcO_4^-). It has high mobility in the environment and high potential for biological uptake. Three years after the discharge of fuel from a reactor, the contribution of ^{99}Tc to the total β-radioactivity is only 0,1% but reaches 30-50% after 300 years. The information about biological influence on technetium is not sufficient.

We were trying to find a new sorbents for purification of low radioactive solutions. In our study we used as pure cultures of microorganisms as bacterial associations and mycelian fungy and their dead biomass, which is a plant waste.

2. MATERIALS AND METHODS

The experiment were carried out with the use of long-lived radionuclides ^{99}Tc and following mixture: cobalt (^{60}Co), cesium (^{137}Cs), protactinium (^{233}Pa), thorium (^{228}Th), americium (^{241}Am), ruthenium (^{106}Ru), antimony (^{125}Sb), europium (^{155}Eu), uranium (^{235}U), cerium (^{144}Ce) (Isotop Joint-Stock Company) cheked for radiochemical purity with β- and α-spectrometry. Final concentration of technetium in experiments was from 12 to 60 mg Tc l^{-1}, which correspond to concentration of this element in low radioactivity waste.

For screaning of cultures which are active for technetium sorption we used meta-stable isotope ^{95m}Tc (half-life is 6,01 h).

Radioactivity was determined with a LKB RackBeta scintillation counter. After separation bacterial cell by centrifugation, 10 or 20 μl of the supernatant was introduced into vials containing ZhS-8 scintillating liquid. The radioactivity of cells was determined on a kapron filter after prolonged washing with water containing 1% NaCl. Specific radioactivity of solution of the mixture of radionuclides was measured either radiometrically or α-spectrometrically.

Our experiments employed the pure cultures of bacteria: *Pseudomonas fluorescens, P.vanadiumreductans, P.chromatophilla, Methylobacterium organofilum, Arthrobacter sp., Thiobacillus ferrooxidans, T.thiooxidans*; bacterial anaerobic consortiums contains sulfatereducing, chromatereducing and chloratereducing bacteria; yeasts: *Pichia wicker, Rhodotorula rubra*; actinomycetes: *Streptomyces galbus, S.scabies*; micromycetes: *Aspergillus niger, Scopulariopsis brevicaulis*; dead biomass of *A.niger*, the waste of citric plant production.

The experiments were carried out in penicillin vials in anaerobic and aerobic conditions. For the creating anaerobic conditions we blue up argon or nitrogen during 5-10 min.

Speciation of technetium in the medium after development of bacteria was carried out by paper chromatography [4] and scintillation counting.

Ferrous and ferric iron was determined by trilonometric titration [5]

We used standart nutrient media [6]

Experiments were also run with the ore of the Degtyarskoe deposit, composed mainly of pyrite and 0,4% chalcopyrite.

3. RESULTS AND DISCUSSION

As it above mention for the screaning we used microorganisms of different taxonomic group (see table 1) and metastable technetium - ^{95m}Tc. The experiments were carried out in penicillin vials, the solution of technetium and microbial suspension in phosphate buffer were added. The duration of experiments was 2-4 h. The mictoorganisms were separated by filtration through kapronic filter. Natural decrease of radioactivity during experiments was taken in consideration. According to Table 1, the most perspective cultures were strains of *Pseudomonas, Methylobacterium organofilum,* yeasts and *A.niger.* It's necessary to note that in anaerobic conditions better results were obtained with pseudomonades.

We suggested that the most perspective such cultures which not only sorb technetium but also have reducing ability. According to this suggestion we chose the following microorganisms: *Scopulariopsis brevicaulis*, which can reduce arsenate and tellurate, *P.vanadiumreductans* and *P.isachenkovii*, reducing vanadate and nitrate and also bacterial consortium which can reduce chromate, chlorate and sulfate (see Table 2). The experiments with living and dead cells of vanadatereducing bacteria show that results, receiving with living culture are higher in 5 fold. The study of cultural liquid by paper chromatography show that pentavalent technetium are formed.

Together with biochemists we isolated from vanadatereducing bacteria enzyme - nitrate reductase, which reduced nitrate and vanadate. We have reason to suggest that the same enzyme have part in reduction heptavalent technetium.

The most part of liquid radioactive waste has acid pH. Regarding this fact we studied influence of acidophilic thionic bacteria *Thiobacillus ferrooxidans* and *T.thiooxidans* on the technetium. Brock and Gustafson [7] were the first to establish the ability of these bacteria to

Table 1. Sorption of metastable ^{95m}Tc by microorganisms of different taxons at neutral pH, t= 20° C. Initial activity of ^{95m}Tc - 0,6 Ci.

Cultures	Uptake of metastable technetium, %
Pseudomonas (3 species)	50
Aspergillus niger	76-78
Methylobacterium organofilum	50
Arthrobacter sp	27-30
Pichia wicker	30-50
Rhodotorula rubra	25-35
Streptomyces galbus	30
S.scabies	30

Table 2. The uptake of Tc-99 from solution by different microorganisms in neutral condition, t= 28-30° C, experiment's duration 5-7 days.

Cultures	Initial concentration of technetium, mg/l	Uptake of technetium, %
Aspergillus niger		58
Scopulariopsis brevicaulis	24	58-65
Pichia wicker		40
Bacterial consortium, reducing		
sulphate	30	30-60
chromate	12	99
chlorate	40	18-20
P. vanadiumreductans	25-40	20-25
P. isachenkovii		

grow anaerobically oxidizing molecular sulphur and using ferric iron as an electron acceptor. In our experiments we used heptavalent technetium as an electron acceptor.

In experiments with *T. ferrooxidans* we used as electron donor not only sulphur, but ferrous iron as well. The experiments took place in anaerobic and aerobic conditions.

Table 3 contains results of 3 series of experiments with 8 replicates. As can be seen, the significant decrease of radioactivity occur, especially in anaerobic conditions. Ferric iron appears after development of bacteria. It is important to note following fact. If after experiment the cultural liquid was frozen, dark-brown colloid was formed. The X-ray microanalysis show that this substance contains iron, sulphur and technetium. We suggest that bacteria reduced heptavalent technetium which appeared in colloid form. Such form is usual for nonsoluble bacterial byproduct. To reveal changes in the valence of technetium under the action of bacteria, paper chromatography was employed. In one of experiments the ratio of valent forms in cultural liquid was follow: Tc^{+4} - 7%, Tc^{+5} - 13%, Tc^{+7} - 80%. In this case the radioactivity decrease only for 20 %. Determination percent ratio of different valent forms of technetium in

Table 3. Bioaccumulation of Tc^{99} by *T. ferrooxidans*. pH 2,5; T= 30° C, t= 7 d.

Conditions	Initial radioactivity, $Bq*10^8$	Final radioactivity, $Bq*10^8$	Radioactivity of colloid, $Bq*10^8$	Uptake of ^{99}Tc, %
Aerobic	1,8	1,3	0,3	25
Anaerobic	3,5	1,9	0,35	46
	1,9	1,3	0,11	30

separated colloid was: Tc^{+4} - 27,5%, Tc^{+5} - 61,6%, Tc^{+7} - 11%. These results confirms our suggestion that reduced technetium forms colloidal particles. The ratio of valent forms differ among experiments. Thus, in another experiment colloid contained no tetravalent technetiun, portion of pentavalent was 91%.

To investigate the involvment of technetium in metabolic processes in cells, we did series of experiments with azide - inhibitor of respiration chain. According to the hypothetical scheme suggested by Pronk [8], 10 μM azide causes 95% inhibition of ferrous iron oxidation by *T. ferrooxidans*, but does not affect the oxidation elemental sulphur at the expense of ferric iron. In our experiments with azide oxidation of ferrous iron by pertechnetate took place in anaerobic conditions. Our results confirmed Pronk's data. Azide did not inhibit this process. We find out that decrease of radioactivity by bacterial influence was the same with or without azide - 62 and 64%, respectively, i.e. azide did not influence the process of anaerobic respiration. It prove that pertechnetate participated in energetic metabolism of *T. ferrooxidans*.

When elemental sulphur was electron donor, the experiments were carry out with both thionic bacteria. Table 4 shows relevant data averaged over six replicates. Thus, the radioactivity of solution, when bacteria oxidaze sulphur decrease about 70%. The speciation of technetium shows formation of Tc^{+4} and Tc^{+5}.

In our experiments, we cultivated *T. ferrooxidans* not only on ferrous iron and elemental sulphur, but also on pyrite ore. The duration of this experiment was one month, during which the pH value decreased to 2,43 on average as compared to 3,2 in the control, and 200 mg/l of ferric iron was formed (only traces were found in the control). Because in this experiment we used vials with large volume (0,5 l), oxygen was removed not complete, so part of ferrous iron was oxidized by oxygen, but not by technetium. After experiment, the culture liquid was centrifuged. The radioactivity of the solution decreased only 9%. However, in this experiment, the sedimentation of large aggregated particles of tetravalent technetium could occur (Foto, left tube). If the mixture was vigorously agitated before centrifugation, the cosedimentation of

Table 4. The removing of technetium-99 from solution by thiobacilli growing on molecular sulphur. Total experimental volume is 4 ml, T= 30° C, t= 7 days

Micro-organism	Conditions	Concentration of ^{99}Tc, mg/l	Remain radioactivity of solution, Bq*10^4	Uptake of ^{99}Tc, %
T. ferrooxidans	anaerobic	50	0,69	57
			0,62	61
Control			1,6	0
T.thiooxidans	aerobic	40	0,36	65
			0,13	87
	anaerobic		0,46	56
			0,35	64,3
Control			1,05	0

Foto 1. Freezing cultural liquid
 after development
T. ferrooxidans on sulfide ore
and technetium-99.
Left tube - experiment,
right tube- control.

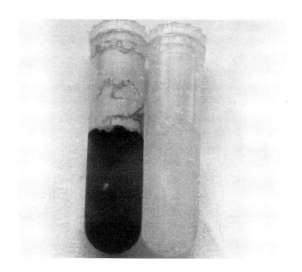

reduced forms of technetium with fine particles of the ore occured and redioactivity of solution decreased 40%.

As we can see from tables 1 and 2, *Aspergillus niger* quite good sorbed of heptavalent technetium - 60-75%. Its depend on high content of chitin in the cell wall of fungy. It is known from literature that chitosan - derivate of chitin, have a good sorption ability. Because chitosan recovery require expenditure, we tried to use dead biomass of *A. niger* - a waste of citric acid production. Penicillin vials contain 10 ml of phosphate buffer (pH 4 and 6,5), 100 mg of fungus biomass and mixture of radionuclides (see "Material and Methods"). All element were removed in different degree (from 30 till 80%), which depends on their chemical properties and concentration (see Fig.1) with the exception of cerium.

4. CONCLUSION

Our experiments shown that different bacteria, yeasts and mycelian fungi can be biosorbents. The advantage of this sorbent is their ability to sorb radionuclides from diluted solutions. This permit to reduce the volume of radioactive waste. The reducing of heptavalent technetium and its transformation to less soluble or nonsoluble forms (Tc^{+4} and Tc^{+5}) lead to decrease of radioactivity of liquid waste.
1. The microorganisms with reducing ability are the most perspective for technetium removal. To them belong sulphate-, vanadate- and chromatereducing bacteria and fungi *Aspergillus niger*, *Scopulariopsis brevicaulis*.
2. It was shown in the first time that autotrophic acidophilic thiobacilli can grow in anaerobic conditions oxidizing molecular sulphur by heptavalent technetium as electron acceptor. *T. ferrooxidans* can oxidize ferrous iron and pyrite ore in the process of anaerobic respiration.
3. In the experiments with vanadatereducing bacteria sorption of technetium by living cells was 5 fold higher than dead.

4. It was proved by paper chromatography reduction of heptavalent technetium to tetra- and pentavalent forms by vanadatereducing and thionic bacteria in neitral and acid conditions, respectively.

5. On the basis of the experiments with dead biomass *Aspergillus niger* we suggested using of fungus chitin without its transformation to chitosan for sorption several radionuclides.

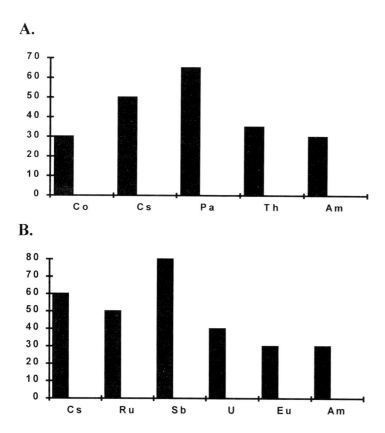

Fig.1. Sorption of different radionuclides by dead biomass *A. niger* , Y-axis - sorption of radionuclides, %. A- pH 4, B- pH 6,5.

ACKNOWLEDGMENTS

We are grateful to Dr. V.Ya. Frenkel, Dr. K.E.Guerman, Prof. V.F. Peretrukhin for the help and discussions our work.

REFERENCES

1. A. Hakajama, T. Horikashi, T. Sakaguchi, Eur. J. Appl. Microbiol. Biotechnol, 16 (1982) 88.
2. L.E. Macaskie, J.D. Blackmore, R.M. Empson, FEMS Microbiol. Lett., 55 (1988) 157.
3. M. Tsezos, R.G.L. MacCready, J.P. Bell, Biotechnol, Bioeng., 34 (1989) 10.
4. S.K. Shukla, J.Chromatogr., 21 (1966) 92.
5. A.A. Reznikov, E.P. Mulikovskaya, I.Yu. Sokolov, Metody analyza prirodnykh vod (Methods for Analyzing Natural Wares), Moscow, Nedra, (1970).
6. Biotekhnologiya metallov: Practicheskoe rukovodstvo (Biotechnology of Metals: A Practical Guide), Moscow TMPGKNT (1989).
7. T.D. Brock, J. Gustafson, Appl. Environ. Microbiol., 32 (1976) 567.
8. J. Pronk. Physiology of the Acidophilic Thiobacilli, PhD Thesis, Delft, 1991.

A novel mineral processing by flotation using *Thiobacillus ferrooxidans*

T.Nagaoka*, N.Ohmura, and H.Saiki

Bio-Science Department, Abiko Laboratory
Central Research Institute of Electric Power Industry (CRIEPI)
1646 Abiko, Abiko-city, Chiba, 270-1194 JAPAN

Oxidative leaching of metals by *Thiobacillus ferrooxidans* has proven useful in mineral processing. Here, we report on a new use for *T. ferrooxidans* whereby bacterial adhesion is used to remove pyrite from mixtures of sulfide minerals during flotation. Under control conditions, the floatabilities of 5 sulfide minerals tested (pyrite, chalcocite, molybdenite, millerite and galena) ranged from 88 to 99%. Upon addition of *T. ferrooxidans*, the floatability of pyrite was significantly suppressed to less than 20%. In contrast, addition of the bacterium had little or no effect on the floatabilities of the other minerals, even when they were present in relatively large quantities: Their floatabilities remained in the range of 70 to 94%. *T. ferrooxidans* thus appears to selectively suppress pyrite floatability. As a consequence, 84 to 95% of pyrite was removed from mineral mixtures, while 73 to 100% of non-pyrite sulfide minerals were recovered. The suppression of pyrite floatability was caused by bacterial adhesion to pyrite surfaces. The number of cells adhering to pyrite was significantly larger than the number adhering other minerals. These results suggest that flotation with *T. ferrooxidans* may provide a novel approach to mineral processing in which the biological functions involved in cell adhesion play a key role in the separation of minerals.

1. INTRODUCTION

Leaching metals from low-grade ores is a well established application of bacteria to mineral processing [1]. For instance, *Thiobacillus ferrooxidans* is able to oxidize pyrite to ferric ions and sulfate thereby contributing to the extraction of many kinds of useful metals from a low-grade mineral. In contrast to this biological processing, physical chemical methods have heretofore been utilized to remove impurities from sulfide minerals. One method, flotation, is based on differences in the surface properties of minerals: Mineral particles with hydrophobic surfaces attach to air bubbles generated at the bottom of a flotation column and float at the top of the column; particles with hydrophilic surfaces do not attach to bubbles and sink to the bottom. Because of its

---

* Corresponding author. E-mail address: nagaoka@criepi.denken.or.jp. Fax: 81 471 83 3347 hydrophobicity, pyrite present as an impurity may contaminate froths containing other more desirable hydrophobic minerals. In many cases, chemical reagents (e.g. cyanide) have been added to flotation columns with the goal of altering the floatability of pyrite. However, success has been limited, and pyrite contamination continues to be a problem.

Recently, much attention has focused on an alternative approach involving the use of bacteria to modify the floatability of mineral particles. Many reports have suggested that certain types of bacteria, including *T. ferrooxidans*, may suppress floatability instead of a chemical reagents [2]. The postulated mechanism of this suppression is an increase in surface hydrophilicity due to adhesion of bacterial cells [3]. If true, the ability of specific bacteria to selectively adhere to and alter the surface properties of specific minerals could be highly useful. In that regard, *T. ferrooxidans* appears able to adhere well to iron-bearing minerals. If the bacterium adheres selectively to iron-bearing pyrite, its addition to a mineral mixture in a flotation column should suppress pyrite floatability without suppressing flotation of the other minerals contained within the mixture.

In this report, the ability of *T. ferrooxidans* to adhere to selected minerals was investigated with the aims of clarifying whether the bacterium selectively adheres to pyrite and determining whether it will effectively remove pyrite from mixtures of sulfide minerals.

2. MATERIALS AND METHODS

2.1. Cell culture

Thiobacillus ferrooxidans ATCC23270 was cultured in 9K basal medium, which contained 44.2g of $FeSO_4 \cdot H_2O$. The pH was adjusted to 2.5 with 6N H_2SO_4. The culture of cell was carried out under aeration at 30℃ for about 3days. The culture medium was passed through filter paper (No.2, ADVANTEC TOYO Co., Tokyo, Japan) to remove the precipitates. The filtrate was centrifuged to collect cells. The obtained cells were washed a few times in sulfuric acid at pH2.0 and utilized in the experiments. Unbound cells were estimated by measuring the optical density at 610nm as a function of the density of the cells

2.2. Preparation of sulfide minerals

The sulfide minerals in this study were museum grade. Each crystal mineral was crushed into fine particle and sized between 53μm and 75μm. The fraction was sonicated for several minutes in acetone to dissociate the ultra-fine particles from the larger one. The detached fine particles were removed by decanting until the supernatant was clear. The washed particles were dried by suction at room temperature and used in subsequent experiment.

2.3. Adhesion experiments

Adhesion experiments were carried out using both individual minerals and mineral mixtures. In the case of the individual minerals, 0.5g of mineral were added to 2ml of cell suspension at the density of 3.2×10^8 cells/ml. In the case of mineral mixtures, 0.2g each of 4 minerals (chalcocite, millerite, molybdenite and galena) plus varying amounts of pyrite (0.25g, 0.5g and 0.8g) were added to 2 ml of cell suspension (1.7 x10^8 cells/ml). The suspension was allowed to stand for 5min after stirring for 1min with vortex mixer, and the optical density of the supernatant was measured. The amount of adherent cells to minerals was determined by subtracting the amount of the cells in the supernatant from that of initially added cells.

2.4. Flotation experiments

The flotation experiments were conducted with a micro flotation column (working volume: 270 ml; height × diameter: 38 × 3 cm). Sulfuric acid solution (pH 2.0) containing methyl isobutyl carbinol (MIBC; 25 μL/L) was used as the flotation liquor. Air bubbles were generated with a porous glass filter situated at the bottom of the column.

The floatabilities of the minerals were measured by adding 0.5g of each sulfide mineral to 2 ml of the cell suspension (3.2 x10^8 cells/ml). The suspension was then shaken for about 1 min and allowed to settle for 5 min. The settled mineral particles were applied to the flotation column, which was then aerated for 10 min at a rate of 100 cm^3/min. The particles that floated to the top of the column were collected as froth, while the particles that sank to the bottom were recovered as tailing. Froths and tailings were dried at 80°C and weighed. The floatability was defined as proportional ratio of a froth weight out of the total weight of a froth and tailing.

The flotation experiments aimed at separating pyrite from a mineral mixture were carried out. The mixtures contained 0.2g each of pyrite and a single sulfide mineral. Prior to flotation analysis, the 0.4g of mineral/pyrite mixture were added to 2 ml of cell suspension (3.7 x10^8 cells/ml) to induce bacterial adhesion. After exposure to the bacteria, the mineral/pyrite mixtures were subjected to flotation. To remove pyrite by flotation, mineral/pyrite mixtures were fed directly into the middle of the column containing 230 ml of flotation liquor and aerated at a rate of 500 ml/min at 1 kgf/cm^2; 10 min of flotation were allowed to complete the separation.

Froths and tailings were recovered as described above and analyzed to determine the distribution of minerals: They were dissolved in nitric and hydrochloric acids, and the elements in the solution were identified and quantitated using an inductively coupled plasma atomic emission spectrometer (model JY48P, Seiko Industry Co., Tokyo, Japan). The absolute amounts of identified elements were converted to mineral weights on the basis of the chemical formula of each mineral. Each mineral's floatability was then calculated from mineral weights in the froths and the tailings as described above.

3. RESULTS

3.1. Adhesion of *T. ferrooxidans* to sulfide minerals.

When the selected sulfide minerals was mixed with cells, the cells floating in the supernatant was decreased as the cells adhere to the surface of mineral. The amount of adhered cells was increased with an increase of added cells as shown Figure 1. However, there was a significant difference of adhesion behavior between minerals. The amount of adhered cells to minerals increased in the order pyrite > molybdenite > chalcocite > millerite > galena. These results showed that *T. ferrooxidans* has an ability adhere more to pyrite than to other sulfide minerals.

The apparent specificity of *T. ferrooxidans* adhesion suggests that the bacterium may selectively adhere to pyrite, even when pyrite is mixed with other minerals. Therefore, the adhesiveness of *T. ferrooxidans* was assessed using mineral mixtures composed of 0.2g each of molybdenite, chalcocite, millerite and galena and selected quantities of pyrite; these findings were compared to pyrite adhesion in the absence of any other minerals. When *T. ferrooxidans* were added to pyrite alone, adhesion increased linearly with increases in added pyrite (Figure 2, dashed line); the number of adherent cells reached 2.75×10^8 in the presence of 0.8g of pyrite. In contrast, in the absence of pyrite, adherence of *T. ferrooxidans* to 0.8 g of mineral mixture was 3.7 times lower. The affinity of *T. ferrooxidans* for pyrite was particularly evident when pyrite was mixed with other minerals. As with pyrite alone, the number of adherent cells increased linearly with the weight of the added pyrite when 0.3, 0.5 or 0.8g of pyrite were added in combination with 0.8g of the mineral mixture (Figure 2, solid line). It was evident from the results that *T. ferrooxidans* selectively adhered to pyrite within mineral mixtures.

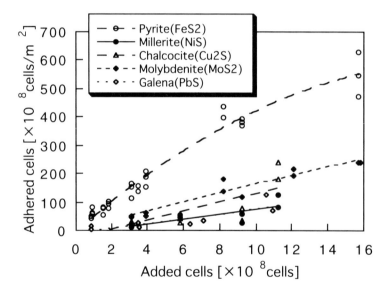

Figure 1. Adhesion of *T. ferrooxidans* to selected minerals. Symbols: ●, pyrite; ○, millerite; △, chalcocite; ◆, molybdenite; ◇, galena. Data points express values obtained from independent experiments.

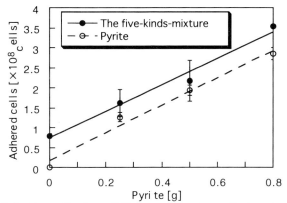

Figure 2. Selective adhesion of *T. ferrooxidans* to pyrite in mixtures of sulfide minerals. Symbols: ●, the number of cells adhering to a mineral mixture; ○, the number of cells adhering to pyrite alone. The mineral mixtures were prepared by blending 0.2 g each of molybdenite, chalcocite, millerite and galena with the indicated quantities of pyrite (x-axis). Data points are means ± SD of triplicate determinations.

3.2. The effects of *T. ferrooxidans* on the floatability of sulfide minerals.

Ohmura et al was reported that *Thiobacillus ferrooxidans* could be utilized as a suppressant of the floatability of pyrite in coal-flotation[3]. It was caused by bacterial adhesion to the surface of pyrite. The effect of the addition of *T. ferrooxidans* on mineral floatabilities was investigated by exposing mineral mixtures to cell suspensions containing 6.5×10^8 cells/ml (Figure 3). In the absence of bacteria, all of the minerals exhibited high floatability (90.8 to 99.0%). Upon exposure to the bacteria, the floatability of pyrite declined dramatically from 95.9 to 19.3%, whereas the floatabilities of the other minerals were unaffected. Thus, pyrite floatability was specifically suppressed by exposure to *T. ferrooxidans*.

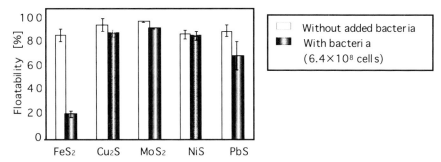

Figure 3. Effect of *T. ferrooxidans* on the floatability of selected minerals. 0.2 g of each mineral were exposed to 2 ml of cell suspension containing 3.5×10^8 cells/ml. Floatability was determined as weight ratio of a froth in total weight of a froth and a tailing. The values were determined by independent triplicate experiments and were expressed as an average with a standard division.

3.3. Pyrite removal from mixtures of sulfide minerals.

The specific ability of *T. ferrooxidans* to suppress pyrite floatability is potentially useful for selectively removing pyrite from mineral mixtures. We initially tested this possibility by blending pyrite (0.5g) with equal quantities of molybdenite, chalcocite, millerite or galena, exposing the mixtures to cell suspensions, and then subjecting them to flotation. In the absence of *T. ferrooxidans*, >90.1% of the pyrite was recovered as froth; i.e. flotation by itself was ineffective for separating pyrite from other minerals. Exposing of mineral mixture to *T. ferrooxidans* dramatically reduced the pyrite content of the froth: 80.5-95.1% of the pyrite in each mixture was now found in the tailings. In contrast, 72.9 to 100% of the other minerals remained in the froth. Consequently, their purity of the other minerals increased from 50% to 84.3% (millerite), 86.0% (galena), 84.4% (chalcocite) and 79.2% (molybdenite), respectively.

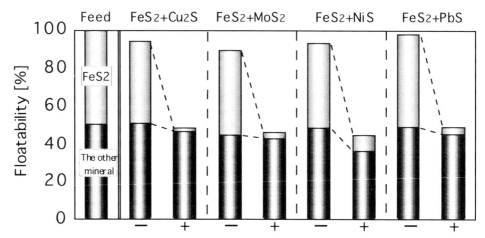

Figure 4. Pyrite removal from sulfide mineral mixtures by flotation with *T. ferrooxidans*. Each mixture contained 0.2g of pyrite and 0.2g of another sulfide mineral (molybdenite, chalcocite, millerite, or galena). The mineral mixtures (0.4g) were exposed to 7.4×10^8 cells. The symbols on vertical axis expressed; (-), without cells exposure and (+), with cells exposure.

4. DISCUSSION

4.1. Selective adhesion of *T. ferrooxidans* to pyrite and its interactions

The selectivity of *T. ferrooxidans* adhesion to pyrite was first suggested when it was observed that the cells adhered to pyrite within coal particles [4,5]. The same phenomenon was observed in iron-rich areas of sulfide ores [6]. Adhesion of *T. ferrooxidans* to pyrite has been compared with adhesion to other minerals [7], but the selectivity was not definitively shown. In contrast, the present study clearly demonstrates that *T. ferrooxidans* will selectively adhere to pyrite, despite the presence of large quantities of one or more other minerals.

Of particular interest to us were the interactions mediating *T. ferrooxidans*' selective adhesion. It is generally understood that bacterial adhesion is governed by physical (e.g. electrostatic or hydrophobic) interactions [8,9,10]. Electrostatic interactions depend on the charges present on the surfaces of the cell and the particle. Therefore, the Zeta-potentials of the minerals used in this study were estimated. All of the minerals had negatively charged surfaces, and the surfaces of *T. ferrooxidans* cells are also negatively charged [11]. Thus, electrostatic interaction would not produce the essential force necessary for *T .ferrooxidans* adhesion; in fact, the negative charges would repulse one another [12].

Adhesion driven by hydrophobic interactions can be modeled thermodynamically taking into consideration the surface tensions of adherent cells, the solid surfaces and the suspending liquid medium [13]. Hydrophobicity can be estimated as the surface tension calculated by the contact angles. The hydrophobicity of the minerals used in this study was reflected by contact angles of 78.4 degrees or greater. In contrast, *T. ferrooxidans* cells have comparatively hydrophilic surfaces with contact angles of 22.7-24.0 degrees [3]. In addition, the surface tension of the sulfuric acid solution that served as the suspending liquid medium in the cell adhesion experiments was calculated to be 71.3 dyne/cm.

The above calculations were incorporated into a thermodynamic model to determine the change of surface free energy (ΔG_{adh}) that occurs when *T. ferrooxidans* adheres to the respective minerals. Because ΔG_{adh} was negative in all cases, bacterial adhesion via hydrophobic interaction with any of the minerals should occur spontaneously; moreover, the more negative the value of ΔG_{adh}, the more likely cells would be to adhere. According to this model, cells are most likely to adhere to molybdenite followed by chalcocite, pyrite, millerite and galena in descending order. With the exception of pyrite, this theoretical determination does approximate the actual order observed experimentally and suggests two conclusions: 1) the adhesion of *T. ferrooxidans* to minerals other than pyrite is governed by hydrophobic interactions between the surfaces of the bacterium and the respective mineral; and 2) the adhesion to pyrite involves specific interactions other than hydrophobic interactions.

Research conducted over the past several years has frequently focused on whether there is a specific mechanism involved when *T. ferrooxidans* adheres to pyrite. For instance, Rojas et al suggested that an organic capsule covering the cell surface is relevant to pyrite adhesion [14,15]. Devacia et al [15] and Arredondo et al [16] both suggested that cell surface proteins play an important role in the adhesion of *T. ferrooxidans* to solid surfaces. Very recently, an apo form of rusticyanin was isolated as a surface protein responsible for *T. ferrooxidans* adhesion to pyrite, which may provide the key to its selectivity [17].

4.2. Application of microbial flotation to mineral processing.

When pyrite is present as a contaminant in mixtures of sulfide minerals, commercial flotation processors most commonly use cyanide to suppress its floatability. Because the suppressive effect of cyanide on pyrite flotation is enhanced in solutions made alkaline by the presence of $Ca(OH)_2$, the separation efficiency of flotation in the presence cyanide + $Ca(OH)_2$ was tested using a pyrite/galena mixture. The pH of flotation liquor was adjusted to 10.0 with $Ca(OH)_2$, and then cyanide (KCN) was added to a concentration of 1 mM. Under these conditions, pyrite rejection and galena recovery were 76% and 68%, respectively; less than the separation efficiency of *T. ferrooxidans* which produced corresponding values of 91% and 91%.

In past years, oxidative metal leaching by bacteria has contributed significantly to mineral processing. It now appears that it may be useful to expand the role played by bacteria in mineral processing to include adhesive control of mineral floatability.

REFERENCES

1. Douglus E Rawlings and Silver Simon, BIO/TECHNOLOGY, 13(1995) 773
2. C.C.Townsley and A.S. Atkins, Process Biochemistry, December(1986) 188
3. N.Ohmura, K.Kitamura, and H.Saiki, Biotechnol. Bioeng., 41(1993) 671
4. R.M.Bagdigian and A.S.Myserson, Biotechnol. Bioeng., 28(1986) 467
5. R.G.L.McCready and B.P.Lc. Gallais, Hydrometallurgy, 12(1984) 281
6. L.E.Murr and V.K.Berry, Hydrometallurgy., 2(1976) 11
7. N.Ohmura, K.Kitamura, and H.Saiki, Appl. Environ. Microbiol., 59, 12(1993) 4044
8. K.C.Marshall, Solid-liquid and solid gas interfaces, In Marshall, K.C.(c.d.), Interfaces in microbial ecology. Harvard University Press, London. (1976) 27
9. H.J.Busscher and A.H.Weekamp, FEMS Microbiol. Rev., 46(1987) 465
10. M.Rosenberg and R.J.Doyle, Microbial cell hydrophobicity: histry, measurement and significance, In Doyle, R.J. and Rosenberg, M.(ed.), American Society for microbiology, Washington DC. (1990) 1
11. R.C.BlakeII, E.A.Shute, and G.T. Howard, Appl. Environ. Microbiol., 60(1994) 3349
12. P.Devancia, K.A.Natarajan, D.N. Sathyanarayana, and G. Ramananda, Rao., Appl. Environ. Microbiol., 59(1993) 4051
13. Darryl R Absolom, Lamberti, Francis V., Policova, Zdenka, Zingg, Walter, Oss, Carel J. van, and Neumann, A. Wilhelm., Appl. Environ. Microbiol., 46, 1(1983) 90
14. J.Rojas, M.Giersig, and Tributsch, H., Arch microbiol., 163(1995) 352
15. J.Rojas, M.Giersig, and Tributsch, H., Fuel, 75(1996) 923
16. R.Arredondo, A. Garcia, and C. A. Jerez., Appl. Environ. Microbiol., 60(1994) 2846
17. N.Ohmura and R.C. BlakeII, IBS BIOMINE '97 conference proceedings, PB1

Sorption of Rare Earth Elements and Uranium on Biomass: a Kinetic Study of Competition Processes

G. Naja[a], C. Peiffert[b], M. Cathelineau[b] and C. Mustin[a]

[a]Centre de Pédologie Biologique. UPR 6831 du CNRS associée à l'Université de Nancy Henri Poincaré, 17 rue Notre Dame des Pauvres, BP n°5, 54501 Vandœuvre-lès-Nancy Cedex, France.

[b]CREGU - UMR 7566 G2R
3 rue Bois de la Champelle, BP n° 23, 54501 Vandœuvre-lès-Nancy Cedex, France.

The quantitative study of the adsorption of metallic ions on both organic and mineral surfaces showed that sorption is strongly dependent on chemical parameters such as pH or competing ions. An increase in pH yields to a greater amount of fixed metals whereas the competition between ions tends to decrease the relative amount of each sorbed element.

Moreover, the behaviour of the adsorption sites and the binding capacity of each material (at different pH) and their selectivities have been studied for complex solutions including up to 7 cations introduced at the same starting concentration. Sorption of cations is very fast and up to 90% of the fixation is achieved in less than 10 minutes. However, fluctuation in the total of sorbed ions is observed for greater times.

Results show that biomasses bound more efficiently uranium than rare earth elements, and that the final amounts of sorbed rare earth elements depends on the nature of the elements, with the following order of increasing sorbed quantities: Ce, Nd, La, Pr and Dy. The same order is found during the sorption of rare earth elements on clays. Finally, kinetics analysis of sorption were done in order to fit experimental results by Freundlich or Langmuir adsorption model which seems more appropriated.

1. INTRODUCTION

The solid-water interface plays an important role in the regulation of the concentrations of the metallic elements in natural water. The contact between aqueous and solid phases constituted by sand, organic matter, clays..., will affect the mobility of the contaminant metals in soils and waters [1-9].

Moreover, sorption of a metallic cation is a surface reaction process recognised for its many industrial application (pollution control and remediation) [10]. Therefore, the study of the surface reactivity, during sorption, is necessary to predict the behaviour of elements in solution [11-14].

Metal sorption involves the competition between protons and metals and the physico-chemical properties of the constituents such as the nature of the functional groups, the surface area...It is accompanied by a relative change of the sorbent's properties (e.g. solubility). Furthermore, the sorption of cations involves formation of bonds with the functional groups of the solid (oxygen atoms) and a release of protons.

This study has been carried out to gain understanding of the sorption mechanisms of cations on different materials and to compare their binding capacities in mining water environments.

A general study carried out on biomass, clays and ferrous oxide showed that biomass bound more efficiently uranium and lead and ferrous oxides bound majority uranium, arsenic and lead whereas clays adsorbed a weak quantity of uranium and have a great capacity to bind rare earth elements. Detailed results presented in this paper concern adsorption on biomass for solution containing uranium and rare earth elements (Dy, La, Nd, Pr, and Ce) which are representative of uranium mine water.

2. MATERIALS AND METHODS

2.1. Experimental solution and solids

The composition of most natural waters interacting with sulphide bearing rocks in subsurface are similar and marked by the prevalence of calcium and sulphate, due to dissolution of the alteration product of sulphides, e.g. gypsum [15].

The experimental solution contains 100mg/l of $CaSO_4$ and H_2SO_4 to fix the pH at 5 or 3.

5 mg/l of uranium, cerium, lanthanum, neodymium, dysprosium and praseodymium have been added to the calcium sulphate matrix.

The ionic strength of the solution is fixed at 0.04M.

The used biomass *Rhizopus arrhizus* (% N=1.61, % C=18.66) was collected from pharmaceutic industry and contains an important quantity of silicium and chlorine which come from the use of diatomite (a filtration adjuvant).

2.2. Experimental procedure

Solid samples (0.5g) were mixed up with an acidified solution (250ml) in a stirred batch reactor at 25°C.

A measure of pH is done as well as an extraction of 2 ml of solution to analyse it by ICP-MS.

Sorption of cations on solid is determined by measuring the change in solute concentration after addition of the solid phase. Samples were collected at: 0.5, 3, 6, 24, 48, 96 and 120 hours.

3. RESULTS AND DISCUSSIONS

The concentration of metallic elements in natural environment or waste mining treatment is determined by the reactions of adsorption of cations having an affinity for the surface of a solid.

Uranium is present in solution as $(UO_2)^{2+}$ for a range of pH between 2.5 and 5.5. The rare earth elements (REE) in solution form trivalent cations with a very small ionic radius ($\approx$100pm) unless for cerium and praseodymium which can form quadrivalent cations.

The ratio of sorption for uranium and REE depends on pH and on the reaction time. To illustrate these observations, the results are presented following two types of curves: the first represents the percentage of the quantity sorbed of the cations and the second type represents the variations of the different concentrations of the sorbed elements (in µg/l), versus the reaction time.

3.1. Biomass

Figures 1 and 2 show that the quantity of the cations sorbed on the surface of the biomass is not equal for all the ions present in solution. It depends on the affinity of each element to sorb as well as on the pH of the solution:

- uranium is the element which is fixed easily at pH=5 and 3. The quantity sorbed is 100% at pH=5 and 90% at pH=3.

- REE are less sorbed. Furthermore, figures 1 and 2 show a strong pH dependence in the case of the uptake of REE. 30% of praseodymium is sorbed at pH=5 and this quantity decreases at 15% for pH=3.

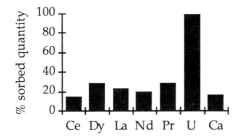

Figure 1. Sorbed quantity at pH=5

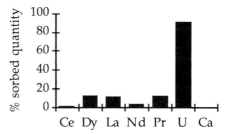

Figure 2. Sorbed quantity at pH=3

The kinetic of sorption of U and Pr (Figures 3 and 4) follows a simple law, however the other cations kinetic uptake (as Ca) is much more complicated. The common property of all these cations is that sorption is very fast and up to 90% of the fixation is achieved in less than 10 minutes.

The decrease of the total sorbed amount (at the end of the reaction time) is similar for all considered rare earth elements (Figures 4 and 5). It suggests redistribution processes, linked to several factors: kinetics of dissolution of the biomass, changes of the surface properties during 5 days,...

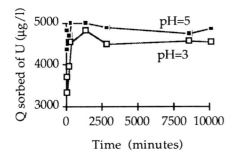

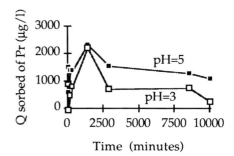

Figure 3. Sorption kinetic of U   Figure 4. Sorption kinetic of Pr

    Biosorption of cations is due to the net negative charge of the surface of the biomass. Indeed the cell wall of a biomass is constituted with carboxylic and hydroxyl groups which confer a global negative surface charge and an anionic ligands property.

    The reaction of biosorption involves the competition between protons and cations and the physico-chemical properties of the organic constituents such as surface functional groups ($-COOH$, $-NH_2$, $-SH$, $-OH$...). These specific sites were found to be involved in the binding processes.

3.1.1. Influence of pH

Because the reactive surface sites are often hydroxyl or carboxylic groups which can coordinate and dissociate protons, one might expect a strong pH dependence of sorption process. As the pH is increased, cation sorption on biomass increases from 0 to 20 percent for Ca and Ce.

The pH of solution affects the sorption of cations which compete with protons to be fixed on the active sites. At pH=3, the available sites on the surface of the biomass are completely protonated, hence the sorption of cations is more difficult (Figure 2). Following the reaction (1), if the concentration of protons increases, the reaction of desorption is favoured (mass law). Moreover the dissociation of the carboxylic groups release their protons which improved the uptake of cations.

$$R\text{-}H + M^+ <=> R\text{-}M + H^+ \tag{1}$$

where M^+ represents a cation of the solution and R-H a functional group of the biomass.

3.1.2. The order of sorption

The order of biosorption is proposed between cations for a pH=3. At this pH the competition cation/proton and cation/cation limits their adsorption on the biomass (sorption of uranium in solution with calcium and in the other hand uranium in presence of the other cations).

The previous results show that the quantity of sorbed U on biomass is more important than the other cations. (All the cations seem to be fixed on the same types of sites). The sorption depends on the degree of affinity or on the

electronegativity of the cations. The electronegativity of REE is around 1.1 and weaker than the other cations (1.2-1.6) which can explain their low sorption.
The sorption kinetics obtained are not easy to explain (Figure 5), however the same type of fluctuations can be distinguished for the five curves. Furthermore, we can point out an order of sorption of REE (at pH=5 and 3) on the biomass Dy ≈ Pr >La ≈ Nd > Ce. The same order of sorption was found for the uptake of these elements on clays and ferrous oxide.

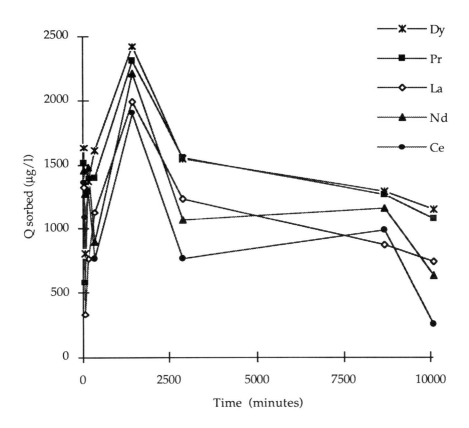

Figure 5. Variation of the sorbed quantity of REE at pH=5. The first point corresponding to t=30 minutes.

3.1.3. The mechanism of sorption
The mechanism of sorption can be considered following 2 steps:
   - the release of protons of the functional groups and the sorption of cations with formation of a bond with the oxygen atom of the carboxylic or hydroxyl groups: this process is very fast and it takes less than 10 minutes for the uptake of uranium (Figure 3). The establishment of a new bond between the cation and the

organic matter is showed by infrared spectroscopy. This step is characterised by a pH decrease,
- the rearrangement of the new molecules to permit to the other cations in solution to be fixed on surface; it is a very long stage which allows the establishment of a physico-chemical equilibrium. At least, an increase of pH is noted.

3.2. Kinetics models

The sorption process for cations in water follows a number of types, including hydrogen bonding, ligand exchange and cation exchange. In this section, sorption models are discussed [16]. The first concept is the relationship of the amount of solute sorbed onto surface as a function of the concentration of the solute. This relationship is an isotherm, which could be described by Langmuir or Freundlich models.

The Langmuir model (Figures 6 and 7) is applied when the solid has a greater affinity for the solute than water. Langmuir hypotheses are: (i) the surface is homogenous, (ii) the sorbed species forms one layer and (iii) there are no interactions between sorbed molecules. The mathematical form of the Langmuir isotherm is shown following equation (2):

$$\mu = \frac{\mu\max \cdot Kads \cdot [A]}{1 + Kads \cdot [A]} \Rightarrow \frac{[A]}{\mu} = \frac{1}{\mu\max \cdot Kads} + \frac{[A]}{\mu\max} \tag{2}$$

where
μ represents the quantity of sorbed element per mass unity (mg/g),
[A] is the concentration of the cation in solution at the equilibrium (mg/l).

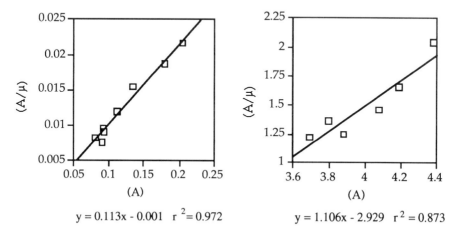

y = 0.113x - 0.001   r^2= 0.972   y = 1.106x - 2.929   r^2 = 0.873

Figure 6. Sorption kinetic of U   Figure 7. Sorption kinetic of Pr

The Freundlich isotherm appears to hold at low concentration of solute and may be used in modelling sorption phenomena. It can be deduced from a multisite Langmuir model. Another hypothesis of Freundlich model is the existence of lateral interactions between adjoining sorbed species. The mathematical formula of the isotherm is:

$$\mu = m[A]^n \Rightarrow \log(\mu) = f(\log[A]) \tag{3}$$

To fit experimental results of cation sorption by Freundlich or Langmuir models, some example of cation sorption Langmuir isotherms are presented in figures 6 and 7 at pH=5. Calculation and determination of the linear regression coefficient were done in order to determine the more appropriate model (Table 1).

Table 1
Application of the kinetic models. Calculation of the different parameters.

	Freundlich			Langmuir	
	n	m	r^2	Qmax sorbed (mg/g solid)	r^2
Uranium	-0.122	7.55	0.188	8.84	0.972
Dysprosium	-0.022	2.74	0	0.941	0.92
Praseodymium	-1.384	18.23	0.055	0.904	0.873
Lanthanum	-2.453	-	0.089	0.42	0.128
Neodymium	-4.995	-	0.553	0.343	0.838
Cerium	-7.852	-	0.649	0.118	0.698

According to the values of the regression coefficient, the Freundlich model is not adequate for the sorption phenomena in comparison with the Langmuir which models the uptake processes. Furthermore, with the Langmuir model, the same order of the maximum sorbed quantity of rare earth elements on biomass is found: Dy > Pr > La > Nd > Ce.

However, these classic models don't involve the competition proton/cation or cation/cation. Hence, they may not describe perfectly the behaviour of an element sorbing on a surface and specially the cationic exchange that may happen on the surface of the solid.

4. CONCLUSION

The mobility of heavy metals and contaminants in soils and waters depends on the chemical properties of the elements and the physico-chemical processes (adsorption, complexation, precipitation...) due to the important surface area of the natural constituents of soils. Therefore, it is important to analyse the kinetic reactions between elements and the functional groups of the surface of solids.

The experimental study of the sorption reactions has shown the effect of pH (competition proton/metal, an increase of pH inducing an increase of the sorbed quantity) and the effect of the competition metal/metal (the presence of a solution containing several cations induce a decrease of the maximal sorbed quantity for a given element) during the uptake of cations on the surface of biomass.

Moreover, it has been shown that the industrial biomass *Rhizopus arrhizus* is efficient for the removal of uranium at pH=5 and 3 whereas clays are able to uptake an important quantity of rare earth elements. Hence, the surfaces of materials have not the same affinity and the same selectivity for cations present in solution.

Future experimental efforts will lead to the development of industrial processes, for chemical remediation, choosing the adequate sorbing material considering the long duration of reactions which can modify the behaviour of the molecular sites and the properties of the materials.

REFERENCES

1. J. Berthelin, Microbial weathering processes. In W.E. (ed) Microbial geochemistry. Blackwell Scientific Publ., Oxford, U.K (1988) 223.
2. M. Tsezos, Biotechnol. Bioeng., 25 (1983) 2025.
3. N. Kuyucak, B. Volesky, Biotechnol. Lett., 10 (1988) 137.
4. L.E. Macaskie and A.C.R. Dean, Env. Technol. Lett., 3 (1982) 49.
5. C. Munier-Lamy, P. Adrian, J. Berthelin and J. Rouiller, Org. Geochem., 9 (1986) 285.
6. E. Guibal and C. Roulph, Journal Francais d'Hydrologie, 21 (1990) 229.
7. J. Berthelin and C. Munier-Lamy, Elsevier Science Publishers, Amsterdam, 3 (1987) 343.
8. C. Munier-Lamy, P. Adrian and J. Berthelin, Toxicol. Environ. Chem., 31-32 (1991) 527.
9. Y. Bizri, M. Cromer, J.P. Scharff, G. Guillet and J. Rouiller, Geochim. Cosmochim., 48 (1984) 227.
10. B.N. Noller, R.A. Watters and P.H. Woods, J. Geochem. Explor., 58 (1997) 37.
11. B. Gueniot, Géologie et géochimie de l'uranium, distribution et mode de fixation, Nancy-I, 1983.
12. A. Manceau, L. Charlet, M.C. Boisset, B. Didier and L. Spadini, In proceedings of Symposium B on clays, Strasbourg (1992) 201.
13. B. Volesky, Biosorption of Heavy Metals, Mc Gill University (ed), Montreal Canada, 1989.
14. C.P. Huang, D. Westman, K. Quirk and J.P. Huang, Wat. Sci. Tech., 20 (1988) 369.
15. M. Cathelineau, A. Guerci, N. Ahamdach, M. Cuney, C. Mustin and G. Milville, J. Pasava, B. Kribek & K. Zak (ed), A.A. Balkema, Rotterdam, (1995) 647.
16 A. Artola and M. Rigola, Biotechnol. Lett., 14 (1992) 1199.

Biosorption of heavy metal ions from aqueous and cyanide solutions using fungal biomass

K.A. Natarajan[a], S.Subramanian[a] and J.M. Modak[b]

[a]Department of Metallurgy, [b] Department of Chemical Engineering
Indian Institute of Science, Bangalore 560012, India

A waste fungal biomass containing killed cells of *Aspergillus niger* was efficiently used in the removal of toxic metal ions such as nickel, calcium, iron and chromium from aqueous solutions. The role of different parameters such as initial metal ion concentration, solution pH and biomass concentration on biosorption capacity was established. The maximum metal uptake was found to be dependent on solution pH and increased with biomass loading upto 10g/L. The adsorption densities for various metal ions could be arranged as Ca> Cr (III) > Ni > Fe > Cr (VI). The effect of the presence of various metal ions in binary, ternary and quaternary combinations on biosorption was also assessed. Ni uptake was significantly affected, while that of Cr (VI) the least, in the presence of other metal ions.

Uptake of base metals from an industrial cyanide effluent was studied using different species of fungi such *as Aspergillus niger, Aspergillus terreus* and *Penicillium funiculosum* and yeast such as *Saccharomyces cerevisiae* which were isolated from a gold mine. Traces of gold present in the cyanide effluent could be efficiently recovered. Among the four base metal contaminants present in the cyanide effluent, zinc was found to be most efficiently biosorbed, followed by iron, copper and lead. The role of both living and dead biomass on biosorption was distinguished and probable mechanisms illustrated.

1. INTRODUCTION

The discharge of heavy metals into aquatic systems has caused world-wide concern since the past few decades. These pollutants are introduced as a result of various industrial operations such as electroplating, chemical manufacturing, leather tanning, petroleum refining, mining and mineral processing, to name a few. The removal of these toxic and heavy metal contaminants from aqueous waste streams and industrial effluents is one of the most important environmental issues facing mankind today. A recent development in environmental biotechnology is the use of microbe-based sorbents for the removal and recovery of strategic and precious metals from industrial waste effluents. Biosorption processes in which microorganisms – live or dead, or their derivatives are employed for detoxification of industrial wastes, offer an attractive alternative to conventional methods of metal removal and recovery. Additionally, the ability of microorganisms to selectively adsorb a specific element in the presence of other elements or species can be used for selective separation and pre-concentration of trace elements. The metal uptake capacities of various microorganisms such as bacteria, algae, fungi and yeasts have received attention in recent times and it has been reported that they can efficiently accumulate heavy and precious metals including radionuclides from their external environment [1-8]. It is also well documented that even when the cells have been killed, the microbial biomass is capable of adsorbing metal ions from aqueous solutions [9,10]. The biosorption of metals using non-living biomass has recently been comprehensively reviewed by Modak and Natarajan [11]. A further advantage of the use of dead biomass is that the problem of toxicity and the economic aspects of nutrient supply and culture maintenance can be eliminated.

In the present investigation, the potential of a waste fungal biomass containing killed cells of *Aspergillus niger* has been assessed for the removal of toxic metal irons such as nickel, chromium, iron and calcium. The role of different parameters such as initial metal ion concentration, solution pH and biomass concentration on the biosorption capacity has been established. The effect of the presence of various metal ions in binary, ternary and quaternary combinations on biosorption has also been studied. The uptake of base metals such as zinc, lead, copper and iron from an industrial cyanide effluent has been investigated using different species of fungi namely, *Aspergillus niger, Aspergillus terreus and Pencillium funiculosum* and yeast such as *Sacchromyces cerevisiae,* isolated from a gold mine. The role of both living and dead biomass on biosorption has been distinguished and probable mechanisms illustrated.

2. EXPERIMENTAL MATERIALS AND METHODS

2.1. Biomass

The biomass samples were obtained from Biocon Ltd., Bangalore. The biomass consisted of *Aspergillus niger* grown on wheat bran in a solid-state fermentation process. After the enzymes were extracted, wheat bran with cells attached to it were autoclaved by the industry before disposal. This waste *Aspergillus* biomass (ABM-1) was used in this investigation. The biomass was in the form of blackish brown granules of $0.16 - 0.2$ mm size and the density was estimated to be 310 g/l. The material was stored in sealed plastic bags at room temperature.

2.2. Metal ions

The metal ions studied were Ca, Fe, Ni, Cr (III) and Cr(VI). Stock solutions of ferrous ammonium sulphate $(NH_4)_2SO_4$ $FeSO_4.6H_2O$ (S.D.Fine Chemicals), nickel sulphate $NiSO_4.6H_2O$ (Merck), chromium nitrate $Cr(NO_3)_3.9H_2O$ (Merck) for Cr(III) and potassium dichromate $K_2Cr_2O_7$ (Merck) for Cr (VI) were prepared. Ca granules (Riedel-de-Haen Seelze, Hannovar) were dissolved in 1:1 HCl and the stock solution prepared. All the reagents were of analytical grade and deionised, double distilled water was used for all experiments.

2.3. Microorganisms

Pure isolates of different species of fungi such as *Aspergillus niger, Aspergillus terreus* and *Pencillium funiculosum* as well as the yeast *Saccharomyces cerevisiae* were obtained from the mine water samples of the Hutti gold mines, located in Raichur district of Karnataka, India. Morphological studies and standard biochemical tests were used to identify the isolates [12].

2.4. Effluent samples

Gold containing cyanide effluent samples were supplied by the Hutti Gold Mines Co. Ltd., Karnataka, India.

2.5. Culturing technique

For the generation of biomass, the fungal and yeast strains were cultured using Sabouraud's media as per the procedure detailed elsewhere [13].

2.6. Biosorption and analytical procedures

A known weight of the ABM-1 biomass was added to a 100 ml solution containing a known amount of metal ion under investigation. The pH was adjusted to a desired value and the suspension was agitated for a desired time in a Remi make incubator shaker at 250 rpm at room temperature (28°C). The slurry after equilibration was filtered through Whatman 41 filter paper. The clear filtrate was analysed for metal concentration using a Thermo Jarrell Ash Video 11 E atomic absorption spectrophotometer in the case of Ca, Fe, Ni and total Cr content using standard procedures [14]. Cr(VI) was determined by a spectrophotometric method using a Shimadzu uv – vis spectrophotometer [15]. Cr (III) concentration was obtained from the difference between the total Cr and Cr (VI) contents. The amount of metal taken up by ABM-1 was calculated as the difference between the initial and final concentrations of metal in solution.

In the case of the effluent samples, both living and non-living biomass were used in the biosorption tests. In the former test, the filtered cells were used as collected. The non-living samples were prepared by boiling the biomass in distilled water for 5 min. The scalded cells were rinsed with distilled water and the excess water was removed. In typical tests, about 4 to 8g (wet weight) of the biomass was contacted with 100 ml of the effluent solution in a 250 ml Erlenmeyer flask and incubated on a rotary shaker at 30°C and 250 rpm. Unless otherwise specified, the contact time was fixed at 24 h to attain sorption equilibrium. The metal-laden biomass was filtered and the supernatant solution analysed by atomic absorption spectrophotometry.

3. RESULTS AND DISCUSSION

3.1. Studies on waste fungal biomass (ABM-1)
3.1.1. Biosorption kinetics

Preliminary tests were conducted to assess the time taken for the biosorption equilibrium to be attained. In these experiments the initial concentration of metal ion was fixed at 100 mg/l, the biomass loading at 10g/l and the pH was 3.5 for Cr(III) and Cr(VI), 4.5 for Ni, 5.4 for Fe and 5.5 for Ca. The results depicted in Figure 1, indicate that the residual concentration decreases with increase in time and a steady state reached at a certain time in each case. It is apparent that equilibrium is attained within an hour for Cr(III) and Cr(VI) while in the case of Ca, Ni and Fe about 1.5 h are required. The decrease in the metal concentration is quite rapid during the first 30 min for the different metal ions studied.

3.1.2. Effect of pH

The influence of pH on the amount of metal ion adsorbed is portrayed in Figure 2. These experiments were performed at a biomass loading of 10g/l and an initial metal ion concentration of 100 mg/l. Based on the kinetic tests, the equilibration time was fixed at 90 min for Ca, Ni and Fe, while for Cr(III) and Cr(VI) it was maintained at 60 min. It is evident from the figure that the amount adsorbed increases with increase of pH upto 4-5 and thereafter remains more or less constant. pH values beyond 7 were not studied to avoid

354

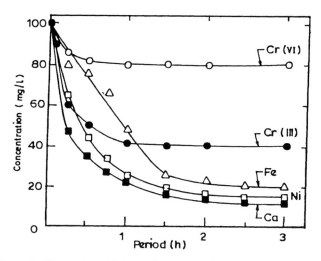

Figure 1. Change in initial metal concentration as a function of time

precipitation of the metal ions. The pH range for higher adsorption is between 5-7 in the case of Ni, 5-6 for Cr(III) and Cr(VI), 4-7 for Ca and 4-5 for Fe. The biosorption process may be considered analogous to an ion-exchange process in the case of dead biomass and thus pH has a significant influence on metal uptake. The cell surface binding sites and the solution chemistry of metal ion are affected by pH. The lesser amount of metal ion adsorbed at highly acidic pH values may be attributed to the fact that protons and the metal cations compete for the binding sites. It has been reported that at low pH values, cell wall ligands are closely associated with H_3O^+ that restricts access to ligands by metal ions as a result of repulsive

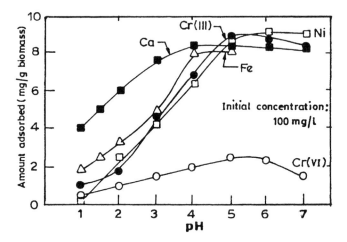

Figure 2. Effect of pH on the amount of metal ion adsorbed

forces [16]. As the pH is increased, more ligand with negative charge would be exposed with subsequent increase in attraction for positively charged metal ions. The decrease in the amount adsorbed beyond the optimum values may be attributed to reduced solubility and precipitation. These findings are in good agreement with the results of other workers [6,17].

3.1.3. Effect of initial concentration of metal ion

The variation in the percentage uptake of metal ion as a function of the initial concentration is shown in Figure 3. In all cases, there is a decrease in the percent uptake as the initial concentration of metal ion is increased. These experiments were conducted at a pH of 5 and the uptake capacity of the biomass for the various metal ions may be arranged in the following sequence of decreasing magnitude:

$$Ca > Cr(III) > Ni > Fe > Cr (VI)$$

It is evident that the biosorption ability is better at lower concentrations of the metal ions, thus facilitating removal of metal ions from dilute solutions.

3.1.4. Effect of biomass loading

The biomass loading is an important parameter governing the extent of metal removal from the solution. The results of the specific uptake capacity (mg/g biomass) as a function of biomass loading depicted in Figure 4 reveal that for all the metal ions there is a steep decrease in the specific uptake capacity upto about 10g/l followed by a gradual decrease beyond that value. In the case of Ca, a fifteen fold decrease is observed, namely from 31.5 mg/g of biomass at 2g/l to about 2 mg/g at 40 mg/l.

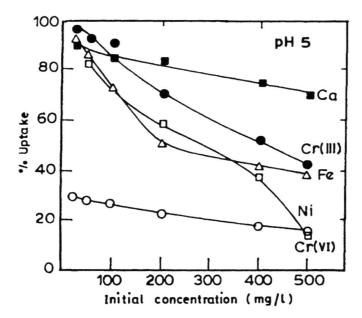

Figure 3. Variation of percent metal uptake as a function of initial concentration

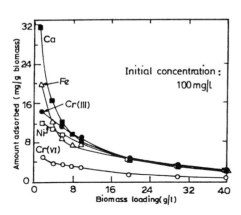

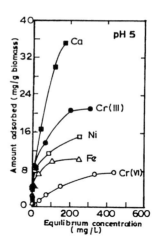

Figure 4. Specific metal uptake as a function of biomass loading

Figure 5. Biosorption isotherms for different metal ions

Similarly for Fe, a seven-fold decrease is observed while for Cr (III) it is about six times, for Cr (VI) about 5 times and for Ni a four-fold reduction is found. It has been observed that at a given equilibrium concentration, the biomass adsorbs more metal ions at low cell densities than at higher loading [18]. The electrostatic interactions between the cells may be a significant factor governing the biomass loading, with a larger quantity of cations being adsorbed when the distance between the cells is more [19]. It has been postulated that high biomass concentration could make a 'screen effect' of the dense outer layer of cells, protecting the binding sites from metal [20].

3.1.5. Biosorption isotherms

The adsorption isotherms for the various metal ions are portrayed in Figure 5. These isotherms were determined at 28°C and at pH 5. At low concentrations, it is evident that there is a steep rise in the adsorption density particularly in the case of Ca, Cr(III), Ni and Fe indicating high affinity behaviour. On a relative basis, the adsorption isotherm for Cr(VI) exhibits a low affinity trend.

All the isotherms were found to follow the Langmuir equation. The adsorption densities may be arranged in the following order for the various metal ions:-

$$Ca > Cr(III) > Ni > Fe > Cr(VI)$$

This follows the trend observed earlier. In the case of dead biomass, as used in this study, the adsorption mechanism is governed by ionic, physical and chemical forces. A variety of ligands located on the fungal cell walls are known to be involved in metal chelation. It has been reported that the cell wall of *Aspergillus niger* comprises of chitin and glucan [21]. These components are suggested to be important sites for metal chelation [22]. An infrared spectrum of the biomass ABM-1 used in this study indicated the presence of carboxyl, amine, hydroxyl and phosphate groups. The biomass was found to be negatively charged in the pH range 3-11. These functional groups assist in the metal binding process apart from the electrostatic forces.

Table 1
Percentage of metal uptake in multi-ion systems

Binary systems	Ca	Fe	Ni	Cr(III)	Cr(VI)
Ca-Fe	53	67	-	-	-
Ca-Ni	70	-	24	-	-
Ni-Fe	-	69	11	-	-
Ca-Cr^{3+}	59	-	-	93	-
Ca-Cr^{6+}	80	-	-	-	26
Ni-Cr^{6+}	-	-	33	-	16
Ni-Cr^{3+}	-	-	3	59	-
Fe-Cr^{3+}	-	69	-	90	-
Fe-Cr^{6+}	-	90	-	-	24
Cr^{3+}-Cr^{6+}	-	-	-	91	21
Ternary Systems					
Ca-Fe-Ni	48	36	5	-	-
Ca-Ni-Cr^{3+}	41	-	0	63	-
Ca-Ni-Cr^{6+}	38	-	19	-	10
Ca-Fe-Cr^{3+}	46	41	-	51	-
Ca-Fe-Cr^{6+}	56	50	-	-	20
Fe-Ni-Cr^{3+}	-	34	0	40	-
Fe-Ni-Cr^{6+}	-	79	6	-	14
Quaternary Systems					
Ca-Fe-Ni-Cr^{3+}	28	19	0	37	-
Ca-Fe-Ni-Cr^{6+}	46	21	11	-	20

3.2. Biosorption studies with multiple metal ion systems

In these investigations, binary, ternary and quaternary combinations of various metal ions namely Ca, Fe, Ni, Cr(III) and Cr(VI) were studied keeping the concentrations of all the ions at 100 mg/l and the pH was fixed at 4 and the biomass loading at 10g/l. The experimental procedure was the same as that used for the individual ion systems. The results are summarised in Table 1.

In general, it can be inferred that the biosorption of each metal ion is inhibited by the presence of various co-ions to a lesser or greater extent, as compared to the situation when they are present alone. For example, in binary combinations the uptake of Ca is decreased in the presence of Fe, Ni and to a lesser extent by Cr(III) and Cr(VI). The uptake of Fe is retarded to almost the same extent by Ca, Ni and Cr(III) while Cr(VI) has a negligible effect. The Ni uptake is significantly inhibited by Cr(III), Fe and Ca and marginally by Cr(VI). It is evident that Ni depresses Cr(III) uptake appreciably, while Ca, Fe and Cr(VI) have a negligible effect. Ni also significantly affects Cr(VI) uptake while Cr(III), Fe and Ca depress Cr(VI) uptake to a lesser extent.

The uptake of a given metal ion is decreased to a greater extent in ternary combinations and furthermore in quaternary systems compared to the binary combinations. In the presence of co-ions in solution, chemical interactions between the ions themselves as well as with the biomass take place resulting in site competition. Many of the functional groups present on the cell wall and membrane are non-specific and different cations compete for the binding sites. Therefore metal uptake from multicomponent systems is lower. It has been reported that metal uptake is increased as the ionic radii of metal cations are increased [7]. The ionic charge and ionic radii of the cations also affect the ion-exchange and adsorption process. The differences in the sorption affinities may also be attributed to differences in the electrode potentials of the various ions, with higher electrode potentials enhancing the affinity for biomass. In multicomponent systems, the complex interplay of several factors such as ionic charge and radii as well as electrode potential account for the difference in the metal uptake capacity of the biomass.

3.3. Studies on uptake of base metals from an industrial cyanide effluent

Biosorption tests were conducted using both living and non-living biomass. The chemical compositions of the industrial cyanide solutions are given in Table 2. Effluent samples 1 and 2, generated at two different stages in the gold processing circuit were used in this study. Effluent-1 was obtained from the zinc boxes after the precipitation of gold and consequently has a higher amount of zinc. Effluent-2 was obtained from the effluent pulp that is discharged to the tailings dumps. The associated minerals such as pyrite, pyrrhotite and chalcopyrite contribute to the iron and copper content in the samples, while lead contamination mainly results from the lead acetate added to the zinc boxes.

The mine water samples on screening revealed the presence of a variety of fungi such as *Aspergillus niger, Aspergillus terreus* and *Pencillium funiculosum*. The yeast species *Saccharomyces cerevisiae* was also isolated. The gold uptake capacity of the different strains is summarised in Table 3. Non-living cells of *Aspergillus terreus* exhibited the highest gold uptake ($\approx$ 60 %) followed by the living cells of *Aspergillus niger* and *Pencillium funiculosum*.

Table 2
Chemical composition of the industrial cyanide solutions

Cyanide solution	Gold (ppm)	Zinc (ppm)	Copper (ppm)	Iron (ppm)	Lead (ppm)	Cyanide (ppm)
Effluent – 1	0.2	23.5	44	23.8	13	350-400
Effluent – 2	0.1	0.4	8.5	20.5	3.4	10-15

Table 3
Gold uptake potential of different fungal strains for living and non-living cells from effluent-1 after 24 h of exposure time

Strain	Gold uptake (%)	
	Living	Non-living
Aspergillus niger	49	30
Aspergillus terreus	33	58
Pencillium funiculosum	49	44
Saccharomyces cerevisiae	33	-

The kinetics of metal uptake by microorganisms is known to be comprised of two main phases. The first phase which is rapid involves metabolism-independent binding to cell walls and other external surfaces and can take place in both living and dead cells [23]. The second phase, which is metabolism-dependent is a slower process that takes place only in live metabolizing cells. The metal uptake in this phase involves an energy dependent intracellular transport across the cell wall and an internal compartmentation and precipitation on the cell surfaces. For a sufficiently long contact time, living cells are normally expected to accumulate higher amounts of dissolved metal ions, since both the above mechanisms of metal uptake may operate simultaneously. The results given in Table 3 attest to these findings.

Microorganisms possess specific transport systems to accumulate metals intracellularly and many are also capable of synthesizing specific intracellular proteins called metallothioneins that can chelate metal ions. Yeasts and certain fungi also have the ability to precipitate metals around the cells as a result of metabolic processes. One or more of these mechanisms could be responsible for the higher gold uptake observed in the living cells. Metabolism-independent uptake essentially involves chemical and physical adsorption processes. Functional groups present on the fungal cell walls are implicated in metal chelation [21].

The uptake of the base metals from effluent-1 was also studied using *Aspergillus niger* and the yeast *Saccharomyces cerevisiae*. These tests were conducted with a solid:liquid ratio of 8:100 (w/v) and for a contact period of 24 h. The results are shown in Table 4.

Among the four base metal contaminants in the gold containing effluent-1, zinc was found to be most efficiently removed using both *Aspergillus niger* and *Saccharomyces cerevisiae*. The upake of copper was better with *Saccharomyces cerevisiae*, while lead removal was around 20% in both cases. In these experiments, the different metal ions present in the effluent sample, compete for the same active sites. The relative preference of the biomass for a particular metal ion depends on various factors such as ionic charge, cation radius, initial concentration of metal ion, solution chemistry and toxicity of metal.

The uptake of gold was also studied using the waste biomass (ABM-1) from the cyanide effluent solutions 1 and 2. The results are summarized in Table 5. As much as 70 to 80% of gold was removed from the cyanide effluents using the waste biomass.

Table 4
Uptake capacities of *Aspergillus niger* and *Saccharomyces cerevisiae* for the base metals from effluent −1

Strain	Metal uptake (%)			
	Zn	Cu	Fe	Pb
Aspergillus niger - live cells	100	8.0	27.3	18.5
killed cells	100	12.7	16.4	19.2
Saccharomyces cerevisiae - live cells	100	35.2	25.1	22.1
killed cells	-	-	-	-

Table 5
Percent uptake of gold by waste biomass (ABM-1) (5% solid content; 24h)

	% Au uptake
Effluent – 1	71
Effluent – 2	83

4. SUMMARY AND CONCLUSIONS

The results obtained in this study clearly demonstrate the potential of a waste biomass (ABM-1) for the removal of Ca, Ni, Fe, Cr and Au from aqueous solutions. The following conclusions can be drawn based on this investigation:

1. The sorption kinetics attain an equilibrium value in about 90 min for Ca, Fe and Ni while 60 min are sufficient for Cr (III) and Cr (VI).
2. The maximum amount of metal ion adsorbed is found to be in the range of pH 4-5 for Fe, pH 5-7 in the case of Ca and Ni, and around pH 6 for Cr (III) and Cr (VI).
3. The percent metal uptake decreases with increase in metal concentration for all the metal ions studied.
4. The specific metal uptake decreases as the biomass loading is increased.
5. The adsorption isotherms exhibit Langmuirian behaviour and the adsorption densities can be arranged in the following series of decreasing affinity:-

$$Ca > Cr (III) > Ni > Fe > Cr (VI)$$

6. The presence of co-ions whether in binary, ternary or quaternary combinations decreases the metal uptake as compared to the case when the metal ions are present alone.
7. The biomaterials tested in this study have shown significantly high values of metal uptake from cyanide effluents particularly for gold and zinc.

REFERENCES

1. P.R. Norris and D.P. Kelly, J.Gen. Microbiol., 99 (1977) 317.
2. J.L. Mowll and G.M. Gadd, J.Gen. Microbiol., 129 (1983) 3421.
3. D.W. Darnell, B. Greene, M.T. Henzl, J.M. Hosea, R.A. Mcpherson, J. Sneddon and M.D. Alexanderm, Environ. Sci.Technol., 20 (1986) 206.
4. N. Kuyucak and B. Volesky, Biotechnol. Bioeng., 33 (1989) 809.
5. R.P. de Carvalho, K.H. Chong and B. Volesky, Biotechnol. Prog., 11 (1995) 39.
6. J.L. Zhou and R.J. Kiff, J. Chem. Tech. Biotechnol., 52 (1991) 317.
7. J.M. Tobin, D.G. Cooper and R.J. Neufeld, Appl. Environ. Microbiol., 47 (1984) 821.
8. E.D. Korobhushkina, G.I. Karavaiko and I.M. Korobhushkin, in Bioengineering of Metals, G.I. Karavaiko and S.N. Groudev (eds.), Moscow, Russia, 1985, 125.
9. A. Nakajima, T. Horikoshi and T.Sagaguchi, Eur. J.Appl. Microbiol. Biotechnol., 12 (1981) 76.
10. M. Tsezos and D.M. Keller, Biotechnol. Bioeng., 25 (1983) 201.
11. J.M. Modak and K.A. Natarajan, Min.Metall. Process., 12 (1995) 189.
12. C.J. Alexopoulos, Introductory Mycology, John Wiley and Sons Inc., New York, 1962, 262.
13. M. Balakrishnan, J.M. Modak, K.A. Natarajan and J.S. Gururaja Naik, Min. Metall. Process., 11 (1994) 197.

14.   Atomic Absorption Methods Manual for Flame Operation, Vol.I, Thermo Jarrell Ash Corp., Waltham, USA.
15.   Z. Marczenko, Spectrophotometric Determination of Elements, Horwood, Chicester, 1976.
16.   R.J. Doyle, T.H. Matthews and U.N. Streips, J. Bacteriol., 143 (1980) 471.
17.   P.O. Harris and G.J. Ramelow, Environ.Sci. Technol., 24 (1990) 220.
18.   G.M. Gadd and C. White, J.Chem. Tech. Biotechnol., 55 (1992) 39.
19.   M. Itoh, M. Yuasa and T. Kobayashi, Plant Cell Physiol., 16 (1975) 1167.
20.   M.P. Pons and C.M. Fuste, Appl. Microbiol. Biotechnol., 39 (1993) 661.
21.   V. Farkas, Microbiol.Rev., 44 (1980) 117.
22.   E.Fourest and J.C. Roux, Appl. Microbiol.Biotechnol., 37 (1992) 399.
23.   B. Volesky, Trends in Biotechnology, 5 (1987) 96.

Development of microbial biosorbents - a need for standardization of experimental protocols

K. M. Paknikar*, P. R. Puranik and A. V. Pethkar

Division of Microbial Sciences, Agharkar Research Institute, G.G. Agarkar Road, Pune 411 004, India

An exhaustive culture isolation and screening program was undertaken in laboratory to obtain highly efficient metal sorbing microorganisms. The isolates belonged to diverse groups viz. i) Gram positive bacteria, ii) Gram negative bacteria, iii) yeasts, iv) fungi and v) algae. It was found that all the microbial cultures tested possessed ability to adsorb metals like lead, cadmium, chromium, copper, zinc and silver from solutions. Bacteria and algae showed higher metal uptake than fungi and yeasts.

Some difficulties arose when an attempt was made to compare the data obtained with those reported in the literature. These difficulties were mainly because of the differences in experimental set-up, protocols and evaluation criteria used as well as the method of expressing results. Such aspects need to be standardized before data from different laboratories is compared. Our studies indicated that (a) there is a need for an exhaustive culture screening for obtaining good metal sorbing cultures and (b) there is an urgent requirement for standardizing the metal biosorption protocols with respect to the experimental parameters used. These parameters could include metal concentration, solution pH, biomass to volume ratio, etc. A standard experimental protocol for metal sorbing microorganisms / biomass in proposed.

1. INTRODUCTION

A lot of work has been done so far on the microbe mediated removal of heavy metals from solutions (1-3). It appears that microbial technology for metal removal would become more attractive in immediate future, considering various limitations of conventional physical-chemical processes and relative advantages of microbial methods. Biosorption i.e. adsorption of metal ions onto microbial surfaces, is now recognized as a valuable technique for metal removal (4). In this process, various mechanisms such as complexation, ion exchange, coordination, adsorption, chelation and microprecipitation operate synergistically or

* Corresponding author, Fax: +91-20-351542, E-mail: paknikar@vsnl.com
PRP is thankful to Council of Scientific and Industrial Research, New Delhi for the award of a Senior Research Fellowship.

independently (5).

It has been a general observation of many researchers that often the results obtained by different groups working on metal biosorption are not comparable (6-8). Results are found to vary widely because of the different criteria used by authors (2). In this paper we have made an attempt to illustrate these variations with the help of our own data. To make realistic comparisons of biosorption data across different laboratories a standard screening protocol is suggested. This simplified protocol would also aid in selection of efficient metal biosorbing microorganisms from a very large number of isolates.

2. MATERIALS AND METHODS

2.1. Microorganisms

A variety of microbial cultures were isolated from diverse habitats such as black cotton soils, acidic soils, acid mine drainage, constructed wetland systems, metal contaminated soils etc. A total of 100 cultures comprising of 32 bacterial, 10 yeast, 52 fungal and 6 algal isolates were screened for metal sorption property.

2.2. Biosorbent preparations

Isolates were grown in bulk quantity in routinely used culture media. Growth was harvested by filtration or centrifugation at 10,000 rpm for 10 minutes using a Sorvall (RC 5B plus, USA) centrifuge. Harvested biomass were washed twice with sterile distilled water and then conditioned to pH 4.0 by repeatedly washing with sodium acetate buffer (0.1 M, pH 4.0). The biomass was dried in an oven at 60°C and if necessary, powdered in a blender.

2.3. Metal solutions

Metal stock solutions (20 mM) were prepared by dissolving appropriate quantities of pure metal powders (lead, cadmium, copper and zinc) in 1% nitric acid or analytical grade salts (silver nitrate and potassium chromate) in deionized distilled water. The pH of the working solutions (1 mM) was adjusted to 4.0 with 0.1 M sodium hydroxide and 0.1 M nitric acid.

2.4. Metal sorption studies

Biosorption of metals by the cultures was determined by a batch equilibrium method. Biomass (50 mg) was contacted with 25 ml metal solutions (1 mM, pH 4.0) in 150 ml Erlenmeyer flasks, which were shaken at 120 rpm and at ambient temperature ($28\pm3°C$) for 30 minutes. Biomass was then separated by filtration or centrifugation (10,000 rpm, 10 minutes). Residual metal in the supernatant was determined on an atomic absorption spectrophotometer (Unicam 929 AA spectrometer, UK). Microsoft-Excel software (MS-Excel, Version 5.0, Microsoft Corporation, USA) was used for statistical analysis.

2.5. Effect of experimental parameters on metal uptake

Biomass of *Nocardia* sp. was conditioned to pH 2.0, 4.0 and 6.0 by washing with sodium acetate buffer of the corresponding pH. The biomass (0.05 g and 0.5 g) was then contacted with zinc and lead solutions (1 mM and 100 mg/l, 25 ml) of the corresponding pH.

3. RESULTS

Screening experiments showed that there were differences in the metal uptake behavior of the isolates. It was observed that among the different types of organisms, bacteria were most efficient metal adsorbers than algae, fungi and yeasts. In order to understand the metal uptake behavior of the isolates, the data was subjected to statistical analysis (Table 1). It could be seen that the difference in mean specific uptake of Gram positive bacteria, Gram negative bacteria and algae was not significant ($p>0.05$). However, mean specific metal uptake by both bacteria and algae was significantly higher ($p<0.05$) than the mean specific metal uptake exhibited by yeasts and fungi. When the sorption data of the cultures was compared with available literature it was seen that such a comparison was not possible due to variations in the experimental methods used by other workers.

The specific metal uptake values of ten most efficient cultures obtained from the above screening program are given in Table 2. It was again seen that the uptake values obtained by us could not be compared with those reported by others because of variations in the experimental design.

4. DISCUSSION

Metal biosorption is a complex phenomenon and is dependent on metal ion chemistry, specific surface properties of the organisms and physico-chemical influence of the environment (9). Gram positive and Gram negative bacteria have significant differences in their cell wall structure and thus, are expected to have distinct metal uptake patterns. Gourdon

Table 1
Statistical analysis of metal uptake data from screening experiments

Type of organisms	Mean specific uptake, X (mg/g biomass)					
	Pb	Cd	Cu	Zn	Ag	Cr
Gram positive bacteria	38.5	14.2	11.3	6.8	35.3	7.8
	(14.5)	(9.9)	(5.1)	(4.2)	(12.5)	(9.6)
Gram negative bacteria	40.1	15.1	9.6	6.0	34.0	8.8
	(19.2)	(11.7)	(4.8)	(4.3)	(15.9)	(13.1)
Yeast	23.9	4.0	3.6	1.9	23.4	1.0
	(10.4)	(2.4)	(1.1)	(0.2)	(5.11)	(0.5)
Fungi	17.8	2.5	2.8	1.9	7.0	3.0
	(5.7)	(1.5)	(1.1)	(4.5)	(10.7)	(3.25)
Algae	43.4	10.7	7.5	5.7	21.3	1.3
	(21.3)	(6.4)	(3.7)	(3.5)	(6.7)	(1.0)
All cultures	35.1	11.5	8.5	5.8	28.0	6.3
	(17.3)	(10.0)	(5.2)	(5.0)	(15.5)	(9.3)

Figures in the parentheses are the standard deviation values

Table 2
Specific metal uptake by selected microbial cultures

Organism	Specific metal uptake, Q (mg/g biomass)					
	Pb	Cd	Cu	Ag	Zn	Cr
Citrobacter sp.[a]	93.1	41.6	21.3	11.8	69.4	49.0
P5[a]	54.4	36.7	26.4	14.9	66.2	23.0
P6[a]	65.1	32.2	15.0	13.3	57.3	29.7
S6[a]	57.0	6.2	8.9	7.8	17.3	1.5
S. cinnamoneum[a]	58.8	18.6	12.4	25.3	49.2	10.4
Aspergillus niger[b]	59.2	31.3	12.2	3.0	8.7	3.3
Mucor haemalis[b]	26.7	4.7	3.9	1.9	4.2	2.1
Saccharomyces cerevisiae[c]	35.5	2.6	2.7	1.9	26.4	0.6
A3[d]	40.2	9.6	5.1	2.1	25.5	0.7
A4[d]	73.7	20.0	12.8	10.1	28.4	2.3

a, bacterium; b, fungus; c, yeast; d, alga

et al. (10) reported that Gram positive cells exhibit approximately 20% more cadmium biosorption than Gram negative cells. Conversely, Morozzi *et al.* (11) reported that Gram negative bacteria tested were able to accumulate higher amounts of cadmium than Gram positive bacteria. In the present study, however, no significant difference was noticed with respect to uptake of cadmium and other metals by Gram positive and Gram negative bacteria (Table 1).

Among various types of microorganisms, fungi and yeasts have been reported to possess relatively higher capacities of metal binding (12). However, the results of our screening experiments showed that there was less uptake of metals by fungi as compared to other types of microorganisms, especially bacteria (Tables 1 and 2). In case of algal cultures, the mean specific uptake of lead and silver was found to be comparable to that of bacterial cultures. However, uptake of other metals was found to be lower than bacterial cultures. These observations suggested that it was difficult to generalize on the biosorption efficiency of different groups of organisms and direct the screening programme for isolation of a particular group. Therefore, a large number of microorganisms from diverse groups need to be screened for obtaining efficient metal sorbing cultures. This task would require a reliable screening method that is capable of weeding out inefficient cultures with minimum testing.

A thorough analysis of the literature showed that difficulties encountered in a screening programme, when a comparative evaluation of the data is desired, may be: a) insufficient information about experimental parameters, such as pH (13-15), temperature (13,16-18), biomass concentration (13,16,19-21), metal concentration (13); b) selection of lower or higher values of pH (22), biomass concentration (7,15,23-26), metal concentration (7,23,27-30) and c) expression of the results in different ways (22,24,27,31). From our experience, there are five important factors which need to be defined in the biosorption experiments. These factors influence metal biosorption in a number of ways and are discussed below.

4.1. pH of the solution

The pH of metal solution influences metal biosorption by changing surface properties of biomass and metal speciation (32-35). In general, metal uptake reduces below pH 2 and above pH 8. Metal sorption capacity of biomass is lowest at its isoelectric pH (36). Most investigators have shown that a pH range of 4.0-8.0 is optimal for metal uptake (37). However, higher uptake of radionucleides (38,39) and precious metals (40,41) have also been reported at alkaline pH values in the range of 8 to 10. Metals such as chromium (20) and gold (54) are sorbed under acidic conditions (pH 2-4). The initial pH of solution is thus an important parameter in biosorption studies.

It is our observation that less attention has been given to the solution pH at equilibrium of metal sorption reaction. In our previous studies, it was observed that pH of the metal solution increased from 4 to 6.5 during the sorption experiment (42). Similar observation was made in case of lead and zinc biosorption by *Rhizopus arrhizus* reflecting an intrinsic ability of the microbial biomass to shift solution pH towards a favorable value (6,17,43). In a majority of reports initial solution pH values are mentioned, which may not be factual unless solution pH is controlled in the desired range by using buffered systems or pH conditioned biomass.

4.2. Temperature

Metabolism dependent metal uptake is affected by changes in temperature (44). Although biosorption is a metabolism-independent process, some reports describe the process as temperature dependent (32,45-47), while others describe it as temperature independent (38,48,49). High temperatures may cause permanent damage to the microbial cells resulting in decreased metal uptake (47).

Some research papers quote room temperature or ambient temperature at which experiments are performed (50,51). However, ambient temperatures may vary considerably depending on the geographical location of laboratories. Therefore, it becomes difficult to compare and interpret data reported from different parts of the world, especially when the process is influenced by temperature.

4.3. Concentration of biomass

At a given equilibrium metal concentration, biomass adsorbs more metal ions at low cell densities as compared to high cell densities (52). Changes in biomass concentration affects electrostatic interactions (53), interactions between binding sites (38,49) and metal to biosorbent ratio (17).

At lower biomass concentrations, it may be difficult to detect changes in metal concentrations after addition of biomass. In such cases, errors could get magnified during calculations. On the other hand, at high biomass concentrations, specific metal uptake values could be underestimated (eventhough metal uptake could exceed 99%) because a large number of free binding sites may still be available on the biomass. Such errors could be minimized by maintaining an appropriate metal to biosorbent ratio.

4.4. Initial metal concentration

It has been observed that amount of metal adsorbed by the biomass increases with the increase in initial metal concentration (35,42,54). In cases where biosorption is efficient at lower metal ion concentrations, biosorbent is quickly saturated and a large number of binding sites may remain free due to non-availability of metal ions. Therefore, calculated specific metal uptake values may be lower than the actual uptake. Such an error in the experimental design could lead to faulty selection of cultures.

When the metal concentrations used are expressed in terms of mg/l, it is difficult to compare sorption capacity of an organism with respect to different metals. Since biosorption is a one-to-one interaction between the metal ions and binding sites, use of equimolar solutions in experiments could easily solve this problem.

4.5. Metal to biomass ratio

Initial metal concentration and biomass concentration in conjunction with the reaction volume (i.e. metal to biomass ratio) determine the net availability of metal for sorption by the biomass (17). Variation in any one of these factors would alter the metal to biomass ratio and have influence on the specific metal uptake value. It is therefore essential to perform experiments with optimum metal to biomass ratio, for proper comparative analysis.

4.6. Expression of the sorption results

Many authors quote metal biosorption in terms of percent efficiency (31,24,51,55). Although this term is simple, it gives a highly unrealistic picture of the biosorbent's metal loading capacity (amount of metal bound per unit weight of biomass), which is important for commercial applications. Various ways of expressing the results have so far been used viz. mg/g (2,5,7,42), mmol/g (14,56), mmol/number of cells (22,27), nmol/cm^2 (27), μg/mg protein (11), etc. Although these different forms of expression may be suitable within a particular group or laboratory, they may often be misinterpreted by others. The following example illustrates this point:

Based on our data (Table 1), a biosorption series based on specific metal uptake (X, mg/g) was found to be:

Zn (5.8) < Cr (6.3) < Cu (8.5) < Cd (11.5) < Ag (28.0) < Pb (35.1)

If mean specific metal values were expressed as mmol/g, the biosorption series would change to:

Zn (0.09) < Cd (0.102) < Cr (0.12) < Cu (0.13) < Pb (0.17) < Ag (0.26)

Variations in the experimental conditions and different ways of expressing results, could lead to numerous types of interpretations of the same experimental data. This point is illustrated with the help of data obtained when a culture of *Nocardia* sp. was used in our laboratory for biosorption of lead and zinc (Table 3). This table indicates that specific uptake of lead and zinc increased with pH of solution. Increase in biomass concentration from 2 g/l to 20 g/l reduced specific uptake of both the metals. However, percent metal removal was more at 20 g/l biomass concentration. Specific uptake of lead was more when 1 mM solution was used; while specific uptake of zinc reduced at 1 mM solution concentration. When 100 mg/l solutions (pH 5.5) were used, specific uptake of zinc (125 μmol/g) was more than that of lead (110 μmol/g) in terms of μmol/g; but was less (8.1 mg/g) than that of lead (22.7 mg/g) in terms of mg/g. Use of equimolar (1 mM) solutions gave lower uptake of zinc (4.3 mg/g or 73 μmol/g) than that of lead (24.7 mg/g or 119 μmol/g) in terms of both mg/g and μmol/g. It is evident from the above example that a standardized protocol for biosorption

Table 3
Effect of different experimental parameters and methods of expressing results on metal uptake

pH	Biomass conc. (g/l)	Initial metal conc.	Specific metal uptake, (Q)				Metal removal	
			(mg/g)		(μmol/g)		(%)	
			Pb	Zn	Pb	Zn	Pb	Zn
2.0	2.0	100 mg/l	2.2	0	10	0	4.6	0
		1mM	2.7	0	13	0	1.3	0
	20.0	100 mg/l	0.5	0	3	0	11.5	0
		1mM	0.9	0	4	0	9.3	0
4.0	2.0	100 mg/l	15.4	2.1	74	32	44.5	4.5
		1mM	17.5	1.9	85	29	20.3	6.5
	20.0	100 mg/l	2.6	1.8	13	27	76.1	38.1
		1mM	7.8	1.2	37	18	91.1	40.7
5.5	2.0	100 mg/l	22.7	8.1	110	125	65.1	17.2
		1mM	24.7	4.3	119	73	28.6	16.4
	20.0	100 mg/l	3.4	3.0	16	47	96.6	64.1
		1mM	8.4	2.0	41	31	97.5	70.3

studies needs to be evolved.

In order to design a protocol which could be used for screening of a variety of organisms across different laboratories, data reported in more than 150 research papers was analyzed statistically and the results obtained are summarized in Table 4. It could be seen that pH values which have been used for metal sorption studies fall in the range 4.9 to 5.5. However, in this range, most of the organisms show little or no sorption of metals such as chromium. Considering the microprecipitation of some metals at pH exceeding 5.5 (2) and adsorption of both cationic (lead, copper, cadmium, zinc and silver) and anionic (chromium) species, pH 4.0 would be appropriate for screening. In order to prevent changes in the solution pH after biomass addition, a suitably buffered system may be used or the biomass may be conditioned to pH 4.0 prior to use.

In general, metal concentrations in the range of 185 to 309 mg/l have been used by most researchers (Table 4). However, the biosorption technology is targeted at removal of metals from dilute (<200 mg/l) solutions. Considering this fact, solutions containing 1 mM concentration of the metals (ca. 50-200 mg/l) would be suitable. Use of molar solutions would also be helpful in comparing biosorption efficiency of an organism for different metals.

Biomass concentrations in the range of 2.2 to 4.1 g/l have been used by most researchers (Table 4). Our calculations reveal that to achieve an optimum metal to biomass ratio, 2 g/l biomass may be used. Metal to biomass ratio in such a system (e.g. comprising of 0.05 g biomass in 25 ml of metal solution) would be 20 mmol metal per gram biomass. At this ratio highest specific metal uptake values obtained would be 104, 56, 54, 33, 32 and 26 mg/g for Pb, Cd, Ag, Zn, Cu and Cr, respectively. If metal to biosorbent ratio is changed, the highest possible specific metal uptake values would be either too high or very low posing

Table 4
Statistical analysis of the literature data on metal biosorption

Metal	Statistical Mean (X)				
	pH	Temp., °C	Biomass, g/l	Co, mg/l	Q, mg/g
Lead	5.0	26	2.3	280	97
	(4.5-5.5)	(25-27)	(1.2-3.4)	(150-409)	(49-144)
Chromium	2.2	25	0.5	360	13
	(1.4-3.0)	(25-26)	(0.3-0.7)	(282-438)	(1-26)
Copper	5.3	26.0	2.4	123	36
	(4.8-5.8)	(25-27)	(1.0-3.8)	(68-178)	(12-60)
Cadmium	6.0	26	2.5	157	36
	(5.5-6.5)	(24-28)	(0.5-4.5)	(79-235)	(8-64)
Zinc	6.1	25	1.7	213	16
	(4.8-7.4)	(23-27)	(1.1-2.3)	(51-375)	(15-17)
Silver	5.1	26	4.9	538	87
	(4.4-5.8)	(25-27)	(0.5-9.3)	(273-804)	(61-112)
All metals	5.2	26.0	3.1	247	52
	(4.9-5.5)	(25-27)	(2.2-4.1)	(185-309)	(37-68)

The values in parentheses are 95% confidence intervals. The analysis is based on data collected from over 150 research papers.

difficulties in proper screening of cultures. In order to allow maximum contact of the biosorbent with metal ions, the system may be incubated on a shake-flask assembly for 30 minutes at 100-120 rpm. Biomass may be separated from the metal solutions by filtration or centrifugation.

Thus, we propose that the following experimental protocol may be followed by researchers for screening of metal sorbing microorganisms / biomass.

4.7. Proposed experimental protocol

Dried and finely powdered biomass may be conditioned to pH 4.0 by repeatedly washing with sodium acetate buffer (0.1 M, pH 4.0). Dried and pH conditioned biomass (0.05 g) may then be contacted with 25 ml metal solution (1 mM, pH 4.0) in a 150 ml Erlenmeyer flask. Flasks may be kept on a shaker (25°C, 120 rpm) for 30 minutes. After contacting, the biomass may be separated by filtration or centrifugation. Filtrates or supernatants may be used for the analysis of residual metal by atomic absorption spectrophotometry and the results may be expressed in terms of mmol metal bound per gram of biomass according to following equation:

$$Q = \frac{0.5 \, (C_0 - C_f)}{AW}$$

where, Q is specific metal uptake (mmol/g), C_0 and C_f are initial and final metal concentrations (in mg/l, as determined by atomic absorption spectrophotometry), and AW is the atomic weight of metal analyzed.

We hope that this protocol would be useful in attaining uniformity in the experimental methods. If adopted, comparative evaluation of the metal biosorption data from different laboratories could be easily carried out. Also, the screening of metal biosorbing microorganisms could be automated in future. Such advancements would greatly facilitate further development of metal biosorption technology.

REFERENCES

1. A. Kapoor and T. Viraraghavan, Biores. Technol., 53 (1995) 195.
2. B. Volesky and Z.R. Holan, Biotechnol. Prog., 11 (1995) 235.
3. C.L. Brierley, Geomicrobiol. J., 8 (1990) 201.
4. S.E. Shumate and G.W. Strandberg, In: M.M. Young, C.N. Robinson, J.A. Howell (eds.), Comprehensive Biotechnology, Vol 4, Pergamon Press, New York, 1986, 235.
5. F. Veglio and F. Beolchini, Hydrometallurgy, 44 (1997) 301.
6. M. Tsezos, E. Remoudaki, V. Angelatou, Int. Biodeter. Biodegrad., (1996) 19.
7. F. Veglio, F. Beolchini and A. Gasbarro, Process Biochem., 32 (1997) 99.
8. M.Z.C. Hu, J.M. Norman, B.D. Faison and M.E. Reeves, Biotechnol. Bioeng., 51 (1996) 237.
9. Y. Sag, D. Ozer and T. Kutsal, Process Biochem., 30 (1995) 169.
10. R. Gourdon, S. Bhende, E. Rus and S.S. Sofer, Biotechnol. Lett., 12 (1990) 839.
11. G. Morozzi, G. Cenci, F. Scardazza and M. Pitzurra, Microbios, 48 (1986) 27.
12. G.M. Gadd, In: H.H. Eccles and S. Hunt (eds.), Immobilization of Ions by Bacteria, Ellis Horwood, Chichester, UK, 1986, 135.
13. A.J. Drapeau, R.A. Laurence, P.S. Harbec, G. Saint-Germain and N.G. Lambert, Sciences et Techniques de l'eau, 16 (1983) 359.
14. J.P.S. Cabral, Microbios, 71 (1992) 47.
15. Z. Tynecka, Z. Gos and J. Zajac, J. Bacteriol., 147 (1981) 305.
16. C. Venkobachar, Water Sci. Technol., 22 (1990) 319.
17. E. Fourest and J.C. Roux, Appl. Microbiol. Biotechnol., 37 (1992) 399.
18. J.S. Chang, R. Law and C.C. Chang, Water Res., 31 (1997) 1651.
19. H. Niu, X.S. Xu, J.H. Wang and B. Volesky, Biotechnol. Bioeng., 42 (1993) 785.
20. M. Nourbakhsh, Y. Sag, D. Ozer, Z. Aksu, T. Kutsal, A. Caglar, Process Biochem., 29 (1994) 1.
21. Z. Aksu, Y. Sag, T. Kutsal, Environ. Technol., 13 (1992) 579.
22. G.W. Garnham, G.A. Codd and G.M. Gadd, Biology of Metals, 4 (1992) 151.
23. M. N. Akthar, K.S. Sastry and P. Maruthi Mohan, Biotechnol. Lett., 17 (1995) 551.
24. K.M. Paknikar, U.S. Palnitkar and P.R. Puranik, In: A.E. Torma, M.L. Apel and C.L. Brierley (eds.), Biohydrometallurgical Technologies Vol. II, The Minerals, Metals & Materials Society, Warrendale, PA, 1993, 229.
25. J.D. McEntee, S.F. Minney and A.V. Quirk, In: Progress in Biohydrometallurgy, Cagliari, 1983, 617.
26. J.E. Sloof, A. Viragh and B. Van der Veer, Water, Air and Soil Poll., 83 (1995) 105.

372

27. B. Wehrheim and M. Wetern, Appl. Microbiol. Biotechnol., 41 (1994) 725.
28. J.A. Scott and S.J. Palmer, Biotechnol. Lett., 10 (1988) 21.
29. A.C.A. Costa and S.G.F. Leise, Biotechnol. Lett., 12 (1990) 941.
30. G.J. Ramelow, Z. Yumo and L. Liu, Microbios, 66 (1991) 95.
31. A.M. Khalid, S.R. Ashfaq, T.M. Bhatti, M.A. Anwar, A.M. Shemsi and K. Akhtar, In: A.E. Torma, M.L. Apel and C.L. Brierley (eds.), Biohydrometallurgical Technologies Vol. II, The Minerals, Metals & Materials Society, Warrendale, PA, 1993, 299.
32. Z. Aksu and T. Kutsal, J. Chem. Technol. Biotechnol., 52 (1991) 109.
33. G.W. Strandberg, S.E. Shumate and J.R. Parrot, Appl. Environ. Microbiol., 41 (1981) 237.
34. M. Tsezos and B. Volesky, Biotechnol. Bioeng., 23 (1981) 583.
35. J.L. Zhou and R.J. Kiff, J. Chem. Technol. Biotechnol., 52 (1991) 317.
36. P.O. Harris and G.J. Ramelow, Environ. Sci. Technol., 24 (1990) 220.
37. K.J. Blackwell, I. Singleton, J.M. Tobin, Appl. Microbiol. Biotechnol., 43 (1995) 579.
38. M.P. Pons and M.C. Fuste, Appl. Microbiol. Biotechnol., 39 (1993) 661.
39. M. Tsezos and D.M. Keller, Biotechnol. Bioeng., 25 (1983) 201.
40. M. Balkrishnan, J.M. Modak, K.A. Natarajan, J.S.G. Naik, Mine and Metall. Proc., 11 (1994) 197.
41. K. Osseo-assare, T. Xue and V.S.T. Ciminelli, In: V. Kudryk, D.A. Corringan and W.W. Liang (eds.) Precious Metals, Kining, Extraction and Processing, TMS-AIME, 1989, 173.
42. P.R. Puranik and K.M. Paknikar, J. Biotechnol., 55 (1997) 113.
43. D. Brady and J.M. Tobin, Enzyme Microb. Technol., 16 (1994) 633.
44. P.R. Norris and D.P. Kelly, J. Gen. Microbiol., 99 (1977) 317.
45. F.T. Awadalla and B. Pesic, Hydrometallurgy, 28 (1992) 65.
46. N. Kuyucak and B. Volesky, Biotechnol. Bioeng., 33 (1989) 809.
47. J.W. Oliver, W.C. Kreye, P.H. King, J. Water Pollut. Control Fed., 47 (1975) 2490.
48. M.L. Failla, C.D. Benedict and E.D. Weinberg, J. Gen. Microbiol., 44 (1976) 23.
49. L. deRome and G.M. Gadd, Appl. Microbiol. Biotechnol., 26 (1987) 84.
50. M.A. Sampedro, A. Blanco, M.J. Llama and J.L. Serra, Biotechnol. Appl. Biochem., 22 (1995) 355.
51. D.K. Sahoo, R.N. Kar and R.P. Das, Bioresource Technol., 41(1992) 177.
52. J.M. Modak and K.A. Natarajan, Min. Metall. Process, (1995) 189.
53. M. Itoh, M. Yuasa and T. Kobayashi, Plant Cell Physiol., 16 (1975) 1167.
54. A.V. Pethkar and K.M. Paknikar, J. Biotechnol., 63 (1998) 121.
55. S.R. Dave and R.A. Patwari, In: A.E. Torma, M.L. Apel and C.L. Brierley (eds.), Biohydrometallurgical Technologies Vol. II, The Minerals, Metals & Materials Society, Warrendale, PA, 1993, 119.
56. B. Volesky, FEMS Microbiol. Rev., 14 (1994) 291.

The 'behaviour' of five metal biosorbing and bioprecipitating bacterial strains, inoculated in a moving-bed sand filter

B. Pernfuß[a], C. Ebner[a], T. Pümpel[a], L. Diels[b], L. Macaskie[c], M. Tsezos[d], Z. Keszthelyi[e], and F. Glombitza[f]

[a]Institut für Mikrobiologie, Universität Innsbruck, Technikerstraße 25, A-6020 Innsbruck, Austria

[b]VITO Vlaamse Instelling voor Technologisch Onderzoek, Boeretang 200, B-2400 Mol, Belgium

[c]School of Biological Sciences, University of Birmingham, Birmingham B15 2TT, United Kingdom

[d]Department of Mining and Metallurgical Engineering, National Technical University of Athens, Heroon Polytechniou 9, GR-15780 Zografou, Greece

[e]A. SKOLNIK GMBH & Co KG, Oberflächentechnik, Lohnergasse 4, A-1210 Wien, Austria

[f]C & E Consulting und Engineering GmbH, Jagdschänkenstraße 52, 09117 Chemnitz, Germany

In the course of a RTD-project[*] we tried to remove nickel from a rinsing water of a plating company in Vienna. To this purpose a moving-bed 'AstraSand' filter was inoculated with a mixture of five well investigated (Brite-Euram 5350) bacterial strains with a high potential to biosorb and bioprecipitate heavy metals. Three of the five bacterial strains (*Pseudomonas mendocina* AS 302, *Arthrobacter sp.* BP 7/26, *Alcaligenes eutrophus* CH 34, *Pseudomonas fluorescens* K 1/8a and *Methylobacillus sp.* MB 127) could be adapted separately to the waste water before inoculation was carried out using a mixed culture. On the basis of morphological and physiological characteristics of the bacteria - four of the strains stain gram-negative, one gram-positive; the strains use different carbon sources and some of them bear heavy metal resistances - selective agar media were collated to re-isolate the strains out of the mixture. With this simple method it could be shown that one of the strains, which could not be preadapted to the rinsing water by separate cultivation, was able to survive in the waste water when it was precultured together with the other strains.

After the inoculation of the non-sterile moving-bed reactor and during it`s operation the proliferation of autochthonous microorganisms could be observed in the waste water in a high density. Periodically the composition of the biofilm, grown on the sand particles, and of the planktonic microorganisms was investigated. The influence of continuous sand regeneration on the growth of the biofilm was documented regarding different operation modes. Eutrophic-, oligotrophic-, Cd-resistant-, Ni-resistant-, As-resistant- and methylo-

---

[*] Brite-Euram BRPR-CT96-0172

trophic bacteria as well as Pseudomonades could be differentiated.

To improve the reliability of the results pure cultures from the stocks of the inoculated strains and colonies from the selective plates, which originated from the biofilm samples, where identified and compared using the 'Biolog'-system for taxonomic identification of bacteria. Often appearing autochthonous strains were also identified.

1. INTRODUCTION

'Biofilms today are extremely active regions in many ecosystems. In some environments, biofilms may be the main source of certain compounds such as methylated metals or newly fixed nitrogen. In other environments, biofilms may be the main locus for processes (such as nitrification, trace metal removal , or photochemical degradation of organic compounds, to name a few) that remove chemical compounds from an ecosystem' [1]. The special properties of biofilms, an already developed bacterial system for biosorption and bioprecipitation [2] and a patented moving-bed filter [3, 4] were combined for the treatment of a nickel containing waste water originating from the chemical nickel line of a plating company in Vienna.

2. MATERIALS AND METHODS

2.1. Bacterial strains

A collection of bacterial strains, which were proved to be able to biosorb and/or bio-accumulate different metal cations, was available. Some of the strains were shown to exhibit plasmid encoded resistances and thus were expected to be able to grow with metal containing waste water.

Pseudomonas mendocina AS 302 [5]: was placed at our disposal by 'VITO' (Vlaamse Instelling voor Technologisch Onderzoek) and shows a non selective and very high metal biosorbing capacity. AS 302 is able to grow with 1 mM arsenic.

Arthrobacter sp. BP 7/26 [6, 7, 8]: was isolated from the sediment of the river 'Inn' and is able to biosorb up to 280 mg of silver per gram dry weight. Investigations with chemical masking of reactive groups in the murein sacculus of the bacterial strain suggested that electropositive amine groups provide the major binding sites for silver, whereas palladium is probably bound exclusively to the carboxyl groups. The biosorption of copper and nickel decreased significantly by the chemical masking of both carboxyl- and amine groups.

Alcaligenes eutrophus CH 34 [9, 10]: was placed at our disposal by 'VITO' and was shown to bear different heavy metal resistances (czc operon). The strain is able to biosorb, bioaccumulate and bioprecipitate heavy metals and can grow with 0.8 mM cadmium and 2 mM nickel.

Pseudomonas fluorescens K 1/8a [2]: was shown to be able to bioreduce heavy metals in a former basic research project funded by the European Commision (BE-5350) and was placed at our disposal by 'University of Birmingham'.

Methylobacillus sp. MB 127 [11]: was isolated from a metal bearing waste water and placed at our disposal by 'C & E' (Consulting und Engineering GmbH) in Germany.

2.2. Inoculation of the moving-bed filter

Three (CH 34, K 1/8a and MB 127) of the five bacterial strains were adapted to the nickel containing waste water (Skol 6) of a plating company (A. SKOLNIK GMBH & Co KG, Skolnik Oberflächentechnik, Wien, Austria), which was supplemented with 2.8 g Na-acetate l^{-1} (pure cultures on agar medium made of waste water Skol 6). The waste water Skol 6 (chemical nickel line) contains nickel and zink (about 10 mg l^{-1}) as well as traces of organic acids (like acetate and lactate) and some (not defined) surfactants.

The stocks of BP 7/26 and K 1/8a were made on nutrient agar (in case of BP 7/26 the medium contained additionally 200 µM of $AgNO_3$).

For the preparation of inoculum baffled Erlenmeyer flasks (500 ml) each filled with 150 ml supplemented waste water were inoculated with surface cultures of the bacterial strains (72 h, 180 rpm, 30°C). After reaching an optical density (OD_{660}) of approximately 0.35 the solution was used to inoculate 10 l of fresh supplemented Skol 6 waste water. Cultivation was done in a stirred, aerated glass reactor. After the transport of the preculture over approximately 500 km, the moving-bed reactor [3, 4] for the cleaning of waste water was inoculated by adding the preculture into 50 l of Skol 6 waste water (non sterile conditions) and aerating and mixing the solution by bubbling compressed air through a perforated tube. After four days start up phase the whole sand filter system (approximately 2 m^3 sand grains, about 680 l between the sand grains, 750 l above the sand) was inoculated by starting continuous filter operation (feed flow 2.4 m^3 h^{-1}; recirculation of filtrate 1.7 m^3 h^{-1}). More detailed information about the pilot plant at 'Skolnik' are presented in a thematically closely allied manuscript, which is also printed in the course of these proceedings [13].

The 'AstraSand' filter is based on the counterflow principle. The waste water flows upwards through the sand bed, prior to discharge through the filtrate outlet. The sand containing the biofilm and the entrapped metals is conveyed from the tapered bottom section by means of the airlift pump to the sand washer. The impurities (abrased biofilm and metals) are discharged through the wash water outlet, while the grains of clean sand are returned to the sand bed.

2.3. Handling of samples and recipes of selective media

Periodically sand samples (ca. 20 cm^3) and samples of waste water were taken and sent to our lab using an insulated cooling device. Samples of sand (5 g fresh weight) and solutions (5 ml of filtrate, washwater or galvanik bath, respectively) were mixed with tetra-sodium diphosphate decahydrate (45 ml SDP - 2.8 g $Na_4P_2O_7$ x $10H_2O$ l^{-1}) and shaken (180 rpm) for approximately 60 min at 30°C. Thereupon the samples were shaken vigorously (Circomix), the sand grains were allowed to settle, and the supernatant was diluted using 0.85% NaCl. Each 100 µl were spread on agar plates of different selective media.

Medium No. 3:   *Basis* - 0.5 g K_2HPO_4 , 1.0 g NH_4Cl, 0.2 g $MgSO_4$ x 7 H_2O, 10 mg $FeSO_4$ x 7 H_2O, 10 mg $CaCl_2$ x 2 H_2O, 16 g agar, 500 ml a.d., pH 7.0 (1 M HNO_3);

                    Trace elements (2 ml) - 10 mg $ZnCl_2$, 100 mg $MnCl_2$ x 4 H_2O, 200 mg $CoCl_2$ x 6 H_2O, 100 mg $NiCl_2$ x 6 H_2O, 20 mg $CuCl_2$ x 2 H_2O, 25% HCl, 1000 ml a.d.;

                    Carbon source - 10 g sodium gluconate, 500 ml a.d.;

                    Addition - 246 mg $CdNO_3$ (final concentration 0.8 mM);

Medium No. 6:   8 g nutrient broth, 16 g agar, 1000 ml a.d.;

Medium No. 8:   *Basis* - 45 g GSP-agar (Merck 10230), 1000 ml a.d.;

Medium No. 22:   *Basis* and *Trace elements* - see medium No. 3;
                  Carbon source: 10 ml methanol, 500 ml a.d.;

Medium No. 23:   *Basis* - see medium No. 8;
                  Addition - 198 mg AsO$_3$ (final concentration 1 mM)

Medium No. 26:   *Basis*, *Trace elements* and *Carbon source* see medium No. 3;
                  Addition - 475 mg NiCl$_2$ x 6 H$_2$O (final concentration 2 mM);

Medium No. 32:   *Basis* - see medium No. 3;
                  Carbon source - 10 g sodium glutamate, 500 ml a.d.;

Medium No. 33:   1000 ml of nickel containing waste water from Skolnik, 16 g agar, 2.8 g
                  sodium acetate;

For each sample three different dilutions and two 'parallels' were spread on the agar plates; incubation was carried out at 30°C and approximately 60% relative humidity. After 48 h (medium No. 6) and 150 h (other media) the number of colony forming units (cfu) was counted. For the evaluation of results (averages of cfu g^{-1} sand fresh weight and cfu ml^{-1}, resp.) agar plates with well separated colonies (50 - 100 cfu) were favoured.

2.4. Taxonomic identification of inoculated strains and re-isolates (Biolog-MicroStation^{TM} System)

From the biofilm out of the moving-bed filter of the plating company, several colonies were re-isolated (on nutrient agar) eight months after the inoculation. The re-isolates were named after the medium they were isolated from (e.g. strain 8 a was isolated from medium No. 8 = selective medium for *Pseudomonas spp.*). Colonies which were counted most frequently on the corresponding selective medium were signed with 'a'; seldom observed colonies with 'c' or 'd'.

After the gram staining, strains were precultured on the media recommended from 'Biolog' for taxonomic identification.

TSA agar: (CASO-agar; Merck 1.05458) with 5% sheep blood for the gram negative
              isolates;

BUGM agar: (Biolog, Inc. Hayward, U.S.A.) with 5% sheep blood for the gram positive
              isolates;

The incubation of precultures was done for 18 - 24 h at 30°C. Inocula for MicroPlates^{TM} (GP-MicroPlates for gram positive, GN-MicroPlates for gram negative strains) were prepared by suspending the biomass in 0.85% NaCl using cotton swabs (OD$_{660}$ ~ 0.15). MicroPlates were incubated at 30°C for 4 h, 24 h, 48 h and 90 h. Bacteria were identified using Biolog's MicroLog^{TM} Software and data bases, which are used to interpret the pattern of purple wells (each of 96 wells is preloaded with a different carbon source along with Biolog's patented redox chemistry in a dry form) in the MicroPlates.

3. RESULTS AND DISCUSSION

3.1. Countings of colony forming units on selective media relative to different filter operation modes

In the **preculture for the inoculation of the moving-bed filter** 1×10^9 cfu ml^{-1} were counted on control plates. The preculture was prepared using the waste water to be treated and a mixed culture of the five described strains (see 2.1.), three of them could be adapted before to the nickel containing waste water. This cell density corresponds with a well developed preculture, the cells were likely to be at the end of the logarithmic phase and thus were well suited for inoculation. The preculture was composed of *Pseudomonas spp.*, methylotrophic bacteria, Ni-resistant- and As-resistant bacteria (Table 2).

Before the inoculation of the filter system one sand sample (Table 1) was tested in order to get an idea of the proportion of autochthonous bacteria present in the moving-bed. Approximately 3×10^6 cfu g^{-1} sand were counted on the control plates (both eutrophic- and oligotrophic bacteria). No Cd-resistant and As-resistant bacteria were found, whereas a relatively high number of Pseudomonades, Ni-resistant- and methylotrophic bacteria could be counted. These results are easily explained in so far as *Pseudomonas spp.* have been found ubiquitously in almost any habitat ever studied concerning bacterial ecology, and as the nickel content of the waste water at 'Skolnik' allowed to expect the presence of autochthonous nickel resistant bacteria. Before inoculation up to 2 mg Ni l^{-1} waste water could be removed within the moving-bed filter [3, 4].

Concerning the filtrate (Table 2) of the moving-bed filter (operation mode: see 2.2.), **24 h after inoculation** approx. 3×10^8 cfu ml^{-1} could be counted on control plates. In contrast to the preculture no planktonic *Pseudomonas spp.* and no As-resistant bacteria were found. Less cfu of methylotrophic and Ni-resistant organisms were counted than in the preculture.

Table 1.
Summary of colony forming units (cfu) per gram sand on control media and selective media in the biofilm samples of the 'AstraSand' filter at Skolnik in Vienna.

	Biofilm bacteria					
	autoch-thonous	1 d after inoculat.	9 d after inoculat.	filter stop for 4 weeks	continuous operation	> oxygen supply
eutrophic	2.8×10^6	3.0×10^9	4.0×10^7	2.0×10^6	0	3.0×10^7
oligotrophic	3.5×10^6	n.d.	2.0×10^7	6.3×10^6	3.3×10^7	0
Cd-resistant	0	n.d.	n.d.	1.5×10^5	2.0×10^6	2.0×10^5
Pseudomonas spp.	1.8×10^6	5.5×10^6	1.2×10^7	1.7×10^6	2.0×10^7	6.2×10^6
methylo-trophic	7.5×10^5	1.5×10^6	8.0×10^6	6.1×10^6	2.0×10^6	8.1×10^6
As-resistant Pseudomonas spp.	0	5.0×10^5	0	0	7.5×10^6	5.0×10^4
Ni-resistant	9.0×10^5	3.0×10^6	6.0×10^6	3.1×10^6	7.0×10^6	6.3×10^6

Table 2.
Summary of colony forming units (cfu) ml⁻¹ on control media and selective media in the
filtrate (1 and 9 d after inoculation) and washwater samples (abrased biofilm and filtrate) of
the 'AstraSand' filter at Skolnik in Vienna. As an additional control the cfu ml⁻¹ Ni-bath (feed
flow) are indicated.

	Planktonic bacteria					
	Ni-bath	Preculture	1 d after inoculation	9 d after inoculation	filter stop for 4 weeks	continuous operation
eutrophic	3.8×10^5	1.1×10^9	2.6×10^8	0	1.0×10^6	2.0×10^7
oligotrophic	4.4×10^5	n.d.	n.d.	7.0×10^5	3.5×10^6	3.5×10^7
Cd-resistant	6.0×10^4	n.d.	n.d.	n.d.	5.0×10^4	0
Pseudomonas spp.	$\sim 2.0 \times 10^5$	2.3×10^7	0	6.0×10^7	1.5×10^5	1.0×10^6
methylotrophic	7.6×10^4	6.3×10^6	1.0×10^6	2.0×10^5	2.0×10^5	0
As-resistant *Pseudomonas spp*	0	5.0×10^5	0	0	0	0
Ni-resistant	2.6×10^5	2.0×10^6	5.0×10^5	0	2.8×10^5	0

Regarding the biofilm sample (Table 1) approximately 10 times more cfu were found than
in the filtrate. Another difference was the appearance of cfu on medium No. 8 and No. 23.
This can be attributed to the effect that *Pseudomonas spp.* took preferably part in the
formation of the biofilm.

Nine days after inoculation of the moving-bed filter and two days after the start of
continuous operation and dosage of sodium acetate less eutrophic cfu (approx. one
hundreth part) were found in the biofilm.

The part of *Pseudomonas spp.*, methylotrophic- and Ni-resistant bacteria increased slightly
whereas the As-resistant bacteria could not be found again. Under these conditions up to
3.7 mg Ni l⁻¹ waste water could be removed.

The 'AstraSand' filter is based on the counterflow principle. The waste water flows
upwards through the sand bed. The sand containing the biofilm and the entrapped metals is
conveyed from the bottom of the filter unit by means of an airlift pump. The impurities
(abrased biofilm and metals) are discharged, while the grains of 'clean' sand are returned to
the surface of the sand bed.

In the course of the operation of the moving-bed filter several biofilm samples (from the
surface of the filter bed) and washwater samples (filtrate and abrased biofilm) were compared
using selective media. By these tests it could be shown that special bacterial groups are
carried out of the system to an increased extent by the regeneration of sand.

Several days **after the start of the continuous filter operation** (Figure 1) preferably
oligotrophic and methylotrophic bacteria as well as Pseudomonades were washed from the

sand particles by sand regeneration. Eutrophic and Ni-resistant bacteria were not found in the wash water samples and seem to take part to a greater extent in the formation of biofilm.

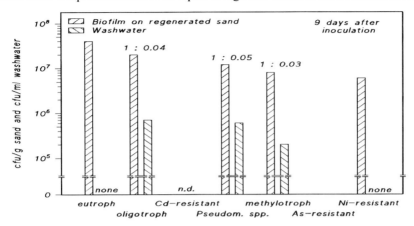

Figure 1. Colony forming units counted on selective agar plates two days after the inoculation of the moving-bed filter. The number of cfu in the biofilm of regenerated sand and in the washwater (filtrate and abrased biofilm) is compared.

After a period under **anaerobic conditions** (filter was out of operation for several weeks) the methylotrophic group of bacteria was affected least by sand regeneration, whereas all other bacteria were carried out to a greater extent: half of the eutrophic-, oligotrophic- and Cd-resistant bacteria, the tenth part of Pseudomonades and Ni-resistant bacteria were washed from the sand particles (Figure 2). With these conditions the highest removal of Ni (up to 5 mg l^{-1}) was documented within the period of observation.

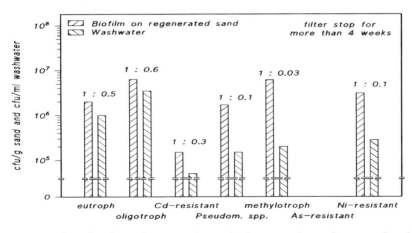

Figure 2. Colony forming units counted on selective agar plates after restarting the moving-bed filter when it lied idled for 28 days. The number of cfu in the biofilm of regenerated sand and in the washwater (filtrate and abrased biofilm) is compared.

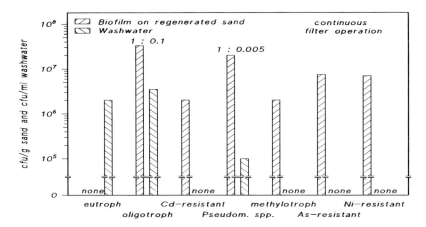

Figure 3. Colony forming units counted on selective agar plates after three weeks of constant filter operation. The number of cfu in the biofilm of regenerated sand and in the washwater (filtrate and abrased biofilm) is compared.

Only a small part of bacteria were washed out under **constant sand filter operation** (Figure 3). No cfu of Cd-resistant-, methylotrophic- As-resistant- and Ni-resistant bacteria were found in the wash water, sampled after four weeks of constant filter operation where the removal of Ni was measured to reach approx. 2.5 mg l^{-1} (Table 2).

3.1. Taxonomic identification of bacteria grown in the biofilm of the moving-bed filter

All bacterial strains for inoculation as well as 20 re-isolated strains were identified using the 'Biolog'-system for taxonomic identification of microorganisms.

One of the most frequently encountered identifications of the 'Biolog' software was '*CDC group IVC-2*'. Also the stocks (medium No. 3) and the adapted pure cultures (medium No. 33) from *Alcaligenes eutrophus* CH 34 were related by the 'Biolog'software to this group of bacterial strains. Colonies of *CDC group IVC-2* were re-isolated from minimal agar (MMS) with gluconate (No. 1), MMS-Cd (No. 3), selective agar for *Pseudomonas spp.* (No. 8), MMS-methanol (No. 22), MMS-Ni and from MMS-glutamate (No. 32). Thus, *Alcaligenes eutrophus* CH 34 seems to build the bulk of the biofilm in the moving-bed filter. The inoculated bacterial strain could assert against the autochthonous flora of the nickel containing waste water.

The 'clear' identification of this bacterial strain out of the biofilm of the moving-bed filter, eight month after the inoculation, is remarkable in so far as it is known that biofilm communities of microorganisms, even those comprising a single species, are genetically heterogeneous [12]. In addition to their responses to various physical and physiological signals, biofilm organisms are likely to undergo genetic changes, where plasmids seem to play a major role.

None of the other inoculated strains (AS 302, BP 7/26, K 1/8a and MB 127) could be identified definitely out of the biofilm, but some of the re-isolated strains (from medium No. 23, which contains arsenic) could be grouped into the genus *Pseudomonas* (like

Pseudomonas mendocina AS 302 and *Pseudomonas fluorescens* K 1/8a), and the family Nocardiaceae (like *Arthrobacter sp.* BP 7/26).

Several re-isolates were related to *Acinetobacter calcoaceticus* BV ALC. Colonies of this species were found on agar plates containing methanol (No. 22) on nickel containing agar plates (No. 26) and, very frequently, on the medium made of the nickel containing waste water (No. 33). Thus, it can be concluded that *Acinetobacter calcoaceticus* BV ALC as an autochthonous strain takes part in the biofilm of the moving-bed filter for the waste water treatment of the plating company 'Skolnik'.

4. SUMMARY

For the waste water treatment (chemical nickel line) of a plating company in Vienna, a moving-bed filter (AstraSand filter) has been inoculated using five bacterial strains with a known biosorption and bioprecipitation capacity. By the use of selective media, which were collated to re-isolate bacteria out of a mixture, it was shown that the biofilm of the filter system contained approximately 10^6 of autochthonous colony forming units before the inoculation. Above all *Pseudomonas spp.* and Ni-resistant bacteria were found to be present in the system (autochthonous organisms). After the inoculation additionally Cd-resistant (*Alcaligenes eutrophus* CH 34) and As-resistant (*Pseudomonas mendocina* AS 302) bacteria could assert. In comparision to the 'non inoculated' filter operation (2 mg Ni l^{-1} removed) the removal of nickel could be increased up to 4.9 mg l^{-1} waste water after the inoculation with selected bacterial strains. By comparing biofilm samples and wash water samples (filtrate and abrased biofilm) it was possible to determine which groups of bacteria took preferably part in the formation of biofilm, or else, which species were discharged with the wash water. By applying this simple methodology the proportions of 'artificial' groups of bacteria could be monitored regarding different modi of filtration and efficiency.

Eight month after the inoculation of the moving-bed filter, the 'Biolog'system for taxonomic identification of bacteria was used to identify some re-isolates from the biofilm with better reliability. It could be shown that *Alcaligenes eutrophus* CH 34 (*CDC group IVC-2*) could assert against the autochthonous flora, and built the bulk of the biofilm. None of the other inoculated strains could be found with certainty, but some of the re-isolates were related to the genus of *Pseudomonas* (like AS 302 and K 1/8a) and to the family of Nocardiaceae (like *Arthrobacter sp.* B 7/26). One of the most frequently found autochthonous re-isolates, was related to *Acinetobacter calcoaceticus* BV.

REFERENCES

1. B. Palenik, J.-C. Block, R.G. Burns, W.G. Characklis, B.E. Christensen, W.C. Ghiorse, A.G. Gristina, F.M.M. Morel, W.W. Nichols, O.H. Tuovinen, G.-J. Tuschewitzki and H.A. Videla, W.G. Charaklis and P.A. Wilderer (eds.) Structure and Function of Biofilms, John Wiley & Sons Ltd, Dahlem, 1989, 351.
2. L. Diels, S. Van Roy, P. Corbisier, S. Nuyts, L. Macaskie, K. Bonthrone, M. Tsezos, E. Remoudaki, V. Angelatou, T. Pümpel, B. Pernfuss, F. Schinner, A. Hummel, L.Eckard, F. Glombitza, D. Boyd and M. Esprit, Modelling of genetic, biochemical, cellular and

microenvironmental parameters determining bacterial sorption and mineralization processes for recovery of heavy- or precious metals, Synthesis Report of BE-5350, Contract N^0: BRE2/0199/C

3. General Patent DynaSand filtration (Holder: Nordic Water Products; representative: Astraco) Patent nr. NL 179/91.

4. Sand Dynameter (holder: Astraco), Patent nr. 0590705.

5. L. Diels, M. Tsezos, T. Pümpel, B. Pernfuss, F. Schinner, A. Hummel, L. Eckard and F. Glombitza, Biohydrometallurgical Processing, C.A. Jerez, T. Vargas, H. Toledo and J.V. Wierz (eds.), University of Chile (1995) 195.

6. T. Pümpel, B. Pernfuss, B. Pigher, L. Diels and F. Schinner, J. Ind. Microbiol. 14 (1995) 213.

7. B. Pernfuss, T. Pümpel and F. Schinner, Biosorption & Bioremediation, T. Macek, K. Demnerova, M. Mackova (eds.), Czech Society for Biochemistry and Molecular Biology, Prague (1995).

8. B. Pernfuss, Biosorption & Bioremediation, T. Macek, K. Demnerova, M. Mackova and J. Kostal (eds.), Czech Society for Biochemistry and Molecular Biology, Prague (1998).

9. L. Diels, Q. Dong, D. van der Lelie, W. Baeyens and M. Mergeay, J. Industr. Microbiol. 14 (1995) 142.

10. L. Diels, S. Van Roy, S. Taghavi, W. Doyen R. Leysen and M. Mergeay, Biohydrometallurgical Technologies, A.E. Torma, M.L. Apel and C.L. Brierley (eds.), Vol II (1993) 133.

11. F. Glombitza, L. Eckhardt and A. Hummel, Bioengeneering 2, (1994) 90.

12. L.J. Stal, E. Bock, E.J. Bouwer, L.J. Douglas, D.L. Gutnick, K.D. Heckmann, P. Hirsch, J.M. Kölbel-Boelke, K.C. Marshall, J.I. Prosser, C. Schütt, Y. Watanabe, Structure and Function of Biofilms, , W.G. Charaklis and P.A. Wilderer (eds.), John Wiley & Sons Ltd, Dahlem (1989) 269.

13. T. Pümpel, C. Ebner, B. Pernfuß, F. Schinner, L. Diels, Z. Keszthelyi, L. Macaskie, M. Tsezos and H. Wouters, Removal of nickel from plating rinsing water by a moving-bed sandfilter inoculated with metal sorbing and precipitating bacteria In: Proceeding of International Biohydrometallurgy Symposium IBS'99 (1999).

Removal of nickel from plating rinsing water with a moving-bed sand filter inoculated with metal sorbing and precipitating bacteria

T. Pümpel[a], C. Ebner[a], B. Pernfuß[a], F. Schinner[a], L. Diels[b], Z. Keszthelyi[c], L. Macaskie[d], M. Tsezos[e] and H. Wouters[f]

[a] Institut für Mikrobiologie, Universität Innsbruck, Technikerstrasse 25, A-6020 Innsbruck, Austria

[b] Vlaamse Instelling voor Technologisch Onderzoek (VITO), Boeretang 200, B-2400 Mol, Belgium

[c] Collini-Skolnik Oberflächentechnik AG, Lohnergasse 4, A-1210 Vienna, Austria

[d] School of Biological Sciences, University of Birmingham, Birmingham B15 2TT, UK

[e] National Technical University of Athens, Heroon Polytechniou 9, GR-15780 Zografou, Greece

[f] Astraco Water Engineering BV, Industriekade 24a, NL-2170 AA Sassenheim, The Netherlands

The MERESAFIN (MEtal REmoval by SAnd Filter INoculation) process presented here was designed to combine the optimum conditions for more than one of the well-known processes of biological metal immobilisation like biosorption and bioprecipitation. The approach makes use of a continuously operated moving-bed AstraSand filter which has been inoculated with a mixed population of metal biosorbing and bioprecipitating bacteria. A pilot plant operating at 1 m^3/h has been erected at a metal plating company in Vienna to treat waste water from an electroless nickel plating line. In addition to several mg/L of nickel the rinsing water also contains some organic acids and inorganic phosphates, which make conventional treatment difficult. The main laboratory experiments as well as preliminary results from the pilot installation are presented.

1. INTRODUCTION

European legislation on waste management in the nineties is based on the principles of 1) minimisation, 2) recycling and valorisation, 3) incineration with energy valorisation, 4) incineration, and 5) landfill in that order. Many attempts have been made to improve points one and two in the plating business too, including regeneration of baths, recycling of metals and acids from waste baths, and introduction of water-saving rinsing technologies. With all those efforts a significant reduction in waste water production has been achieved, but waste water cannot be avoided completely within the next few years. The need to develop more effective and cheaper technologies therefore remains.

Conventional physico-chemical treatment of metal-bearing waste waters may not always prove successful due to the high costs of processing effluents of high volume and low contamination or because the treated water does not meet certain legal standards e.g. it contains complexing organic matter. Biotechnological approaches can succeed in those areas and are designed to cover such niches. Certain micro-organisms are able to remove heavy metals from diluted solutions by a variety of biological (e.g. uptake), biologically mediated (e.g. bioprecipitation) and physico-chemical (e.g. biosorption) mechanisms. Through the synergistic effort of ten partners from industry and research, comprising engineers, chemists and biologists, supported by the Brite-Euram program of the CEC, the MERESAFIN system presented here has been developed.

2. MATERIALS AND METHODS

2.1. Laboratory experiments

2.1.1. Selection of micro-organisms

Bacteria with the potential to sorb and precipitate heavy metals or to degrade certain organic compounds typically found in industrial waste waters were supplied by four of the ten project partners. After identification of the waste waters of interest by the industrial partners the collection of micro-organisms was screened for the following features:

- risk classification (database of the Robert Koch Institut, Germany; Internet: *www.rki.de*).
- growth on economically acceptable nutrients (2.1.2.)
- tolerance to the metals present and to the waste waters (2.1.3.)
- biosorption/ bioprecipitation of the metals of interest (2.1.4./2.1.5.)
- degradation of organic substances present (2.1.5.)

The finally selected strains were adapted to the waste waters by cultivation on waste water agars.

2.1.2. Screening for nutrients

The application of the 'BIOLOG' system for identification of bacteria (BIOLOG Inc., USA) yielded an overview of the oxidation of 95 defined carbohydrates by the individual bacteria. (The system makes use of the reduction of tetrazolium violet to a red form by the electron transfer chain. The reactions occur in the 96 inoculated wells of a microtiter plate, each well containing a single carbon compound and one blank respectively. The intensity of colouring is a measure of the utilisation of the specific substrate and is read with a microtiter-plate photometer.) The procedure was carried out according to the manufacturer's instructions.

Further, the growth of all the preselected strains was assessed on agars prepared from molasses, green syrup, sweet whey, waste waters from paper mill, brewery, dairy, fruit juice and cigarette production, as well as supplementary mixtures thereof.

2.1.3. Determination of tolerance for waste water

Waste water was supplemented with the components of a minimal medium, enabling growth of all selected strains. The pH was adjusted to 7, and the solution sterile filtered and diluted 1:2 with sterile minimal medium in 4 subsequent steps. These solutions and a blank without any waste water were transferred to glass tubes and inoculated with suspensions of the individual strains. After 24 h of incubation growth was determined by measuring the optical density at 660 nm.

2.1.4. Biosorption capacity of biomass in real waste water

Individual bacteria were cultivated in 500 mL conical flasks on a gyratory shaker. After sufficient growth the biomass was harvested by centrifugation, washed with 0.6% $NaNO_3$ solution, and resuspended in fresh $NaNO_3$ solution. The dry weight content of the bacterial suspension was determined gravimetrically.

Biosorption tests were carried out in acid-leached centrifugation tubes (Polyallomer) using five different concentrations of biomass with the real waste water (Table 1). The suspensions were thoroughly mixed on a Circo-Mix and allowed to equilibrate. After 30 min the metal laden bacteria were sedimented by centrifugation (10 min, 10,000 rpm), the supernatants decanted and acidified prior to residual metal analysis (2.3.).

The five resulting data pairs (dry weight/residual metal concentration) are subjected to mathematical fitting (first order exponential decay). The calculated slope (first derivative) of the regression line at zero dry weight gives a good estimation of the equilibrium biosorption capacity at the original metal concentration (Figure 1; [1]).

2.1.5. Integrated evaluation of waste water treatment

Experiments with the real waste water samples were carried out under non-sterile conditions. In 100 mL conical flasks water samples were supplemented with the different nutrients under investigation. After adjustment of the pH value either individual bacterial strains or mixtures of them were inoculated from adapted cultures on waste water agar. In each test the non-inoculated control was used to assess the potential of autochthonous micro-organisms from the water samples. The flasks were incubated on a gyratory shaker at room temperature. After 24 to 48 h the following analyses were performed: pH, optical density (660 nm); after centrifugation: concentrations of metals and organic acids (2.3.).

2.2. The pilot plant

2.2.1. Concept and physical dimensions of the moving-bed sandfilter

In the moving-bed Astrasand filter (Astraco Water Engineering BV) the water flows upwards through a sand bed in a cylindrical vessel. Inoculated bacteria grow on the sand grains forming a biofilm, which traps trace elements from the feed water. From the bottom of the filter the laden sand is moved upwards by an internal airlift and the sand grains are cleaned by attrition. On the top of the filter, particles of metal containing biofilm are extracted from the system with some wash water. The sand grains retain a residual biofilm and fall to the top of the bed, where bacteria start to grow again and to immobilise the metals [2, 3].

Technical data of the pilot filter (DST-06-D):

sand filling:	1.7 m^3	total height:	4.5 m
filter bed area:	0.6 m^2	feed flow rate:	1-8 m^3/h
effective bed height:	2 m		

386

Table 1
Experimental pattern for biosorption tests

Tube-#	A [ml]	B [ml]	C [ml]	A+B+C
1	0	2	6	8
2	0.5	1.5	6	8
3	1	1	6	8
4	1.5	0.5	6	8
5	2	0	6	8

A: bacterial suspension
B: 0.6% $NaNO_3$ solution
C: waste water or synthetic metal solution

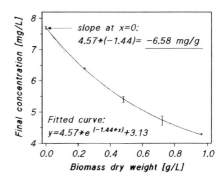

Figure 1. Example of the calculation of biosorption capacity.

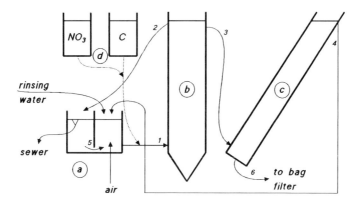

Figure 2. Simplified flow diagram of the pilot plant. Feed **1**, filtrate **2**, wash water **3**, separator effluent **4**, recirculated filtrate **5**, sludge path **6**.

2.2.2. Experimental set-up

The sandfilter forms the core of a water treatment system which consists of a buffer tank (**a**) to smooth concentration spikes, the filter itself (**b**), a lamella separator (**c**) to separate the laden biomass from the wash water, bag filters (not shown) to dehydrate the biomass, dosing equipment to supply nutrients (**d**), and on-line measuring devices (Figure 2).

2.2.3. Inoculation of the filter

The five selected bacteria were grown in acetate-supplemented waste water Skol6 up to the 50 l stage in airlift reactors, which was then used to inoculate the sand of the pilot plant.

2.3. Analytical methods

Sample preparation: centrifugation in Polyallomer tubes (10 min, 10,000 rpm).

Dissolved metals: Flame atomic absorption spectrometry in acidified (to 1% HNO_3) supernatants, standard conditions of manufacturer.

Dissolved organic acids (acetic, lactic, malic): HPLC; Column: Biorad's Aminex HPX87H, 40°C; eluent: 4 mN H_2SO_4, 0.6 mL/min; detection: UV, 210 nm.

Nitrate, method A: 2x5 mL of each sample and standard are acidified with 200 µL of 10% H_2SO_4 each. Add one copper-plated zink granule to one of the parallels for reduction of nitrate. Read UV-absorbance after 20 to 24 h and use the calculated differences of absorbencies between the parallels to construct a calibration curve and to determine the nitrate concentrations. Linear calibration range: approx. 20 mg/L nitrate-N [4].

Nitrate, method B: HPLC; Column: LiChrospher 100RP18 5 µ, ambient temperature; eluent: 0.1 M NaCl with 1.3 mL/L n-octylamine, pH 3.5 with o-phosphoric acid, 1 mL/min; detection: UV, 210 nm [5].

3. RESULTS AND DISCUSSION

3.1. Selection of bacterial strains and waste water

According to the selection criteria (2.1.1.) the following five bacterial strains were chosen to form the consortium for detailed laboratory investigations as well as for the inoculation of the pilot plant (the '5-Mix' consortium).

Pseudomonas mendocina AS302: risk class 1, high biosorption capacity, grows on lactic and acetic acid, metal resistant (Sb, Hg, As, Tl) [6, 7]

Arthrobacter sp. BP7/26: risk class 1, high biosorption and probably bioprecipitation, metal resistant (Ni, Pb) [7-9]

Alcaligenes eutrophus CH34: risk class 1, good bioprecipitation (metal carbonates), grows on lactic and acetic acid, cleaves some metal-lactate complexes, alkalises the medium (promotion of hydroxidic precipitation), also grows under micro-aerophilic conditions, metal resistant (Cd, Zn, Ni, Co, Cr, Tl, Pb, Tl) [7, 10, 11]

Pseudomonas fluorescens K1/8a: risk class 1, good bioprecipitation, grows on lactic acid

Methylobacillus sp. MB127: risk class 1, high biosorption capacity, metal resistance (U), grows on acetic acid [7, 12, 13]

Based on the toxicity tests for the tested waste waters (Figure 3 shows the average growth of selected bacteria in the rather toxic MB3 and relative non-toxic Skol6 water), and on analyses of their constituents, the rinsing water of an electroless nickel plating plant was selected for process development (waste water 'Skol6'). Besides a few (2-10) mg/L of Ni the water mainly contains inorganic chloride, sulphate, phosphate and ammonium, organic acids (lactic, malic, acetic), and some unknown surface active compounds in varying concentrations. The pH-value fluctuates around the neutral point.

3.2. Biosorption of nickel by selected strains

As already known from previous work the sorption capacity of bacterial biomass for nickel cations generally is rather low, compared with other transition elements. The reported values range around a few milligrams of Ni per gram dry weight, at an equilibrium Ni-concentration of 50 to 100 mg/L [7, 14-16]. In addition to the expected low biosorption capacity, the presence of complexing organic acids, phosphates, and other unknown organic compounds in the waste water was thought to further decrease the biosorptive potential. With regard to the nickel and organic acids only, the PHREEQE-code for modelling solution equilibria indicated the nickel to be fully complexed with lactate at neutral pH ($Ni(lac)_2$).

Experiments with a waste water sample containing 10 mg/L of Ni ('Skol6/1') revealed that the measured biosorption capacities deviated only marginally from the isotherm predictions determined in pure Ni-sulphate solutions (Figure 4). There was practically no negative effect caused by the waste water matrix. The best biosorbents were MB127, AS302, and BP7/26 with 3.5, 2.9, and 2.7 mg Ni/g dry wt. respectively.

A waste water from a nickel sulphate production plant ('MB5') was also investigated. Although it does not contain organic matter, nickel biosorption was much lower than predicted (Figure 4). The chemical composition of the water matrix could not yet be analysed in detail.

The low biosorptive power of bacteria for nickel showed that a technology based on simple biosorption with dead bacterial biomass would never lead to a working system. Moreover, the COD content of the waste waters would require an additional water treatment stage.

3.3. The integrated approach to water treatment

In contrast to conventional chemo-physical and biosorptive methods active micro-organisms open up the possibility to develop a single-stage process for the cleavage of organo-metallic complexes, degradation of organic compounds, immobilisation of dissolved and fine-dispersed metallic and metalloid elements, and to some extent also the removal or degradation of inorganic ions like ammonium, nitrate, and phosphate. All these classes of substances form the matrix of a typical waste water from the plating industry, like the selected water 'Skol6'.

Concentrations of the organic acids in 'Skol6' were too low, on average, for sufficient biomass production by the selected bacteria. So an additional carbon source had to be found. Making use of the 'BIOLOG' system only a few substrates were found to be oxidised by most bacteria (in principle the system aims at producing a very distinct pattern for each organism). Also, a range of complex substrates was tested on normal agar plates (2.1.2.).

Finally, acetate was selected as the substrate of choice for further process development. Acetate supports good growth of most of the selected bacteria, it does not complex metal cations, it is economically acceptable, and it can be accurately traced in the process with the usual analytical methods (HPLC). The complex sources are much cheaper, but they are never completely degraded and therefore contribute to COD in the waste water. This fact is of special importance for re-use of the treated water, if discharge limits have to be met, or ecotaxes are to be paid. With regard to both the pilot-scale and future full-scale operations, safe and risk-free storage and handling of a substrate must also be considered.

With the chosen waste water the only supplementation required is carbohydrate as the growth and energy source. Nitrogen is available from ammonia, added for pH control. Phosphorous is present in inorganic phosphates, and trace elements come from the water (Ca, Mg) as well as from the plating process (etched base metals and all their alloying constituents).

With regard to the prerequisites mentioned above laboratory experiments were performed in order to optimise supplementation and the experimental set-up. Under aerobic conditions (agitated conical flasks) the addition of 50 to 200 mg/L of acetate-C resulted in sufficient growth of the mixed bacterial population '5-Mix' to reduce the nickel concentration from 10 to below 0.5 mg/L. Substrate overdosing is not only a financial problem; it also led to decreased nickel removal (Figures 5 and 6). Excess acetate, although not complexing the metal, may compete with anionic sites on the bacterial cell walls for Ni cations.

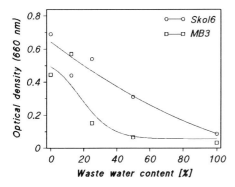

Figure 3. Toxicity testing for selected bacteria with waste waters Skol6 and MB3.

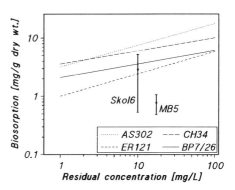

Figure 4. Biosorption capacities of selected bacteria for nickel in synthetic solution (isotherms) and real waste waters (points).

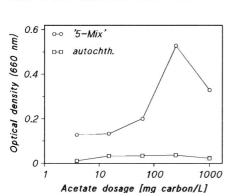

Figure 5. Growth of mixed culture '5-Mix' and autochthonous micro-organisms on waste water Skol6 with acetate.

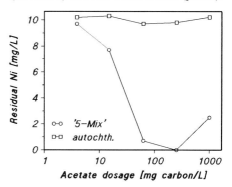

Figure 6. Removal of Ni from waste water Skol6 with acetate by '5-Mix' and autochthonous micro-organisms.

The estimated Ni-load of the biomass (approx. 50 mg Ni/g dry wt.) by far exceeded the potential of simple biosorption, and raised hopes for successful process development. These results clearly indicate the predominance of biologically mediated processes of nickel immobilisation over passive biosorption.

Further experiments with individual bacterial strains revealed great differences in their patterns of growth and Ni-removal. As the behaviour of the strains in the non-sterile pilot plant could not be foreseen it was decided to keep using the '5-Mix'. In order to observe the surveillance of each of the inoculated strains, microbiological differentiating tests based on selective media were developed. The results will be presented elsewhere [17].

3.4. The pilot plant

3.4.1. Basic considerations

After long discussions about the pros and cons of the various techniques applied in biological waste water treatment (e.g. packed and fluidised bed reactor, rotating disc reactor, trickling reactor) the concept of Astrasand's moving-bed sand filter promised to fulfil most requirements. Sand is a reliable carrier for micro-organisms, and the continuous regeneration of the sand allows for steady removal of metal-bearing biomass. It was further evident that a biofilm reactor (like a sand filter) would offer some advantages over reactors with suspended biomass: (i) Biofilms are known to withstand extreme conditions which may arise in industrial waste water (e.g. spikes of extreme pH or high metal concentration). (ii) Biofilms provide a micro-environment which could be very beneficial for metal precipitation (e.g. high pH, high concentration of CO_2). (iii) Biofilms contain a lot of microbial exopolymers which may also help to entrap dispersed solids, as well as to biosorb dissolved metals.

The only problem of the sand filter concept, which was recognised before starting the first plant, is the limited availability of oxygen. With the original design (for physical filtration) oxygen input is limited to its solubility in the feed water. With the maximum of 8 to 10 mg/L of dissolved oxygen under atmospheric pressure, only 9 to 11 mg/L of acetate-C can be oxidised (determined in laboratory experiments with the bacterial culture '5-Mix'). As at least 50 to 100 mg/L of acetate-C were shown to be necessary for the complete removal of 10 mg/L of Ni (Figure 6) an additional electron acceptor had to be identified. Nitrate was chosen due to its wide-spread microbial utilisation and its non-complexing properties.

3.4.2. Start-up

A clear strategy was followed to start-up and operate the pilot installation.

Phase 1:

In order to gain experience with the system and to analyse the parameters of the selected feed flow, the filter was run without inoculation and supplementation and without any modification of the water. The pH of the rinsing water varied around neutral and needed no correction. The concentration of dissolved oxygen fluctuated greatly and had to be stabilised by installing an air supply to the buffer tank. Due to the presence of organic acids some autochthonous micro-organisms emerged in the filter, but the removal of nickel was marginal.

This phase was completed with intensive flushing of the plant in order to remove most of the grown biomass.

Phase 2:

After successful inoculation with the selected bacteria ('5-Mix') and initialisation of the dosing of the additional carbon source, the filter system worked as expected.

3.4.3. Results of aerobic operation

So far the feed water has been aerated. The nutrient doses have been varied and the effects analysed. It could be shown that the removal of nickel correlates strictly to carbon consumption and therefore also to the growth rate of the biomass. Within the reported period the carbon dose has been increased step by step up to 15 mg carbon/L, resulting in the removal of around 1 mg Ni/L (Figures 7 and 8). This is somewhat less than measured with the flask experiments (Figure 6), but residence time was two days compared with less than one hour in the filter. Additionally the pH in the flasks rose to around 8.5, whereas it fluctuated between 7 and 8 in the filter.

As expected from the theory it was necessary to compensate for the lack of oxygen for the oxidation of more than 10 mg/L of acetate-C. Nitrate was shown to work, but analysis on the precise stoichiometry of substrate oxidation is yet to be performed.

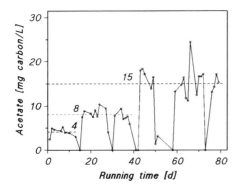

Figure 7. Dosage of carbon source to the filter feed.

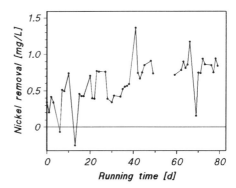

Figure 8. Differences in Ni-concentration between feed and filtrate.

HPLC-analysis of organic acids showed their complete degradation, including the part of the lactate which was calculated to be complexed by nickel.

The bio-sludge produced during the reported period contained about 20 mg/g Ni in the dry matter. With the Ni-concentration typically ranging between 2 - 5 mg/L in the rinsing water, the concentration factor therefore reaches 10,000. The analysed Ni- content in the sludge exceeds the biosorptive capacity of the biomass by a factor of ten (Figure 4), which is clear evidence of additional biological or biologically mediated processes. X-ray diffraction patterns revealed no crystalline structures in the sludge. Besides precipitation of Ni-hydroxides by the observed rise in pH, the formation of Ni-carbonates may also occur. Further chemical analysis of the precipitates is to start immediately in order to provide better understanding and optimisation of the whole process.

4. CONCLUSIONS AND OUTLOOK

If the preliminary calculations can be confirmed, the cost of installation and operation of the sand filter is already below that of competing technologies. From the results to date, the MERESAFIN system could prove successful in the integrated treatment of continuously flowing (waste) water streams containing both metals or metalloids and organic compounds.

ACKNOWLEDGEMENT

The authors are grateful for financial support from the Brite-Euram program of the CEC (Contract BRPR-CT96-0172).

REFERENCES

[1] Pümpel, T. and Schinner F. Biosorption & Bioremediation Symposium. Macek, T., Demnerova, K., Mackova, M. (Eds.), Czech Society for Biochemistry and Molecular Biology, Prague, (1995), 1.

[2] Kramer, J.P. and Wouters, J.W., J. Water SRT ,42 (1993) 97.

[3] Diels, L., Spaans, P.H., Van Roy, S., Hooyberghs, L., Wouters, H., Winters, J., Macaskie, L., Pümpel, T. (1999) Heavy metals removal by sandfilters inoculated with metal sorbing and precipitating bacteria. Submitted to IBS99.

[4] Schinner, F., Öhlinger, R., Kandeler, E. and Margesin, R. Bodenbiologische Arbeitsmethoden, 2nd edition, Springer-Verlag, 1993.

[5] Doblander, C. and Lackner, R., Biochim Biophys Acta 1289 (1995) 270.

[6] Diels L., Tsezos, M., Pümpel, T., Pernfuss, B., Schinner, F., Hummel, A., Eckard, L. and Glombitza, F. Biohydrometallurgical Processing. Jerez, C.A., Vargas, T., Toledo, H. and Wiertz, J.V. (Eds.), University of Chile, (1995) 195.

[7] Tsezos, M., Remoudaki, E. and Angelatou, V. Int Biodeterioration, (1995) 129.

[8] Pernfuß, B., Pümpel, T. and Schinner, F, Biosorption & Bioremediation Symposium, Macek, T., Demnerova, K., Mackova, M. (Eds.), Merin. Czech Society for Biochemistry and Molecular Biology, Prague (1995).

[9] Pernfuß, B., Biosorption und Bioakkumulation von Silber durch Bakterien: Weiterentwicklung einer Screeningmethode und grundlegende Untersuchungen zu den Anreicherungsprozessen und Bindungsstellen von *Arthrobacter* BP 7/26; Thesis, Leopold-Franzens Universität Innsbruck, (1997).

[10] Diels, L., Sadouk, A. and Mergeay, M., Toxicol Environ Chem 23 (1989) 19.

[11] Mergeay, M., Nies, D., Schlegel, H.G., Gerits, J. and Van Gijsegem, F., J Bacteriol 162 (1985) 328.

[12] Iske, U., Glombitza, F.: Verfahren zur Reinigung von toxischen Abwässern, Patent DD 239 197 A1.

[13] Glombitza, F. Hummel, A., Eckardt, L., Cleaning of an uranium mining drainage water by means of biosorption. VAAM Jahrestagung, Hannover (1994).

[14] Kutsal, Y.S.T., Chem Eng J., 60 (1995) 181.

[15] Savvaidis, I., Hughes, M.N. and Poole, R.K., FEMS Microbiol Lett., 92 (1992) 181.

[16] Scott, J.A., Karanjkar, A.M. and Rowe, D.L., Minerals Eng., 8 (1995) 221.

[17] Pernfuß, B., Ebner, C., Pümpel, T., Diels, L., Macaskie, L., Tsezos, M., Keszthelyi, Z. and Glombitza, F., The behaviour of five metal biosorbing and bioprecipitating bacterial strains, inoculated in a moving-bed sandfilter. Submitted to IBS-99.

Adhesion of microorganism cells and jarosite particles on the mineral surface

Z. Sadowski

Department of Chemical Engineering and Heating Equipment
Wroclaw University of Technology, ul. Wybrzeze Wyspianskiego 27, 50-370 Wroclaw,
Poland

It has been know that during both bioleaching and biomodification of the mineral surface the microbial cell should be tenaciously adhered to the mineral surface. The main goal of this research was evaluate the effect of polysaccharides (dextrine) on both the bacterial cells and precipitated jarosite particles deposition onto the mineral surface. The determination of the free energy of solid surface was med. by means of the thin-layer wicking technique. It was found that the quartz particles which were covered by the dextrine film have value of $\gamma^{LW}=$ 121.44 mN/m (without dextrin γ^{LW} = 62.18 mN/m). The adsorption of dextrin caused a decrease of the γ^- component of the free energy from 171.90 to 123.19 mN/m. The treatment of both quartz and gold refractory ore by polysaccharides caused an increase of the adhesion of microbial cells and jarosite colloid particles. The deposition of jarosite on the surface of gold ore stopped by the dispersing reagent addition.

1. INTRODUCTION

Although, bioleaching operation of sulphide minerals and biooxidation of refractory gold ores are well established commercial processes[1-4], a special emphasis should be placed on the adhesion of both microorganism cells and bioleaching products onto the mineral surface. Deposition of both small particles and microbial cells onto the mineral is important in bioleaching areas. In direct bioleaching conditions, the deposition of microorganism cells on the mineral surface should promoted this process [5]. Whereas, the adhesion of jarosite particles and other products should be minimised. [6]. To improve understanding of the mechanisms influencing on the both mineral cells and colloid particles attachment will be a target of this research. Bacterial cells adhesion is difficult to quantify at a fundamental level. Hydrophobic interaction, macromolecular bridging, hydration forces, electrical double-layer interaction and dispersion forces created a very complicate picture. The rule of both an electrical double-layer and dispersion forces in the particle deposition has been discussed by Matijevich [7,8]. The DLVO theory has been used with a success to the interpretation of obtained results [9]. Most bacteria are negative charged because of the predominance of the anionic groups present within the cell wall. The interaction of microbial cell with a strong negatively charged of the collector surface results in an energy barrier. It means the adhesion of microbial cells can be realised when the energetic barrier is overcome [10]

The second approach to the microbial adhesion is based on the surface thermodynamics of interfacial interactions when the role of surface free energy is discussed [11]. The adsorption of organic macromolecules (polysaccharides) influences the surface free energy [12]. Hence, the presence of polysaccharides changes the mineral surface properties and it should be effected on the adhesion of both microbial cells and colloid particles.

The purpose of this work is to investigate the changes in surface free energy of the quartz surface, which are due to adsorption of dextrin on the surface.

2. EXPERIMENTAL

2.1. Materials and reagents

Quartz was supplied from WARD'S Natural Science Establishment Inc., (New York). The quartz sample was ground using Fritsch planetary-type ball mill and then sieved to obtain two fraction size. First size fraction between 0.4-0.1 mm and belong 0.04 mm. The surface of quartz was cleaned by hydrochloric acid.

Jarosite particles were prepared following the procedure described by Baron and Palmer [13]. The obtained particles had a mean diameter of 12.76 μm. A Phillips diffractomer was used to the analysis of precipitated solid. The X-ray diffractogram of the prepared sample showed that the particles compound with jarosite. The material was cleaned by repeated cycles of centrifugation and redispersion in double distilled water.

Gold refractory ore: The gold ore was from Zloty Stok (Poland). The lumps were hand-sorted and ground in laboratory mill. Particle in the size range –320 +125 μm were used to both adsorption and adhesion experiments. The gold ore sample assayed 51,1% SiO_2 , 1.6% Al_2O_3, 7.9% Fe_2O_3, 7.1% MgO and 5.7% CaO according to Dr. Luszczkiewicz from Mining Department [14]

Reagents: Potato dextrin and antrone obtained from Sigma Chemical Co., and natrium hexametaphosphate purchased from Benckiser-Knapsack GMBH (Germany).

2.2. Bacterial Culture

A pure strain of *Nocardia amarae* provided by Dr. Maliszewska from Institute of Biochemistry, Biotechnology and Organic Chemistry was used. The *Nocardia* was grown in a medium having the composition [g/l] : 20 glycerin, 0.2 K_2HPO_4, 0.2 $MgSO_4$ H_2O, 0.2 NaCl, 0.1 K_2SO_4. The culture was maintained at 28°C. Steady state was attained after three days of growth. After this period of time the biomass was harvested and used for the adhesion experiment.

2.3. Experimental procedures

The thin-layer wicking technique. This technique was applied by Chibowski and Holysz [15,16]. Details of the experimental procedure and theoretical background was described elsewhere [17].

The surface free energy of minerals is expressed as a sum of apolar Lifshitz – van der Waals and polar acid-base components. The acid-base component results from interaction of electron donor (Lewis base) and electron acceptor (Lewis acid).

The polymer (dextrin) adsorption. Adsorption of dextrin onto both quartz and gold refractory ore was performed at various polymer concentrations. A known amount (1g) of the mineral sample was mixed with 100 ml of dextrin solution. The flasks with such prepared mineral suspensions were agitated in a mechanical shaker for 8 h. After that, the mineral suspension was centrifuged and the supernatant was collected for the determination of the non-adsorbed dextrin. For the polysaccharides concentration the colorimetric method which was described by Dubois et al. [18] was applied.

The sedimentation and adhesion test. The sedimentation experiments were carried out with a special glass cylinder (35 cm long and 50 ml volume). The initial suspensions were prepared using both quartz and gold ore samples. The time of sedimentation of these suspensions were tested at different conditions of pH. Then, the mixed suspension of jarosite and gold ore was tested. The mixed suspensions were vigorously shaken by 3 min. In order to obtain information on the adhesion of both small jarosite particles and microbial cells the turbidity (light absorbance at 450 nm) of the suspensions were determined at a constant depth of cylinder. If jarosite particle or microbial cells were deposited on the mineral surface the concentration and turbidity decreased.

The adhesion of Nocaria cells onto the quartz surface was measured using a method which was previously used by Clayfiels and Lumb [19] and the author [20].

3. RESULTS AND DISCUSSION

The knowledge of the surface free energy of solids has a significant important in the field of adhesion, leaching, wetability and adsorption. The surface free energy describes the physical interaction potential of the solid surface. In Table 1 the surface free energy components of quartz as obtained from thin-layer wicking technique are given. For comparison, the values for quartz treated with dextrin are also given.

Table 1
Effective Pore Radius, R, and Surface Free Energy Components: Lifshitz-van der Walls, γ^{LW}, Electron Acceptor,γ^{+} and Electron Donor,γ^{-},of untreated quartz and treated quartz with dextrin

Material	$R \times 10^{-4}$ [cm]	γ^{LW} [mN/m]	γ^{+} [mN/m]	γ^{-} [mN/m]
Untreated quartz	1.69	62.18	0	171.90
Quartz with dextrin	3.25	121.44	0	123.19

The adsorption of dextrin on the solid surface should be accompanied by changes of the surface free energy of quartz. It can be seen that the effective pore radius increases for the quartz layer when quartz was treated with dextrin. The γ^{LW} component also causes an essential increase from 6.218 to 123.19 mN/m. Simultaneously, the adsorption of dextrin caused a decrease of the electron donor γ^{-} component from 171.90 to 123.19 mN/m. The relative high value of the γ^{-} component for the quartz surface is due to the presence of the hydrophilic silanol group. The adsorption of polysaccharide covered these groups.

The shape of adsorption isotherms for quartz and gold refractory ore suggests a simple Langmuir type adsorption behaviour. The saturation adsorption densities was obtained for gold refractory ore. It can suggest a high affinity of dextrin to the surface of ore particles. For quartz the adsorption isotherm is not of the high affinity type and indicates weak interaction.

Adhesion of jarosite particles on the ore particles is presented at Fig. 2. The results provide a further evidence for the role of polysaccharides in the adhesion of colloidal particles. The pre-treatment of ore particles with dextrin solution causes an increase of jarosite adhesion. This effect is more evident at pH 7.5.

On the other hand, the addition of dispersing reagent (natrium hexametaphosphate to the suspension give an opposite effect. These results need further studies.

The deposition of jarosite particles onto the surface of gold ore particles is analogous to heterocoagulation and therefore can be treated theoretically of the same way [7]. The total interaction forces between a sphere (microbial cell or jarosite particle0 approaching to the collector surface is given by

$$F_T = F_{vdW} + F_{elec} + F_{steric} + F_H \qquad (1)$$

Where F_{vdW} and F_{elec} are referred to the classic van der Waals and the electrostatic forces, F_{steric} is the special steric interaction and F_H is a hydrophobic interaction. In the case of the dextrin layer on the mineral surface the steric interaction dominates.

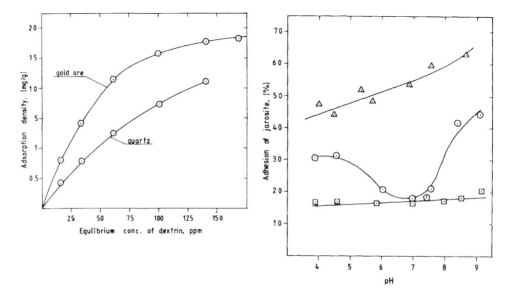

Figure 1. Adsorption isotherm of dextrin on quartz and gold refractory ores.

Figure 2. Adhesion of small jarosite particle onto the gold refractory ore, (Δ) gold ore treated by dextrin, (0) pure gold ore and (□) pure gold ore and dispersant reagent.

Figure 3 shows the effect of pre-treatment of the quartz surface with different dextrin solutions on the adhesion of *Nocardia* cells.

It is evident that an increase of the dextrin concentration the adhesion of microbial cells increases. This suggests a strong affiliation of microorganism cells to the mineral surface when it was covered by a polysaccharide layer.

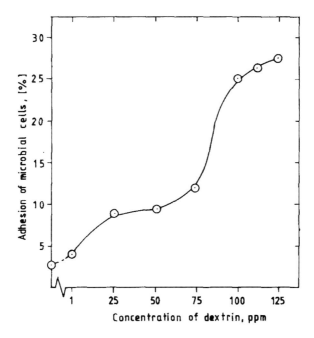

Figure 3 . Adhesion of Nocardia cells onto the quartz surface

The adsorption of dextrin at the mineral surface was broadly investigated [21-23]. Depending on the nature of the mineral surface, the adsorption of polymer macromolecules can take place through tree different ways [21]. From Figure 3 it is evident that polymer layer had a subsequent effect on the microorganism cell adhesion. It will be interesting to compare the structure of polymer layer with the results of microbial cell adhesion.

4. CONCLUSIONS

On the basis of the results obtained with the thin-layer wicking measurements, adsorption experiments and adhesion data, the following conclusions can be formulated:

1. Adsorption of polysaccharides onto the mineral surface affected the surface free energy components.
2. Adsorption isotherms of dextrin on the different mineral surface followed the Langmuir model.
3. The presence of adsorbed polysaccharides caused an increase of jarosite adhesion onto the surface of gold refractory ore.
4. The addition of a dispersing reagent prevented the adhesion of jarosite particles
5. Dextrin interacted with microorganism cells through both steric and hydrophobic interaction.

ACKNOWLEDGEMENTS

The author thanks Dr. Maliszeska for the microorganism cells culture and Dr Luszczkiewicz for both valuable information and a sample of refractory gold ore. This work was support by grant No. 9 T12B 028 14 from KBN (Committee of Scientific Research), Poland.

REFERENCES

1. H.L.Ehrlich and C.L.Brierley, Microbial Mineral Recovery, McGraw-Hill Publishing, New York, 1991.
2. G.Rossi, Biohydrometallurgy, McGraw-Hill Book Company GmbH, Hamburg, 1990.
3. R.Poulin and R.W.Lawrence, Minerals Engineering, 9, (8) (1996) 799.
4. D.S.Holmes and R.W.Smith (Eds.) Minerals Bioprocessing II, Publication of TMS, 1995.
5. R.T.Espejo and P.Ruiz, Biotechnology Bioengineering, 30 (1987) 586.
6. R.W.Smith and R.Misra (Eds.), Minerals Bioprocessing, Publication of TMS, 1993.
7. E.Matijevic, Progr. Colloid Polym. Sci., 101 (1996) 38.
8. A.Zelenev and E.Matijevic, Colloids Surfaces, 125 (1997) 171.
9. R.L.Qiao, Z.Li, and C.S.Wen, J. Colloid Interface Sci., 202 (1998) 205
10. S.E.Truesdail, J.Lukasik, S.R.Farrah, D.O.Shah,and R.B.Dickinson, J. Colloid Interface Sci., 203 (1998) 369.
11. J.Skvaria, J.Chem. Soc. Faraday Trans., 89, (15) (1993) 2913.
12. A.Ontiveros-Ortega, M.Espinosa-Jimenes, E.Chibowski and F.Gomzalez-Caballero, J. Colloid Interface Sci., 199 (1998) 99.
13. D.Baron and C.D.Palmer, Geoch. Cosmoch. Acta, 60, (2) (1996) 185.
14. A.Luszczkiewicz and A.Muszer, Physicochemical Problems of Mineral Processing, 31 (1997) 197.
15. E.Chibowski and L.Holysz, Langmuir, 8 (1992) 710.
16. E.Chibowski and L.Holysz, Langmuir, 8 (1992) 717.
17. E.Chibowski, J. Adhesion Sci. Technol., 6, (9) (1992) 1069.
18. M.Dubois and K.A.Gilles, Anal. Chem., 28, (3) (1956) 350.
19. E.J.Clayfield and E.C.Lumb, Discuss. Faraday Soc., 42 (1972) 413.
20. Z.Sadowski, in Minerals Bioprocessing II (D.S.Holmes & R.W.Smith, Eds.) Publication of TMS, (1995) 279.
21. G.A.Nyamekye and J.S.Laskowski, J. Colloid Interface Sci., 157 (1993) 160.
22. G.B.Raju, A.Holmgren and W.Forsling, J. Colloid Interface Sci., 193 (1997) 215.
23. J.Addai-Mensah, J.Dawe, R.Hayes, C.Prestidge and J.Ralston, J. Colloid Interface Sci., 203 (1998) 115.

Multi-component biosorption of lead, copper and zinc ions on *R. arrhizus*

Y. Sag, A. Kaya and T. Kutsal

Hacettepe University, Chemical Engineering Department, 06532, Beytepe, Ankara, Turkey

The biosorption of three divalent metal ions-namely lead(II), copper(II) and zinc(II)- on *R. arrhizus* has been studied for three single-component, three binary and one ternary systems. The ability of the fungal biomass to bind three metal simultaneously is shown as a function of the number of metals competing for binding sites, metal combination and levels of metal concentration. The mono and multi-component sorption phenomena have been expressed by the non competitive and competitive Freundlich adsorption models developed for optimum adsorption conditions.

1. INTRODUCTION

One of the difficulties in describing the adsorption of metal ions from wastestreams is that wastewaters contain not one, but many metal ions. When several components are present, interference and competition phenomena for adsorption sites occur and lead to a more complex mathematical formulation of the equilibrium. Several isotherms have been proposed to describe competitive adsorption equilibrium for such a system. These isotherms range from simple models related to the individual isotherm parameters only, to more complex models related to the individual isotherm parameters and to correction factors [1,2].

Several multisolute isotherms have been suggested for solutes that individually obey the Freundlich expression [3]:

$$q_{eq} = a^0 \, C_{eq}^{b^0} \tag{1}$$

and the equation may be linearised by taking logarithms:

$$\ln q_{eq} = b^0 \ln C_{eq} + \ln a^0 \tag{2}$$

Therefore a plot of $\ln q_{eq}$ versus $\ln C_{eq}$ enables the constant a^0 and exponent b^0 to be determined.

One of models related to the individual isotherm parameters and to correction factors is an empirical extension of the Freundlich model, restricted to binary mixtures [4]:

$$q_{1,eq} = \frac{a_1^0 \cdot C_{1,eq}^{b_1^0 + b_{11}}}{C_{1,eq}^{b_{11}} + a_{12} \cdot C_{2,eq}^{b_{12}}} \tag{3.a}$$

$$q_{2,eq} = \frac{a_2^0 \cdot C_{2,eq}^{b_2^0 + b_{22}}}{C_{2,eq}^{b_{22}} + a_{21} \cdot C_{1,eq}^{b_{21}}} \tag{3.b}$$

where the a_i^0 and b_i^0 are derived from the corresponding individual Freundlich isotherm equations. This isotherm requires six new parameters to be derived experimentally in bisolute adsorption tests.

A Freundlich type multi-component adsorption isotherm was also derived and employed successfully to describe adsorption data of various bicomponent and three-component systems [5,6]. The adsorption isotherm for component i in a k-component system expressed in terms of weight of sorbate, is written in the form:

$$q_{i,eq} = a_i^0 \, C_{i,eq} \left(\sum_{j=1}^{k} a_{ij} \, C_{j,eq} \right)^{b_i^0 - 1} \tag{4}$$

The pre-exponental coefficient a_i^0 and the exponent b_i^0 can be determined from the monocomponent systems. The competition coefficients a_{ij} describe the inhibition to the adsorption of component i by component j, and can be determined from experimental data of multi-component systems.

2. MATERIALS AND METHODS

2.1. Microorganism, growth conditions and preparation of the microorganism for biosorption

Rhizopus arrhizus, a filamentous fungus, was obtained from the US Department of Agriculture Culture Collection. Batch experiments were conducted with 100 ml cultures in 250 ml Erlenmeyer flasks. The flasks were agitated on a shaker for 96 h at 150 rpm and at a temperature of 30°C. The growth medium contained malt extract (17.0 g l^{-1}) and soya peptone (5.4 g l^{-1}). The pH was adjusted to 5.4-5.6 with H_2SO_4. After the growth period, *R. arrhizus* was washed twice with distilled water, inactivated using 1% formaldehyde and then dried in an oven at 60°C for 24 h. For biosorption studies, a weighed amount of dried cells was suspended in 100 ml of distilled water and homogenized for 20 min in a homogenizer at 8000 rev min^{-1}.

2.2. Preparation of biosorption media containing single metal ions, binary and ternary metal mixtures

Pb(II), Cu(II) and Zn(II) solutions were prepared by diluting 1.0 g l^{-1} of stock solutions of lead(II), copper(II) and zinc(II), obtained by dissolving anhydrous lead(II) nitrate, copper(II) nitrate trihydrate and zinc(II) nitrate hexahydrate in distilled water, respectively. The range of concentrations of prepared metal solutions varied between 5 and 250 mg l^{-1}. For the determination of adsorption characteristics of the dominant metal in binary metal mixtures, the initial concentrations of the dominant metal were varied between 20 and 210 mg l^{-1} while the competing metal ion concentrations in each biosorption medium were held constant over the range 25-110 mg l^{-1}. For the determination of adsorption characteristics of Pb(II), Cu(II) and Zn(II) ions in ternary metal mixtures, the initial concentrations of the metal ions were varied simultaneously between 25 and 200 mg l^{-1}. The pH of the biosorption media was adjusted to the desired value for the biosorption of Pb(II), Cu(II) and Zn(II) ions with 1 mol l^{-1} of HNO_3. The fungal suspension (20 ml) was mixed with 180 ml of the desired metal solution in an Erlenmeyer flask. The flasks were agitated on a shaker at 25°C for 48 h, which is sufficiently long for adsorption equilibrium. The samples were centrifuged at 6030 g for 3 min and the supernatant liquid was analysed for metal ions.

2.3. Analysis of heavy metal ions

The concentrations of unadsorbed Pb(II), Cu(II) and Zn(II) ions in the sample supernatant were determined using an atomic absorption spectrophotometer (ATI-UNICAM 929) with an air-acetylene flame. ATI-UNICAM lead, copper and zinc hollow cathode lamps were used. Pb(II), Cu(II) and Zn(II) were measured at 217.0, 324.8 and 213.9 nm, respectively.

3. RESULTS AND DISCUSSION

The kinetic results are given as the initial adsorption rate, r (mg g^{-1} min^{-1}), equilibrium results as the adsorbed metal ion quantity, q_{eq}, per unit weight of dried biomass (mg.g^{-1}), and the unadsorbed metal ion concentration, C_{eq} (mg l^{-1}), in solution at equilibrium. The initial adsorption rate was obtained by calculating the slope of a plot of the adsorbed metal ion quantity, q, per gram of dried microorganism (mg g^{-1}) versus time (min) at t=0.

3.1. Biosorption of Pb(II), Cu(II) and Zn(II) on *R. arrhizus* in single component and binary systems

The biosorption of single species of Pb(II), Cu(II) and Zn(II) ions on *R. arrhizus* was studied with respect to adsorption pH and initial metal ion concentration in order to determine the optimum conditions for heavy metal removal. Pb(II), Cu(II) and Zn(II) ions are more effectively adsorbed to the biomass at pH in the range 4.0-5.0. The optimum pH for the single-component biosorption of Pb(II), Cu(II) and Zn(II) ions on *R. arrhizus* was determined to be 5.0, 4.0 and 5.0, respectively. At pH values higher than 5.5, Pb(II) and Cu(II) ions precipitated. At pH values higher than 6.5, Zn(II) ions also precipitated and adsorption studies at these pH values could not be performed. The simultaneous biosorption of Pb(II), Cu(II) and Zn(II) from binary mixtures was investigated at both pH 4.0 and 5.0. Biosorption of binary mixtures of Pb(II)-Cu(II), Pb(II)-Zn(II) and Cu(II)-Zn(II) on *R. arrhizus* was studied firstly at pH 5.0, the optimum pH value for Pb(II) and Zn(II). At the first stage of biosorption, rapid equilibrium is established between adsorbed metal ions on the fungal cell and unadsorbed metal ions in solution. This equilibrium can be represented by adsorption isotherms. Experimental data for the binary metal systems were tested for various proposed models [2,7]. It was found that none of the suggested models, except the empirical multi-component Freundlich model restricted to binary mixtures, could represent the entire data set properly.

The Freundlich adsorption isotherms for the simultaneous biosorption of Pb(II), Cu(II) and Zn(II) ions on *R. arrhizus* from the binary mixtures in the presence of the other metal ion at a constant concentration of 50 mg l^{-1} and at pH 5.0 are given in Figures 1-3 and are compared with the single-metal situations. In all the figures where metal uptake equilibrium data are shown, the model profiles are presented as solid and dashed lines and the experimentally obtained values are given by open symbols. The error bars indicated in the figures represent 5 per cent of the error between the predicted and the experimental values when these values exceed the dimensions of the symbols. The individual Freundlich constants, a_i^o and b_i^o, were determined from the intercept and slope of the linearized equilibrium equation, respectively. a_i^o and b_i^o can be considered as indicators of adsorption capacity and adsorption intensity, respectively. The individual Freundlich constants, a_i^o and b_i^o for Pb(II), Cu(II) and Zn(II) ions were found to be 22.939 and 0.161; 3.804 and 0.254; 1.886 and 0.336, respectively. To determine competition coefficients, the empirical bicomponent Freundlich model given by Equations (3a) and (3b) was written for both the components and solved simultaneously by using an MS Excel 7.0 computer program. The competition coefficients for the binary mixtures of Pb(II) and Zn(II) at pH 5.0 were determined as $b_{11}= 0.797$, $a_{12}= 1.648$, $b_{12}= 0.609$, $b_{22}= 0.010$, $a_{21}= 0.482$, $b_{21}= 0.506$. The competition coefficients for the binary mixtures of Cu(II) and Pb(II) at pH 5.0 were found to be $b_{11}= 0.010$, $a_{12}= 0.136$, $b_{12}= 0.733$, $b_{22}= 0.761$, $a_{21}= 1.002$, $b_{21}= 0.780$. The competition coefficients for the binary mixtures of Cu(II) and Zn(II) at pH 5.0 were determined as $b_{11}= 0.928$, $a_{12}= 2.263$, $b_{12}= 0.717$, $b_{22}= 1.216$, $a_{21}= 0.285$, $b_{21}= 0.010$. The relative capacities in the binary mixtures were in the order Pb(II)>Cu(II)>Zn(II), in agreement with the single-component data.

To obtain selectivity for Cu(II) ions, the pH of the binary mixtures was adjusted to 4.0, the optimum pH value for Cu(II) ions. The Freundlich adsorption isotherms for the simultaneous biosorption of Cu(II) ions on *R. arrhizus* from binary mixtures of Pb(II)-Cu(II) and Cu(II)-Zn(II) in the presence of the other metal ion at a constant concentration of 50 mg l^{-1} and at pH 4.0 are given in Figure 4 and are compared with the single-Cu(II) situation. The individual Freundlich constants, a_i^o and b_i^o for Pb(II), Cu(II) and Zn(II) ions at pH 4.0 were found to be 7.335 and 0.279; 8.325 and 0.257; 0.279 and 0.657, respectively. The competition coefficients for the binary mixtures of Cu(II) and Pb(II) at pH 4.0 were found to be $b_{11}= 0.410$, $a_{12}= 1.168$, $b_{12}= 0.555$, $b_{22}= 0.485$, $a_{21}= 0.347$, $b_{21}= 0.749$. The competition coefficients for the binary mixtures of Cu(II) and Zn(II) at pH 4.0 were determined as $b_{11}= 0.984$, $a_{12}= 3.120$, $b_{12}= 0.743$, $b_{22}= 0.523$, $a_{21}= 0.212$, $b_{21}= 0.856$. In the binary mixtures, the adsorbed Cu(II) ion quantity per unit weight of dried biomass at equilibrium at pH 4.0 increased, compared with pH 5.0.

3.2. Biosorption of Pb(II), Cu(II) and Zn(II) on *R. arrhizus* in ternary systems

The competitive biosorption of Pb(II), Cu(II) and Zn(II) on *R. arrhizus* as three metal-ion systems at pH 5.0 was investigated as a function of different combinations of initial metal ion concentrations. The initial adsorption rates of Pb(II), Cu(II) and Zn(II) ions from the ternary mixtures are given in Table 1. The initial adsorption rates of the metal ions increased with increasing metal ion concentrations and the ratio of the metal ion concentration to the other metal ion concentrations and/or to the total metal ion concentration. Although the adsorbed

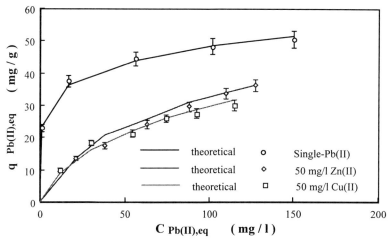

Figure 1. Comparison of the amounts of Pb(II) experimentally adsorbed on *R. arrhizus* and calculated using the single-component and the empirical bicomponent Freundlich models in the binary mixtures of Pb(II)-Zn(II) and Pb(II)-Cu(II) at pH 5.0.

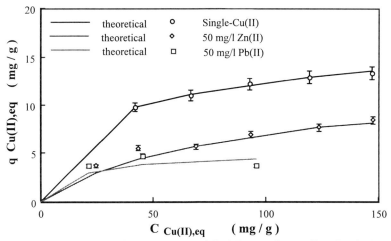

Figure 2. Comparison of the amounts of Cu(II) experimentally adsorbed on *R. arrhizus* and calculated using the single-component and the empirical bicomponent Freundlich models in the binary mixtures of Cu(II)-Zn(II) and Pb(II)-Cu(II) at pH 5.0.

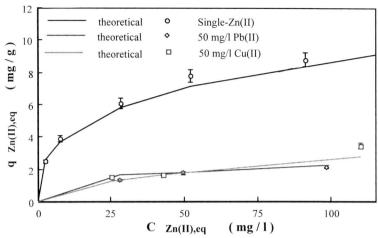

Figure 3. Comparison of the amounts of Zn(II) experimentally adsorbed on *R. arrhizus* and calculated using the single-component and the empirical bicomponent Freundlich models in the binary mixtures of Cu(II)-Zn(II) and Pb(II)-Zn(II) at pH 5.0.

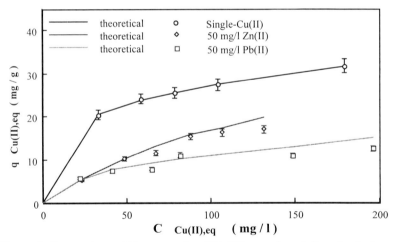

Figure 4. Comparison of the amounts of Cu(II) experimentally adsorbed on *R. arrhizus* and calculated using the single-component and the empirical bicomponent Freundlich models in the binary mixtures of Cu(II)-Zn(II) and Pb(II)-Cu(II) at pH 4.0.

lead(II) quantities at equilibrium decreased, compared with the binary systems, lead(II) ions were also adsorbed selectively from the ternary metal mixtures. The combined inhibition effect of Cu(II) and Zn(II) ions on the biosorption of Pb(II) was greater than the inhibition effect observed in the situation where Cu(II) and Zn(II) ions were the ' sole competing metal ' in solution. The selectivity of the fungus for Cu(II) ions in the ternary systems increased in comparison with the binary systems when the Cu(II) ion concentration and/or the ratio of the Cu(II) ion concentration was increased with respect to the total metal ion concentration. On the other hand, zinc biosorption decreased substantially (Figure 5). The competition between metal ions to bind to active components of the cells can result in both synergistic and antagonistic responses depending on metal combination, number of metals competing and metal concentrations. The greater the total metal concentration, the most effective was the competitive metal uptake because of a strong driving force or a large difference in concentration between adsorbent surface and metal solution. The three components were also found to obey the multi-component Freundlich model. The competition coefficients given by Equation (4) were estimated from the competitive adsorption data of Pb(II), Cu(II) and Zn ions by using an MS Excel 7.0 computer program (Table 2). The average percentage errors between the experimental values and the predicted values using the multi-component Freundlich model for the entire data set of Pb(II), Cu(II) and Zn ions were 14.8%, 9.3% and 17.4%, respectively.

Table 1
Initial adsorption rates of Pb(II), Cu(II) and Zn(II) ions on *R. arrhizus* from ternary metal mixtures

Data No.	$C_{Cu(II),i}$ mg l^{-1}	$C_{Pb(II),i}$ mg l^{-1}	$C_{Zn(II),i}$ mg l^{-1}	$r_{Cu(II),i}$ mg (g-min)$^{-1}$	$r_{Pb(II),i}$ mg (g-min)$^{-1}$	$r_{Zn(II),i}$ mg (g-min)$^{-1}$
1	53.91	22.41	53.77	3.64	1.75	0.27
2	27.72	48.02	54.52	2.13	4.00	0.29
3	54.05	46.53	50.41	2.79	2.98	0.14
4	75.24	39.80	77.22	3.68	1.79	0.24
5	41.18	75.80	74.88	2.36	4.98	0.28
6	74.25	73.10	75.60	3.54	4.01	0.18
7	103.46	51.68	96.57	3.65	2.54	0.34
8	53.10	101.75	96.30	2.28	5.12	0.27
9	103.37	101.11	96.75	2.95	4.13	0.17
10	153.09	73.44	149.70	4.13	3.15	0.28
11	77.69	148.10	151.30	2.98	6.80	0.33
12	155.93	152.02	152.00	3.28	4.80	0.27
13	203.22	98.70	205.48	3.96	2.63	0.29
14	104.78	198.70	200.56	3.17	5.76	0.38
15	201.97	203.60	201.53	3.96	4.94	0.21

406

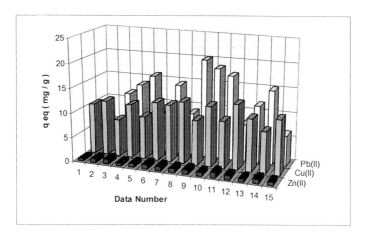

Figure 5. Comparison of the adsorbed Pb(II), Cu(II) and Zn(II) quantities per gram of dried biomass at equilibrium from ternary metal mixtures.

The adsorption isotherms for the simultaneous biosorption of Pb(II) ions on *R. arrhizus* in the presence of increasing concentrations of Cu(II) and Zn(II) ions in the range 50-200 mg l⁻¹ are given in Figure 6. In all the figures where metal uptake equilibrium data are shown, the model profiles are presented as full lines while the open symbols denote experimentally obtained values. The adsorption isotherms for the simultaneous biosorption of Cu(II) ions on *R. arrhizus* in the presence of increasing concentrations of Pb(II) and Zn(II) ions in the range 50-200 mg l⁻¹ are also given in Figure 7. The experimental equilibrium uptake values and the predicted values using the multi-component Freundlich model for Zn(II) ions are given in Table 3.

Table 2
Individual Freundlich constants based on single-component data and the competition coefficients based on multi-component data

i	Solute	a_i^o	b_i^o	a_{i1}	a_{i2}	a_{i3}
1	Copper	3.804	0.254	1.000	0.100	0.100
2	Lead	22.939	0.161	1.580	1.000	1.627
3	Zinc	1.886	0.336	34.207	19.678	1.000

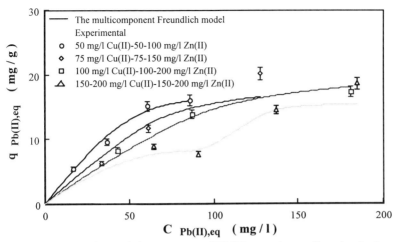

Figure 6. Comparison of the amounts of Pb(II) experimentally adsorbed on *R. arrhizus* and calculated using the multi-component Freundlich model in the ternary mixtures of Pb(II), Cu(II) and Zn(II).

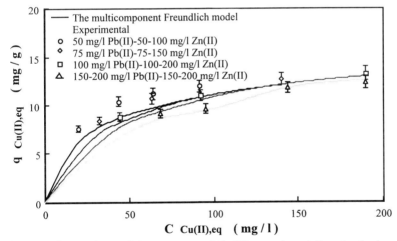

Figure 7. Comparison of the amounts of Cu(II) experimentally adsorbed on *R. arrhizus* and calculated using the multi-component Freundlich model in the ternary mixtures of Pb(II), Cu(II) and Zn(II).

Table 3
For competitive Zn(II) biosorption by *R. arrhizus* from ternary mixtures, the comparison of experimental and predicted $q_{Zn,eq}$ values.

Data No.	$C_{Cu(II),i}$	$C_{Pb(II),i}$	Experimental $C_{Zn(II),i}$	Theoretical $q_{Zn(II),eq}$	$q_{Zn(II),eq}$
1	53.91	22.41	53.77	0.74	0.68
2	27.72	48.02	54.52	0.97	0.80
3	54.05	46.53	50.41	0.52	0.56
4	75.24	39.80	77.22	0.72	0.72
5	41.18	75.80	74.88	0.88	0.80
6	74.25	73.10	75.60	0.57	0.63
7	103.46	51.68	96.57	0.96	0.72
8	53.10	101.75	96.30	0.83	0.83
9	103.37	101.11	96.75	0.56	0.64
10	153.09	73.44	149.70	0.90	0.85
11	77.69	148.10	151.30	0.93	0.99
12	155.93	152.02	152.00	0.90	0.74
13	203.22	98.70	205.48	0.88	0.95
14	104.78	198.70	200.56	1.12	1.05
15	201.97	203.60	201.53	0.79	0.82

ACKNOWLEDGEMENT

The authors wish to thank TÜB‹TAK, the Scientific and Technical Research Council of Turkey, for the partial financial support of this study (Project No: YDABÇAG-525).

REFERENCES

1. M. Sheintuch and M. Rebhun, Wat. Res., 22 (1988) 421.
2. J.C. Bellot and J.S. Condoret, Process Biochem., 28 (1993) 365.
3. J. M. Smith, Chemical Engineering Kinetics, 3rd edn., McGraw-Hill, New York, 1981.
4. W. Fritz and E.U. Schluender, Chem. Eng. Sci., 29 (1974) 1279.
5. C. Sheindorf, M. Rebhun and M. Sheintuch, J. Colloid Interface Sci., 79 (1981) 136.
6. C. Sheindorf, M. Rebhun and M. Sheintuch, Water Res., 16 (1982) 357.
7. M. Sheintuch and M. Rebhun, Water Res., 22 (1988) 421.

Heavy metal ions removal by biosorption on mycelial wastes

L. Stoica and G. Dima

University "Politehnica" Bucharest, Dept. Inorganic Chemistry, 1 Polizu St.
7000, Bucharest, Romania

Mycelial wastes of Penicillium resulted in great quantities from pharmaceutical industry were studied regarding their ability for heavy metal ions (Cd(II), Pb(II)) removal from synthetic aqueous solutions. Different preliminary treatments were applied to mycelial wastes in order to use them during biosorption experiments. The main parameters of the biosorption process were studied: pH solution, heavy metal ions and biomass concentrations and the type of metallic salt anion. Equilibrium sorption isotherms of Cd(II) and Pb(II) follow the typical Langmuir adsorption model.

The loaded inactive biomass was separated from the aqueous solutions by dissolved air flotation and consequently was submitted to elution for heavy metal ions recovery.

The following parameters were studied for the biosorption-flotation process: pH, sorbent-metallic ion contact time and metallic ion concentration in order to obtain an effective separation of the loaded biomass from the aqueous solution.

1. INTRODUCTION

Heavy metals can be removed from aqueous solutions by different physico-chemical processes according to the real pollution context. Some toxic metals (i.e. cadmium, lead) must be removed up to very low concentrations, less than 0.1 mg/l. These severe limits imposed for the effluents discharging require advanced purification processes for heavy metals removal.

The last ten years researches demonstrated that heavy metals removal from aqueous systems could be successfully accomplished by using active or inactive biomasses, the process being already known as "biosorption". Active and inactive biomasses have similar biosorption capacities (1). Hence, the mycelial residues produced during many industrial fermentation processes like enzymes, flavors or antibiotics production, may be used as biosorbents at least for economically reasons.

Fungal mycelial residues were found to accumulate metals and radionuclids by physico-chemical and biological mechanisms including their binding by metabolites and biopolymers or specific polypeptides (2). Biosorption on fungal cell walls was studied for many metallic species (3), the possibilities of subsequently metals desorption and biomasses reuse being also investigated.

If the biosorption process is operated in stirred tanks using a suspended biomass (4), a subsequently solid/liquid separation stage is required. The specific characteristics of this kind of sorbat/sorbent system make difficult the separation by filtration (the process needs more time and may face filter blocking problems especially in the case of fine or ultrafine particles), centrifuging (apparent more expensive) or sedimentation (relatively slow process inadequate to biological materials which are usually of low density. Some flotation techniques were applied for microorganisms separation (5) and the possibility of combining biosorption and flotation were also studied (6). Thus, flotation became of great interest among the bioseparation processes.

Our work presents a study on heavy metal ions (Cd(II), Pb(II)) recovery from synthetic aqueous solutions by biosorption on mycelial Penicillium residues resulted from pharmaceutical industry and also on loaded biomasses separation by dissolved-air-flotation (DAF). The optimal values founded for the critical processes parameters allowed us to reduce the Cd(II) and Pb(II) concentrations under the limits of effluents discharging imposed by international legislation.

2. EXPERIMENTAL

The biosorption experiments were carried out in batch system. The samples of Cd(II) or Pb(II) solutions were shaken together with corresponding quantities of biomass in 50 ml flasks on a Braun shaker at a low constant rate (150 rpm) for 24 hours (the time we considered for reaching the equilibrium). A mycelial Penicillium residue, resulted from the fermentation process in pharmaceutical industry, was used as biosorbent. The biomass was previously treated by repeated washing and drying up to constant weight; the size particles was less than 100µm.

Cadmium and lead solutions of different initial concentrations were prepared using p.a. reagents: $Cd(NO_3)_2.H_2O$, $Pb(NO_3)_2$; $CdCl_2.5/2H_2O$, $CdSO_4$.

Chemical analyses by AAS of the remaining solutions were used in order to assay the unremoved cadmium or lead. For pH regulation we used a Philips pH-meter and 0.1 M HCl or NaOH solutions.

The separation by the flotation process was carried out in a dissolved-air-flotation apparatus (7). The stirring of synthetic mixtures (300 ml), containing the cadmium or lead solution and the biomass was carried on by an electrical stirrer.

3. RESULTS AND DISCUSSIONS

Biosorption of metallic ions from aqueous solutions on inactive biomass occurs by sorption mechanisms. In figure 1, equilibrium isotherms for Cd(II) and Pb(II) biosorption on inactive Penicillium biomass, obtained without pH controlling, are presented. It is obviously that Langmuir type isotherms describe well the biosorption process. The respective equation can be written as:

$$q_e = \frac{bq_m C_{eq}}{(1 + bC_{eq})}$$
(1)

where C_{eq} represents the free metallic ions concentration at equilibrium, and b, q_m are empirical constants, characterizing the metal-biomass interactions.

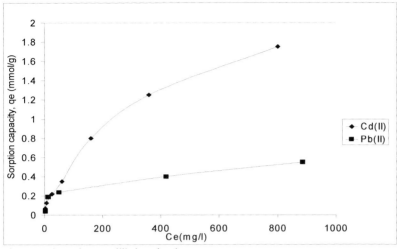

Figure 1. Sorption equilibrium isotherms
        solution pH = 5.5; biomass amount=1g/l; sorption time=24h

Sorption capacities obtained for the two metallic ions were expressed in terms of mmol M(II)/g biosorbent for a good comparison of the results. The respective values increase along with the increasing of the metallic ions initials concentrations.

In the range of low concentrations (under 100 mg/l), the biomass presents similar sorption capacities for the two metallic species (figure 1) but in the range of high initials concentrations (over 500 mg/l) we observed a significant lowering of Pb(II) retention on the Penicillium biomass.

3.1. Biosorption Parameters

The solution pH is an important parameter of the process.

Sorption experiments of Cd(II) on Penicillium biomass showed a great influence of pH solution on biosorption process (figure 2), our experimental results indicating a significant drop in biosorption capacity values for pH = 6.

Cd(II) ions are present in aqueous solutions as stable aqueous complexes $[Cd(OH_2)_4]^{2+}$ in the domain of low pH values. Up to pH=7 the aqueous system contains (even for very low concentrations) only positively charged hydroxospecies ($CdOH^+$) and Cd^{2+}, after this pH value being also present negatively charged hydroxospecies ($[Cd(OH)_3]^-$ and $[Cd(OH)_4]^{2-}$). Considering that the biosorption process involves positives species of Cd(II) having different sizes and charges we can explain why the effectiveness of the biosorption process increase continuously with the solution pH increasing. Starting with pH = 7, for an aqueous solution with initial concentration of 100 mg Cd(II)/l, the metallic ion precipitation begins so that Cd(II) recovery occurs not only by biosorption but also by precipitation. It is interesting to note that, at the end of the biosorption process, aqueous solution pH values are the same (pH = 5,5).

The amount of biomass added to the aqueous solution of Cd(II) represent also an important parameter. The experimental results obtained (figure 3) for Cd(II) sorption on different amounts of biomass show, as we expected, that the recovery efficiency of Cd(II) from aqueous solutions (C_i=100mg/l and C_i=50mg/l) is greater when the amounts of biomass added increase (from 1g/l to 4g/l). In the same time the sorption capacity decreases accordingly.

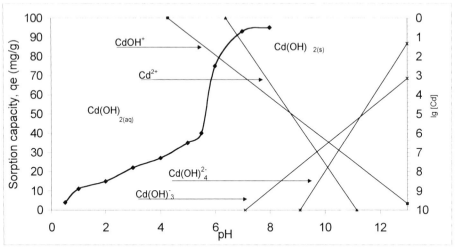

Figure 2. Influence of solution pH on biosorption process; correlation with the structure of Cd(II) species in pH range of 0-13.

C_i = 100 mg Cd(II)/l, sorption time = 24h, biomass amount = 1g/l

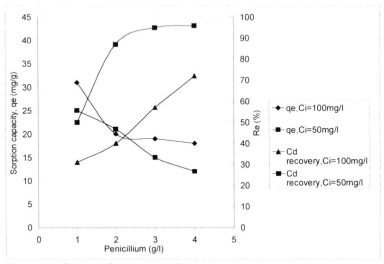

Figure 3. Influence of biomass addition
solution pH = 5.5, biosorption time=24h

A considerable increase of Cd(II) recoveries from the solutions it was observed for a biomass amount of 2g/l (64% for the solution with Ci =100 mg/l and 87% for the solution with Ci =50 mg/l), so that a supplementary addition of biomass to Cd(II) solutions does not seem to be justified from economic point of view.

The time of biosorption also was varied during sorption experiments (figure 4). For two initials concentrations of Cd(II) solutions we observed similar behavior which means that the process is rapid , the first 30 minutes of contacting being enough to reach the equilibrium sorption capacity (considered for 24 hours of contacting).

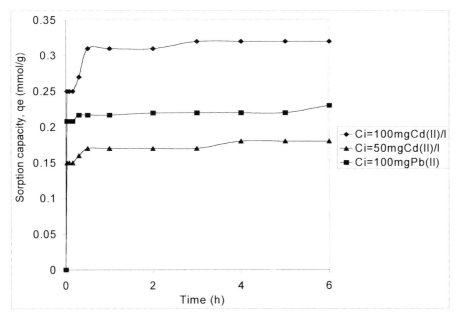

Figure 4. Influence of biosorption time
    solution pH = 5.5, biomass amount =1g/l

The process is rapid for Pb(II) sorption, too (the results for an aqueous solution of $Pb(NO_3)_2$ of C_i=100 mg/l are presented), being in accord with other values reported (8, 9).The study of process parameters was carried on simple dried Penicillium biomass.

3.2 The influence of metallic salt anion

The sorption isotherms for three metallic salts, with corresponding anions sulphate, nitrite and, chloride and the same cation Cd(II) are presented in figure 5. We observed no significant differences between the obtained sorption capacities, the values being closed along the whole range of initials concentrations of Cd(II) solutions.

414

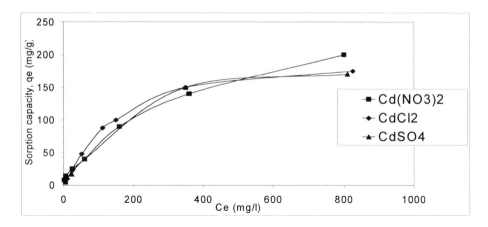

Figure 5. The influence of metallic salt anion
    solution pH = 5.5, biosorption time=24h, biomass amount=1g/l

3.3 The influence of pretreatment on biomass sorption capacity

The residues of Penicillium were previously treated for sorption experiments development.
The biomasses obtained by two different drying ways present distinguishes sorption capacities
and recovery efficiencies. We obtained better results (figures 6 and 7), in terms of sorption
capacity or R (%), for the acetone dried biomass, especially in the range of high initials
concentrations both for Cd(II) and Pb(II).

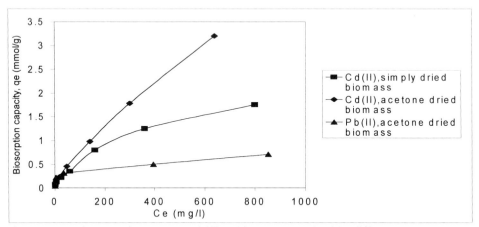

Figure 6. Sorption experiments on Penicillium biomasses obtained by different pretreatments
    solution pH = 5.5, biomass amount=1g/l, sorption time=24h

The explanation for these results may be the nondestructive effect of acetone drying process
(occurring probable by osmotic pressure modification).

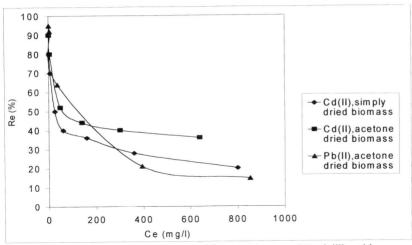

Figure 7. Cd(II) and Pb(II) recovery by different pretreated Penicillium biomass
solution pH = 5.5, biomass amount=1g/l, biosorption time=24h

3.4 The biosorption-flotation as a separation process

As the organic sorbent is difficult to separate from the liquid phase, particularly by filtration, we experimented the separation of loaded biomass by flotation, the biosorbent presenting a natural flotation tendency. Cd(II) solutions were contacted with the biosorbent, under stirring at different pH values. After the biosorption stage the suspensions were transferred in a flotation cell and submitted to the DAF process. Water saturated with air in the saturator and kept under pressure of 4.10^5 N/m^2 was introduced to the base of the cell. When releasing water to the atmospheric pressure, fine air bubbles were generating, appropriated for solid/liquid separation.

Table 1
%R= f(pH) dependence in Cd(II) recovery by sorption and biosorption-flotation processes;

	Sorption				Sorption-flotation			
C_i (mg/l)	sorption time (min)	sorption pH	C_f (mg/l)	Cd(II) R(%)	pH Sorpt-flot	C_f (mg/l)	Cd(II) R(%)	Biomass R(%)
10	10	6	2.8	72	6	1.9	81	95
10	10	7	0.6	94	7	0.5	95	95
10	10	8	0.85	91.5	8	0.65	93.5	95
10	10	9	0.9	91	9	0.8	92	95
10	20	6	0.9	91	6	0.85	91.5	95
10	20	7	0.5	95	7	0.4	96	95
10	20	8	0.8	92	8	0.6	94	95
10	20	9	1	90	9	0.8	92	95

Table 2
%R= f(sorption time) dependence in Cd(II) recovery by sorption and biosorption-flotation processes ;C_i=10mg/l; pH=7.

| Sorption time(min) | Sorption | | Sorption-flotation | | |
	C_f(mg/l)	R(%)	C_f(mg/l)	R(%)	Biomass R(%)
10	0.6	94	0.5	95	95
20	0.5	95	0.2	98	95
30	0.8	92	0.4	96	95

Table 3
%R= f(C_iCd(II)) dependence in Cd(II) recovery by sorption and biosorption-flotation processes ;sorption time=20 min.;pH=7.

| C_i(mg/l) | Sorption | | Sorption-flotation | | |
	C_f(mg/l)	R(%)	C_f(mg/l)	R(%)	Biomass R(%)
100	17	83	16	84	95
50	5	90	4.5	91	95
10	0.5	95	0.2	98	95

The dilution ratio (sample volume: water volume) was of 3:1. Cd(II) concentration and pH value were determined for the effluents. The biosorbent was the acetone dried Penicillium biomass. A simple scheme for the whole process is presented in figure 8:

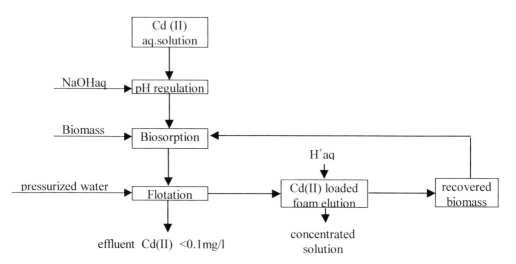

Figure 8. Biosorption process scheme

We can assume that, in fact, the pH flotation correspond to the pH value resulted at the end of biosorption step. We must note that the efficiency of the flotation process is a promising one for this particular case: Cd(II) loaded biomass separation and pH = 5.5.

4. CONCLUSIONS

Biosorption may be applied for Cd(II) recovery from aqueous solutions, the efficiency being dependent on process optimal parameters. Even for more concentrated solution, the acetone pretreated biomass may ensure, in optimal work conditions, high values for Cd(II) recovery. The difficulties of separation the Cd(II) loaded sorbent from the liquid phase may be avoided if DAF technique is applied. This separation process offer the possibility of higher Cd(II) removals (comparing with the simple biosorption) and also high loaded biomass removals (over 95%, the remaining being mechanical losses). What appear of great importance from our laboratory experiments is the fact that flotation does not need the subsequently change of optimal sorption parameters for the loaded biomass separation to occur very well.

REFERENCES

1. C.L. Brieley, J.A. Brieley and M.S. Davidson in Metal Ions and Bacteria, Wiley, Chichester, UK (1989) 359.
2. A.I. Zouboulis and K.A. Matis, An innovation in reclamation of toxic metals, 1996.
3. G.M. Gadd, Biotechnology, H.J. Rehm ed., vol. 6b, 1988.
4. P.J. Jackson, A.P. Torres and E. Delhaize, U.S. Patent 5,120,441 (1992).
5. G.M. Gadd and C. White, J. Chem. Tech. Biotechnol, 55 (1992) 39.
6. R.W.Smith, Min. Process. Extract. Metallic Rev., 4 (1989) 277.
7. L. Stoica, Ion and Molecular Flotation, Did. and Ped. Ed., Bucharest, Romania, 1997.
8. K.A. Matis, A.I. Zouboulis, I.Ch. Hancock, Separation Science and Technology, 29 (8) 1055.
9. E. Fourest and J.C. Roux, Appl. Microbiol. Biotechnology, 37 (1992) 399.

Cadmium removal by the dry biomass of *Sargassum polycystum*

S. Srikrajib[a], A. Tongta[b*], P. Thiravetyan[b], K. Sivaborvorn[a]

[a]Department of Sanitary Engineering, Faculty of Public Health, Mahidol University, 240/1 Rajvithi Rd., Bangkok, 10400, Thailand

[b]Division of Biotechnology, School of Bioresources and Technology, King Mongkut's University of Technology Thonburi, Bangmod, Thungkru, Bangkok 10140, Thailand

This study was concerned with a biosorption by biomass of the brown marine alga *Sargassum polycystum* in a batch system. The optimum condition for cadmium removal was investigated by determination of cadmium uptake capacity of alga which were dried at different temperatures; 80 and 100 °C. In each case of algal preservation, three different conditions of pH; 4.0, 4.6, and 5.6 were studied. A sodium acetate buffer was used in a system to maintain constant pH. The results showed that *S. polycystum* had the highest cadmium uptake at pH 4.0 at both different drying temperatures. The cadmium uptake capacity estimated from Langmuir isotherm were 103.36 and 95.49 mg Cd/g biomass for the alga dried at 80 and 100 °C, respectively. The biomass which were dried at 80 °C had a cadmium uptake capacity of 71.39 mg Cd/g biomass at pH 4.6 and 70.66 mg Cd/g biomass at pH 5.6. The cadmium uptake capacity of alga dried at 100 °C in the conditions of pH 4.6 and 5.6 were 56.78 and 56.67 mg Cd/g biomass respectively. Alga dried at 80 °C was found to have slightly higher cadmium uptake capacity per gram dry weight than that of alga dried at 100 °C.

The recovery of cadmium loaded alga with 0.2 M HCl resulted in no changes of the algal cadmium uptake capacity through five cycles of regenertion process. The dry biomass of *S. polycystum* exhibited high swelling volume. The values of distention index were 13.8 and 11.1 ml/g for alga dried at 80 and 100 °C, respectively. These results indicated the low stability of biosorbent. However, its high cadmium uptake capacity and abundant availability indicated that *S. polycystum* can be used for the development of biosorbent for heavy metal removal from wastewater.

---

* Corresponding author : Fax : 662-427-9623, e-mail : ianangta@cc.kmitt.ac.th

1. INTRODUCTION

The contamination of heavy metal in the water is a significant environmental problem because of accumulation in microorganisms and other components of the benthos [1]. The larger aquatic animals and human can accumulate heavy metals from these organisms through the food chain [2,3].

The main industrial uses of cadmium are protective plating on steel, stabilizer for PVC, pigments in paint and plastic, electrode material in nickel-cadmium batteries, and components of alloys [1,4]. Chemical precipitation and ion exchange are the most commonly used procedures for metal removal. Biosorption, the passive uptake of metals by dead biomass has been well documented [2,3,5-12]. The interest has particularly focused on brown marine algae as a cheap and efficient material for biosorption.

In this study, the biosorption of cadmium by biomass of brown marine alga *Sargassum polycystum* was investigated in a batch system. *S. polycystum* is a marine alga found abundantly in the Gulf of Thailand and grows in the intertidal region [13].

2. MATERIALS AND METHODS

A fresh sample of brown marine alga *S. polycystum* was harvested from Sattaheep, Chonburi and sundried on the beach. The biomass was washed with deionized water and dried at two different temperatures; 80 and 100 °C overnight. Dry materials were ground in a laboratory blender and sieved (20-30 mesh).

2.1. Cadmium Uptake

Solutions of 10-1200 mg/l cadmium in sodium acetate buffer were prepared at various pH values of 4.0, 4.6, and 5.6. The contact equilibrium experiments were performed in 125-ml Erlenmeyer flasks applying 100 mg of dry biomass to 50 ml of buffered cadmium solution of known initial concentration. The suspension was agitated on rotary shaker at 250 rpm for 3 hours at room temperature. Metal-free and biosorbent-free blanks were used as controls. The sorbents were separated from the solution by filtration through a 0.45 μm membrane filter into the syringe.

The cadmium uptake was evaluated from the following equation:

$$q = V(C_i - C_f)/M \tag{1}$$

where q is the cadmium uptake (mg Cd/g of biomass), V is the volume of solution in the flask (ml), C_i is the initial concentration of cadmium in the solution (mg/l), C_f is the equilibrium concentration of cadmium in the solution (mg/l), and M is the mass of biomass (g). The concentrations of cadmium in solution were determined by flame atomic absorption spectrophotometry (GBC 932) at the wavelength of 228.8 nm.

Langmuir sorption model [14,15] was chosen for the estimation of maximum cadmium uptake:

$$q = q_{max} bC_f / (1 + bC_f) \tag{2}$$

where b is the Langmuir constant, ratio of the adsorption/desorption rates and q_{max} is the maximum theoretical metal uptake.

2.2. Regeneration

After the metal sorption experiment, the cadmium-loaded biomass was regenerated with 0.2 M HCl solution until no cadmium was detected in the eluant. The regenerated biosorbents were then used in the second sorption process. The sorption-desorption process was consecutively performed up to five cycles.

2.3. Swelling Characteristics

The swelling characteristics (distention index, swelling ratio, and volume of adsorbed solvent) were obtained from the weights and volume of dry and swollen particles. Dry particles were swollen in cylinder with deionized water for 12 hours. The weight and volume of particles were measured before and after the swelling.

Distention index (DI) was calculated from:

$$DI = V_s/W_d \tag{3}$$

The swelling ratio (Q) was calculated from:

$$Q = W_s/W_d \tag{4}$$

The volume of adsorbed solvent (VAS) was calculated as the following ratio:

$$VAS = (W_s - W_d)/W_d \tag{5}$$

Where W_d is the weight of the dry particles, W_s is the weight of swollen particles, and V_s is the volume of the particles after swelling.

Multiplication q values by the reciprocal value of DI was used to transform the metal uptake values based on weight (q) to based on volume (q_v).

3. RESULTS AND DISCUSSION

3.1. Cadmium Uptake by S. polycystum

The determination of cadmium uptake for *S. polycystum* at different conditions of pH; 4.0, 4.6, and 5.6 was shown in Figures 1-3. The optimum drying temperature was investigated by comparing the cadmium uptake capacities of alga which dried at 80 °C with that dried at 100 °C. Results in Figures 1-3 suggested that the biomass dried at 80 °C had the cadmium uptake capacities slightly higher than that of the biomass dried at 100 °C. The similar drying temperature effect has been reported for the cadmium uptake by the dry biomass of *Ascophyllum nodosum* [2]. These results showed that the higher drying temperature could cause more damage on the metal binding sites of *S. polycystum* (i.e., carboxyl group). This result may also be the effect of swelling characteristics in which the biomass dried at 80 °C had higher swelling volume per gram. The swelling characteristics of *S. polycystum* is presented later in this report. If we consider only the adsorption capacity per mass of the alga, the preferred temperature for drying the marine alga *S. polycystum* should be at 80 °C.

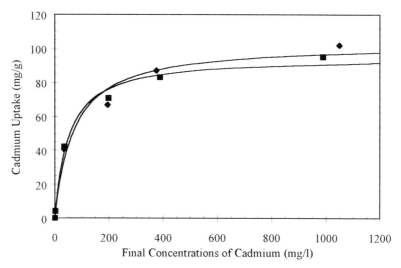

Figure 1. Cadmium sorption isotherm at constant pH 4.0 for *S. polycystum* dried at different temperatures : (◆) 80 °C; (■) 100 °C; (——) theoretical results.

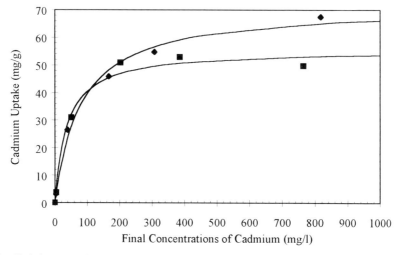

Figure 2. Cadmium sorption isotherm at constant pH 4.6 for *S. polycystum* dried at different temperatures : (◆) 80 °C; (■) 100 °C; (——) theoretical results.

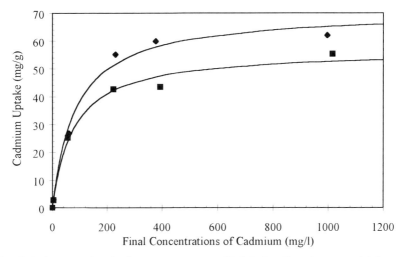

Figure 3. Cadmium sorption isotherm at constant pH 5.6 for *S. polycystum* dried at different temperatures : (◆) 80 °C; (■) 100 °C; (—) theoretical results.

The values of maximum metal uptake and Langmuir constant of Equation 2 for the experiments at various pH values and drying temperatures are presented in Table 1 together with the correlation coefficient (R^2) of the fittings. A comparison of the adsorption capacity of several algae and other biomass is presented in Table 2.

Table 1
Langmuir isotherm constants for cadmium biosorption by *S. polycystum*

Experimental conditions		Langmuir isotherm constants		R^2
pH	Drying temperatures (°C)	q_{max} (mg/g)	b (l/mg)	
4.0	80	103.36	0.014	0.983
	100	95.49	0.020	0.993
4.6	80	71.39	0.012	0.994
	100	56.78	0.018	0.989
5.6	80	70.66	0.012	0.992
	100	56.67	0.013	0.991

The results in Table 1 showed that the optimum pH for the adsorption of cadmium on *S. polycystum* was at pH 4.0. It was shown in several reports that the adsorption capacities were decreased at lower pH values [2,3,7,11]. At low pH, protons compete with metal ions for the binding sites resulting in the reduction of metal uptake [11].

The use of buffer is a good method to maintain constant pH. However, it leads to a multicomponent system with sodium, hydronium and metal ions in sodium acetate buffer possibly competing for binding sites. The results of metal uptake in this work showed that the metal uptake was decreased as the pH of the buffer solution was higher than 4.0 since the amount of sodium in sodium acetate buffer is higher at higher pH. The sodium concentration about 18, 49 and 91 mM for sodium acetate buffer pH 4.0, 4.6, and 5,6 respectively. It had been found in several reports that the metal uptake capacities were reduced when the sodium concentrations increased [10,16].

Table 2
Comparison between maximum cadmium uptake for *S. polycystum* and other dry biomass

Biomass type	Size (mm)	Drying temperatures (°C)	pH of cadmium soluiton	q_{max} (mg/g)	References
S. polycystum	0.50-0.84	80	4.0	103.36	
S. polycystum	0.50-0.84	100	4.0	95.49	
S. fluitans	0.50-0.84	80	4.5	108.39	Leusch & Volesky, 1995
S. fluitans	0.50-0.84	60	4.5	102.37	Fourest & Volesky, 1996
S. fluitans cross-linked with Glutaraldehyde	0.84-1.00	80	3.5	120	Leusch et al., 1995
S. natans	N/A	N/A	3.5	132	Holan et al., 1993
Fucus vesiculosus	N/A	N/A	3.5	73	Holan et al., 1993
Ascophyllum nodosum	N/A	N/A	3.5	133	Holan et al., 1993
A. nodosum	N/A	N/A	4.9	215	Holan et al., 1993
A. nodosum cross-linked with Formaldehyde	N/A	N/A	4.9	149	Holan et al., 1993
Glutaraldehyde	N/A	N/A	4.9	138	Holan et al., 1993
Seccharomyces cereviseae	N/A	N/A	4.5	28	Volesky et al., 1993

N/A ≡ Not available

3.2. Regeneration of alga biosorbent

Cadmium loaded *S. polycystum* was tested for desorption and regerneration with 0.2 M HCl [2]. Due to its highest cadmium uptake capacity, the algal biosorbent dried at 80 °C and adsorbed with cadmium ions at pH 4.0 was regenerated with 0.2 M HCl. The regenerated algal biosorbents were used in the second cadmium adsorption process. The results of cadmium uptake for the five cycles of sorption-desorption process are presented in Table 3. The cadmium uptake of *S. polycystum* in all five cycles were not different significantly (the differences were less than 7%). However, the desorption of cadmium with 0.2 M HCl resulted in 30% loss of mass of alga due to some components dissolved from alga. The 0.2 M HCl solution could damage the composition of algal cell wall. Furthermore, the debris of cells could block the flow of liquid if the biosorption is operated in a column system.

Table 3
Comparison of cadmium uptake of *S. polycystum* for five cycles of regenerations

Cycle	Cadmium uptake (mg/g)	Different from the 1st cycle (%)	Dry weight of *S. polycystum* (mg)	Loss weight (%)
1	105.86	-	100	-
2	101.33	-4.28	80.3	19.7
3	112.89	6.64	76.0	24.0
4	99.46	-6.05	74.4	25.63
5	102.67	3.01	69.8	30.19

3.3. Swelling characteristics and metal isotherm based on swollen biomass

Brown marine algae contain alginate in cell wall about 10-47 % by dry weight [17]. This compound, like gel, can swell in the water. The swelling characteristics of *S. polycystum* dried at different temperatures are compared in Table 4. *S. polycystum* which dried at 80 °C had higher swelling characteristics (*DI*, *Q*, and *VAS*) than the alga dried at 100 °C. This result indicated that the higher drying temperature could damage the alginate in cell wall of alga.

Table 4
Swelling characteristics of *S. polycystum*

Drying temperatures (°C)	Dry weight	Swollen weight	Swollen volume	Distention index	Swelling ratio	Volume of absorbed solvent
	W_d (g)	W_s (g)	V_s (ml)	DI (V_s/W_d)	Q (W_s/W_d)	VAS [(W_s-W_d)/W_d]
80	0.5502	3.9390	7.6	13.8	7.1	6.1
100	0.6290	3.9191	7.0	11.1	6.2	5.2

Generally, the metal uptake is based on weight (mg metal per g biomass). However, it is useful to express the metal uptake per reactor volume (mg metal per ml swollen biomass) for the sorption column applications. This can be done by multiplying q values by the reciprocal value of *DI*.

Figure 4 demonstrates the differences of cadmium sorption isotherm between the alga dried at 80 °C and 100 °C based on swollen volume. Based on volume, the alga dried at 100 °C had higher metal uptake capacity than that of the alga dried at 80 °C. The calculated maximum cadmium uptake per swollen volume were 7.49 and 8.60 mg/ml for *S. polycystum* dried at 80 and 100 °C respectively.

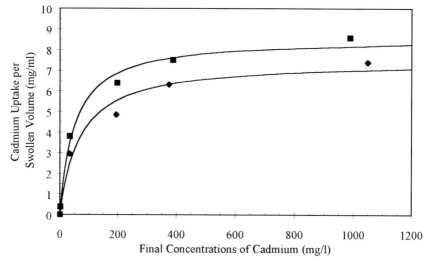

Figure 4. Cadmium sorption isotherm based on swollen volume at constant pH 4.0 for *S. polycystum* dried at different temperatures : (◆) 80 °C; (■) 100 °C; (—) theoretical results.

S. polycystum biomass had higher swelling characteristics than other biomass reported [2,7]. These characteristics are inappropriate for use in sorption column, but modification of the biomass may prevent the swelling of this alga and enhance the stability of biomass [2,5]. It should be noted here that ions in solution could influence the swelling characteristics of the alga. Binding of cations with carboxyl groups of alginate causes the alginate to gain a different structure. Protons and cadmium ions make a lower volume of alginate structure than sodium ions [11], thus alginate and cadmium alginate show a lower swelling. The swelling decreased when the concentration of proton or cadmium increased, because there were more cations binding with alginate [16].

4. CONCLUSIONS

Biosorption of cadmium by dry brown marine alga, *Sargassum polycystum* biomass was studied at different pH values. The effect of the drying temperature on the adsorption capacity was studied. The cadmium uptake capacity per gram dry weight of the alga dried at 80 °C was slightly higher than that of alga dried at 100 °C. The optimum pH of the buffer solution was found to be at pH 4.0. Langmuir adsorption isotherm was used to describe the equilibrium adsorption of cadmium onto the dry algal biomass. The results from the regeneration experiments with 0.2 M HCL showed that the cadmium uptake capacity remained stable for five cycles of regeneration. However, 0.2 M HCl may not be the appropriate eluent, due to the loss of biomass after the wash by the eluent. The high swelling characteristics may reduce the cadmium uptake capacity per volume if used in a column system. An abundant availability of *S. polycystum*, the high cadmium uptake, and the stability of the uptake capacity after regeneration are the advantages of using this biosorbent.

REFERENCES

1. L. Friberh, C. G. Elinder and Kjelström, Environmental Health Citeria 134 Cadmium, Finland , 1992.
2. Z. R. Holan, B. Volesky and I. Prasetyo, Biotechnol. Bioeng., 41 (1993) 819.
3. J. T. Matheickal and Y. Qiming, Wat. Sci. Tech., 34 (1996) 1.
4. M. Borsari, Cadmium: Inorganic & Coordination Chemistry, R. B. King (eds), Encyclopedia of Inorganic Chemistry, Chichester, England, 1994.
5. N. Kuyucak and B. Volesky, Biotechnol. Bioeng., 33 (1989) 809.
6. B. Volesky, H. May and Z. R. Holan, Biotechnol. Bioeng., 41 (1993) 826.
7. Z. R. Holan and B. Volesky, Biotechnol. Bioeng., 43 (1994) 1001.
8. B. Volesky and I. Prasetyo, Biotechnol. Bioeng., 43 (1994) 1010.
9. A. Leusch, R. Zdenek, Z. R. Holan and B. Volesky, J. Chem. Tech. Biotechnol., 62 (1995) 279.
10. A. Leusch and B. Volesky, J. Biotechnol., 43 (1995) 1.
11. S. Schiewer and B. Volesky, Env. Sci. Technol., 29 (1995) 3049.
12. E. Fourest and B. Volesky, Env. Sci. Technol., 30 (1996) 277.
13. L. Khanhanapai and O. Hiaso, Common Seaweeds and Seagrassess of Thailand, 1995.
14. T. Vermeulen, M. D. LeVan, N. K. Hiester and G. Klein, Adsorption and Ion Exchange, H. Robert and P. Dongreen (eds), Perry's Chemical Engineers's Handbook, 6th edition, USA, 1984.
15. D. M. Puthven, Principles of Adsorption Process, USA, 1984.
16. S. Schiewer and B. Volesky, Env. Sci. Technol., 31 (1997) 2478.
17. J. P. Clinton, Marine Botany, USA, 1981.

Modelling of fixed bed biosorption columns in continuous metal ion removal processes. The case of single solute local equilibrium

A. Hatzikioseyian [a] * , F. Mavituna [a], and M. Tsezos [b]

[a] Department of Chemical Engineering, University of Manchester Institute of Science and Technology, (UMIST), Manchester, PO Box 88, United Kingdom

[b] National Technical University of Athens, Department of Mining and Metallurgical Engineering, Environmental Engineering Laboratory,
Heroon Polytechniou 9, 157 80 Zografou, Athens, Greece

A modelling approach for a fixed bed biosorption column is presented. The developed model includes the solute bulk movement through the void space of the bed, a solute dispersion term for simulating cases of non ideal flow, and sorption terms expressed by the appropriate sorption isotherm. The main assumption of the model is that biosorption equilibrium is rapid and no mass transfer resistances exist in the liquid and solid phase. The resulting second order non-linear partial differential equation describes the performance of a column filled with biosorbent material and used to remove single solute metal ions from dilute solutions. The model predicts the optimum expected operation of the bed by simulating breakthrough and concentration profile curves, under different operating conditions. Sensitivity analysis of the operating parameters, revealed that the most important of them are the sorption capacity and the sorption intensity of the biosorbent material as these terms are expressed through the Freundlich's isotherm coefficients.

1. INTRODUCTION

Adsorption is well-known and widely used industrial separation process in which chemical molecules are selectively bound on the surface of a solid phase. Traditional sorbent materials such as zeolites, granular activated carbon (GAC), silica and alumina have been used extensively in industrial scale gas and liquid separation processes. In the light of searching for inexpensive, renewable, alternative natural sorption materials the use of metabolically active or inactive microbial biomass as sorbent material has been proposed[1,2]. The term biosorption has been adopted to describe the phenomenon of sequestering metal ions from aqueous solutions by different microorganisms such as bacteria, fungi, algae and yeast[3,4]. Biosorption, has been extensively documented in the literature, over the last three decades, for many types of

---

* Current address: National Technical University of Athens, Department of Mining and Metallurgical Engineering, Environmental Engineering Laboratory, Heroon Polytechniou 9, 157 80 Zografou, Athens, Greece

microbial biomass. In the last few years, the industry has shown also an increasing interest in applying biosorption as the key component of a waste water treatment technology, for achieving the effluent horizon values imposed by the increasingly more stringent environmental effluent standards. Biosorption is among the candidate alternative technologies for achieving low metal effluent concentrations at reasonable cost, especially for high volume - low concentration applications. Pilot scale units installed at different industrial sites have provided with promising results[5].

Most of the experimental data reported in literature refer to results from small scale batch experiments (shaking flasks), of single, binary or multi-element synthetic metal ion solutions, on different biomass types, correlating the metal uptake capacity through an isotherm equation. In few cases, small-scale laboratory systems have been set up, (usually fixed bed columns), with attempt to correlate their performance to the design and scale up of an integrated metal decontamination process. The reasons for the limited mathematical modelling work on biosorptive systems reported in the literature are primary the following:

- The studies on biosorption have focused for many years mostly on the microbiological aspects of process, whereas the engineering view for developing an applied novel waste water treatment technology has been overlooked.

- Biosorption process itself is complex. Predictive modelling of biosorption in micro scale molecular level, with sorption or ion exchange reactions between metal ions and cell components is difficult. The mechanism of such interactions is not well elucidated, the microbial surface is highly heterogeneous, and the metal sequestering also depends on several other parameters such as metal ion chemistry, the nature of the microbial biomass, solution pH, anion or cation co-ion effects, presence of organic molecules etc. In addition, other phenomena such as bioprecipitation, bioreduction/biooxidation, may also contribute significantly to the metal uptake mechanism when metabolically active biomass is used. The complex matrix of the waste water in industrial applications, may also interfere significantly in the sequestering process, in a way that can not be predicted beforehand and almost inevitably by reducing the metal uptake capacity as compared to that estimated from pure, single element, chemically defined metal solutions.

- Modelling of adsorption as a separation process is complex itself. In many cases empirical models are used for design and the process scale up is based on experimental data obtained from similarly configured lab scale units, (usually packed beds). Complete models and detailed design equations for the most common process configurations such as batch fixed beds and fluidised beds are complicated due to the fact that the sorption process is a non steady state operation concerning the solute concentration on the solid and liquid phase. Mathematical modelling of such systems although is more often based on simple material balances and mass transfer phenomena from the liquid to the solid phase, usually results in a system of non-steady state partial differential equations, in the general case of non linear form. These equations, can only be solved by applying advanced numerical analysis methods. Therefore, as the solution is not provided in simple explicit form, (except for some simple cases), the manipulation of the model is difficult. In addition, the parameters involved in modelling continuous fixed bed columns, (axial dispersion coefficient, mass transfer coefficients in liquid and solid phase, kinetic constants, etc.), are difficult to be measured experimentally from batch or continuous systems. Therefore, reliable parameter values should either be calculated from generalised expressions, or be estimated by fitting the model equations to data obtained experimentally. This multivariable fitting procedure is mathematically cumbersome, many times arbitrary, and the values obtained are not always easily scaleable to large systems.

Development of reliable process design and modelling tools for biosorption technology is important for simulating the results obtained from currently running systems. In addition, model sensitivity analysis will reveal the significance of each process parameter, leading to optimisation of process efficiency. The scale up criteria and wastewater treatment costs should also be considered for any industrial scale applications. Principles from chemical reactor design, separation processes, applied environmental biotechnology, and numerical analysis methods should be combined for successful modelling.

So far, biosorption modelling in batch mode systems has been studied thoroughly for the case of immobilised biomass particles of *Rhizopus arrhizus*,[6,7,8,9]. The outcome of the model is a kinetic profile curve (predicted or fitted to experimental data), which converges to the sorption equilibrium values. Although, batch systems are not likely to be used in industrial applications, the manipulation of batch models reveals the model mathematical complexity, even for the simplest configuration. Nevertheless, it provides the basis for tackling more complex systems.

A summary of possible contacting configurations for biosorptive systems has been presented elsewhere[10]. Continuous contacting configurations, such as fixed bed columns (canisters), similar in operation to activated carbon and ion exchange resins, are more likely to be used in industrial applications. Metabolically active or inactive microbial biomass can be immobilised by different techniques (encapsulation, biofilm formation, etc.) to formulate biosorbent particles, similar in the behaviour to ion exchange resins. The performance of such columns is described through the concept of the breakthrough curve, which is the concentration profile of the sorbate at the column exit as a function of time. The time of breakthrough appearance and the shape of the breakthrough curve are very important characteristics for determining the operation of a sorption column. Mathematical models for fixed beds, originate mainly from activated carbon sorption processes and ion exchange or chromatographic applications. Although, models for ordinary sorption columns have been studied extensively[11,12,13,14], modelling of continuous biosorption contacting configurations has not been examined in depth. Only recently, experimental breakthrough data for Cd biosorption have been treated theoretically[15].

The present paper aims to contribute to the mathematical modelling of biosorption process design, for a packed bed column configuration. The model examines the limited case described by the concept of rapid equilibrium. Under this concept, the model predicts the maximum anticipated "life duration" or best performance for a biosorption column used for sequestering a single solute from a solution. Although mass transfer limitations, both in liquid and solid phase, exist in most sorption processes, the developed model provides new information about the importance of key process operating parameters.

2. FIXED BED DESIGN

2.1. Principles

2.1.1. The breakthrough concept

In the simplest type of adsorption processes in which an adsorption column is used to remove a trace impurity from a process stream or waste water, the main requirement for rational design is an estimate of the dynamic or breakthrough capacity of the bed. In such systems the adsorbable impurity is strongly adsorbed with a favourable isotherm and the concentration profile therefore rapidly approaches constant-pattern form. The constant-pattern assumption provides the basis of a very simple design method, which permits reliable scale-up from small-scale laboratory experiments. Thus, a breakthrough curve and in particular the width of the sorption zone are important characteristics for describing the operation of a fixed bed biosorption column. A typical sketch of a breakthrough curve is presented in Figure 1, (for downflow mode of operation). The width of the sorption zone is shown in grey colour. At the early stages of the operation the solute is retained at the top of the column, and the most of the sorbent material is unsaturated, (white area). As the operation continues, a sorption zone is developed (grey area). This reaction zone moves downward, while saturated material is left behind (black area). At the beginning of breakthrough, at which the lower end of the sorption zone touches the bottom of the column, the total volume treated is represented by V_b. From this point afterward, the concentration of the solute at the exit increases constantly. The operation of the column can be continued until the exhaustion point (t_e, or V_e), at which the sorption zone has reached the exit of the column.

2.1.2. Length of unused bed (LUB)

The length of the unused bed (LUB) is defined by the equation,[16] :

$$LUB = \left(1 - \frac{t_1}{t_2}\right) L \tag{1}$$

where L is the length of the fixed bed and the time parameters t_1 and t_2 are defined by the integrals, (see also figure 1):

$$t_1 = \int_0^{t_b} \left(1 - \frac{C}{C_0}\right) dt = dark\ area \tag{2}$$

$$t_2 = \int_0^{\infty} \left(1 - \frac{C}{C_0}\right) dt = dark + light\ area \tag{3}$$

The time t_b is the breakthrough time at which the effluent concentration reaches its maximum permissible discharge level. This level is usually considered to be between 1 to 5% of the feed solute concentration. In some cases time parameters t_1 and t_2 are substituted by t_b, t_e respectively.

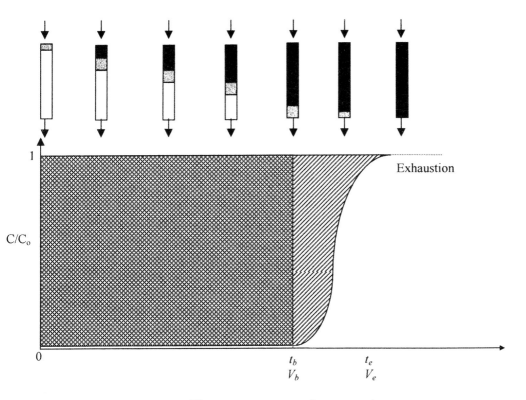

$$C/C_o$$

Time or waste water volume treated

Figure 1. A typical breakthrough curve showing the movement of the adsorption zone, breakthrough and exhaustion time.

3. DESIGN BASED ON EXPERIMENTAL DATA

The length of sorption zone can be calculated from the experimental breakthrough curve, for the system being studied, by applying material balance equations and implementing the maximum solute uptake capacity[17]. It should be noted that the total mass of pollutant entered the column from the beginning to the completion of breakthrough is $(V_b-V_e)C_o$. At the same period, the solute that escaped removal can be calculated by the integral below the breakthrough curve from the points V_b to V_e. Thus the solute removed (M_r) in the reaction zone is :

$$M_r = (V_e - V_b)C_0 - \int_{V_b}^{V_e} CdV \qquad (4)$$

M_r is also given by the product $A_s \, \delta \, \rho_p \, q_o$, where A_s is the superficial area of the bed, ρ_p the fixed density of the bed, q_o the maximum solute uptake capacity, and δ the width of reaction zone. Combining with equation 4 and solving for δ gives:

$$\delta = \frac{(V_e - V_b)C_0 - \int_{V_b}^{V_e} CdV}{A_s \rho_p q_o} \tag{5}$$

The integral in equation 5, should be calculated by numerical approximations from experimental breakthrough data.

Empirical models[18] applied for the results obtained from ion exchange columns, resulted the equation 6 which describes the breakthrough curve in terms of t_b, t_m and β,

$$\frac{C}{C_0} = 1 - \exp\left[-\left(\frac{t-t_b}{t_m}\right)^\beta\right] \tag{6}$$

where t_b is the experimental breakthrough time, t_m is the time when $C/C_o = 0.632$, and β is a curve shape factor. The parameter β can be estimated by fitting a set of experimental data to the linear logarithmic form of the equation. Application of the equation 6 requires experimental breakthrough data, which in many cases might not be available. In addition, such mathematical expression does not predict the performance of the column when various operating parameters are changed, thus no information for the scale up of the process are provided. Further details on designing of fixed bed columns are given elsewhere[19,20].

Dimensionless numbers and correlation equations applied for the design of fixed bed columns

Two different velocities can be defined for the liquid flowing into the column. The superficial velocity based on the total volume of the column (or based on the empty volume), and the interstitial velocity based on the void volume of the packed bed, defined by the equations:

$$u_{superficial} = \frac{F}{A} \tag{7} \qquad\qquad u_{interstitial} = \frac{F}{A\,\varepsilon} \tag{8}$$

where F is the volumetric flow rate (cm^3/min), A is the cross sectional area of the column (cm^2) and ε is the porosity of the bed. The velocity mentioned in the ADR equation in the following paragraphs is the interstitial velocity.

Three forms of particle Reynolds numbers have been used in correlation equations for a liquid flow through a fixed bed[21]:

$$N_{Re} = \frac{D_p G}{\mu} \tag{9} \qquad N_{R'e} = \frac{D_p G}{\mu\,\varepsilon} \tag{10} \qquad N_{R''e} = \frac{D_p G}{\mu\,(1-\varepsilon)} \tag{11}$$

where G is the superficial mass flow rate for the solution, (mass per time per surface area, e.g gr/min cm^2), ε is the bed porosity, and μ is the absolute viscosity of the flowing liquid, (gr/cm min) and D_p, the particle diameter (cm).

The axial dispersion coefficient for the liquid flowing in the fixed column can be calculated by the correlation coefficient [22]:

$$\frac{D_h \; \rho_l}{\mu} = \frac{N_{Re}}{0.20 \; + \; 0.011 \; N_{Re}^{0.48}} \qquad 10^{-3} < N_{Re} < 10^3 \tag{12}$$

Finally, the pressure drop in the column can be calculated by the equation [16]:

$$f = (\frac{2 \, R_p}{L}) \, \frac{\Delta P}{\rho_f \; (\varepsilon u \,)^2} \tag{13}$$

where εu is the superficial fluid velocity (cm/min), L is the bed length (cm), and ΔP is the bed pressure drop (gr/cm min^2).

The f coefficient can be calculated by the Ergun equation:

$$f = (\frac{1 - \varepsilon}{\varepsilon^3}) \left(\frac{150(1 - \varepsilon)}{N_{Re}} + 1.75 \right) \tag{14}$$

In equations 12 and 14 the Reynolds number used is the one calculated from equation 9.

4. MATHEMATICAL MODELLING OF FIXED BEDS

A design approach based on a mathematical model preliminary relies on the solution of the material balance equations for the solute being adsorbed by the solid phase. What is required, is to determine the expected performance of a fixed column through the prediction of the breakthrough curve. For the case of a biosorption column the sorbent material is composed of biological material. The biosorbent particles can be native or immobilised biomass particles similar to the particles developed in previous works[9,23].

Models for sorption phenomena, and material balance equations have been summarised and reported in an excellent review paper[24]. The movement of a solute in a column type reactor is generally described by the three dimensional generalised advection-dispersion-reaction (ADR) equation[24,25].

$$\frac{\partial C}{\partial t} = div \, (D_h \; grad \; C) - v \; grad \; C + \left(\frac{\partial C}{\partial t} \right)_r + S(C) \tag{15}$$

The subscript r denotes reaction that affects the solute concentration. The reaction under consideration for the case of biosorption is solute retention by the biomass solids, a heterogeneous-phase, mass transfer process.

The above equation is a generalised three dimensional non steady state material balance equation for a solute being transported by diffusion, bulk flow, and depleted from the solution by a sorption reaction. For generality, a solute production term is also included. The solution of this equation describes the performance and process dynamics of a sorption column.

Simplification of the general ADR equation to one spatial axial dimension (z) for a single solute, subject only to sorption without any other fluid-phase reaction or source term, yields the following equation.

$$\frac{\partial C}{\partial t} = D_h \frac{\partial^2 C}{\partial z^2} - u_z \frac{\partial C}{\partial z} - \frac{\rho_s(1-\varepsilon)}{\varepsilon}\left(\frac{\partial q}{\partial t}\right) \qquad (16)$$

The most important term of this material balance equation, is the rate of the solute uptake by the biosorbent, $\partial q/\partial t$. Various methods can be used to characterise this term. All methods consist of describing two general components:

(i) the aqueous-solid equilibrium phase distribution relationship, (i.e. the sorption isotherms), and

(ii) the rate at which this equilibrium is approached.

4.1. Local equilibrium models

The most simplistic approach to model sorption phenomena, is to assume that the time scale associated with the microscopic processes of mass transfer to the sorption site and subsequent sorption is very much smaller than that associated with the macroscopic processes of fluid transport. This effectively assumes that the equilibrium prevails locally, and is approached rapidly. Putting that in other words, this concept implies, that for the case of a fixed column, the rate of change of the sorbed phase concentration , q, at any point z, is instantaneously reflected by the rate of change of the solution phase concentration, C, at that point. Therefore, it is assumed that mass transfer limitation in the liquid and solid phase are negligible and the sorption reaction is rapid. This approach yields the so called local equilibrium models, (LEM), which are presented in table 1. The general equation (16a), which describes the local equilibrium model, is a special case of equation (16).

The most simple local equilibrium model, assumes that the equilibrium distribution between the solid phase and the fluid phase is linear. In this case a linear isotherm equation is applied and the partial derivative $\partial q/\partial C$ can be substituted by the partitioning coefficient K_p (The partial derivative is equal to the derivative dq/dC as q is only a function of C in this model).

$$q = K_p C \qquad (17) \qquad\qquad\qquad \frac{dq}{dC} = K_P \qquad (18)$$

The linear local equilibrium model (LLEM), shown in equation 23, is a very popular modelling method for column configurations, due in part to the simplicity of its solution, which can be obtained analytically, and differs simply by a constant from any solution to the conservative form of the advection-dispersion equation. Although the LLEM version of ADR equation has been widely employed for describing solute retardation by sorption in subsurface systems, it has become increasingly apparent that this model frequently fails to provide adequate representation of the effect of sorption processes on solute transport. Inclusion of more sophisticated non-linear equilibrium models, such as the Freundlich or Langmuir isotherms, often provides better representation of sorption phenomena, especially over extended equilibrium concentration ranges[24].

If the Freundlich isotherm is used to describe the sorption term, then a local equilibrium model is obtained by substituting the term $\partial q / \partial C$ by the equation (20):

$$q = K_F C^n \qquad (19)$$

$$\frac{dq}{dC} = K_F \; n \; C^{n-1} \qquad (20)$$

Similarly, for the case of Langmuir isotherm the term $\partial q / \partial C$ can be substituted by the derivative of Langmuir isotherm:

$$q = Q^o \frac{C}{1 + bC} \qquad (21)$$

$$\frac{dq}{dC} = \frac{Q^o b}{(1 + bC)^2} \qquad (22)$$

A nonlinear equilibrium isotherm generally precludes analytical solution of equation 16a, thereby complicating model solution and application. In addition in many models the term of the axial dispersion coefficient is neglected before proceeding to the solution. A detailed review of the analytical solutions of the ADR equation for many special cases, is available elsewhere[16].

Table 1
Local equilibrium models for fixed column reactor

LOCAL EQUILIBRIUM MODELS	
FIXED COLUMN REACTOR CONFIGURATION	
$\dfrac{\partial C}{\partial t} = D_h \dfrac{\partial^2 C}{\partial z^2} - u_z \dfrac{\partial C}{\partial z} - \dfrac{\rho_s (1 - \varepsilon)}{\varepsilon} \dfrac{\partial q}{\partial C} \dfrac{\partial C}{\partial t}$	(16a)
LINEAR ISOTHERM $\qquad$ $\dfrac{\partial C}{\partial t} = D_h \dfrac{\partial^2 C}{\partial z^2} - u_z \dfrac{\partial C}{\partial z} - \dfrac{\rho_s (1 - \varepsilon)}{\varepsilon} K_p \dfrac{\partial C}{\partial t}$	(23)
FREUNDLICH ISOTHERM $\qquad$ $\dfrac{\partial C}{\partial t} = D_h \dfrac{\partial^2 C}{\partial z^2} - u_z \dfrac{\partial C}{\partial z} - \dfrac{\rho_s (1 - \varepsilon)}{\varepsilon} K_F \; n \; C^{n-1} \dfrac{\partial C}{\partial t}$	(24)
LANGMUIR ISOTHERM $\qquad$ $\dfrac{\partial C}{\partial t} = D_h \dfrac{\partial^2 C}{\partial z^2} - u_z \dfrac{\partial C}{\partial z} - \dfrac{\rho_s (1 - \varepsilon)}{\varepsilon} \dfrac{Q^o b}{(1 + bC)^2} \dfrac{\partial C}{\partial t}$	(25)

The model developed and solved by the authors, has the advantage that non linear isotherms can be incorporated in the model and there is no need to eliminate the term of axial dispersion coefficient for the solution of the model[26].

4.2.   Rate Models

The assumption that the equilibrium is attained instantaneously, (local equilibrium models), is not always valid. There are cases, such as the sorption of hydrophobic organic compounds on soil[27,28], and the sorption of Uranium[9] by *Rhizopus   Arrhizus* where the equilibrium is attained after several hours.

For all these cases, the term $\partial q/\partial t$ should be approximated by rate models. The most significant rate models have been summarised elsewhere[24,27]. Two of the most popular used are the dual resistance diffusion model and pore diffusion model[29].

5. SOLUTION OF THE ADR EQUATION FOR THE CASE OF LOCAL EQUILIBRIUM

The ADR equation is a partial differential equation of parabolic type. The implementation of a non-linear sorption isotherm precludes its analytical solution. Thus numerical methods should be used.

Problem description

The effluent solute concentration of a fixed bed biosorption column, as a function of time and the other operating parameters, is to be known, (breakthrough curve). At the beginning of the column operation the sorbent material of the column is assumed to be fresh or completely regenerated, for the entire column length. At time *t=0* waste water containing a single solute is pumped introduced in a downward mode through the column as shown in figure 1.

Model assumptions

The main model assumptions are the following:

(i) The concept of rapid equilibrium has been applied in all cases.

(ii) Operation is isothermal, which is valid for the cases of sorption from dilute solutions. Wastewater treatment by biosorption is applicable mainly to large volume low concentration solutions. Thus, this assumption is almost always valid. Indirectly, isothermal operation, implies that the sorption parameters, (coefficients of sorption equilibrium equation), are constant along the column length and duration of column service life.

(iii) Treatment of single solute wastewater. The model can be extended to describe multi-solute sorption phenomena, assuming that the appropriate sorption equilibrium equations are provided and competition effects among the species can be expressed analytically through a multi-solute isotherm equation.

(iv) Uniform bed packing. Uniform bed porosity throughout the column length. Hydraulic defects such as short-circuiting or channelling effects are not considered.

(v) Solute dispersion to the radial direction is negligible,

(vi) No radial velocity gradient. Uniform u_z in any cross section of the bed and along the column length.

Equations

The one dimensional local equilibrium cases of ADR equation have been used (equations 23 or 24 or 25) with the appropriate initial and boundary conditions.

Initial and boundary conditions

Solving the ADR equation (equation 16a), requires the definition of one initial and two boundary conditions which can be formulated as follows[28,30,31]:

Initial condition	$t \leq 0$	$C_i = 0$	$0 < z < L$	(26)
1st boundary condition	$t > 0$	$C = C_{initial}$	$z = 0$	(27)
2nd boundary condition	$t > 0$	$\dfrac{\partial C}{\partial z} = 0$	$z = L$	(28)

The second boundary condition denotes that at lengths longer than L, solute transfer does not takes place because no sorbent material is present.

Solution method

The authors have developed a solution method based on the application of the finite differences method for the solution of the material balance equation[26]. The methods of Euler or Crank-Nicolson have been used for the discretisation of the space and time derivatives[32].

It is important to notice that at any point of the column, the concentration of the solute in the liquid phase is both a function of time and the position of it along the column length, because the system is at non steady state concerning the liquid and solid phase solute concentrations.

6. RESULTS AND DISCUSSION

The column dynamic response for the case of Cu biosorption by immobilised *Rhizopus arrhizus* assuming rapid equilibrium has been studied. The Freundlich isotherm has been used to describe Cu equilibrium distribution between the liquid and solid phase. The operating conditions summarised in Table 2 have been selected as a "typical" data set, around which model sensitivity analysis has been performed.

The column length and the internal diameter, are representative of a small scale laboratory column. Longer columns could also be selected. The bed porosity depends on the packaging arrangement and on the radius of the biosorbent particles used. Biosorbent particles are likely to have a density value slightly higher than that of water, (as described[33] by the BIO-CLAIM[TM] system), whereas the immobilized particles developed by the research group of Tsezos have a density value of about 0,773 gr/cm^3. Wastewater, at a flow rate of 1 lt/min, loaded with 100 ppm Cu is assumed to be pumped downflow through the column. The mathematical expression to present the sorption isotherm is the Freundlich equation with the appropriate constants[34]. The axial dispersion coefficient D_h has been estimated by the correlation equation 12 for the conditions described above, while pressure drop ΔP and f factor, have been calculated from equations 13 and 14 respectively.

Preliminary runs of the computer program had the aim to identify the conditions that would allow increased accuracy and solution stability. Although, no significant difference on the solution accuracy has been observed, between the Euler and Crank-Nicolson methods, the later has been used in all cases.

A broad range of sensitivity analysis has been performed by changing each time one of the values of the operating parameter or sorption characteristics of a fixed column and monitoring the breakthrough response of the system. The following tables and figures summarise the results obtained from the sensitivity analysis of the model. Length of unused bed (LUB) has been calculated from equation 1 after numerical calculation of t_1 and t_2 from equations 2 and 3 respectively.

440

Table 2
Operational data of a supposed laboratory scale column (base point).

Column length (L)	20 cm
Column internal diameter (ID)	2 cm
Column length Internal diameter (L/ID)	10/1
Bed porosity (ε)	0.25
Biosorbent particles density (ρ_s)	1.1 gr /cm^3
Fluid flow rate	1000 cm^3/ min
Initial solute concentration (C_o)	100 mg/l
Isotherm type	Freundlich
Metal Ion	Cu^{2+}
Freundlich exponent (n)	0.782
Freundlich coefficient (K_F)	0.991 mg/gr dry biomass
Immobilised Particles Diameter(D_p)	0.1 cm
Flow density (ρ_l)	1000 gr/l
Absolute viscosity (μ)	0.6 gr/min cm
Interstitial velocity	1273 cm/min
Superficial mass flow (G)	31830 gr/min cm^2
$N_{Re}=D_pG/\mu$	5300
D_h (estimated from eq. 12)	3 10^{-3} cm^2/min

The parameter, which affects most significantly the breakthrough of the column, has been shown to be the exponent of the Freundlich equation. This value is characteristic of the sorption intensity. At values $<<1$, a highly favourable equilibrium pattern is observed, which means a preference of the solute to be sorbed at the solid phase at concentrations multiple of that of the liquid phase. Thus, the length of the sorption zone is narrow. At values of the exponent near but less than 1, the sorption zone is wider as the sorption intensity is weaker. For constant Freundlich coefficient value, the increase at the value of the Freundlich exponent, means indirectly an increase of the available sorption capacity for low solute concentrations. Thus, later breakthrough and exhaustion times are observed. A non-uniform behavior is observed at the exponent value of 0.5, (Table 3 and Figure 2).

The coefficient of the Freundlich equation is the second important parameter. It indicates the sorption capacity of the biosorbent material. The higher this value (for constant exponent), the higher is the capacity of the biosorbent material. Thus, as it was expected increased values of Freundlich's coefficient result in longer breakthrough and exhaustion times, without affecting the width of the sorption zone, (Table 3 and Figure 3).

The effect of the axial dispersion coefficient is well known from the studies of its effect on the reaction efficiency of plug flow chemical reactors. From the engineering point of view, a small axial dispersion coefficient is required for higher column performance. The ideal limit would be plug flow pattern. High dispersion coefficient values result in flattening of the breakthrough curve as shown in figure 4. In the simulated results, as the axial dispersion coefficient decreases, the breakthrough curves becomes steeper and the length of unused bed shorter, (Table 4 and Figure 4).

Table 3
Effect of isotherm parameters on breakthrough curves

Parameter	Breakthrough time (min)	Exhaustion time (min)	% Length of unused bed	$(1-t_b/t_e)*100\%$
Exponent of Freundlich equation				
0.1-0.3		Oscillation		
0.4	8.0	8.1	0.2	1.2
0.5	7.7	7.8	0.5	1.3
0.6	9.1	9.3	0.9	2.2
0.9	30.7	35.7	5.3	14.0
Coefficient of Freundlich equation				
0.01		Oscillation		
0.50	9.5	10.1	2.1	5.9
2.00	37.3	39.9	2.2	6.5
5.00	92.9	99.6	2.2	6.7

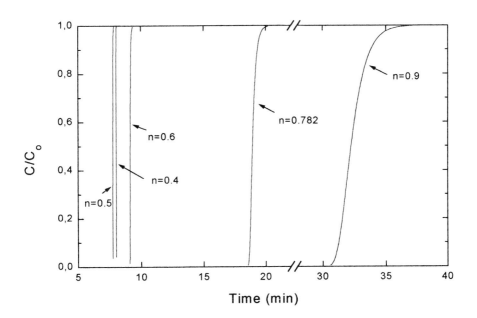

Figure 2. Effect of Freundlich exponent on breakthrough curves.

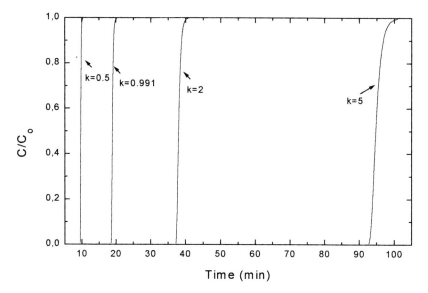

Figure 3. Effect of Freundlich coefficient on breakthrough curves.

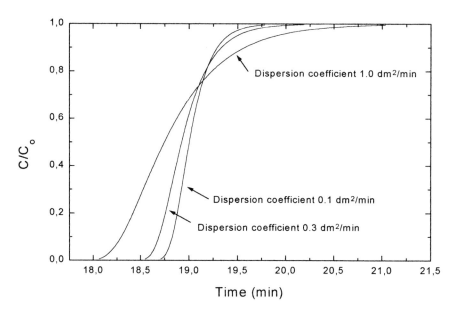

Figure 4. Effect of axial dispersion coefficient on the breakthrough curve.

Table 4
Effect of various parameters on the operation of the column.
Results obtained by ADR model for rapid equilibrium

	Breakthrough time (min)	Exhaustion time (min)	% Length of unused bed	$(1-t_b/t_e)*100\%$
Base point (set values of Table 2)	18.6	19.9	2.2	6.5
Parameter variation				
Bed porosity (dimensionless)				
0.10	22.4	23.6	1.7	5.1
0.35	16.0	17.4	2.5	8.0
0.50	12.3	13.5	3.0	8.9
0.65	8.6	9.6	3.3	10.4
Axial dispersion coefficient (dm^2/min)				
0.1	18.7	19.7	1.6	5.1
1.0	18.1	20.7	4.0	12.6
Flow rate (lt/min)				
0.5	36.6	40.3	3.0	9.2
5	4.0	4.1	1.6	2.4
10	2.2	2.3	1.1	4.3
Solute concentration (ppm)				
50	21.6	23.1	2.1	6.5
200	16.0	17.1	3.0	6.4
500	13.1	14.0	4.4	6.4
Sorbent density (gr/cm^3)				
0.9	15.2	16.3	2.0	6.7
1.0	16.9	18.1	2.1	6.6
1.2	20.2	21.7	2.2	6.9
Column length (dm)				
1	9.3	10.2	3.0	8.8
5	46.4	48.9	1.7	5.1
Internal column diameter (dm)				
0.1	4.9	5.1	1.6	3.9
0.5	109.1	134.9	6.6	19.1

The change of bed void volume revealed that as bed porosity increases, the breakthrough and exhaustion time decrease, because less sorbent material is present in the bed, the void volume of the column is higher, therefore, the sorption front moves faster toward the exit of the column. This behaviour has been quantified in table 4. The length of unused bed also increases as the porosity of the bed increases.

The increase of the flow rate with constant solute concentration decreases the useful column operation time (for constant bed capacity). Thus, as the fluid flow rate increases the breakthrough and the exhaustion times decrease. Increase at the column flow rate reflects also to increase of the interstitial velocity of the fluid in the column. This might affect the axial dispersion coefficient, resulting possibly in the flattening of the breakthrough curve. In our work increase of flow rate has not been interrelated to simultaneous changes of the axial dispersion coefficient values.

For constant column sorption capacity it is expected that the saturation of the column will occur earlier as the solute concentration increases, and thus an earlier breakthrough curve is expected. This behaviour has been verified.

According to the simulated results obtained, the density of the biosorbent material is also significant to the breakthrough time. An increase on the biosorbent density results in an increase in breakthrough time due to the presence of more sorbent material in the column and thus the higher sorption capacity of the system.

It is obvious that the increase of column length, results to higher sorption capacity of the bed thus to a later breakthrough. A linear relationship between breakthrough time and column length has been observed. This observation has been verified experimentally in the biosorption literature[15].

A comparative way of presenting sensitivity analysis data is shown in Figure 5. The relative importance of the various parameters can be identified by plotting the relative change of the value of interest (i.e. breakthrough time) against the relative change in the value of the parameter under investigation. The percentage relative changes have been calculated from the formula:

$$\% \frac{dx}{x} = \frac{x_{new} - x_{base\ point}}{x_{base\ point}} \, 100\% \tag{29}$$

The origin of the axis correspond to the base point selected previously (see Table 2). The breakthrough time at the base point is 18.56 min. The x axis shows, the percentage deviation from the base point of the variable under consideration. In y axis the percentage change of the breakthrough time from the base point breakthrough time is presented. From Figure 5, it is clear that the most important operating parameter is the sorption intensity and the sorbent uptake capacity, as expressed by the Freundlich exponent and coefficient respectively. The effect of column length is the same as of the Langmuir coefficient, which is in agreement with the common sense that doubling the column length is equivalent to doubling the maximum uptake capacity. Presenting sensitivity analysis data in such normalising way reveals the relative significance of the parameters in the model. Nevertheless, the plot may be misleading because the easiness of changing one parameter by 10% might not be as easy as for the other parameters. For example 10% change of the flow rate may represent a flow disturbance, whereas 10% change of the Freundlich equation coefficient might require the development of a new sorbent material with higher uptake capacity.

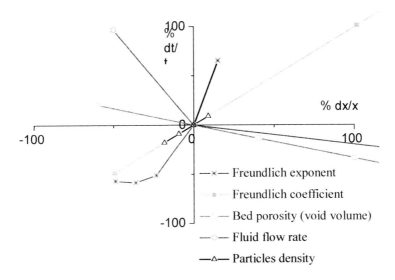

Figure 5. Sensitivity analysis plot showing the relative importance of various parameters on the breakthrough time of a fixed bed column.

7. CONCLUSIONS

A modelling approach to simulate the operation of a fixed bed biosorption column has been presented. Although the model is restricted for the case described by the concept of local equilibrium, it simulates the optimum performance expected from a sorption column under different operating conditions. The breakthrough curves predicted are steep because the overall sorption rate in the model is considered infinite. This is actually an ideal situation for maximum exploitation of the sorption capacity of a column. The importance of the sorbent high uptake capacity has been demonstrated. In addition, the flattening effect of the axial despersion on the breakthrough curves has been also observed. Experimental validation of the results predicted from the model is difficult to be obtained, as in most of the cases mass transfer resistances in solid and/or liquid phase determine the overall sorption rate. In cases where the sorbent particles are fine, highly porous and the flow regime is turbulent, the operation of a column is believed to approach that predicted from the model, without in any case exceeding that. Therefore, the assumption of rapid equilibrium provides one of the two extremes for the operation of sorption columns. The other extreme condition would be flow without sorption reaction.

The presented model, provides the background for developing more complex models. The core ADR equation remains the same, whereas it should be completed with the equations describing liquid and solid phase mass transfer phenomena. Sorption reaction kinetic coefficients should also be used if the sorption time scale is comparable to the diffusion time scale. Preliminary results from extension of the model to implement mass transfer resistances in the liquid and solid phase have revealed much more earlier breakthrough curves with more flattening pattern concentration profiles.

NOMENCLATURE

A_s	Superficial area of the bed (cm^2)
b	Langmuir isotherm parameter (cm^3/mg)
C,	Solution phase solute concentration (mg of solute / cm^3 of solution)
C_o, $C_{initial}$	Initial solute concentration at the feed (mg/l)
D_h	Hydrodynamic dispersion coefficient (cm^2/min),
ε	Bed porosity, void volume per unit total volume (dimensionless)
F	Volumetric flow rate (l/min)
f	f coefficient
G	Superficial mass flow rate (gr/cm^2 min)
K_p	Partitioning coefficient
L	Column length (cm)
LUB	Length of unused bed (cm)
M_r	Solute removed in the reaction zone
N_{Re}, $N_{Re'}$, $N_{Re''}$	Reynolds number (dimensionless)
q	Volume averaged sorbed-phase solute mass per solid-phase mass (mg solute/gr sorbent).
q_o, Q^o	Maximum solute uptake capacity of the solid phase
r	subscript denotes reaction that affects the solute concentration
R_p	Biosorbent particle radius (cm)
ρ_s	Biosorbent material bulk density (gr/cm^3),
$S(C)$	Fluid-phase solute source term (mg of solute / cm^3 min)
t	Time (min)
t_1	Time parameter until the breakthrough point
t_2	Time parameter until the exhaustion point
t_b	Breakthrough time (min)
t_e	Exhaustion time (min)
t_m	Time when $C/C_o = 0.632$, (min)
u_z	One dimensional fluid phase interstitial velocity (cm/min),
v	Pore velocity vector (cm/min)
z	Space direction (cm)
β	Curve shape factor (dimensionless)
δ	Width of reaction zone (cm)
ΔP	Column pressure drop (gr/cm min^2)
μ	Absolute viscosity of the wastewater (gr/cm min)
$\rho_p = \rho_s(1-\varepsilon)$	Fixed density of the bed (biosorbent mass per unit bed volume, gr /cm^3)
ρ_s	Biosorbent material density (g/cm^3)

REFERENCES

1. B. Volesky. Biosorption of Heavy Metals. CRC Press Boca Raton, 1990.
2. J. Wase and C. Forster. Biosorbents for Metal Ions. Taylor and Francis, 1997.
3. G.M. Gadd. Microbial control of pollution, Symposium 48, J.C. Fry, G.M. Gadd, R.A. Herbert, C.W. Jones and I.A. Watson-Craik(eds.), (1992) 59.
4. M. Tsezos and B. Volesky. Biotechnology and Bioengineering 24 (1982) 385.
5. L. Diels, H. Wouters, M. Esprit, F. Glombitza, L. Macaskie, T. Pumpel, and M. Tsezos. Pollutec 96 - Separative Techniques and Environment. Eurexpo-Lyon., (1996).
6. M. Tsezos. Microbial Mineral Recovery, H.L. Ehrlich and C.L. Brierley (eds.), McGraw-Hill, New York, 1990, 325.
7. M. Tsezos and A.A. Deutschmann. J. Chem. Tech. Biotechnol., 48 (1990) 29.
8. M. Tsezos and A.A. Deutschmann. J. Chem. Tech. Biotechol., 53 (1992) 1.
9. M. Tsezos, S.H. Noh and M.H.I. Baird. Biotechnology and Bioengineering, 32 (1988) 545.
10. M. Tsezos, E. Remoudaki and A. Hatzikioseyian. Fifth International Symposium on Environmental Issues and Waste Management in Energy and Mineral Production, SWEMP 98, Ankara, Turkey, 1998.
11. M.W. Balzli, A.L. Liapis and D.W.T. Rippin. Trans IChemE, 56 (1978) 145.
12. N.S. Raghavan and D.M. Ruthven. AIChE Journal, 29(6) (1983) 922.
13. W.J.Jr. Weber and J.C. Crittenden. Journal of Water Pollution Control Federation, 47(5) (1975) 924.
14. Q. Yu, and N.H.L. Wang. Computers Chem. Engng., 13(8) (1989) 915.
15. B. Volesky and I. Prasetyo. Biotechnology and Bioengineering, 43 (1994) 1010.
16. D.M. Ruthven. Principles of adsorption and adsorption processes. John Wiley & Son, 1984.
17. A.P. Sincero and G.A. Sincero. Environmental Engineering. A Design Approach. Prentice - Hall Inc., New Jersey, 1996.
18. M.S. Doulah. Ion Exchange for Industry, British Society of Chemical Industry. M. Streat (eds.), Elis Horwood Publishers, Chichester U.K., (1988).
19. W.L. McCabe, J.C. Smith and P. Harriot. Unit operations of chemical engineering. McGraw Hill. Fifth edition, 1993.
20. R.T. Yang. Gas separation by adsorption processes. Butterworths series in chemical engineering, 1987.
21. P.N. Dwivedi and S.N. Upadhyay. Ind. Eng. Chem., Process Des., 16 (1977) 157.
22. S.F. Chung and C.Y. Wen. AIChE Journal, 14 (1968) 857.
23. M. Tsezos, M.H.I. Baird and L.W. Swemilt. Hydrometallurgy, 17 (1987) 357.
24. W.J.Jr. Weber, P.M. McGinley and L.E. Katz. Wat. Res., 25 (1991) 499.
25. S.V. Patankar. Numerical heat transfer and fluid flow. McGraw Hill Inc, 1980.
26. A. Hatzikioseyian. Design and Economic Analysis of a Biosorption Process for the Removal of Low Level Metal Ion Contaminants from WasteWater. M.Sc. Dissertation. University of Manchester Institute of Science and Technology, UMIST, United Kingdom, (1992).
27. W.J.Jr. Weber and C.T. Miller. Wat. Res., 22 (1988) 457.
28. W.J.Jr. Weber and C.T. Miller. Wat. Res., 22 (1988) 465.
29. W.T. Weber and R.K. Chakravorti. AIChE Journal, 20 (1974) 228.
30. S.H. Lin. J. Chem. Tech. Biotechnol., 51 (1991) 473.
31. S. Veeraraghavan, L.T. Fan and A.P. Mathews. Chemical Engineering Science, 44 (1989) 2333.

32. S. Nakamura. Applied numerical methods with software. Prentice-Hall International Editions, 1991.

33. J.A. Brierley. Biosorption of Heavy Metals, B. Volesky (eds.), Boca Raton: CRC Press , 1990, 305.

34. R. Ileri. An experimental and theoretical study of the biosorption of copper ions by immobilised dead *Rhizopus arrhizus*. Ph.D. thesis. University of Manchester Institute of Science and Technology, UMIST, United Kingdom, (1992).

Mechanism of palladium biosorption by microbial biomass. The effects of metal ionic speciation and solution co-ions

E. Remoudaki, M. Tsezos, A. Hatzikioseyian and V. Karakoussis

National Technical University of Athens, Department of Mining and Metallurgical Engineering, Environmental Engineering Laboratory, Heroon Polytechniou 9, 157 80 Zografou, Athens Greece

Biosorptive palladium uptake capacities have been quantitatively evaluated through a systematic experimental approach for six selected strains of microbial biomass. Soluble ionic species are of interest for interaction with microbial biomass to quantitatively assess palladium removal by biosorption mechanisms. The ionic species $Pd(NH_3)_4^{2+}$ and $PdCl_4^{2-}$ are the main soluble palladium ions present in the process and/or waste waters. Two sets of biosorption experiments have been conducted corresponding to the above soluble palladium ionic complex species. The concentrations of other ions in the contact solutions have been chosen appropriately to define known initial ionic compositions. The obtained sorption uptake capacity values revealed that biosorption can be successfully used for palladium sequestering from aqueous solutions. The role of the solutions pH on the biomass characteristics is extensively discussed. Palladium species hydrolysis behavior, chemical coordination, stereochemical characteristics of the metal complexes and the knowledge of solution ionic composition have been proven necessary for the elucidation of the observed palladium biosorption uptake capacities.

1. INTRODUCTION

Palladium is a precious metal of the platinum metals group with important industrial applications[1]. Besides environmental concern regarding the presence of metals in water streams, recovery of palladium from process and/or industrial waste waters is of economic interest.

Sorption has been one of the processes on which much attention has been focused for metal sequestering from process and/or waste water streams primarily in response to increasing concern over their fate in the environment. Sorptive materials of high performance such as activated carbon are usually expensive to use and as such, is not always the viable option[2]. Because of this, attention has turned to the sorptive capabilities of biological materials. The use of microorganisms for the removal of metal ions from process and/or waste waters may represent a potential alternative to the existing technologies[3,4,5]. This is especially true in the cases of high volume, low concentration, complex solution matrix applications.

Sequestering of metals by microbial biomass, taking place independently of the cell metabolism, is termed biosorption. The detailed mechanism underlying biosorption processes are largely unknown. A variety of physical, chemical and biological mechanisms may be involved, including adsorption, precipitation, complexation and reduction of the targeted metals[6,7]. The elucidation of the biosorption mechanisms and the evaluation of the parameters influencing the metal uptake capacities are necessary for successful applications of biosorptive processes. Significant efforts for the elucidation of the biosorption mechanisms have been reported in the literature[4,8,9].

The present paper reports experimental results on biosorption equilibrium, reversibility and overall kinetics for two different soluble species of palladium : $Pd(NH_3)_4^{2+}$ and $PdCl_4^{2-}$. These are the main soluble palladium ions present in process and/or waste waters. Six microbial species have been selected for testing the biosorption of both species. The present effort to throw some light to palladium biosorption mechanism was based on (1) simultaneous considerations of palladium species hydrolysis behavior, chemical coordination and stereochemical characteristics of the above metal complexes and (2) biomass surface charge behavior as a function of pH.

2. MATERIALS AND METHODS

2.1. Microbial strains

The strains tested for biosorption in the present work have been collected after a systematic microbiological work for characterization and selection of a highly sorbing species from a large number of different heavy metal resistant microbial strains from several sites around the world, contaminated by industrial or mining activities [10,11,12]. Table 1 summarizes the microbial species and their origin. Detailed work on the genetic characterization, the culturing conditions, biochemical (i.e. metal binding proteins, secondary metabolites) and cellular parameters (i.e. cell charge, functional groups, exopolymers production) have been reported elsewhere [10,11,12].

Table 1
Microbial strains examined and their origin.

Microbial strain	Origin
Alcaligenes eutrophus CH34	Sediment of a zinc decontamination tank, Belgium
Alcaligenes eutrophus ER121	Zinc contaminated soil, Belgium
Pseudomonas mendocina AS302	Copper mine in Shituru, Zair
Pseudomonas stutzeri EM77	Rhodium Madeleine Mines, Canada
Arthrobacter sp. BP7/26	Sediment of the river Inn, Austria
Arthrobacter sp. BP7/15	Silver mine, Austria

Samples from each microbial species were received from collaborating Institutes [10,13,14,15]. Cultivation was carried out under well defined conditions [10,11,12] followed by freeze drying for the preparation of dry biomass which was used for the biosorption, desorption and kinetic experiments. Prior to use in experiments, each biomass was treated (washed and dried) to remove possible residual from the culture media used, according to the procedure already described elsewhere [13].

2.2. Biosorption isotherms

Two different ionic species of palladium were used for the preparation of the biosorption contact solutions : $Pd(NH_3)_4^{2+}$ and $PdCl_4^{2-}$. The contact solutions corresponding to the first species were prepared by dissolving $Pd(NH_3)_4Cl_2$ in deionized water. The initial pH of these solutions was 8. The corresponding biosorption experiments were carried out without any further pH adjustment. Solutions of $PdCl_4^{2-}$ (12.8 g Pd/l) in 0.25 N HCl were used for the preparation of the biosorption contact solutions corresponding to the second palladium species tested. The final pH of these solutions was adjusted to 3.

Biosorption equilibrium isotherms were determined by contacting palladium solutions of various concentrations with different quantities of biomass of the strains. The initial solution metal concentrations extended over the range of 10^0 to 10^3 mg/l. The reagents used were of analytical grade. The detailed experimental procedure for the experimental definition of the biosorption isotherms is described elsewhere [13,14]. The results of apparent (overall) kinetic experiments, presented in the following paragraph, for all the strains tested, have shown that a contact time of 24 h was sufficient for the attainment of equilibrium. A 24 h contact time was selected for the execution of the biosorption experiments.

The concentrations of palladium were determined by flame atomic absorption spectrometry using a Perkin-Elmer 2100 Atomic Absorption Spectrophotometer. Mean relative errors of palladium solutions concentrations have been estimated to be less than 5%. The reported experimental values of the biomass biosorption metal uptake capacities (q) are the results of duplicate experiments and the reproducibility of the reported values is better than 10% of the stated value.

2.3. Desorption experiments

Desorption experiments were conducted immediately after the biosorption experiments. Each amount of biomass, separated following the biosorption experiment, was resuspended in a volume of 100 ml deionized water of appropriate pH (8 for $Pd(NH_3)_4^{2+}$ and 3 for $PdCl_4^{2-}$). The suspensions were agitated in the orbital shaker at 250 rpm for 24 h. After that period, the biomass was separated from the contact solutions by centrifugation and filtration. Palladium concentrations were determined in the solutions. The detailed procedure for desorption experiments and the formula used for mass balance calculations for carry over corrections have been presented elsewhere [13].

2.4. Apparent overall kinetic experiments

The apparent overall kinetic experiments were carried out by setting up biosorption experiments using contact flasks which contained identical biomass and metal ion quantities for each metal species and strain. Each biomass sample was kept in contact with the metal ion solution for different time intervals ranging between 1 and 24 h, in order to determine the overall rate of biosorption and contact time necessary for attainment of equilibrium.

3. RESULTS AND DISCUSSION

3.1. Biosorption isotherms

The biomass biosorption equilibrium uptake capacity was calculated for each sample according to the mass balance already presented in Tsezos et al[13]and using the Freundlich model equation.

3.1.1. Biosorption of Pd(NH₃)₄²⁺

3.1.1. Biosorption of $Pd(NH_3)_4^{2+}$

Figures 1 and 2 illustrate examples of the obtained linearized isotherms corresponding to $Pd(NH_3)_4^{2+}$ solutions for the strains CH34 and BP 7/15 respectively. The values of the Freundlich isotherm model parameters corresonding to each strain tested are summarized in Table 2.

Table 2
Experimentally determined Freundlich parameters corresponding to each strain tested for biosorption of $Pd(NH_3)_4^{2+}$ at pH 8.

Strain	k	1/n
BP 7/26	63	0.02
BP 7/15	55	0.07
CH 34	53	0.11
ER 121	43	0.13
AS 302	44	0.17

The high values of k reported in Table 2, for all strains, show good adsorbent capacity for $Pd(NH_3)_4^{2+}$ species.

In order to evaluate the possible binding mechanism(s) we will examine solution and metal characteristics such as pH, metal solubility, hydrolysis, stereochemical and coordination chatracteristics.

The initial solution pH, was measured immediately after the preparation of the palladium tetra-ammonium complex $Pd(NH_3)_4^{2+}$ solutions, as already mentioned above. The resulting pH of the solutions was at values about 8 and biosorption experiments were carried out without any further pH adjustment. The initial solution pH is an important parameter which may influence metal biosorption[3,16].

The concentration of H^+ ions, expressed by the pH, influences the cells surface charge. Determined biomass electrophoretic mobility values (EPM), for the strains tested are shown in Figure 3 which is reconstituted from results reported by Glombitza et al.[17] . From this figure we observe that EPM values present negative maxima towards neutral pH values. This means that the surface charge by the shear plane of the biomass particles is negative at pH = 8.

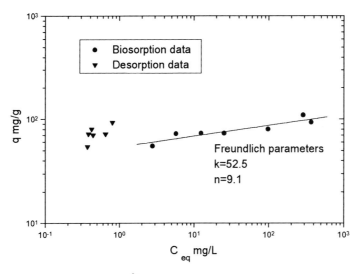

Figure 1. Linearised $Pd(NH_3)_4^{2+}$ biosorption isotherm, strain CH34. Desorption data are also reported.

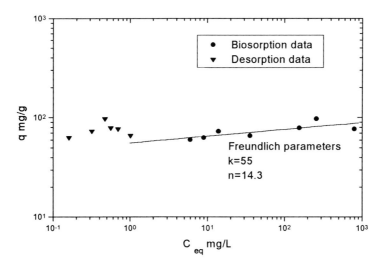

Figure 2. Linearised $Pd(NH_3)_4^{2+}$ biosorption isotherm, strain BP 7/15. Desorption data are also reported.

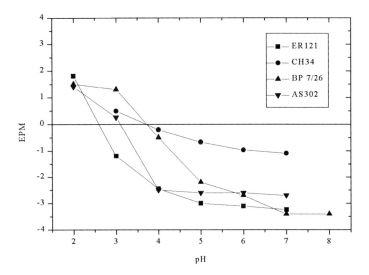

Figure 3. Electrophoretic mobility values (EPM) as a function of pH for the strains ER 121, CH 34, BP 7/26, AS 302.

Regarding the speciation of the metal at that pH value, palladium is present in solution as $Pd(NH_3)_4^{2+}$. This complex does not readily hydrolyze[1]. Palladium species in solution ($Pd(NH_3)_4^{2+}$ positively charged) and the surface charge by the shear plane of the biomass (negatively charged) suggest that the initial approach of the metal complex to the biomass is favored by the coulombic interaction.

The palladium tetra-ammonium complex ion $Pd(NH_3)_4^{2+}$ is a square planar complex [1,18]. Interaction of this complex ion with other ligands presented by the biomass is expressed through the substitution of ammonia by another ligand. The substitution reactions of this complex are characterized by the "trans effect". According to this effect, the position at which an incoming ligand might effect a substitution does not depend on the substituting or substituted ligand but on the nature of the ligand trans to that position. Ligands can be arranged in the following order indicating their relative abilities to labilize ligands trans to themselves[1]:

$$F^-, OH^-, H_2O, NH_3 < Cl^- < Br^- < I^-, -SCN^-, NO_2^-, C_6H_5^- < S=C(NH_2)_2, CH_3^- < H^-, PR_3, AsR_3 < CN^-, CO, C_2H_4$$

This order is related to the nucleophilic character of a ligand. The ligands exhibiting strong nucleophilic character are placed in the beginning of the classification and nucleophilicity is reduced from the left to the right. The ligands exerting the strongest trans effect are those whose bonding to a metal is thought to have most π acceptor character (e.g. PR_3, $-COO^-$, CO, C_2H_4) and which therefore remove most π electron density from the metal. This reduces the electron density at the coordination site directly opposite, i.e. trans, to that ligand and is there that further nucleophilic attack is most likely. This interpretation is not directly concerned with labilizing a particular ligand but rather with encouraging the attachment of a further ligand [1].

Regarding the attachment of palladium to biomass binding sites, it is probable that after the substitution of the first molecule of ammonia by an available biomass ligand, for example B, the second ligand will be placed in a position which depends on the relative classification of B in comparison to the ammonia NH_3 on the series reported above. The resulting structure can be trans or cis as illustrated by the following reactions (1) and (2) :

trans substitution of NH_3 by ligand B

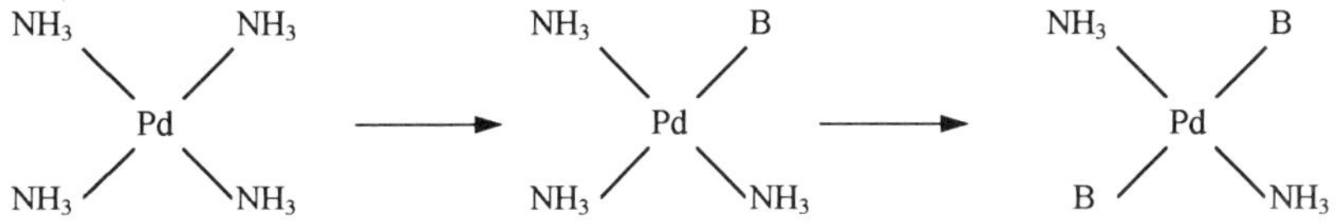

(1)

cis substitution of NH_3 by ligand B

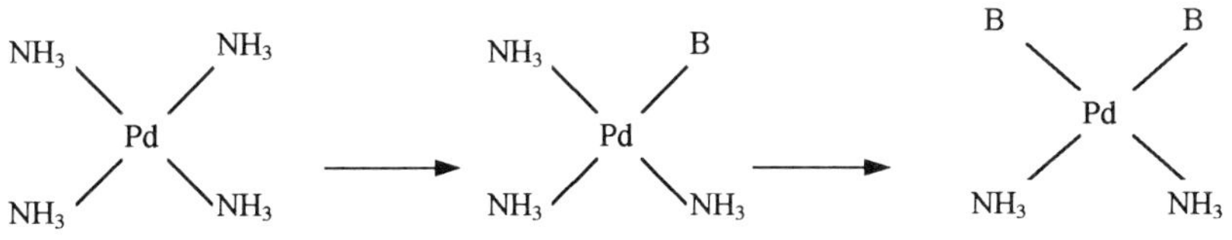

(2)

Regarding the nature of ligand B, indirect information is available from the elecrophoretic mobility measurements (EPM). According to Figure 3, the strains tested exhibit positive charge at acidic pH values. Above pH 3, the charge is negative and decreases almost linearly as pH increases to 5-6, above which the negative value stabilizes to an almost constant value up to pH 8.

James[19] showed that the response of a bacterial surface to changes in the pH of the suspension media is related to the chemical origin of its surface charges. Experimental data from different types of bacteria can be grouped in to four different types of EPM-pH responses [4,19]. On the basis of this classification[19], the strains tested in the present work belong to type (group) 4 which is characterized by the presence of mixed amino-carboxyl groups on the microbial surface such as those of *B. subtilis* [4]. Experimental evidence shows that carboxyl groups are present in excess over amino groups in these bacterial surfaces and thus carboxylic groups dominate the electrokinetic behavior of the cells. The above studies have demonstrated the important influence of solution pH on the surface charge properties of the bacterial surface, which as a consequence, can affect their metal sorption capacity. Beveridge and Murray [20], have shown that the most apparent site is the carboxyl group of the glutamic acid of the peptidoglycan in the cell walls of *Bacillus subtilis*. Amino groups contribution appears to be not significant in binding Au(III) as well as other metals [20].These authors suggest that even though the glycan strands are firmly rigid, their constituent cross-linking groups have a high degree of motional freedom, which would allow growth of metal aggregates. Soluble metal could easily penetrate the peptidoglycan and come into contact with the highly reactive -COO$^-$ of glutamic acid.

It is therefore suggested that the ligands provided by the microbial biomass for palladium binding may be mainly carboxyl and to a lesser extend amino-groups. If most of the B binding sites (reactions (1), (2)) are carboxyl groups, they are less nucleophilic than ammonia. The first substitution of an ammonia molecule by a carboxyl group will then be likely followed by a second substitution to a position trans to the carboxyl group leading to a trans structure. This means that as carboxyl groups have been proposed as mostly involved in binding of Pd by the biomass, they would lead the second substituting ligands to trans positions. Formation of trans structures (reaction (1)) is then expected to result to a likely stable binding of palladium to the biomass.

From TEM and EDAX spectra recorded for three of the strains tested in the present study (BP 7/26, AS 302 and ER 121), it has been shown that biosorbed palladium is mostly found in the form of electron dense clusters mainly detected in intracellular locii of the cells suggesting the formation of elemental palladium[15]. The passive reduction of gold from Au^{3+} to Au^o has also been observed and reported in isolated cell walls of *Bacillus subtilis*[21] and non metabolizing whole cells of *Bacillus subtilis*[22]. Beveridge and Murray[20] have suggested that after the contact of gold (in the form of AuCl$_4$$^-$) with the highly reactive -COO$^-$ of glutamic acid, the metal forms deposits in the vicinity of the cross-linking chains.In both studies microscopic crystals of elemental gold were observed. It is suggested that gold is bound to sites on and within the cell wall and these sites act as nucleation points for the reduction of the gold and growth of crystals.

In a study on the influence of other metal co-ions in the biosorption of palladium by two of the strains tested in the present work BP 7/26 and AS302, it has been shown that a strong competition exists between PdCl$_4$$^{2-}$ and AuCl$_4$$^-$ during biosorptive uptake, translating into a competition of both complexes for the same biomass binding sites [14]. This competition has been attributed to the similarities of the Pd aned Au complexes in terms of chemical coordination behaviour and stereochemical characteristics. Standard reduction potential values of the palladium and gold complexes and the corresponding reduction reactions (3) and (4) are shown below :

$$PdCl_4{}^{2-} + 2e^- \rightarrow Pd + 4Cl^- \hspace{4cm} E^o = 0.623 \text{ Volts (3)}$$

$$AuCl_4{}^- + 3e^- \rightarrow Au + 4Cl^- \hspace{4cm} E^o = 0.994 \text{ Volts (4)}$$

These values mean that the reduction of both complexes is thermodynamically favored supporting the proposed mechanism hypothesis.

From the above discussion, it is suggested that palladium biosorption mechanism is a two-step process : (1) Initiation of the uptake at discrete points by chemical bonding. (2) Reduction of palladium as a second step of the biosorption process. The proposed mechanism is in agreement with the strongly favorable isotherms obtained in our biosorption experiments. The obtained results on palladium desorption confirm the irreversibility of palladium biosorption. This mechanism agrees well with the one proposed by Beveridge and Murray [20,21] for gold biosorption by *B. subtilis*.

3.1.2. Biosorption of $PdCl_4^{2-}$

Palladium species predominance diagram[18] indicates that two parameters influence the speciation of palladium chloride complexes: chloride ion concentration and solution pH. In the range of Pd concentrations used in our biosorption experiments, soluble palladium ionic species are predominant in acidic pH values. The pH of the contact solutions used in the present work was adjusted to 3 in order to maintain soluble palladium ionic species in solution.

Regarding chloride concentration, palladium species predominance diagram[18] indicates that, at chloride concentrations higher than 0.1 M, $PdCl_4^{2-}$ is only present in solution. At chloride concentration less than 0.1 M, other hydrolysis species of palladium are also present (i.e. PdOHCl, Pd (OH)$_2$).

For our biosorption experiments, two experimental sets, at different values of chloride concentration, were carried out at an initial pH=3. The first experimental set was carried out without any adjustment of chloride concentration ([Cl⁻]=0.002 M), and the second with at a chloride concentration of 0.1 M, added as KCl.

Figure 4 presents an example of the obtained isotherms corresponding to $PdCl_4^{2-}$ solutions at two different chloride concentrations for the strain EM 77. From this figure, it is suggested that Pd uptake capacities are higher at 0.002 M chloride concentration as compared to those obtained at 0.1 M chloride concentration. This can also be observed by comparing the values of k reported in Table 3. The values of k are higher for the case of [Cl⁻]=0.1 M.

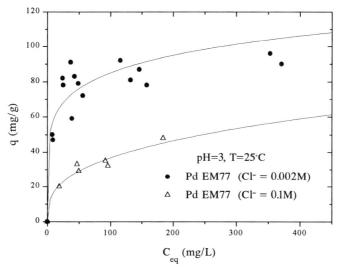

Figure 4. Comparative sorption isotherms of $PdCl_4^{2-}$, at two different chloride concentrations for the strain EM 77.

458

The low values of 1/n indicate a favorable sequestering pattern independent of chloride concentration (Table 3). The values of k corresponding to [Cl⁻]=0.1 M are lower than those corresponding to [Cl⁻]=0.002 M. The only species present in solution in the case of 0.1 M chloride concentration is $PdCl_4^{2-}$. Therefore, this is the only palladium species interacting with the microbial biomass, and palladium biosorption is represented by the corresponding isotherm in Figure 3. For the case of [Cl⁻]=0.002 M, two processes are active : biosorption and hydrolysis. These processes are additive and responsible for the observed increased palladium uptake capacity (Figure 4).

Table 3
Experimentally determined Freundlich parameters corresponding to each strain tested for biosorption of $PdCl_4^{2-}$ at pH 3 and two different chloride concentrations

Strain	[Cl⁻]=0.002 M		[Cl⁻]=0.1 M	
	k	1/n	k	1/n
BP 7/26	-	-	28	0.09
EM 77	40	0.14	8	0.34
CH 34	52	0.11	30	0.13
ER 121	40	0.14	25	0.12
AS 302	-	-	31	0.14

At pH=3, the electrophoretic mobility (EPM) values are positive except for strain ER 121 which presents a negative value of EPM (Figure 3). Palladium complex $PdCl_4^{2-}$ is negatively charged thus, the initial approach of the metal complex to the biomass does not appear to be hindered by coulombic interactions.

Figures 5 and 6 illustrate biosorption isotherms corresponding to $PdCl_4^{2-}$ solutions at chloride concentrations 0.1 M for the strains BP 7/26 and AS 302 respectively. In the same figures the biosorption isotherms corresponding to $Pd(NH_3)_4^{2+}$ species are reported. The palladium biosorption uptake capacities observed are higher in the case of $Pd(NH_3)_4^{2+}$ than those corresponding to $PdCl_4^{2-}$.

The stereochemical characteristics of both complexes are very similar since both are square planar. The substitution of chloride by ligands of the biomass will follow reactions (1) and/or (2). If carboxyl groups are the dominant binding sites, trans structures (reaction (1)) are expected to be more favored because the nucleophilicity of the carboxyl group is lower than that corresponding to chloride ions. However, at pH = 3 the availability of the carboxyl groups decreases due to their protonation. The observed EPM values at pH=3 (Figure 3) indirectly evidence this protonation. The ionic equilibrium of glutamic acid as a function of pH can be described as an example of the protonation of the groups of this amino-acid as the pH decreases (reaction (5)):

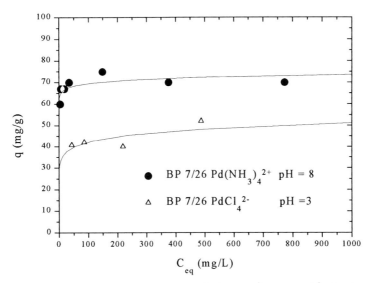

$$
\begin{array}{c}
\text{COOH} \\
\text{CHNH}_3{}^+ \\
\text{CH}_2 \\
\text{CH}_2 \\
\text{COOH}
\end{array}
\underset{-\text{H}^+}{\overset{+\text{H}^+}{\rightleftarrows}}
\begin{array}{c}
\text{COO}^- \\
\text{CHNH}_3{}^+ \\
\text{CH}_2 \\
\text{CH}_2 \\
\text{COOH}
\end{array}
\underset{-\text{H}^+}{\overset{+\text{H}^+}{\rightleftarrows}}
\begin{array}{c}
\text{COO}^- \\
\text{CHNH}_3{}^+ \\
\text{CH}_2 \\
\text{CH}_2 \\
\text{COO}^-
\end{array}
\underset{-\text{H}^+}{\overset{+\text{H}^+}{\rightleftarrows}}
\begin{array}{c}
\text{COO}^- \\
\text{CHNH}_2 \\
\text{CH}_2 \\
\text{CH}_2 \\
\text{COO}^-
\end{array}
\tag{5}
$$

From the above reaction, it is suggested that protonation begins with H^+ binding by the amino groups. Protonation of the carboxyl groups is accomplished at more acidic pH<3.22 = pI (I : isoelectric point). At pH=8 ($Pd(NH_3)_4{}^{2+}$ biosorption experiments) the carboxyl groups are non protonated and available for nucleophilic substitution. At pH=3, ($PdCl_4{}^{2-}$ biosorption experiments), substantially less carboxyl groups are available, since protonation takes place extensively. The result is that at pH = 3 the number of available biomass binding sites is likely considerably lower than that corresponding to pH=8. It can therefore be proposed that this is the reason for which palladium uptake capacities in the case of biosorption of $PdCl_4{}^{2-}$,at pH=3, are lower than those observed for $Pd(NH_3)_4{}^{2+}$ at pH=8.

Figure 5. Comparative biosorption isotherms of $Pd(NH_3)_4{}^{2-}$ and $PdCl_4{}^{2-}$, for the strain BP 7/26.

460

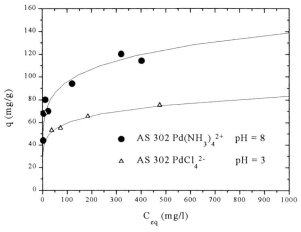

Figure 6. Comparative biosorption isotherms of $Pd(NH_3)_4^{2-}$ and $PdCl_4^{2-}$, for the strain AS 302.

3.2. Desorption results

The experimentally obtained desorption data for $Pd(NH_3)_4^{2+}$ are reported in Figures 1-2. The results reported in these figures show that very little of the metal sorbed, is released back at desorption experiments. This applies for all the strains examined and for both palladium species tested. The reason for the irreversibility of Pd biosorption is obvious on the basis of the biosorption mechanism proposed earlier.

3.3. Biosorption apparent overall kinetic results

Figure 7 presents two examples of palladium apparent overall kinetic profiles, showing that biosorption is practically accomplished within the first hour of contact and therefore reported results refer to biosorption in equilibrium conditions.

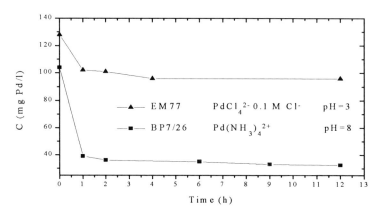

Figure 7. Examples of apparent overall kinetic biosorption profiles.

4. CONCLUSIONS

The following conclusions can be drawn based on the results of this study :

(1) Biosorption of palladium is favorable for both ionic species $Pd(NH_3)_4^{2+}$ and $PdCl_4^{2-}$ and for all the strains tested. Higher biosorption capacities systematically correspond to the tetra-ammonium palladium complexes.
(2) Solution pH is an important parameter influencing palladium biosorption. Besides the influence of pH on the hydrolysis of the metal complexes, biomass surface charge is also modified as a function of pH, reflecting the protonation - deprotonation of biomass sites in the cell surface.
(3) The consideration of chemical coordination and stereochemical characteristics have been shown to be significant for the understanding of the observed biosorptive behavior.
(4) Biosorption of palladium square planar complexes starts with ligand substitution reactions. After the first substitution by a biomass ligand, a second ligand is most likely placed to a position trans to the first, leading to a trans structure. The growth of microscopic elemental crystals has been observed through a subsequent Pd reduction step.
(5) In the case of the palladium chloride complex, chloride concentration should be considered for the determination of biosorptive uptake capacities in order to avoid the overlapping of hydrolysis and precipitation with biosorption.

REFERENCES

1. Greenwood N. & Earnshaw, A., Chemistry of the Elements. Pergamon Press, Oxford, 1993.
2. Forster C. F. and Wase D. A. J., Biosorbents for metal ions, Wase D. A.. J., and Forster C. F., Eds, Taylor and Francis, 1997.
3. Garnham G. W., Biosorbents for metal ions, Wase D. A.. J., and Forster C. F., Eds, Taylor and Francis, 1997.
4. Urrutia M. M., Biosorbents for metal ions, Wase and Forster Eds, Taylor and Francis, 1997.
5. Kapoor A and Viraraghavan T., Biosorbents for metal ions, Wase D. A.. J., and Forster C. F., Eds, Taylor and Francis, 1997.
6. Gadd, G., M., Biosorption, Chemistry and Industry, 13 (1990) 421.
7. Volesky, B. Biosorption of heavy metals, Ed CRC Press, Boca Raton, FL, 1990, 27.
8. Tsezos M., and Volesky B., Biotechnol. Bioengng, 24 (1982) 385.
9. Tsezos M., and Volesky B., Biotechnol. Bioengng, 24 (1982) 755.
10. Diels L., L. E. Macaskie, M. Tsezos, T. Puempel and F. Glombitza, (1995a), Modelling of genetic biochemical, cellular and microenvironmental parameters determining bacterial sorption and mineralization processes for recuperation of heavy or precious metals, Final technical report, Brite Euram Programme.
11. Diels L., M. Tsezos, T. Puempel, B. Pernfuss, F. Schinner, A. Hummel, L. Eckard, F. Glombitza, Pseudomonas mendocina AS 302, a bacterium with a non selective and very high metal biosorption capacity, Biohydrometallurgical Processing, C.A. Jerez, T. Vargas, H. Toledo, and J. V. Wiertz (Eds.), University of Chile, (1995).

12. Diels L., P. Corbisier, L. Hooyberghs, F. Glomitza, A. Hummel, M. Tsezos, T. Pumpel, B. Pernfuss, F. Schinner and M. Mergeay, Heavy metal resistance and biosorption in Alcaligenes Eutrophus ER121, Biohydrometallurgical Processing, C.A. Jerez, T. Vargas, H. Toledo, and J. V. Wiertz (Eds.), University of Chile, (1995).

13. Tsezos M., E. Remoudaki and V. Angelatou, International Biodeterioration and Biodegradation, 35 (1995) 129.

14. Tsezos, M., E., Remoudaki and, V., Angelatou, International Biodeterioration and Biodegradation, 19 (1996)29.

15. Tsezos, M., E., Remoudaki and, V., Angelatou, Comp. Biochem Physiol., 118 (1997) 478.

16. Greene B. and Darnall D.W., Microbial Mineral Recovery, H. L. Ehrlich and C. L. Brierley Eds, Mc Graw-Hill, Inc., 1990.

17. Glombitza F., A. Hummel and L. Eckardt, Modelling of genetic biochemical, cellular and microenvironmental parameters determining bacterial sorption and mineralization processes for recuperation of heavy or precious metals, 4th Intermediate report, Brite Euram Programme (1995).

18. Baes, C.F. & Mesmer, R.E. ,The Hydrolysis of Cations. Wiley, New York, 1976.

19. James, A. M., Advances in colloid and Interface Science, 15 (1982) 171.

20. Beveridge, T. J., and Murray, R. G. E., J. Bacteriol., 141 (1980) 876.

21. Beveridge, T. J., and Murray, R. G. E., J. Bacteriol., 127 (1976) 1502.

22. Gee, A. R., and Dudeney, A. W. L Biohydrometallurgy, Science and Technology Letters, P. R. Norris and D. P. Kelly Eds, Kew Surrey, U. K., (1988) 437.

Biosorption of toxic metals by immobilised biomass and UF/MF membrane reactor

F. Veglio'[a], F. Beolchini [a], R. Quaresima [a] and L. Toro [b]

[a] Dipartimento di Chimica, Ingegneria Chimica e Materiali, Facolta' di Ingegneria, Universita' degli Studi dell'Aquila, 67040 Monteluco di Roio, L'Aquila, Italy (e-mail veglio@ing.univaq.it)

[b] Istituto di Chimica Industriale, Universita' ''La Sapienza'' di Roma, P.le A. Moro, Rome, Italy

A study on the biosorption of toxic metals (copper, nickel and lead) by a species of *Arthrobacter* sp. bacterium is reported in this work. *Arthrobacter* sp was tested as a biosorbing material free in solution and immobilised in a polymeric matrix of poly-hydroxoethylmethacrylate. The equilibrium of the process was in all cases well described by the Langmuir isotherm. In the case of immobilised biomass, the Shrinking Core Model has been used for the fitting of experimental data. A good fit has been found in the case of controlling intraparticle diffusion in all experimental trials. A membrane process represents an alternative to immobilisation in polymeric matrices, which might be too expensive. A simulation has been performed by combining equilibrium data obtained in equilibrium biosorption trials and material balances in a membrane module. The simulation showed the technical feasibility of the biosorption process where biomass is confined inside the reactor by an appropriate membrane.

1. INTRODUCTION

Industrial activities generate an intensification of environmental pollution with the accumulation of several pollutants such as heavy metals. A growing attention is therefore given to the potential health hazard presented to the environment by heavy metals. Mining and metallurgical wastewater are considered to be the major sources of heavy metal contamination, and the need for economical and effective methods for the removal of metals has resulted in the development of new separation technologies [1-4]. The research on new technologies, involving the removal of toxic metals from wastewaters, has directed attention to *biosorption,* based on the metal binding capacities of various biological materials. Algae, bacteria, fungi and yeast have proved to be potential metal sorbents [1-7].

Microbial biomass consists of small particles of a low density, poor mechanical strength and little rigidity. It is not practical to contact large volumes of metal bearing aqueous solutions with microbial biomass within conventional unit processing operations, largely because of solid/liquid separation problems. The immobilisation of the biomass in solid structures creates a material of the correct size, mechanical strength, rigidity and porosity, necessary for use in columns. Immobilisation can also yield beads or granules that can be stripped of metals, reactivated and reused in a manner similar to ion exchange resins and activated carbons. The possibility of using the biosorbent material through adsorption-desorption cycles would also improve substantially the economics of the biomass technical applications. The economics of the process are also improved employing waste biomass instead of purposely-produced biomass [1,6,7].

The aim of this work is to study the toxic metals (copper, nickel and lead) removal from solutions by the bacterium *Arthrobacter* sp., isolated from a natural environment. *Arthrobacter* sp. was chosen as a test biosorbent material. *Arthrobacter* sp. was tested as a biosorbing material in *free* conditions (i.e. suspended in solution) and in immobilised conditions (trapped in a polymeric matrix of polyhydroxyethylmethacrylate). A simulation study has been also carried out to evaluate preliminarily the performance of biosorption processes obtained in tangential flow filtration systems (ultrafiltration and microfiltration).

2. MATERIALS AND METHODS

2.1. Microorganism
Arthrobacter sp. harvested from natural waters collected near L'Aquila (Italy) was kindly supplied by the *Dip. di Biologia di Base ed Applicata* (L'Aquila University). Further details about cell cultivation, harvesting and use can be found in elsewhere [1,6-8]. Further details about the preparation of immobilised biomass can be found in elsewhere [6].

2.2. Analytical determinations
Nickel, lead and copper concentration determinations were carried out with a Perkin-Elmer Model 2380 atomic absorption spectrophotometer. The total amounts of biomass entrapped were estimated as already reported [6].

2.3. Data analysis
The metal specific uptake (q) was determined as follows:

$$q = \frac{\left(C_0 - C_{eq}\right)}{X} \tag{1}$$

where q (mg of metal per g of biomass) is the metal specific uptake, C_0 (mg/L) is the initial metal concentration, C_{eq} (mg/L) is the metal equilibrium concentration, X (g/L) is the biomass concentration in solution.

The Langmuir sorption model was chosen for the fitting of experimental data obtained with biomass *free* in solution (not immobilised):

$$q = \frac{q_{max} b C_{eq}}{(1 + b C_{eq})} \tag{2}$$

where q_{max} (mg/g) is the maximum metal specific uptake and b (L/mg) is the Langmuir constant, ratio of the adsorption/desorption rates, i.e., related to energy of adsorption through the Arrhenius equation [5].

For the fitting of experimental data, the model was linearized as shown in the following:

$$\frac{1}{q} = \frac{1}{q_{max}b} \frac{1}{C_{eq}} + \frac{1}{q_{max}} \tag{3}$$

Parameters were estimated by using the least squares method modified by Mezaky [9]. In fact, the linearization of the Langmuir's model requires the introduction of an appropriate weight function to transform the experimental error variance as a constant: this is required by the least square method hypothesis [9].

In the case of the immobilised biomass the process is not as rapid as the one by free biomass, since mass transfer resistances inside the particle take place. Hence, in addition to equilibrium models described in the case of free biomass, such as Langmuir model, a kinetic model has to be developed, in order to estimate mass transfer characteristic parameters. The overall rate of binding (diffusion plus reaction) depends primarily on diffusivity [10]. Many investigators have calculated the diffusivities of heavy metals in biopolymers using the shrinking core model (SCM). This model was applied to fluid-particle chemical reactions by Levenspiel [10] and it was subsequently modified by Rao and Gupta [11] for the application to sorption of heavy metals in ion exchange resins. The model is based on the observation that when examining a cross section of a partly reacted solid particle, one often finds an unreacted core of material surrounded by an outer layer of reacted material. The dynamic picture is one of metal ions diffusing through a transformed shell material to a core of unreacted material that is progressively shrinking. Such a picture, though obviously simplified, allows for an exact mathematical solution for the kinetics of the process. A necessary condition for the model to be applicable is that the reaction between the metal and the microorganism is irreversible. Otherwise the reacted shell would not be inert and would continue to react as long as the metal concentration changes.

In the case of a process controlled by the diffusion of metal ions through the liquid film (*film diffusion control*), the extent of the biosorption process as a function of time will be given by the following expression (detailed derivation can be found in [11]):

$$\chi = \frac{3D}{\delta \cdot R \cdot C^0} \int_0^t C dt \tag{4}$$

where χ, the extent of the biosorption process, is the following:

$$\chi = \frac{C_0 - C}{C_0 - C_{eq}} \tag{5}$$

and C^0 is calculated from the total amount of metal bound over the course of an experiment and the total volume of biosorbent material used (mol/m^3) [10]. Consequently, if the film diffusion is controlling, a plot of χ vs $\int_0^t C dt$ yields a straight-line relationship.

If the process is controlled by the diffusion through the reacted shell (*particle diffusion control*), the SCM is described by the following equation (detailed derivation can be found in Rao and Gupta, 1982):

$$F(\chi) = 1 - 3(1-\chi)^{\frac{2}{3}} + 2(1-\chi) = \frac{6D_e}{R^2 C^0} \int_0^t C dt \tag{6}$$

In the case of particle diffusion control a plot of function $F(\chi)$ vs $\int_0^t Cdt$ will give a straight-line relationship and the effective diffusivity in the resin-biomass complex could be obtained from the slope of such a plot [10] as follows:

$$D_c = [slope] \, C^0 \, \frac{R^2}{6} \qquad (7)$$

3. RESULTS

An ionic characterisation of the biomass was performed in order to show the cell wall active groups which are able to capture metals. The experimental results are reported elsewhere [7,8,12]. It is observed that the *Arthrobacter* sp. cell wall has two main functional groups. In fact the titration curve presents two equivalence points. The first group has a pK_A equal to about 4.7 , and the second one has a pK_A around 9. Regarding the group with a pK_A equal to 4.7, it is probably a carboxylic group. In fact, carboxylic acids have pK_A values of about 4.7 (CH_3COOH has a pK_A equal to exactly 4.7) [12]. As regards the second group it may even be an aminic group. In this case, the pK_A observed is quite smaller with respect to the pK_A of simple amines (CH_3NH_2 has a pK_A equal to 10.7) and it is more similar to ammonia pK_A (that is equal to 9.2). It may be that there are other groups on the cell wall which mask aminic group behaviour. The hypothesis that *Arthrobacter* sp. cell wall mainly consists of carboxylic groups and aminic groups finds confirmation in the literature [13].

3.1. Biosorption equilibrium

Arthrobacter sp. was tested as a biosorbing material suspended in aqueous solutions in order to study the equilibrium of the biosorption process. The experimental results (here not shown) are well fitted by equation (2). It is seen that the highest value experimentally observed for copper specific uptake was of 46 mg/g (at C_{eq} = 210 mg/L), for nickel specific uptake it was of 13 mg/g (at C_{eq} = 150 mg/L) and for lead specific uptake it was of 130 mg/g (at C_{eq} = 250 mg/L). The Langmuir model parameters found from the fitting of experimental points are shown in Table 1. As reported in the table, the standard error of parameter b is high both for nickel and lead biosorption. For a better estimation of this parameter, an investigation in the linear zone of the Langmuir model (C_{eq} < 50 mg/L) should be performed.

The biosorbent concentration in solution during biosorption trials was demonstrated to be an important factor in the biosorption process [3,4]. The effect of biomass concentration in solution on q_{max} (Langmuir model parameter) values obtained in the case of copper accumulation was also investigated.

Table 1
Langmuir model parameters estimated from the fitting of experimental points of copper, nickel and lead biosorption (pH=4.0)

metal accumulated	q_{max} [mg Cu/g cells d.w.]	b [L/mg]
copper	55 ± 6	0.029 ± 0.005
nickel	12.7 ± 0.6	0.14 ± 0.07
lead	125 ± 8	1.4 ± 0.9

Results suggest that copper specific uptake decreases when the biomass concentration rises and the phenomenon seems to be restricted to the lowest values of the biomass concentration. Gadd and White [3] explained such a behaviour hypothesising that an increase of biomass concentration leads to interference between binding sites. This hypothesis was invalidated by Fourest and Roux [4] attributing the responsibility of the specific uptake decrease to metal concentration shortage in solution. The explanation of this behaviour, not considered by a simple Langmuir adsorption, could be related to the different concentration gradient between the solution and the inner side of the microbial cell. In fact, in the presence of an high biomass concentration there is a very fast superficial adsorption on the microbial cells that produces a lower metal concentration in solution with respect to the case where the cell concentration is lower. In this second case the concentration gradient between the solution and the inner side of the microbial cell is much higher and it could be possible to reach a metal concentration threshold in which the mechanism of biosorption is due also to the metal penetration through the cell wall, thus resulting in an higher metal uptake [14].

3.2. Immobilised biomass

Arthrobacter sp. was entrapped inside of a macro- and microporous matrix build up with polyhydroxoethylmethacrylate (poly-HEMA) cross-linked with trimethylolpropanetrime-thacrylate (TMPTM). The resin-biomass complex (RBC) was prepared with different characteristics according to a factorial experiment. Fig.1 shows the immobilised biomass in the resin. Factors investigated were: crosslinker (TMPTM) molar fraction, biomass concentration in the solid and particles granulometry [6,12].

Fig. 2 shows copper specific uptake vs time profiles during biosorption trials performed using RBC with two levels of microorganism's concentration inside the resins.

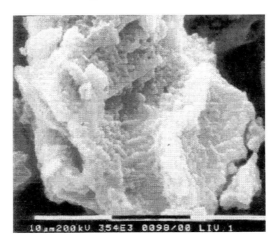

Figure 1. Typical picture of the resin-biomass complex (RBC)

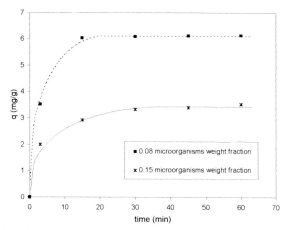

Figure 2. Experimental (points) and calculated through SCM (lines) values of copper specific uptake (mg Cu/g resin) at two levels of cells concentration inside the resins (0.02 cross-linker weight fraction; 425-750 μm granulometry; copper initial concentration: 25 mg/L) [12].

It is evident a negative effect of biomass concentration on the metal specific uptake as it was observed with free biomass. The Shrinking Core Model was used for the fitting of experimental data, considering external film diffusion and intraparticle diffusion control. The kinetic control of the sorption reaction was not considered because preliminary trials performed with not immobilised biomass showed the reaction to be very fast. Figs. 3 and 4 show results obtained considering external film diffusion control (Fig. 3) and intraparticle diffusion control (Fig. 4) in a typical trial. It is seen that a good fit was found in the case of controlling intraparticle diffusion. Other profiles, not shown here, were similar. Copper diffusion coefficient in RBC was estimated from the slope of the regression lines. The obtained values did not differ from each other with respect to the estimation error. An average copper diffusion coefficient of about $3 \ 10^{-6} \ cm^2 \ s^{-1}$ was found. This value is obviously lower than the copper diffusion coefficient in aqueous solutions [8].

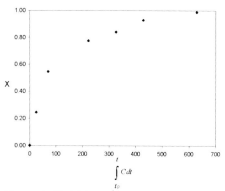

Figure 3. Shrinking Core Model: external film diffusion control

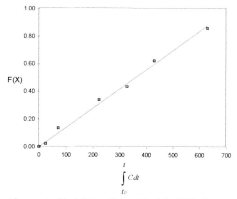

Figure 4. Shrinking Core Model: diffusion through the shell control

After filtration and washing with water the material was suspended in EDTA 0.01 M for possible regeneration [6]. Subsequently the RBC was resuspended in copper solution and four further biosorption-desorption cycles were performed. It was seen (data here not shown) that the moderate biosorption ability of RBC was essentially maintained after five runs [12].

3.3. Biosorption in UF/MF membrane systems

An alternative way to confine biomass in a reactor may involve the use of tangential flow membrane filtration (ultrafiltration or microfiltration – UF and MF respectively). This system is widely used in biotechnological separation processes (up and downstream processes) -in particular for cell recoveries [15]-, for its particular advantages linked to the possibility of low pressure conditions applied on the membrane, coupled with large flux rates. Several authors have shown the potential application of this system on biosorption processes [15]. In this case cells can be used without any immobilisation procedure, such as the one investigated in this work. Fig.5 shows a typical configuration for biosorption processes in CSTR coupled with UF/MF membrane systems. The major advantage of this system with respect immobilised cells is due to the absence of the immobilisation procedures. In fact these are often characterised by high costs. Furthermore biosorption kinetics are also improved: in the case of cells immobilised inside of a solid support, the metal diffusivity through the polymeric matrix plays an important role in the kinetic limiting step.

Considering the configuration shown in Fig.5, the model of the biosorption process can be evaluated coupling the material balance for the toxic metal in transitory regime with the equilibrium equation for a particular metal. In particular, for a system with one toxic metal:

$$V \cdot \frac{dC}{dt} = F \cdot Co - F \cdot C - \frac{dq}{dt} \cdot X \tag{8}$$

Biosorption by not immobilised cells was demonstrated to be very fast (results not shown) so that it can be considered at equilibrium. Consequently, introducing the equation (2) into the equation (8) and rearranging, the following differential equation can be obtained:

$$\frac{dC}{dt} = \frac{(Co - C)}{\tau \cdot \left[1 + \dfrac{q_{max} \cdot Ks \cdot X}{V \cdot (Ks + C)^2} \right]} \tag{9}$$

where:

C = actual metal concentration in the reactor (mg/L);
Co = metal concentration in the feeding stream (mg/L);
t = time (min);
t = residence time (V/F) (min);
V = reactor volume (L);
X = biomass concentration (g/L);
Ks = 1/b (mg/L);
q_{max} = maximum metal uptake (mg/g);

470

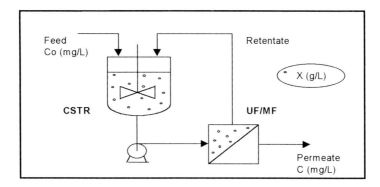

Figure 5. Typical configuration for biosorption processes in CSTR coupled with UF/MF membrane systems.

Fig.6 and 7 shown the results of the simulation studies obtained integrating numerically equation (9) in different operative conditions. It is possible to observe how the biomass concentration improves, as expected, the breakthrough conditions. By using 5 g/L of biomass concentration 1200 L of wastewater (with 10 mg/L of toxic metal concentration) per kg of biomass (as dry weight) can be processed in 30 h, with a metal cumulative efficiency of 95-98%. Obviously these results must be confirmed by experiments and the simulation study will be used to verify the developed mathematical model. However, the qualitative behaviour of the results shown in Fig.6 can be well compared with the experimental results reported by Duncan et al. [15].

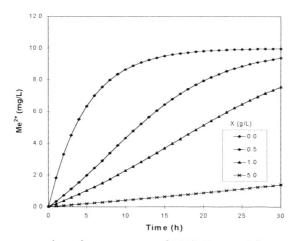

Figure 6. Metal concentration after one stage of UF/MF: F=2 L/h; V=10 L; Ci=10 mg/L; qmax=50 mg/g; Ks=5 g/L: effect of the biomass concentration.

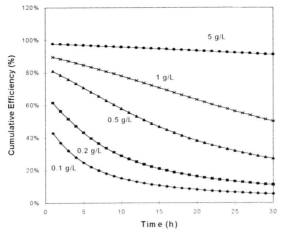

Figure 7. Metal cumulative efficiency after one stage of UF/MF: F=2 L/h; V=10 L; Ci=10 mg/L; qmax=50 mg/g; Ks=5 g/L: effect of the biomass concentration.

CONCLUSIONS

The equilibrium of the metal biosorption process was described by the Langmuir isotherm in the free biomass experiments. The highest values experimentally observed for the metal specific uptake were 46 mg Cu/g, 13 mg Ni/g and 130 mg Pb/g. In the case of biomass entrapped inside of a poly-HEMA matrix, the shrinking core model was used for the fitting of experimental data and the effective diffusivity of copper inside the polymeric matrix was estimated at about $3 \cdot 10^{-6}$ cm^2 s^{-1}. Entrapped biomass can be regenerable and usable in subsequent sorption/desorption cycles. A simulation study of biosorption performances by using UF/MF systems has shown the potential technical feasibility of the process. With this last system experimental runs are in progress to validate the mathematical model.

ACKNOWLEDGEMENTS

The authors are grateful to Dr. Angela Gasbarro, Dr. Deborah Simone and Mr. Marcello Centofanti for their helpful collaboration in the experimental work. They are also indebted to Prof. Benedetto Corain for his precious suggestions about the use of micro- and macroporous resins. The work was carried out with the financial support of MURST 60%.

REFERENCES

1. Veglio', F. & Beolchini, F., Hydrometallurgy, 44 (1997) 301.
2. Schiewer, S., Fourest, E., Chong, K.H. and Volesky, B., In: Biohydrometallurgical Processing , C.A. Jerez, T. Vargas, H. Toledo and J.V. Wiertz Eds, University of Chile, (1995) 219.

3.   Gadd, G.M. & White, C., Biotechnology & Bioengineering, 33 (1989) 592.
4.   Fourest, E. & Roux, J.C., Appl. Microbiol. Biotechnol., 37 (1992) 399.
5.   Holan, Z.R., Volesky, B., Prasetyo, I., Biotechnology & Bioengineering, 41 (1993) 819.
6.   Veglio', F. Beolchini, F., Gasbarro, A., Lora, S., Corain, B., Toro, L. Hydrometallurgy, 44 (1997) 317.
7.   Veglio', F., Beolchini, F. & Gasbarro, A., Process Biochemistry, 32, 2: (1997) 99.
8.   Veglio', F., Beolchini, F., Toro, L., Ind. & Eng. Chem. Res., 37, 3 (1998), 1107.
9.   Himmelblau, D.M., Process Analysis by Statistical Methods. John Wiley & Sons Inc, 1970.
10.   Chen, D., Lewandowski, Z., Roe, F. & Surapaneni, P., Biotechnology and Bioengineering, 41 (1993) 755.
11.   Rao, M.G. & Gupta, A.K, The Chemical Engineering Journal, 24 (1982) 181.
12.   Beolchini, F., Veglio', F., Gasbarro, A., Toro, L., Ubaldini, S., In: Innovations in Mineral and Coal Processing (Atak, S., Onal, G., Celik, M.S., Eds.), Balkema Publisher, Rotterdam, The Netherlands (1998) 787.
13.   Pelczar, M.J., Reid, R.D., Chan, E.C.S., Microbiologia, Zanichelli, Italy 1982.
14.   Peng, T.Y., Koon, T.W., Microb. Util. Renewable Resour., 8 (1993) 181.
15.   Duncan, J.R., Brady, D., Stoll, A., Wilhelmi, B., In: Biohydrometallurgical Processing , C.A. Jerez, T. Vargas, H. Toledo and J.V. Wiertz Eds, University of Chile, (1995) 237.

Biosorption of Cd and Cu by different types of *Sargassum* biomass

B. Volesky[a]*, J. Weber[a] and R. Vieira[b]

[a]Department of Chemical Engineering , McGill University
3610 University Str., MONTREAL, Canada H3A 2B2

[a]Labomar, the Federal University of Ceara, Fortaleza, Brazil

Sargassum seaweed biomass has been established as an excellent biosorbent material for heavy metals.   Three different species of non living *Sargassum* biomass were compared for their equilibrium Cd and Cu uptake from aqueous solutions using experimental sorption isotherms which fitted well the Langmuir sorption model. Uptakes of Cd at the optimum pH 4.5 ranged from q_{max} = 87 mgCd/g for *S. vulgare,* 87 mgCd/g for *S. fluitans* and 74 mgCd/g for *S. filipendula*. The amount of Cu uptake at pH 4.5 was q_{max} = 59 mgCu/g for *S. vulgare*, 56 mgCu/g for *S. filipendula* and 51 mgCu/g for *S. fluitans.*

Potentiometric and conductometric titrations carried out with the biomass of the three *Sargassum* species revealed a high degree of similarity among these materials. A separate contribution to the metal binding by the strong and weak acidic groups in the biomass could not accurate due to the similarities in their dissociation constants. Estimates of 0.3 mmol/g of strong and 1.5 mmol/g of weak acidic groups were obtained for *Sargassum* biomass types studied.

1. INTRODUCTION

Heavy metals are well known for their ecological hazard. While persistent in the environment, their circulation and eventually accumulation in the food chain is the consequence [1]. This accumulation presents a serious threat to human health.

While the acute toxicology of Cu is debated [1], Cd is one of the most dangerous heavy metals. The RCRA (US Resource Concentration and Recovery Act) regulations permit the maximum concentration for Cd of 0.01 mg/L in sources of potable water [2].

Biosorption process based on metal-sequestering properties of non-viable biomass provides a basis for the development of a new approach to remove heavy metals particularly when they occur at low concentrations [3]. Due to the high uptake capacity and the very cost-effective source of raw material, recent studies focused on marine algae, seaweeds [4]. The performance of the predominant ion exchange mode of binding mechanism in *Sargassum*

---

* Corresponding author. E-mail: boya@chemeng.Lan.mcgill.ca/
http://mcgill.ca/biosorption/biosorption.htm

seaweed depends on the chemical composition of the cell wall and on the solution-chemistry of the metal [5, 6, 7]. Species of the brown seaweed *Sargassum* are excellent metal-sorbers.

2. MATERIALS AND METHODS

2.1 Biomass and its Preparation:

Beach-dried *S. vulgare* was harvested from Natal, Brazil. Two different batches of beach-dried *Sargassum* were collected along the coast near Naples, Florida, containing *S .filipendula* and *S. fluitans*, respectively. The biomass was prepared by cutting it with a knife into irregular shaped particles between 1 mm and 4 mm in size. It was then washed twice with distilled water and dried in an oven at $45^{0}C$ overnight.

Analytical grades of $Cd(NO_3)_2 \bullet 4H_2O$ (Fisher Scientific) and $CuSO_4 \bullet 5H_2O$ (ACP Chemicals) were used for sorption experiments. Metal solutions were prepared by dissolving the salts in distilled water. Batch equilibrium sorption experiments were performed in 125mL Erlenmeyer flasks, containing 50mL of metal solution of a known initial concentration and 100mg of dry biomass. The initial concentrations of cadmium were 10, 50, 100, 150, 200, 250 and 450 mg/l; for copper from 10 to 250 mg/l in the same steps. The suspension was mildly agitated on a rotary shaker (New Brunswick Scientific) at 2.5 Hz, at room temperature ($22^{0}C$) for 6 hrs. The pH was adjusted periodically to pH 4.5 by adding 0.2M HCl or 0.1M NaOH as required. By the end of the third hour of contact, the pH remained relatively constant at pH 4.5 (± 0.3). The separation of the supernatant liquid from the biomass was achieved by decanting. Metal-free and biosorbent-free blanks were used as controls. Metal content of sample supernatant was analyzed by an atomic absorption spectrometer (AAS, Thermo Jarrel Ash, model Smith-Hieftje II). Atomic absorption standards (1000mg/L) were obtained from Fisher Scientific.

2.2 Titration of protonated biomass

The same cut biomass as for sorption experiments was used. It was protonated by soaking 2g of it in 100 mL of 0.1M HCl twice. Following a distilled water wash it was dried at $60^{0}C$ for 4 h. 500 mg of biomass used for each titration was resuspended in 100 ml of 0.1M HCl again and rinsed with dist. water until constant conductivity (measured by the Cole Palmer model 4070). Wet biomass was eventually suspended in 1mM NaCl. 0.5 mL of exactly 0.106M NaOH were added stepwise until the pH and conductivity values stabilized.

2.3 Equilibrium biosorption evaluation and models

Biosorption metal uptake (q) is determined from the sorption system mass balance:

$$q = \frac{V \cdot (c_i - c_f)}{M} \qquad (1)$$

In the present work, the Langmuir sorption model was applied to evaluate the isotherms since it gives a good fit to the data points.

$$q = \frac{q_{max} \cdot c_f}{b^{-1} + c_f} \qquad (2)$$

While the model provides only a mechanistic representation of the data, it is convenient in that it uses q_{max} and b as parameters which can be easily practically interpreted.

3. RESULTS

In this work, seaweeds *Sargassum vulgare*, *Sargassum fluitans* and *Sargassum filipendula* were used (class of *Phaeophyceae*, order of *Fucales*). For biosorption their two-layer structure of the cell wall is important since the metal binding occurs there. The inner fibrillar skeleton layer is basically made up of cellulose. The amorphous matrix of outer layer is attached by extracellular polysaccharides. Both contain alginic acid (~40%) and fucoidan (~20%).

3.1. Solution chemistry of cadmium- and copper-species

The uptake capacity is strongly affected by the pH. Usually, the sorption capacity increases with increasing pH, since the competition between protons and metal-cations decreases. However, heavy metals, especially copper-species, precipitate at higher pH-values. The chemical equilibrium program MINEQL$^+$ was used to get information about the solubility of Cu- and Cd-species as a function of pH at a certain concentration. In Figure 1, this relationship is shown for the highest initial concentration used in these experiments; that is 250 mg/L for Cu (CuSO$_4$) and 450 mg/L for Cd (Cd(NO$_3$)$_2$).

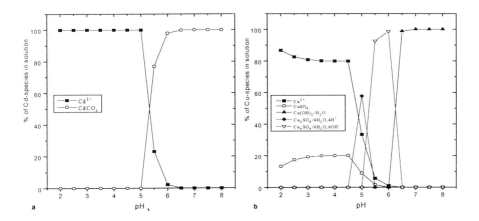

Figure 1: Solubility of cadmium and copper ions in water with respect to pH, calculated by MINEQL$^+$,

a) for Cd(NO$_3$)$_2$ at c_{Cd}=250mg/L,   b) for CuSO$_4$ at c_{Cu}=450mg/L.

Figure 1a shows the limit of solubility for cadmium cations to be approximately at pH 5. Solid Otavite (CdCO$_3$) is the main compound at pH above pH 5. The solubility of copper cations is more than 80% at pH below 4.5, the residual amount consist of the soluble CuSO$_4$-complex. Above pH-values of 4.5, copper exists as a solid in the form of Tenorite (Cu(OH)$_2$/H$_2$O), Anterlite (Cu$_3$SO$_4$/ 4H$_2$O,4H$^+$) and Brachantite (Cu$_4$SO$_4$/6H$_2$O,6OH$^-$). To operate at comparable conditions for both metals, the sorption experiments were performed at pH 4.5, the highest possible value for Cu without precipitation at an initial concentration of Cu 250mg/L.

3.2 Cadmium and copper isotherms

The sorption properties of *Sargassum* species studied were evaluated from the experimentally obtained biosorption isotherms. Figures 2 and 3 show the sorption isotherms for *S. vulgare, S. fluitans* and *S. filipendula* brought in contact with Cd and Cu. For all three biomass types the pH had to be adjusted only with acid to keep it at pH 4.5 during the first three hours of the sorption process. Especially at lower metal concentrations, for this raw, non-protonated biomass, the pH had a tendency to increase to pH 5.5 between pH adjustments.

Metal uptakes at the final concentrations $c_f=10$ mg/L and $c_f=200$ mg/L are usually selected for comparison of sorption behavior, representing the capacity of a sorbent at arbitrarily selected a 'low' and a 'high' residual concentrations. They were calculated from the Langmuir sorption model which gave a good fit for all isotherms obtained in these experiments: all curves are confident with a probability of P=99%. Table 1 summarizes the q_{10}-, q_{200}- values and the corresponding Langmuir parameters, q_{max} and b.

S. vulgare had the highest Cd uptake in the low as well as in the high metal concentration ranges (Fig. 2). For Cd, the q_{10}-value of *S.vulgare* was 29.3mg/g; 16% more than for *S. fluitans* and 25.4% more than for *S. filipendula*. The q_{200}- value of *S. vulgare* (78.96mg/g) was about 10% higher than that for *S. fluitans*. The difference between *S. fluitans* had q_{200} 15.3% higher than *S. filipendula*. A high value of Langmuir parameter *b* means a steep desirable beginning of the isotherm which reflects a high affinity of the biosorbent for the solute. *S. vulgare* had the highest value of b=0.0509 L/mg for Cd. The values for *S. fluitans* and *S. filipendula* were almost identical, 0.0464 L/mg and 0.0643 L/mg, respectively.

With regard to copper, *S. vulgare* shows best sorption behavior in the entire investigated concentration range. *S. vulgare* differs from *S. filipendula* in the q_{10}- and q_{200}- values by 41.4% and 18.4%, respectively. The former value is also confirmed by the Langmuir constant *b* that is very high for *S. vulgare*, yielding 0.143 L/mg. This means that at $c_f = 50$ mg/L, already 90% of the saturation capacity is achieved. The differences between *S. fluitans* and *S.filipendula* for q_{10} and q_{200} are very low amounting to 4.5% and 8.7%, respectively.

Comparing the Cd and Cu uptakes for each biomass shows higher uptake for Cd with regard to the metal-mass, especially at high concentrations. However, referring to the number of adsorbed metal-cations, fewer ions of Cd were sequestered. That is due to the higher molecular weight of Cd ($M_{Cd}= 112.4$ g/mol) in relation to Cu ($M_{cu}= 63.55$ g/mol). Consequently, for comparison of different metals, the uptake expressed on the molar or electrical equivalents basis is more appropriate. Table 1, summarizes values of all the characteristic sorption parameters q_{max} in mmol/$g_{biomass}$ and *b* in L/mmol.

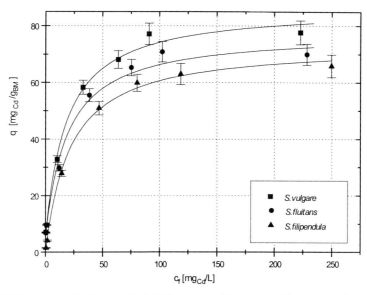

Figure 2: Cd biosorption isotherms for 3 different raw *Sargassum* biomas species at pH 4.5.

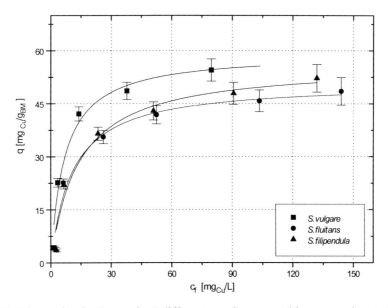

Figure 3: Cu biosorption isotherms for 3 different raw *Sargassum* biomas species at pH 4.5.

Table 1

Biosorption isotherm parameters for Cd and Cu: different *Sargassum* seaweeds

Cadmium	q_{10} [mg/g]	q_{200} [mg/g]	q_{max} [mg/g]	$b \bullet 10^{-2}$ [L/mg]	q_{max} [mmol/g]	b [L/mmol]
S. vulgare	29.25	78.96	86.72	5.09	0.79	5.33
S. fluitans	25.17	71.69	79.41	4.64	0.71	5.19
S. filipendula	23.32	66.51	73.69	4.63	0.66	5.20
Copper						
S. vulgare	34.97	57.53	59.44	14.26	0.93	9.16
S. fluitans	24.73	48.5	51.09	9.38	0.80	5.96
S. filipendula	23.67	52.73	56.37	7.24	0.89	4.16

While the differences in q_{max} (mmol/g$_{BM}$) between Cd and Cu by *S. vulgare* is 17.7% in favour of Cu uptake, and 12.7% in favor of Cu for *S fluitans*, the differences in metal uptakes by *S. filipendula* are more distinct, revealing 35% higher uptake of Cu. Likewise, each biomass shows higher affinity for Cu than for Cd, particularly in the case of *S. vulgare* where the affinity is 1.7 fold higher for Cu.

3.3 Titration of protonated biomass

Potentiometric and conductometric titration provide information about the amount of strong and weak acidic functional groups in the biosorbent. With the knowledge of the basic mechanisms of heavy metal complexation, this information may permit predictions for the heavy metal uptake. Carboxyl groups (COO$^-$) of alginate (~45% of biomass dry weight) and, to a lesser extent, sulphonate groups (SO$_3^-$, ~20% of biomass) contribute to heavy metal bisorption. whereas sulphonate groups are found in fucoidan (up to dry weight).

The result of the conductometric titration is nearly identical for each biomass examined (Figure 4). At the beginning of the titration, the conductivity decreases sharply due to the neutralization of free protons from strong acidic groups, probably sulphonate groups. The neutralization reaction releases water and a salt; thus it may be written:

$$A^- + H_3O^+ + Na^+ + OH^- \rightarrow A^-Na^+ + 2H_2O \qquad \text{The symbol A represents an acid.}$$

It should be mentioned that not only the number of charges, but also the motility of the ions contributes to the conductivity. Protons have the highest mobility, thus their removal yields lower conductivity. Moreover, a further contribution to the decreasing conductivity is the repression of dissociation of weak acidic groups. With increasing NaOH-values the slope of the conductometric curve also increases, since all strong acidic groups are neutralized and the weaker groups dissociate more and more. The transition between both slopes should correspond to the first equivalence point of the potentiometric titration (Figure 4). For *S. fluitans* and *S. filipendula* the equivalent point, which is more of an equivalent range, is approximately 0.3 mmol/g$_{BM}$ (±0.1mmol/g$_{BM}$) whereas the conductivity curve of *S. vulgare* shows an indistinct transition that may be estimated as 0.5 (±0.1mmol/g$_{BM}$).

The potentiometric curve, however, reveals no clear sign of this equivalent point. On the other hand, a second equivalent point occurs for all three biomasses due to the neutralization of the weak acidic groups, mainly carboxyl groups from alginate. The number of equivalents of weak acids, obtained by subtracting the equivalents of strong acid, amounts to approximately $(2-0.5) = 1.5$ mmol/g_{BM} for *S. vulgare*, $1.8-0.3=1.5$mmol/g_{BM} for *S. fluitans* and $(1.9-0.3) = 1.6$ mmol/g_{BM} for *S. filipendula*.

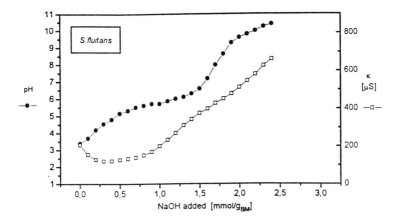

Figure 4: Potentiometric and conductometric titration of protonated *Sargassum fluitans* biomas.

The conductivity turns out not to be quite appropriate for determination of these weak groups. Theoretically, weak acidic groups are neutralized at the point where a steep increase in the slope is observed. At this point, the excess NaOH added for titration determines the conductivity by leaching out alginate at pH-values higher than at pH~5 which increases the conductivity, and therefore gives a blurring intersection. The leaching of alginate could be observed during the experiments. The solution was clear at the beginning of the titration and became more and more brown to the end. Table 2 summarizes the data, including the Langmuir parameter q_{max} for metal uptake obtained by the isotherms.

The correlation between the metal uptake capacity and the amount of acidic is related to the number of carboxylic groups (weak acid) occurring in alginate, since these groups play the predominant role for metal binding. For *S. fluitans* this relationship is very good, resulting in a 6.6% difference in Cu uptake and a 5.3% difference in Cd uptake. The difference for *S. vulgare* is 24% for Cu and 17.7% for Cd, whereas the results for *S. filipendula* also show a large deviation for the Cd-uptake (17.5%); the Cu-uptake, however, differs much less (11.3%).

4. DISCUSSION

The present experiments characterize, in general, the metal uptake performance of different species of *Sargassum* seaweed. The equilibrium data obtained in batch experiments are also necessary for the flow-through biosorption column experiments to follow. In comparison with other biomass types such as fungi, bacteria and other algae types, brown algae, especially the *Sargassum* used here, show very good metal sequestering capabilities. For instance, the biosorbent *Penicillium chrysogenum* reveals a cadmium uptake of

$q_{100}=39$mg/L [8]. *S. filipendula*, binding the least amount of Cd in the present experiments, sorbed 61 mg$_{Cd}$/g$_{BM}$, 56% more. Among bacteria, *Pseudomonas testosteroni* bound 41.6 mgCd/g$_{BM}$ [9] and *Bacillus subtilis* sequestered 79.2 mgCu/g biomass treated with alkali to enhance biosorption properties. The untreated biomass bound only 9.2 mgCu/g$_{BM}$ [10]. Chemical modification with glutaraldehyde, for instance, can easily increase the uptake capacity of raw *Sargassum*-species from 55-60 mg$_{Cu}$/g$_{BM}$, to 100 mg$_{Cu}$/g$_{BM}$ [11].

Table 2
Acidic group content and heavy metal uptake capacity in three *Sargassum* species.

	strong acidic groups [mmol/g$_{BM}$]	weak acidic groups [mmol/g$_{BM}$]	weak acidic groups $\bullet$ 0.5 [mmol/g$_{BM}$]	q_{max} cadmium [mmol/g$_{BM}$]	q_{max} copper [mmol/g$_{BM}$]
S. vulgare	0.5	1.5	0.75	0.79	0.93
S. fluitans	0.3	1.5	0.75	0.71	0.80
S. filipendula	0.3	1.6	0.80	0.66	0.89

While potentiometric and conductometric titrations can characterize the cell wall in terms of acidic groups, this approach did not prove to be sufficiently sensitive to evaluate the biomasses accurately because of insufficient differences in the dissociation constants of the acidic groups. Indistinct equivalent points obtained could not show a separate contribution of strong and weak acidic groups. Fourest and Volesky [12] determined 0.275 mmol/g$_{BM}$ strong acid groups and 1.95mmol/$_{gBM}$ weak acid groups in raw *S .fluitans*, corresponding very well to the amount of sulphonates and alginate obtained by chemical analysis. In the present experiments, the *Sargassum* metal uptake capacity was predicted with an error-range from 5.3%. to 17.5%.

4.1 Sources of errors
In order to estimate the accuracy of sorption isotherms, error bars for each data point were calculated (Figures 2 and 3), quantifying the experimental errors by including concentration, mass and volume measurements. The basis for this calculation is the law of Gauss yielding the maximum possible error for the metal uptake. When applied to equation (1), it leads to the following expression:

$$\Delta q = \left| \frac{\partial q}{\partial V} \right|_{M,c_{i,f}} \cdot \Delta V + \left| \frac{\partial q}{\partial M} \right|_{V,c_{i,f}} \cdot \Delta M + \left| \frac{\partial q}{\partial c_i} \right|_{M,V,c_f} \cdot \Delta c_i + \left| \frac{\partial q}{\partial c_f} \right|_{M,V,c_i} \cdot \Delta c_f \tag{3}$$

Upon derivation of partial differential quotients in equation (3), the following is obtained:

$$\Delta q = \frac{(c_i - c_f)}{M} \cdot \Delta V + \frac{V \cdot (c_i - c_f)}{M^2} \cdot \Delta M + \frac{V}{M} \cdot \Delta c_i + \frac{V}{M} \cdot \Delta c_f \tag{4}$$

Thus, Δq represents the absolute error of the uptake and may be expressed as: $q = q_{exp} + \Delta q$. Applying equation (4), the error for each parameter has to be estimated. The volumetric error is estimated to be $\Delta V = \pm 1 mL$, as labeled on the metric cylinder; the mass related error concerning the accuracy of the balance can be assumed to be $\Delta M = \pm 1 mg$. To determine Δc, a constant value for each metal is presupposed, corresponding to the sensitivity of the AAS. This was for Cd ± 0.1 mg/L and for Cu ± 0.3 mg/L using the less sensitive adjustment of the AAS, whereby at the same time the dilution error of the samples decreases. However, this error can not be ignored, especially at high concentrations where the dilution factor is large. A dilution error of $\pm 1\%$ could be established, and is taken into account when the linear range of the AAS is exceeded and a dilution is inevitable. For Cd, the limit is 20 mg/L and for copper 40 mg/L. Eventually, these errors where added to the sensitivity error of the AAS yielding the total concentration error Δc. The resulting absolute error Δq by applying equation (4) for each data point are additionally illustrated as error bars in Figures 2 and 3. At low concentrations (<15mg/L) the deviation does not exceed $\pm 1 mg/g_{BM}$ and is negligible. In the high concentration range, however, the error becomes more considerable, adding up to approximately ± 5 mg/g_{BM} for each isotherm.

5 NOMENCLATURE

b	Langmuir constant [L/mg, L/mmol]
c	metal concentration [mg/L]
c_i	initial or influent metal concentration [mg/L]
c_f	final metal concentration [mg/L]
κ	conductivity [μS]
m_{ad}	metal mass adorbed to the biomass [mg, g]
M	dry weight of biomass [mg, g]
q	metal uptake capacity of biosorbent [mg$_{metal}$/g$_{biomass}$]
q_{max}	maximum metal uptake capacity of biosorbent [mg$_{metal}$/g$_{biomass}$]
$q_{10,200}$	metal uptake at the final metal concentration of 10 and 200 mg/L, respectively [mg$_{metal}$/g$_{biomass}$]
V	volume of metal solution [mL, L]

REFERENCES

1. J.W. Moore and S. Ramamoorthy, Heavy Metals in Natural Waters, Springer-Verlag, New York 1984, pp. 28, 77, 182.
2. M.F. Harry, Standard Handbook of Hazardous Waste Treatment and Disposal, (1989) Ch.12, 56.
3. B. Volesky, in Biosorption of Heavy Metals, B. Volesky (ed.), CRC Press, Boca Raton, FL, 1990, pp. 3-6.
4. D. Kratochvil and B. Volesky, Trends Biotechnol., 16 (1998) 291.
5. E. Fourest and B. Volesky, Appl Biochem Biotechnol, 67 (1997) 33.

6. S. Schiewer and B. Volesky, Environ. Sci. Technol., 31 (1997) 2478.

7. S. Schiewer and B. Volesky, Biosorption by marine algae, F. Roddick, M. Britz and N. Robbins (eds.), Kluwer, Dordrecht, The Netherlands, (in press, 1999).

8. B. Volesky, FEMS Microbiol. Rev., 14 (1994) 291.

9. E. Rudolph, Untersuchungen zur Biosorption und zur Toxizitaet von Schwermetallen auf Bakterien, U-GH 1993.

10. B. Volesky, ed., Biosorption of Heavy Metals, CRC Press, Boca Raton, FL 1990.

11. A. Leusch, Z.R. Holan and B. Volesky, J. Chem. Tech. Biotechnol., 62 (1995) 279.

12. E. Fourest and B. Volesky, Environ. Sci. Technol., 30 (1996) 277.

Removal and concentration of uranium by seaweed biosorbent

Jinbai Yang and Bohumil Volesky[1]

Department of Chemical Engineering , McGill University
3610 University Street, MONTREAL, Canada H3A 2B2

While only weakly radioactive, traces of natural uranium in surface waters represent a danger because of high toxicity. Mine drainage brings enough uranium into the environment as to cause concern. Acid washed, protonated, non-living biomass of *Sargassum* seaweed sequestered uranyl ions from solution extremely effectively. At pH 4.0, pH 3.2 and pH 2.6, the maximum uranium uptake values were 560 mg/g, 330 mg/g and 150 mg/g, respectively. The uranium biosorption mechanism was affected by the solution pH through the hydrolysis of uranyl ions in aqueous solution. At low pH value, the uranium is present in the solution mainly in the form of free UO_2^{2+} ions, and it was competing with protons for the binding sites on the biomass. The high uranium biosorption at higher pH was attributed to the ion exchange between the hydrolyzed uranyl ions, UO_2OH^+, $(UO_2)_3(OH)^{5+}$, $(UO_2)_2(OH)_2^{2+}$ and protons. Experimental sorption isotherms could be reliably fitted by conventional Langmuir and/or Freundlich models. A flow-through biosorption column demonstrated a high overall column uranium sorption capacity of 105 mgU/g at pH 2.5. About 36 bed volumes of 238 mg/L uranium solution was purified before the breakthrough (at 1 mgU/L in the column effluent). The regeneration of the column by elution with 0.1 N HCl resulted in a very narrow peak of the elution curve reflecting a high efficiency in uranium recovery.

1. INTRODUCTION

Uranium is one of the most seriously threatening heavy metals mainly because of its high toxicity, not so much radioactivity. Uranium contamination poses a threat in some surface and groundwaters [1, 2]. Activities associated with the nuclear industry have brought excessive amounts of uranium into the environment [3]. Biosorption has been proposed as a potential alternative for removing toxic heavy metals. Various biosorbents based on non-living biomass have demonstrated an excellent uranium adsorption performance.   For example, filamentous fungi, yeast, bacteria, actinomycetes and fresh-water algae, such as *Chlorella*, have been reported binding uranium in excess of 150 mg/g of dry biomass [4, 5, 6, 7, 8, 9, 10].   Marine algae represent an important biosorbent type.   They proliferate ubiquitously and abundantly in the litoral zones of world oceans often posing environmental nuisance. Metal biosorption capacities of non-living seaweed biomass were summarized by Kuyucak and Volesky [11].   The brown alga *Sargassum fluitans* has been is particularly effective in binding heavy metal ions of gold, cadmium, copper, zinc, etc. [12].

---

[1] Corresponding author. e-mail: boya@chemeng.Lan.mcgill.ca/
http://www.mcgill.ca/biosorption/biosorption.htm

The high sorption capacity, easy regeneration and low-costs make this biomass of special interest for purification of large volumes of wastewater with lower concentration levels of metal toxicity to be removed [12, 13, 14, 15]. This is either difficult and/or expensive to accomplish by conventional metal-removal processes.

Although the fact that marine algae are capable of biologically concentrating radionuclides such as radium, thorium and uranium has been known for a long time [16], the biosorption of uranium by non-living marine algae has not been reported. The present work extended investigations of metal biosorption by the protonated *Sargassum* alga biomass to the uranium removal and recovery. In this work, basic parameters of equilibrium uranium biosorption were determined and the biosorption-desorption in a flow-through packed column was examined.

2. MATERIALS AND METHODS

2.1. Preparation of sorbent.

Beach-dried *Sargassum fluitans*, collected in Naples, FL, was chopped up in a homogenizer and sieved to different fraction sizes. The batch of dry biomass with particle size (1.0 - 1.4) mm was selected for subsequent protonation pretreatment aimed at standardizing the biomass by eliminating the light metals Ca^{2+}, Mg^{2+} etc. The protonation wash using 0.1 N HCl (10 g biomass / L) resulted in some biomass weight loss. After 3 hours of contacting with acid, the biomass was rinsed with deionized water in the same volume many times until a stable wash solution pH 4.0 was reached. The biomass was then dried in an oven at 40-60 $^{\circ}$C overnight. So prepared biomass was stored for later use.

2.2. Metal concentration analysis.

Dissolved uranium and lithium concentrations in filtered or centrifuged sample supernatant solutions were assessed simultaneously by an inductively coupled plasma atomic emission spectrophotometer (ICP-AES, Thermo Jarrel Ash, Model TraceScan). The ICP analyses were conducted at wavelengths of 409.014 nm and 367.8 nm for uranium and lithium, respectively. Trial tests confirmed that no sorption of dissolved metals occurred during processing of samples.

2.3. Sorption dynamics experiments.

In order to determine the contact time required for the sorption equilibrium experiments, the sorption dynamics experiments were conducted first. 0.1 g of biomass was mixed with 50 mL of 200 mg/L $UO_2(NO_3)_2$ solution in a magnetically stirred vessel with standard baffles. The pH value of the solution was controlled by a computer-driven autotitrator assembly (PHM82 pH meter, TTT80 Titrator and ABU80 AutoBurette, Radiometer, Copenhagen, Denmark). The autotitrator was set in the end-point titration mode, maintaining the pH value of the reacting solution at the level of the designed end-point. The 0.05 N LiOH solution was added into the titrated system by the internal high-speed pump and a burette. The pH value and the volume of the alkali *vs.* contact time were recorded by the controlling computer. A series of 0.2 mL samples of solution were removed from the vessel at pre-defined time intervals. After appropriate dilution, the samples were analyzed by the ICP-AES for metal concentrations.

2.4. Sorption equilibrium experiments.

With control samples containing no biomass, a series of uranium nitrate solution concentrations (50 mL) were mixed with 0.1 g biomass in 100 mL Erlenmeyer flasks which were shaken on a rotary shaker at 3Hz and room temperature. The pH was adjusted with 0.05 N LiOH or 0.05 N HNO$_3$ during the sorption process and the volume of LiOH added was recorded. After 3 hours of contact (according to the preliminary sorption dynamics tests), the sorption equilibrium was reached and the solution was filtered or centrifuged. The biomass was then cleaned by soaking and rinsing it with deionized water several times (no biosorbed metal loss occurred) before drying it at 40-60°C in an oven overnight. The dried metal-loaded biomass was used in desorption experiments later. The supernatant was diluted with D-H$_2$O for uranium and lithium concentration analyses by the ICP-AES.

The initial uranium concentrations C_i correspond to the control samples, and the final uranium concentrations C_f were from the supernatant solution. The uranium uptake was calculated by the concentration difference method that is based on the mass balance as follows:

$$q = (C_i - C_f) \, V \, / \, W \qquad (1)$$

with V being the solution volume, and W being the mass of biosorbent.

In the desorption experiments, 0.1 g of metal-loaded biomass was mixed with 50 mL 0.1 N HCl in a 100 mL Erlenmeyer flask. The remaining procedure was the same as that in the sorption equilibrium experiments except that no pH adjustment was required. The eluted biomass metal content could be calculated directly from the amount of metals desorbed into the HCl solution as follows:

$$q_{des.} = C_{des.} \, V \, / \, W \qquad (2)$$

with $q_{des.}$ being eluted metal content per gram of biomass, and $C_{des.}$ being the metal concentration in the HCl eluent solution.

2.5. Sorption and desorption column experiments.

The column (Diameter/Length: 3/45 cm) was uniformly packed with 22.64 g (dry basis) of protonated biomass. During the column sorption operation, an aqueous solution containing 238 mg/L (1 mM) uranium at pH 2.5 was pumped upward through the column at a constant flowrate (340 mL/h) continuously. The samples, collected from the outlet of the column by a fraction collector (Gilson 205) at pre-set time intervals, was analyzed for the uranium concentration by the ICP-AES. The pH value of the outlet solution was recorded by the controlling computer. After the biomass in the column became saturated, the column was washed at the same flowrate by D-H2O for several hours, before a subsequent uranium elution with 0.1 N HCl acid. The outlet sample collection and analysis was the same as that used in the biosorption uptake run.

3. RESULTS

3.1. Sorption dynamics and isotherms at different pH values in the batch system.
In order to determine the minimum contact time for the equilibrium experiments, the sorption dynamics was examined first. The profiles of dimensionless uranium concentration *vs.* contact time are plotted in Figure 1.

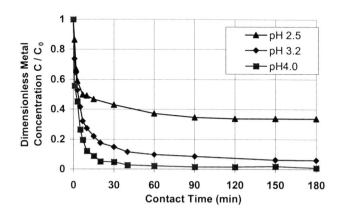

Figure 1: The uranium concentration decreases with the contact time for different pH values, $C_0 = 190.5$ mg/L.

The uranium biosorption rate was strongly influenced by the sorption system pH value, the uranium solution concentration in the solution decreased with contact time faster at higher pH values. At various pH values, approximately 70 - 80% of the uranium present originally in the solution was sorbed onto the biomass in about 15 minutes after the start of biosorption and the equilibrium could be reached within 3 hours. This provided a guide for the biosorption contact time to be used in the following equilibrium experiments.

The equilibrium of the uranium biosorption on *Sargassum* biomass was expressed in resulting biosorption isotherms for pH 2.6, pH 3.2 and pH 4.0 (Figure 2).

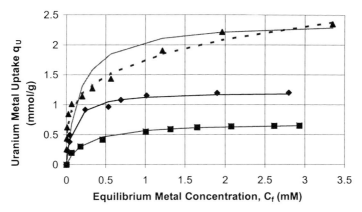

Figure 2:   Uranium sorption isotherms experimental data and Langmuir model regression,
(■) pH 2.6; (◆) pH 3.2; (▲) pH 4.0;   (—) Langmuir model; (---) Freundlich
model

The lines in Figure 2 refer to model-calculated values and the points are for experimental uranium uptakes. At pH 2.5 and pH 3.2, the isotherms could be represented well by the Langmuir sorption model, $q = q_m * C_f / (K + C_f)$, while the Freundlich model, $q = k*(C_f)^n$, could represent a better regression than the Langmuir model at pH 4.0.

The model parameters, the maximum uptake capacities q_m and the equilibrium constants K in the Langmuir model as well as k and n in the Freundlich model, were regressed from the experimental data at various pH values and are listed in Table 1:

Table 1 shows that q_m and K were largely dependent on the final solution pH values. The fact that q_m increased while the K decreased for higher solution pH values indicated that the sorption affinity of uranium for the biomass was enhanced at higher solution pH values. It is worth noting that the q_m value at pH 4.0 was close to the amount of the biomass binding sites, 2.25 mmol/g, as determined from titrations [17]. The control samples demonstrated that the very high uranium uptake could not be attributed to micro-precipitation.

Table 1

Regressed Langmuir and Freundlich sorption isotherm model parameters

	pH 2.6	pH 3.2	pH 4.0
K (mmol/L)$^{-1}$	0.233	0.084	0.1695
q_m (mmol/g)	0.701	1.215	2.40
k (Freundlich Model)			1.756
n (Freundlich Model)			0.249

488

3.2. Desorption of uranium by HCl

The uranium-loaded biomass was eluted by various elutants, $NaHCO_3$, $(NH4)_2SO_4$ and by mineral acids. It was established that the diluted mineral acids, such as H_2SO_4, HNO_3 and HCl, were effective in uranium desorption and that no significant biomass damage resulted after several sorption-desorption cycles. The experimental results for the elution of the uranium-loaded biomass with various initial uranium loading by 0.1 N HCl are presented in Figure 3.

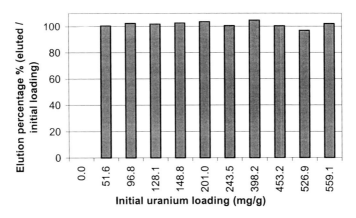

Figure 3: Comparison of the initial uranium loading and the uranium elution with 0.1 N HCl

The Y axis in Figure 3 stands for the percentage of the eluted uranium metal over the initial uranium loading on the biomass. It could be noticed that the elution percentage values are close to unity within 4% error range, indicating that the elution with 0.1 N HCl was complete. The biomass weight loss during the acidic desorption process was less than 5%. The biomass was also protonated at the same time and it was thus ready for the next cycle of uranium biosorption uptake.

3.3. Uranium sorption and elution in a packed biomass column.

In the biosorption of uranium in the column packed with *Sagarsum* biomass, approximately 10 L of 238 mg/L uranium solution was processed before the column breakthrough point occurred which was arbitrarily established at 1.0 mg/L of uranium in the column outlet. In this case, the column residence time was approximately 49 minutes and the total amount of 2,380 mg of uranium was accumulated on 22.64 g (dry) biomass. That gives the column overall uranium biosorption capacity of 105 mg U/g (dry biosorbent), including the only partially saturated portion of the dynamic sorption zone still inside at the column breakthrough (outlet 1 mgU/L). The results are illustrated in Figure 4 where the uranium concentration in the column outlet is plotted vs. the number of the column volumes that passed through the column. The column breakthrough took place at 36.5 bed volumes passing through.

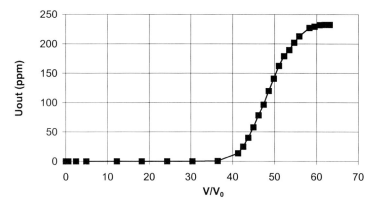

Figure 4: Biosorption column breakthrough curve. Feeding uranium concentration C_0 = 238mg/L, flowrate F = 340 mL/h, volume of empty column V_{bed} = 280 mL, biomass weight W = 22.64 g (dry), pH 2.5.

The elution curve for the column acid wash and recovery of uranium is shown in Figure 5. The uranium concentration in the elution acid (0.1N) at the outlet of the column was plotted against the volume of the elutant passing through the column.

The narrow peak, about 6000 mg/L average concentration in 400 mL volume, and the low residual uranium concentration (< 1 ppm) indicated a highly efficient and complete column elution. After one month of continuous sorption-regeneration operation (5 cycles), no significant damage in biomass structures was observed.

4. DISCUSSION

4.1. The effect of pH on uranium biosorption mechanisms and the maximum sorption capacity. For algal biomass, ion exchange has been considered as a main mechanism responsible for metal sequestering [18, 19, 20]. The ion exchange mechanism for uranyl ion binding to the biomass is complicated by the fact that the uranium cation UO_2^{2+} is hydrolyzed in aqueous solutions within the pH range of the sorption system studied here. Partioning of the hydrolysed uranium species depends on the solution pH and on the total uranium concentration in the solution. In the range of acidic to near neutral pH values, four major hydrolysed complex ions, UO_2^{2+}, $(UO_2)_2(OH)_2^{2+}$, UO_2OH^+, $(UO_2)_3(OH)_5^+$ and a dissolved solid schoepite exist in the solution [21]. The hydrolysis equilibrium constants are pK = 5.8 for UO_2OH^+, pK = 5.62 for $(UO_2)_2(OH)_2^{2+}$ and pK = 15.63 for $(UO_2)_3(OH)_5^+$[22]. The equilibrium coposition calculations could be carried out by the computer program MINEQL+ [21].

490

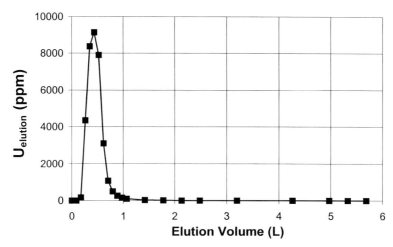

Figure 5: Elution of uranium with 0.1 N HCl from the biosorption column. Flowrate F = 340 mL/h, volume of empty column V_{bed} = 280 mL, biomass weight W = 22.64 g (dry)

At pH 4.0, all hydrolyzed ions UO_2OH^+, $(UO_2)_3(OH)_5^+$ and $(UO_2)_2(OH)_2^{2+}$ existed in the solution within all the present experimental concentration ranges of uranium. The percentage of the free ion UO_2^{2+} decreased while that of $(UO_2)_2(OH)_2^{2+}$ increased with the increase in the total uranium concentration. The two monovalent ions took about 10 - 15 % of the total in all concentration ranges. According to Collins and Stotzky [23], the hydrolyzed species can apparently be sorbed better than the free hydrated ions. Particularly the monovalent ions, compared with the divalent hydrolyzed ions, have even higher affinity to the biomass in ion exchange with protons because they could replace single protons on separate binding sites in the biomass. The binding of the hydrolyzed ions onto the biomass would drive the hydrolysis equilibrium toward the formation of hydrolyzed complex ions when the hydrolyzed proton ions, H_3O^+s, were neutralized by the added LiOH to the system to maintain the constant solution pH 4.0. Eventually, the uranium would be sorbed on the binding sites in the form of hydrolyzed ions. When the hydrolyzed ions exchanged with protons, the ion exchange stoichiometry would be $U/H^+ = 1 : 1$ for UO_2OH^+, $U/H^+ = 3 : 1$ for $(UO_2)_3(OH)_5^+$ and $U/H^+ = 2 : 2$ for $(UO_2)_2(OH)_2^{2+}$, comparing with a $U/H^+ = 1 : 2$ exchange ratio for the free ion UO_2^{2+}. In another word, the hydrolyzed uranyl ions have a higher binding capacity on the biomass than the free ions. When the hydrolysed ions become predominant in the ion exchange, the maximum molar uranium uptake could thus become close to or even higher than the value for the total binding capacity in the biomass, i.e. 2.25 meq/g. The extremely high experimental q_m value of 2.4 mmol/g at pH 4.0 (Table 1) may be the demonstration of this case.

With decreasing system pH, the percentage of UO_2^{2+} in the solution increased accordingly. The lower pH suppressed the enhancement of uranium biosorption occurring normally because of the hydrolyzed ions. When the pH became low enough, for example at pH 2.6, the divalent free UO_2^{2+} became the dominant ion form in the solution for a wide uranium concentration range from 0.3 to 1000 mg/L. Apart from a less preferable ion exchange ratio of U / H$^+$ = 1 : 2, some binding sites which are far away from other sites may not be available for the divalent UO$_2$$^{2+}$ which needs to exchange with two protons. This leads to an even lower uranium binding capacity at low pH.

On the other hand, the non-ionic dissolved solid schoepite starts appearing in the solution when the pH is too high. The uranium sorption may be hindered by the decrease in ion concentration in this situation. For example, Guibal et al. [5] observed a decrease in uranium uptake by filamentous fungus biomass at pH 6.0.

In summary, the biosorption of uranium on *Sargassum* biomass is a ion exchange process between the uranium ions and protons introduced to the biomass binding sites during the acid pre-treatment and biomass regeneration process. The hydrolysis of uranium ions, which is dependent on the solution pH, increased the uranium uptake by forming monovalent hydrolyzed complex ions. The number of available binding sites in the biomass for hydrolyzed ions was twice or more that for the divalent free UO2^{2+} ions. Correspondingly, at pH 4.0, the maximum uranium uptake was as high as 566 mg/g or 2.38 mmol/g, which is quite close to the total amount of biomass binding sites.

The present work demonstrates that uranium can be very effectively removed from the uranium-containing solution by the continuous-flow biosorption process. The uranium-laden *Sargassum*-based biosorbent can be conveniently eluted from the sorption column with a small volume of an HCl (0.1 N) wash which concentrates the metal. The high efficiency of biosorption and elution, low biomass damage and stability over a prolonged operation time make the new biosorption process an effective alternative for uranium pollution control which is coupled with the possibility of a feasible uranium metal recovery.

5. GLOSSARY

C_0 Column feeding concentration (mmol/L)
C_i, C_f Initial and final metal concentration (mmol/L)
$C_{des.}$ Metal concentration in the eluant solution (mmol/L)
F Flowrate passing through column (mL/hr)
q Metal uptake (mmol/g)
$q_{des.}$ Eluted metal content per gram of biomass (mg/g)
V Solution volume (L)
V_b Column empty bed volume (mL);
W Biomass weight (g)

REFERENCES

1. S.K. White, J. Am. Water Works Assoc., 75 (1983) 374.
2. J.C. Laul, Radioanal. Nucl. Chem. Articles, 156 (1992) 235.
3. B. Benedict, T.H. Pigford and H.W. Levi, Nuclear Chemical Engineering, McGraw-Hill, New York 1981.
4. B. Volesky and M. Tsezos, Separation of Uranium by Biosorption, U.S. Patent 4 320 093 (1981); Canadian Patent 1 143 007 (1983).
5. E. Guibal, C. Roulph and P. Le Cloirec, Water Res., 26 (1992) 1139.
6. L.E. Macaskie, R.M. Empson, A.K. Cheetham, C.P. Grey and A.J. Skarnulis, Science,

257 (1992) 782.

7. N.D.H. Munroe, J.D. Bonner, R. Williams, K.F. Pattison, J.M. Norman and B.D. Faison, Binding of dissolved uranium by *Pseudomonas aeruginosa* CSU, in Abstracts, Amer. Soc. Microbiol. Ann. Meeting, Washington, DC 1993

8. M.Z.-C. Hu, J.M. Norman, N.B. Faison and M. Reeves, Biotechnol. Bioeng., 51 (1996) 237.

9. T. Horikoshi, A. Nakajima and T. Sakaguchi, Agric. Biol. Chem., 332 (1979) 617.

10. J.J. Byerley, J.M. Scharer and A.M. Charles, Chem. Eng. Journal, 36 (1987) B49.

11. N. Kuyucak and B. Volesky, in BookBiosorption by algal biomass, B. Volesky (ed.), CRC Press, Boca Raton, FL, 1990.

12. B. Volesky and Z.R. Holan, Biotechnol. Prog., 11 (1995) 235.

13. N. Kuyucak and B. Volesky, Biorecovery, 1 (1989) 189.

14. A. Leusch, Z.R. Holan and B. Volesky, J. Chem. Tech. Biotechnol., 62 (1995) 279.

15. I. Aldor, E. Fourest and B. Volesky, Can. J. Chem. Eng., 73 (1995) 516.

16. D.N. Edgington, S.A. Gorden, M.M. Thommes and L.R. Almodovar, Limnol. Ocean., 15 (1970) 945.

17. E. Fourest and B. Volesky, Environ. Sci. Technol., 30 (1996) 277.

18. R.H. Crist, K. Oberholser, D. Schwartz, J. Marzoff, D. Ryder and D.R. Crist, Environ. Sci. Technol., 22 (1988) 755.

19. D.R. Crist, R.H. Crist, J.R. Martin and J. Watson, in BookIon exchange system in proton-metal reactions with algal cell walls, P. Bauda (ed.) , Societe Francaise de Microbiologie, France, 1993.

20. S. Schiewer and B. Volesky, Environ. Sci. Technol., 29 (1995) 3049.

21. W.D. Schecher, MINEQL+ : A Chemical Equilibrium Program for Personal Computers, Users Manual Version 2.22, Environmental Research Software, Inc., Hallowell,   ME 199).

22. C.F.J. Baes and R.E. Mesmer, The Hydrolysis of Cations, Wiley-Interscience, John Wiley & Sons, New York 1976.

23. Y.E. Collins and G. Stotzky, Appl. Environ. Microbiol., 58 (1992) 1592.

Enhancement of gold-cyanide biosorption by L-cysteine

Hui Niua[a], Bohumil Volesky[a]* and Newton C. M. Gomes[b]

[a]Department of Chemical Engineering , McGill University
3610 University Street, MONTREAL, Canada H3A 2B2

[a]Instituto de Microbiologia Prof. Paulo de Goes, CCS
Universidade Federal do Rio de Janeiro, Rio de Janeiro, Brazil[1]

The presence of L-cysteine increased gold-cyanide biosorption by protonated *Bacillus subtilis*, *Penicillium chrysogenum* and *Sargassum fluitans* biomass at pH 2 by (148 ~ 250)%. The respective Au uptake by these biomass types was 20.5 μmol/g ,14.2 μmol/g and 4.7 μmol/g of Au. Au-loaded biomass can be eluted with 0.1M NaOH. The elution efficiencies exceeded 90% at pH 5.0 with the Solid-to-Liquid ratio S/L = 4. Biosorption of anionic $AuCN_2^-$ complex involved ionizable protonated cysteine-loaded biomass functional groups. An adverse effect of increasing solution ionic strength ($NaNO_3$) was explained by a triple layer surface complexation mechanism. The NO_3^- anion competed with $AuCN_2^-$. Results confirmed that certain waste microbial biomaterials are capable of effectively removing and concentrating gold from solutions containing residual cyanide.

1. INTRODUCTION

Recent experimental results demonstrated that *Bacillus subtilis, Penicillium chrysogenum* and *Sargassum fluitans* biomass could extract Au from cyanide solution [1]. The main mechanism of Au biosorption involved anionic $AuCN_2^-$ species adsorption onto N-, P-, or O- containing functional groups on biomass through ion-pairing (H^+- $AuCN_2^-$). However, the capacities for Au biosorption by *Bacillus subtilis, Penicillium chrysogenum* and *Sargassum fluitans* biomass were not encouraging .

Proteins are known to be capable of complexing with metal ions. Cysteine, which figures prominently in discussions of metal ion binding to proteins, has three possible coordination sites, namely sulfhydryl, amino and carboxylate groups [2]. Hussain attributed the protection of isolated human lymphocytes from silver toxicity to cysteine through the formation of Ag-thiol complexes [3]. The complexation of Cu-cysteine was ascribed to the complexing of Cu to thiol as well as amino groups [4]. These results showed that cysteine had a tendency to combine well with metals. However, the behavior of L-cysteine in Au-cyanide complex biosorption have never been examined.

---

* Corresponding author. E-mail: boya@chemeng.Lan.mcgill.ca/
http://www.mcgill.ca/biosorption/biosorption.htm

The objectives of this work are to investigate the effect of L-cysteine on Au biosorption from cyanide solution by dead *Bacillus subtilis*, *Penicillium chrysogenum* and *Sargassum fluitans* biomass. The mechanism of Au-cyanide biosorption under these unconventional conditions was also examined.

2. MATERIALS AND METHODS

2.1. Biosorbent preparation

Waste industrial biomass samples of *Bacillus subtilis* and *Penicillium chrysogenum* were collected from Sichuan Pharmaceutical Company, Chengdu, P. R. China. *Sargassum fluitans* seaweed biomass was collected beach-dried on the Gulf Coast of Florida. Biomass was ground into particles around (0.5-0.85) mm in diameter, then washed with 0.2N HNO_3 for 4 hrs and rinsed with distilled water to pH~4.5. Finally, the biomass was dried in the oven at 50^0C for 24 hrs to a constant weight.

2.2. Acidification of gold cyanide solution

A solution of $AuCN_2^-$ was prepared by dissolving solid $NaAuCN_2$ in NaOH solution at pH 11 to simulate the industrial gold cyanide leach solution. Metal biosorption by biomass is usually taking place at pH values less than pH 6 since some biomass types or their constituents could seriously hydrolyze at elevated pH levels [5]. In order to make a full use of the biomass potential to concentrate gold from the cyanide solution, the pH of the gold cyanide solution needs to be adjusted for adequate biosorption. Since toxic hydrogen cyanide gas released at pH lower than 9.3 [6] is extremely dangerous, the conventional AVR process (Acidification, Volatilization and Reneutralization of cyanide) was employed [7]. The only difference from the standard AVR process in the current experiments was stripping of cyanide gas by nitrogen instead of air to avoid any oxidation of HCN. Total cyanide analysis was done by standard cyanide distillation followed by the titrimetric method for free CN^- in the alkaline solution [8].

2.3. Cysteine adsorption by biomass

Approximately 40 mg dried protonated biomass contacted with 20 ml cysteine solution with certain initial cysteine concentration 0~ 1.2 mmol/l in 150 ml Erlenmeyer flasks. The solution was mixed and left to equilibrate for 4 hrs. The cysteine uptake was determined from the difference of cysteine concentrations in the initial and final solutions. Cysteine was analyzed by a UV-visible spectrophotometer (Cary 1).

2.4. Equilibrium sorption experiments

Approximately 40 mg dried protonated biomass was combined with 20 ml sodium gold cyanide solution with or without L-cysteine in 150 ml Erlenmeyer flasks. The solution was gently mixed and equilibrated for 4 hrs. Uptakes of chromium were determined from the difference of metal concentrations in the initial and final solutions. The pH of the solutions before and during the sorption experiments was adjusted with 0.1M NaOH or HNO_3. The ionic strength was controlled by adding $NaNO_3$. All reagents were ACS reagent grade quality. Au concentration was determined by a sequential inductively-coupled plasma atomic emission spectrometer (Thermo Jarrell Ash, Trace Scan).

3. RESULTS AND DISCUSSION

3.1. Effect of L-cysteine on Au biosorption

The effect of L-cysteine on Au biosorption by *Bacillus* , *Penicillium* and *Sargassum* biomass was examined by varying L-cysteine concentration in the Au cyanide solution from 0 ~ 0.6 mmol/l at pH 2.0, with the initial Au concentration of 0.1015 mmol/l. No cyanide was released during the process. The results are shown in Figure 1. The final cysteine concentration around 0.5 mmol/l enhanced Au uptakes by *Bacillus* , *Penicillium* and *Sargassum* biomass up to 250% , 200% and 148%, respectively.

L-cysteine biosorption isotherms for *Bacillus* , *Penicillium* and *Sargassum* biomass in Figure 2 show encouraging uptakes by *Bacillus* and *Penicillium* biomass, while *Sargassum* biomass sorbed very little. Under the experimental conditions, the sequence for the cysteine uptake by the three biomass types is *Bacillus* > *Penicillium* > *Sargassum* , which agreed with the sequence of increased Au uptake in the presence cysteine. Enhancement of Au biosorption in the presence of cysteine apparently relates to the "bridging" function provided by cysteine between the Au-cyanide complex and biomass. The main active sites on the cysteine molecule are sulfhydryl, amino and carboxyl groups [4]. The dissociation constants (pK) of those groups are respectively 8.12, 10.36 and 1.90 [9]. At low pH 2.0, the carboxyl group is more active than the other two groups of cysteine and tends to combine with positively charged groups on biomass.

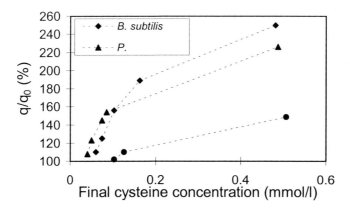

Figure 1: Effect of cysteine on Au-cyanide uptake
   0.04 g biomass, initial Au concentration 0.1015 mmol/l, 20 ml solution, pH2.0,
   4hrs, room temperature

496

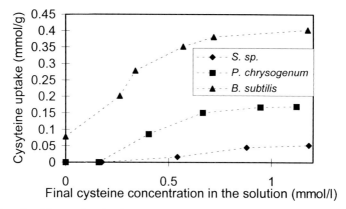

Figure 2: Cysteine biosorption isotherms
0.04 g biomass, 20 ml solution, pH2.0, 4hrs, room temperature

Bacillus cell walls contain as much as 70% of the dry weight as teichoic acid. This polymer (2-D -glucopyranosyl glycerol phosphate) is covalently linked to peptidoglucan [10], which mainly contain weak base groups like amine, phosphate and hydroxyl. *Penicillium* cell walls contain up to 40% of chitin which is linked to glucan [11]. This complex contains acetylamine and hydroxyl groups. All of these groups on these two biomass types could be positively charged by protons at pH 2.0 which would make them amenable to combining with the carboxyl moiety on cysteine.

Sargassum cell walls contain up to 70% polysaccharides , 40% of which is alginate containing abundant carboxyl groups. Since carboxyl groups on biomass cannot combine with the same groups on cysteine, *Sargassum* cannot effectively bind cysteine. The low cysteine binding by *Sargassum* may be due to the smaller amount of phenolic groups also present in the cell wall. Results revealed that the presence of cysteine did increase the Au-cyanide uptake by biomass and the increased Au uptake was related to the cysteine uptake by biomass.

3.2. Effect of pH

The effect of pH on Au biosorption in the presence of cysteine was examined by varying pH from 2~6. The initial ratio of Au : cysteine was 1:5. During the process of acidifying the Au cyanide solution and Au biosorption equilibration, there was no cyanide released from the solution. Figure 3 shows that in the presence of cysteine Au adsorption by *Bacillus*, *Penicillium* or *Sargassum* biomass was strongly affected by pH. The equilibrium uptakes of Au at pH 2 were greater than those at pH>2. The pH had a tendency to increase during the equilibration, hence 0.1 N HNO_3 was used to adjust the pH. This observation is opposite to that reported for biosorption of Zn, Cd, and $Pb(NO_3)_2$ by cysteine alone [4]. Divalent ions of these metals complexed with the cysteine sulfhydryl ($-S^-$) and amino groups. During the adsorption process, the hydrogen of sulfhydryl was dissociated, accounting for the pH drop. However, in the present case of $AuCN_2^-$ uptake, it was hard to dissociate Au from the cyanide complex at room temperature. The present Au cysteine-aided biosorption probably still involved anionic $AuCN_2^-$ complex adsorption. A similar general behavior was reported for biosorption of anionic Cr(VI) [12], whereby lowering of the equilibrium pH from neutral to acidic yielded an increase in Cr uptake by *Sargassum*. This was also the case for $AuCN_2^-$ uptake by *Bacillus*, *Pencillium* and *Sargassum* biomass without cysteine addition [1]. Basically, sulfhydryl and amino groups on cysteine behaved like weak bases which could be protonated at low pH, the same as those found in biomass. When cysteine is encountered in aqueous solution, there exists a surface charge on ionizable functional groups. As the concentration of protons is increased at pH 2.0, more and more weak base groups either on cysteine or in biomass become protonated and many acquire a net positive charge. These charged sites become available for binding anionic $AuCN_2^-$. Meanwhile, some carboxyl groups on cysteine may still be dissociated as the solution pH was higher than the dissociation constant (pK=1.9) of the carboxyl group on cysteine. This allows binding on biomass through the combination with some positively charged biomass function groups. As biomass bound cysteine, the amount of weak base groups of $AuCN_2^-$.on biomass increased resulting in enhanced Au uptake.

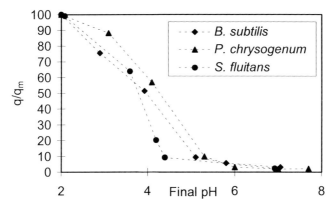

Figure 3: Effect of pH on Au uptake
0.04g biomass, 20 ml solution, initial Au concentration 20 mg/l, initial cysteine concentration 75 mg/l, 4hrs, room temperature

While cysteine presence definitely enhanced the Au biosorption uptake, the results were still lower than those observed for cation biosorption [13, 14, 15, 16, 17, 18, 19, 20, 21, 22]. This may be because the sites responsible for anionic gold cyanide complex binding need to be protonated and positively charged, while sites for cations binding are all those that could just be protonated.

3.3. Ionic strength effect
Ionic strength of experimental solutions was changed from 0.005 M to 0.15 M by adding $NaNO_3$ at pH 2. During the process, there was no cyanide released from the solution just as the case was in the pH effect experiments, indicating that sodium nitrate can not assist in

dissociating the gold-cyanide complex. The effect of ionic strength is illustrated in Figure 4. Increasing ionic strength reduced the Au biosorption. As the concentration of $NaNO_3$ increased to 68mM, the uptake of Au by *Bacillus* and *Penicillium* biomass was respectively reduced to 70% and 50% of that without $NaNO_3$ in the solution. The Au uptake by *Sargassum* decreased almost to zero at 20mM $NaNO_3$. Changing ionic strength (i.e. the background electrolyte concentration) influences adsorption in at least two ways:
(a) by affecting the interfacial potential and hence the activity of electrolyte ions and adsorption;
(b) by affecting electrolyte ions and adsorbing anions competition for available sorption sites.

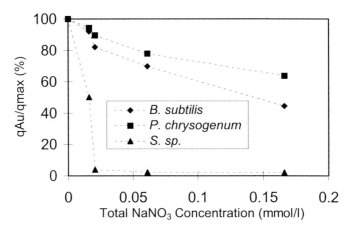

Figure 4: Effect of ionic strength on Au uptake
0.04 g biomass, 20 ml solution, pH 2.0 , initial Au concentration 20 mg/l, cysteine concentration 75 mg/l, 4hrs, room temperature

Ions such as metal cations and inorganic anion species present in aqueous solution (either in free or complex forms) often display the tendency toward preferential adsorption on ionizable function groups [23]. Although the extent of adsorption can be described by conventional adsorption isotherm expressions such as the Langmuir or Freundlich equations [24], the nature of ion adsorption is more chemical than physical and it is more appropriate to consider ion sorption through mechanisms based on chemical reactions or surface complexation. Hayes et al. [25] provided an explanation for the effect of ionic strength on anionic ion sorption from liquid to solid phase by considering the triple-layer mechanism (TLM).

In cysteine-loaded biomass, the main part for biosorption is on the cysteine and the cell wall [26], therefore the terminology "surface" used here includes surface on cysteine and all micro-surface throughout the cell wall where sites responsible for Au binding are situated .

According to the triple-layer surface complexation mechanism , ion adsorption is the formation of surface complexes at certain sites of biomass (Figure 5). The surface charge is assumed to be caused by the ionization of discrete identifiable site groups (FH) on cysteine or in biomass or, conversely, from the adsorption of charge-determining ions, which influences the distribution of nearby ions in the aqueous solution, leading to the formation of an electrical double layer. This electrical double layer is assumed to be composed of three parts demarcated by the site surface plane (denoted by o); an outer Helmholtz plane (denoted by d) indicating the closest distance of approach of hydrated ions or the start of the diffusive double layer;

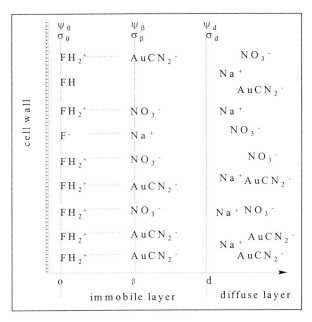

Figure 5: Schematic representation of the structure of Three Layer Mechanism of aurocyanide adsorption on cysteine-loaded biomass in the presence of electrolyte $NaNO_3$.

and the inner Helmholtz plane (denoted by β), which indicates centers of ions (electrolyte ions, Na^+ and NO_3^-) that form complexes with the surface groups F^- or FH_2^+ (Figure 5) [27]. There are two analogs of the TLM: the outer-sphere or the inner-sphere analog. As the electrolyte ion concentration exerted a significant influence on Au-cyanide adsorption which was involved in the outer-sphere analog of the TLM, meaning that $AuCN_2^-$ was placed on β-plane and $AuCN_2^-$ was adsorbed on biomass by ion-pair sorption ($FH-H^+ - AuCN_2^-$). The inner-sphere model analog, assuming the replacement of negative functional groups on cysteine or biomass by Au-cyanide complex and having no direct dependence on the β-potential , would be less influenced by ionic strength changes[25]. It seems that the inner-sphere analog of the TLM could be neglected in the $AuCN_2^-$ biosorption system.

In the ($FH- H^+ - AuCN_2^-$) complex, weak base functional groups FH_2^+ are placed on o-plane, and $AuCN_2^-$ is placed on β-plane. NO_3^- undergoes the same reaction as $AuCN_2^-$ and was placed on β-plane. Na^+ may exchange the proton on the FH moiety and was placed on either o-plane or β-plane. A site charge by proton is occurring on $-F^-$ and $-FH$ on o-plane.

At a certain controlled pH, an increase of $NaNO_3$ concentration caused the variation of surface potential (ψ_o, potential on o-plane; and ψ_β, potential on β–plane) leading to a decrease in activity coefficients as well as charge density (σ_o, surface charge on o-plane; σ_β, surface charge on β–plane and σ_d, surface charge on d-plane). Meanwhile, the increase of NO_3^- led to the competition with $AuCN_2^-$ for the binding sites on biomass. As a result, Au uptake was reduced. The relatively low Au biosorption uptake, as compared to free metal biosorption, corresponded to the outer-sphere analog. That was because the inner layer capacitance between o-plane and β-plane is much higher than the outer layer capacitance between β-plane and d-plane [25]. Cations tend to be adsorbed on both inner layer (o-plane) and outer layer (on β-plane) [27] and, therefore, their uptakes are generally higher. A similar phenomenon was found in the anion ion-exchange process whereby a weak base resin would have a relatively low binding capacity with anions indirectly attached onto active sites through proton bridges [28].

3.4. Desorption of Au-loaded biomass

The possibility of desorbing Au from biomass with sodium hydroxide was examined by first sorbing Au onto biomass in the presence of cysteine at pH 2 and then desorbing Au with 0.1 M NaOH at pH 3, 4 or 5. The initially Au-loaded *Bacillus* biomass contained 20.5 μmol Au /g of dry biomass, *Penicillium* biomass 14.2 μmol/g, and *Sargassum* biomass 4.7 μmol/g. The percentage of Au recovery, represented by the ratio of the amount of Au released per gram of the biosorbent during desorption and the equilibrium sorption uptake, was calculated for desorption experiments lasting for 4 hours. Figure 6 shows that more than 90% of Au was recovered at pH 5 with the solid-liquid ratio S/L=4 for all of these three biomass types, which indicated that Au could be easily eluted. These results further confirmed the outer-sphere complexation ($FH-H^+-AuCN_2^-$) postulated. When the "bridge" was broken up by OH- combining with H+, Au was dissociated from solid phase. Basically, the outer-sphere complex is not stable probably because of the relatively long distance between sorbent functional groups (on o-plane) and the sorbate (on b-plane). From the view point of adsorption reaction equilibrium, increasing the pH pushed the reaction of anionic gold cyanide complex adsorption to the left, allowing the Au elution. However, in the case of biomass, the use of concentrated NaOH leads to massive leaching of a variety of compounds from the biomass and to the destruction of the biomass cellular structure. Therefore, Au elution was limited to pH no higher than 6.

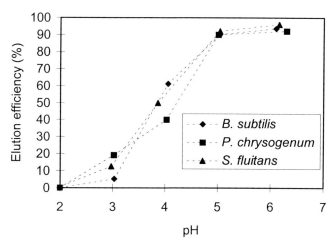

Figure 6: Effect of pH on Au elution efficiency
0.02 g biomass, 5 ml solution, initial Au loading was 20.5 mmol/g *Bacillus* biomass, 14.2 mmol/g *Penicillium* biomass and 4.7 mmol/g *Sargassum* biomass; 4h, room temperature.

4. SYMBOLS:

F :   functional group on biomass or cysteine

H: dissociable hydrogen on function group, the binding of which make cysteine-loaded biomass neutral

σ_o: surface charge on o-plane

σ_β :   surface charge on β–plane

σ_d :   surface charge on d-plane

ψ_o: potential on o-plane

ψ_β: potential on β–plane.

REFERENCES

1. H. Niu and B. Volesky, J. Chem. Technol. Biotechnol. (in press, 1999)
2. H. Shindo and T.L. Brown, J. Amer. Chem. Soc., 87 (1965) 1904.
3. S. Hussain and B. Volet, In Vitro Toxicol., 8 (1995) 377.
4. N.C. Li and R.A. Manning, J. Amer. Chem. Soc., 77 (1955) 5225.
5. E. Percival and R.H. McDowell, Chemistry and Enzymology of Marine Algal Polysaccharides, pp. 157-176. Academic Press, London, U.K. (1967).
6. J. Marsden and L. House, Chemistry of Gold Extraction,. Ellis Horwood, Hartnoll, UK 1993.
7. P.A. Riveros and R. Molnar, CIM Bulletin, 89 (1996) 153.
8. A.D. Eaton, L.S. Clesceri and A.E. Greenberg (eds.), in Standard Methods for the Examination of Water and Wastewater, Amer. Pub. Health Assoc., Washington, DC 1995, pp.4.19.

9. D.D. Perrin, Stability Constants of Metal-Ion Complexes, Pergamon Press, Oxford,UK1979.

10. T.J. Beveridge and R.G.E. Murray, J.Bacteriol., 127 (1976) 1502.

11. F.A. Troy and H. Koffler, J. Biol. Chem., 244 (1969) 5563.

12. D. Kratochvil and B. Volesky, Environ. Sci. Technol., 32 (1998) 2693.

13. B.J. Fen, C.J.Daughney, N.Yee and T.A. Davis, Geochim.Cosmochim.Acta, 61 (1997) 3319.

14. H. Niu, X.S. Xu, J.H. Wang and B. Volesky, Biotechnol. Bioeng., 42 (1993) 785.

15. N. Kuyucak and B. Volesky, Biorecovery, 1 (1989) 189.

16. E. Percival and R.H. McDowell, Chemistry and Enzymology of Marine Algal Polysaccharides, Academic Press, London, U.K. 1967, pp. 83, 99, 127, 157.

17. S. Schiewer and B. Volesky, in Metals: Mine Drainage, Removal and Toxicity, F. Roddick, M. Britz and N. Robbins (eds.), Kluwer, Dordrecht, The Netherlands (in press 1999).

18. D. Kratochvil, E. Fourest and B. Volesky, Biotechnol. Lett., 17 (1995) 777.

19. M.Z.-C. Hu, J.M. Norman, N.B. Faison and M. Reeves, Biotechnol. Bioeng., 51 (1996) 237.

20. J.P. Huang, J. Westman, K. Quirk and J.P. Huang, Water. Sci. Technol., 20 (1988) 369.

21. C.-P. Huang, C. Huang and A.L. Morehart, Water. Res., 24 (1990) 433.

22. N.A. Yakubu, Biosorption of Uranium, Ph.D. Thesis, University of London, U.K.1986.

23. C.Tien, Adsorption Calculation and Modeling, Butterworth-Heinemann, Boston 1994, 29.

24. A.J. Rubin and D.L. Mercer, Adsorption of Organics at Solid/liquid Interface, Ann Arbor Science, Ann Arbor, MI 1981, 245.

25. K.F. Hayes, J. Coll. Interf. Sci., 125 (1988) 717.

26. A. Rothstein and R. Meier, J.Cell.Comp. Physiol., 38 (1951) 245.

27. K.F. Hayes, J. Coll. Interf. Sci., 115 (1987) 564.

28. J.A. Marinsky, Ion Exchange, Marcel Dekker, New York 1966, 228.

Multimetal biosorption in a column using *Sargassum* biomass

M. M. Figueira[a], B. Volesky,[b1] K. Azarian[b] and V. S. T. Ciminelli[a]

[a] Department of Metallurgical Engineering, Federal University of Minas Gerais, Belo Horizonte, MG, Brazil

[b] Department of Chemical Engineering , McGill University, 3610 University Street, MONTREAL, Canada H3A 2B2

Removal of Cu, Cd and Zn from multicomponent mixtures by biosorption was studied in equilibrium biosorption systems and in a flow-through column packed with *Sargassum* algal biosorbent pre-saturated with potassium. The elution order of Zn, Cd and Cu from the column and their concentration overshoots at the column exit, were due to the differing affinities to the biosorbent. Low-affinity Zn broke through the column faster than Cd with a sharp favorable breakthrough curve due to the pretreatment of the biomass with K which features the lowest affinity of all the ionic species examined. K-form biomass gives the most effective utilization of the column bed. An overshoot of the Zn exit concentration was observed and explained by the sequential ion exchange between Cd and Zn during the operation of the column bed. The same effect was observed for both Zn and Cd when Cu was present in the feed solution. sorbing onto K-*Sargassum* biomass. A sharp Zn exit concentration overshoot was followed by the one of Cd that lasted until the later breakthrough of Cu was complete. In this system, the overshoot of Zn was caused mainly by its exchange for Cd ions in the biosorbent. While the Zn and Cd exit concentrations overshoots were close together, the time interval between Zn and Cu breakthroughs was much longer. This behavior is all based on the magnitude of relative affinities of Zn, Cd and Cu for the K-biomass determined from batch equilibrium experiments. The relative affinity coefficients were: $K_{Zn/K}$= 1.96, $K_{Cd/K}$= 3.70 and $K_{Cu/K}$= 16.51.

1. INTRODUCTION

Toxic heavy metals pose an environmental threat and need to be removed from usually large volumes of industrial effluents. Biosorption is a process which utilizes inexpensive dead biomass to sequester heavy metals from aqueous solutions. *Sargassum* seaweed-based biosorbents can accumulate metal concentrations that can be orders of magnitude higher than those in the liquid phase. Biosorption processes can serve as an

---

[1] corresponding author e-mail: boya@chemeng.Lan.mcgill.ca

attractive cost-effective alternative to reduce volumes of dilute metal-bearing and toxic industrial effluents. The resulting small volume of high-concentration metal-bearing by-product wash liquid could be economically reprocessed for metal recovery [1]. As compared to conventional methods for removing toxic metals from industrial effluents, such as precipitation with lime, and/or ion exchange, the biosorption process offers low operating costs and high efficiency in detoxification of relatively dilute effluents. These advantages serve as the primary incentives for developing full-scale biosorption processes to clean-up heavy metal pollution [2].

Biosorbents are prepared from the naturally abundant and/or waste biomass of algae, moss, fungi, or bacteria which is inactivated and usually pretreated by washing with acids and/or bases before final applications [3]. While simple cutting and/or grinding of the dry biomass may yield stable biosorbent particles [4], some types of biomass have to be either immobilized in a synthetic polymer matrix [5] and/or grafted on an inorganic supporting material such as silica in order to yield particles with the required mechanical properties [6]. Biosorbent particles can then be packed in sorption columns which are perhaps the most effective device for the continuous removal of heavy metal [1, 7].

Certain biosorbent materials can be viewed as natural ion-exchange materials that primarily contain weakly acidic and basic groups [8, 9, 10]. Relatively well-developed knowledge of ion exchange can be applied in the study of biosorption, to describe both the mechanism(s) involved as well as to predict the performance of the biosorption process on a large scale. That was the approach used by Kratochvil and Volesky [11] when studying the biosorption of Cd, Cu, Zn and Fe from multicomponent mixtures in a flow-through column packed with *Sargassum* biomass-based algal biosorbent in the Ca-form. The biosorption column operation resulted in a demonstration and explanation of concentration "overshoots" in the column effluent assigned to the differences in ion affinities to the biomass sorption bed. The biomass used was pretreated and saturated with Ca which had a higher affinity than Zn in the column feed. It was suggested that the biomass should be pretreated by loading it with a cation with a lower affinity for the biomass than the other influent ions, thus avoiding the development of undesirable premature effluent concentration overshoots.

Evaluation of the Zn biosorption performance of *Sargassum* biomass pretreated with several cations, Mg, Na and particularly K needs to be carried out. At least an approximate comparison with the earlier Ca-biomass results is desirable. Moreover, the selected K-biomass form will be studied in a flow-through biosorption column using a mixture of metals often found in metal-bearing industrial effluents.

2. MATERIALS AND METHODS

2.1. Preparation of the biomass

The brown seaweed *Sargassum* was collected in Naples (Florida) in March. The raw, sun-dried biomass was treated with a 0.2N HCl solution in a batch for 3 h at the biomass concentration of 10g/L, followed by a rinse with distilled water until the pH of the solution reached the value of 4.5, and finally dried overnight in the oven at 60°C. The conversion of the protonated biomass to K-biomass was carried out using a 20 mM solution of KOH. The pH of the spent wash solution was about 5.5. Separated biomass was dried overnight (60°C) and used in the experiments.

2.2. Metal binding experiments

20 mg of biomass was put in contact with 10 mL of a heavy metal (Cd, Zn and Cu) sulfate solution in a test tube for 6 h (1 to 10 mM of the heavy metal, pH adjusted to 5.0). Mixing was promoted by mild bubble aeration. Samples were taken of the initial metal solution, and of the supernatant solution after the sorption equilibrium was reached. Blanks, represented by either biomass in distilled water or metal solutions without biomass, were run as required. The samples were analyzed for Cd, Zn, Cu and K content using inductively coupled plasma atomic spectrometer (ICP-AS Thermo Jarel Ash, Trace Scan). The metal uptake was determined from the difference of metal concentrations in the initial and final solutions. No significant change in the pH values was observed throughout the equilibrium contact experiments.

2.3. Column experiments

Dry K-biomass was packed into a 50 cm long column of 2.5 cm in diameter, yielding an approximate packing density of 100 g/L. The column was then slowly flooded with distilled water from the bottom. The metal solutions were fed into the column from the top, at a rate of approximately 8 mL/min, and samples of the column effluent were collected every 30 minutes from the bottom by means of a fraction collector (FC203 Gilson). The pH in the effluent as well as the pressure at the column inlet were continuously monitored using a flow-through cell and a pressure transducer (PX602, Omega), respectively. The data acquision card (PCL 711S, Omega) was used in combination with a PC 286 computer to convert the analog signals corresponding to the effluent pH and the pressure to a digital format and to store them on a computer hardisc. The computer also served for the remote control of the fraction collector. The samples obtained from the column effluent were analyzed for K and the heavy metals used in the feed solution.

In order to compare the different systems studied, the dimensionless concentration (C/C_0) was plotted against the dimensionless time (T), calculated according to the equation:

$$T = \frac{C_0 Ft}{\rho_b QV_c} \tag{1}$$

where C_0 is the total normality of the solution (meq/L), F is the volumetric flowrate (mL/h), t is time (h), ρ_b is the packing density of the dry biomass in the packed-bed (g/m), Q is the concentration of binding sites in the biosorbent (meq/g) obtained from the titration of the biomass against a NaOH solution [10] and V_c is the volume (L) of the packed-bed in the column.

2.4. Theoretical approach: the multicomponent Langmuir model

A mathematical model of biosorption by *Sargassum* biomass has been developed and tested by Schiewer and Volesky [10]. The model defines equilibrium binding constants k_i, and k_j for cations I and J, respectively, sorbing onto binding sites in the biomass B according to the following reactions:

$$I^+ + B \leftrightarrow \quad IB \qquad\qquad\qquad K_i = q_i / C_i C_B \tag{2}$$

$$J^{2+} + 2B \quad \leftrightarrow \quad 2BJ_{0.5} \qquad\qquad Kj = q_j^2 / C_j C_B^2 \qquad\qquad (3)$$

where q_j and q_j are the equilibrium uptakes of I and J, respectively, and C_i and C_j represent the equilibrium concentrations of I and J in solution, respectively. C_B denotes the concentration of free, unoccupied, binding sites in the biomass.

By rearranging equations (2) and (3), the following general expression can be obtained for equilibrium uptake of metal I in multimetal systems:

$$q_i = q_{max} * \frac{(K_i * c_j)^{1/\gamma i}}{1 + \sum_{i=1}^{n} (K_i * c_j)^{1/\gamma i}} \qquad\qquad (4)$$

where q_{max} represents the maximum uptake of metals by *Sargassum* biomass (calculated by the model) and γ is the ionic charge of the element. This equation represents the explicit sorption isotherms whereby uptakes of each metal is expressed as explicit functions of the equilibrium composition in the liquid. With the above equation it is possible to determine the equilibrium bindings constants K of metal ions as a function of their final concentrations.

2.5. Fitting the models

Equilibrium data sets for Zn, Cd and Cu sorbing onto sorption K-biomass provided the experimental basis for the fitting of the biosorption models. Using the binary systems data, the multicomponent Langmuir model was fitted and the constants K_{Zn} $K_{Cd,}$ K_{cu} and K_k were obtained by minimizing the objective function F_M represented by equation (5):

$$F_M = sum \left\{ \frac{q_{Cd}^{exp} - q_{Cd}^{theor}}{q_{Cd}^{exp}} \right\}^2 \qquad\qquad (5)$$

3. RESULTS AND DISCUSSION

3.1. Modeling of the batch equilibrium data

Ion exchange was previously demonstrated as the main mechanism involved in heavy metal uptake by *Sargassum* biosorbent in protonated [12] or Ca-form [11]. This can be confirmed in the present work through ion mass balances drawn for each set of batch equilibrium for biosorption of Zn, Cd,and Cu by K-biomass form.

The calculated equilibrium constants K obtained by fitting the experimental data with the multicomponent Langmuir model are presented in Table 1. The model, developed from the Langmuir sorption isotherm model, assumes that all sites are initially free and does not consider any reverse reaction of a displaced ion with the site. The value of unity in the denominator of equation (4) is related to the amount of free sites still available in the biomass at a given equilibrium concentration of the metal. By dividing the values obtained for individual K_M of each metal by those for K_K which was initially present in the biomass one can determine the relative equilibrium constants for bi-component sorption systems, K_M/K_K.

The value of F_M objective function reveal that the model fitted the experimental data well within a deviation expected for this type of non-uniform sorption systems [10].

Table 1
Model parameters for multicomponent Langmuir model: equilibrium constants K (L/meq), K_M/K_K ratios:

Metal ion	K^*	K_M/K_K
K	2.16	1.00
Zn	4.26	1.96
Cd	8.02	3.71
Cu	35.35	16.51

*: function error F_M for the system $= 0.092$

The relative affinities of all the metals tested toward the K-form biomass were higher than that of K, confirming the feasibility of biomass K-pretreatment for sorption of those metals. While the relative affinity of Zn and Cd were relatively close to each other ($K_{Zn/K} = 1.96$ and $K_{Cd/K} = 3.71$), the one for Cu was much higher ($K_{Cu/K} = 16.51$).

3.2. Dynamic biosorption in the fixed bed

The most common and most efficient arrangement of a sorption process is the packed bed flow-through column. It combines the highest sorbent density with a high concentration difference driving force throughout the apparatus. The biosorbent is equilibrated at the high incoming concentration of the metal bearing solution ensuring high uptake values, whereas the low-concentration effluent encounters still fresh and powerful sorbent material.

The order of metals as they start leaving the column upon break-through, and the peak overshoot concentrations of the toxic species in the column effluent eventually determine the overall efficiency of the biosorption water treatment process. These kinds of results were in the focus of experiments carried out with in a flow-through column packed with K-*Sargassum* biomass treating a multi-metal feed. In order to compare results of different biosorption column runs, identical operating conditions were maintained for these experiments.

Figure 1 shows the concentrations of Zn and K in the columns effluent as a function of dimensionless time (T) (throughput) defined above. The sorption uptake of Zn from the feed was accompanied by the elution of K ions from the packed-bed. At approximately T=0.2, the concentration of Zn in the column effluent reached the level in the feed, i. e. X_M (or C/C_0) = 1. The mirror image formed by the Zn breakthrough and the K elution curves indicates that K in the biomass was being stoichiometrically exchanged for Zn.

The performance of a continuous-flow biosorption column receiving an equimolar mixture of two metals (Cd and Zn, 1. 5 meq/L) is displayed in Figure 2. A higher affinity of selectivity of the *Sargassum* K-form biomass for Cd over Zn is well exhibited in the resulting breakthrough curves. The affinity of Zn which is lower than that of Cd for the biomass caused Zn to break through the column faster than Cd (at T approx. 0.75). A sharp favorable breakthrough curve observed for Zn is the result of preloading the biomass with K whose affinity for the biomass is lower than that of Zn. This leads to a more effective utilization of

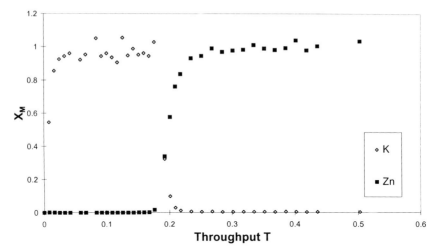

Figure 1: Experimental breakthrough curve of Zn sorbing onto K-*Sargassum* biomass in a continuous flow-through column

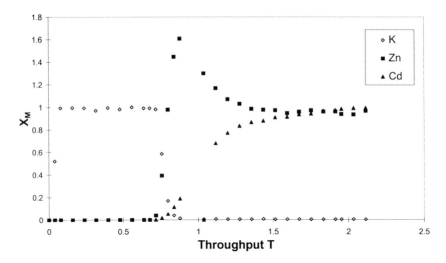

Figure 2: Experimental breakthrough curves for Zn and Cd sorbing onto K-*Sargassum* in a column

the biosorbent material inside the column. At $T = 0.8$, the concentration of Zn in the column effluent plotted in Figure 2 reached the level of Zn in the feed and started overshooting. That can be explained by the ion exchange between Cd and Zn, whereby Cd from the solution was displacing the "faster-running" Zn already bound to the biosorbent. Since no more Zn was

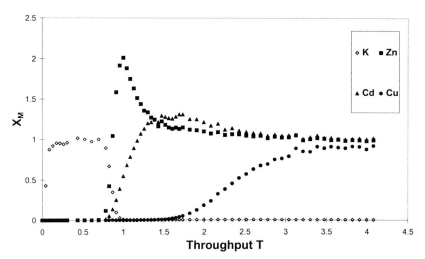

Figure 3: Experimental breakthrough curves for Zn, Cd and Cu sorbing onto
K-*Sargassum* in a column

being sorbed from the liquid beyond this point, the released Zn increased the overall concentration of Zn in the effluent liquid above the feed level (up to 1.6 times the feed concentration). This phenomenon, referred to as effluent concentration "overshooting" and was well demonstrated by Kratochvil and Volesky [11] in a multicomponent flow-through column system. The overshooting of Zn is caused by a "chromatographic effect", as the high affinity Cd desorbs the low affinity Zn which had previously sorbed onto the biosorbent in the bed. The current results show a generally better biosorption column performance due to the K-form biosorbent used as opposed to the Ca-form alternative [11]. Eventually, when the effluent concentrations of Zn, as well as Cd, reached the feed concentrations (around T > 1.3), the column feed was in equilibrium with both Zn and Cd ions sorbed on the biomass and the column was completely saturated.

Similar effluent concentration overshoot effect can be seen in Figure 3, representing the breakthrough curves of Zn, Cd and Cu sorbing onto K-*Sargassum* biomass in a biosorption column fed with equimolar concentrations of the metals (1.0 meq/L). Even greater overshoot ($X_M > 2$ at T > 1) resulted for Zn followed by a slower increase and another overshoot in the effluent Cd concentration, peaking at $C/C_o = 1.25$, which slowly subsided when the breakthrough of Cu occurred. Obviously, Zn was was being exchanged for mainly Cd ions which, eventually, were displaced by Cu on the binding sites of the biomass in the bed. While the Zn and Cd breakthroughs were close together, the breakthrough of Cu came much later. This is the reflection of the relative affinities of Zn and Cd being much closer to each other ($K_{Zn/K} = 1.96$ and $K_{Cd/K} = 3.70$) than to that of Cu ($K_{Cu/K} = 16.51$).

The present results with a K-biomass column fed with equimolar mixture of Cu, Cd and Zn (1 mM/L each) (Figure 3) could be compared with those obtained by Kratochvil and Volesky [2] with Ca-biomass. Although the order of elution of the metals in each experiment was the same (Zn, Cd, Cu), the magnitude of effluent concentration overshooting differed significantly. The K-biomass column gave steep and high Zn overshoots (twice C_o lasting

~0.5 T), comparing with ~1.3 C_o and prolonged overshoots (1.5 T) reported for the Ca-biomass column. Moreover, the overshoot of Cd in the Ca-biomass column was even higher than that of Zn (~1.9 the feed concentration) as compared to the overshoot of Cd in the K-biomass system: 1.25 C_o. This results from the relative affinities of the heavy metals towards Ca-biomass quoted as [2]: $K_{CuCa} = 2.01$, $K_{CdCa} = 0.67$ and $K_{ZnCa} = 0.45$. These affinity values resulted in the overshoots of both Zn and Cd due to their simultaneous exchange with Cu being fed in the column.

According to Kratochvil and Volesky [11], the condition under which overshoots occur can be formulated as follows. Low-affinity species present in the feed overshoot in the column effluent only if the species with the highest affinity in the feed is bound to the biosorbent more strongly than the species with which the biosorbent had been presaturated. Consequently, a metal species may or may not overshoot in the column effluent depending on the ionic form of the biosorbent. This is the explanation for the overshoots of both Zn and Cd in K-*Sargassum* column: with K being the species with the lowest affinity, all the metals in the feed solution would overshoot in the presence of Cu, the species with the highest affinity.

4. FINAL REMARKS

Very much like synthetic ion-exchange resins, some biosorbents can be prepared in different ionic forms such as protonated (H-form) or saturated with Ca, Mg, Na, etc., by washing the biomass with mineral salts and/or bases. The relative positions of heavy metals and light metals in the affinity series determine the suitability of the biosorbent ionic forms and regenerants for the optimal performance of the biosorption column being considered for the possibility of industrial application. The present work established not only the affinity sequence of heavy metals toward the pre-loaded K-*Sargassum* biomass, but also examined the biosorption column performance with a new modification of the biosorbent used.

The service time of the column in practical applications is determined by the toxic metal in the feed that has the lowest affinity (the key element), since this is the species that flows out of the column first. When the concentration of the key toxic species in the column effluent exceeds the regulatory limit, the operation of the sorption column has to be usually stopped. The extent of the biosorbent bed utilization increases with steeper breakthrough curves. The present study points out the advantage of the K-form biosorbent which does not give any breakthrough before K is almost completely replaced in the column bed by the feed heavy metals (Zn, Cd, Cu). The breakthrough of Zn in the Ca-biomass column occurred even before the complete exchange of Ca inside the bed with heavy metals [11]. For certain types of heavy metal combinations, the K-form biosorbent enables a more complete use of the sorption column before the first metal breaks through. This is the main advantage of the use of K-biomass over Ca-biomass for the heavy metal biosorption system studied in this work.

ACKNOWLEDGEMENTS

The Brazilian CAPES Scholarship support for M.M.F. is gratefully acknowledged. NSERC of Canada funded the project with a partial contribution from the Brazilian PADCT project (940/95).

REFERENCES

1. Kratochvil, D.; Volesky, B. Trends Biotechnol. No. 16 (1998) 291.
2. Kratochvil, D., PhD Thesis, McGill University (1997).
3. Volesky, B. In Volesky, B. (ed.), Biosorption of Heavy Metals CRC Press, Boca Raton, 1990.
4. Fourest, E. and Roux, J.C. Appl. Microbiol. Biotechnol. No. 37 (1992) 399.
5. Jeffers, T.H. and Corwin, R.R. In Torma, A.E., Apel, M.L. and Brierley, C.L., (eds.) Biohydrometallurgical Technologies, Proceedings: International Biohydrometallurgy Symposium. The Minerals, Metals and Materials Society, Warrendale, 1993.
6. Mahan, C.A. and Holcombe, J.A. Anal. Chem. No. 64 (1992) 1933.
7. Volesky, B. Trends Biotechnol. No. 5 (1987) 96.
8. Crist, D.R., Crist, R.H., Martin, J.R.and Watson, J. In Bauda, P. (ed.) Metals-Microorganisms Relationships and Applications, FEMS Symposium Abstracts, Societe Francaise de Microbiologie: Paris, 1993.
9. Fourest, E. and Roux, J.C. FEMS Microbiol. Rev. No. 14 (1994) 325.
10. Schiewer, S. and Volesky, B. Environ. Sci. Technol. 1995, 29, 3049.
11. Kratochvil, D.and Volesky, B. Water Res. (1998) (in press)
12. Schiewer, S., Fourest, E., Chong, K.H.and Volesky, B. In Jerez, C.A., Vargas, T., Toledo, H.and Wiertz, J.V. (eds.).Biohydrometallurgical Processing: Proceedings of the International Biohydrometallurgy Symposium; University of Chile, Santiago, 1995.

Biosorption of free and complexed cadmium ions by *Aspergillus niger*

L.H. Rosa[a], P.F. Pimentel[a], M.M. Figueira[b], L.C.S. Mendonça-Hagler[c] and N.C.M. Gomes[ac]*

[a]Department of Biotechnology and Chemical Technology – CETEC, Belo Horizonte, MG, Brazil

[b]Graduate Program in Metallurgy and Mining, Federal University of Minas Gerais, Brazil

[c]Institute of Microbiology Prof. Paulo de Góes, Federal University of Rio de Janeiro, Brazil

Metal ion adsorption by microorganisms is commonly denoted as biosorption and does not depend on cellular metabolic activity. In this work the biomass of the fungus *Aspergillus niger* was tested for the capacity to accumulate cadmium in the presence and absence of an organic ligand (citrate). An overall improvement on the Cd sorption capacity of at least 15% was achieved with caustic treatment. The present work elucidates, to some extent, the effects of caustic treatment on the biosorption of Cd by *A. niger* in terms of improvement of the metal uptake as well as some changes in the biomass sites due to both the treatment and the presence of the metal.

1. INTRODUCTION

Rapid industrialization has led to increased disposal of wastes containing pesticides, organic and inorganic compounds and high concentrations of heavy metals. Cadmium, for example, is a heavy metal that ranks with lead and mercury as one of the top three metals with respect to the hazards that they pose to humans and the environment (1). In contrast to the decline in use of lead and mercury, cadmium use in industrial processes such as pigment manufacture and metal plating has been steadily increasing. Thus, the study of cadmium removal from industrial waste using biosorption to facilitate its subsequent recovery is of high interest (2). In general, treatment of effluents contaminated with metals involves physicochemical processes of flocculation and/or precipitation, electrolysis, crystallization and adsorption. However, these processes can be expensive and/or lead to production of new contaminants. Biosorption of heavy metals on fungi occurs as a result of ionic interactions and complex formation between metal ions and functional groups present on the fungal cell surface. Functional groups such as carboxyl, phosphate and amino groups were observed to be responsible for biosorption of heavy metals (3). In the present work, the capacity of *Aspergillus niger* to accumulate cadmium in free and complexed forms was evaluated.

---

* CX Postal. 2306, 31.170-000, Belo Horizonte, MG, Brazil. e-mail: gomesncm@tulipa.cetec.br

2. MATERIALS AND METHODS

2.1. Microorganism

The fungus *Aspergillus niger* was isolated from samples obtained from a gold extraction plant as described previously (4). Spores were inoculated in shake flasks containing Sabouraud liquid medium (0.5% yeast extract, 1% peptone and 2% glucose) and incubated at 150 rpm and at 30°C for 72 hours. The biomass was separated from the medium by filtration, washed with distilled deionized water and then dried at 60°C overnight.

For acidic treatment, *A. niger* biomass was prepared by washing with 0.2 M H_2SO_2 and then rinsing it with distilled water (5).

For caustic treatment, the acid-treated biomass was treated with 200 ml of 0.75 M NaOH at 70-90°C for 10 -15 min.(6).

Dried *A. niger* biomass was added to 25 ml of distilled deionized water to a final concentration of 1 mg (dry weight) ml^{-1} in 250 ml Erlenmeyer flasks. Free and complexed cadmium ions, Cd^{2+} and Cd-citrate⁻ were added to the suspensions to the desired final concentration. The flasks were incubated for 3 h (the pH of the samples was adjusted to 7.0 every 30 min. with 1 N HCl or NaOH). The biomass was then centrifuged (22,100 g, 10 min.), and the supernatant was retained for cadmium analysis. The amount of metal removed was determined by the difference in metal content between flasks containing no biomass (control) and flasks containing the biomass (test):

$$\text{Metal uptake } q = (C_i - C_f)\, V/M \tag{1}$$

where C_i and C_f are the initial and equilibrium (final) metal ion concentrations (mgl^{-1}) respectively, V is the volume of sample solution (L), and M is the dry weight of the biomass added (g).

2.2. Analytical methodology

Metal analyses were performed using an inductively coupled plasma emission spectrophotometer (model Optima 3000 according to standard methods (7).

Infrared (IR) spectra of cadmium loaded and unloaded *A. niger* biomass with and without caustic treatment were recorded on a Perkin Elmer FTIR Spectrometer PARAGON 1000 PC. Cadmium unloaded biomass was previously protonated (acidic treatment) to release any metal ion from *A. niger* thus revealing the acidic, non-loaded groups in the biomass (8). Ten milligrams of biomass was encapsulated in 90 mg of KBr. Translucent disks (> 20mm) were obtained by pressing the ground material with the aid of a bench press (10,000 Kg for 10 min).

3. RESULTS AND DISCUSSION

In the present work the ability of the *Aspergillus niger* biomass, with and without chemical treatment, in scavenging cadmium was investigated in the presence and absence of citrate. The biomass without chemical treatment was able to sorb Cd^{2+} and Cd-citrate$^{(-1)}$ at different pH values, with the highest rate occurring at the pH 7.0 (data not shown). According to thermodynamic assumptions of the composition of different solutions (Table 1), at pH 7.0 most of the Cd present in Cd-citrate solution would be as Cd-citrate anion (96.2%), while the inorganic metal salt would be present in approximately equivalent concentrations in the forms of mono and divalent cations ($CdCl^+$ and Cd^{2+}). However, one should note that the species distribution obtained by the calculations through MINEQL program is that of the solution in equilibrium and thus does not consider the kinetics of the reactions involved.

Table 1
Speciation of cadmium at pH 7.0, 25°C, calculated by the thermodynamic data using the program MINEQL+ (9)

Solution	Metal species	% of total Cd
	Cd^{2+}	57.6
CdCl$_2$	$CdCl^+$	42.0
	$CdCl_2$	1.4
Cd-citrate	Cd^{2+}	2.4
	$CdCl^+$	2.4
	Cd-citrate [(-1)]	96.2

Note: Cd-citrate solution was prepared by adding $CdCl_2$ to a citrate solution, which is why part of the Cd was present as $CdCl^+$ in this solution.

The caustic treatment of the biomass improved the Cd uptake capacity of *A. niger* both in the presence and absence of citrate. In general, alkaline substances may increase the metal ion uptake capacity of some microorganisms (6). The inhibition of metal biosorption by complexing substances can be attributed to the strength of subsequent metal-complex generated and/or due rejection of the anionic charged microbial cell surface for anionic metallic complexes such as Cd-citrate[(-1)].

Figure 1 represents the sorption of free and citrate-complexed Cd ions by *A. niger* before and after caustic treatment. In both biomass types, a relatively higher maximum uptake of free Cd can be observed (~ 8 mg.$^{-1}$ and 10 mgg^{-1} for non-treated and treated biomass, respectively). The treatment of the biomass with NaOH led to an increase of the maximum uptake of both free and complexed Cd of 15 to 20% when compared to the non-treated biomass. This proves the advantage of using this type of treatment in improving the capacity of *A. niger* biomass.

Infrared spectra of *A. niger* biomass before and after caustic treatment were obtained to evaluate the possible effects of caustic treatment on the cell surface, as well as to identify the chemical groups involved in the cadmium biosorption (Figure 2). Comparing the spectra of the untreated biomass with that after caustic treatment, one can conclude that no major changes of carboxylic sites have happened after the treatment. However, shifts of troughs at 852 and 899 cm^{-1} to lower frequencies (816 and 855 cm^{-1}) were observed after the caustic treatment of the biomass. These results suggest that caustic treatment cause changes in the structure of the sulfate groups of the biomass. Alkaline substances are commonly able to dissolve proteins and hydrolyze lipids bound to active sites in biomass, thereby increasing the capacity of cell wall to sorb metals from solutions (10).

516

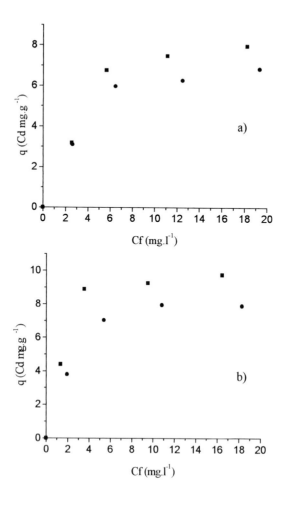

Figure 1. Representative adsorption isotherms of Cd (■) and Cd-citrate (●) by *Aspergillus niger* biomass before (a) and after caustic treatment (b) at 27°C, pH 7.0.

Both spectra of unloaded biomass (Figure 2 a and b) display troughs at 1740 cm^{-1}, corresponding to the stretching of the free carbonyl double bond from the carboxyl functional group. After contacting the caustic treated biomass with $CdCl_2$ and Cd-citrate solutions (Figure 2 c and d, respectively), the biomass exhibited spectra with clear shifts of the carbonyl stretching band to lower frequencies (1653 and 1652 cm^{-1}, respectively). This shift is typical for the complexation of the carbonyl group by dative coordination (11).

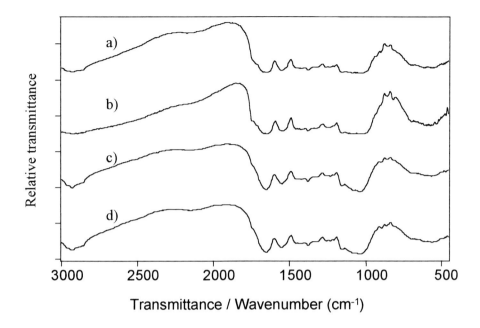

Figure 2. FTIR spectra of *Aspergillus niger* biomass before (a) and after (b) caustic treatment, and the loaded with Cd (c) and Cd-citrate (d).

Transmittance troughs also are shown around 850 cm^{-1} in both cadmium unloaded biomass types, which corresponds to S=O bounds. These were shifted to higher frequencies (860 and 890 cm^{-1}) after free and complexed cadmium sorption, suggesting that sulfate groups may also participate in the process of cadmium biosorption. These functional groups were also found important in the uptake of Cd and Fe in other biosorbent prepared with the seaweed *Sargassum* biomass (8, 12).

It is not totally clear whether the sorption of Cd-citrate would be due to the participation of carboxyl and sulfate groups of the biomass, since repulsion would be expected between anionic complexes and those groups. However, the citrate solution composition for Cd species presented in Table 1 is only applicable for equilibrium state. An explanation for the similar changes observed in the FTIR spectra of the biomass exposed to both free and complexed Cd would be that, when in the experimental conditions, the solutions would not be completely in equilibrium. In this case, there would still be cationic forms of Cd present in solution which would bind to negatively charged groups in the biomass such as the carboxyl and sulfate groups. This explanation should be, however, investigated in terms of chemical analysis of the solution as well as other methods that could determine the metal species adsorbed onto the biomass, such as XPS (X-ray photoelectron spectroscopy) or the surrounding environment of the metal sorbed, like the EPR (electron paramagnetic resonance) technique.

4. CONCLUSIONS

Biosorption of metals by fungi has been previously described by several authors, and the initial analysis of the changes in functional groups in the biomass depending on the treatment used has been performed with the use of FTIR technique (3). However, there is no consistent work correlating the metal uptake capacity and the functional groups involved in the sorption process with the different treatment types used. The present work elucidates, to some extent, the effects of caustic treatment on the biosorption of Cd by *A. niger* in terms of improvement of the metal uptake as well as some changes in the biomass sites due to the treatment and the presence of the metal.

Relatively scant information is available on the biosorption of metals in the presence of ligands in solution (4). Industrial solutions commonly contain several organic substances that form complexes with metals at relatively high stability. The present work describes the differences on the metal sorption by the biomass due to the presence of ligands such as citrate. Although the overall yield of metal sorption was not as high as that of some other types of biosorbents, such as those produced with seaweed biomass (13), the potential use of *A. niger* biomass for treating metal-bearing solutions resides in its capacity of sorbing metal ions both in the presence and absence of metal ligands. While an overall improvement on the sorption capacity of at least 15% was achieved with caustic treatment, other suitable treatment of this type of biomass, which can be easily obtained from fermentation processes, should be further investigated in order to apply this technique in large scale.

ACKNOWLEDGEMENTS

The Brazilian FAPEMIG Scholarship to L.H.R. and CNPq Scholarship support for N.C.M.G. are gratefully acknowledged. This work was funded by the PADCT project (430/95) and Fundação de Amparo à Pesquisa do Estado de Minas Gerais – FAPEMIG, in Brazil.

REFERENCES

1. B. Volesky (ed.), Biosorption of Heavy Metals, CRC Press, Boca Raton, 1990.
2. I. Aldor, E. Fourest and B. Volesky, Can. J. Chem. Eng., 73 (1995) 516.
3. A. Kapoor, T. Viraraghavan, Biores. Technol., 61 (1997) 221.
4. N.C.M. Gomes and Linardi, V.R., Rev. Microbiol., 27 (1996) 218.
5. M. M. Figueira, B. Volesky and V. S. T. Ciminelli, 54 (1997) 344.
6. Y. Lu, and E. Wilkins, J. Hazard. Mater., 49 (1996) 165.
7. American Public Health Association, APHA, Washington, 1992.
8. M. M. Figueira, B. Volesky and Mathieu, H.J., Environ. Sci. Technol., (1998) (submitted).
9. W.D., Schecher, Chemical Equilibrium Management System. MINEQL+ ver. 3.01b. Environmental Research Software, 1994.
10. T.R. MuraleedHaran and C. Venkobachar, Biotechnol. Bioeng., 35 (1990) 320.
11. K. Nakamoto (ed), Infrared and Raman Spectra of Inorganic and Coordination

Compounds - B. Application in Coordination, Organometallic, and Bioinorganic Chemistry; John Wiley and Sons, New York, 1997.

12. E. Fourest and B. Volesky, Environ. Sci. Technol., 30 (1996) 277.
13. Z. R. Holan, B. Volesky and I. Prasetyo, Biotechnol. Bioeng., 41 (1993) 819.

Specific metal sequestering acidophilic fungi

C. Durán, I. Marín and R. Amils

Centro de Biología Molecular Severo Ochoa, Universidad Autónoma de Madrid, Cantoblanco, Madrid 28049, Spain

Sixty acidophilic fungi isolated from the Tinto River (pH between 2 and 2.5 and high concentration of heavy metals) selected on the basis of their taxonomic diversity were exposed to increasing concentrations of selected metals (Ag^+, Hg^{2+}, Cu^{2+}, Zn^{2+}, As^{5+}, Cr^{3+}, Ni^{2+} and Cd^{2+}) to determine their resistance and sequestering profiles. Some of the isolates were extremely sensitive to the different metals assayed, while others were capable of growing in the presence of metal concentrations as high as 0.4 M. In general, the acidophilic fungi from the Tinto River are more resistant to heavy metals than the reference systems obtained from type collections, exhibiting a characteristic polyresistance profile which is extremely variable. Some of the fungi can sequester heavy metals with rather high efficiency. Metal sequestering is normally associated with metal resistance. In several cases specific heavy metal sequestering was observed. Preliminary data obtained with P34, a *Penicillium* isolate, suggested that the mechanism of specific copper sequestering (33% at 100 mM of Cu^{2+}) depends on active cell growth, involving metal transport and formation of cellular inclusions. The amount of metal absorbed in non growing or disrupted cells was rather low. The copper sequestering specificity of P34 was maintained in the presence of complex metal mixtures of industrial bioleaching solutions. The potential applications of these fungi in biohydrometallurgy are discussed.

1. INTRODUCTION

The presence of relatively high concentrations of toxic heavy metals in industrial waste waters constitutes a complex environmental issue since the difficulties involved in separating its different metal components impedes recycling and makes costly, environmentally safe disposal methods necessary. Different separation techniques, such as differential precipitation or ionic exchange have been described in detail, but most of them are not feasible on an industrial scale, because they require strict control of the physico-chemical parameters of the solutions to be treated, or simply because they are too expensive, as in the case of ionic exchange.

The discovery that many biological systems are not victims of heavy metal toxicity, but active agents of metal mobilization, dramatically changed this picture and opened up the opportunity to explore the importance of biological-mineral interphases, and, of course, their biotechnological potential. Metals are not only the source of energy for an important number of microorganisms (chemolithotrops), but they are also an absolute requirement for many enzymatic activities which use metals as cofactors, a probable memory of its inorganic origin.

Interestingly enough, the same metallic cations that are required for biological activity can be extremely harmful if the cells are exposed to rather high concentrations of the very same metals. Biological systems control the mobilization of appropriate toxic heavy metals. How then do the chemolithotrophic microorganisms, that obtain energy from the oxidation of metallic sulfides, overcome the toxicity produced by the toxic products of these reactions ? One possibility is that they induce metal resistance mechanisms. Another might be that they precipitate or sequester the metal inside or outside of the cell.

In this context, it has to be mentioned that the cell walls of many microorganisms are negatively charged, a property that allows their ionic interactions with cationic metals, a phenomena known as biosorption. This mechanism is based on the same principles as ionic exchange, and in general is rather nonspecific and can be used to remove metallic cations from diluted solutions (1,2). On the other hand, some plants accumulate high concentrations (up to three orders of magnitude) of specific toxic heavy metals, an operation than requires an active transport system in addition to the correspondent resistance mechanisms. These hyperaccumulating plants are considered very useful to bioremediate industrial soils contaminated with heavy metals (phytoremediation) (3).

With the exception of the metal specificity associated with mechanisms of energy transduction (bioleaching), the most studied metal-biological system interactions in biohydrometallurgy are those related with biosorption, which, as mentioned, are rather nonspecific. Special attention has been devoted recently to the search for specific metal bioaccumulators, to facilitate metal recovery and recycling from industrial wastewaters.

The microbial characterization of the Tinto River, an extreme acidic environment with high concentration of heavy metals, allowed the isolation of an important number of fungi (over 400) which thrive in the extreme conditions found in the river (4). This property strongly suggests that they are active members of the heterotrophic community of the habitat rather than casual external resistant forms (5,6). Due to the extreme conditions of pH and the correspondent metal content present in the river, we decided to study the metal resistance for different members of the Tinto microbial community, in order to generate models of how they deal with this important ecological problem. As a consequence a screening program was devised to gather information about the metal resistance and sequestering abilities of selected acidophilic fungi from the Tinto River (6).

2. MATERIALS AND METHODS

For the resistance and sequestering screening, the selected fungal isolates were grown in the liquid media described in (7), supplemented with required metals obtained from concentrate stock solutions (1 M Ag_2SO_4, 0.1 M $HgSO_4$, 0.5 M $CuSO_4.5H_2O$, 1 M $ZnCl$, 1 M $Na_2HSO_4.7H_2O$, 1 M $CrCl_3.6H_2O$, 1M Ni $SO_4.6H_2O$ and 1 M $CdSO_4.8/3\ H_2O$). The initial metal concentrations were 0.1 mM for Ag, Hg, Ni and Cd, and 1 mM for Cu, Zn, As, and Cr, in a final volume of 100 ml. The flasks were inoculated with selected fungi grown in solid media (aliquots of 1/100). Flasks were shaken at 30°C at 150 rpm. Cells growth was followed for a week. Fungi able to grow in the presence of a given metal concentration were assumed to be resistant to it. In this case an inoculum (1/100) was transferred to a flask with a higher metal concentration in order to determine the limit of metal tolerance. To evaluate the metal sequestering associated to the growth of different fungi, aliquots of the metal solution before the fungal inoculation and after fungal growth were collected and their metal concentrations determined by atomic absorption and by TXRF (Total X-ray Diffraction). Control experiments were performed in the absence of fungal inoculation to follow any non biological metal disappearance promoted by the components present in the complex growth medium or transformed during the long lasting incubations. The metal biomass retention was measured after digestion of dried cells with nitric acid as described in (7). Scaning and transmission electron microscopy equipped with EDAX probes were used to analyze fungal biomass exposed to different heavy metal solutions. The molecular taxonomy used for the identification of the fungi was based on the amplification by PCR of ITS regions of the rRNA operons using ITS4 and ITS5 fungal primers.The PCR products and their restriction digestion fragments (*EcoR*I, *Sau*3A and *Rsa*I) were separated by conventional electrophoresis in the conditions described in (8).

3. RESULTS AND DISCUSION

3.1 Metal resistance and sequestering profiles

Sixty different acidophilic fungal isolates from the Tinto River were selected according to their taxonomic status, to ensure an appropriate diversity in the resistance and sequestering screening. Eight metals (Ag^+, Hg^{2+}, Cu^{2+}, Zn^{2+}, As^{5+}, Cr^{3+}, Ni^{2+} and Cd^{2+}) were selected for this study based on their toxicity, economic interest and presence in the Tinto habitat. Table 1 presents a selection of metal resistance profiles for different fungal isolates.

The screening for metal sequestering associated to cell growth showed that an important number of the fungal isolates could reduce the amount of metal in solution, some of them with an interesting level of specificity. Table 2 presents a selection of specific metal sequestering efficiencies for different acidophilic fungal isolates.

Due to the type of screening performed only those microorganisms capable of growing in the presence of a given concentration of metal have the opportunity to sequester the metal. Using this approach we focused our screening on the selection of metal sequestering mechanisms related with cell growth, that is to say, those which were dependent on cellular metabolism. Nonspecific absorption in the negative charged cell envelopes, although not excluded, was not favored. As a first approximation we did not try to differentiate the diverse mechanisms that could promote the metal removal from solution. Under the term sequestering it has to be considered: absorption to the negative charges of cell envelops, production of quelating agents, change in the metal redox potential and metal transport. The reproducible levels of specific metal removal from solution observed for several of the Tinto fungal isolates, allowed us to consider that metal transport might be involved in some of the observed metal sequestering.

As expected, a strong correlation was found between the ability to sequester a specific cation and the correspondent level of resistance. The reverse, however, was not always true, many isolates exhibited important levels of metal resistance which did not correspond to significant values of metal sequestering. Most of the fungal isolates from the Tinto River exhibited rather important levels of metal resistance (polyresistance), which underlines the ecological significance of this property (4,5). In general fungal isolates from the Tinto River are much more resistant to different heavy metals than the reference systems obtained from

Table 1 : Metal resistance values for different acidophilic fungal isolates

Fungal strain	higher resistance values	other resistances
Alternaria sp.I1 9	---	1 mM Cd
Alternaria sp.I14	1mM Ag	no significative
Aspergillus sp.P51	100 mM Cr	no significative
Aspergillus sp.P37	---	50 mM Cu, As and Cr
Bahusakala sp.O66	400 mM As, 1 mM Ag and Hg	10 mM Ni and Cd
Bahusakala sp.O62	1 mM Ag	no significative
Cladosporium sp.I18	400 mM Zn and Cr	50 mM Cu
Cladosporium sp.P72	400 mM As	100 mM Zn, 200 mM Cr
Hormonema sp.I12	400 mM As,1 mM Ag	200 mM Cr, 50 mM Cu and Zn
Hormonema sp.I17	1 mM Ag	50 mM Zn
Nodulisporium sp.V56	400 mM As,10 mM Ni	50 mM Cr, 10 mM Zn
Nodulisporium sp.V58	400 mM As	100 mM Cr
Penicillium sp.P54	1 mM Ag, 400 mM As	50 mM Cu, 100 mM Cr, 10 mM Cd
Penicillium sp.V82	1 mM Ag and Hg	10 mM Cr
Penicillium sp.P34	200 mM Cu,1 mM Ag	50 mM As, 100 mM Cr,
Scytalidium sp.O64	1 mM Hg, 400 mM As	200 mM Cr
Scytalidium sp.O74	1mM Ag, 400 mM As	10 mM Cu and Cr
Trichoderma viride O6	1 mM Ag, 10 mM Ni and Cd	200 mM As
T. viride O8	1 mM Ag	50 mM Zn
T. viride CECT 2423	---	no significative

Table 2. Percentage of metal removal after fungal growth

Fungal isolate	Metal concentration	% of metal sequestered
Alternaria sp.I14	0.1 mM Hg^{2+}	95
Hormonema sp.I12	200 mM Cr^{3+}	35
	10 mM As^{5+}	72
Penicillium sp.I25	200 mM Zn^{2+}	93
Penicillium sp.I28	1 mM Ni^{2+}	90
Penicillium sp.P34	100 mM Cu^{2+}	35
	10 mM As^{5+}	68
Scytalidium sp.P65	10 mM Cd^{2+}	90
Bahusakala sp.O66	1 mM Ag^{+}	66

type collections. Only a few isolates were unable to grow at the lowest concentration used in the screening, which might mean that they do not belong to the system, or that they can develop in structures that protect the sensitive microorganisms from the high metal concentration found in solution (biofilms).

3.2 Metal resistance and sequestering induction.

Different experiments were designed to learn more about the mechanism involved in the specific metal sequestering promoted by acidophilic fungi. The first was to determine whether the metal resistance/sequestering observed at rather high concentrations was already present in the isolates or required induction. Four isolates were chosen for this experiment according to their metal sequestering specificity. The maximum metal resistance value obtained by stepwise increase of the metal content in the growth medium was used to test the capacity of unexposed cells to grow by direct exposure to the correspondent metal concentration. Table 3 shows the different behavior exhibited by the selected fungal isolates for different metallic cations.

Table 3. Constitutive and inducible metal tolerance in selected acidophilic fungi

Fungal isolate	Maximum stepwise metal resistance	Direct metal resistance
Penicillium sp. P3	200 mM Cu^{2+}	none
Nigrospora sp. V12	1 mM Ag^+	1mM Ag
Penicillium sp. Y22	400 mM Zn^{2+}	none
Cladosporium sp. Y18	400 mM Cr^{3+}	none

As can be seen, the direct exposure to high concentration of Zn, Cr and Cu did not allow the growth of the correspondent fungal isolates, although resistance to these metal concentrations could be achieved through induction by exposing the cells to increasing concentrations of the metals. In the case of Ag it could be that the induction mechanism was not observed due to the difficulties of obtaining high concentrations of this metal in solution in the presence of the different components of the growth medium.

3.3 A study case: specific copper sequestering using Penicillium sp. P34

The specificity and efficiency for copper sequestering shown by the fungal isolate Penicillium sp. P34 not only in the presence of solutions of copper sulfate, but also with industrial bioleaching solutions from chalcopyrite concentrates, were strong arguments for its selection as a model case for further characterization of the mechanisms involved. To further validate that cell growth was required for an efficient Cu sequestering, 1 g of Penicillium sp. P34 cells grown in the absence of copper were exposed to a 100 mM Cu solution for 24 h in conditions in which cell growth was not favored (lack of energy source). An equivalent amount of cells from the same source were ground with sterile sand, to destroy cell structure and their viability, and were also exposed for 24 h to a 100 mM Cu solution, in identical ionic and pH conditions than for intact cells. Finally an aliquot of induced *Penicillium* sp. P34 cells was used as a diluted inoculum (1/100) to grow them for three and seven days in the presence of a 100 mM Cu solution in reach medium (YEPD). The results presented in Table 4 clearly show that growth is required for an efficient Cu sequestering. An efficiency of copper sequestering of 173 mg per gram of fungal biomass was obtained in these experimental conditions. Non growing or disrupted cells are only able to sequester a small amount of copper, probably by non specific absorption mechanisms.

Table 4. Cu sequestering efficiencies of *Penicillium* P34 depending on the growth conditions

cells status	biomass	exposure time	medium conditions	% of Cu sequestered
non growing cells	1 g	24 h	100 mM Cu	0.8
active growing cells	0.3 g	72 h	100 mM Cu + YEPD	10.7
active growing cells	1.1 g	7 d	100 mM Cu + YEPD	35
disrupted cells	1g	24 h	100 mM Cu	3.3

Penicillium sp. P34 cells were also able to sequester copper with similar specificity and efficiency when they were grown in the presence of a bioleached solution from a chalcopyrite concentrate. Using an industrial effluent with an initial concentration of 7000 ppm of Cu an efficient cell growth was obtained starting with a diluted inoculum (1/100) of P34, which removed 33% of the copper present in the solution. The remaining copper solution (after elimination of the cells by centrifugation) was complemented with concentrated YEPD medium, and a second cycle of P34 growth was started with a diluted inoculum. In this condition the efficiency of copper removal was 51% . The total efficiency of extraction using two cycles was 67%. The introduction of a new cycle allowed the removal of 86% of the total copper present in the complex industrial solution. If we used a concentrated inoculum (1/2) instead of a diluted (1/100) one a removal efficiency of 47% could be obtained in less time.

Table 5. Cu removal from an industrial bioleaching solution by *Penicillium* sp. P34.

number of cycles	type of inoculum	incubation time	removel efficiency
one cycle	1/100	6 d	33%
two cycles	1/100	6 d + 6	67%
three cycles	1/100	6 d + 6 d + 6 d	86%
one cycle	1/2	2 d	47%

The results presented in Table 5 outline the design of different strategies for Cu removal from complex industrial solutions, which should be tested at pilot plant scale. It can be concluded from these experiments that sequential growth cycles with an appropriate dilution factor could be used for an efficient removal of cooper from a complex industrial metal mixture.

The transmission electron microscopy observation of *Penicillium* sp. P34 grown in the presence of high concentrations of Cu showed the formation of vesicular structures with dense inclusions in the cells exposed to copper (Figure 1).

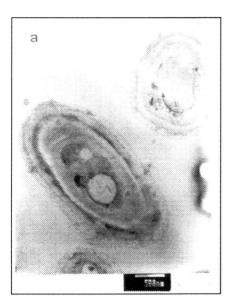 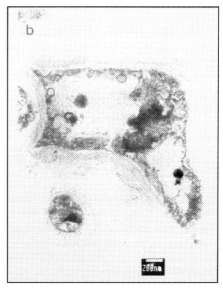

Figure 1 . Transmission electron microscopy of *Penicillium* P34 cells grown a) in the absence of heavy metals, and b) in the presence of 3500 ppm of Cu solutio.

The analysis of the transmission electron microscopy preparations by an EDAX probe showed that the dense particles that appear when the fungal cells are grown in the presence of copper have an intense diffraction signal corresponding to this cation which is absent in the control cells, suggesting that an active transport system followed by cytoplasmic compartamentalization might be responsible of the specific copper sequestering properties detected in the acidophilic *Penicillium* isolate P34.

528

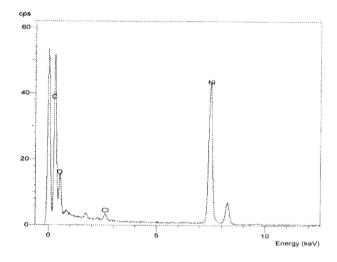

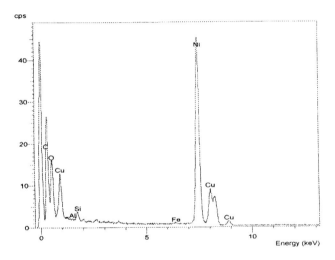

Figure 2. EDAX analysis of the transmission electron microscopic preparations shown in Figure 1. Panel a) corresponds to dense regions of unexposed control cells and panel b) to the dense regions of *Penicillium* sp. P34 cells exposed to 3500 ppm of industrial copper solution.

In contrast to the important amount of information concerning bioabsorption, the references related to metal sequestering associated with fungal growth are rather limited (for a review see reference 2,). Although, it has been reported that yeasts can accumulate grater amounts of metals by transport than by biosorption, for many filamentous fungi it appears than biosorption accounts for a major portion of total metal uptake (2) As mentioned in the introduction, one of the main goals of our study was to understand the mechanisms by which the acidophilic fungi from the Tinto River can cope with the high concentrations of heavy metals found in its waters. The metal resistance profile measured for a representative selection of the main taxonomic genus characterized so far in the river showed an outstanding level of polyresistance, which is probably a necessary adaptation to the habitat characteristics. The low levels of resistance found in the type collection reference systems underlines this ecological interpretation. In general, most of the isolates were more sensitive to Cu, As and Ni, than to the other cations measured, although we found several isolates that grow in the presence of high concentrations of these metals. Only a few isolates were unable to grow in the lowest concentrations of metals used. The resistance/sensitivities to different metals do not show any preferential taxonomic distribution, they are scattered throughout the different representatives of the taxonomic biodiversity of the habitat. In relation to this point is important to underline that conventional morphological fungal taxonomy has serious limitations.

An important effort is being made in this area to provide adequate molecular taxonomical tools which will ensure an appropriate level of taxonomic confidence to the screening (7).The specific copper sequestering exhibited by the *Penicillium* P34 isolate seems to be related with specific transport and intracellular storage in vesicular structures. We still do not know what are the cellular components involved in the bioaccumulation mechanism, but its reproducibility and efficiency opens up an important area of applications not only in bioremediation but also in metal purification. Different metal sequestering specificities have been also observed for other acidophilic fungal isolates from the Tinto River, which might be also of biotechnological interest. The close phylogenetic proximity of fungi and plants allow us to suggest that the high level of specific metal accumulation found in members of both kingdoms might be related.

4. CONCLUSIONS

- In general the acidophilic fungi isolated from the Tinto River are resistant to high concentrations of heavy metals.
- The resistance profile is quite variable and peculiar to each isolate
- Some acidophilic fungi show important metal sequestering properties
- Some of the isolates exhibit biotechnologically interesting metal sequestering specificity and polyresistance profile.
- One of those, *Penicillium* sp. P34 isolate, can efficiently sequester copper from complex industrial solutions. The copper sequestering seems to involve active transport and the formation of cytoplasmatic inclusions which concentrate this metal.
- A pilot plant is being designed to test the possible applications of P34.

ACKNOWLEDGEMENTS

This work was supported by grant AMB97-0547-C02-02 from the Comisión Interministerial de Ciencia y Tecnología and 07M/0180/1997 from the Comunidad Autónoma de Madrid (CAM), and by an institutional grant to the Centro de Biología Molecular Severo Ochoa from the Fundación Areces. C.D. is a posdoctoral fellow the C.A.M.

REFERENCES

1. Volesky, B. 1990. Bisorption of heavy metals, CRC Press, Florida (1990).
2. Gadd, G.M. Biotechnology, Rehm H.J., Reed, G. (eds), VCH, Weinheim, vol 6b (1988), 401.
3. Brooks, R.R (ed), Plants that hyperaccumulate heavy metals, CAB International, Wallingford, (1998)
4. López-Archilla, A.I., Marín, I., Amils, R. Geomicrobiol. J., 11 (1993) 223.
5. López-Archilla, A.I., Marín, I., Amils, R. Bbiohydrometallurgical Processing, Jerez, C.A., Vargas, T., Toledo, H., Wiertz, J.W. (eds), University of Chile, Santiago (1995) 63.
6. Durán, C., Marín, I., Amils, R. Proceedings of the International Biohydrometallurgy Symposium 97, Australian Mineral Foundation, Glenside (1997), E-ROM2.1
7. López-Archilla, A.I., Marín, I., González, A., Amils, R. Fungal identification techniques, Rossen, L., Rubio, V., Dawson, M.T. and Frisvad, J. (eds), European Commission, Brussels (1996) 202.
8. Boysen, M., Borja, M., del Moral, C., Salazar, O., Rubio, V. Curr. Genet., 29 (1996) 174

Chapter 5

Bioremediation

Microbial Leaching in Environmental Clean-up Programmes

K. Bosecker

Federal Institute for Geosciences and Natural Resources (BGR),
Stilleweg 2, D 30655 Hannover, Germany

Microbial leaching is a simple and effective technology for extracting valuable metals from low-grade ores and mineral concentrates. Besides the industrial application for raw materials supply, microbial leaching has some potential for remediation of mining sites, treatment of mineral industrial waste products, detoxification of sewage sludge and for remediation of soils and sediments contaminated with heavy metals. There is no routine treatment for toxic metals dispersed in solid materials, and autotrophic and heterotrophic leaching processes may be considered for environmental clean-up programmes. The problems of bioremediation for heavy metal-contaminated sites are very different from those of bioremediation for organic pollution, but intensive interdisciplinary collaboration in basic and applied research in this economically important field is expected to be very beneficial in the near future. It would be ideal if the bioremediation system maximised the extent and rate of degradation of waste materials, simultaneously minimising the level of toxic substances during the operation.

1. INTRODUCTION

As a consequence of technological and industrial development many industrial sites are contaminated with heavy metals and/or organic compounds. The main problem of this kind of anthropogenic waste is that it is toxic to organisms of any kind and particularly to human beings. Therefore it is important to remove or minimise these metal and organic contaminants to minimise the danger to health when consuming food and breathing the air and to prevent contaminants being dissolved and distributed by surface and groundwater.

Organic contaminants may be degraded biologically and CO_2 and water are the final products [1-3]. In contrast to organic compounds heavy metals cannot be decomposed, either biologically, chemically or physically. Metals may be solubilised, they may be changed in valence or in chelating state, they may be immobilised, but they are still metals. At low concentrations many metals are essential parts of metabolic processes but at high concentrations they are toxic and careful clean-up programmes are necessary to save our environment.

Microbial leaching is a simple and effective technology used for metal extraction from low-grade ores and mineral concentrates [4-6]. Metal recovery from sulphide minerals is based on the activity of chemolithotrophic bacteria, mainly *Thiobacillus ferrooxidans*, *T.thiooxidans* and *Leptospirillum ferrooxidans*, which convert insoluble metal sulphides into soluble metal sulphates. Non-sulphide ores and minerals can be treated by heterotrophic bacteria and fungi. In these cases metal extraction is due to the production of organic acids and chelating and complexing compounds excreted into the environment. In addition, metals may be solubilised or immobilised by chemical or biochemical changes in oxidation states. Besides the industrial application for the raw materials supply, microbial leaching has some potential for the environmental clean-up of mining sites, treatment of mineral industrial waste products, detoxification of sewage sludge and for remediation of soils and sediments contaminated with heavy metals.

As reported by A. O. Summers [7] the basic problem of metal bioremediation is "...the occurrence of metals in soil, rock or sediment, or in water, at concentrations too dilute to be worth mining (if the metals are valuable) or sufficiently concentrated to be an environmental concern (if the metals are toxic). The process employed for metals that are more valuable than toxic is called biorecovery and the hope of cheaper and more efficient strategies for their enrichment and re-use is the driving force for interest in this area. For metals that are toxic but not intrinsically valuable the processes are referred to as bioremediation and the driving forces are increasingly strict standards for water and air quality."

2. BIOREMEDIATION FEASIBILITY

2.1. Mining sites

Intensive mining and ore processing have produced billions of tonnes of waste all over the world, a continuing process. Depending on the mineral composition of the mine tailings and waste rocks, and affected by the environmental geological and climatic conditions, mining sites represent a serious hazard to the environment. Tailings containing heavy metal sulphides are oxidised by weathering processes accelerated by naturally occurring sulphur and iron-oxidising bacteria, and the dissolved metals will contaminate ground and surface waters. This phenomenon is recognised world wide as acid mine or acid rock drainage (AMD/ARD) and several technologies have been developed for the treatment of the acid and metal containing mine waters. Treating the mine water however, does not mean removing the root of all evil. For this reason techniques for deactivating the pollution source and inhibition of metal dissolution are developed (neutralisation, revegetation). On the other hand heavy metal-contaminated mine sites can be remediated by using adapted, sophisticated heap and dump leaching technologies where optimum conditions for the growth of the leaching micro-organisms are kept constant and any seepage of the leachate is prevented. In this way both processes, remediation of mining sites and recovery of valuable metals, may be achieved simultaneously [8].

2.2. Mineral industrial waste materials

Mineral industrial waste products (fly ash, slag, incineration cinders) often contain substantial amounts of toxic metals, which in the case of inadequate disposal may be mobilised and may cause hazardous environmental problems if they reach the soil or the groundwater. Meanwhile, special waste disposal sites exist and special waste treatment plants have been established in several countries. Very often waste disposal is subject to strong governmental restrictions. Therefore the costs of waste removal continuously increase, and environmental protection laws may create economic problems.

In the case of industrial mineral residues conventional bioleaching may fail, since most metals are present mainly as oxides rather than as sulphides. Metal oxides in such residues can be leached by microbial acid production, e.g. sulphuric acid generated by *T. thiooxidans*. In some cases, chemical acid leaching is easier. Bioleaching with *T. thiooxidans* may be advantageous when in chemical leaching high costs would arise for the transport of the acid and on the other hand, sufficient sulphur for bacterial acid production is cheaply available. Another advantage is that as a consequence of sulphuric acid production during the growth of *T. thiooxidans* the pH in the leach suspension falls only gradually, so that the metals pass into solution at different rates corresponding to their solubilities, and can be separated from the leaching suspension selectively. Residues containing metal carbonates and silicates may be remediated by organic acids produced by heterotrophic bacteria and fungi. In these cases one of the main problems is in searching for cheap organic substrates necessary for the growth of heterotrophic microorganisms. [9,10].

2.3. Sewage sludge

Sewage sludge has been used as fertiliser on farmland for a long time. But with the realisation that sewage sludge might be seriously contaminated by toxic metals this common practice of waste management has become no longer acceptable or has even been banned. Since the work of Tyagi and Couillard [11,12] who isolated chemolithotrophic thiobacilli tolerant to organic compounds, previous detoxification of sewage sludge by bioleaching seems to be a practicable alternative. In the meantime an appropriate technology is being developed [13].

2.4. Aquatic sediments

Disposal of sediments from rivers and docks causes major environmental problems because most sludge contains high amounts of toxic metals. When exposed to air, due to oxidation processes, the sludge turns acid and heavy metals are mobilised. At present remediation techniques are being developed based on accelerating naturally occurring bioleaching by activating the autochthonic thiobacilli. The addition of elemental sulphur increases acidification, followed by an increase in metal extraction [14,15].

2.5. Soils

Because the most important metal leaching bacteria (*Thiobacillus ferrooxidans* and *T. thiooxidans*) have been found to be sensitive to even low concentrations of organic substances [16] only a few efforts have been made to apply bioleaching techniques to remediation of contaminated soils. In preliminary studies we were able to show that metals mobilising bacteria could be isolated from various heavy metal-contaminated soils. The enrichment cultures solubilised toxic metals to various extents, depending on the enrichment medium, the soil sample, and the type of contaminant. In some cases the heavy metal contamination was reduced to such an extent that threshold values were reached, recommended for almost unrestricted use of the soil [8,17].

2.6. Radioactive waste

Radionuclides are present in soils, ores and residues mainly as oxides, usually in crystalline form, insoluble and often co-precipitated with iron oxides. Direct dissolution is possible by enzymatic reduction from higher to lower oxidation state, indirect mobilisation occurs via solubilisation by microbial metabolites such as organic acids and chelating agents. Uranium contaminated soil was successfully remediated by bacterial leaching with *T. ferrooxidans* and at a laboratory scale 99% of uranium was extracted within 17 days. [18,19].

536

3. CONCLUSIONS

Bioleaching has serious potential for remediation of heavy metals contaminated materials. There is no routine treatment and leaching processes using autothrophic or heterotrophic microorganisms may be considered for environmental clean-up. As with other bio-hydrometallurgical techniques remediation by bioleaching is of great economic advantage because biohydrometallurgical processes are low in capital and energy costs, they show high flexibility, they may be used on site and do not cause environmental pollution. As a rule of thumb, bioprocesses are one-third to one-half the cost of conventional chemical and physical remediation technologies [8]. Although the problems of bioremediation for heavy metal-contaminated sites are very different from those with organic pollution, intensive inter-disciplinary collaborations in basic and applied research will be very beneficial in the near future [7]. Genetic improvement of metal solubilising microorganisms, whether by mutation and selection or by genetic engineering, will allow the bioremediation processes to be improved. It would be ideal if the bioremediation system yielded maximum degradation of waste materials and minimised the hazardous risk potential at the same time[8].

REFERENCES

1. H.D. Skippers and R.F. Turco (eds.), Bioremediation - Science and Applications, SSSA Special Publ. No. 43, Madison, 1995.
2. G.A. Lewandowski and L.J. De Filippi, Biological Treatment of Hazardous Wastes, John Wiley & Sons, Inc., New York, 1998.
3. B.S. Schepart (ed.), Bioremediation of Pollutants in Soil and Water, ASTM Publ. Code No. 04-012350-48, Philadelphia, 1995.
4. K. Bosecker, FEMS Microbiol. Rev. 20 (1997) 591.
5. G. Rossi, Biohydrometallurgy, McGraw-Hill Book Company GmbH., Hamburg, 1990.
6. A.E. Torma, in: H.J. Rehm and G. Reed (eds.), Biotechnology Vol. 6 B, 367, VCH Verlagsgesellschaft, Weinheim, 1989.
7. A.O. Summers, Current Opinion in Biotechnol., 3 (1992), 271.
8. G.A. Torma, P.-Ch. Hsu and A.E. Torma, Proc. 2nd. Internat. Symp. on Extraction and Processing for the Treatment and Minimization of Wastes, Scottsdale, Arizona, Oct. 27-30, 1996.
9. K. Bosecker, Acta Biotechnol., 7 (1987) 487.
10. C. Brombacher, R. Bachofen and H. Brandl, Appl. Environ. Microbiol., 64 (1996) 1237.
11. R.D. Tyagi and D. Couillard, Process Biochem., Aug.(1987) 114.
12. D. Couillard and S. Zhu, Water, Air, Soil Pollut., 63 (1992) 67.
13. T.R. Sreekrishnan and R.D. Tyagi, Environ. Technol. 15 (1994) 531.
14. H. Seidel, J. Ondruschka and U. Stottmeister, in: W.J. van den Brink, R. Bosman and F. Arendt (eds.), Contaminated Soil'95, 1039, Kluwer Acad. Publ., Netherlands, 1995.
15. D. Couillard, M. Chartier and G. Mercier, Rev. Sci. Eau, 7 (1994) 251.
16. J.H. Tuttle and P.R. Dugan, Can. J. Microbiol., 22 (1976) 719.
17. M. Bock and K. Bosecker, in: DECHEMA Monographs Vol. 133, Biodeterioration and Biodegradation, 639, VCH Verlagsgesellschaft, Weinheim, 1996.
18. A.J. Francis, J. Alloys Compd., 213/214 (1994) 226.
19. R.G. Reddy, S. Wang and A.E. Torma, New Remediation Technology in the Changing Environmental Arena, B.J. Scheiner, T.D. Chatwin, H. El-Shall, S.K. Kawatra, A.E. Torma (eds.), 157, SME, Littleon, Colorado, 1995.

Influence of bacteria and sulphite ions on the transformation of pyritic tailings: shake flask tests

C. García[a], A. Ballester[b], F. González[b] and M.L. Blázquez[b]

[a]Departamento de Tecnología Industrial. Universidad Alfonso X el Sabio. Madrid. Spain

[b]Departamento de Ciencia de los Materiales. Facultad de Ciencias Químicas. Universidad Complutense. 28040 Madrid. Spain

Mining and metallurgical tailings, once dumped, are exposed to changing weather conditions. The action of natural agents such as water, oxygen, bacteria and other complex factors, transforms and, at the same time, modifies the composition and causes deterioration of the tailings. In the present study, the affect of two factors on the transformation of pyritic tailings from the flotation of a complex sulphide ore is analyzed. Shake flask tests were carried out to determine the behaviour of the tailings under the influence of bacterial catalysis and the reducing conditions induced by the presence of sulphite ions in the system. As expected, the presence of lithotrophic bacteria in the system catalyzed the tailings (pyrite) oxidation. On the other hand, during the flotation of these ores, the sulphite ion is used to condition the medium in which the solid is concentrated resulting reducing conditions. The presence of sulphite ion influenced the subsequent bacterial action and controlled the final transformation of the solid residue.

1. INTRODUCTION

An abandoned mine, a mine waste, or any heap of ore exposed to the changing environmental conditions suffers important transformations giving rise to a phenomenon named as weathering. When the residue is a sulphide ore these transformations produce acid mine drainages (AMD). To study these transformations different tests have been designed to describe the changes concerning an abandoned residue using controlled laboratory conditions. To design the weathering tests it is necessary to know previously several geological and climatic factors as the rain frequency, temperature fluctuations, etc. with the aim of deciding the appropriate and specific tests to follow the deterioration of a specific residue. Although there are no concrete rules defining this experimentation, different tests have been proposed which lay down the guideline and determine, in general, the capacity of a residue either to produce acid or to neutralize it. Among them, several static and kinetic tests are included.

From the beginning of the studies related to the generation of acid waters, it has been shown that different acidophilic bacteria of the genus *Thiobacillus* (*T. ferrooxidans* and *T. thiooxidans*) and *Leptospirillum* (*L. ferrooxidans*) are the responsible for the oxidation of pyrite and other metal sulphides with the consequent production of acidity. Another important group of microorganisms related to the transformation of the compounds appearing in the mine waters are sulphate reducing bacteria (SRB) which, at the same time, reduce the sulphate of the medium to sulphide, increase the pH and, as a consequence, facilitate the precipitation of the dissolved metals.

In the present work, a study was carried out of different factors involved in the transformation when mining solid residue derived from a polymetallic sulphide flotation plant, was placed in a pond. As starting point, the final consequences of these transformations were known but the different factors influencing the system and the mechanism behind the observed phenomena were unknown. For this, several factors influencing this type of system were studied to prove, in comparison with the real data, their implication on the process. In this sense, two static weathering tests were used: the ABA test and the shake flask test. The ABA (Acid-Base Accounting) test was chosen because it produces quick results giving a first idea of the water quality after the residue is abandoned. On the other hand, the shake flask test permits performance of controlled experiments to determine the influence of particular variables such as temperature, microbiological activity, etc. In this case, the influence of two factors, the presence of bacteria and the sulphite ions, was studied. It is necessary to underline that the tests presented in this paper are a part of a global study of the system in which kinetic weathering tests and column experiments were carried out.

2. MATERIALS AND EXPERIMENTAL METHODS

2.1. Site

The site studied was a real system consisting of a pond of pyritic residues belonging to a mining company from the pyritic belt in the Southwest of Spain. The residues came to the pond from a flotation plant producing three differential concentrates from a polymetallic complex sulphide.

2.2. Sampling

To characterize the system, the pond was sampled as a whole. The samples from the bottom were collected using a special device named Ballcheck KB. All the samples were taken in duplicate and introduced into sterile flasks which were completly filled in to avoid the presence of oxygen and the possibility of future oxidation.

2.3. Chemical and mineralogical analysis

Solid samples

For the characterization of the solids contained in the samples, X-ray diffraction (Philips X'Pert-MPD) and granulometric analysis (Microtac FRA) were carried out. The chemical composition was determined after acid digestion of the samples; the subsequent analysis of the metals in solution was performed by Atomic Absorption Spectrophotometry (AAS) (Perkin-Elmer 1100B). The sulphur content was determined using an automatic analyser from Leco.

Liquid samples

The pH evolution was followed using a Crison 2001 electrode. For sulphate analysis a photocolorimeter (Metrohm 662) was used which determined the turbidity of $BaSO_4$ precipitates formed by reaction of sulphate ion with barium chloride (1). The concentration of sulphite ion was measured using an automatic titroprocessor from Metrohm and registered the potential at which the reaction between sulphite ion and iodine took place (1,2).

The concentration of metals in the samples taken in the water pond and from the shake flask tests was measured by means of AAS.

2.4. Microbiology

Samples taken from the different points of the pond as well as the solid sample used to carry out the weathering experiments were microbiologically characterized. For microorganism isolation the following culture media were used:

Aerobes

9K medium without iron: $(NH_4)_2SO_4$, 3 $g.L^{-1}$; KCl, $0.1g.L^{-1}$; $K_2HPO_4.3H_2O$, 0.618 $g.L^{-1}$; $MgSO_4$, 0.5 $g.L^{-1}$; $Ca(NO_3)_2.4H_2O$, 0.013 $g.L^{-1}$.

- Heterotrophs (bacteria, yeasts, and fungi):

Medium A: 9K medium without iron at pH 3 with 10 $g.L^{-1}$ glucose; 5 $g.L^{-1}$ yeast extract; and 15 $g.L^{-1}$ agar.

Medium Y: 9K medium without iron at pH 3 with 0.5 $g.L^{-1}$ bactotryptona; 1 $g.L^{-1}$ malt extract; 10 $g.L^{-1}$ glucose; and 15 $g.L^{-1}$ agar.

- Chemolitotrophs

9K medium containing one of the following compounds as energy source: iron, tetrathionate, thiosulphate or elemntal sulphur. pH was adjusted to 2, 3 or 6 with sulphuric acid.

Anaerobes

Postgate's C medium was used (ASTM D 4412-84). The same medium but with sodium molybdate was also used as a control test.

All the cultures were maintained at 30ºC. Bergey's Manual of Determinative Bacteriology was used for the identification of the isolated microorganisms. To count the microorganisms the most probable number (MPN) method was used (1,3).

2.5. Experimental tests

ABA Test

The ABA test was carried out in duplicate in accordance with the Method 1312 introduced by EPA (Acid-Base Accounting, (4).

Shake flask tests

To perform these tests 25 g of ore, previously dried and ground, were put in an Erlenmeyer flask together with 250 mL of either distilled water or 9K medium at pH 2 tests in the presence of bacteria. To carry out the sterile tests, in addition, 20 mL of a 2 % thymol solution were added. The cultures were maintained during one month with orbital agitation and at 35ºC. Periodically a 2 mL sample was taken out to determine pH, metals (Fe, Cu and Zn) in solution and sulphate concentration. Furthemore, in each sample, the number of bacteria was checked. In the tests carried out in the presence of sulphite ion, 0.1 g/L of this were added as sodium sulphite was added.

In addition the behaviour of the pulp leaving the flotation plant was also studied. In this case 100 g of sludge from the plant were placed in the 250mL flask as initial ore.

3. RESULTS AND DISCUSSION

3.1. Chemical and microbiological characterization of the tailings

The residue discharged from the flotation plant was a basic pulp of pyrite with 30% of solids. The chemical composition of the water contained in this pulp is shown in Table 1 in the column "Initial discharge". From a mineralogical point of view the sample was a pyritic ore with the following chemical analysis: 0.5% Cu; 2.83% Pb; 0.05% Zn; 38.8% Fe; and 37.6 % S. The mineralogical composition of the solid contained in the pulp was: chalcopyrite, 1.4%; sphalerite, 0.8%; galena, 0.5%; pyrite, 63.4%; carbohydrates and silicates, 34.0%. The particle size of the solid was between 25 and 30 μm. The number of microorganisms contained in the pulp in the initial conditions was: 18,300 cells/mL of sulphur-oxidizing lithotrophic bacteria and 8,060 cells/mL of total heterotrophs. Anaerobic SRB and iron-oxidizing lithotrophic bacteria were not detected.

During the storage of the pulp in the pond two different phenomena took place. On the one hand, the discharged basic pulp transformed producing a new residue, an acid water in which the concentration of metals and sulphate ion had appreciably increased (Table 1, column "Final AMD"). Only one ion, the sulphite ion, virtually disappeared during the transformation. On the other hand, a chemical and microbiological gradient appeared when the system was tested from the surface to the bottom. This gave rise to a new distribution of chemical species (Table 2) and a microbiological partitioning as a function of these new conditions (Table 3). As the data show, in the pond surface the conditions were clearly oxidizing with a potential of 405 mV (versus SHE) which favoured the decrease of the pH in the medium up to 3 and the consequent release of metals to the aqueous solution.These conditions were uniform in the surface layer of pond water.

Iron requires a special attention because the ferric/ferrous relationship can be a reference to the bacterial activity. The iron contained in the surface samples of the pond water was basically ferric ion: of the 50 mg/L of total iron contained in the water sample only, 1.7 mg/L corresponded to ferrous iron. This reveals the possible microbiological contribution to the surface oxidation in the pond water, because the chemical oxidation of ferrous ion in an acid medium is a very slow process.

At the same time a microbiological selection took place which was characterized by the presence of acidophilic lithotrophic bacteria in the surface water: *T. ferrooxidans* and *T. thiooxidans* were identified. In the samples taken from the pond bottom, SRB were found. A number of heterotrophs, including fungi and bacteria, in different parts of the system. (5). The microorganism distribution in the pond is shown in Table 3. To carry out the experiments in shake flasks a pyrite pulp from the discharge point in the flotation plant was used.

Table 1
Chemical analysis of the initial tailings discharge and the final residue (mg/l)

	Initial discharge	Final AMD
pH	9-10	2,5-3,5
Eh (versus SHE, mV)	(-225) - (+215)	425 - 555
Sulphates	500-650	1800-2000
Sulphites	88-120	2-3
Copper	0,1-0,2	0,4-0,8
Iron	0,4-2	50-55
Zinc	0,1-2	30-50
Lead	0,1-0,2	5-6
Calcium	400-450	400-450
Magnesium	15-20	50-55

Table 2
Chemical gradient in the pond (mg/l)

	Surface AMD	Bottom sludge
pH	2,5-3,5	6-7
Eh (versus SHE, mV)	425-555	(-250)-(-208)
Sulphates	1800-2000	400-500
Copper	0,4-0,8	0,3-0,6
Zinc	30-50	0,1-0,3
Iron	50-55	3,3-5

Table 3
Microbiological gradient in the pond (MPN.g^{-1})

Location	Lithoautotrophic aerobic bacteria	Anaerobic bacteria	Total heterotrophs
Surface water	$1,1x10^2-7,0x10^3$	0	$1,0x10^3$
Bottom sludge	$2,5x10^1-9,5x10^2$	$1,7x10^2- 2x10^3$	$6,0x10^1$

3.2. Weathering tests on the tailings

3.2.a. ABA test

For the digestion of the samples, 80 mL of 0.5N HCl were used with orbital agitation for 48 hours. The volume of NaOH consumed by the sample was 20.65 mL. The initial and final pH values were 0.6 and 8.3, respectively.

The calculated neutralizing potential of the samples was 12.22 and the acid potential of the samples obtained from the total contained sulphur was 1,175. Therefore, the net neutralizing potential (NNP) was -1,163 kg CaCO$_3$/t of sample, that is to say, negative and the conclusion was that the sample had capacity to produce acidity. This first approximation was later confirmed by means of other static and dynamic tests which showed the conditions and rates these acid producing samples were attacked. It is important to take into account that the interpretation of these types of results requires experience and care, because, samples showing a positive NNP did not always exhibit this behaviour either in the kinetics tests or in the field

542

experiments. However, this does not mean that under certain conditions these samples could not produce some acidity.

3.2.b. Bacterial catalysis

To study the influence of bacteria on pulp weathering three different experiments were prepared in which pulp was used as starting solid residue. In one test, sludge alone was used. In a second test, 10 mL of an inoculum of iron-oxidizing bacteria isolated from the pond were added. In a third test, a sludge with 20 mL of thymol was used with the aim of knowing the chemical evolution of the solid in the absence of bacteria. The inoculum (10 mL) was taken from a *T. ferrooxidans* culture which was previously centrifuged to avoid incorporating iron into the experiment.

The pH evolution in the three tests (Figure 1) proved that the presence of bacteria influences acid generation. Chemical factors decreased the pH to about 4, but microbiological activity was necessary to bring the pH closed to 2. The pH behaviour of the pulp only and the pulp with bacteria was similar. The explanation for this is that bacteria, indigenous to the sample, grew. Growth did not occur when a bactericide was added.

The changes in the pulp among the three tests were very evident when the behaviour of the total iron in solution was analyzed (Figure 2). In the presence of the bactericide practically no iron was solubilized, whereas in the presence of bacteria the iron concentration increased remarkably both when the bacteria populations were natural and when adapted bacteria were inoculated.

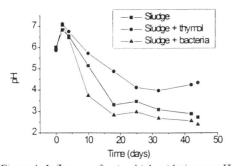

Figure 1: Influence of microbial oxidation on pH

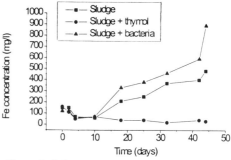

Figure 2: Influence of microbial oxidation on soluble iron concentration

The presence of bacteria associated with the initial ore was proven during system characterization and microorganism isolation. This test work determined whether the bacteria adhered to the solids or if they were free in the pulp water, and ascertained the bacterial activity in each case. Duplicate experiments were performed using dry ore obtained from the pond pulp. The ore, once dried, was finely ground before beginning the test; 25 g of ore in 250 mL of distilled water were used. Five different experiments were prepared: (a) ore only (O); (b) ore plus thymol (OT); (c) ore in 9K medium at pH 2 with 33.3 g/L of ferrous sulphate and a 10% inoculum of a *T. ferrooxidans* culture (OB); (d) ore in 9K medium at pH 2 with iron and thymol (OBT); and (e) ore in 9K medium at pH 2, 33,3 g/l, ferrous sulphate and a 10%

inoculum from a mixed culture of bacteria grown from a pulp of the same ore (OMC). The inocula were centrifuged before addition to the flasks to avoid the addition of ferric ion to the solutions. The pH in the test carried out with only ore (test O) evolved to acid values. This is difficult to explanation if only chemical factors are considered (Figure 3). In inoculated tests, pH change paralleled cellular growth, which was monitored by bacterial counts. Cell numbers appreciably increased as the conditions became more acid, hence, more suitable for the bacterial growth. This sharp decrease in the pH was not observed when a bactericide was used (test OT), although with thymol evaporation, some acidification was detected (Figure 3). Thymol was not replaced during the test. Bacterial monitoring using optical microscopy was carried out in the test with thymol. No appreciable cell groth was observed. In test O 10^6 cells/mL were counted; this number increased with time. The catalytic effect of bacteria on the ore oxidation was clearly detected. This effect was more evident in the test carried out in the presence of a mixed culture as well as an iron-oxidizing culture (Figure 3). The pH decrease was much more evident in the tests carried out in the presence of bacteria (Figure 3) than in the sterile test. The main difference between the experiment OMC, using a mesophilic mixed culture, and the experiment OB, using a *T. ferrooxidans* culture, was that pH values were much lower in the first case than those reached in the presence of *T. ferrooxidans*. The reason could be that in a mixed culture it is possible to find bacteria oxidizing both iron and sulphur. *T. thiooxidans* is common and can reduce the pH of the medium to 0.6 (6). In the test OMC pH values of 1.0 were reached.

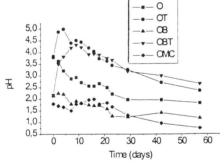

Figure 3: pH variation in shake flask tests. Legend: ore (O); ore + thymol (OT); ore + T. ferrooxidans + thymol (OB); ore + + mixed culture (OMC)

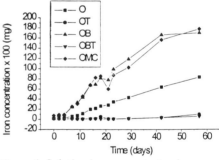

Figure 4: Solution iron concentration in shake flask tests. Legend: ore (O); ore + thymol (OT); ore + T. ferrooxidans + thymol (OB); ore + + mixed culture (OMC)

The variation of the total iron concentration observed in these tests reflected an important bacterial activity both in the inoculated tests (OB and OMC, Figure 4) and in the tests where indigenous bacteria grew (test M). Solubilization of iron in test M was paralleled to the bacterial growth curve. This trend was much more evident in the case of the tests inoculated with *T. ferrooxidans* (test OB, Figure 4). In tests performed in the presence of thymol (tests OT and OBT), oxidation of the ore was not detected. Sulphate concentration in the tests showed the same trend as iron (Figure 5). A higher concentration of sulphate in solution was observed in the test inoculated with *T. ferrooxidans*. Based on soluble sulphate, the tests carried out in the presence of thymol again showed the absence of ore leaching. In test O, in which the ore bacteria evolved freely, the sulphate concentration also increased appreciably in comparison with the test where the bacterial activity was non-existent although the oxidation level was always lower than that detected in the experiment inoculated with *T. ferrooxidans*.

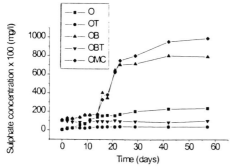

Figure 5: Changes in sulphate ion concentration in shake flask test. Legend: ore (O); ore + thymol (OT); ore + T. ferrooxidans + thymol (OB); ore + + mixed culture (OMC)

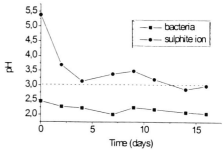

Figure 6: Influence of bacteria and sulphite ion on pH during weathering

3.2.c. Sulphite ion affects

The affect of sulphite ion in the pond waters was reproduced by performing different experiments in shake flasks. A new test was designed whereby changes in the pulp in the presence of sulphite ion were compared with changes in the pulp in the presence of bacteria. For this, two flasks each containing 150 g of partialy leached pulp were prepared. Fresh solid was acid conditioned for 18 days at pH 3. The reason for using this type of raw material was to avoid the adaptation phase of the microorganisms to the extremely basic conditions of pulp fresly discharged from the flotation plant. In the first flask 10 mL of an inoculum of *T. ferrooxidans* were added to 150 g of acid conditioned oulp; in the second flask 0.1 g/L of sulphite ion as sodium sulphite was aded to the acidic pulp. Changes in pH and total iron concentration in solution as a consequence of ore weathering are shown in Figures 6 and 7. Sulphite ion diminished (oxidation), as evidenced by a higher pH (Figure 6) and less iron being solubilized (Figure 7). In the presence of bacteria the final pH was 2 (Figure 6) and more iron was released (Figure 7). These results strongly indicate that sulphite affected the global chemical condition of the system because the final pH of the tailings pond was similar to the final pH of about 3 reached in the laboratory test with sulphite ion (Figure 6). The moderately reducing character of sulphite minimizes metals dissolution as demostrated by iron behaviour ilustrated in Figure 7.

4. CONCLUSIONS

The affect of microbiological oxidation of solid residue in the tailings pond is an important factor; however the final state of the system depended on all the factors (physical, chemical and microbiological) acting jointly. In laboratory, the bacteria oxidized the ore and produced waters with pH lower than 2; nevertheless pH values measured in the pond were higher. In the tailings pond there was a concominant relationship between the reducing conditions imposed by the presence of sulphite ion, and the microbiological activity. In the absence of sulphite ion the bacterial activity occurred and was responsible for both low pH values and the increase of metals and sulphate ion in solution. In the presence of sulphite ion moderately reducing conditions resulted in higher pH values in the tailings, decreased metals dissolution and minimized pyrite oxidation.

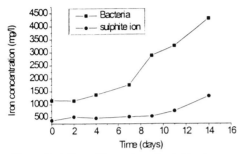

Figure 7: Affect of bacteria and sulphite ion
on iron solubilization during residue weathering

ACKNOWLEDGEMENTS

The authors would like to thank to the *Comisión Interministerial de Ciencia y Tecnología* (Spain) for supporting this research.

REFERENCES

1. APHA, Standard Methods for the Examination of Water and Wastewater, 17th edition, American Public Health Association, USA, Washinhton D.C., 1989.
2. K. Bhaskara and G. Rao, Analytica Chimica Acta, 13 (1955) 313.
3. J. Pochon and P. Tardiex, Techniques d´analyse en microbiologie du sol. Collection techniques de base, Editions de le Tourelle, France, Seine, 1974.
4. A. Sobek, W. Schuller, J.R. Freeman and R.M. Smith, Field and laboratory methods applicable to overburdens and mine soils. US Environmental Protection Agency, EPA 600/2-78-054, USA, Cincinnati, 1978.
5. C. García, A. Ballester, F. González, M.L. Blázquez and M. Acosta, Minerals Engineering, 9 (1996) 1127.
6. Bergey´s Manual of Systematic Bacteriology, Willians & Wilkins, USA, Baltimore, 1989.

Development of a bioremediation process for mining wastewaters

P. Blumenroth[a] , K. Bosecker[a], A. Michnea[b], A. Varna[b] and N. Sasaran[b]

[a]Federal Institute for Geosciences and Natural Resources (BGR), Stilleweg 2, 30655 Hannover, Germany

[b]Mining Research and Design Institute, 62 Dr. V. Babes St., 4800 Baia Mare, Romania

Various bacterial and fungal strains capable of degrading cyanides or having a high potential for adsorbing heavy metals were isolated from water and sediment samples from a mining wastewater deposit in Baia Mare. Some of the isolates were characterised and identified. For those strains displaying the most rapid degradation or the greatest adsorption capacity, important parameters were analysed in synthetic media and in process waters from the Baia Mare pond and the Sasar cyanidation plant. Optimal pH and temperature levels, maximum tolerable cyanide and metal concentrations, as well as possible additional substances required, such as hydrocarbon sources or phosphate were determined. Disruptive factors inhibiting the activity of the organisms and making the complex analysis of the cyanides more difficult were found. With a view towards the development of a pilot plant the immobilization of microorganisms was tested on different support materials.

1. INTRODUCTION

Industrial wastewater containing cyanide, for example from ore leaching, is usually treated chemically, e.g. with alkaline hypochlorite [1; 2]. Purification is necessary because aquatic organisms in particular are affected by the toxic effects of the cyanide emitted, even in small concentrations (at the ppb-ppm level). In the current joint project the chemical treatment of the wastewater will be replaced by the use of microorganisms that can degrade cyanide. This ability has already been shown at several laboratories for different microbes [3 - 7]. The only commercial plant for cyanide biodegradation currently known, treats wastewater at the Homestake Mine, USA [8; 9]. The pilot plant scheduled for Baia Mare in a cubic metre range has to be adapted to local conditions. Therefore we enriched and isolated several microorganisms from wastewater and sediment samples and performed degradation and sorption tests with local wastewaters [10; 11]. Depending on the origin of the wastewater from ore mining heavy metal contamination also exists. Heavy metals can be removed from the contaminated water by active uptake or passive attachment to the cell surface of filamentous fungi or bacteria (biosorption) [12 - 15].

2. MATERIALS AND METHODS

2.1. Water samples and analyses

Water and sediment samples were collected at the Bozinta dam in Baia Mare, Romania (wastewater A). Temperature and pH were measured in the laboratory. For the analysis of heavy metals, water samples were filtered (0.45 µm) and stabilized by adding concentrated nitric acid (< pH 1). For the stabilization of cyanides, pH was raised to 11 and ascorbic acid was added (method DIN 38405-D13-1). Sediment and water samples for the isolation of microorganisms were left untreated. Further water samples were taken irregularly from the Bozinta basin over several months and from the cyanidation plant (wastewater B) and were analysed for their respective heavy metal and cyanide level, for anions and total cell count. Heavy metals were determined by ICP-AES; cyanide contents were measured photometrically with barbituric acid/pyridine (DIN 38405-D14), either directly (for the detection of free cyanide) or after destillation of the samples (total cyanide concentration). Copper-cyanide complexes were analysed by HPLC. Total cell counts (CFU) were achieved by plating serial dilutions of the untreated samples on nutrient agar (see below) and incubation at room temperature followed by colony counting.

2.2. Microorganisms and growth media

Selection of cyanide-degrading bacteria was carried out by enrichment cultures and subsequent isolation on selective agar. For enrichment, 10% of sediment (w/v) or wastewater (v/v) were incubated at 30°C in minimal medium M9 with filter sterilized potassium cyanide (KCN) as the sole N-source in a concentration range of 0.5 to 2 mmol/l, supplements (2.5 ml per liter M9 medium) and 0.4% glucose. KCN solution and glucose were added after autoclaving. The procedure was repeated three times with supernatant aliquots as inoculum for fresh medium. Composition of M9 medium was as follows: Na_2HPO_4 7.0 g, KH_2PO_4 3.0 g, NaCl 0.5 g, add H_2O_{dest} 1 l, pH 7.0-7.2. Supplements comprised 25 ml of a 1 M $MgSO_4$ solution, 25 ml of a 36 mM filter sterilized $FeSO_4 \cdot 7H_2O$ solution and 50 ml of a stock solution of trace elements. This stock solution contained: $MgCl_2$ 10.75 g, $CaCO_3$ 2.0 g, $FeSO_4 \cdot 7H_2O$ 4.5 g, $ZnSO_4 \cdot 7H_2O$ 1.44 g, $MnSO_4 \cdot 4H_2O$ 1.12 g, $CuSO_4 \cdot 5H_2O$ 0.25 g, $CoSO_4 \cdot 7H_2O$ 0.28 g, $H_3BO_3 \cdot 7H_2O$ 0.06 g, HCl $_{conc.}$ 51.3 ml, make up volume to 1000 ml with $H_2O_{dest.}$.

Single colonies were achieved by repeated striking out or plating aliquots of the enrichment cultures on M9 agar (18 g/l M9 medium). KCN was applied externally on sterile filter papers and put in the lid of the plate. Plates were sealed with Parafilm™ and incubated "upside down" in a glass chamber. Fresh KCN was added every 3-4 days until colonies were detectable. Cultivation of isolates was done in M9 medium with supplements as described before. For non-selective conditions KCN was substituted by NH_4Cl (1 g/l M9 medium). Other carbon sources like molasses, whey and residues from beer and juice production were also tested. For this, particles and precipitations were removed by paper filtration, filtrate was sterilised by fractionated sterilisation and added as 1-4% v/v to the medium. For total cell counts in wastewater nutrient FP agar, diluted 1:10, and R2A agar [16] were used. FP agar was routinely used for purity checks. Composition of FP agar: casein peptone 2 g, meat peptone 8 g, yeast extract 0.5 g, NaCl 2 g, agar 18 g, add 1 l of dest. water, pH 7; add glucose 4 g after autoclaving.

Selection of fungi was focused on cyanide tolerance and adsorption ability for heavy metals. Sediment extract agar was prepared for each sampling point (modified after Martin, in [17]).

100 g of sediment together with 200 ml of water were boiled for 30 min, filtered after cooling and filled up to 200 ml with water. The extract was amended with 1% glucose and 0.1% $NaNO_3$ and KH_2PO_4 each. Agar (4 g) was added and pH was adjusted to 5 before autoclaving. Aliquots of the corresponding water samples were plated in the absence and the presence of KCN to look for cyanide tolerance of fungal strains. To inhibit bacterial growth, 50 µg/ml nalidixic acid was added. Heavy metal tolerance was tested in shake flasks, containing medium with increasing amounts of copper, zinc and iron. Growth was looked at in comparison to growth without the addition of heavy metals. Further cultivation was carried out in minimal medium Duff (2 g/l yeast extract, 4% glucose or other carbon sources, see above, pH 4.6) or malt medium (malt 40 g, casein peptone 5 g, $(NH_4)_2HPO_4$ 1 g, add $H_2O_{dest.}$ 1 l; pH 6.0, fractionated sterilisation).

2.3. Degradation and adsorption studies

Kinetic studies for the biodegradation of cyanide and its derivatives were carried out with i) synthetic solution, consisting of minimal medium M9 and different concentrations of KCN (1-50 mmol/l), ii) artifically contaminated wastewater A (KCN 1-5 mmol/l) and iii) diluted and undiluted wastewater from the cyanidation plant (wastewater B). Cells were grown to late log phase and harvested by centrifugation, washed twice with sterile NaCl (0.9%) and resuspended in an appropriate volume of the solution/wastewater to be tested to get an optical density of about 0.8 to 1.0. Flasks with a small neck were sealed with a silicon stopper to avoid evaporation; they were incubated at 30°C on a rotary shaker. Aliquot samples were taken periodically by the help of a sterile syringe with a sterile needle and analysed for their cyanide content.

Metal adsorption studies with fungi - After sufficient growth, fungal mycelium was harvested via filtration and rinsed with sterile NaCl (0.9 %). Wet weight was determined and equal amounts were used for inoculation of shake flasks containing test solutions i) Duff medium, contaminated with different amounts of metals (50-200 mg/l copper, zinc and iron (II), or ii) wastewater A with added metals or iii) wastewater B, diluted or undiluted. To stabilize iron ions in the medium, citric acid was added. Aliquots were taken from the supernatant immediately after inoculation, acidified with concentrated nitric acid and analysed for their metal content. After several days of incubation (1 to 7) at 30°C and agitating on a shaker, the experiment was stopped by filtration of the suspension. Again, wet weight was determined and the biomass was dried at 60°C until weight remained constant. Metal concentration was determined in the filtrate. Adsorption rates (as % of control) as well as specific adsorption (mg metal/g biomass dry weight) were calculated. Flasks without biomass were used as abiotic control.

2.4. Immobilization of bacteria

For the technical application of bacteria immobilization seems to be an appropriate method to avoid wash-out of the biomass and to possibly enhance degradative action due to surface enlargement [18 - 20]. Two types of zeolites were tested, with zeolite type 1 being tuff material from Italy with the main component philippsite (and heulandite) and zeolite type 2, also volcanic tuff (main component clinoptilolite), from Zalau near Baia Mare, Romania. Both zeolites had a grain size of about 3-5 mm. Prior to use the material was sterilized at 150° C for 90 minutes. Afterwards zeolite material was added to M9 medium (about 10% w/v) with glucose and 2 mmol/l cyanide and inoculated with a bacteria suspension. Incubation was done at 30° C on a rotary shaker (115 rpm) for 3-5 days. Grown zeolites were collected afterwards in a sieve, carefully rinsed with sterile NaCl (0.9%) and used for degradation kinetics. Zeolites without grown bacteria were used as a control. Attachment of the bacteria was controlled by cell staining with acridine orange (0.1 g dissolved in 100 ml acetate buffer 0.1 mol/l, pH 4.66, filter sterilized, 3 minutes incubation) and subsequent microscopic analysis (fluorescence filter 450-490 nm).

3. RESULTS

3.1. Water analysis

Wastewater A from the storage basin is a highly diluted mixture (cyanidation leach, acid mine drainage, tailings). Levels fluctuate depending on the processing plant in Baia Mare. At 2.7-12.8 mg/l the cyanide level was relatively low. The sulphate level was between 679-1640 mg/l. The water was hardly contaminated at all with organic material. The main components in heavily contaminated water from the cyanidation plant (wastewater B) were cyanide (approx. 1200 mg/l) as well as copper and zinc. Both wastewaters have a high level of calcium (> 200 mg/l) because the pH level is controlled by lime during cyanidation. A list of some of the data is given in Table 1.

Table 1: Composition of the water sample taken from the Bozinta basin (A) and the cyanide leaching in the Sasar mine (B)

Parameters (all [mg/l] besides pH and cell numbers)	Wastewater A (mean values)	Wastewater B (mean values)
pH value	7.2	11.6
Cyanide, total	7.4	1160
Chloride	8.9	36.8
Nitrate	4.4	6.1
Sulphate	846	280
Phosphate	0.24	< 0.1
Copper	5.0	209
Iron	< 0.1	1.3
Zinc	2.5	269
Calcium	310	218
Cell numbers/ml (medium R2A/FP)	$1.4 \cdot 10^4/3.8 \cdot 10^4$	0/0

3.2. Isolates

Various bacteria and fungi were isolated under selective conditions (cyanide as the single nitrogen source, tolerance towards cyanide and heavy metals, heavy metal adsorption) from water and sediment samples from the Bozinta basin. The best bacterial degraders and the strongest metal adsorbing fungi were identified (DSMZ, Braunschweig). The most important bacteria were *Pseudomonas spec.* and *Burkholderia cepacia*, both common aquatic organisms. The fungal isolates were *Trichoderma koningii, T. harzianum, Epicoccum nigrum* and *Aspergillus fumigatus* as well as an unidentifiable strain named isolate 1. Because best results were obtained with *Burkholderia cepacia, Aspergillus fumigatus* (also named isolate 2 in the following) and isolate 1, the further experimental work focussed on these organisms.

3.3. Cyanide degradation
3.3.1. Synthetic medium

The bacterial isolates were analysed for their effectiveness where the use of cyanide as sole nitrogen source is concerned. The quickest growing strains were used in further experiments. Two very good degraders from the species *Burkholderia cepacia* were found. Kinetics with different cyanide concentrations (Fig. 1) were carried out at 30°C or 15°C. Complete reduction took just a few hours, as expected more quickly at 30°C than at 15°C.

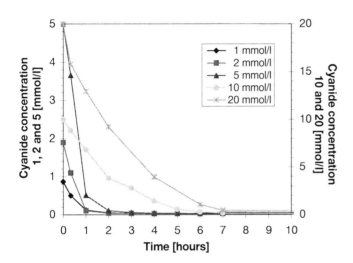

Fig. 1: Degradation of different cyanide concentrations at 30°C by *Burkholderia cepacia*

Comparable degradation rates were achieved at pH 7 and 9 in batch experiments. The maximum tolerable and degradable cyanide concentration was between 20 and 30 mmol/l CN^-. Ammonium appeared during degradation as a metabolite and was subsequently used.

High concentrations of sulphate, as encountered in the wastewater, did not influence the degradation. The degradation of cyanide did not require a hydrocarbon source but general metabolism of the organisms did (to gain energy). As well as glucose as the classical carbon source, cheap alternatives such as molasses, whey and liquid residues from beer and juice production were examined for suitability. The addition of juice and beer residues (4%; v/v) achieved successful growth results compared to the glucose concentration of 0.4 % used otherwise.

3.3.2. Wastewater

The autochthonous bacteria present in the wastewater from the mine basin did not contribute to the degradation of cyanide because of the minimal concentration of cells. As the cyanide concentration was very small, wastewater A was artificially contaminated for the majority of experiments. Phosphate as a nutrient was vital for the degradation of cyanide by inoculated bacteria in wastewaters A and B. The rate of degradation depended on the level of phosphate present (Fig. 2). Calcium was bound with the help of EDTA to prevent precipitation with phosphate. Without the addition of EDTA degradation was slown down considerably.

The maximum metal tolerance was ascertained in various dilutions of wastewater B; according to the isolate tested levels are between 5 and 43 mg/l copper or 9 and 64 mg/l zinc (when present simultaneously). These results can only provide an approximate tolerance area because the metals mentioned are present not just as ions but also as different complexes and for this reason they have a lower toxicity.

When wastewater B was diluted differently according to the tolerance levels calculated (factors of 20, 10, 5 according to about 55, 110 and 220 mg/l CN⁻), the highest percentual reduction was seen in degrading experiments with the lowest initial concentrations of cyanide and metals (91% > 81% > 77%) (Fig. 3). Comparing these degradation rates with those from synthetic solutions, where concentrations up to 520 mg/l of CN⁻ were degraded totally, one might have expected equal degradation rates in wastewater B, too. A reason for the different results could be a higher rate of dissociation of complex compounds like $[Cu(CN)_4]^{3-}$ in diluted solutions and consequently a higher level of degradable free cyanide.

Degradation was however not yet complete: Where the different residual amounts of cyanides according to the initial concentration are concerned, it was on the one hand a copper-cyanide complex and on the other probably thiocyanate that has not yet been identified or quantified.

These two cyanide compounds are however partially degradable in a synthetic medium under certain conditions. Even in a degradation test of cyanide and derivatives in wastewater using mixed cultures (cyanide and thiocyanate-using isolates) no improvement was made, i.e. the residue levels of cyanide compounds that fake free cyanide in analysis could not be reduced any further in short time experiments. This again clearly shows the complexity of wastewater and the difficulty of transferring laboratory data, based on tests with defined media, "into the field". More recent tests showed that the levels of cyanide and their complexes with copper can be reduced also by the fungal isolates used for biosorption.

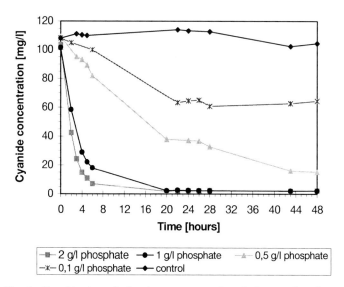

Fig. 2: Cyanide degradation in wastewater in relation to phosphate addition (*B. cepacia*)

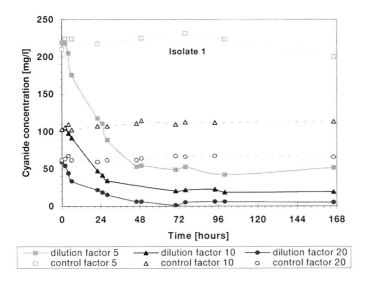

Fig. 3: Cyanide degradation in different dilutions of wastewater B (*B. cepacia*)

3.3.3. Influence of glucose

When glucose was added to experiments with wastewater without any biological influence a continual reduction in the cyanide concentration present was observed. In control tests without glucose no losses were observed. The phenomenon was also seen in synthetic media, but only to a significant extent at high cyanide concentrations (> 10 mmol/l) and glucose amounts greater than 8 g/l (Fig. 4).

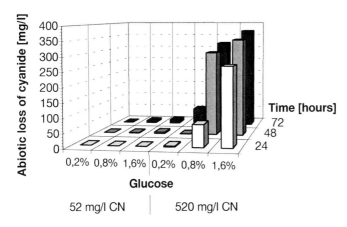

Fig. 4: Abiotic loss of cyanide depending on glucose/cyanide concentration

It is assumed that so-called glucocyanohydrine had formed (an addition reaction of aldehyde). The by-product of this compound is ammonium, which was quantified in appropriate tests in synthetic media and the amount of which corresponded to the abiotic loss of cyanide. The reaction is largely dependent on the cyanide and glucose concentration. Tests with the alternative carbon sources in wastewater B mentioned above already showed a lower abiotic cyanide reduction.

3.4. Immobilization of bacteria

Apart from commercially available plastic support materials, the growth on zeolite rock was also looked at. The use of zeolites has been reported for wastewater treatment [18; 21; 22] and provides a cheap alternative, as the zeolites can be obtained near the processing plant. Immobilization on the plastic carrier was not successful, but the bacteria did attach themselves to zeolite. Cell density was controlled via cell staining with acridine orange. Investigations with an electron microscope are to follow.

When zeolite type 1 was used in kinetic tests for the biodegradation of cyanide, decrease followed at the same speed as in suspended cells (< 1 mg/l after 4 hours). On the other hand growth on zeolite type 2 was poorer and the reduction in cyanide correspondingly slower too (< 1 mg/l after 29 hours; Fig. 5).

Recent long-term column experiments (3 weeks) with grown zeolite type 2 and a repeated addition of glucose (0.2%) showed total degradation of all cyanides in wastewater B (10-fold dilution, pH 8).

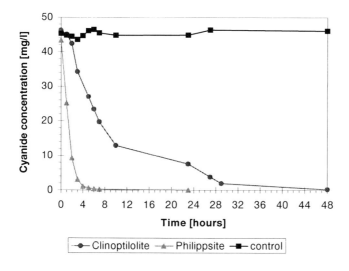

Fig. 5: Degradation of cyanide by immobilised cells of *Burkholderia cepacia*

3.5. Metal biosorption by fungi

Those fungal strains isolated from sediment samples from the Bozinta pond were subjected to a screening for metal and cyanide tolerance and the isolates then proving to be resistent were selected for metal sorption tests in batch culture. Artificially contaminated synthetic medium (Duff) or deionised water served as a matrix for the sorption tests. Metals were used singly or as a mixture of iron, copper and zinc in various concentrations (e.g. 25, 50 or 100 mg/l each) and sorption capacities were calculated under different conditions.

It could be shown that the different fungi did have preferences where the sorption of a particular metal (iron, copper or zinc) is concerned, but sorption rates were however dependent on whether metals were offered separately or in a mixture. The optimal pH levels for the sorption of the individual metals were not taken into consideration, because the wastewater has to be treated at alkaline pH values in any case to avoid massive formation of gaseous hydrogen cyanide. Just one single pH level was given in all experiments right at the beginning and any changes were noted during the course of the experiment.

Sorption capacity further depended on the composition of media and the glucose level in the pre-culture. The fungal mycelium that grew from the residue from juice production adsorbed the metals better than biomass cultivated with molasses or glucose. ("juice" 1% > molasses 1% > glucose 0.4% > glucose 4%; see figure 6).

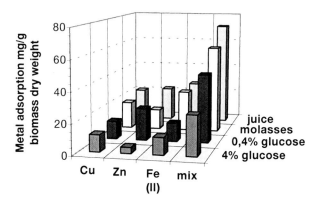

Fig. 6: Adsorption levels depending on preculturing conditions (*A. fumigatus*)

Sorption rates of living (resting) biomass were far greater than those of dead biomasses. Dried mycelium showed the lowest effect. The percentual amount of sorbed metals increased with greater initial use of fresh biomass, whereas the specific adsorption, expressed as mg metal per gram dry weight, was greater when less biomass was used. The highest sorption rates amounted for iron to 39 mg/g dry weight, for copper to 28 mg/g dry weight and for zinc to 36 mg/g dry weight in wastewater A with an artificial contamination of 50 mg/l of each metal; the sorption for the total metal mixture was between 77 (isolate 2) and 106 mg/g dry weight (isolate 1; see Fig. 7). The sorption process took time until the saturation of the resting

556

biomass. The incubation time was a maximum of seven days, but after two to four days the majority of the metal was adsorbed.

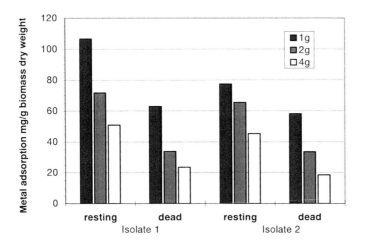

Fig. 7: Comparison of metal sorption (mix) by different amounts of dead and resting biomass

Sorption experiments with wastewater B were carried out with different metal concentrations produced by step-wise dilutions and an initial biomass of 4 g. Results showed that isolate 2 adsorbed better in this matrix than isolate 1, especially at the highest concentrations of contaminants (Fig. 8).

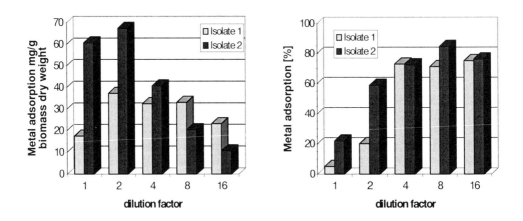

Fig. 8: Adsorption rates at pH 9 at different wastewater concentrations

The amount of percentually adsorbed metal increased as metal concentration decreased and remained roughly constant at a dilution factor of 4 (72-79 %). At this dilution factor metal concentrations for zinc and copper approximately corresponded to those in the experiments with artificially contaminated wastewater (52 or 55 mg/l). The sorption rate for isolate 2 with about 41 mg/g dry weight was comparable to that of wastewater A (45.4 mg/g dry weight), whereas isolate 1 adsorbed only 32.5 mg/g dry weight in wastewater B (compared to 51.0 mg/g dry weight in wastewater A). Absolute sorption amounts in undiluted wastewater B amounted to 60 to 81 mg/g dry weight according to the pH level used (7 or 9). Regarding each isolate, there was no correlation between sorption capacity and the pH value used.

4. CONCLUSIONS

Results achieved under defined laboratory conditions are been transmitted to real mining wastewaters. It is shown, that our isolates are able to remove high concentrations of cyanide and heavy metals from alkaline wastewaters. Factors like preculturing conditions or nutrient supply were discovered that are important for the strains to work efficiently and leading to an improved purification capacity of the organisms in highly contaminated wastewaters. Further improvements like, for example, the choice of treatment type (trickling filter, sand bed filter, etc.) and optimal retention time will follow as part of the development of a pilot plant on a technical scale.

ACKNOWLEDGMENTS

The project is supported as part of the scientific-technical cooperation by the BMBF (German Federal Ministry for Education and Technology). We gratefully acknowledge the technical assistance of I. Stapelfeldt and D. Spier.

REFERENCES

1. A. Smith and T. Mudder, Chemistry and Treatment of Cyanidation Wastes. Mining Journal Books Ltd., London, England ,1991.
2. A. Smith and T. Mudder, Mining Environmental Management, June (1995) 4
3. K.D. Chapatwala, G.R.V. Babu, O.K. Vijaya, K.P. Kumar, J.H. Wolfram, J. Ind. Microbiol. & Biotechnol. 20 (1998) 28.
4. C. Boucabeille, A. Bories, P. Ollivier, G. Michel, Environ. Pollut. 84 (1994) 59.
5. S.K. Dubey and D.S. Holmes, World Journal of Microbiology & Biotechnology 11 (1995) 257.
6. M.M. Figueira, V.S.T. Cifninelli, V.R. Linardi, In: C.A. Jerez, T. Vargas, H. Toledo and J.V. Wiertz: Biohydrometallurgical Processing, University of Chile (1995) 333.
7. S.A. Raybuck, Biodegradation 3 (1992) 3.
8. J.L. Whitlock, T.I. Mudder, In: Lawrence, R.W.; Branion, R.M.R. and Ebner, H.G. (eds.): Fundamental and Applied Biohydrometallurgy, Elsevier, Amsterdam; Proceedings 6th Internat. Symp. Biohydrometallurgy, Vancouver (1986) 327.

9. J.L. Whitlock, Geomicrobiol. J. 8 (1990) 241.

10. P. Blumenroth, H. Dettmers, K. Bosecker, Proceed. International Biohydrometallurgy Symposium IBS´97 - Biomine ´97, 4-6 August, Sydney Australia; Australian Mineral Foundation, Glenside, South Australia (1997).

11. P. Blumenroth and K. Bosecker, Tagung junger Geochemiker, 21.-23. Mai 1998, GUG-Schriftenreihe „Geowissenschaften und Umwelt", Springer Verlag (in press).

12. G.M. Gadd, New Phytol. 124 (1993) 25.

13. A. Stoll and J.R. Duncan, Biotechnol. Lett. 18(10) (1996) 1209.

14. M. Tsezos, Microbial Mineral Recovery, Chapter 14. Eds.: H.L. Ehrlich and C.L. Brierley, McGraw Hill, Inc. USA, 1990.

15. B. Volesky, FEMS Microbiol. Rev. 14 (1994) 291.

16. D.J. Reasoner and E.E. Geldreich, Appl. Environ. Microbiol. 49(1) (1985) 1.

17. O. Fassatiova, Progress in industrial microbiology, Vol. 22, Elsevier, 1986.

18. M.-C. Dictor, F. Battaglia-Brunet, D. Morin, A. Bories, M. Clarens, Environ. Poll. 97(3) (1997) 287.

19. J. Klein and H. Ziehr, BioEngineering 3 (1987) 8.

20. D.M. White and W. Schnabel, Wat. Res. 32(1) (1998) 254.

21. S.I. Marton and A. Liebmann, 2. Internat. Symp. and Exhibition on Environm. Contam. in Central and Eastern Europe (20.-23.9.94). Proc. Budapest (1994).

22. Y.-J. Suh, J.M. Park, J.-W. Yang, Enzyme Microb. Technol. (June) 16 (1994) 529.

Biological treatment of acid mine drainage

J. Boonstra, R. van Lier, G. Janssen, H. Dijkman and C.J.N. Buisman

PAQUES Bio Systems, P.O. Box 52, 8560 AB BALK, The Netherlands

In this paper experience obtained with THIOPAQ technology treating Acid Mine Drainage is described. THIOPAQ Technology involves biological sulfate reduction technology and the removal of heavy metals as metal sulfide precipitates. The technology was developed by the PAQUES company, who have realised over 350 high rate biological treatment plants world wide. 5 plants specially designed for sulfate reduction are successfully operated on a continuous base (1998 status).

At Budelco, a zinc refinery in the Netherlands, an acid groundwater stream is effectively treated since 1992, removing metals and sulfate.

At Kennecott Utah Copper (USA) a demo plant is in operation since 1995. An acid groundwater flow is treated to remove sulfate and metals, whereas the excess sulfide is used to selectively recover copper economically.

Early 1998, a demonstration project was executed at the Wheal Jane mine in Cornwall, UK. In this demonstration project it has been proven that THIOPAQ technology can effectively be used to treat the Wheal Jane Acid Mine Drainage. Relative to lime dosing technology, very high removal efficiencies of all heavy metals (including cadmium and arsenic) can be obtained.

1. INTRODUCTION

Acid Mine Drainage (AMD) is a world wide recognised problem. It is generally accepted that these diluted streams of sulfuric acid, contaminated with metals, have to be treated. Currently, treatment involves mostly neutralisation using lime or other alkalinic components. This results in the precipitation of sulfates and metals as gypsum and metal hydroxides respectively, which have to be landfilled. The operating costs of this process are high while sulfate and metals removal efficiencies are relatively low (1500 mg/l for sulfate and 0.5-5 ppm for the metals). In addition, all valuable metals are lost in the sludge.

The recently developed THIOPAQ technology on the other hand, produces sulfur and metal sulfides. Compared to neutralisation, the sludge volume is 6 to 10 times lower and the toxic metals are removed to a 1-100 ppb level. Also valuable metals like Cu, Zn and Co can be recovered.

The biological treatment consists of two main process steps which take place in separate reactors. First, sulfate is converted into sulfide; subsequently, the sulfide is converted into sulfur. Due to the production of alkalinity during the conversion of sulfide into sulfur, influent neutralisation can be achieved by recirculation of this stream eliminating the need to add alkaline chemicals.

The metals precipitate as sulfides, which are separately removed. It is possible to selectively recover different metals from the contaminated water, based on the principle of different solubilities of metal sulfides at different pH's. In this way it is possible to selectively recover, for example, Cu^{2+} and Zn^{2+}. Compared to precipitation of metals with lime, much lower effluent concentrations can be achieved (ppb level) by using biotechnology due to the much lower solubility of metal sulfides. Basically, removal efficiencies comparable to the EC directives, shown in table 1, can be obtained.

Table 1
EC Dangerous Substances Directives (1996) for some metals

Compound	EC-directive (mg/l)
Iron	1.0 (D)
Zinc	0.5 (T)
Arsenic	0.05 (T)
Cadmium	0.001 (T)
Copper	0.028 (D)

(T) = total metal; (D) = Dissolved metal

The end products of the biological installation are suitable for reuse. The sulfur can either be produced as a sulfur cake, 60% dry solids with a purity of 95% S^o, or as pure liquid sulfur. The sulfur cake can be used for the production of sulfuric acid. An other option is to use it as soil amendment. The metal sulfides can be recovered in e.g. zinc or copper refineries/smelters.

2. TECHNOLOGY DESCRIPTION

The process of removing both sulfate and metals consists of four main process steps. These include two biological reactors and one or two separation steps for the produced solids, namely sulfur and metals sulfides. The general overall reactions that take place can be simplified as follows:

$$H_2SO_4 + 4H_2 + \frac{1}{2}O_2 \rightarrow S^o + 5H_2O \tag{1}$$

$$Me^{2+} + SO_4^{2-} + 4H_2 \rightarrow MeS + 4H_2O \tag{2}$$

The above equations show that the effluent stream is completely neutralised by producing elemental sulfur and metal sulfides. If for a certain application no sulfate discharge standards apply, the system can be operated to produce the stoichiometric amount of sulfide required for metal precipitation. This can be achieved by adjusting the dosing rate of reductant (in the above equations expressed as hydrogen gas).

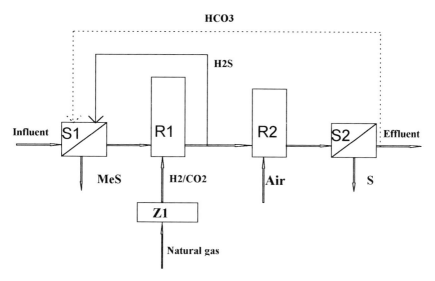

Figure 1. Process for sulfate and metals removal using hydrogen as electron donor.

In unit S1 (see Figure 1) the influent is mixed with two recycle streams. The first recycle, anaerobic effluent, contains the necessary sulfide to precipitate the metals. The metal sulfides, which are separated in unit S1, can be further dewatered by means of a decanter centrifuge. The second recycle is from the effluent of the tilted plate separator (unit S2) where the sulfur is separated. This recycle stream is shown dotted in Figure 1 since it is, in general, only necessary for concentrated streams or if the influent pH needs to be increased.

In the first reactor (R1) biological sulfate reduction takes place under anaerobic circumstances. Depending on the sulfate load (the amount of sulfate reduced per hour) there are different chemicals available to serve as reductant or electron donor for the biomass. For high sulfate loads, hydrogen gas is most economic. This gas can be produced (if not available at site) by cracking methanol or by means of a natural gas reformer (Z1). In both cases a mixture of H_2 and CO_2 is produced. The carbon dioxide is used as a carbon source by the sulfate reducing bacteria.

The main reaction that occurs in the anaerobic reactor using hydrogen is:

$$SO_4^{2-} + 4H_2 + H^+ \rightarrow HS^- + 4H_2O \tag{3}$$

For lower sulfate loads, ethanol or a concentrated organic waste stream can best be used as electron donor. In this case a cracker or reformer (Z1) is obviously not required. For ethanol the main reaction is:

$$3SO_4^{2-} + 2C_2H_5OH \rightarrow 3HS^- + 3H_2O + 3HCO_3^- + CO_2 \tag{4}$$

The sulfate reducing bacteria (Oude Elferink, 1998) that grow in the reactor agglomerate into tiny granules that have good settling characteristics. Due to the growth, excess of biomass must be periodically removed from the reactor. This biomass can be used to start up new reactor systems or can be disposed of.

The second biological reactor, the aerobic reactor (R2), contains Thiobacillus micro-organisms (Visser, 1997) that oxidise the sulfide formed in the previous reactor into elemental sulfur:

$$HS^- + \frac{1}{2}O_2 \rightarrow S^O + OH^- \tag{5}$$

It is clear from the above reaction that the pH increases. The produced alkalinity can be used for influent neutralisation, as pointed out earlier.

Obviously only the excess sulfide which was not precipitated with metals is oxidised in the second reactor. Relative to the sulfate reducing reactor, this aerated reactor can therefore have a very small volume in most cases.

The produced sulfur is separated by means of a tilted plate separator. From here, two options are possible:

1. The sulfur can be dewatered in a decanter centrifuge to about 60% dry solids; the sulfur will have a purity of about 95% on dry basis.
2. For large amounts, the sulfur can be melted to produce pure liquid sulfur.

Depending on metal- and sulfate concentrations and the options considered for reuse, metal sulfide precipitates can either be removed up front the anaerobic reactor or along with the produced sulfur. If removed up front, the sulphide rich effluent stream from the anaerobic reactor is recycled to unit S1 in the scheme 1. For other applications, unit S1 can be excluded and all particles can be removed in S2. In the last case the metal sulfides are formed in the anaerobic reactor.

Selective recovery of metals is based on the different solubilities of metal sulfides at different pH-values. By adjusting the pH in the precipitation tank to a certain value, it is possible to form a specific metal sulfide precipitate. This principle is being used at the Kennecott demo plant, where copper is recovered from a leach water stream.

3. EXPERIENCE

3.1. Full scale experience at the Budelco Zinc refinery

Budelco B.V. (Netherlands) has been operating a zinc refinery since 1973. Annual zinc production exceeds 200,000 metric tons. More than 100 years of zinc refining by various companies on the same site has resulted in considerable heavy metals and sulfate contamination of soil and groundwater underneath the plant. To prevent the pollution of spreading further, a hydro-geological containment system was developed. The water balance of the area (including rainfall and influx through aquifers) requires that some 5000 m3/d is extracted from a combination of twelve shallow and deep wells.

Various technologies for the treatment of this groundwater stream have been considered. After extensive laboratory and pilot scale testing, Budelco decided in 1990 to install a commercial scale biological water treatment plant in which sulfate is reduced to hydrogen sulfide, a major part of the hydrogen sulfide is used for precipitating the zinc and other metals, and the remaining hydrogen sulfide is converted to elemental sulfur in a second biological reactor.

PAQUES was awarded the turn-key contract. The treatment plant installed by PAQUES consisted of four main components: an Upflow Anaerobic Sludge Blanket (UASB) reactor for the sulfate reduction step using ethanol as electron donor, including a gas handling system; a Submerged Fixed Film (SFF) reactor for the aerobic conversion of sulfide present in the UASB effluent to elemental sulfur; a tilted plate settler for the removal of solids; and a continuously cleaned sand bed filter as a solids polishing step before discharge. The design criteria for the commercial scale treatment plant are shown in Table 2.

Table 2
Design criteria of the Budelco biological metal and sulfate removal plant

Component	Unit	Influent	Effluent
Flow	m^3/h	300*	
Zinc	mg/l	100	<0.3
Cadmium	mg/l	1	<0.01
Sulfate	mg/l	1000	<200

*The plant is now being expanded to handle a flow of 400 m^3/hour.

The treatment plant was commissioned in May of 1992. Typical effluent concentrations are comparable with the results obtained in the pilot plant (Scheeren et al., 1993). This means that the effluent zinc concentration is < 0.05 mg/l and the effluent sulfate concentration is well below 200 mg/l. Cadmium is normally not detectable in the effluent (<1 ppb). Both the metal sulfides and elemental sulfur are returned to the roaster. The metals are recovered and the sulfur is converted to sulfuric acid in the acid plant. No new solid waste streams in need of disposal are created. More than 6 years of operational experience at Budelco show that the biological treatment system operates very reliable.

3.2. Demo plant at Kennecott's Bingham Canyon, Utah Copper Mine (USA)

Kennecott operates a large open pit copper mine in Bingham Canyon, Utah, producing over 300,000 tons of copper per year. Open pit mining started in 1906. Since 1995 Paques and Kennecott have been co-operating in the development of technologies for treatment of water with elevated metal and sulfate concentrations. A demo plant is in operation to asses the THIOPAQ sulfate and metal removing technology using hydrogen as electron donor. This plant contains several process steps and can be divided in to two sections as shown in Figure 2.

1. Biological H_2S production
In this part of the demo plant, sulfate reducing bacteria, using hydrogen as electron donor, convert the sulfate to hydrogen sulfide. Part of this hydrogen sulfide is oxidised to form elemental sulfur and alkalinity. Acid Mine Drainage containing high levels of sulfate and metals is treated. A metal separation unit and sulfur separation unit are installed to remove metal sulfides and elemental sulfur.

2. Copper recovery from leach water
The H_2S containing gas leaving hydrogen sulfide producing bioreactor is directed to a gas/liquid contactor. In this contactor the hydrogen sulfide reacts with the copper in a leach water stream to form copper sulfide. The produced copper sulfide is removed from the water in a lamella clarifier.

564

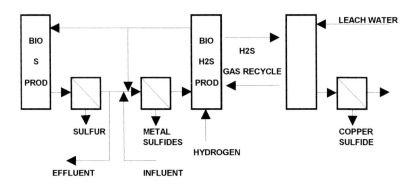

BIOLOGICAL H2S PRODUCTION   COPPER RECOVERY

Figure 2. Simplified flow diagram of the pilot plant used at Kennecott Utah Copper

Operation of the pilot plant started in November 1995. Water containing high levels of sulfate and metals is used as influent at a typical flow rate of 0.2 m^3/h. Metals and sulfate are reduced to low levels without addition of alkaline chemicals. Part of the produced hydrogen sulfide is used for selective recovery of copper from a leach water stream (up to 5,5 m^3/h). Copper is recovered at very high efficiencies up to 99.9 %. Preliminary results are presented in Table 3 and Table 4.

Table 3
Influent and effluent concentrations of the high-sulfate water treated biologically for H$_2$S production in the Paques pilot plant @ Kennecott Utah Copper

Parameter	Influent (mg/l)	Effluent (mg/l)
pH	2.5	8.5
Sulfate	30,000	<500
Aluminium	2,200	<2
Calcium	480	50
Copper	60	<0.1
Iron	675	<0.3
Magnesium	4,500	1,950
Manganese	350	0.3
Zinc	65	<0.1

Table 4
Influent and effluent concentrations of the leach water treated for selective recovery of copper at the Paques pilot plant @ Kennecott Utah Copper

Parameter	Influent (mg/l)	Effluent (mg/l)
pH	2.6	2.2
Copper	180	< 0.3
Iron	380	379
Zinc	200	199

3.3. Demo plant at the Wheal Jane Mine (Cornwall, UK)

An international consortium of four major companies has executed a THIOPAQ technology demonstration project for the treatment of the heavily polluted acid mine drainage (AMD) at the former Wheal Jane mine near Truro, Cornwall. Partners in the consortium are ACWa Services (UK-contractor), BP-chemicals (electron donor supplier), NUPAQ (Dutch own & operate company) and PAQUES Bio Systems (technology supplier).

Around 700 m^3/h of contaminated minewater from the former mine requires treatment prior to discharge. The contamination is caused by the presence of a variety of heavy metals such as zinc and the highly toxic cadmium and arsenic. To avoid these dangerous substances entering the environment, treatment of the stream is required before it is discharged into the Carnon River. Currently, a contingency plant is being operated in which the AMD is treated by lime dosing, followed by settling of the resulting metal oxides and metal hydroxides in the mine tailings dam.

In order to demonstrate the applicability of THIOPAQ technology at the Wheal Jane mine, a pilot plant has been operated from the end of November 1997 till the end of February 1998. Typically, a flow of 1 - 1,5 m^3/h was treated continuously. The main objective of the project was to demonstrate the stability and robustness of both biological processes (sulfate reduction and sulfide oxidation) under the specific Wheal Jane conditions, in order to show that inhibition of the biomass is no issue.

During the trial, it was demonstrated that the Wheal Jane acid mine drainage can be successfully treated using THIOPAQ technology. The biomass has proven to be able to treat the water without encountering any problems. BP's Aquaguard was tested as electron donor for the sulfate reducing bacteria. This compound has proven to be an effective reductant; high sulfate reducing activity could be obtained at relative low costs (Aquaguard is for example about 30% cheaper than ethanol).

The demo project has resulted in a proposal to treat the entire AMD stream. The heavy metals concentrations that are guaranteed for the effluent of the plant are very low when compared to lime dosing technology, which is believed to be the most seriously considered treatment alternative. Specifically for this project, the following main advantages of THIOPAQ relative to lime dosing have been recognised:

- Effluent metals concentrations can be obtained which are comparable with EC-directives
- A sludge product is obtained which is more easy to dewater and lower in volume
- Simultaneous removal of sulfate (better reuse possibilities for drinking water production or irrigation)
- Possible recovery of base metals

Economical figures are strongly dependent on local circumstances, but in general economics for both technologies are more or less comparable for the treatment of AMD.

When looking at the annual loading of some heavy metals to the Carnon river and the Fal Estuary, a significant difference can be observed between the two technologies considered. The dissolved metals concentrations in the following table are based on tentative results from the lime dosing contingency plant and on expected effluent concentrations of THIOPAQ technology applied at Wheal Jane. An average flow of 685 m3/h is accounted for.

Table 5
Comparison of discharge of lime dosing technology and THIOPAQ technology treating Wheal Jane AMD

	Effluent Lime (mg/l)	Effluent Thiopaq (mg/l)	kg/year discharged less (Thiopaq)
Iron	0.2	< 0.1	> 600
Zinc	1.7	< 0.05	> 9900
Copper	0.03	< 0.01	> 120
Aluminium	0.7	0.1	3600
Manganese	2.1	0.5	9600
Cadmium	0.004	< 0.0005	21
Sulfate	*850*	*200*	*3.895.000*

Although the differences in effluent concentrations seem small for most compounds, the annual metals loading to the river is seriously reduced applying THIOPAQ.

It should be noted that the 21 kg/year cadmium decrease is 10% of the current total annual Carnon river loading. Arsenic is not included in this table. Removal efficiencies applying lime dosing are not accurately known. THIOPAQ technology can reduce Arsenic below 1 ppb.

4. CONCLUSIONS

THIOPAQ technology can be applied effectively to treat Acid Mine Drainage streams. Over the past 6 years, full scale experience has proven that very high metal removal efficiencies can be obtained relative to conventional technologies, which might be considered as an alternative. Further, considerable advantages are recognised with respect to aspects like re-use, sludge disposal and water quality.

The technology is in a mature stage and available for metallurgical operations and abandoned mine sites now.

REFERENCES

Gabb,P.J., Howe,D.L., Purdie,D.J., Woerner,H.J. The Kennecott smelter hydrometallurgical impurities process, Proceedings of COPPER 95- COBRE 95 International Conference Volume III, 1995.
Vegt de, A.L., Bayer,H.G., Buisman,C.J., Biological sulfate removal and metal recovery from mine waters, SME Annual Meeting Denver Colorado, 1997, 93.

Houten van,R.T, Biotechnology and Bioengineering, 44 (1994) 586.

Buisman,C., Post,R., Yspeert,P., Geraats,G. and Lettinga,G., Acta Biotechnol. 9 (1989) 255.

Buisman, C., P. Yspeert, G. Geraats and G. Lettinga, Biotechnology and Bioengineering, 35 (1990) 50.

Hammack, R.W. , Dvorak, D.H. , Edenborn, H.M. ,Selective metal recovery using biogenic hydrogen sulfide, Extraction and processing for treatment and minimization of wastes, The Minerals, Metals and Materials Society, 1993.

Oude Elferink, S.J.W.H., Sulfate-reducing Bacteria in Anaerobic Bioreactors, PhD Thesis, Wageningen Agricultural University, The Netherlands, 1998.

Peters, R.W. , Ku, Y. , AIChE Symp. Ser. 81 (1985) 9.

Scheeren, P.J.H., R.O. Koch and C.J.N Buisman. Geohydrological Containment System and Microbial Water Treatment Plant for Metal-Contaminated Groundwater at Budelco, International Symposium - World Zinc '93, Hobart 10-13 October: 373, 1993.

Vegt de, A.L., H. Dijkman and C.J.N. Buisman. Hydrogen sulfide produced from sulfate by biological reduction for use in metallurgical operations, TMS conference, San Antonio (February 1998) and SME conference, Orlando, 1998.

Visser, J.M., Sulfur compound oxidation and sulfur production by Thiobacillus sp. W5, PhD Thesis, Delft University of Technology, The Netherlands, 1997.

Computer-munching microbes: Metal leaching from electronic scrap by bacteria and fungi

H. Brandl[*], R. Bosshard and M. Wegmann

University of Zurich, Institute of Environmental Sciences,
Winterthurerstrasse 190, 8057 Zurich, Switzerland

We applied microbiological processes to mobilize metals from electronic waste materials. Bacteria (*Thiobacillus thiooxidans, T. ferrooxidans*) and fungi (*Aspergillus niger, Penicillium simplicissimum*) were grown in the presence of electronic scrap. The formation of inorganic and organic acids caused the mobilization of metals. Initial experiments showed that above a concentration of 1% (w/v) of scrap in the medium microbial growth was inhibited. However, after a prolonged adaptation time, fungi grew also at concentrations of 10% (w/v). Both fungal strains were able to mobilize Cu and Sn by 65%, and Al, Ni, Pb, and Zn by >95%. At scrap concentrations of 0.5 % *Thiobacilli* were able to leach >90 % of the available Cu, Zn, Ni, and Al. Pb precipitated as $PbSO_4$, Sn probably precipitated as SnO. For a more efficient metal mobilization a two-step leaching process is proposed where biomass growth is separated from metal leaching.

1. INTRODUCTION

Relatively short lifetimes of electrical and electronical equiments (EEE) are leading to an increased amount of waste materials. In Switzerland, approx. 110,000 t of electrical appliances have to be disposed yearly, in Germany ten times more (1.5 mio t). Specialized companies are responsible for recycling and disposal. EEE is dismantled and manually sorted (Figure 1). The resulting material is subjected to a mechanical separation process. Dust-like material is generated by shredding and other separation steps during mechanical recycling of electronic wastes: approx. 4% of the 2400 t of scrap treated yearly by a specialized company is collected as fine-grained powdered material. Whereas most of the electronic scrap can be recycled (e.g. in metal manufacturing industries), the dust residues have to be disposed in landfills or incinerated.

[*] phone 0041 1 635 6125, fax 0041 1 635 57 11, email: hbrandl@uwinst.unizh.ch. Financial support was provided by the Swiss National Science Foundation within the Priority Program Environment. Electronic waste was provided by IMMARK AG, Kaltenbach, Switzerland.

However, these residues contain metals in concentrations which might be of economical value: 24% (w/w) Al, 8% Cu, 3% Zn, 2% Sn, 2% Pb and 2% Ni (Table 1). In addition, precious metals (Ag, Au) are also present in small amounts. Provided a suitable treatment and a recovery process, this material might serve as a secondary metal resource.

Table 1
Amounts of selected elements in dust obtained from recycling of electronic equiment. Concentrations were determined by ICP-MS (rapid screening).

Element		Content (g/kg)
Aluminum	Al	237
Copper	Cu	80
Lead	Pb	20
Nickel	Ni	15
Tin	Sn	23
Zinc	Zn	26

Biohydrometallurical techniques allow the cycling of metals by a process close to natural biogeochemical cycles [1-3]. Using biological techniques, the recovery efficiency can be increased where thermal or physico-chemical methods alone are less successful as shown in copper and gold mining where low-grade ores are biologically treated to obtain metal values which are not accessible by conventional treatments [4, 5].

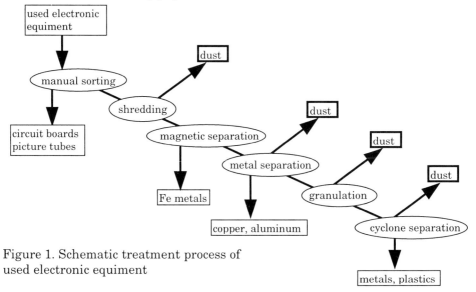

Figure 1. Schematic treatment process of used electronic equiment

The objective of the project is to apply a bacterial leaching process ("bioleaching") for the mobilization of metals from different metal-containing solid waste such as fine-grained electronic waste. Leached and recovered metals might be recycled and re-used as raw materials by metal-manufacturing industries.

Only very few data are available on the bioleaching of metals from electronic waste materials [6]: Heterotrophic bacteria and fungi (*Bacillus* sp., *Saccharomyces cerevisiae*, *Yarrowia lipolytica*) have been used to mobilize Pb, Cu, and Sn from printed circuit boards [7]. A *Sulfolobus*-like organism was able to leach gallium arsenide from semi-conductors [8].

2. MATERIAL AND METHODS

2.1. Material

Dust collected from shredding processes was fractionated by sieving and dried at 80°C over night. Size distribution is shown in Table 2. The fraction "<0.5 mm" was used throughout all bioleaching experiments.

Table 2
Size distribution of dust-like material from electronic scrap recycling process

Fraction (mm)	Percent	Description
<0.5	95.3	homogeneous dust
0.5 - 1	2.0	fluffy material
1 - 2	1.0	fluffy material
> 2	1.7	Al-coated capacitor foils

2.2. Organisms and culture conditions

Thiobacillus thiooxidans and *T. ferrooxidans* were purchased from the German Collection of Microorganisms and Cell Cultures, Braunschweig, Germany (strain DSM 622 and DSM 2392, respectively). Organisms were cultivated under non-sterile conditions in baffled 250-ml Erlenmeyer flasks and incubated at 30°C on a rotary shaker at 150 rpm or in aerated and stirred 1- and 5-l flasks. A mixed culture was grown on a medium containing (in g/l) KH_2PO_4 (0.1); $MgSO_4 \bullet 7H_2O$ (0.25); $(NH_4)_2SO_4$ (2.0); KCl (0.1); $FeSO_4 \bullet 7H_2O$ (8.0). 1 % (w/v) elemental sulfur was added and the pH was adjusted with sulfuric acid to 2.5-2.7. Growth was monitored following pH (Hamilton pH-Minitrode), cell counts (Neubauer counting chamber), and redox potential (Hamilton redox electrode PT 4805). For leaching experiments, different quantities (25, 50, or 100%, respectively) of the stock culture were used for inoculation.

Penicillium simplicissimum was purchased from Centralbureau voor Schimmelcultures, Baarn/Delft, The Netherlands (strain CBS 288.53).

Aspergillus niger was obtained from Institute of Plant Biology of the University of Zürich, Switzerland. Organisms were cultivated on a sucrose medium in baffled 250-ml Erlenmeyer flasks and incubated at 30°C on a rotary shaker at 150 rpm. The growth medium contained (in g/l) sucrose (100); $NaNO_3$ (1.5), KH_2PO_4 (0.5); $MgSO_4 \bullet 7H_2O$ (0.025); KCl (0.025); and yeast extract (1.6). Growth was monitored following pH (Hamilton pH-Minitrode) and the formation of metabolites by high pressure liquid chromatography. Fungal cultures were grown in the presence of different amounts of electronic scrap.

2.3. Analytical methods

Metal analyses were performed by inductively coupled plasma atomic absorption spectroscopy (Spectroflame-ICP-AES, Spectro, Analytical Instruments, Kleve, Germany) at the following wavelengths (nm): Al (396.2), Cd (228.8), Cr (267.7), Cu (324.8), Fe (261.2), Mn (294.9), Ni (352.5), Zn (206.2). Prior to analysis, the samples were centrifuged at 23700 g for 15 min., acidified with 5 drops of conc. HNO_3 per 30 ml aqueous solution, passed through a glass fiber filter (Whatman GF/C) to guarantee particle-free suspensions, and stored at 4 °C.

As an alternative method, Cu and Pb were analyzed by colorimetric tests (Merck Spectroquant 14767 and 14833, respectively). Commercially available tests were modified to be applied for sample volumes of 1 ml.

High pressure liquid chromatography was used to analyze culture liquids for metabolites (sugars and organic acids in fungal cultures). Compounds were separated at 25°C on an Aminex HPX-87H column (Biorad, Glattbrugg, Switzerland) using 5 mM H_2SO_4 as eluent (0.4 ml/min). Injection volume was 20 µl. Two serially connected detectors were used for identification: UV-detector at 210 nm for acids, refractive index detector for sugars.

3. RESULTS AND DISCUSSION

3.1. Metal solubilization by bacteria

Growth (pH as a measure for growth) of *a mixed* culture of *Thiobacillus ferrooxidans* and *T. thiooxidans* on different amounts of electronic scrap is shown in Figure 2a. Due to its alkalinity, the addition of the scrap led to an increase of the initial pH. Organisms grew at scrap concentrations of <1 % (w/v). Obviously, electronic scrap shows a certain toxicity. At concentrations of 0.5 % *Thiobacilli* were able to leach >90 % of the available Cu, Zn, Ni, and Al (Figure 2b). Pb precipitated as $PbSO_4$, Sn probably precipitated as SnO. Pb and Sn were, therefore, not detected in the leachate. Only for aluminum was a correlation between the size of the inoculum and the leaching efficiency observed (Figure 2b). Leaching of Ni and Zn were independent from the size of the inoculum.

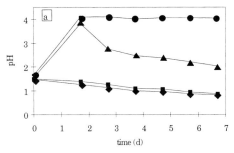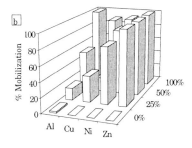

Figure 2. (a) Growth (measured as pH decrease) of a mixed culture of *Thiobacillus ferrooxidans* and *T. thiooxidans* on different amounts of electronic scrap (% w/v). ◆ 0 (control); ■ 0.1; ▲ 0.5; ● 1. Points represent mean values of duplicates. (b) Mobilization of different elements in the presence of 0.5 % (w/v) electronic scrap after 7 d incubation when inoculated with 25, 50, or 100% (v/v) of a mixed culture. 0 % represents leaching with distilled water.

3.2. Metal solubilization by fungi

Growth (pH as a measure for growth) of *Aspergillus niger* and *Penicillium simplicissimum* on different amounts of electronic scrap is shown in Figure 3. Initial experiments demonstrated that at a concentration of >1% (w/v) of scrap in the medium microbial growth was inhibited. However, after a prolonged adaptation time of six weeks and longer, fungi grew also at concentrations of 10% (data not shown). Inhibiting compounds were not identified.

During growth on sucrose various organic acids (citrate, gluconate, oxalate) were formed. After an incubation period of 21 d, *A. niger* formed 3 mM oxalate and 180 mM citrate, whereas *P. simplicissimum* formed 5 mM oxalate and 20 mM citrate in the same period. Acid formation was influenced by the strength of the phosphate buffer applied (data not shown).

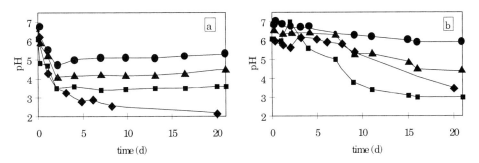

Figure 3. Growth (measured as pH decrease) of *Aspergillus niger* (a) and *Penicillium simplicissimum* (b) on different amounts of electronic scrap (% w/v). ◆ 0 (control); ■ 0.1; ▲ 0.5; ● 1. Points represent mean values of duplicates.

Differences can be seen in the leaching pattern of the two fungal strains applied (Figure 4). In general, *P. simplicissimum* was able to mobilize more metals as compared to *A. niger* under the same conditions. However, *A. niger* has a certain preference to mobilize copper. Both fungal strains were able to mobilize Cu and Sn by 65%, and Al, Ni, Pb, and Zn by >95%.

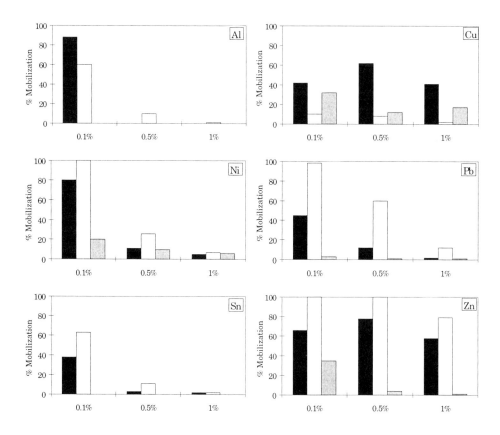

Figure 4. Leaching of metals from electronic scrap by *Aspergillus niger* (■) and *Penicillium simplicissimum* (☐) after 21 d at 30°C on different amounts of electronic scrap (% w/v). Sterile non-inoculated medium was used as control (☐). Bars represent mean values of duplicates.

From the results obtained it is obvious that for a commercially interesting process a direct growth of organisms in the presence of electronic scrap is poorly suited. Therefore, a two-step process seems appropriate to increase leaching efficiencies (e.g. for an industrial application): In a first step organisms are grown in the absence of electronic scrap followed by a second step where the formed metabolites are used for metal solubilization. This has already been suggested for the metal mobilization from fly ash for both bacteria and fungi [9, 10]. There are several advantages: (i) biomass is not in direct contact with metal-containing waste and might be recycled; (ii) waste material is not contaminated by microbial biomass; (iii) acid formation in the absence of waste material can be optimized; (iv) higher waste concentrations can be applied as compared to the one-step process resulting in increased metal yields.

Preliminary experiments were performed to increase the amount of waste material to be leached. Commercially available gluconic acid (Naglusol™, 2.5 M) produced by *A. niger* was used as leaching agent (Table 3). Concentrations of up to 10% electronic scrap could easily be treated resulting in an almost complete solubilization of available Cu, Pb, Sn, and Zn. Al is present in the electronic scrap in relatively high concentration allowing a solubilization of only approximately 40 %. At higher scrap concentration, Cu precipitated from solution as red sludge.

Table 3
Percent (%) metal mobilization when Naglusol™ (4 fold diluted) was used as leaching agent. Values represent means of duplicates.

Element		Scrap concentration (% w/v)			
		0.1	1	5	10
Aluminum	Al	62	57	42	43
Copper	Cu	85	86	70	8
Lead	Pb	100	92	99	97
Nickel	Ni	100	100	100	100
Tin	Sn	100	100	100	100
Zinc	Zn	100	100	100	100

In addition to the two-step process, microorganisms might be selected by adaptation experiments to tolerate electronic scrap at elevated concentrations and to leach metals selectively from electronic waste materials.

REFERENCES

1. K. Bosecker, In: P. Präve et al. (eds.) Handbuch der Biotechnologie. Oldenbourg, München, pp. 835, 1994.
2. D.E. Rawlings (ed.) Biomining. Springer, Berlin, 1997.
3. A.E. Torma, In: H.J Rehm H.J. and G. Reed (eds.) Biotechnology. Vol. 6b, Verlag Chemie, Weinheim, pp. 367, 1998.
4. D.E. Rawlings, J. Ind. Microbiol. Biotechnol., 20 (1998) 168.
5. G. Rossi, Biohydrometallurgy. McGraw-Hill, New York, 1990.
6. W. Krebs, C. Brombacher, P.P. Bosshard , R. Bachofen and H. Brandl, FEMS Microbiol. Rev. 20 (1997) 605.
7. M.S. Hahn, S. Willscher, G. Straube, In: A.E. Torma et al. (eds.) Biohydrometallurgical Technologies. The Minerals, Metals & Material Society, Warrendale, pp. 99, 1993.
8. G. Bowers-Irons, R. Pryor, T. Bowers-Irons, M. Glass, C. Welsh, R. Blake, In: A.E. Torma et al. (eds.) Biohydrometallurgical Technologies. The Minerals, Metals & Material Society, Warrendale, pp. 335, 1993.
9. P.P. Bosshard, R. Bachofen, H. Brandl, Environ. Sci. Technol. 30 (1996) 3066.
10. C. Brombacher, R. Bachofen, H. Brandl, Appl. Environ. Microbiol. 64 (1998) 1237.

Acid Mine Drainage (AMD) treatment by Sulphate Reducing Bacteria

C.M. Estrada Rendon[a], G.Amara[b], P.Leonard[c], J.Tobin[c], J.Roussy[d], J.R.Degorce-Dumas[d]*

[a] Universidad Nacional de Columbia, Medellin (Columbia).

[b] Università di Firenze, Dpt. Ingegneria Ambiente e Territorio, Florence (Italy).

[c] Dublin City University, School of Biological Sciences, Dublin (Ireland).

[d] Ecole des Mines d'Alès - Laboratoire Génie de l'Environnement Industriel, 6, avenue de Clavières 30319 Alès cedex (France).

This study was carried out on the AMD been emitted from the tailings dam at Carnoules (Gard, France), an old polymetallic mine (Zn, Pb). The tailing dam is located at the end of a valley where the tributary Reigous originates. Consequently, the water source is continuously polluted with AMD producing acidification and high sulphate and metal solubilization in the tributary Reigous. Fe, Pb, Zn, Mn and As ions concentrations were shown to fluctuate throughout the year.

Batch tests were conducted to determine if biogenic H_2S could be used to eliminate soluble As and metals ions as insoluble sulphides. Dynamic treatment in pilot scale units using cheap, reliable mineral and organic materials was investigated. Promising removal levels, especially of arsenic, were achieved.

1. INTRODUCTION

The first activities in exploitation the polymetallic mine of Carnoulès (blend, galene, pyrite) date from the beginning of the XIX century and was finally abandonned as recently in 1963. Millions of tons of waste materials including sulphides ores accumulated during the period of activity of the mine.

The tailing dam of Carnoulès is located at the end of a valley where the tributary Reigous originates. Consequently, the water source is continuously polluted with AMD, producing acidification and metal solubilization in the tributary Reigous. The acidity which arises from the subsurface oxydation of sulphides, causes the water of the Reigous tributary to gain acidity (pH 2-4) with the resulting solubilisation of arsenic and heavy metals such as Fe, Zn, Pb, Mn (see Table 1) It is well known[1-4] that such pyritic mine tailings leach acid mine drainage originating in a large part from the metabolic activity of *Thiobacillus* sp. and hosts. As a visible result of the contamination of the Reigous, the aquatic ecosystem is abiotic and a yellow-orange-reddish coloured deposit "yellow boy" (iron hydroxide and related compounds such as jarosite) is present along the 2 km long tributary and continues even after the confluence with the Amous river.

That natural attenuation is a major abatement process for soluble metals (Fe, Zn, Pb, Mn) and As was demonstrated by determining their concentration in water along the tributary[5]. The majority of the metal and arsenic removal was shown to occur within the upper section of the tributary emerging from the dam, where AMD streams in a small basin which acts as an efficient passive treatment system containing numerous aerobic and acidophilic iron-oxidizing bacteria such as *T. ferrooxidans* and protozoa.

An other potential alternative treatment of AMD involves the precipitation of metals as insoluble sulphides[6-10]. In the presence of sulphides (HS^-, H_2S in particular), iron and many other metal ions would be expected to precipitate[7].

The aim of this work was to study the feasibility of an anaerobic biotreament of AMD including Sulphate-Reducing Bacteria (SRB) to eliminate the heavy metals and arsenic ions under as insoluble sulphide forms.

Table 1 : AMD characterization and variation along the years 97-98. Water was sampled in november 1997, march may, and june 1998.

	pH	SO_4^{2-}	Fe as Fe^{2+}	As	Pb	Zn	Mn
Limit values mg/L	2-4	2-3000	720-1480	95-230	1-2	15-35	5-15

2. MATERIAL AND METHODS

2.1. Analytical methods

In order to chatacterize the AMD from the Reigous tributary, samples were taken in 500 mL plastic bottles at different periods of time and analysed as follows : 20mL aliquots were filtered using 0.2 μm membrane filters (GelmanSciences), acidified with 3 drops of concentrated HCl to maintain metal solubility and stored at 4°C until ICP analysis (Jobin Yvon JY36 ICP spectrometer). Prior to analysis the samples were diluted using a laboratory diluter (Hamilton, micro lab 1000). Standard concentrations were used to calibrate the spectrometer before usage. Results are given as total metal concentration. The same procedure was used to anlayse the samples from the batch tests and pilot units experiments (inlet and outlet sampling). Fe(II) was analysed using the phenantroline colorimetric method.

2.2. Batch tests

In vitro experiments using SRB indigenous to activated sludges (from the local urban wastewater treatment plant) as inoculum were performed with different AMD ratios both in 200 mL flasks and 20 mL test tubes, incubated at 30°C. Lactate was used as energy source. The influence of pH on SRB activity was studied in order to determine the pH limits for operation of continuous treatment in the pilot units.

2.3. Pilot scale treatment unit

The AMD stream was pumped upflow through the pilot units which consisted of two columns A and B installed in series (Figure 1). The first one was filled with Siporex granular material (0.5-1.5 cm). This alcaline mineral product ("Béton cellulaire") consists of an hydrated calcium silicate with a macroporous and microporous (2-5 μm) structure, a water absorption coefficient of 140% and a density of $0.35g/cm^3$ (with a particle size around 1.6 to 0.5 mm). This material was used because of its general characteristics and notably its acidic water neutralizing potentiality which is an important factor governing the SRB activity. Then, as described in Figure 1, the upflow AMD stream entered the column B which contained straw previously inoculated with SRB. The operating conditions of the columns are given in Table 2. Water samples were taken at entry and exit of the columns A and B and analysed by ICP after filtration, as outlined in section 2.1, to determine their respective efficiency.

Table 2 : Operating conditions of siporex and straw columns.

	Column A	Column B
Diameter	2.5 cm	10cm
Packing height	1.8 m	1 m
Flowrate	0.1 - 0.2L/h	0.1 - 0.2L/h
Flow velocity	0.05- 0.1m/h	0.013 - 0.026m/h
Residence time	36 / 18h	77 / 38.5h

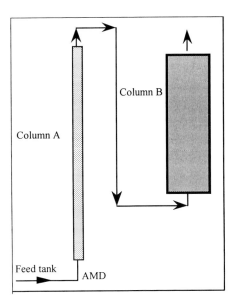

Figure 1 : Schematic diagram of the pilot scale units.

3. RESULTS AND DISCUSSION

3.1. Sulphate reduction assesment

3.1.1 SRB Inoculum.

Assesment of activated sludges (AS) as a source of SRB was obtained from qualitative fermentation studies using decanted activated sludges suspension as a source of SRB.Anaerobic fermentation experiments were conducted in 20mL tubes. The liquid medium consisted of : activated sludges suspension (20% v/v), sodium lactate (6g/L) and 25mg/L of ferrous iron. Sodium sulphate (100mg/L) was added as sulphate source ; the final volume of medium was 20mL and pH was 6.5. Experiments were performed with and without sulphate. As shown in Table 3, sulphate reduction occured in the presence of sulphate as electron acceptor ; this was indicated by blackening of the medium (by production of FeS), accompagnied by a foul odour characteristic of H_2S production. These simple and qualitative experiments showed that SRB can be readily found in activated sludges.

Table 3 : Sulphides production with AS as inoculum. + indicates blackening of the medium.

Days	1	2	3	4	8
+ Sulphate	-	+	+	++	+++
- Sulphate	-	-	-	-	-

3.1.2 Evidence of sulphate reduction and metal abatement in the presence of AMD.

The objective was to determine the feasibility of sulphate reduction using AS as a source of SRB and AMD as a source of metals and suphate. 400 mL of the acidic water AMD, pH 3.1 was mixed to 500mL of AS suspension and lactate (1g/l) yielding a final pH was 5.9 and incubated anaerobically at 30°C. After 8 days sulphate reduction was indicated by the presence of a distinct dark black precipitate. This was confirmed by the analysis of the residual soluble metals in the medium (Figure 2).

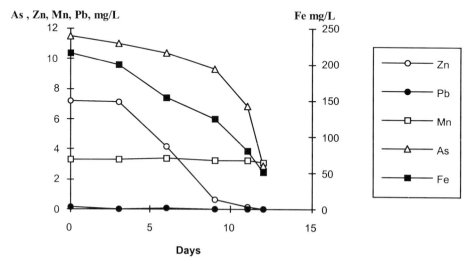

Figure 2 : Soluble metal and arsenic concentration during sulphate reduction experiment.

As shown, the concentration of soluble Fe, As, Zn decreases with time. After 12 days, the percentages of the removal efficiency for Fe, As, Zn, Pb and Mn were 76%, 75%, 99%, 93% and 5%, respectively. Mn ions stayed in soluble form and were not precipitated by sulphides. Due to its low initial concentration (0.13mg/L), precise calculation for Pb removal yield is difficult.

3.1.3 Influence of pH and AMD concentration on sulphate reduction.

The effect of pH and dilution factor of AMD on performance of sulphate reduction were investigated in 20 mL tubes containing lactate (1g/L). The dilution of AMD was in the range 20% to 80% v/v, and pH varied from 4.5 to 6.5. Results are given in Table 4. At low pH value (pH = 4.5), sulphate reduction was not observed after 7 days. In the same conditions, it was shown that sulphate reduction occured at pH 5.2 but less rapidly than at pH 6.5. Dilution of AMD is another important parameter which also needs consideration. The influence of this factor can be readily observed by examination of the kinetics of the sulphate reduction. At high AMD concentration, ie 80%, sulphate reduction occured but slowler than at 50% or 20%. In the absence of a carbon source such as lactate, no reaction was observed.

Table 4 : Influence of pH and AMD dilution on sulphate reduction. Colour profile ranging from totally black ++++ to no colour -.

Days	AMD pH =4.5			AMD pH = 5.2				AMD pH = 6.5		
	No lactate,	20% ,	50%	No lactate,	20%,	50%,	80%	No lactate,	20%,	80%
1	-	-	-	-	-	-	-	-	-	-
2	-	-	-	-	+++	-	-	-	+++	-
3	-	-	-	-	++++	++	+	-	++++	++
4	-	-	-	-	++++	++	+	-	++++	+++
7	-	-	-	-	++++	+++	++	-	++++	+++

3.2. Treatment of AMD on pilot scale units

3.2.1 Neutralisation of AMD using siporex material

As expected, the batch tests showed that pH is a critical factor governing the eventual success of any SRB treatment. Neutralisation with the conventional alkaline substances such as sodium hydroxide is not viable at field scale where abundant and cheap neutralising materials such as crushed limestone are of more interest. In this study, siporex material was used to neutralise the acidic AMD. Continuous neutralisation of influent AMD (with initial pH ranging from 2 to 4 depending on the sampling period of time) was operated under a flowrate ranging from 0.1 L/h to 0.2 L/h (see material and methods). As seen from Figure 3, the neutralising effect of the material was good during the first operating days but tended to decrease with time (after 3 weeks the pH of the effluent AMD was ca. 4.5). On days 34 and 64, the packing material was changed and consequently the pH of the effluent AMD increased. Depending on the initial pH value of AMD, the pH of the effluent was low as ca. pH 3, which is too low for SRB activity. An orange-reddish precipitate around the material particle appeared in the first days and increased with time.

Soluble metal abatement was followed as presented in Figure 4. As shown, the iron removal efficiency which was ca. 50% initially but decreased to 10% after one operating month. Over the same time, As ions removal efficiency was higher and remained stable at between 60-80%. Replacing the packing material had a strong effect on pH of the effluent as discussed

582

before, and also on the metal removal especially for iron. As expected, trends in pH variation and iron removal were closely related. On the other hand, As abatement which fluctuated between 60% and 98% was less dependent on the pH.

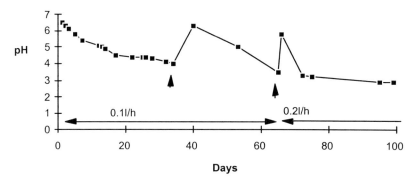

Figure 3 : Evolution of the pH after AMD treatment on siporex column. Arrows indicate the changes of the packing material.

Removal efficiency %

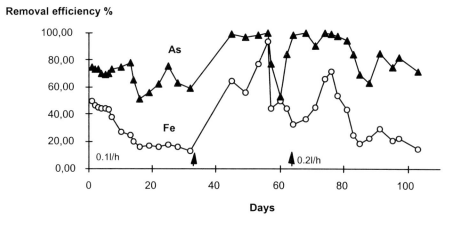

Figure 4 : Removal of As and Fe on siporex column. Arrows indicate the changes of packing material.

3.2.2   Siporex / straw linked process.

The feasibility of a two step process including neutralisation of AMD on siporex material and a finishing treatment process on straw material where sulphate reduction occurs was studied for 3.5 operating months. Soluble iron and arsenic ion analysis was performed to determine their respective elimination performance (Figure 5). From a general point of view, the global Fe abatement curve (after the second step) shows the same trends that were obtained in the first step (neutralisation). The Fe abatement curve shows 3 peaks corresponding to each change of siporex material and can be superimposed on to the pH curve. Though the pH of the treated AMD from the straw column was slightly higher than that of siporex column (4.2<pH>6.5) (it depended on the pH of the efflent at the entry of the straw column), values as low as 4.2 were measured. Such pH values are low for SRB activity. The finishing step did not ameliorate significantly the performance of Fe elimination.

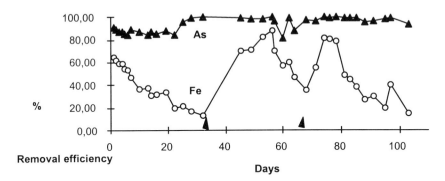

Figure 5 : Removal of As and Fe on linked siporex/straw columns. Arrows indicate the changes of packing material.

On the other hand, the second stage of the treament (on straw) increased efficiency of As ions removal leading to a global and regular 90% - 100% elimination yield. Overall the specific capacity of the straw column allowed a further 50-100% elimination of the remaining As with respect to the siporex exit and only about 0-25% on average for iron.

Concerning the elimination of the other soluble metals (Zn, Pb and Mn) present in AMD, their respective removal efficiency corresponding to columns A, B and A + B, obtained after three operating months, are given in Table 5.

Specific and global percentages are expressed as the range of the limit values of the removal efficiency percentages obtained for each metal ion.

Table 5 : Removal efficiency of Zn, Pb and Mn. * with respect to the influent metal concentration.

Removal efficiency	Column A (siporex)	Column B (straw)*	Linked process (Global efficiency)
Zn	15-80%	25-95%	38-98%
Pb	60-100%	80-100%	90-100%
Mn	4-7%	2-15%	5-9%

It was apparent that the elimination of zinc (and to a lesser extent, elimination of lead), were closely related to the pH trends. Acidic pH usually corresponded to the lowest removal efficiency. Nevertheless, good elimination of Pb was achieved. Elimination of manganese was considered as negligible.

3.2.3 Mechanisms involved

Initially, only physical and chemical reactions were though to be involved in the first neutralising step with the siporex material. It was observed that the precipitation on the packing particles of iron under an insoluble reddish precipitate (ferric form) occured more drastically at neutral pH. However on further examination, it appears that microbial processes are occuring in this step. Effluent AMD from siporex columnn was sampled and observed under the microscope. About 4.10^4 bacteria/ml was numbered besides flagellated protozoa. The siporex material which was characterised by weak aerobic conditions as revealed by its low dissolved oxygen level (1-2 ppm O_2 in the influent AMD contained in the feed tank) contained numerous motile, gram-negative, regular rod shaped bacteria. On a morphological basis, no neutral iron bacteria (ie. members of Gallionella, Leptothrix, Siderocapsa...) were observed in the microbial consortium. Taking into account morphological criteria and the pH conditions of this ecological "niche" the majority of these bacteria (1-2.5 µm) was identified as members of the genus Thiobacilli which are known to catalyse the oxidation of iron. On the other hand, T.ferrooxidans which is generally assumed to be an obligate aerobe, can be grown anaerobically on elemental sulphur as energy source, using ferric ion as electron acceptor. In the present work, this hypothesis could be verified by the appearance, in the upper section of the column of a green-dark coloured precipitate which was considered as a ferrous precipitate. In order to facilitate aerobic process into the siporex column, downstream percolation of the AMD onto the mineral material would be preferable.

On straw material, adsorption and precipitation phenomena were considered to occur allowing a polishing treatment, especially for arsenic and to a lesser extent for zinc and lead ions. But unsurprisingly, the treated AMD from this column also contained numerous bacteria $(6.10^4$ cells/ml) and a flagellated protozoa similar as that observed in the siporex effluent. These bacteria were mostly curved, rod shaped and identified as SRB taking into account the anaerobic conditions, the presence of dark coloured precipitates appearing into the straw column and the weak sulphide odour found in the exit water. Nevertheless, sulphate reduction, which was probably limited when pH was too low, was not sufficient enough to allow complete metal removal (Fe, Zn in particular) as sulphides.

Finally, the presence of bacteria in both columns indicates that biological as well as physical and chemical processes were occuring; these microbial consortia are suspected to play an important role in the elimination of soluble metals especially for arsenic. Further studies are required in order to elucidate the metal speciation and to precisely identify the microbial populations involved in such a linked process.

4. CONCLUSIONS

In vitro, batch experiments showed that sulphate reduction could occur even in the presence of high initial metal concentrations and pH as low as 5.2. Fe, As, Zn and Pb were precipitated or co-precipitated as insoluble sulphides. In the same conditions, Mn ions remain in solution.

Continuous soluble metal elimination was achieved using a two step process including neutralisation of AMD on an alkaline mineral material (siporex) and treatment on an organic support (straw). The columns were operated in series under anaerobic conditions and low flow velocity, allowing total elimination of arsenic and to a lesser extent lead, zinc and iron. Manganese was not eliminated.

Besides physical, chemical reactions occuring at each step of the process, biological reactions are believed to play a crucial role for the metal abatement. But, before scaling up, further study is required to fully understand the complex mechanisms involved.

REFERENCES

1. A.P. Harrison, Annu. Rev. Microbiol. 38, 265-292, (1984).
2. H.L.Ehrlich, Geomicrobiology, Marcel Decker New York, 1990.
3. V.P.Evangelou, Pyrite oxidation and its control, CRC Press, Boca Raton, 1995.
4. G.Rossi, Biohydrometallurgy, McGraw-Hill, Hamburg, Germany, 1990.
5. P.Leonard et al this proccedings.
6. L.P.Miller, Contrib. Boyce Thompson Inst. plant. Res. 16, (1950) 85.
7. H.L.Ehrlich, Geomicrobiology, Marcel Decker New York, 1990.
8. N.Wakao, T.Takahashi, Y.Sakurai, H.Shiota, J.Ferment.Technol., 57, (1979) 445.
9. J.S.Whang, D.Young, M.Pressman, Environ.Prog., 1, (1982) 110.
10. R.W.Hamack, H.M.Edenborn, Appl.Microbiol. Biotech., 37, (1992) 674.

Natural attenuation study of the impact of acid mine drainage (AMD) in the site of Carnoulès

P.Leonard[a], C.M.Estrada Rendon[b], G.Amara[c], J.Roussy[d], J.Tobin[a], J.R.Degorce-Dumas[d]*

a Dublin City University, School of Biological Sciences, Dublin (Ireland).

b Universidad Nacional de Columbia, Medellin (Columbia).

c Università di Firenze, Dpt. Ingegneria Ambiente e Territorio, Florence (Italy).

d Ecole des Mines d'Alès - Laboratoire Génie de l'Environnement Industriel,
      6, avenue de Clavières 30319 Alès cedex (France).

This study was carried out on the AMD being emitted from the tailings dam at Carnoules (Gard, France), an old polymetallic mine (Zn, Pb). The tailings dam is located at the end of a valley where the tributary Reigous originates. Consequently, the water source is continuously polluted with AMD producing acidification and metal solubilisation in the tributary Reigous.

The fact that natural attenuation is a major abatement process for soluble metals (Fe, Zn, Pb, Mn) and As was demonstrated by determining their concentrations in water along the tributary. After 750m, percentage abatements as high as 85%, 80%, 98%, 72% and 94% respectively were observed. These values were shown to fluctuate during the year. The majority of the metal and arsenic removal is achieved within the first part of the tributary emerging from the dam (first 60 meters). In order to explain the rapid abatement which can not be satisfactorily explained by the local physical and chemical properties of the water (pH, O_2, temperature or metal levels) biological data were collected.

Biological studies showed that this short distance was also the most populated in terms of bacteria (i.e. Thiobacilli) and protozoa, as revealed by adenosine triphosphate (ATP) determinations, microscopic observations and counts.

1. INTRODUCTION

Mining generating metals or coal puts great stress on the environment, especially in terms of Acid Mine Drainage (AMD), water quality and biodiversity of aquatic organisms. The first activities at the polymetallic mine of Carnoulès (blend, galena, pyrite) date from the beginning of the19th century and work continued until the mine was finally abandoned in 1963. Millions of tons of waste materials accumulated during the period of activity of the mine.

Waste materials (especially those containing metal sulphides) from mining activities discharged as tailings are subject to bioleaching[1-3] (via direct and/or indirect mechanisms[4]) as well as chemical and physical weathering, leading to acid mine drainage pollution. The tailing dam of Carnoulès is located at the end of a valley where the tributary Reigous originates. Consequently, the water source is continuously polluted with AMD, producing acidification and metal solubilisation in the tributary Reigous. The acidity which arises from the subsurface oxidation process of sulphides, causes the water of the Reigous tributary to gain acidity (pH 2-4) and solubilises arsenic and heavy metals such as Fe, Zn, Pb and Mn.

It is well known that such pyritic mine tailings generate acid mine drainage which result due to the metabolic activity of *Thiobacillus ferrooxidans* and similar species.

As a visible result of the contamination of the Reigous, the aquatic ecosystem is abiotic and a yellow-orange-reddish coloured deposit known as "yellow boy" (iron hydroxide and related compounds such as jarosite) is present along the 2 km long tributary and even continues after its confluence with the Amous river.

In order to characterise the site and the local phenomena involved, chemical, physical and biological studies were performed. The analysis, realized at different periods of time during the years 97-98, revealed that pollution of the Reigous water emerging from the dam, fluctuated and contained respectively (mg/L) : Fe, 720 to 1480 (as Fe^{2+}); Pb, 1 to 2 ; Zn, 15 to 35 ; Mn, 5 to 15 ; As, 95 to 230 and sulphate 2000 to 3000.

Data were collected to establish the profile of the Reigous. They showed that the majority of the metal and arsenic removal was achieved within the first part of the tributary emerging from the tailings (i.e. 60 meters) where numerous bacteria (in particular *T. ferrooxidans*) and two species of protozoa were present. A close relationship between the evolution of the chemical and biological studied parameters has been identified.

2. MATERIAL AND METHODS

2.1. Samples

In order to determine the characteristics of the Reigous tributary, samples were taken in 500mL plastic bottles from different points (Table 1) along its 2 km course. Eight samples were analysed for the following metals, Fe, Zn, Pb, Mn and for As. For the biological tests, deposit and water were taken. After overnight settling, water samples were taken from the solid/liquid interface before ATP or microscopic determinations.

Table 1 : The different sampling points on the Reigous .

Pt 0	Pt 1	Pt 2	Pt 3	Pt 4	Pt 5	Pt 6	Pt 7
Source*	30m	35m	45m	60m	500m	750m	1,800m

(*distance is measured from the tailings dam ie. source)

2.2. Analytical methods

The following procedure was used to analyze the metals contained in the water ; 20 mL aliquots were filtered using 0.2 μm membrane filters (Gelman Sciences), acidified with 3 drops of concentrated HCl to maintain metal solubility and stored at 4°C until ICP analysis (Jobin Yvon JY36 ICP spectrometer). Prior to analysis the samples were diluted using a lab diluter (Hamilton, micro lab 1000). Standard concentrations were used to calibrate the spectrometer before usage. All samples were analysed for their degree of acidity or dissolved oxygen and temperature along the profile of the drainage course using portable pH or dissolved oxygen meters. (WTV equipment). ATP analysis (Lumac Biocounter M2500) was carried out on 20 μL water samples after dilution with 80 μL phosphate buffer pH 7 and addition of 100 µL of NRB solution and 100 µL of Lumit solution. Microscopic observations and counts were performed using a Leitz Wetzlar microscope connected to a camera and a video copy processor.

3. RESULTS AND DISCUSSION

3.1. Physical and chemical characterisation of the profile of the Reigous.

3.1.1. Metals
  Studies of AMD at Carnoules have shown that Natural Attenuation (NAT) is a major factor in metal abatement. As can be seen in Figure 1 the soluble iron concentration decreases with distance. This process seems to be occuring rapidly, within the first 60 meters. It can be clearly observed that the same phenomenon is occuring in November (1997) as in May (1998) and to a lesser extent in March (1998), but the metal concentrations were fluctuating (944 mg/L, 1480 mg/L and 719 mg/L respectively in November, March and May).
  These results reinforce the finding that NAT is a major abatement process showing respectively a 78%, 50% and 85% abatement in soluble iron concentration in November, March and May over a distance of 750 m. The final iron concentrations were respectively 210, 730 and 104 mg/L.
  The seasonal conditions (monthly averaged temperatures and rainfalls) in November 1997, March 1998 and May 1998 were respectively : 15°C and 286mm, 3.2°C and 7 mm, 17.9°C and 254mm. The concentration of iron in the Reigous tributary appears to be related to the proportion of water infiltrating from the surface (rainfall). Thus in dry month (March 98) values are relatively low but after a wet period (as in November 1997 and May 1998) the iron concentrations increase in the Reigous due to the elution of elements by infiltrating waters. Such phenomenon has already been observed[7].

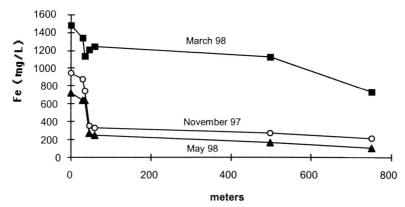

Figure 1. Variation of the soluble iron concentration with distance from the tailings dam at different periods of time.

  Concentration profiles of the other metals present in the acidic water with respect to distance are shown in Figure 2. From 0 to 750 m, As, Mn and Zn show similar abatement profiles to that of iron in Figure 1. Pb is low (0.7 - 0.5 mg/L in the first section (60m) and tended to decrease after 60 m (0.3 - 0.1 mg/L). After 750 m, percentages of abatement for As,

Zn, Mn and Pb are respectively 94%, 80%, 98% and 72%. After 1800m, the final concentrations are (mg/l) As : 3, Zn : 12, Mn : 5.4, Pb : 0.4. The metal removal is achieved by precipitation as revealed by the yellow-orange-reddish coloured deposit, occuring via, in particular, the co-precipitation of As and the metals with oxidized iron. These results agree well with the literature which reports that oxides of iron strongly adsorb As compounds [6].

The chemical Fe(II) oxidation rate depends on several factors such as initial Fe(II) concentration, pH, temperature, O_2 concentration and is considered to occur slowly[5]. In nature, especially in the case of aerobic and acidic environments (pH < 4), this Fe(II) oxidation process is greatly enhanced because of the metabolic activity of *T. ferrooxidans* leading to the ferric hydroxide precipitate. The rapid iron precipitation which is observed here, suggests that complex chemical and biological processes occur in the upper section of the tributary.

Figure 2 shows that an increase in Zn, Pb, and Mn occured in the Reigous between points 6 and 7 (750 and 1800 m), respectively : 3 to 12 mg/L; 0 to 0.4 mg/L and 2.1 to 5.3 mg/L. As demonstrated by the data in Table 2, this augmentation in metal level is due to the presence of three small tributaries (containing these metals) which join the Reigous between these points.

Depending on the season and more particularly on the rainfalls (which can be sudden and torrential), the total flowrate of these subtributaries must be considered as mostly irregular and sometimes can be larger than the Reigous flowrate. Data from Table 2 are given as an example of their chemical composition as it was determined on the 5/5/98.

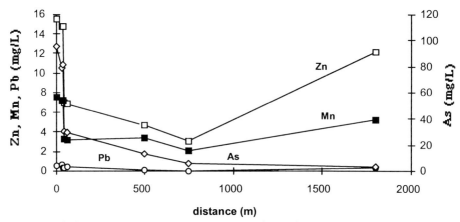

Figure 2. Variation of As, Mn, Zn and Pb concentrations with distance from the tailings dam on the 5/5/98.

Table 2. Subtributary pH and metal concentration on the 5/5/98.

Subtributary	Distance	Zn	Pb	As	Fe	Mn	pH
1	850m	11.5	1.6	0.3	1.3	1.5	5.5
2	1200m	19	0.5	0.8	57	4	2.5
3	1400m	11	0.8	0.3	34	3.6	2.6

3.1.2 pH

The variation of pH along the Reigous profile is given in Figure 3. As can be seen by the graph, the profile is not regular.

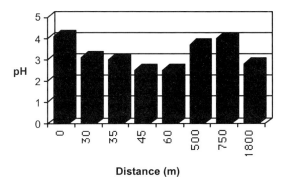

Distance (m)

Figure 3. pH profile of the Reigous on the 5/5/98.

The drop in pH between point 0 and point 4 (0 to 60 m) is consistent with the metal precipitation which is occuring in this section according the following equations:

$$Fe^{2+}_{(aq)} + 1/4\ O_{2(g)} + H^+_{(aq)} \text{-------}> Fe^{3+}_{(aq)} + 1/2\ H_2O_{(l)} \tag{1}$$

$$Fe^{3+}_{(aq)} + 3\ H_2O_{(l)} \qquad \text{--------}> Fe(OH)_{3(s)} + 3\ H^+_{(aq)} \tag{2}$$

Geological factors such as changing surface rock and soil type can also play a role in altering the pH of the polluted water as well as the joining with other acidic subtributaries (see subtributaries 2 and 3 outlined above) which explains the sharp drop in pH between points 6 (750) and 7 (1800 m).

3.1.3. Temperature and dissolved oxygen.

Dissolved oxygen and temperature profiles of the Reigous were determined on the same day in order to better characterise the ecosystem (Figure 4). It can be observed that dissolved O_2 at the source (water exit at the tailings dam) is low (1.8 mg/L) as expected but quite rapidly increases up to 8.5 mg/L after a few meters and then stays high due to the natural oxygenation in the environment. The temperature varies along the tributary from 13.5°C (source) up to 20°C as determined by the water temperature of joining subtributaries and the extent of foliage cover at the sections sampled.

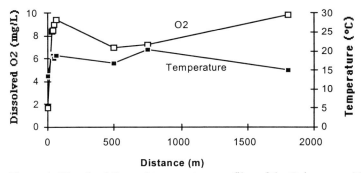

Distance (m)

Figure 4. Dissolved O_2 and temperature profiles of the Reigous on 5/5/98.

3.2. Biological characterisation of the profile of the Reigous.

3.2.1 Microscopic observation

Besides physical and chemical determinations, microscopic observations and adenosine tri-phosphate (ATP) measurements were performed for a better characterisation of the Reigous. No macrophytes or metazoa were observed in the tributary. The observation of deposited sediment samples from the points 1 to 7 were carried out under 40X or 100X magnification. Cell counts were used to quantify the biomass using an hematimeter type. As shown in Figure 5, samples from points 1 to 4 were found to be the most heavily populated. Sample 1, in particular, showed considerable biomass abundance in comparison with the other points including bacterial cells, and two morphological types of protozoa.

The bacteria were found to be rod shaped, from 0.6 to 1 μm in length, mostly motile and gram negative. As these bacteria grow on mineral media (pH 2.5) containing sulphate and ferrous ions leading to a reddish ferric hydroxide precipitate, they were identified as *T. ferrooxidans*. Other rod shaped bacteria (1 to 3 μm) were observed in the AMD sample but were not identified. Possible heterogenous microbial population must be considered to exist and thrive in such acidic conditions and would need more detailed microbiological work.

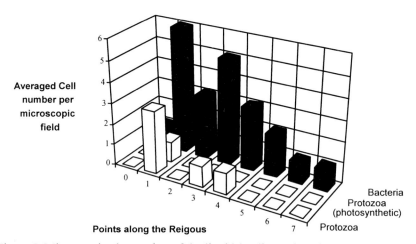

Figure 5. Microscopic observation of the liquid / sediment interface of samples 0 to 7. Average of 10 values.

Two different types of protozoa were observed and their identification is ongoing : one (100 μm in length) is photosynthetic containing green globular vesicules (algae?); it presents a very distinct red pigment at one end of its cytoplasm and is able to move in an amoeboid motion as a member of the group Sarcodina. The other, smaller (5-10 μm) organism, seems more like a flagellate. These organisms seem to be well adapted to this specific acidic environment. The green coloured organism is colonizing sections of the surface of the "yellow boy" precipitate, in particular during the summer period, corresponding to the warmest conditions and the lowest flowrate of the tributary. It was noticed that in cold seasons, this colonization was limited to the 30-35 m zone.

3.2.2 ATP measurements

The data (Figure 6) were obtained from sampled water for two different dates, 24/4/98 and 6/5/98. The ATP profiles of the Reigous are close but show that the water samples are not equivalent in term of biomass concentration, confirming the microscopic data.

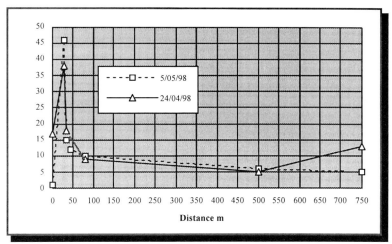

Figure 6. Profile of the Reigous using the ATP response (given as % of total ATP response).

The ATP response (given in arbitrary units) varies at point 0 and is highest for points 1, 2 and 3. After point 3 (45m) the response seems rather constant, ranging from 5% to 13% of the total response for both data. Taking into account that the reagent used for cell lysis is specific for bacterial cells (manufacturer's instruction) it is not possible to assess that eukaryotic cells are disrupted using this protocol.

The Fe-As rich and acidic waters of Carnoulès were previously studied[8]. Due to the local pH-Eh conditions, Leblanc *et al.*, referring to the stability domain calculated from the ferrous and arsenite ions and eventhough Fe^{2+} should be the more stable ionic form for Fe (especially in the upper part of the Reigous), oxidation of Fe^{2+} occurs, leading to the precipitation of iron under Fe^{3+} - arsenate and $Fe(OH)_3$ forms. The yellow accretions observed in the upper part of the Reigous were found particularly Fe-As rich (22 to 49% Fe_2O_3 and 9 to 20% As respectively) and contain numerous microbial structure (bacterial stromatolites) as revealed by the SEM microphotographies. Our results confirm this previous work, showing that the biological activity is probably a major phenomenon occuring in the upper part of the Reigous.

Living forms of *Thiobacillus* sp. and their hosts, more specifically colonize the upper section of the tributary where most of the metal abatement is observed. Among different hypotheses to explain such a specific ecological "niche" for microorganisms, it can be proposed that the shallow water depth (few centimeters), the low flowrate and velocity of the water due to the widening of the bed tributary forming a small basin at this section where steady state conditions, high oxygen concentration and light accessibility occur, allow better bacterial growth and photosynthesis activity which also favour the microbial aerobic metabolism.

4. CONCLUSIONS

The collected physical, chemical and biological data demonstrate that complex reactions occur in the Reigous tributary, especially near its origin. These reactions lead to rapid metal and arsenic precipitation. Undoubtely, the intrinsic toxicity of As and its high concentration encountered in the Reigous is an acute problem. Nevertheless, NAT was shown to greatly minimize the impact of the As toxicity due to its high removal efficiency (94% and 97% respectively at 750 m and 1800 m). In this context, the iron-oxidizing bacteria that thrive in acid pH waters induce precipitation of iron and other metals such as arsenic, zinc, manganese, lead as oxidized minerals that appear yellow or yellow-orange. A high degree of depollution is achieved due to the presence of a natural basin acting as a passive treatment sytem. Studies are ongoing to determine the interactions between metals (including their speciation) and the microorganisms involved in the oxidation-reduction processes.

Nervertheless, from an ecological point of view, it must be considered that due to the sudden, torrential rainfalls which occur in the Mediterranean climate, certain precipitated metals (or arsenic compounds) are subject to resuspension and possible downstream resolubilisation when joining the Amous river.

REFERENCES

1. A.P. Harrison, Annu. Rev. Microbiol. 38 (1984) 265.
2. H.L.Ehrlich, Geomicrobiology, Marcel Decker New York, 1990.
3. V.P.Evangelou, Pyrite oxidation and its control, CRC Press, Boca Raton, 1995.
4. G.Rossi, Biohydrometallurgy, McGraw-Hill, Hamburg, Germany, 1990.
5. C.S.Kirby, H.M.Thomas, G.Southam and R.Donald, Applied Geochemistry, 6 (1998).
6. W.M.Mok, C.M.Wai, Wat. Res., 23 (1989) 1.
7. C.A.Johnson, I.Thornton, Wat. Res. 21 (1987) 3.
8. M.Leblanc, B.Achard, D.Ben Othman, J.M.Luck J. Bertrand-Sarfati, J.Ch.Personné, Appl.Geochem., 11 (1996) 541.

Bioremediation of cyanide leaching residues

X. Díaz and R. Caizaguano

Escuela Politécnica Nacional, Instituto de Investigación Tecnológica,
P.O.Box 17-01-2759, Quito - Ecuador

Bioremediation is an economic and effective technique to detoxify cyanide leaching residues. This treatment technique was investigated to determine its applicability to effluents and solid wastes from Ecuadorian cyanidation plants. Tailings from different cyanide leaching plants that utilized different types of ore were used to isolated native strains of bacteria using several nutrient media. These tests were carried out on a laboratory scale in Erlenmeyer flasks with a solid/liquid ratio of 1/10 in a rotating shaker at 150 rpm and 30°C. The pH of the pulp and the bacterial population were measured daily by standard plate counting. The obtained colonies were characterized for *Pseudomonas*-type bacteria. The obtained cultures were progressively conditioned for higher concentrations of free cyanide at an initial pH of 11 in nutrient media. *Pseudomonas* population, pH and free cyanide were measured daily. The results show that bacteria levels higher than 10^4 cells/ml were able to resist a concentration of 100 ppm free cyanide, while 10^8 cells/ml were required to resist a concentration of 750 ppm free cyanide in the media. At concentration levels higher than 750 ppm free cyanide, the toxicity of the media did not allow bacterial survival. The cultures adapted to 750 ppm of free cyanide were able to detoxify a fresh cyanide residue containing more than 3000 ppm of total cyanide. Blank tests (without bacteria) were also carried out. The results showed a reduction of total cyanide content between 200 and 390 ppm/day, in the first 3-4 days for the solid and liquid residue, respectively. The total cyanide remediation rate in the blank controls was constant at approximately 45 ppm/day.

1. INTRODUCTION

An important source of environmental contamination in metallurgical processes for gold recovery is cyanidation wastes. They contain heavy metals, as well as cyanide and very stable cyanide complex compounds [1]. Mining and metallurgical operations in Ecuador are artisanal. Currently, no control of effluent discharge exists, resulting in detrimental conditions in the water and soil due to the presence of toxic compounds. Technical solutions to treat effluents, particularly in mining and metallurgy, must be efficient, economic and easy-to-implement. Biological treatment of cyanide wastes meets all these requirements.

An almost complete biodegradation of weak and strong cyanide complex compounds using native bacteria has been reported [2]. The process is simple in concept and application, no toxic byproducts are generated, and operating costs are low [2,3].

While the basic process, including the type of bacteria to be used, is similar for different sites, each one presents site-specific problems and issues that must be addressed before this biotreatment process can be used for cyanide remediation. Some of the issues include mineralogical composition of the waste material, microflora composition, bacteria production and application of this technology in situ [3].

The basic problems associated with the use of cyanide are the presence of free cyanide and heavy metals, as well as the formation of stable complexes in solid and liquid residues [4]. These complexes foul the cyanide solution such that it can not be recycled to the gold recovery process, and introduce heavy metals into the soil [2]. According to some estimations, approximately 600 tons of cyanide and cyanide toxic compounds are discharged to the environment each year in Ecuador [5].

The cyanide biodegradation by *Pseudomonas* remediates soil and waters from free cyanide and its complex compounds in a natural way [3,4]. The development of this process may provide a technical solution in Ecuador for the protection of the environment from the detrimental effects of cyanide residues. The objective of this research study is to investigate the biodegradation of cyanidation residues using native bacteria isolated from three different industrial cyanidation tailings. This bacterial culture was tested for free cyanide tolerance and adapted to different cyanide concentrations

2. EXPERIMENTAL

2.1 Bacterial isolation

Three industrial cyanidation tailings from Ecuadorian plants were used to isolate cyanide-tolerant bacteria. The general characteristics of the materials are described in Table 1. The first material (Tails 1) had been dumped for over five years; the others (Tails 2 & Tails 3) had been in the tailings dams for less than two months. Four different nutrient media were used, which are specific for autotrophic bacteria isolation [6], as shown in Table 2.

A 20-g tail sample was added to 200 ml of nutrient medium in a 300 ml Erlenmeyer flask, and the pH was adjusted to 11 with 1N NaOH. The flasks were placed in a thermostated, rotating shaker at 30°C and 150 rpm. A daily count of the bacterial population was performed by inoculating 0.1 ml of the liquid sample from the relevant culture in Mac Conkey agar, followed by incubation at 35°C.

2.2 Characterization of isolated bacteria

Several biochemical tests [7,8] were carried out on liquid samples from the isolation media as well as on colonies from Mac Conkey agar control plates kept in 1 g/l beef extract. The cultures were tested by microscopic examination with both direct and ultraviolet light, Gram staining reaction, citrate agar growth, oxidase reaction for detection of cytochrome oxidase, oxidative mechanism by Hugh Leifson medium, fermentative test in TSI (triple sugar iron) medium, nitrate agar motility, and urea agar reaction. Pure cultures of *Pseudomonas putrefasciens spp.* and *Pseudomonas pseudoalcaligenes spp.* were used for comparative characterization of Pseudomonas-type bacteria. Similar tests were conducted with bacteria after adaptation to 400 ppm of cyanide.

2.3 Adaptation of the isolated bacterial cultures to cyanide

The mixed culture from Tails 1 and medium M2 was selected as the bacteria to be further studied because it demonstrated the highest bacterial population and the fastest growth kinetics in the isolation step. The nutrient medium M4 was chosen because it does not contain $FeSO_4$, which may interfere with the results by consuming any free cyanide present to form iron stable complex compounds.

2.3.1 Bacterial tolerance to cyanide concentrations

Three hundred-ml flasks containing 100 ml of nutrient medium M4 were inoculated with different amounts of the mixed culture selected before. The initial populations of bacteria ranged between 5.5×10^3 up to 1.2×10^6 cells/ml. The pH was adjusted to 10.5 with 1N NaOH, and sodium cyanide was added at different concentrations, ranging from 50 to 250 ppm of free cyanide. The temperature was maintained at 25°C. The bacterial populations were determined daily by colonies counting on Mac Conkey agar.

The bacterial culture which survived and grew at the higher amount of cyanide was selected as the most tolerant bacteria

Table 1
Physical characteristic of materials used for bacterial isolation

Material	Density (g/cm^3)	Total cyanide content (mg/kg)	Humidity (%)	pH	d_{80} (μm)
Tails 1	2.2	< 10.0	< 0.1	3.4	162.3
Tails 2	2.7	18.1	8.8	8.5	162.8
Tails 3	2.5	< 10.0	13.3	4.9	196.8

2.3.2 Procedure for bacterial adaptation to cyanide

The isolated culture with the most tolerance to cyanide was used in the adaptation step. Six hundred ml of nutrient medium M4 were inoculated with 400 ml of the most tolerant isolated bacteria. The total volume was divided over four flasks each with 250 ml of medium M4. The pH was then adjusted to 12 with 1N NaOH. The cultivation flasks were placed on a rotating shaker at 150 rpm and 25°C for four to seven days, until the bacterial culture began its exponential growth phase. At this point, the first stage of adaptation was started by adding sodium cyanide was to adjust the free cyanide concentration to 200 ppm. Bacterial population, pH and free cyanide were monitored periodically. At the end of the exponential phase (around the 12th day), a suitable inoculum was taken from the medium, and a new culture was inoculated with fresh medium. The procedure was followed as before, adding a higher amount of sodium cyanide to the medium each time, such that for the second and third stages of adaptation concentrations of 400 and 750 ppm of free cyanide were used, respectively.

2.4 Biodegradation of cyanidation residues

The degradation of total cyanide by the adapted culture was studied in both effluents and solid wastes of cyanide leaching tests, which had been prepared in the laboratory. The bacteria adapted to 750 ppm of free cyanide were used in the nutrient medium M4.

2.4.1 Biodegradation of effluents

Six hundred ml of cyanidation effluent were mixed with 400 ml of the adapted bacterial culture in theM4 nutrient medium in 2000 ml flasks. The pH was adjusted with 1N NaOH or 1N H_2SO_4 as required. Two hundred-fifty ml of this solution were placed in 500 ml flasks and agitated at 150 rpm in an orbital, thermostated shaker.

Table 2
Nutrient media used for bacterial isolation

Medium No.	Description	Reference [6]
M1	For autotrophic bacteria related to genus of *Pseudomonas*	Medium C
M2	For general autotrophic bacteria	Medium A
M3	For soil autotrophic bacteria	Medium B
M4	Medium M2 without $FeSO_4$	

Blank solutions were incubated under the same conditions, using distilled water instead of the bacterial culture and 0.1% bactericide (Tego) in solution. These blank solutions represented the natural cyanide degradation. Daily controls monitored the total cyanide content by distillation and Liebig titration, the bacterial population and the pH. Temperature and initial pH were the variables studied in this part. pH was not adjusted during the test. Total cyanide concentrations of 1500 and 3400 ppm were used in these studies.

2.4.2 Biodegradation of solid wastes

Ninety ml of M4 nutrient medium were added to 250 ml-Erlenmeyer flasks and inoculated with 10 ml of the bacterial culture. They were placed on a rotating, thermostated shaker at 150 rpm and 25°C. The populations of bacteria were measured until the start of the exponential growth phase. Then 10 g of cyanidation tailings were added to the culture. Ten flasks were prepared following the previously described procedure. Daily, one flask was analyzed. The bacterial population and the pH were measured. The solid residue was analyzed for total cyanide by distillation and Liebig titration. Tests were done at several pH values. As the original pulp had a pH 10, the pH was controlled the addition of H_2SO_4. The cyanidation residue used in these tests had a total cyanide concentration of 700 mg/kg of total cyanide.

3. RESULTS AND DISCUSSION

3.1 Bacterial isolation

The isolation tests were carried out during a 19-day period. The results are presented in Table 3. Figure 1 shows that the Tails 1 sample, which was the most oxidized material, produced the largest bacteria population from the onset of the experiments and also showed the most rapid bacterial growth kinetics. Figure 2 shows the influence of the nutrient media on the isolation of *Pseudomonas*. The nutrient media M2, M1 and M4 was favored, in this order, for the best bacterial population growth.

3.2 Characterization of isolated bacteria

The isolated bacteria are Gram negative. They are rod-like in shape and do not fluoresce under an ultraviolet light. The microorganisms grew in Mac Conkey agar, forming bright red and brown colonies. The citrate agar growth test was positive. The colonies displayed a positive oxidase reaction.

Table 3
Results of bacterial isolation

Medium	Tails 1 Maximum growth			Tails 2 Maximum growth			Tails 3 Maximum growth		
	Population [cells/ml]	Day	pH	Population [cells/ml]	Day	pH	Population [cells/ml]	Day	pH
M1	2.3×10^{10}	7	7.8	1.1×10^{10}	9	8.2	2.3×10^{10}	9	8.1
M2	7.8×10^{10}	7	7.7	2.1×10^{10}	9	8.3	5.3×10^{10}	10	7.9
M3	6.6×10^{8}	8	7.5	6.5×10^{8}	13	8.0	6.6×10^{8}	10	7.8
M4	9.8×10^{9}	7	7.9	1.2×10^{9}	10	8.1	9.8×10^{9}	9	8.2

The positive reaction in Hugh Leifson medium indicated an oxidative metabolism, while the negative reaction in TSI media confirmed a non-fermentative metabolism. The agar nitrate test for motility was positive, while the urea agar test indicates ureasa-positive microorganisms. The results from all of these tests confirms the presence of a mixed *Pseudomonas*-type culture [7,8]. The results were similar for both the liquid media sample and the colonies grown using Mac Conkey agar cultivation plates. The pure *Pseudomonas* cultures responded similarly to the biochemical tests.

600

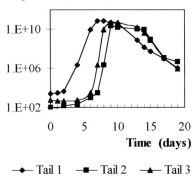

Population (cells/ml)

Time (days)

——◆—— Tail 1 ——■—— Tail 2 ——▲—— Tail 3

Figure 1. Isolation of bacterial from mineral
tailings from three different locations.
Medium M2, 30°C, 150 RPM

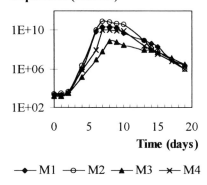

Population (cells/ml)

Time (days)

——◆—— M1 ——○—— M2 ——▲—— M3 ——×—— M4

Figure 2. Influence of nutrient medium
on bacterial isolation. Tail 1,
30°C, 150RPM.

3.3 Bacterial adaptation to cyanide

3.3.1 Bacterial resistance to cyanide concentrations

An culture with an initial population of 1.2×10^4 cells/ml resisted 100 ppm of free cyanide, while 1.0×10^6 cells/ml were required to survive a maximum free cyanide concentration of 200 ppm. Lower initial populations were unable to grow in the presence of free cyanide. The results indicate that the bacteria tolerance to free cyanide is influenced by the initial concentration of bacteria.

Table 4
Bacterial adaptation to cyanide

Initial conditions

NaCN [ppm]	Population [cells/ml]	pH	Degradation time [days]	Maximum population [cells/ml]
200	4.0×10^5	11.4	6	7.0×10^8
400	5.0×10^5	11.6	9	5.5×10^8
750	9.5×10^7	11.4	10	5.0×10^8

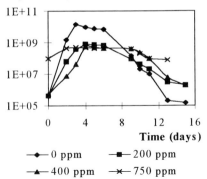

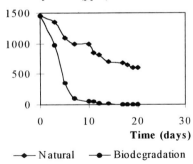

Figure 3. Bacterial growth kinetics in the
adaptation tests. Medium M4, 25°C, initial
pH of 11 and 150 RPM.

Figure 4. Biodegradation and natural
degradation in cyanidation effluents
at 25°C and an initial pH of 8.

3.3.2 Bacterial adaptation to cyanide

Table 4 reports the results of the adaptation of the bacteria to cyanide. Bacterial cultures
were able to continuing growing with up to 750 ppm of free cyanide in solution. These mixed
cultures were readily adapted to different free cyanide concentrations while they are in the
exponential phase. However, the maximum bacterial population achieved was less than that
obtained in the isolation step. Again, the influence of the initial bacterial concentration was
noticed during the tests. A population of 4.0×10^5 cells/ml and 5.0×10^5 cells/ml was needed to
resist 200 and 400 ppm of free cyanide, respectively. An initial population of 9.5×10^7 cells/ml
was required for the 750 ppm cyanide concentrations. Bacterial cultures were unable to grow
at higher cyanide concentrations.

Table 5
Influence of pH on the biodegradation of cyanidation effluents at 25 °C

Initial conditions				Bacterial degradation	Natural degradation
Total cyanide [ppm]	Population [cells/ml]	pH	Time [days]	Total cyanide [ppm]	Total cyanide [ppm]
1500	1.0×10^5	11.5	30	100	1250
1500	1.0×10^5	9.2	15	20	950
1500	1.5×10^5	8.0	14	< 10	700
3400	1.3×10^5	8.0	12	< 10	1200
3400	1.3×10^5	9.2	12	450	2100

Table 6
Influence of temperature on biodegradation of cyanidation effluents at pH 8

Initial conditions				Bacterial degradation	Natural degradation
Total cyanide [ppm]	Population [cells/ml]	Temperature [°C]	Time [days]	Total cyanide [ppm]	Total cyanide [ppm]
1500	2.5×10^5	20	30	400	1200
1450	1.5×10^5	25	14	< 10	700
1450	1.5×10^5	30	20	400	500

Figure 3 shows the bacterial growth kinetics in the adaptation tests. It appears that the length of the stationary growth phase increased with increasing free cyanide concentration, indicating that the bacterial cultures became more resistant to the toxic, free cyanide.

It is interesting to note that while the cultures exposed to cyanide never reached the maximum level of bacteria populations in the cultures without cyanide, the growth rate in the presence of cyanide was still similar to that without cyanide.

3.4 Biodegradation of cyanidation residues

3.4.1 Biodegradtion of effluents
The results in Table 5 show that a rapid biodegradation of total cyanide concentration from 3400 ppm to 10 ppm was achieved at an initial pH 8 in 12 days.

Over the same period, natural degradation reduced the total cyanide concentration from 3400 ppm to 1200 ppm. The kinetics of biodegradation and natural degradation of total cyanide is given in Figure 4.

The pH is an important factor in the detoxification process. At a high pH (11.0) a longer time was needed, as the process proceeded at a lower rate as shown in Figure 5. Both Figures 5 and 6 show that for an initial pH of 8 and 9, initially there is a very rapid decrease in the cyanide concentration until the 5th and 7th day for an initial concentration of 3400 and 1500 ppm, respectively. After that, a substantial reduction in the degradation rate occurred.

The temperature was a crucial parameter as shown in Table 6 and Figure 7. Increasing the temperature from 20°C to 25°C resulted in increased biodegradation of the cyanide. A further increase of temperature to 30°C reduced biodegradation.

This reduction in cyanide biodegradation is thought to be due to a high bacteria mortality, which is presumably caused by an increased toxicity in the medium with increased temperature [4]. The optimal biodegradation for the conditions studied was achieved at 25°C.

3.4.2 Biodegradation of solid wastes
The results indicate that the optimal conditions for biodegradation of cyanide are similar for cyanidation effluents and solid residues. Table 7 shows that the best biodegradation was obtained at pH 8 in less time (8 days). Figure 8 shows that at an initial pH of 10, the rate of

biodegradation was much lower than for initial pH's of 8 and 9, but after 12 days a similar final cyanide concentration in the solid residues was attained.

Table 7
Effect of pH on biodegradation of cyanide in solid residues at 25°C.

Initial conditions				Bacterial degradation	Natural degradation
Total cyanide [ppm]	Population [cells/ml]	pH	Time [days]	Total cyanide [ppm]	Total cyanide [ppm]
700	4.0×10^5	8.0	8	< 10	200
700	4.0×10^5	9.0	11	< 10	200
700	4.5×10^5	10.0	12	< 10	560

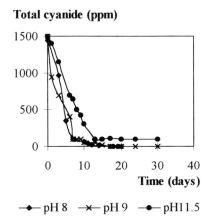

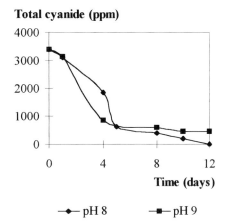

Figure 5. Effect of initial pH on biodegradation Cyanadation effluents at an initial cyanide Concentration of 1500 ppm at 25°C, 150 RPM

Figure 6. Effect of initial pH on of biodegradation of cyanadation effluents at an initial cyanide concentration of 3400 ppm at 25°C, 150 RPM

604

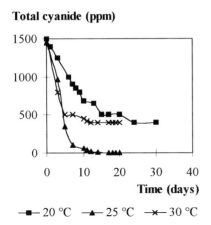

Total cyanide (ppm)

—■— 20 °C   —▲— 25 °C   —✕— 30 °C

Total cyanide (mg/kg)

—▲— pH 8   —■— pH 9   —◆— pH10

Figure 7. Effect of temperature on
Biodegradation of cyanadation effluents
At an initial pH 8, medium M4, 150 RPM

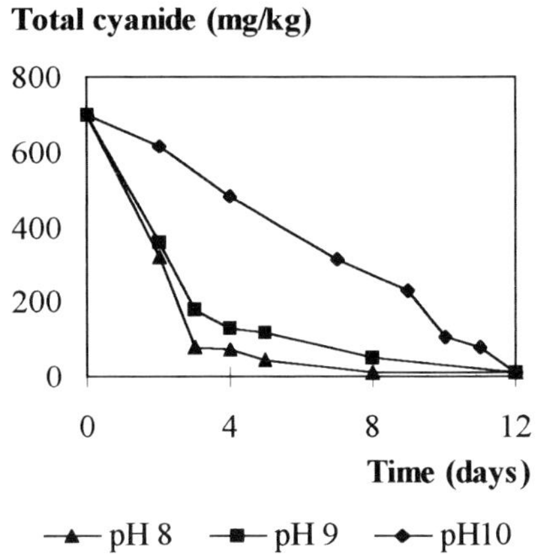

Figure 8. Effect of pH on biodegradation
of in cyanadation solid residues. At 25°C
medium M4, 150 RPM

4. CONCLUSIONS

Native bacteria cultures were isolated from industrial cyanidation tailings in a standard nutrient media for autotrophic organisms. The isolated bacteria was readily adapted to 400 ppm free cyanide concentrations and, in 75% of the tests, survived up to 750 ppm. In the adaptation tests, a maximum population of $5x10^8$ cells/ml was obtained in 4 days. The best conditions for the bio-detoxification of effluents and solid wastes were achieved at 25°C and an initial pH 8. These conditions are within the optimum growth range of the *Pseudomonas* bacteria [2]. However, the cultures were not able to survive at 30°C, presumably due to a higher toxicity of the cyanide in the solutions, which resulted in a reduction of the biodegradation process. Under the best conditions, initial biodegradation rates of 230 and 420 mg total cyanide / day were achieved in the effluents during seven to five days with initial total cyanide concentrations of 1500 and 3400 ppm and residual total cyanide concentrations of 100 ppm and 400 ppm, respectively. Longer times were required to achieve total detoxification due to a reduction in the biodegradation rate. This change in degradation rate could be due to the increased difficulty of biodegradation of the strong cyanide complex compounds, while the initial high kinetics could be due to the biodegradation of WAD cyanide complex compounds.

In the solid residue, a maximum degradation rate of 204 mg total cyanide/day was obtained. The average rate of natural degradation was 45 mg total cyanide/day. The same kinetic behavior was observed as in the effluents.

ACKNOWLEDGMENTS

This research is part of the project BID - FUNDACYT No. 099. The authors are grateful to Dr. Irma Paredes for her technical assistance in microbiology, and to Dr. Saskia Duyvesteyn for her kindly review of the paper.

REFERENCES

1. Malhotra D. and Tuka R., Proceedings of the Symposium on Emerging Process Technologies for a Cleaner Environment., S. Chander (Editor), Society for Mining, Metallurgy and Exploration, Inc., Littleton, CO, (1992) 169.
2. Marsden J. and House I., The Chemistry of Gold Extraction. Ellis Horwood, New York, 1992.
3. Thompson L.C., Proceedings of the Symposium on Emerging Process Technologies for a Cleaner Environment, S. Chander (Editor), Society for Mining, Metallurgy and Exploration, Inc., Littleton, CO, (1992)197.
4. Smith A. and Mudder T., The Chemistry and Treatment of Cyanide Wastes. Mining Journal Books Limited, London., 1991.
5. De la Torre, E., Díaz, X., Castro, L, Memorias Seminario Internacional La Metalurgia del Oro en el Ecuador, (1996) 127.
6. Noel D., Fuerstenau M. and Hendrix J., Proceedings of the Conference Mineral Bioprocessing, R. Smith and M. Misra (Editors), The Mineral, Metals and Materials Society, Warrendale, PA, (1991) 355.
7. Cowan, S.T. and Steel, K.J., Manual for the Identification of Medical Bacteria. Cambridge University Press, Cambridge, 1974.
8 Escobar, M. B., Manual de Técnicas y Procedimientos. Trabajo Práctico No. 35: Estudio Bioquímico de Pseudomona y Otros Bacilos Gram Negativos no Fermentadores. Universidad de Antioquia, Medellín, (1993) 251.

Heavy metals removal by sand filters inoculated with metal sorbing and precipitating bacteria

L. Diels[a], P. H. Spaans[b], S. Van Roy[a], L. Hooyberghs[a], H. Wouters[b], E. Walter[b], J. Winters[c], L. Macaskie[d], J. Finlay[d], B. Pernfuss[e], T. Pümpel[e]

[a] Vlaamse Instelling voor Technologisch Onderzoek (VITO), Environmental Technology, Boeretang 200, B-2400 Mol, Belgium

[b] Astraco Water Engineering bv, T. de Boerstraat 24, NL-8561 EL Balk, The Netherlands

[c] Union Minière nv, UM Research, Kasteelstraat 7, B-2250 Olen, Belgium

[d] School of Biological Sciences, University of Birmingham, Birmingham B15 2TT, UK

[e] Institut für Mikrobiologie, Technikerstrasse 25, A-6020 Innsbruck, Austria
e-mail:dielsl@vito.be

Large volumes of wastewater containing metals like Cd, Zn, Cu, Pb, Hg, Ni or Co are mainly treated by precipitation processes. However the metals concentration does not always reach the regulatory standards and in many cases ecotaxes must be paid on the heavy metals load in the discharged water. Therefore a second polishing treatment is often necessary. In order to be economically acceptable the technology must be cheap and adapted to the treatment of large volumes. The use of sandfilters inoculated with heavy metal biosorbing and bioprecipitating bacteria fulfils these objectives.

The system is based on a moving bed sandfilter. A biofilm is formed on the sand grains after inoculation with heavy metal resistant bacteria able to biosorb or to bioprecipitate heavy metals. Passage of the wastewater over these biofilms leads to the binding of the metals to the biofilm and consequently the removal of the metals from the wastewater. The metal laden biofilm is removed from the sand grains in a sand washer created by an airlift for the continuous movement of the filterbed. The metal loaded biomass is separated from the sand in a labyrinth on the top of the sand washer. Nutrients and a carbon source are provided continuously in the system in order to promote the regrowth of the biofilm on the sand grains.

The reactor can be used for the removal of heavy metals and some COD. The obtained biosludge contains heavy metals at concentrations of more than 10% of the dry weight.

1. INTRODUCTION

Heavy metal cations are well recognized as environmental problems. Many wastewaters are contaminated by metals like Cd, Zn, Cu, Pb, Hg, Ni or Co. Metal recycling or processing companies (non-ferrous) as well as surface treatment companies (galvanization) produce metal containing wastewaters. These industries have, in most cases, a large range of wastewater treatment facilities mainly based on precipitation technology. However due to the strengthening of environmental standards and the ecotaxes that must be paid for the discharge of metals into the sewer or surface waters, problems arise with the removal of trace concentrations (a few ppm) sometimes complexed, or in the presence of organic contaminants. Surface treatment companies have some problems with the removal of metals due to the complexing properties of some organics. The closing and flooding of mines in several European countries generates large volumes of metal contaminated (U, Sn, Cu, As, Fe) water. The old metal processing sites are distributed throughout Europe and lead to major problems of contaminated groundwater. One solution is to pump and treat the low level contaminated waters.

The need for safety and protection of groundwater and/or drinking water resources results in an increase of the environmental protection standards with the requirements for lower concentrations of heavy metals in wastewaters. The European directive 96/61/EEC requires member states to achieve an Integrated Prevention Pollution and Control (IPPC) of pollution arising from different industrial activities. Therefore it is important to look for suitable and cheap cleaning and treatment methods for wastewaters, groundwaters and landfill leachates.

Some specific biomasses can be used for the removal of heavy metals from wastewater by biosorption via the binding of metals to functional groups on the cell surface polymers [1-3] and bioprecipitation by crystallization of metals at the cellular surface [4-5] processes. Several industrial processes [6-8] based on this behaviour have been developed. Testing of such immobilized biomass revealed the following problems. The metal removal efficiency is dependent of the concentration of biomass. After saturation of the biomass, a regeneration with weak acids is necessary. It could be indicated that the necessary regeneration time increased with the number of biosorption cycles. In addition full capacity was not reached on rechallenge following regeneration. After some cycles the metals binding capacity decreased due to biomass destruction or incomplete metal removal.

In order to overcome these problems a special method, based on the use of viable biomass which can be grown during the metal sorbing or bioprecipitation process, was developed. The final objective was also to generate a sludge which can be processed in an extant pyrometallurgical process. The MERESAFIN (MEtal REmoval by SandFilter INoculation) system described here is based on the inoculation of a sand filter with metal biosorbing or bioprecipitating bacteria.

2. MATERIALS AND METHODS

2.1. MERESAFIN concept

Bacteria, able to biosorb or bioprecipitate heavy metals, grow in a biofilm on a supporting material. During contact with heavy metal-containing wastewater the biofilm adsorbs the metals. Subsequently the metal-loaded biomass is removed from the supporting material and the resting biomass residual on the substratum can be reused, after regrowth, for a subsequent treatment cycle.

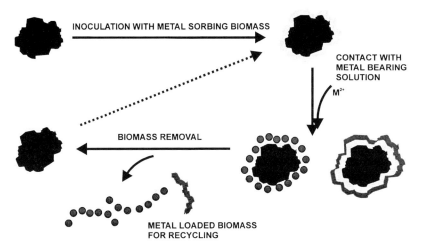

Figure 1. Concept of sand particles inoculated with metal biosorbing or bioprecipitating biofilms.

The supporting material can be sand or other materials retained within a moving bed sand filter which is based on a counterflow principle (figure 2). The water to be treated is admitted through the inlet distributor (1) in the lower section of the unit and is cleaned as it flows upward through the sand bed, prior to discharge through the filtrate outlet (2) at the top. The sand containing the heavy metals bound to the biofilm is conveyed from the tapered bottom section of the unit (3) by mean of an airlift pump (4) to the sand washer (5) at the top. Cleaning of the sand starts in the pump itself, in which metal-loaded biofilms are separated from the sand grains by the turbulent mixing action. The contaminated sand spills from the pump outlet into the washer labyrinth (6), in which it is washed by a small flow of clean water. The metal loaded flocs are discharged through the washwater outlet (7), while the grains of sand with a partly removed biofilm are returned to the sand bed (8). As a result, the bed is in constant downward motion through the unit. In this concept water purification and sand washing both take place continuously, enabling the filter to remain in service without interruption.

The complete water treatment system is shown in figure 3. Wastewater is pumped through the Astrasand filter and purified. The washwater, containing the metal-loaded biomass, is drained to a lamella separator or setling tank. The water, coming from the thickener, is reintroduced in the sand filter. The sludge coming from the thickener is treated further in a filter press or a decanting centrifuge after addition of lime or flocculants respectively. The filtercake obtained in this way, containing the metals, is recycled in a pyrometallurgical treatment facility (shaft furnace) of a non-ferrous company.

610

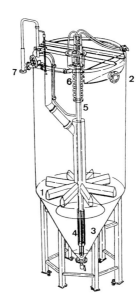

1. Inlet distributor
2. Outlet
3. Dirty sand
4. Air-lift pump
5. Sand washer
6. Washer labyrinth
7. Wash water outlet
8. Cleaned sand

Figure 2. Moving bed sand filter concept

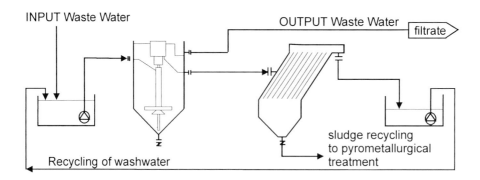

ASTRASAND filter   Lamella Separator

Figure 3. Flow sheet of the wastewater treatment

2.2. Sand filter

An Astrasand (Astraco Water Engineering – Paques, Balk, The Netherlands) moving bed sand filter AS-30-40 with a diameter of 3.00 m, a filter bed height of 4.00 m and a filter bed volume of 12 m³ was filled with sand with a grain size diameter between 0.8 and 1.2 mm. The normal filtration velocity was between 6 – 7 m/h and a feed flow between 20 and 45 m³ /h was applied. The washwater flow was between 3 and 5 m³/h and the sand circulation speed between 0.5 and 1.0 cm/min. The empty bed contact time was about 36 minutes.

2.3. Bacterial inoculation of the sand filter

The heavy metal resistant and bioprecipitating *Ralstonia eutropha* CH34 (former name *Alcaligenes eutrophus*) [9] was used together with the metal biosorbing bacteria *Pseudomonas mendocina* AS302 [10] and *Arthrobacter sp.* BP7/26 [11].

An inoculum of the three strains was prepared and added in the filtrate room. The feed pump was bypassed and only the airlift was operated, ensuring that the sand bed was moving in a certain volume of water and filtrate and wash water were continuously mixed. The final concentration of the bacteria in the sand filter water was approximately 10^8 cfu/ml. After 4 days wastewater was added into the system and the normal operation was started.

2.4. Wastewater

Wastewater from a non-ferrous company was used after a neutralisation and precipitation step. This water contained 1.5 mg Co/l, 0.7 mg Ni/l, 0.2 mg Zn/l and 0.1 mg Cu/l all analysed by ICP. The COD was nearly 20 mg O_2/l and no nitrates, ammonia or phosphates analysed by standard methods were present. The input oxygen concentration was, depending on the weather conditions, between 0 and 6 mg O_2/l.

2.5. Sampling

Daily samples of the input, filtrate and wash water were taken with an Endress Hauser automatic sampler. The heavy metals (Co, Ni, Zn, Cu) concentrations were measured as well as COD, NO_3-N, NH_4-N and PO_4-P. Also water samples from different bed heights (0.7, 1.7 and 3.7 m) were taken. Oxygen concentration, pH and E_h were also measured.

3. RESULTS AND DISCUSSION

3.1. Bacterial inoculation

After the inoculation the metals-bearing wastewater was introduced into the sand filter and nutrients (carbon, nitrogen and posphorous source) were added. After a while a steady-state was reached. The bacterial concentrations in the water were counted at different levels in the filter bed. These results are presented in table 1. It seems that more than 10 % of the bacteria can grow under anoxic (denitrifying) conditions. Most of the bacteria were counted as *Pseudomonas mendocina* AS302 (selective growth on GSP-medium with 20 mM arsenate). The CH34 bacteria (Ni-resistant) were counted on minimal medium plates containing 2 mM Ni. These bacteria seemed to be present for at least more than 1 % of the total concentration. The low Ni concentration in the wastewater would not give a sufficient selective pressure (i.e. advantage) to the Ni-resistant CH34.

Table 1. Bacterial countings in the water of the sand filter

Bed height	Total cfu	Anoxic	AS302	Zn-resistant	CH34
0.7 m	2.3E+07	2.4E+06	1.8E+07	7.5E+04	2.7E+05
1.7 m	2.9E+07	1.1E+06	2.4E+07	1.5E+04	2.6E+05
2.7 m	3.3E+07	2.9E+07	2.1E+07	1.5E+04	5.0E+05

Concentrations are expressed in cfu/ml water (cfu = colony forming units)

The bacterial concentrations and proteins on the sand particles showed approximately 10^8 cfu/g wet weight of sand and 1000 µg protein/g wet weight of sand. The bacteria in the biofilm showed a similar distribution as in the water : around 10^7 cfu AS302 /g wet sand, 10^4 cfu Zn-resistant/g wet sand and 10^5 cfu CH34/g wet sand.

During the inoculation the ambient air temperature fell to below 0°C. The water temperature was 7°C. Afterwards during the operation the water temperature was always between 17 and 22°C.

Scanning electron microscopy showed a thin dense biofilm of around 10 µm as a basal layer and concentrated into crevices in the sand grains. On top of this biofilm a more open biofilm structure of again 10 µm was observed. This seems to be the biofilm layer which is removed every cycle in the sand washer. Samples from the sand washer showed the biofilm released from the basal layer. Also this layer seems to be the active biofilm.

The nutrients consumption is presented in table 2. The COD, NO_3-N, NH_4-N and PO_4-P are presented for the influent (Input), after addition of nutrients (Input + nutrients), at different bed heigths of the filter (filter + 0.7 m, + 1.7 m and + 3.7 m) and in the filtrate. Table 2a and 2b present the values for two different nutrient additions. Oxygen was consumed the first 10 cm of the filterbed together with the NH_4-N. In the following anoxic zone NO_3-N was consumed with the carbon source (COD) as electron donor. Only a minor part of the phosphate was consumed.

Table 2a. Nutrient consumption in the filter bed

Sample	COD mg O_2/l	NO_3-N mg/l	NH_4-N mg/l
Input	15.8	0.24	0.69
Input + nutrients	36.4	0.52	6.20
Filter + 0.7 m	21.0	0.58	4.0
Filter + 1.7 m	13.9	0.31	1.7
Filter + 2.7 m	14.4	0.32	0.9
Filtrate	11.9	0.31	< 0.23

Table 2b. Nutrient consumption in the filter bed

Sample	COD mg O_2/l	NO_3-N mg/l	NH_4-N mg/l	PO_4-P mg/l
Input	18.8	0.92	0.47	< 0.05
Input + nutrients	65.7	6.24	1.17	0.87
Filter + 0.7 m	51.2	< 0.23	0.35	0.74
Filter + 1.7 m	31.6	< 0.23	0.34	0.69
Filter + 3.7 m	19.1	0.71	0.03	0.57
Filtrate	23.4	0.35	0.23	0.74

3.2. Heavy metal removal with the MERESAFIN concept

Figure 4 and 5 show the metals concentrations in the input and the filtrate for the period between 17-3-98 and 10-4-98 for Cu and Zn and for the period between 6-8-98 and 20-8-98 for Ni and Co. This shows that under all the conditions tested the Cu and Zn were removed completely. Under optimal conditions Ni and Co could be removed to an extent of more than 95 %. No additional COD, NO_3-N or NH_4-N were measured in the filtrate.

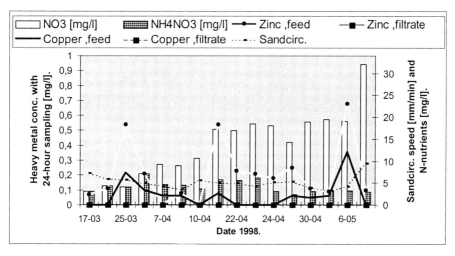

Figure 4. Zinc and copper concentrations in input and filtrate

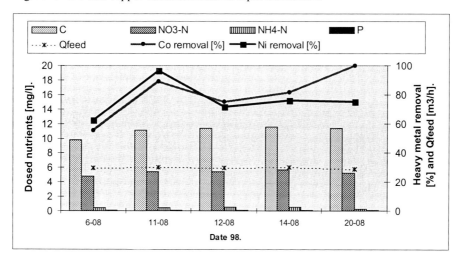

Figure 5. Cobalt and Nickel removal percentages.

From separate analysis (not shown) it was observed that most of the metals were removed in the anoxic zone of the filter bed.

3.3. Sludge dewatering and pyrometallurgical treatment

The sludge obtained from the setling process of the washwater was dried and analysed. This sludge typically showed 0.53 % Al, 0.23 % As, 2.51 % Co, 0.40 % Cr, 0.80 % Cu,

614

20.00 % Fe, 1.01 % Ni, 0.20 % Pb and 2.66 % Zn. This product could be treated in a graphite furnace at 1100 °C.

To the sludge coming from the thickening process in the settler 50 % (weight to weight) of lime was added and further dewatered in a filter press. The obtained filter cake (50% dry weight) was sent to a metal processing company (Montanwerke Brixlegg) for treatment in a graphite furnace.

Other tests

In other laboratory scale experiments the removal of Total Organic Carbon (TOC) and a metal like Tl was observed. Table 3 presents the results of treatment of a real waste water. The TOC was reduced from 65.6 mg/l to 13.4 mg/l. The Tl concentration was reduced from 694 µg/l to 240 µg/l.

Table 4 presents the removal of silver cyanide from a surface treatment plant. The cyanide is mineralized and the Ag is precipitated. Two *Pseudomonas* strains were used for the cyanide degradation. But only in the presence of *Alcaligenes eutrophus* CH34 was Ag could be removed by a bioprecipitation process.

Table 3. TOC biodegradation and Tl removal in a lab scale sand filter.

Time (hours)	TOC (mg/l)	Tl (µg/l)
0	65.6	694
17	-	365
41	-	292
65	26.9	271
137	17.3	212
161	-	218
185	13.4	240

Table 4. Silver cyanide removal in a lab scale sand filter.

Time (days)	Ag (mg/l)	CN⁻ (mg/l)
0	4.5	4.2
1	2.0	2.2
2	1.4	-
3	0.3	0.2

4. CONCLUSIONS

It was shown that the use of a moving bed sand filter inoculated with heavy metal biosorbing and bioprecipitating bacteria, forms a new reliable technology.

The removal of Zn and Cu was between 95 and 100 %. Co removal was between 80 and 90 %. Fe was removed for 60 – 80 % and other metals e.g. Al, Ag, Cr, As and Se were removed at least for more than 80 %.

In separate experiments removal of TOC (65 mg O_2/l to 13 mg O_2/l) was demonstrated in addition to the removal of metals like Tl (625 µg/l to 235 µg/l).

In addition some COD, NO_3-N and NH_4-N were removed. The technology can therefore be used as a cheap polishing step. It allows metal using companies to reach the new environmental standards and hence to reduce ecotaxes on the discharged metals and nutrients.

The moving bed sand filter technology allows a homogeneous treatment of waste water containing heavy metals. Due to the continuous washing system the filter bed resistance and flow rate are always stable and self regulating, providing a stable output and steady-state operation.

The sludge released in this washing process could be thickened into a filter cake of 50% dry weight. This cake could be recycled in a shaft furnace and in that way reintroduced in the process. So no extra waste products are released from the process. The use of active bacteria in a biofilm allows the system to remove the heavy metals not only by biosorption but also by bioprecipitation processes. The bioprecipitation is induced by the presence of functional groups and crystallization foci together with the physico-chemical microenvironment created by the biofilm. The biofilm generates steep pH gradients at the cell surface which allow the crystallization to occur.

The biofilm is further studied now by the use of molecular biology techniques. These include PCR and DGGE in order to identify the microbial population and the presence of the inoculated strains. After one year of operation about 9 different bacterial species could be observed in the biofilm. The use of different primers in the PCR will allow to distinguish the bacteria from each other. This information will be combined with the results obtained from Biolog and metal resistance patterns. Complementary results are presented by Pümpel et al. [12].

The technology is now protected (patent pending) and will allow an economically acceptable treatment of large waste water flows containing heavy metals. These waste waters include waste water from metal processing and surface treatment companies, mine water, groundwater and landfill leachate.

Experiments are indicating that the presence of the specific bacteria is of high importance for a successful metal removal.

ACKNOWLEDGEMENTS

The European Community is acknowledged for their support via the Brite-Euram project BRPR-CT96-0172.

REFERENCES

1   F. Glombitza and U. Iske, Biohydrometallurgy, J. Salleys, R.G.L., McReady, P.L., Wichlacz (Eds.), Jackson Hole. (1990) 329.
2   M. Tsezos, E. Remoudaki and V. Angelatou, Int. Biodet. Biodegrad. (1995) 129.
3   T.J. Beveridge and R.G.E. Murray, J. Bacteriol., 141 (1980) 876.
4   L.E. Macaskie, K.M. Bonthrone and D.A. Rauch, FEMS Microbiol. Lett. 121 (1994) 141.
5   L. Diels, Biohydrometallurgy, J. Salleys, R.G.L., McReady, P.L., Wichlacz (Eds.). Jackson Hole. (1990) 369.

6   T.H. Jeffers, C.R. Ferguson and D.C. Seidel, Biohydrometallurgy, J. Salleys, R.G.L., McReady, P.L., Wichlacz (Eds.). Jackson Hole. (1990) 317.

7   D.W. Darnall, R.M. McPherson, J. Gardea-Torresday, Biohydrometallurgy, J. Salley, R.G.L., McReady, P.L., Wichlacz (Eds.). Jackson Hole. (1990) 341.

8   C.J. Brierley and J.A. Brierley, Biohydrometallurgical Technologies, ed. by A.E. Torma, M.L. Apel, C.L. Brierley, the Mineral, Metals & Material Society, (1993) 35.

9   L. Diels, Q. Dong, D. van der Lelie, W. Baeyens, M. Mergeay, J. Ind. Microbiol. 14 (1995) 142.

10 L. Diels, M. Tsezos, T. Pümpel, B. Pernfuss, F. Schinner, A. Hummel, L. Eckard and F. Glombitza, Biohydrometallurgical Processing, Jerez, C.A., Vargas, T., Toledo, H. and Wiertz, J.V. (Eds.),. University of Chile. (1995) 195.

11 B. Pernfuss, T. Pümpel and F. Schinner, Biosorption & Bioremediation Symposium, Macek, T., Demnerova, K., Mackova, M. (Eds.), Merin. Czech Society for Biochemistry and Molecular Biology, Prague (1995).

12 T. Pümpel, C. Ebner, B. Pernfuss, F. Schinner, L. Diels, Z. Keszthelyi, L. Macaskie, M. Tsezos, H. Wouters, Removal of nickel from plating rinsing water by a moving-bed sandfilter inoculated with metal sorbing and precipitating bacteria. Submitted to IBS99 (1999).

Removal of Iron from Silica Sand: Integrated Effluent Treatment by Sulphate Reduction, Photochemical Reduction and Reverse Osmosis

A.W.L. Dudeney, A. Narayanan and I.I. Tarasova

T.H. Huxley School of Environment, Earth Science and Engineering,
Imperial College of Science Technology and Medicine, South Kensington,
London SW7 2BP, United Kingdom

Following several years of development as part of a multinational EEC project, a full material balance has been developed for the continuous treatment of effluent from heap or column leaching of silica sand with oxalic acid to remove iron-bearing impurities. The paper gives the results of leaching, photochemical precipitation of iron from the leachates as iron(II) oxalate, microbiological precipitation of residual iron in an upflow anaerobic sludge blanket (UASB) reactor and water recycle via reverse osmosis (RO). RO facilitates the simultaneous production of purified water for sand washing and concentrated bicarbonate for pre-neutralisation of the acidic leachate feed to the UASB system. Data have been produced for scale-up and industrial application.

1. INTRODUCTION

Efficient water management and waste product minimisation are general objectives in hydrometallurgical and biohydrometallurgical processing. In this context the potential advantages of organic acids (particularly oxalic acid) over mineral acids in leaching industrial minerals to remove iron have received considerable attention (e.g., 1-4). The advantages centre on the formation during leaching of soluble complexes (e.g., the *tris*oxalatoiron(III) anion) which are both photochemically and microbiologically degradable. In the case of oxalic acid the main reactions occurring may be represented as in Eqns 1-3.

Leaching $\quad\quad\quad\quad$ $6H_2C_2O_4 + Fe_2O_3.H_2O = 2H_3[Fe(C_2O_4)_3] + 4H_2O$ $\quad\quad\quad\quad\quad$ [1]

Photodegradation $\quad$ $2H_3[Fe(C_2O_4)_3] + 2H_2O + h\upsilon = 2FeC_2O_4.2H_2O + 3H_2C_2O_4 + 2CO_2$ [2]

Biodegradation $\quad$ $Fe(C_2O_4)_3^{3-} + 1.5C_2O_4^{2-} + SO_4^{2-} + 4H_2O = FeS + CO_2 + 8HCO_3^{-}$ [3]

The previous work cited indicated that the photochemical reaction involves an electron transfer from oxalate to iron resulting in the precipitation of iron(II) oxalate (in the form of a well-crystallised dihydrate), the evolution of carbon dioxide and the release of half of the original oxalic acid for recycle to leaching. Photochemical degradation (Reaction 2) does not go to stoichiometric completion and the residual iron and oxalate concentrations are suited to biodegradation by anaerobic sulphate reduction (Reaction 3), with the formation of ferrous sulphide, carbon dioxide and bicarbonate. In principle, the bicarbonate formed can be recycled for partial neutralisation of oxalic acid prior to biodegradation. Such neutralisation is necessary because the bacterial systems employed cannot tolerate the low pH generated in leaching and photodegradation, and the use of recycled reagent rather than added reagent is clearly advantageous. However, an efficient means of bicarbonate concentration is required both to recover process water and avoid unduly diluting the biodegradation feed.

Further potential advantages of the oxalic acid system arise from the nature of the sludges produced (ferrous oxalate and ferrous sulphide). Studies aimed at converting these to added-value products will be reported elsewhere.

The present paper deals with the treatment of the iron-bearing leachate resulting from oxalic acid leaching of silica sand (actually leaching of the surfaces of the sand grains, to which a large proportion of the iron impurities adhere). Representative configurations and material balances are described for the overall process of heap leaching and effluent treatment, and (in more detail) for the integration of reverse osmosis with biodegradation to concentrate bicarbonate (and thus facilitate liquor pre-neutralisation) while at the same time purifying process water for recycle to sand washing. The objective is to demonstrate the feasibility of this novel system to produce sand suitable for clear white glass bottle production. In order to avoid a green coloration the iron(III) oxide content of the sand must not exceed 0.033%.

2. EXPERIMENTAL

2.1. Leaching and photodecomposition

Sand containing >95% (dry weight) quartz grains (nominal particle size range +0.16-0.40 mm), 0.044% iron(III) oxide and 4% w/w moisture, received from Fife Silica Sands Ltd., Scotland, was leached in laboratory columns, thus simulating heap leach operations. Typically, 1 kg samples were leached with set concentrations of 1-4% w/v oxalic acid under acid cure conditions, i.e., sufficient acid was added in one volume (normally 100 ml) to wet the sand surfaces and allowed to react without disturbance for set periods of time (24-72 hr) before being washed from the sand particles with the minimum required quantities of water (normally two volumes of 50 ml) under suction. Details of the methods employed have been provided previously (2).

Leach liquors prepared as above were used for photodecompostion (and biodegradation) studies. As relatively large volumes were required, synthetic solutions - prepared by dissolution of freshly precipitated and washed hydrated iron(III) oxide in oxalic acid (4) - were also used. No significant differences were observed between the properties of the two types of liquors. Photodecomposition was carried out under batch conditions in daylight (illumination >2500 Lux), as described previously (3), and continuous conditions in 20 l trays simulating ponds of 20 cm depth and residence times of 2-3 days (2). Starting solutions contained typically 80-240 mg/l Fe and an excess of oxalate. The effluents from photodecomposition contained 80-100 mg/l Fe and up to 2 g/l oxalate (nominal pH of 2.0).

2.2. Biodegradation and reverse osmosis

Biodegradation was carried out under upflow anaerobic sludge blanket (UASB) conditions. The UASB system described previously (2, 4) - essentially vertical glass columns each containing approximately 0.7 l anaerobic sludge and through which solutions typically containing 80 mg/l Fe and 1.5 g/l oxalate were fed at approximately 1.44 l/d per column - was integrated for continuous steady state operation with a Purite reverse osmosis (RO) / ion exchange water purification unit. The unit was modified (in collaboration with the Purite company) by disconnecting the ion exchange system and connecting a recirculation pump to the RO system.

Preliminary assessment of the operating characteristics of the RO system were carried out using a synthetic UASB effluent (25 1 0.04 M sodium bicarbonate containing 55 mg/l sulphate as sodium sulphate). The solution was fed to the membrane at approximately 1 l/min (at 3 bar). One liter of permeate and the corresponding volume (about 180 l) of reject were collected over about 5-6 minutes and sampled for analysis. The reject was recycled to the membrane and a new permeate (1 l) and reject collected. This procedure was repeated a further 14 times with each RO feed comprising the previous reject and 1 l permeate being collected and analysed.

Under integrated continuous conditions the UASB effluent was similarly passed over the RO membrane (but via a stocktank containing previously prepared 0.1 M sodium bicarbonate) with the permeate recycling as purified water for sand washing and the reject recycling (once again via the stocktank) as a bicarbonate concentrate for pre-neutralisation of the UASB feed. In order to produce a homogeneous feed to the UASB system, a second stocktank was employed to mix the recycle stream with the incoming photodecomposition effluent.

3. RESULTS AND DISCUSSION

Figure 1 shows an example of a material balance based on experimental data obtained from column leaching, photodecomposition and biodegradation. A small size type-face has been adopted to show the overall scheme on a single page. The data were used to simulate the treatment (digestion) of 400 tonne conical heaps of sand with 1% w/v oxalic acid for 48 hours at a liquid:solid ratio 1:10. These data indicated that approximately 30% of the iron would be solubilised under the conditions employed (thus reducing the iron oxide content from 0.044% to 0.031%, which is sufficient for clear white bottle production).

It was assumed that the leach liquor could be removed from the sand particles to about 4% moisture with the aid of suction manifolds underneath each heap - as is common practice in the industrial minerals industry - and piped to a shallow (<0.5 m deep) open pond having large enough surface area (about 8 m x 20 m) to remove up to 80% of the iron as ferrous oxalate by photochemical reduction. According to Figure 1 some 68 kg of the material would be produced per 40 tonne heap of sand. The overflow from this pond would receive pH adjustment and oxalic acid make-up before utilisation with the next heap of sand.

The film of leachate remaining on the sand particles after suction would be removed by washing (and suction) with recycled water equivalent in volume to that of the original leachant. Assuming the continued applicability of simple linear scale-up, the laboratory results predicted that the iron(III) oxide, oxalate and moisture contents of the washed sand would be 0.032, 0.008 and 4%, respectively, while the washings would contain approximately 275 mg/l Fe and 2 g/l oxalate (pH approximately 2.0). The washings would be provided with bacterial mineral nutrients as necessary (mainly ammonium, sulphate and phosphate ions) and appropriate neutralisation and/or dilution before biodegradation in the UASB system. Experiment showed that the minimum permissible feed pH was 3.0 (below which the columns became 'soured'). At pH 3.0 a feed rate of 1.44 l/day per column (0.7 l sludge volume) gave essentially complete precipitation of iron (equivalent to 12 kg ferrous

620

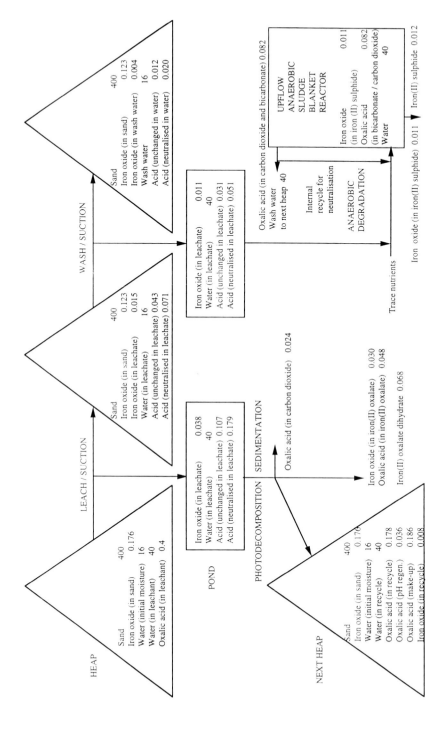

Figure 1. Material balance for oxalic acid leaching of silica sand and effluent treatment by photodecomposition and biodegradation

sulphide per 40 tonne heap) and biodegradation of oxalate. Greater feed rates could be employed at higher feed pH values (approximately an order of ten greater at neutral pH).

This model analysis did not take account of other factors, e.g., possible evaporation and iron-catalysed photochemical decomposition of oxalic acid (3), but provided a sound basis for on-site pilot scale studies. The details of biological effluent treatment are given below.

Figure 2 shows an example of an integrated material flow balance for the UASB and RO systems, including recycle of bicarbonate for pre-neutralisation of the UASB feed. The equipment configuration and flows shown were developed on the basis of previous experience and trial and error. Once the conditions had been established, the equipment was operated at steady state for 2 weeks in conjunction with a synthetic photodecomposition effluent.

It was established as a pre-requisite that the RO system would settle at a suitable steady state at a relatively high concentration of salts (e.g., 0.1 M bicarbonate and 200 mg/l sulphate) as much lower concentrations would unduly dilute the UASB feed. Figure 3 shows the results obtained. As expected, the bicarbonate permeate and reject concentrations increased with each cycle. The figure indicates smooth curves over 16 cycles from 0.001-0.014 M (permeate) and from 0.040-0.107 M (reject). Above about 0.1 M the permeate leakage began to increase greatly. The time (and therefore the volume of feed as recycled reject) required to collect 1 l permeate also increased with the salt concentration. Some 8 cycles were necessary to double the salt concentration in the reject (with 0.64/.086 = 7.4% leakage to the permeate) and, after 16 cycles the concentration had increased some 2.5 fold (with 13% leakage). The pooled permeate concentration was 0.0043 M and the total recovery of bicarbonate was 100x60.3/61.0 = 98.9%. The sulphate distribution was similar and the final RO reject concentration was 145 mg/l (a concentration factor of 2.6).

Similar experiments carried out with a clear UASB effluent (bicarbonate 0.022 M, sulphate 45 mg/l and iron <0.5 mg/l) collected over several days. The bicarbonate and sulphate concentrations were increased to 0.1 M and 195 mg/l, respectively. The concentration factors were about 4.5. The conclusion was that the concentration of salts in the RO reject was large enough for pre-neutralisation purposes and the leakage to the permeate small enough to avoid undue contamination of the wash water.

The subsequent design work included matching (i) the photodecomposition effluent inflow to Stocktank 1 to the permeate outflow from RO (X l/d, Figure 2) (ii) the recycle stream to Stocktank 1 (Y l/d) to a UASB inlet pH 3 and (iii) the UASB sludge bed volume (i.e., the number of laboratory reactors operating in parallel) to the combined flow (X+Y l/d) and loading from Stocktank 1. The work also included determining the feed rate (X+Z l/d) to RO, the rate of reject recycle (Z l/d) and the reject:permeate ratio (R = Z/X).

Despite significant variability in the permeate flow (and consequent bicarbonate variation from 0.082-0.104 M), acceptable numerical balances were established. As part of the process of determining these balances, it was found that every 4 l synthetic photodecomposition effluent required approximately 1 l recycled bicarbonate to raise the pH to 3 or greater: thus the nominal feed:recycle ratio was 4. The feed rate to a single UASB reactor was 1.44 l/day and the rate of RO permeate production was approximately 8 l/day. Therefore, the number of reactors required was 8/(0.8 x 1.44) = 7. The RO reject:permeate ratio was 179.

The UASB results obtained are given in Table 1 (in which each line of data represents two days of operation) and indicated satisfactory continuous operation. Oxalate was undetectable in the effluents and the iron concentrations were consistently << 1 mg/l. The large sulphate

622

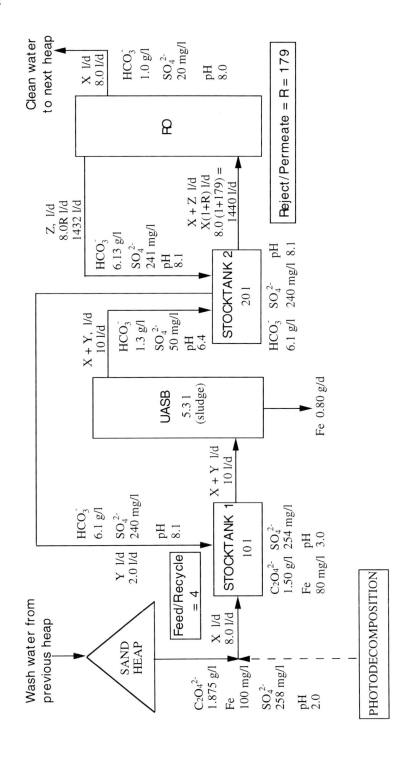

Figure 2.   Material balance: washing cycle

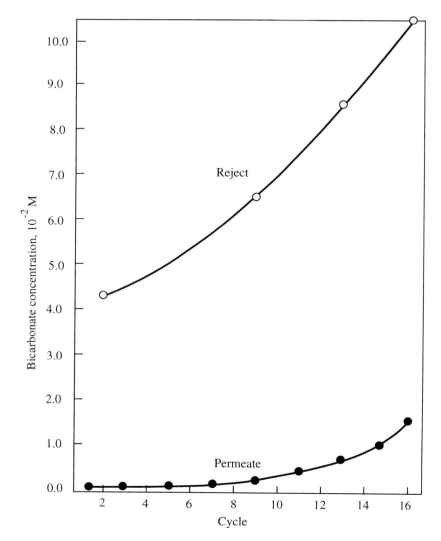

Figure 3. Bicarbonate analyses of reverse osmosis permeate and reject

624

utilisation and increase in pH indicated strong bacterial activity despite the low influent pH. Effluent bicarbonate concentrations were always about 0.02 M.

Table 1
UASB Feed (pre-neutralised) and effluent analyses for a two-week period

Influent				Effluent			
Oxalate g/l	Iron mg/l	Sulphate mg/l	pH	Iron mg/l	Bicarbonate g/l	Sulphate mg/l	pH
1.514	74	194	3.5	0.30	1.403	43	6.4
1.487	71	230	3.2	0.25	1.342		
1.501	80	271	3.1	0.40	1.312	43	6.4
1.501	78	271	3.0	0.40	1.342	43	
1.478	81	256	3.1	0.22	1.305	57	6.3
1.478	81	250	3.1	0.40	1.305	63	6.3
1.496	80	256	3.0	0.40	1.293	43	6.4

4. CONCLUSIONS

Although all of the techniques discussed are well known in different sectors of industry, their particular use and integration in the ways described are substantially novel. Leaching with organic acids is not currently established in the mineral industry and oxalic acid is relatively costly (about £1/kg). Nonetheless, the work has shown that, with use of low leachate:solid ratios (i.e., acid cure conditions), the method becomes feasible for heap leaching. The photochemical reactions of iron(III) oxalates have been studied for more than 40 years but were not until the current initiative proposed as an alternative method of precipitating iron from leachates. UASB systems have been used for 30 years or so for treating organic wastewaters but only in recent years have they been studied for mineral processing applications Reverse osmosis, although commonly employed for water purification, has rarely been considered as a means of simultaneous purification and concentration.

Overall, the results indicate that the leaching and effluent treatment system described is technically feasible. The next step should be on-site pilot plant testwork to assess the likely economics of the integrated system. An earlier outline economic comparison (5) indicated that upgrading silica sand by oxalic acid leaching (with effluent treatment by photodecomposition and UASB biodegradation, but excluding RO) should be competitive with the traditional method of sulphuric acid leaching. Other circuit configurations - such as routing part of the wash water to photodecomposition (to reduce the load on the UASB system) are possible and should be tested.

The results also demonstrated the fundamental viability of reverse osmosis in utilising the inherent alkalinity of UASB effluents to counteract the inherent acidity of oxalic acid leach liquors, while at the same time solving the water conservation problems which would otherwise exist in the circuit. However, for economic use the loading rates of both the UASB sludge and the RO membrane need to be substantially increased. The early indications are that significant progress should be possible.

ACKNOWLEDGEMENTS

The work in this paper was carried out under CEC contracts BRE2-CT92-0215 and BRPR-CT96-0156. The support of the European Union is gratefully acknowledged. Thanks are also due to the contract partners who provided advice and assistance.

REFERENCES

1. R. Chiarzia and E.P. Horwitz, Hydrometallurgy 27 (1991) 339.
2. A.W.L. Dudeney, A. Narayanan and I.I. Tarasova. Proceedings of the 5th International Symposium on Environmental Issues and Waste Management in Energy and Mineral Production - SWEMP '98, Eds. A.G. Pasamehmetoglu and A. Ozgenoglu, Ankara, Turkey (1998) 695.
3. A.W.L. Dudeney and I.I. Tarasova, Hydrometallurgy 47 (1998) 243.
4. A.W.L. Dudeney, A. Narayanan, I.I. Tarasova, J.E. Teer and D.J. Leak, Biohydrometallurgical Processing, Volume II, Proceedings of the International Biohydrometallurgy Symposium IBS '95, Eds. C.A. Jerez, T. Vargas, H. Toledo and J.V. Wiertz, Universidad de Chile, (1995) 277.
5. C.F. Bonney (Ed.), Removal of Iron from Industrial Materials - final report of the CEC Contract MA2M-0014 (1991-3) Removal of Iron from Industrial Minerals, obtainable from the Commission of The European Communities, Directorate General XII, Science, Research and Development, Rue Montoyer 75, B-1049 Brussels.

Bioremediation of a soil contaminated with radioactive elements

S.N. Groudev [a], P.S. Georgiev[a], I.I. Spasova[a] and K. Komnitsas[b]

[a] - Department of Engineering Geoecology, University of Mining and Geology, Studentski grad - Durvenitza, Sofia 1100, Bulgaria

[b] - Laboratory of Metallurgy, National Technical University of Athens, Zografos 15780, Greece

Samples of a soil heavily polluted with radioactive elements and toxic heavy metals were subjected to bioremediation in lysimeters containing 50 kg (dry weight) of soil each. The soil treatment was connected with the solubilization of the pollutants located in the upper soil layers (mainly in the horizon A) and their removal from the soil by drainage waters percolating through the soil mass. The solubilization was carried out by the soil microflora whose activity was enhanced by suitable changes in the levels of some essential environmental factors such as water, oxygen and nutrient contents in the soil. The leaching of radioactive elements and of some toxic metals was efficient and their contents in the soil were decreased below the relevant permissible levels within 10 months of treatment. It was also possible to transfer portions of the pollutants into the deeply located soil layers (mainly in the horizon B_2) where they were immobilised mainly as a result of the activity of the anaerobic sulphate-reducing bacteria.

1. INTRODUCTION

Pollution of soils by radionuclides may be of different kinds. An important type of radionuclides originates from the emission to the atmosphere, e.g. nuclear explosions (^{3}H) or reactor operations (^{85}Kr). Subsequent fallout of radionuclides with precipitation and infiltration causes pollution of different aqueous and terrestrial ecosystems. Such antropogenic pollution is sufficiently well documented worldwide.

Natural radioactivity is due mainly to ^{40}K and the members of the natural ^{238}U decay chain. Natural radionuclides are released to the surface and ground waters from rocks and ores by dissolution and desorption, or by diffusion or atomic recoil, during radioactive decay. Different chemolithotrophic and heterotrophic microorganisms are able to leach uranium and other radioactive elements from different mineral raw materials in both acidic and alkaline water solutions (1-4). The intensity of these microbial processes in the virgin uranium ore deposits is usually limited by the shortage of oxygen, water and/or dissolved organic compounds. The artificial disturbance of the ore bodies as a result of human activities largely enhances the leaching of radioactive elements.

In Bulgaria for a long period of time uranium was leached commercially in a large number of deposits using different in situ technologies. Most of these commercial-scale operations were connected with the acid leaching of uranium due to the presence of pyrite and the negative net neutralization potentials of the relevant uranium ores. Several years ago all commercial-scale operations for uranium leaching in the country were stopped due to a complex of different political, economical and environmental reasons. Regardless of some preventive and remedial actions during the uranium recovery, many natural ecosystems were heavily polluted with radioactive elements, mainly through the seepage of acid drainage waters. Such waters are still a persistent environmental problem at many abandoned mine sites. Soils around the water flowpath are polluted with radioactive elements and are unsuitable for agricultural use. In some deposits alkaline leaching of uranium was applied, using solutions containing carbonate and hydrocarbonate ions. This process also was connected with a heavy pollution of the soils. In some cases air transportation is also an essential factor for the soil pollution.

Different methods for remediation of soils contaminated with radioactive elements are known but only few of them have been applied under large-scale conditions. The excavation and transportation of the heavily polluted soils to specific depositories is still a common practice in most countries.

Recently, a biotechnological method for detoxification in situ of soils contaminated with toxic heavy metals and arsenic was developed and applied in some agricultural lands in Bulgaria (5, 6). This method is connected with the transfer of the soil contaminants into the deeply located soil horizons where the soluble contaminants are turned into the relevant insoluble sulphides. Microorganisms related to different physiological groups, mainly acidophilic iron- and sulphur-oxidising chemolithotrophic bacteria and anaerobic sulphate-reducing bacteria, respectively carry out both the transfer and precipitation of the contaminants.

The ability of different microorganisms to dissolve uranium in waters with a slightly alkaline pH as well as the ability of the sulphate-reducing bacteria to precipitate the dissolved uranium by reducing it to the tetravalent form are well known (1, 7, 8). For that reason, a study was carried out to establish the possibility to apply the above-mentioned remediation method for treatment of soils polluted with uranium and the other radioactive elements from its decay chain. In this paper some data about this study are presented.

2. MATERIALS AND METHODS

A sample of soil heavily polluted with radioactive elements (uranium, radium, thorium) and toxic heavy metals (copper, cadmium, lead) was used in the experimental work. The sample was excavated in a way preserving the natural soil profile. Portions of the sample were used in the bioremediation experiments within 48 h after the excavation to preserve the viable indigenous microflora of the soil.

Elemental assays in the different soil horizons were performed by digestion and measurement of the ion concentration in solution by ICP spectrophotometry. Mineralogical analysis was carried out by X-ray diffraction techniques. The main geotechnical characteristics of the soil were measured by methods described elsewhere (9). The bioavailable fraction of the toxic elements was determined through leaching soil samples by DTPA and EDTA (10). The speciation of the pollutants with respect to their mobility was determined by the sequential extraction procedure (11). The toxicity of soil samples was determined by the EPA Toxicity Characteristics Leaching Procedure (12).

The freshly excavated soil samples were subjected to bioremediation in plastic rectangular (39 cm x 19 cm) lysimeters containing 50 kg (dry weight) of soil each. The soil in the lysimeters was arranged in a way reflecting the natural distribution of the different soil horizons. The soil profile in the lysimeters was 80 cm high (horizon A 25 cm, horizon B 55 cm, and 10 cm from the horizon C). The soil treatment was connected with the solubilization of the pollutants and their removal from the soil by drainage waters percolating through the soil mass. Leach solutions containing carbonate ions and some dissolved organic compounds were used for this purpose. The upper soil layers were ploughed up periodically to enhance the natural aeration. The pregnant soil effluents were treated by a passive system consisting of an anoxic cell containing a mixture of organic substrates (spent mushroom compost, sawdust, horse and cow manure) and a community of different metabolically interdependent microorganisms. The barren waters from the anoxic cell were recycled to the lysimeters.

In some lysimeters water solutions of dissolved organic compounds were injected through vertical pipes to the deeply located soil layers (horizon B_2) to enhance the activity of the anaerobic sulphate-reducing bacteria. The dissolved metal concentrations were determined by ICP spectrophotometry. Uranium concentration was measured photometrically using the arsenazo III reagent. Radium was measured radiometrically. Sulphate concentrations were determined photometrically.

The isolation, identification and enumeration of the microorganisms inhabiting the lysimeters were carried out by methods described previously (13, 14).

3. RESULTS AND DISCUSSION

Data about the chemical composition and some essential parameters of the soil are shown in Table 1. The concentrations of contaminants were higher in the upper soil layers (mainly in the horizon A) (Table 2). Considerable portions of the contaminants were present as the relevant inert fractions, which are refractory to solubilization. Regardless of this, the leaching of the contaminants in the upper soil layers was very efficient (Table 2) and was undoubtedly connected with the activity of the indigenous soil microflora. Thus, in a control lysimeter treated by leach solutions containing microbial inhibitor ($HgCl_2$) the leaching of contaminants was negligible.

The analysis of the soil microflora revealed that it included a rich variety of microorganisms (Table 3). In the upper soil layers the aerobic heterotrophic bacteria were the prevalent microorganisms. Their total number was higher than 10^8cells/g dry soil. Bacteria related to the genera Pseudomonas and Bacillus were the most numerous in these microbial communities. The fungi were low in number. Some chemolithotrophic bacteria able to oxidise S^0 and soluble inorganic sulphur compounds at neutral and alkaline pH values were also present. Thiobacillus thioparus was the main species among these chemolithotrophs but Thiobacillus neapolitanus, the anaerobic Thiobacillus denitrificans and some mixotrophic bacteria (mainly such related to the species Thiobacillus novellus) were also present.

Table 1
Characteristics of the soil used in this study

Parameters	Horizon A (0 - 25 cm)	Horizon B (26 – 70 cm)
Chemical composition (in %)		
- SiO_2	60.6	61.1
- Al_2O_3	14.3	14.0
- Fe_2O_3	12.5	5.90
- CaO-	5.72	2.35
- MgO	3.21	1.31
- K_2O	3.92	3.81
- Na_2O	0.88	1.94
- S total	0.28	0.21
- S sulphidic	0.14	0.10
- humus	3.7	1.4
Bulk density, g/cm^3	1.51	1.63
Specific density, g/cm^3	3.14	3.43
Porosity, %	44	41
Moisture capacity, %	42	39
Permeability, cm/s	5×10^{-2}	5×10^{-2}
pH (H_2O)	7.54	7.65
Net neutralization potential, kg $CaCO_3$/t	53	+ 215

The activity of this microflora was enhanced by suitable changes in the levels of some essential environmental factors such as water, oxygen and nutrient contents in the soil. This was achieved by regular ploughing and irrigation of the soil. The optimum soil humidity was about 45-50% from the moisture capacity of the soil but periodic flushing with leach solutions was needed to remove the dissolved contaminants. Zeolite saturated with ammonium phosphate was added to the soil (in amounts in the range of 2-5 kg/ton dry soil) to provide the microorganisms with ammonium and phosphate ions. The presence of zeolite enhanced to some extent the rate of leaching and decreased considerably the inefficient consumption of the nutrients. The slow-release solid zeolite made nutrients available to the microorganisms over a period of weeks rather than all being immediately water soluble and subject to rapid washout. This was connected with the maintenance of relatively constant concentrations of ammonium and phosphate ions in the soil solution due to the existence of a stable equilibrium between the dissolved and adsorbed forms of these ions. Furthermore, the zeolite improved the physicomechanical properties of the soil.

The temperature during the treatment was in the range of 16-20°C. However, using control lysimeters it was found that the process was efficient even at temperatures as low as 9-10°C but practically stopped at temperatures lowers than 3-4°C.

Table 2.
Toxic elements in the horizon A of the soil before and after the bioremediation

Parameters	Uranium	Radium	Copper	Cadmium	Lead
Content of toxic elements, ppm:					
- before treatment	35	15	648	7	275
- after treatment	9	5	262	3	114
Bioavailable fraction, ppm:					
a. by DTPA leaching:					
- before treatment	2.60	2.35	268	0.8	0.9
- after treatment	0.09	0.05	41	0.1	0.1
b. by EDTA leaching:					
- before treatment	0.10	0.12	97	< 0.04	1.4
- after treatment	<0.01	0.01	14	< 0.004	0.15
Easily leachable fractions – exchangeable + carbonate, ppm:					
- before treatment	2.7	1.9	28	0.2	0.7
- after treatment	0.5	0.3	1.4	<0.04	0.08
Inert fraction, ppm:					
- before treatment	12	6.2	125	4.1	161
- after treatment	8	4.1	102	2.8	97
Toxic elements solubilized during the toxicity test, ppm:					
- before treatment	0.35	0.14	1.92	0.01	0.08
- after treatment	0.05	0.03	0.23	<0.004	<0.004

Note: The contents of radium are given in $\mu g/t$ and for the radium solubilized during the toxicity test in - $\eta g/l$.

Under optimum conditions portions of the contaminants were removed from the upper soil horizons and their residual concentrations, with the exception of that of the lead, were lowered below the relevant permissible levels within 10 months of treatment. The leaching rate markedly depended on the presence of organic compounds and carbonate ions in the leach solutions. Uranium was leached efficiently in these ecosystems with a slightly alkaline pH probably by two different mechanisms. (4). The first mechanism was connected with the microbial production of peroxide compounds, which turn the tetravalent uranium to the hexavalent state. The U^{6+} is then solubilized as uranyl carbonate or as complexes with some organic compounds. The second mechanisms was connected with the microbial secretion of some organic metabolites, mainly organic acids, which form complexes with this metal.

The toxic heavy metals were solubilized mainly by means of microbially secreted organic acids. Iron and manganese were, however, solubilized also as a result of an enzymatic

Table 3
Concentration of various physiological groups of microorganisms in the soil during the treatment

Microorganisms	Horizon A (0 – 25 cm)	Horizon B (26 – 70 cm)
	Cells/g dry soil	
Aerobic heterotrophic bacteria	$10^7 - 10^8$	$10^5 - 10^6$
Oligocarbophiles	$10^4 - 10^6$	$10^3 - 10^4$
Cellulose-degrading microorganisms	$10^4 - 10^6$	$10^2 - 10^4$
Nitrogen-fixing bacteria	$10^3 - 10^5$	$10^2 - 10^3$
Nitrifying bacteria	$10^2 - 10^5$	$10^1 - 10^3$
Chemolithotrophic sulphur-oxidising bacteria	$10^3 - 10^6$	$10^2 - 10^5$
Chemolithotrophic iron-oxidising bacteria	$1 - 10^1$	$0 - 10^1$
Anaerobic heterotrophic bacteria	$10^3 - 10^5$	$10^4 - 10^7$
Denitrifying bacteria	$10^4 - 10^5$	$10^3 - 10^5$
Anaerobic bacteria fermenting carbohydrates with gas production	$10^3 - 10^5$	$10^3 - 10^6$
Sulphate- reducing bacteria	$10^3 - 10^4$	$10^4 - 10^7$
Fe^{3+} - reducing bacteria	$10^2 - 10^3$	$10^2 - 10^4$
Mn^{4+} - reducing bacteria	$10^1 - 10^2$	$10^2 - 10^3$
Methanogenic bacteria	$0 - 10^1$	$1 - 10^2$
Streptomycetes	$10^3 - 10^5$	$10^2 - 10^3$
Fungi	$10^3 - 10^6$	$10^2 - 10^4$
Total cell numbers	$1 \times 10^8 - 5 \times 10^8$	$1 \times 10^6 - 3 \times 10^7$

reduction of the Fe^{3+} and Mn^{4+} to the relevant bivalent forms. Toxic metals were solubilized, although at low rates, even from the relevant sulphide minerals. This was due to the activity of some chemolithotrophic bacteria, mainly such related to the species Thiobacillus thioparus and Thiobacillus neapolitanus. These bacteria enhance the oxidation of sulphide minerals by removing the passivation films of S^o deposited on the mineral surface as a result of different chemical, electrochemical and biological processes (1). Some of the metals, mainly the copper, were solubilized also as complexes with the ammonium produced from the biodegradation of organic matter in these systems.

In some experiments the dissolved contaminants were removed from the lysimeters by the pregnant effluents. The treatment of these effluents in the anoxic cell resulted in an efficient removal of the contaminants. The toxic heavy metals were precipitated mainly as the relevant insoluble sulphides by the hydrogen sulphide produced by the sulphate-reducing bacteria inhabiting the cell. Uranium was precipitated as uraninite (UO_2) and most of the radium was adsorbed on the solid organic substrates in the cell. The role played by microbial biomass, both viable and dead, in the uptake of these contaminants from aqueous solutions also has been well documented (1).

In other experiments it was possible to immobilise the contaminants in the deeply located soil layers (mainly in the horizon B_2) as a result of the activity of the indigenous sulphate-reducing bacteria. This activity was enhanced by the water solutions of soluble organic compounds injected to this soil horizon. The precipitated contaminants were further immobilised by their sorption on the clay minerals present in the horizon B_2.

The field application of the above-mentioned treatment is connected with a detailed characterisation of the subsurface geologic and hydrogeologic conditions of the relevant site and with the construction of an effective collection system to prevent soil effluent migration and pollution of surface and ground waters. Furthermore, some conventional remediation procedures such as grassing of the treated soil, addition of certain fertilisers and animal manure as well as with periodical ploughing, liming (in the case of acidic soils) and irrigation are needed to restore completely the physical, water and biological properties of the soil.

ACKNOWLEDGEMENTS

The authors would like to greatly acknowledge the financial support of the European Commission under the Copernicus project entitled "Marine pollution in the Black Sea due to mining activities: risk assessment, development of preventive and remedial actions, "Contract No : ERB-ICI5-CT96-0114.

This paper is dedicated to the memory of Professor Antonios Kontopoulos of National Technical University of Athens, who suddenly passed away on April 26, 1998 at the age of 53.

REFERENCES

1. G.I. Karavaiko and S.N. Groudev (eds.), Biogeotechnology of Metals, GKNT International Projects, Moscow, 1985.
2. J.E. Zajic, Dev. Ind. Microbial., 11 (1969) 413.
3. L. Fekete, B. Czegledi, K. Czako-Ver and M. Kecskes, in: Use of Microorganisms in Hydrometallurgy, pp. 43-47, Hungarian Academy of Sciences, Pecs, 1980.
4. S.N. Groudev, A.Kontopoulos, I.I. Spasova, K.Komnitsas, A.T. Angelov and P.S. Georgiev, in: M.Herbert and K.Kovar (eds.), Groundwater Quality: Remediation and Protection, pp. 249-255, IAHS Press, Wallingford, UK, 1998..
5. S.N. Groudev, Mineralia Slovaca, 28 (1996) 335.
6. S.N. Groudev, in: Proc. XX Int. Min. Proc. Congr. Aachen, Germany, Vol. 5 (1997) 729.
7. D.R. Lovely and E.J.P. Phillips, Appl. Environ. Microbiol., 58 (1992) 850.
8. E.J.P. Phillips, D.R. Lovely and E.R. Landa, J. Ind. Microbiol., 14 (1995) 203.
9. DOE, A Guide to Risk Assessment and Risk Management for Environmental Protection, 1995.
10. A.A. Sobek, W.A. Schuller, J.R. Freeman and R.M. Smith, U.S. EPA 600/2-78-054, Cincinnati, Ohio, 1978.
11. A.Tessier, P.G.C. Campbell and M. Bisson, Anal. Chemistry, 51 (1979) 844.
12. U.S. Environmental Protection Agency, Characterics of EP Toxicity, Paragraph 261.24, Federal Register 45(98), 1990.

13. V.I.Groudeva, I.A. Ivanova, S.N. Groudev and G.C. Uzunov, in: A.E.Torma, H.L. Apel and C.L. Brierley (eds.), Biohydrometallurgical Technologies, vol. II, The Minerals, Metals & Materials Society, Warrendale, Pennsylvania, (1993) 349.
14. G.I. Karavaiko, G.Rossi, A.D. Agate, S.N. Groudev and Z.A. Avakyan, Biogeotechnology of Metals. Manual, GKNT International Projects, Moscow, 1988.

Effect of flooding of oxidized mine tailings on *T. ferrooxidans* and *T. thiooxidans* survival and acid mine drainage production: a 4 year restoration-environmental follow-up

R. Guay [a], P. Cantin [a], A. Karam [b], S. Vézina [c] and A. Paquet [d]

[a] Dept Microbiology, [b] Dept of Soils and Agrifood Eng., Laval University, Canada, G1K 7P4,

[c] CAMBIOR Inc., 800 Blvd René-Lévesque W., Montréal (Qc), Canada H3B 1X9,

[d] Enviromine Inc., 1398 Jacques Bureau, Ste-Foy (Qc), Canada G2G 2M1

A pilot-scale study on the effect of flooding unoxidized and oxidized Cu/Zn tailings demonstrated the technical feasability of this technology to remediate a mining site where over 3 million tons of tailings were impounded. Full-scale flooding of the tailing pond with free running water was undertaken after the construction of an impervious dam; approximately 2 million m^3 of surface water at pH 7,4 completely covered the tailings after 16 months. The minimal water column over the tailings was established at 1,20 m and reached 4,5 m, depending on the site topography. Water and tailings samples were collected from 9 different locations from the surface of the man-made lake using a specially designed borer and were analyzed for pH, conductivity, iron- and sulfur-oxidizing bacteria activity and numbers as well as the sulfate reducing bacteria (SRB) population. We showed that over a four year period of flooding, the overall population of iron-oxidizers decreased considerably; their numbers drastically fell from 1×10^6 to 1×10^2 active cells per g of oxidized tailings while the SRBs increased from 10^1 to $10^5/g$. The pH of the influent, the reservoir and the effluent water remained fairly constant between 6,9 up to 7,4 over the entire period. During this time, interstitial water pH increased from 2,9 to 4,3 in flooded tailings where lime could not be incorporated in the first 20 cm of tailings; elsewhere, the pH of the tailings suspensions remained fairly constant around neutral values (pH 7,0). Dissolved oxygen was measured at fixed intervals and remained also constant between 6 and 7.5 mg/L while water temperatures fluctuated below freezing point to +20C respectively in winter and summer season.

Corresponding author: Roger Guay, Dept of Microbiology, School of Medicine, Laval University, Ste-Foy, (Qc), Canada, G1K 7P4. E-mail: mcbrog@hermes.ulaval.ca

1. INTRODUCTION

Microorganisms which actively contribute to the production of acid mine drainage are members of lithotrophically growing genera *Thiobacillus* and *Leptospirillum* , especially *Thiobacillus* (*T.*) *thioparus* , *T. thiooxidans* , *T. ferrooxidans* and *Leptospirillum ferrooxidans.* Once the iron oxidizers have colonized sulfidic mine tailings, it is almost impossible to eradicate them. Many attemps were made in the past to prevent the sequential colonization of mine tailings, it usually start with heterotrophic *Bacillus, Pseudomonas* and *Alcaligenes* strains that utilize organic chemicals present in the slurried tailings during their deposition. This step is rapidly followed by mild thio-oxidizers (*T. thioparus* and related species) which slowly decrease the pH interstitial water confined to the first 5 to 10 cm below the surface, to a 4 to 5 values. Then *T. thiooxidans*, *T. ferrooxidans* and *Leptospirillum ferrooxidans* take over from that point and their sulfide and iron oxidizing activities contribute to the rapid decrease (less than 3,0) of the tailings pH. Within 18 months, the unsaturated sulfidic mine tailings are heavily colonized by strong iron and sulfur oxidizing strains.

Therefore, measures such as the use of encapsulated slow-release surfactants and detergents were developed to prevent colonization of mine tailings by the iron oxidizers (1). Subaqueous disposal of tailings has also be considered as an effective method to interfere and possibly prevent the bacterial oxidation of metal sulfides (2-4). Microbiological oxidation of most of the copper, iron and zinc sulfides unrecovered during the hydrometallurgical processing of the ores is highly dependent on oxygen availability, acidic pH and high Eh. It is thought that any physico-chemical measure that would interfere with one of these parameters would also stop the bacterial activity in tailings.

This study presents evidence that flooding of oxidized copper and zinc tailings considerably retarded their further oxidation and decreased the viability and oxidative activity of the *Thiobacilli* in the submerged tailings over a period of 4 years.

2. MATERIALS AND METHODS

2.1. Sampling of flooded tailings

The initial sampling of the site was carried out before flooding and lime application using grab sampling method; the samples were aseptically collected and placed in sterile bags. Collection of tailings specimens in flooded areas was done from a row-boat using a home-made borer device. The sampler was made of a 30 cm long CPVC tube section, 1,75 cm inside diameter; this borer-type sampler was fitted on 1 to 3 screwed-together 2,0 m sections of copper tubing in order to reach the flooded tailings. Approximately 50 g of consolidated tailings (~10 cm long tailings cores) were recovered with the borer. The sampler was designed to retain, above the tailings sample, about 40 mL (~15 cm) of the water column sitting right on top of the solid sample as outlined below on Figure 1.

The water from the bottom of the man-made lake was recovered and analyzed *in situ* for pH, temperature and conductivity measurements. The solid core was aseptically pushed from the sampler using a sterile CPVC rod in sterile plastic bag and stored on ice. A second tailings core was recovered in a screw-cap test tube prefilled with a transport medium (sterile NaCl 0,15% solution), and left to overflow in order to preserve the sample anoxic conditions.

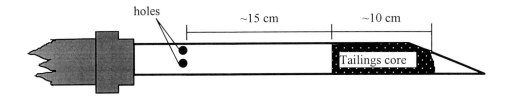

Figure 1. Schematic view of the CPVC tailings borer/sampler

2.2. Physico-chemical analysis

The pH of the surface water and of the water recovered from the sampler was measured using a portable Cole Parme pH meter (model 612) calibrated before the sampling. Water temperature was measured at the surface and the bottom of the man-made lake using a precision thermometer. Interstitial water pH, obtained from the tailings core, was measured at the laboratory within three hours after collecting the specimens.

2.3. Microbiological analysis

2.3.1. Determination of bacterial iron- and sulfur oxidizing activities

A 5 g sample of tailings was suspended in 9K (5), BS-4 and BS-7 (6) growth media adjusted respectively at pH 2.3, 4.0, and 7.0. The suspension was diluted 1:100 and used to inoculate 100 mL of the three culture media distributed in 300 ml Erlenmeyer flasks. The incubation was carried out on a gyrotory action shaker, model G-53 (New Brunswick Sci.Co.), at 28°C for two to three weeks. Iron oxidation activity in the 9K medium was followed by volumetric analysis of the remaining ferrous iron (Fe^{2+}) and sulfur oxidation was evaluated by periodic pH measurements on samples withdrawn during the active growth on BS-4 and BS-7 media.

2.3.2. Enumeration of iron- oxidizers

A quantity of 0,5 g of tailings was placed in 4,5 mL of modified 9K growth solution (MPN medium) and agitated vigorously for 1 min to detach bacteria from the solid mineral particles. A modified MPN procedure (7) was used to enumerate the iron-oxidizers after 10 days of incubation at 30°C.

A similar methodology was used to enumerate the sulfur-oxidizing bacteria: 0,5 g of tailings were placed in 4,5 mL of BS-7 growth medium and diluted with the same medium using standard MPN procedure. Positive activity in the reaction mixture was revealed with a pH indicator after microbiological oxidation of thiosulfate, used as the energetic substrate, to sulfuric acid.

2.3.3. Detection and enumeration of sulfate-reducing bacteria (SRB)

A sample of tailings (0,5 g) that has been maintained in anoxic conditions is suspended in 4,5 mL of a de-aerated saline solution and diluted using the Postgate medium. An MPN procedure is then used to enumerate the SRB after the oxygen was microbiologically removed from the medium.

638

3. RESULTS

The Solbec-Cupra tailings impoundment site after processing some 4,2 Millions tons of $CuFeS_2$, Cu_5FeS_4, $Cu_{12}Sb_4S_{13}$, FeS_2, PbS, and ZnS minerals occupied a 66ha area (2 468 000 m^3) prior to its flooding. Two impervious dams were built in 1994 and the site was left to flood at a minimal depht of 1,3 m by running water; more or less 2 million m^3 accumulated over the tailings in less than 2 years. No special measures were taken to prevent the resuspension of the tailings but a recommandation to spread agronomic lime at a rate of 125 t/ha was followed where practically feasable to neutralize the accumulated H_2SO_4 and solubilized Cu, Fe, and Zn heavy metals.

Sampling of the tailings was carried out before, during and after total flooding of the tailings area following the pattern outlined in Figure 2. The sampling of tailings was carried out tree times a year, in May, August, and November, from 1995 to 1998, a total of 108 tailings specimens were collected during this period. The sampling was carried out in the same area on the artificial lake but not at the very same location at each occasion.

3.1. Initial iron- and sulfur- oxidizers populations

According to the recorded iron and sulfur oxidation activity of the bacteria isolated from the nine sampling points on the site prior to the flooding, we were able to divide the 45 ha uncovered tailings area in three distinct sections in which iron- and sulfur-oxidizers represented different proportions of the *Thiobacilli*. The acidophilic iron bacteria *T. ferrooxidans* and *L. ferrooxidans* were located mainly in the upper portion of the site (where the tailings are under 1,0 m of water), whereas the mid-section of the site was occupied by a mixed population of the acidophiles *T. thiooxidans* and *T. ferrooxidans*. The lowest area of the tailings pond has always been maintained soaked near the marsh and this is where we characterized a strong neutrophilic sulfur oxidizing *T. thioparus* population.

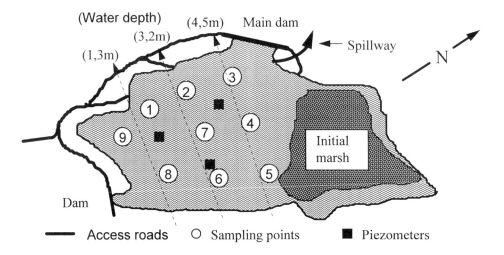

Figure 2. Artificial lake after flooding the Solbec-Cupra tailings site showing the positions of the sampling points numbered from 1 to 9.

It seems that the cycles of wet and dry areas in the tailings site had some influence on the type of sulfide (S^{2-}) or ferrous iron (Fe^{2+}) substrate preferred by the *Thiobacilli*. This observation may influence the restoration procedures in that, the need for neutralization of the already oxidized sulfidic tailings may be different from a microbiological perspective : it may not be as important to eliminate the neutrophilic *T. thioparus* as to eradicate the strong acidophilic microorganisms.

3.2. Evolution of the iron-oxidizers over the years

Figure 3 clearly shows that the flooding of millions of tons of tailings with 2 millions of m^3 of fresh water affected the iron-oxidizing activity of the *T. ferrooxidans* and *L. ferrooxidans* populations in the Solbec-Cupra site. Almost in every sampling point the overall population of iron-oxidizers fell from record numbers (1×10^7 viable cells/g) to barely detectable levels ($< 1 \times 10^2$ viable cells/g). It was further shown that the microbiological oxidizing activity could not be maintained under water, at least not at a noticeable level, because none of the water samples recovered from the borer did demonstrate any decreasing pH when compared to the incoming waters. No mixing of the tailings interstitial waters and the waters sampled a few centimeters above the submerged tailings was observed, the pH remained well above 6,0 over the entire artificial lake.

These results are attributed in part to flood water infiltration through the tailings, which was estimated to be around 10 to 15%, but it cannot explain the drastic decrease of both bacteria viability and oxidizing activity.

When first sampled, the unflooded tailings sustained a fairly large iron-oxidizing bacteria population in the first 20 cm below the surface. The iron oxidizing activity was maximal : the normal Fe^{2+} content of the 9K growth medium was completely exhausted after 72 h of culture of the isolates, the lag phase of the iron-oxidizers was measured at 24 to 48 h and within a day, the 9 00 ppm of ferrous iron was converted to ferric state. Two years only after flooding, submerged tailings samples still contained abundant numbers of iron-oxidizers but their growth on ferrous sulfate showed extended lag phase, up to 10 days before Fe^{2+} was completely oxidized to Fe^{3+}.

3.3. Evolution of the sulfur-oxidizing bacteria populations

The sulfur-oxidizing activity of *T. ferrooxidans* and *T. thiooxidans* was also affected by flooding but to a lesser extent than was the iron oxidation. Before the development of an appropriate enumerating methods for the sulfur-oxidizing bacterial cells, we had to rely on growth parameters evolution to evaluate the global microbiological sulfur-oxidizing activity in the tailings. In the mean time, we used a modified MPN methodology to enumerate neutrophilic sulfur-oxidizing bacteria during the last 27 months of sampling. Figure 4 shows fairly rapid decrease in the numbers of sulfur-oxidizing neutrophiles in the deepest part of the artificial lake (more than 4,2 m in depth). A similar observation was also made in Fig. 3 where the iron-oxidizing bacteria rapidly fell to low values. The numbers of *T. thioparus*-like bacteria remained relatively constant in submerged tailings under 1,0 to 3,2 m of water.

3.4. Evolution of the sulfate-reducing bacteria

Two years after the flooding was initiated, the tailings were completely covered with water, and we noticed that in areas where vegetation had been buried with the oxidized minerals, gas bubbles erupted from the shallow bottom of the artifical lake.

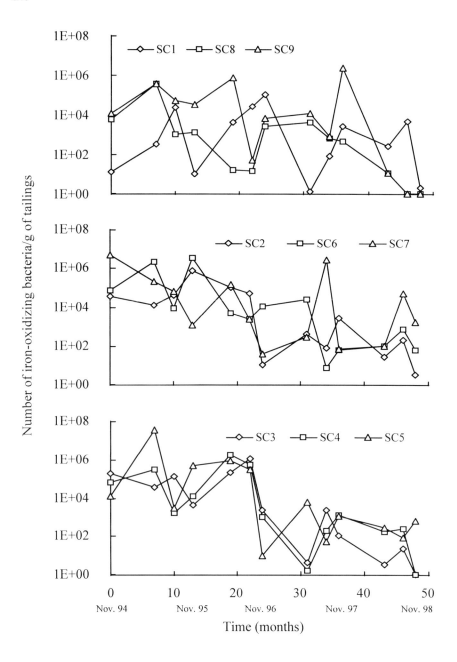

Figure 3. Fluctuations in the populations of acidophilic iron-oxidizing bacteria recovered from the submerged tailings in 9 sampling areas (SC1 to SC9), over a period of 48 months after flooding of the Solbec-Cupra tailings site.

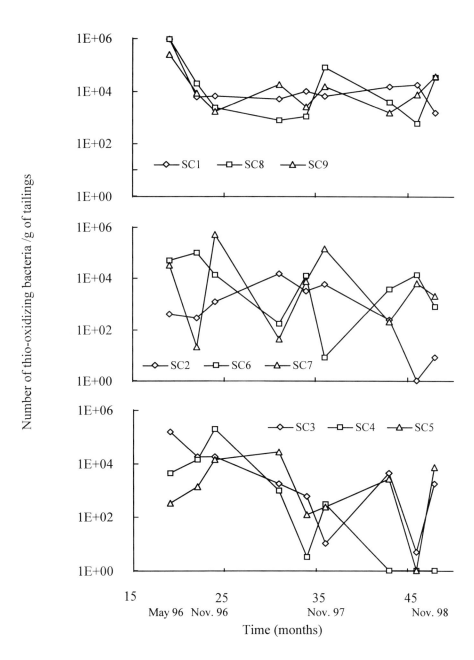

Figure 4. Fluctuations in the global population of neutrophilic sulfur-oxidizing bacteria recovered from the submerged tailings in 9 sampling areas (SC1 to SC9), over a period of 29 months after flooding of the Solbec-Cupra tailings site.

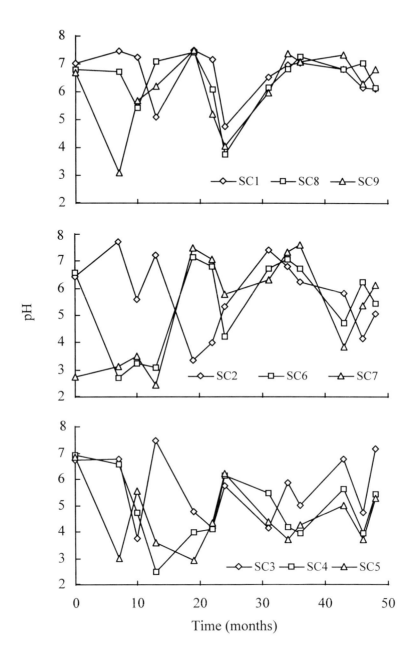

Figure 5. Evolution of pH of interstitial water recovered from the submerged tailings sampled at different locations (SC1 to SC9) over a period of 48 months after flooding of the Solbec-Cupra tailings site.

We then collected more samples from each sampling point to determine the eventual presence of sulfate-reducing bacteria (SRB). The SRB counts were first recorded early in 1996 and gave an average of 1×10^4 viable cells per gram of tailings; these numbers varied from 10 to 10^5 SRBs/g of solid since then but did not increase significantly. The overall concentration of organic carbon present in the flooded tailings was measured at 1,75% (W/W) and we estimated that the organic carbon content was probably not sufficient to support sulfate-reducers for a long period of time.

4. CONCLUSIONS

We conclude that the absence of suitable growing conditions, such as low pH, CO_2 availability, and a prolonged deficiency in their energy supply is likely to cause the loss of the iron-oxidizing activity of *T. ferrooxidans* and *L. ferrooxidans*, and considerably affect their survival in submerged habitat. This conclusion is supported by the results shown in Figure 5 where pH of the interstitial water is following a constant rise. Ths tendency was observed in tailings samples recovered from an area where flood water varies from 1 to 3 m in depth. Tailings sampled from deeper environments exhibited a very slow increase of their intersttial water pH, and even under these conditions, iron- and sulfur-oxidizing bacterial populations fell to low numbers. This situation probably reflects the low oxygen availability in the flooded tailings at this depth in addition to the already altered environment for these autotrophs.

ACKNOWLEDGMENTS

The joint financial assistance of the Canadian Centre for Mineral and Energy Technology (CANMET), and the Ministry of Natural Resources of the Quebec Province, through the MEND Program, is gratefully acknowledged.

REFERENCES

1. J. Ondrushka and F. Glombitza. Proc. Internat. Conf. Contaminated Soils, Kluwer Academic Publishers, Dordrecht, (1993) 1195.
2. M. Kalin. Treatment of Acidic Seepages Employing Wetland Ecology and Microbiology, the ARUM Process. MEND project 3.11.1 (Canadian Centre for Mineral and Energy technology), 1993.
3. Centre de Recherche Minérale, Groupe Roche and Monterval Inc. Experimental flooding of Solbec-Cupra Tailings Pond. MEND Project 2.13.2a (Canadian Centre for Mineral and Energy technology), 1993.
4. R. Guay, A.E. Et-Touil and A. Karam. Bench scale Experimental Flooding of Solbec-Cupra Tailings : Chemical and Microbiological Aspects. MEND Project 2.13.2c (Canadian Centre for Mineral and Energy technology), 1994.
5. M.P. Silverman and D.G. Lundgren. J. Bacteriol., 77 (1959) 642.
6. L.L. Barton and J.M Shively. J. Bacteriol., 95 (1968) 720.
7. R. Lafleur, E. Desjardins-Roy, D. Couillard and R. Guay. Biohydrometallurgical Technologies, II. Fossil Energy Materials Bioremediation, Microbial Physiology. (A.E. Torma, M.L. Apel, and C.L. Brierley), vol 2 (1993) 433.

Remediation of phosphogypsum stacks. Field pilot scale application

K. Komnitsas[a], I. Paspaliaris[a], I. Lazar[b], I.G. Petrisor[b]

[a]National Technical University of Athens, Lab. of Metallurgy,
GR-157 80, Athens, Greece. E-mail:komni@central.ntua.gr

[b]Institute of Biology of the Romanian Academy, Dep. of Microbiology,
Spl. Independentei 296, 79651 Bucharest, Romania

At Navodari, 20 km north of Constanta in Romania, the treatment of phosphate ores for the production of fertilisers has resulted in the production of over 3,000,000 m^3 of phosphogypsum which have been disposed of in three stacks over an area of 40 ha.

For the assessment of the risks resulting from these tailings an environmental characterisation study and a risk analysis based on a source-pathway target principle were performed. These studies revealed that phosphogypsum stacks are characterised by residual acidity and elevated concentration of sulphates, toxic elements and radionuclides. In addition, the lack of a cover favours aerial transportation of fine particles and solubilisation and migration of toxic and radioactive elements causing contamination of surrounding soils, surface and ground waters.

Previous laboratory glasshouse studies conducted have shown that modification of phosphogypsum with the addition of amendments and nutrients leads to the establishing of a substrate suitable for development of a vegetative cover consisting of perennial species and/or bushes. Based on these laboratory data, a field pilot scale study was carried out in order to examine in situ the potential of phosphogypsum stacks revegetation.

In this paper, data derived from the field tests are presented and discussed. The critical parameters which play the most important role for the development of a vegetative cover are soil pH, presence of sufficient amounts of micronutrients in the tailings, sufficient water holding capacity of the substrate and selection of species which show a potential for growth in such environments.

Data derived after a seven month field trial period indicate that a number of perennial species and bushes can be used for the development of a vegetative cover which will minimize future environmental risks and improve aesthetics in the area.

1. INTRODUCTION

Navodari is located 20 km north of Constanta and 3 km west of the Romanian Black Sea coast. The 15 ha chemical complex produces fertilisers by treating phosphates ores using the wet method. This method includes treatment of pyrites for the production of sulphuric acid and subsequent attack of phosphate ores. The plant which commenced operation in 1959 has resulted in the production of 3,000,000 m^3 of phosphogypsum which have been disposed of in two stacks over an area of 30 ha. The new tailings discharge facility which operates since 1996 covers an area of 10 ha.

Phosphogypsum (PG) is a calcium sulphate dihydrate, $CaSO_4.2H_2O$, a by-product of phosphoric acid production from phosphate rocks and is characterised by pH values ranging between 2 and 4 (7) and inherent low level radioactivity associated to Ra-226.

The main pathways for the exposure of humans to radioactivity associated with PG are direct gamma irradiation, radon emission in air, infiltration of rain water into the tailings mass causing solubilization of toxic and radioactive elements and therefore contamination of surface and ground waters, and finally aerial transfer and subsequent deposition of fine particles on the surrounding soils resulting in increased levels of Ra and U uptake in crops.

Therefore use of PG for agricultural, construction and other purposes has been prohibited in most countries, and PG stacks have to be reclaimed in an environmentally acceptable manner. The application of a vegetative cover seems a viable solution for minimising most of the abovementioned risks bearing in mind that the following undesirable characteristics of PG will be altered:

- residual acidity (mainly due to phosphoric and sulphuric acids)
- nutrient deficiency and low nutrient holding capacity
- tendency for caking and crust formation

Previous studies on vegetative covers for PG dumps reclamation (8,9) have indicated that before applying vegetation, acidity should be reduced to acceptable levels and nutrients should be added in sufficient quantities to support vegetation.

An environmental characterisation study performed recently (5) has shown that abandoned Navodari PG stacks are characterised by:

- pH values ranging between 4.9-5.5. The higher values are reported in the upper parts of the dumps due to leaching with rainwater. It has to be mentioned that paste pH of PG in the operating discharge facility varies between 4.05 and 5.0.
- elevated concentration of sulphur (17.7%) and radionuclides (Ra-226: 475 Bq/kg).
- solubilisation and transportation of toxic elements and radionuclides by several physicochemical and mobilisation mechanisms, favoured by local weather conditions.
- high soluble or readily available fractions of some heavy metals, as measured by sequential extraction (in mg/kg): Fe:424, Al: 64, Cu: 25.8 and Zn: 8.7.
- isolation of a great number of microbial communities involved in solubilisation and biosorption processes.
- low TCLP toxicity.

A risk assessment study performed on a source-pathway-target basis (4) has shown that the major risks emanating from PG stacks for the area of Navodari are:

- inhalation of dust by residents as well as deposition of dust on agricultural lands.
- solubilisation of heavy metals from PG stacks causing contamination of surface streams, surrounding soils and groundwater abstraction for potable supply.

Dry ash analysis of plants spontaneously grown on the slopes of older stacks has shown elevated concentration of heavy metals: Pb: 500-5,000 ppm, Zn:1,400-3,000 ppm, Cu: 700-3,000 ppm and Ra-226: 50-370 Bq/kg.

Additional environmental risks for the area result from the discharge of decant waters, containing high levels of cadmium, sulphates, phosphates and thorium, from the operating discharge facility to the Black Sea, and the solubilisation of heavy elements from the cinders dump located in the vicinity of the PG stacks. The environmental impact resulting from these sources of pollution is beyond the scope of this work.

Based on environmental characterisation and risk analysis data, the application of a vegetative cover on PG dumps has been considered as a viable and cost effective remediation scheme. The main advantages of such cover for mitigating environmental risks are (5):

- isolation of PG from the environment and elimination of wind erosion and dusting
- prevention of oxygen diffusion into the tailings mass
- temporary storage of rain water and subsequent return to the atmosphere via evaportanspiration
- inhibition of acidic water generation
- minimisation of radon emmissions in air, and plant uptake in toxic elements
- potential use of PG dumps as a habitat
- improvement of aesthetics

Prior to establishing an *in situ* vegetative cover glasshouse tests were conducted in pots for a period of 12 months. In these tests, tailings were modified so that they exhibited the proper characteristics for maintaining plant growth such as optimum pH, optimum moisture and nutrient content and an adequate bacterial population (5). The results obtained from this study indicated that a number of perennial species and bushes show sufficient tolerance of the growth conditions and therefore can be used for the establishment of a vegetative cover. The species with the best performance were used for the development of the *in situ* pilot scale cover.

The field application was conducted in an area of 200 m^2, over a period of seven months, on a slope of an older PG dump. Modification of the tailings surface included the addition of several amendments such as dolomite, kaolin, sewage sludge and clean soil.

Addition of dolomite is common practice for reclaiming acidic mine spoils. Dolomite provides nutrients such as calcium and magnesium in the substrate and can neutralize any potential acidity produced. Kaolin was added in order to provide nutrients to the system and to increase PG water holding capacity. Sewage sludge is generally characterised by a near neutral pH and with its organic content assists in the establishment of a soil microbial community, thus enabling the development of the vegetative cover. Soil was mainly used for the development of an overburden cap and as a source of nutrients.

The objective of this paper is to examine the possiblity of applying a vegetative cover, which is considered a low cost and easy to maintain system, for the *in situ* remediation of PG stacks. Critical parameters such as modified tailings pH, availability of nutrients, soil microbial activity and intensity of plant growth were recorded on a monthly basis. Data derived after a seven month period, as presented in the following sections, indicate that the development of a vegetative cover on Navodari PG stacks seems a feasible rehabilitation scheme which will minimise most of the environmental risks for the area.

2. EXPERIMENTAL PROCEDURE

Two series of field tests were carried out for a period of seven months over an area of 200 m^2 on a slope of the dump. PG tailings were mixed with amendments such as dolomite, $CaMg(CO_3)_2$ (D), kaolin $Al_2O_3.2SiO_2.2H_2O$ (K), sewage sludge (SS) and clean soil (CS) in various configurations, so that proper soil pH and nutrients content conditions were established *in situ*.

In each series PG and amendments were mixed in plots (1x2x0.25m) as follows:

P1 : PG (no amendments)

P2 : PG + D (2 g/kg) + K (15 g/kg) + SS (30 g/kg)

P3 : PG + SS (30 g/kg)

P4 : PG + CS (25 cm overburden cap)

Series 1

The perennial herbaceous species *Agropyron (Elymus) repens (Salt couch grass), Artemisia absintium (Salt warmwood), Hierochloe repens (Holy grass), Bromus inermis (Smooth brome grass), Cynodon dactylon (Bermuda grass), Cardaria draba (Hoary grass) and Galium humifusum (Bedstraw)* were planted in each plot. The initial height of the perennial species varied between 1 and 5 cm.

Series 2

The bushes *Acer negundo (Box elder), Ailanthus altissima (Tree of heaven), Robinia pseudacacia (Acacia), Hippophae rhamnoides (Sea buckthorn), Populus alba and Populus nigra (Simon's popular) and Eleagnus angustifolia (Oleaster)* were also planted in each plot. The initial height of the bushes planted, having no branches, varied between 50 and 150 cm.

The contents of each plot were moistened using deionized water to which fertilizers were applied as solutions. N was applied as NH_4NO_3 (50 mg/kg of substrate), K as K_2SO_4 (20 mg/kg) and Mg as $MgSO_4$ (12 mg/kg).

Prior to planting as well as at several predefined time intervals after planting, representative samples (mixtures of tailings and amendments) from each plot were collected and analysed for pH and electrical conductivity (EC) using a 2:1 water:soil ratio. Melich I solution was used to determine the extractable elements in the substrates (2). The amendments used were analyzed for HCl-extractable elements before application (10).

Elemental analysis for PG tailings and for all amendments used was performed with digestion techniques and measurement of ion concentration in solution by Atomic Absorption Spectrophotometry. Plant aerial biomass was harvested, dried, digested and analyzed for plant uptake in heavy metals and toxic elements.

The soil microbial activity, which is an indicator of soil fertility and subsequently plant growth, was determined in soil samples by recording dehydrogenase activity (1,3).

3. RESULTS AND DISCUSSION

3.1. Substrate characteristics

Prior to conducting the field tests, a complete chemical analysis of the tailings and amendments used was carried out. These findings (in mg/kg) are seen below:

PG tailings, $S(SO_4)$: 177,300; Ca:175,500; Mg: 77.5; Fe: 2,350; Cu:55; As:50; Zn:75; Al:22, Pb:34.5, Cd:4.5; P_2O_5: 22.9; F:1100; Ra-226:475 Bq/kg, Pb-210:442 Bq/kg; Th-232: 12.4 Bq/kg; Th-234: 53.4 Bq/kg.
Dolomite, Ca: 160,000; Mg: 75,000; Fe: 590; Cu: 59; Na: 287; K: 220
Kaolin, Ca: 33,029; Mg: 9,663; Fe: 5,465; Na: 1,907; K: 10,277; Al:104,752
Sewage Sludge, Ca: 24,335; Mg: 1,508; Fe: 3,406; Na: 682; K: 1,095; Al:1,310; Cu:531

Table 1 presents paste pH, electrical conductivity and extractable elements for raw and amended tailings.

Table 1
Paste pH, electrical conductivity (mS/cm) and extractable elements for raw and amended tailings (mg/kg)

Substrate	Paste pH	EC	Mn	Zn	Cu	Ca	Mg	Fe	K
PG	4.9	1.9	1.1	3.0	1.6	1530	18	10.8	4.9
PG + D (2 g/kg) + K (15 g/kg) + SS (30 g/kg)	6.5	1.3	3.1	12.3	3.9	1960	55	32.0	28.3
PG + SS (30 g/kg)	5.7	1.5	1.3	4.4	3.1	1840	36	23.3	19.5
Soil cap	7.2	0.5	0.2	0.4	4.0	1320	94	46.7	34.3

n.d: not detected

Analysis of these data shows that:
- PG tailings have a pH of 4.9 and contain elevated concentrations of sulphates, cadmium, fluorine and Ra-226.
- The modified substrates contain, to a greater or lesser extent, sufficient amounts of extractable elements. Soil supplies sufficient amounts of Mg and K.
- Addition of amendments has a beneficial effect in raising substrate pH. The highest pH value recorded was 6.5 when dolomite, kaolin and sewage sludge were mixed with PG. The pH of the soil was 7.2.
- Electrical conductivity values for all substrates decreased to a greater or lesser extent when amendments were added. As stated in the literature, decreased EC values along with a favourable pH(>4.5) contribute to germination and growth of plants (8).

It is therefore concluded that modification of substrates with the addition of amendments results in the establishment of an environment which favours the development of a vegetative cover. The critical parameters related to plant growth are soil pH, electrical conductivity, content in micronutrients and accumulation of nutrients and toxic elements in plants.

3.2. Plant growth characteristics

Prior to planting the selected perennial species were analyzed for their content of several elements and micronutrients (Table 2).

Table 2
Concentration of elements and nutrients in plants prior to planting

Species	Mn (mg/kg)	Zn (mg/kg)	Cu (g/kg)	Ca (g/kg)	Mg (g/kg)	Fe (g/kg)	Na (g/kg)	K (g/kg)	Cd (mg/kg)	Pb (mg/kg)
Cynodon dactylon	19.9	76.6	8.9	0.7	0.5	0.1	0.7	6.4	n.d	n.d
Galium humifusum	139.4	905.9	771.5	14.9	3.1	35.9	1.1	13.4	5.97	n.d
Agropyron repens	64.5	135.1	110.6	4.9	0.9	4.7	0.9	4.0	n.d	74.4
Bromus inermis	84.6	87.6	16.7	7.9	1.6	2.2	0.8	23.4	n.d	n.d
Hierocloe repens	49.7	84.0	11.4	3.8	1.2	0.4	0.6	20.9	n.d	n.d
Cardaria draba	54.7	164.5	37.8	9.9	4.8	3.1	2.1	25.3	2.5	n.d
Artemisia absinthium	82.2	96.2	41.1	3.8	2.1	0.7	0.6	12.7	2.1	n.d

n.d: not detected

It is seen from Table 2 that species *Galium humifusum, Agropyron repens* and *Cardaria draba*, which grew sparsely on older PG dumps, show a higher plant uptake in toxic elements and heavy metals (Mn, Zn, Cu, Cd) compared to other plants grown on natural clean soil. These species are expected to exhibit higher tolerance for growth since they are adapted to the local conditions.

As was mentioned previously, plant growth intensity, as measured by recording the height of the planted species, the number of branches and leaves, as well as plant uptake in heavy metals and nutrients were recorded at specified time intervals. At the same periods, all other characteristics of the substrates, including paste pH, electrical conductivity and remaining amounts of extractable elements were also recorded. These data, obtained after a test period of seven months, are presented in Tables 3, 4 and 5. The height of the plants in the *in situ* tests as recorded after 7 months is presented in table 6. In this table literature data quoting the average height of the same species grown in clean soil are also given. Similar data showing the growth characteristics of the bushes (height of the plant, number of new branches and average length of the branches) are shown in table 7.

Table 3
Paste pH and electrical conductivity values in substrates after a period of 7 months

Plot	Substrate	Paste pH		EC (mS/cm)	
		Before planting	After 7 months	Before planting	After 7 months
P 1	PG	4.9	4.5	1.9	1.5
P 2	PG + D (2 g/kg) + K (15 g/kg) + SS (30 g/kg)	6.5	5.9	1.3	1.1
P 3	PG + SS (30 g/kg)	5.7	5.1	1.5	1.2
P 4	Soil cap (25 cm)	7.2	7.0	0.5	0.4

Table 4
Extractable elements in substrates (mg/kg) after a period of 7 months

Plot	Substrate	Mn	Zn	Cu	Ca	Mg	Fe	K
P 1	PG	0.5	2.9	1.4	1340	12	6.5	3.4
P 2	PG + D (2 g/kg) + K (15 g/kg) + SS (30 g/kg)	3.0	5.5	3.8	1620	50	6.9	5.2
P 3	PG + SS (30 g/kg)	1.3	4.0	3.0	1340	20	6.6	2.0
P 4	Soil cap (25 cm)	0.1	0.2	0.1	1280	70	0.3	18

Table 5
Plant uptake after a period of 7 months (mg/kg)

Plot	Plant	Mn	Zn	Cu	Cd	P	Fe	F
P 3/P 4	Cynodon dactylon	27/59	75/59	12/15	4/5	7/17	469/560	3100/3260
P 3/P 4	Artemisia absinthium	111/125	152/160	26/29	12/13	27/25	275/380	6200/6280
P 3/P 4	Hierochloe repens	79/94	76/91	20/33	4/3.5	6/10	750/3320	3650/4000
P 3/P 4	Elymus repens	64/114	59/83	27/31	4/3.5	10/11	650/1500	4100/4220
P 3/P 4	Agropyron repens	40/44	123/64	15/30	5/5	5/7	965/1200	6400/6450

Table 6
Comparison of plant height for plants grown in the field and in clean soil (in cm)

Plot	Cynodon dactylon	Galium humifusum	Agropyron repens	Agropyron repens (from stack)	Bromus inermis	Hierochloe repens	Artemisia absinthium
P 1	32	30	21	42	32	85	85
P 2	42	45	45	67	70	40	98
P 3	37	32	37	64	86	86	95
P 4	40	30	41	75	47	75	90
Literature	10-50	10-50	20-150	20-150	30-150	20-70	20-90

Table 7
Growth characteristics for bushes (height, number of branches, length of branches), in cm

Plot	Acer negundo	Ailanthus Altissima	Robinia pseudacacia	Hippophae rhamnoides	Eleagnus angustifolia
P 1	100-11-12	167-2-13	41-7-6	120-11-9	48-14-6
P 2	88-15-15	175-2-25	78-13-18		68-9-15
P 3	60-7-20	165-2-10	115-10-30		80-27-34
P 4	57-4-33	147-1-40	100-6-23	74-7-15	

From all these data, the following conclusions concerning the development of the vegetative cover can be extracted:

- Perennial species *Agropyron repens* and *Galium humifusum* grown already sparsely on PG as well as *Cynodon dactylon* and *Artemisia absinthium* from Series 1 tests show the best growth intensity in almost all different substrates. Optimum substrates are those resulting from the addition of sewage sludge and the application of a soil cap. Sufficient growth was also recorded when these species were planted directly on PG. This was probably due to the initial presence of sufficient amounts of nutrients in PG stacks, but in this case growth is not expected to last for a long period, because nutrients are gradually washed away with surface run off and infiltrating rain water.

- The beneficial effect of sewage sludge addition on plant growth is due to the enrichment of the substrate with organic material that enables the establishment of an active soil microbial community. The establishment of microbial communities in a soil system is considered as a critical parameter for the development of a self-sustaining vegetative cover on reclaimed mining areas.

- Bushes *Acer negundo*, *Ailanthus altissima*, *Robinia pseudacacia* and *Eleagnus angustifolia* were the species showing the best growth intensity when planted in substrates formed either by mixing PG with sewage sludge or by the application of a soil cap. Very limited growth was observed when bushes were planted directly on PG, with the exception of *Acer negundo* and *Ailanthus altissima*, which exhibited a noticeably higher growth potential.

- The height of both perennial species and bushes grown *in situ* can be considered as promising and can be compared with literature data quoting the height of these species when grown in clean soil. In some cases recorded heights were higher than those quoted in literature. Treatment of sewage sludge may be required in some cases, in order to avoid the development of wild herbaceous species competing with the planted species and therefore hindering their growth.

- Substrate pH and electrical conductivity was maintained in all plots at levels suitable for plant growth. Soil pH values had decreased slightly after a period of seven months, but this decrease never exceeded 0.6 units and in all cases soil pH was maintained at values higher than 4.5.

- Sufficient amounts of extractable elements, as calculated after treatment with Mehlich I solution, remained in the substrates after a period of 7 months. These remaining amounts of micronutrients indicate that plant growth will not be hindered in the following months.

- Preliminary measurements of plant uptake in toxic elements indicate extremely high accumulation of fluorine in perennial plants. Heavy metals accumulated in relatively low levels. It has to be determined however to which extent all these toxic and heavy elements accumulate in the various parts of the plants after an additional period of 12 months and how this concentration affects plant growth.

- Finally, measurements of the microbial activity of the substrates, as measured by the quantity of formazan produced (300-1200 µg/g of substrate), indicate that most of the substrates used provide sufficient soil fertility.

The evaluation of all these experimental data concerning the field application leads to the conclusion that the conditions created *in situ* favour the establishment of a vegetative cover on

PG tailings. A continuation of these tests for an additional period of 12 months should provide useful information concerning suitability of the substrates and tolerance of the plants for growth in such environments.

3.3. Cost of proposed rehabilitation scheme

For the estimation of the cost required of the selected rehabilitation scheme, the prices of raw materials purchased from nearby sites, including processing and transport, were taken into account. These prices given in ECU (based on the exchange rate of Romanian Lei to ECU in September 1998, 1ECU=9700 Lei) are for dolomite 15 ECU/t, for kaolin 30 ECU/t, for sewage sludge 10 ECU/t and for clean soil 15 ECU/t.

The cost for rehabilitating of an area of 1000 m^2 with all options examined is presented in the following table 8.

Schemes P2, P3 and P4 seem more feasible for the establishment of a vegetative cover since direct planting on PG must be excluded for reasons mentioned previously. The application of a soil cap offers additional advantages, such as development of a more suitable substrate for plant growth, increase of water holding capacity, minimisation of the potentially produced acidity and elimination of dusting. A final decision will be made after an additional period of *in situ* testing and will depend on the success of each proposed scheme and availability of funds.

Table 8
Cost for proposed rehabilitation schemes (in ECU/1000 m^2)

	P 1	P 2	P 3	P 4
Purchase of amendments	0	240	115	5625
Modification of PG surface	0	100	100	100
Purchase of plants / planting	200	200	200	200
Maintenance costs	500	500	500	500
Total	700	1040	915	6425

4. CONCLUSIONS

The application of a vegetative cover on PG dumps in Navodari, Romania seems a feasible rehabilitation scheme in order to eliminate future environmental risks in the area associated with solubilisation of toxic elements and radionuclides, emission of radon and dusting.

In order to study the feasibility of this scheme, based on previous glasshouse studies experiments, pilot scale field tests were carried out over an area of 200 m^2, testing a number of perennial herbaceous species and bushes. In order to assist plant growth by creating suitable substrates, PG tailings were modified with the addition of several amendments such as dolomite, kaolin, sewage sludge and clean soil.

The results obtained after an initial trial period of 7 months indicate that several perennial species, namely *Agropyron repens (Salt couch grass), Artemisia absinthium (Salt warmwood), Galium humifusum (Bedstraw)* and *Cynodon dactylon (Bermuda grass),* and bushes, namely *Acer negundo (Box elder), Ailanthus altissima (Tree of heaven), Robinia pseudacacia (Acacia)* and *Eleagnus angustifolia (Oleaster)* show an excellent growth potential. This potential was defined by the growth of the planted species, taking into consideration the height of the species, the number and the length of the new branches and the toxic element uptake.

Optimum growth characteristics were recorded in substrates established by mixing PG with sewage sludge as well as when a soil cap of 25 cm was applied. The application of these amendments resulted in the addition of micronutrients, the increase of water holding capacity and the addition of organic matter in PG, parameters which are considered important for the development and maintenance of a vegetative cover.

The final decision for the selection of the optimum rehabilitation scheme will be taken after an additional trial period and will depend on the ability of the scheme to minimise potential environmental risks and on availability of funds.

ACKNOWLEDGMENTS

The authors would like to acknowledge the financial support of the European Commission under the Copernicus project entitled "Marine pollution in the Black Sea due to mining activities: risk assessment, development of preventive and remedial actions", Contract No: ERB-IC15-CT96-0114.

REFERENCES

1. L.E.jr. Casida, D.A. Klein and T. Santoro, Soil Sci., 98 (1964) 371.
2. E.A. Hanlon and J.M. De Vore, FL Coop. Ext. Serv., Circ. 812, Univ. FL, Gainesville, (1989).
3. J.A. Harris and P. Birch, Proceedings of the 1990 Mining and Reclamation Conference and Exhibition, Charleston, West Virginia (1990).
4. A. Kontopoulos, K. Komnitsas et al., Proceedings EC 1st European Thematic Network EUROTHEN Workshop, A. Kontopoulos and G. Katalagariannakis (eds.), Athens, Greece, (1998) 221.
5. K. Komnitsas, I. Lazar and I. Petrisor, Minerals Engineering, 2 (1999).
6. A. Kontopoulos, K. Komnitsas and A. Xenidis, Proceedings of the III International Conference on Clean Technologies for the Mining Industry, M.A. Sanchez, F. Vegara and S.H. Castro (eds.), Concepcion-Chile (1996) 391.
7. A. May and J.W. Sweeney, Proceedings of the International Symposium on Phosphogypsum, D.P. Morris and P.W. Moody (eds.), Bartow, Florida, (1980) 481.
8. S.K. Patel, J.B. Sartain and S.G. Richardson, Proceedings of the International Land Reclamation and Mine Drainage Conference and the Third International Conference on the Abatement of Acidic Drainage, Pittsburgh, PA, (1994) 139.
9. S.G. Richardson, Proceedings of the International Land Reclamation and Mine Drainage Conference and the Third International Conference on the Abatement of Acidic Drainage, Pittsburgh, PA, (1994) 184.
10. D.A. Whitney, North Dakota Agric. Expt. Stn. Bull., 499 (1988) 20.

Selection of remedial actions in tailings disposal sites based on risk assessment studies. Two case studies

C. Hallett[a], M. Cambridge[a], K. Komnitsas[b]

[a] Knight Piésold Ltd., Kanthack House, Station Road, Ashford, TN23 1PP, Kent, UK

[b] National Technical University of Athens, Laboratory of Metallurgy, GR 157 80, Athens, GR

Intensive mining and ore processing activities concerning phosphate and polymetallic sulphidic ores and concentrates have generated throughout Europe millions of tonnes of hazardous wastes which contain high residual concentrations of heavy elements and radionuclides in mobile forms. Under the action of several physicochemical mechanisms, toxic and radioactive elements contained in the wastes are mobilised, migrate to the surroundings and cause severe and widespread contamination of soils, surface and ground waters. In order to assess the risk posed by each source of pollution at each affected area and to select a viable remedial action, a risk assessment study based on a source-pathway-target basis is considered necessary. This study utilises all available data concerning wastes, probable transfer routes and target groups affected and defines the magnitude of risk for each case. In this paper, two areas of Eastern-Europe, Vromos Bay in Bulgaria and Navodari in Romania are used as case studies. At Vromos Bay, several million tonnes of flotation tailings containing heavy and radioactive elements have been disposed at a tailings dam and by the Black Sea coast. At Navodari, several million tonnes of phosphogypsum tailings and pyritic cinders also containing heavy and radioactive elements have been disposed at tailing dumps. Besides, in the latter case, decant waters often without neutralisation are discharged with a pipeline to the Black Sea. For both areas, and in order to define a viable remediation scheme, a risk assessment study was undertaken, based on a complete environmental characterisation of the pollution sources. Finally, the rehabilitation schemes being developed are briefly presented and discussed.

1. INTRODUCTION

Intensive mining and ore processing activities over the last fifty years, in coastal areas in Bulgaria and Romania, have resulted in the production of millions of tonnes of mining wastes and tailings that are characterised as radioactive, toxic and hazardous. These wastes, which have been deposited in coastal areas often without pre-treatment, do not comply with the environmental standards for safe deposition. This has resulted in widespread contamination of the neighbouring soils and the Black Sea itself due to a number of mobilisation and transportation mechanisms.

The main sources of pollution that are currently affecting the sites of Vromos Bay in Bulgaria and Navodari close to Constanta in Romania were identified and characterised. The purpose of this paper is to outline how the results of this environmental characterisation were used to undertake a risk analysis in order to assess the level of risk posed by each source for each case study area. Rehabilitation schemes were then examined for the two sites, aiming at the development of techniques to deactivate the pollution sources and rehabilitate the contaminated areas with preventive and remedial actions.

Vromos Bay is located approximately 12 kilometres E-SE of Burgas city, in Bulgaria, and forms a secondary bay within the larger Burgas Bay at the western shore of the Black Sea. The beach, with a maximum width of 200 metres, extends over approximately 3 kilometres in E-W direction, covering an area of approximately 200.000 m^2. This area has been contaminated by approximately 8 million tonnes of copper tailings from the nearby Rossen flotation plant between 1954 and 1977. Since that period, tailings were discharged in a constructed tailings dam located close to the flotation plant at a distance of 2 km from the bay. These tailings contain, besides copper and iron minerals, other heavy minerals containing uranium, radium, thorium and other radionuclides.

Navodari is situated on an isthmus at the mouth of Casimcea valley between Lake Siutghiol and Lake Tasaul close to Constanta, in Romania. The chemical complex which operates at Navodari, consists of two plants: one for the production of sulphuric acid and the other for the production of fertilisers. The distance between the chemical plant and the Black Sea is 3 km. The treatment of raw materials has resulted in the production of millions of tonnes of phosphogypsum and pyritic cinders. Phosphogypsum has been deposited in three tailings dams; two of them were formed during the periods 1960-1975 and 1975-1996, covering an area of approximately 40 ha and are no longer operational, being partially covered with spontaneous vegetation. The third dam is under operation since 1996 covering an area of 10 ha. Pyritic cinders, resulting from the processing of pyrite to sulphuric acid, have been deposited on a single unvegetated dump, with a volume of 975,000 m^3, also in the vicinity of the plant. In addition, decant water from the operating dam, sometimes without neutralisation, is discharged with a pipeline to the Black Sea.

2. ENVIRONMENTAL CHARACTERISATION

The methodology which was applied for the environmental characterisation of the pollution sources, tailings, soils and effluents in the two sites under study was based on: a proper sampling strategy, chemical, mineralogical and radiometric analyses, measurements of the geotechnical characteristics of tailings, determination of the Net Neutralisation Potential and toxicity of tailings and soils, evaluation of the metals speciation, determination of the bioavailable fraction of heavy metals and monitoring of the level and quality of pore water [1].

2.1. Results from Vromos Bay
Analysis of the data obtained from the environmental characterisation of Vromos Bay indicated the following [2]:
• Discharge of tailings into Vromos Bay has resulted in a disturbance of the ecological balance of the littoral zone due to the deposition of sand and silt over an extended area.

- The tailings at the tailings dam contain low levels of residual sulphides. The specific activity of the contained radionuclides (especially U-238 and Ra-226) is considered high. The bioavailable fraction of Pd, Zn and Cu is substantial. Tailings potential for generating acidity and their toxicity according to the Toxicity Characteristics Leaching Procedure (TCLP test) are low. Pore water analysis within the tailings shows minimum dissolution of heavy elements.
- The tailings at the beach show a normal size distribution due to the leaching action of the sea currents and they contain elevated concentrations of residual sulphides and heavy metals. The specific activity of the radionuclides shows elevated values. Due to the sea action, their potential for generating acidity, their toxicity according to the TCLP test and the bioavailable fraction of Pd, Zn and Cu are considered low.
- Soils between the tailings dam and the beach are characterised by concentrations of heavy metals and radionuclides which sometimes exceed the Bulgarian standards set for agricultural use of soils (up to 4 times higher than natural background values).
- The two surface streams which transfer mainly all leachates from the tailings dam to Vromos bay are characterised by neutral-alkaline pH values, low concentration of heavy metals, high concentration of anions (chlorides and sulphates) and radionuclides and elevated concentrations of dissolved O_2 and total dissolved solids.
- The quality of the water in the natural wetland existing close to the bay and receiving part of the water of the streams is better and this is mainly attributed to the physicochemical reactions occurring in a wetland which aid in water purification.

2.2. Results from Navodari

Analysis of the data obtained from the environmental characterisation of Navodari indicated the following [3]:

- Phosphogypsum tailings are characterised by low-neutral pH values and contain elevated concentrations of sulphur, lead, cadmium and radionuclides. The main concern from the radioactivity associated to Ra-226 is the emanation of radon-222 gas. Although their toxicity according to the TCLP test and their bioavailable fractions are considered low, solubilisation of heavy elements and radionuclides is favoured by local weather conditions. In addition, a great number of microbial communities involved in metals solubilisation and biosorption processes was isolated from phosphogypsum and cinders dumps and from decant waters.
- Due to the lack of a vegetative cover and the particle size of the phosphogypsum tailings aerial transportation and deposition of fine particles occurs on surrounding soils. Dry ash analysis of plants grown spontaneously on the dumps showed elevated concentration of heavy metals (ppm): Pb:500-5000, Zn:1400-3000, Cu:700-3000 and Ra-226:50-370 Bq/kg.
- Solubilisation of heavy elements occurs also from the cinders dump and causes pollution of the surrounding soils.
- Decant waters which are very often discharged to the Black Sea without any neutralisation are characterised by acidic-neutral pH values and contain elevated concentrations of cadmium, fluorine, sulphates, phosphates and Th-234.

3. RISK ASSESSMENT

A risk assessment was performed [2] to assess the environmental risks associated with mining and chemical wastes at each case study site in Bulgaria and Romania. The aim of the risk assessment was to identify the targets and objectives for research into preventive and remedial methods for waste rehabilitation.

The risk assessment was based on the site characterisation data summarised above. When data was not considered to be sufficiently comprehensive for a full evaluation of the risks, a qualitative and subjective risk assessment was undertaken based on the available information, and where necessary using stated assumptions. It was also intended that the risk assessment exercise be used to identify where important gaps in environmental characterisation exist, to enable updating and review as additional data became available.

The approach to risk assessment [2] was based on U.K. guidelines using the terminology and methodology outlined in the following Tables 1 and 2.

Table 1
Risk Assessment Terminology

Term	Definition
Hazard	A property or situation that in particular circumstances could lead to harm.
Consequences	The adverse effects of harm as a result of realising the hazard which causes the quality of human health or the environment to be impaired in the short or longer term.
Probability of consequences	The assessment of the probability that an identified hazard will reach a target in sufficient concentration (or leading to sufficient exposure) to cause harm.
Magnitude of consequences	The prediction of the significance of an impact in terms of its magnitude should the hazard reach the target.
Risk	A combination of the probability, or frequency of occurrence of a defined hazard and the magnitude of the consequences of the occurrence.
Source	A contaminant or potential pollutant which constitutes a hazard.
Receptor (target)	A human, living organism, ecological system or property (including crops or livestock) that may potentially be affected by a contaminant. The assessment of risk to employees (occupational risk assessment) has been excluded from the assessment.
Pathway (pollutant linkage)	Routes or means by or through which the receptor is being or could be exposed to or affected by a contaminant, such as wind, groundwater and surface water.
Harm	Harm includes disease, serious injury or death to humans; irreversible or other adverse change in the functioning of an ecological system; substantial damage to or failure of buildings, plant or equipment; or disease, physical damage or death of livestock or crops.

Hazards, or potential sources of contamination which may lead to some form of harm on the local environment were identified and evaluated. This was carried out by establishing whether a plausible source-pathway-target relationship existed, such that targets could be exposed to or affected by a hazard.

For each plausible pollutant pathway identified, a risk assessment was undertaken. The risk was evaluated by assessing the probability of the hazard reaching the target as well as assessing the significance of the impact should the hazard reach the target.

The probability that a contaminant will reach a target in sufficient concentration (or leading to sufficient exposure) to cause harm was assessed according to a qualitative scale [2] as: *high* (certain or near certain to occur), *medium* (reasonably likely to occur), *low* (seldom likely to occur) and *negligible* (never likely to occur).

Table 2
Methodology for the Risk Assessment

Stage	Method
Hazard identification and assessment	Identify whether a *plausible source-pathway-target relationship* exists (i.e. the identification of a *hazard* source migrating along a *pathway* and causing potential *harm* to a *target*).
Risk assessment	Once a plausible *pollutant pathway* has been established, the risk is defined by assessing the probability that the hazard will reach the target and the significance of the impact should the hazard reach the target.
Recommendations for further data collection	Recommendations for additional data collection in order to further define the risk.
Risk management	Use the assessed risk to identify appropriate controls.

Similarly the significance of an impact in terms of its magnitude, should a hazard reach a target was assessed as [2]: *severe* (human fatality, major injury or illness causing long term disability; a significant change in number or eradication of one or more species including beneficial, endangered or key species; irreparable damage to non-living structures), *moderate* (human injury or illness causing short term disability; a significant change in species population densities but not total eradication of a species (no effect on endangered or beneficial species); damage to structures which are present in limited numbers e.g. Grade II listed buildings, *mild* (other human injury or illness, some change in species population densities with no negative effects on ecosystem function; damage to commonplace present day structures which could be repaired) and *negligible* (nuisance rather than harm to humans, no significant changes in any species populations or in any ecosystem functions; very slight damage to structures).

The level of risk associated with each pollutant pathway is then assessed by the combination of this probability of consequence with magnitude of consequence as shown in the following Table 3.

Table 3
Risk Assessment

Probability of consequence	Magnitude of Consequences			
	Severe	Moderate	Mild	Negligible
High	High	High	Medium/Low	Near Zero
Medium	High	Medium	Low	Near Zero
Low	High/Medium	Medium/Low	Low	Near Zero
Negligible	High/Medium/Low	Medium/Low	Low	Near Zero

3.1. Results for Vromos Bay

The risk assessment undertaken for each of the pathways for Vromos Bay is summarised in Tables 4 and 5.

From this analysis it is concluded that the major risks for Vromos Bay are:

- increased radiation levels for humans and livestock through the radioactive decay
- surface drainage which deteriorates surface water quality and affects humans, crops and livestock as well as freshwater ecosystems and the Black Sea
- wind erosion which causes deposition of fine particles containing heavy metals and radionuclides on crops
- solubilisation of heavy metals and radionuclides which affect surface water quality and crops

3.2. Results for Navodari

The risk assessment undertaken for each of the pathways for Navodari is summarised in Table 6.

It is therefore concluded that the major risks for Navodari are:

- inhalation of dust by residents as well as deposition of dust on agricultural lands
- contamination of groundwater abstraction for potable supply and contamination of ecosystems in local streams and the Black Sea
- solubilisation of heavy metals and anions from phosphogypsum stacks and pyritic cinders dump and contamination of surface streams and surrounding soils

4. PROPOSED REHABILITATION SCHEMES

The results of the above risk assessment were used to define the scope of research efforts in both areas for the development of rehabilitation schemes by focusing on high and medium risk pathways as priority areas.

Research efforts focused on appropriate, practical, consistent, cost-effective and technically defensible technologies aiming at protection of human health and the environment.

Table 4
Summary Risk Assessment for the Operational Tailings Facility, Vromos Bay

Source - Pathway – Target				Risk Assessment		
Source	Hazard	Pathway	Target	Probability	Magnitude	Risk
		Wind erosion → increased atmospheric dust levels → inhalation	Humans	Low	Mild	Low
		Wind erosion → increased atmospheric dust levels → deposition on agricultural land	Crops	Medium	Moderate	Medium
Operational	Elevated conc. of	Wind erosion → increased atmospheric dust levels → deposition on agricultural land → bioaccumulation	Humans and livestock	Medium	Mild	Low
Tailings	selected metals and	Seepage → groundwater quality → potable water abstraction	Humans	Medium	Moderate	Medium
Facility	radionuclides	Surface drainage → surface water quality	Freshwater ecosystems	Medium	Moderate	Medium
		Surface drainage → surface water quality → abstraction for crops and livestock	Crops and livestock	Medium	Moderate	Medium
		Radioactive decay → increased radiation levels	Humans	Low	Severe	High/ medium
		Seepage and surface drainage → groundwater and surface water → Black Sea quality	Marine ecosystem	Low	Mild	Low
		Seepage and surface drainage→ groundwater and surface water → Black Sea quality → bathing	Humans	Low	Negligible	Near Zero

4.1. Research at Vromos Bay

For the development of an integrated remediation strategy laboratory research has been focused on techniques aiming to decontaminate tailings and soils and clean up effluents.

In order to study solubilisation of different pollutants (heavy metals and radionuclides) from flotation tailings, leaching tests were conducted in: lysimeters, with the use of effluents from the main stream as leach solutions; mechanically stirred reactors, with the use of chemical acids and; shake flasks, with the use of chemical acids, nutrients and a bacterial inoculum. Preliminary data, although they are promising, indicate that large scale decontamination of tailings and soils with the application of solubilisation techniques may not be considered as a viable remediation technique. The main problems related to these applications are the high costs required for reagents and the disposal of the resulting sludges.

For such large scale remediation the application of a soil cover may prove a better option. This application besides the initial high cost required is characterised by low maintenance costs and eliminates aerial transfer of fine particles and therefore contamination of areas which are mainly used for agricultural activities. Furthermore, anaerobic conditions within the tailings could be maintained, resulting in the retention of uranium in the less mobile tetravalent state.

In order to study the clean up of effluents from the main stream a large scale laboratory passive treatment system, operating under continuous flow conditions and consisting of an anaerobic cell, an aerobic cell and a filter system was constructed and used. Preliminary data indicate the high efficiency of this scheme for the clean up of effluents, in terms of heavy metals, anions and radionuclides.

Table 5
Summary Risk Assessment for the Beach Tailings, Vromos Bay

Source - Pathway - Target				Risk Assessment		
Source	Hazard	Pathway	Target	Probability	Magnitude	Risk
Beach Tailings	Elevated conc. of selected metals and radio-nuclides	Wind erosion → increased atmospheric dust levels → inhalation	Humans	Low	Mild	Low
		Wind erosion → increased atmospheric dust levels → deposition on agricultural land	Crops	Medium	Moderate	Medium
		Wind erosion →increased atmospheric dust levels → deposition on agricultural land → bioaccumulation	Humans and livestock	Medium	Mild	Low
		Radioactive decay → increased radiation levels	Humans	Low	Severe	High/ medium
		Erosion and leaching → Black Sea quality	Marine ecosystem	Medium	Moderate	Medium
		Erosion and leaching → Black Sea quality → bathing	Humans	Medium	Moderate	Medium

4.2. Research at Navodari

The application of a vegetative cover on phosphogypsum dumps has been considered as a viable remediation option in order to isolate tailings from the environment, eliminate wind erosion and dusting, prevent oxygen diffusion and rain water infiltration into the tailings mass, inhibit acidic water generation and finally improve aesthetics in the area.

For the study of the potential establishment of a vegetative cover on PG dumps, glasshouse and field tests have been carried out testing a number of herbaceous species and bushes.

Table 6
Summary Risk Assessment for the Navodari site

Source – Pathway – Target				Risk Assessment		
Source	Hazard	Pathway	Target	Probability	Magnitude	Risk
		Wind erosion→increased atmospheric dust levels→ inhalation	Humans	Medium	Moderate	Medium
		Wind erosion→increased atmospheric dust levels→ deposition on agricultural land	Crops	Medium	Moderate	Medium
		Wind erosion→increased atmospheric dust levels→ deposition on agricultural land→bioaccumulation	Humans	Medium	Mild	Low
		Radioactive decay→ increased radiation levels	Humans	Low	Mild	Low
		Seepage→groundwater quality→portable water abstraction	Humans	Medium	Moderate	Medium
Phospho-gypsum and pyrite cinders	Elevated conc. of selected metals and radionuclides	Surface drainage→surface water quality	Freshwater ecosystems	Medium	Moderate	Medium
		Surface drainage→surface water quality→ bioaccumulation	Humans	Low	Negligible	Near zero
		Surface drainage→surface water quality→abstraction for portable supply	Humans	Low	Negligible	Near zero
		Surface drainage→surface water quality→abstraction for crops and livestock	Crops and livestock	Low	Negligible	Near zero
		Seepage and surface drainage→groundwater and surface water→Black Sea quality	Marine ecosystem	Low	Negligible	Low
		Seepage and surface drainage→groundwater and surface water→Black Sea quality→bathing	Humans	Low	Negligible	Near zero

In order to assist plant growth by creating suitable substrates, PG tailings were modified with the addition of several amendments such as dolomite, bentonite, kaolin, sewage sludge and clean soil. Laboratory and field data indicate that several perennial species and bushes show an excellent growth potential and can be used for the establishment of a vegetative cover.

5. CONCLUSIONS

Risk analysis based on the source-pathway-target principle identified and characterised all pollution sources which directly or indirectly affect humans, soils and freshwater ecosystems in the areas of Vromos Bay, Bulgaria and Navodari, Romania. This study utilised all available data concerning wastes, probable transfer routes and target groups affected and defined the magnitute of risk for each case.

Based on the above findings, a rehabilitation strategy was designed aiming at deactivating the pollution sources and rehabilitating the areas with preventative and remedial actions. Research efforts focused on appropriate, practical, consistent, cost-effective and technically defensible technologies aiming at protection of human health and the environment.

Encouraging experimental data led to the application of *in situ* remediation technologies in a pilot scale.

ACKNOWLEDGMENTS

The authors would like to acknowledge the financial support of the European Commission under the Copernicus project entitled "Marine pollution in the Black Sea due to mining activities: risk assessment, development of preventive and remedial actions", Contract No: ERB-IC15-CT96-0114.

REFERENCES

1. A. Kontopoulos, K. Komnitsas, A. Xenidis and N. Papassiopi, Minerals Engineering, 10 (1995) 1209.
2. A. Kontopoulos, K. Komnitsas et al., Proceedings EC 1st European Thematic Network EUROTHEN Workshop, A. Kontopoulos and G. Katalagariannakis (eds.), Athens, Greece, (1998) 221.
3. K. Komnitsas, A. Kontopoulos, I. Lazar and M. Cambridge, Minerals Engineering, 12 (1998) 1179.

The application of sulphate-reducing bacteria in hydrometallurgy

Alena Luptáková[a] and Mária Kušnierová[a]

[a]The Institute of Geotechnics of the Slovak Academy of Sciences[1], Košice, Slovakia

The occurrence, characteristics, scientific and practical significance of sulphate-reducing bacteria (SRB) are discussed in this article. On the basis of the natural biological activities of SRB, i.e., their isolation and cultivation, and their chemistry of autotrophic and heterotrophic reduction of sulphates, a process utilizing SRB in Cu-removal from solution is described. Some practical applications in hydrometallurgy are also examined.

1. INTRODUCTION

The environment, mainly its basic elements - water and air - are currently contaminated by large amounts of gaseous, liquid and solid wastes each year. This leads to a very negative influence on the life of human beings and on nature as a whole.

Some detrimental substances, to which heavy metals belong, can be taken up in the food chain and accumulated in amounts that later may cause acute or chronic diseases or even death [1]. Heavy metals belong to the environmental pollutants and are mainly of anthropogenic origin [2]. Industrial waste waters are their main source. Many technologies for heavy metals removal from waste water exist. Biological processes, so-called biotechnologies, belong to those in which there has recently been increasing interest as a means to clean up the environment and protect it.

Biotechnology is the controlled utilization of suitable microorganisms or their products for technological purposes [3]. In principle, it involves the harnessing of microbial activities, which have been part of nature since its creation [4]. The sulphate-reducing bacteria (SRB) are microorganisms that can be exploited in this way. They form part of the biosphere, and with their vital metabolic processes, they influence the biological sulphur cycle in nature. Promoting this cycle is the dominant role in their life on the Earth (Figure 1) [5]:

[1] Acknowledgment: The authors are grateful to the Slovak Grant Agency for Science (Grant No. 2610399) for the financial support of this work.

666

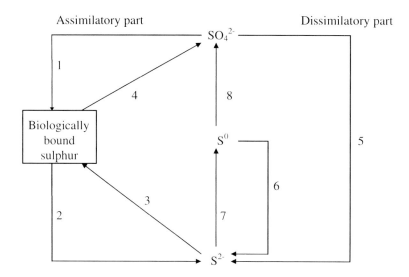

Figure 1. The biological sulphur cycle (modified from TRÜPER, 1984) [5].
1 assimilatory sulphate reduction by plants, fungi and bacteria; 2 death and decomposition by fungi and bacteria; 3 sulphide assimilation by bacteria and some plants; 4 excretion of sulphate by animals; 5 dissimilatory sulphate-reducing bacteria; 6 dissimilatory sulphur-reducing bacteria; 7 phototrophic and chemotrophic sulphide-oxdizing bacteria; 8 phototrophic and chemotrophic sulphur-oxidizing bacteria.

1.1 Occurrence and characteristics of sulphate-reducing bacteria

SRB can by found virtually everywhere: in soil, thermal sulphur springs, mining waters from sulphide deposits, oil deposits, stationary fresh waters, sea and ocean beds, and in the intestines of insects and human beings [6].

The SRB represent a group of chemoorganotrophic, strictly anaerobic and gram-negative bacteria, which exhibit great morphological and physiological diversity. Characteristics of some contemporary SRB genera [based on date from Widdel and Hansen (1991)] are shown in Table 1 [7].

Despite their considerable morphological variety, they have one property in common, which is the ability to utilize sulphates and other partially oxidized forms of sulphur as electron acceptors, which are reduced to sulfide during anaerobic respiration. The electron donors in these processes are organic substrates as lactate, malate, alcohols, etc., or gaseous hydrogen.

The typical species of SRB is *Desulfovibrio desulfuricans* (*D. desulfuricans*), which may reduce sulphate autotrophically with hydrogen gas, as shown in reaction (1) as well as heterotrophically with an organic electron donor like lactate, as shown by reaction (2):

$$4 \, H_2 + SO_4^{2-} \xrightarrow{\text{SRB}} H_2S + 2 \, H_2O + 2 \, OH^- \tag{1}$$

$$\text{2 C}_3\text{H}_5\text{O}_3\text{Na} + \text{MgSO}_4 \xrightarrow{\text{SRB}} \text{2 CH}_3\text{COONa} + \text{CO}_2 + \text{MgCO}_3 + \text{H}_2\text{S} + \text{H}_2\text{O} \qquad (2)$$

SRB can produce a considerable amount of hydrogen sulphide, which reacts easily in aqueous solution with the cations of heavy metals, forming metal sulphides that have low solubility. Reaction (3) summarizes this type of reaction:

$$\text{Me}^{2+} + \text{H}_2\text{S} \xrightarrow{\hspace{2cm}} \text{MeS} + \text{2 H}^+ \qquad (3)$$

in which Me^{2+} is the cation of a particular metal.

Table 1.
Characteristics of some contemporary sulphate-reducing bacterial genera [7]

Genus	Morphology	Cell wall/membrane	Organic compound oxidation	Spore formation	Temperature effects
Desulfovibrio	Vibrio	Gram-, Euba	I	No	M
Desulfotomaculum	Rod	Gram+, Euba	I, Co	No	M, T
Desulfomicrobium	Oval/rod	Gram-, Euba	I	No	M
Desulfobulbus	Oval	Gram-, Euba	I	No	M
Desulfobacter	Oval/vibrio	Gram-, Euba	Co	No	M
Desulfobacterium	Oval	Gram-, Euba	Co	No	M
Desulfococcus	Sphere	Gram-, Euba	Co	No	M
Desulfosarcina	Oval (aggregrates)	Gram-, Euba	Co	No	M
Desulfomonile	Rod	Gram-, Euba	Co	No	M
Desulfonema	Multicellular filamentes	Gram-, Euba	Co	No	M
Desulfobotulus	Vibrio	Gram-, Euba	I	No	M
Desulfoarculus	Vibrio	Gram-, Euba	Co	No	M
Thermodesulfobacterium	Rod	Gram-, Euba	I	No	T
Archaeoglobus	Sphere	Archaebacteria	Co	No	T

Reprinted from: F. Widdel and T. A. Hansen, The dissimilatory sulphate- and sulphur-reducing bacteria, Spinger-Verlag, New York, 1991, p. 583.
Notes: Gram+/-, Euba = typical eubacterial Gram-negative or -positive cell wall and membrane; I = incompletely oxidize organic compounds to acetate; Co = completely oxidize organic compounds to CO_2; M = mesophilic strains; T = thermophilic strains.

SRB are promoters of many processes that can be used in industrial technologies, including hydrometallurgy, in which the focus is on removal of cations of heavy metals from solution. Several procedures have been developed up to now. They are based on the natural biological activities of SRB described by reactions (1), (2) and (3). The following examples can by cited:
1. Removal of Fe^{2+} and SO_4^{2-} from waste waters in hydrometallurgic copper production [8].

2. Removal of SO_4^{2-} from mine waters, waters from coal desulphurization and metallurgic plants [9].
3. Removal of Cu^{2+} from waste waters [10].
4. Removal of heavy metals from water and soil [11].

In order to remove heavy metals from waste waters, e.g., from galvanizing plants, mine waters (Smolník locality) and metallurgic plants (e.g., Krompachy) by use of the activity of SRB, mixed strains were isolated, cultivated, and their to produce hydrogen sulphide assessed. The cultures were then tested for ability to precipitate copper from a model solution.

2. MATERIAL AND METHODS

2.1. Isolation
Bacteria were isolated from water samples from three localities:
1. (VSŽ) – water from a waste water collection tank that been used for washing machinery in Východoslovenské železiarne (works), pH 7.5-8, strong H_2S odour, heavily polluted, black, oily;
2. (Gj) – potable mineral water (Gajdovka spring), pH 7-8, spring water, hygienically acceptable, strong odour of H_2S;
3. (J) – potable mineral water (Jánovce spring), pH 8, spring, hygienically acceptable, slight odour of H_2S.

A selective nutrient medium DSM-63 (Desulfovibrio medium, DSM – catalogue 1989) was used for isolation. Agar in the amount of 1.5% was added to his medium in the preparation of solid medium for plating. The isolation was performed by the modified dilution method [12].

2.2. Cultivation
Cultivation was carried out under following conditions:
- Statically
- Incubation temperature 30 °C
- Nutrient medium DSM-63, pH 7.5
- Anaerobic
- Conditions (ANAER-cult, Diagnostic Merck Art. Nr.5 13807 for Petri disches).

2.3. Eliminating Cu from solution
Microbial copper precipitation was studied in a discontinuous reactor in the thermostat at 30 °C and run for a period of 8 days under anaerobic conditions. Anaerobiosis was generted after inoculation, by introducing an inert gas (Ar) into the reactor. The reactor was then hermetically closed. A concentration of Cu of 20 mg L^{-1} was chosen for the test solution because that was the average Cu concentration in mine waters or waste waters from hydro-metallurgical processing of copper concentrates. The solution was prepared with $CuSO_4$. $5H_2O$. Residual copper in solution was measured by atomic absorption photometry (VARIAN spectrometer) in samples taken from the reactor at 12-h intervals.

2.4. Microscopic observations

After heat-fixation of a smear from a culture on a microscope slide, the bacteria were gram stained [12] and then observed under a NIKON ECLYPSE 400 light microscope, using oil-immersion at 1000-fold magnification.

3. RESULTS AND DISCUSSION

SRB were isolated in a selective, liquid nutrient medium from all samples tested. They were recognized by a typical smell of H_2S and by intensive blackness of the nutrient medium after 7 to 9 days, probably caused by the generation of secondary sulphides. The SRB isolates were then purified by the modified plating method, which was repeated several times. Based on morphology as determined from gram stains and other microscopic observations, we identified the following genera:

1. Sample (VSŽ) – mixture of *Desulfovibrio* and *Desulfotomaculum* (Figure 2).
2. Sample (Gj) – mixture of *Desulfovibrio* (Figure 3), *Desulfotomaculum* (Figure 4) and *Desulfosarcina* (Figure 5).
3. Sample (J) – *Desulfovibrio* (Figure 6).

The isolated cultures were then tested for their ability to precipitate Cu at a concentration of 20 mg L^{-1} from solution. The objective of the experiment was to determine the activity of the isolates, their effectiveness and the kinetics of Cu^{2+} precipitation by them. The results are shown in Figure 7. They indicate high activity of the bacteria in that they precipitated 98-99% of the copper in solution.

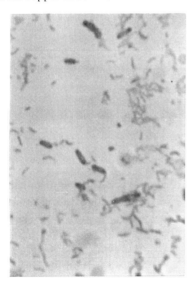 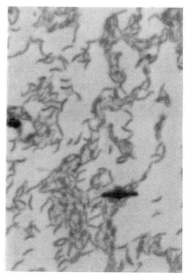

Figure 2. Sample (VSŽ), (1000x) by oil immersion, *Desulfovibrio* - spiral shaped bacteria, *Desulfotomaculum* - rod shaped bacteria.

Figure 3. Sample (Gj), (1000x) by oil immersion, *Desulfovibrio* - spiral shaped bacteria.

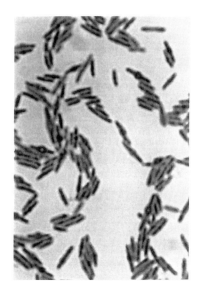

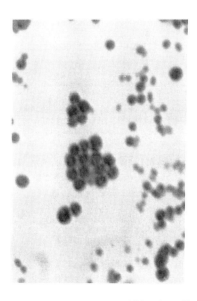

Figure 4. Sample (Gj), (1000x) by oil immersion, *Desulfotomaculum* - rod shaped bacteria.

Figure 5. Sample (Gj), (1000x) by oil immersion, *Desulfosarcina* - oval (aggregrates) bacteria.

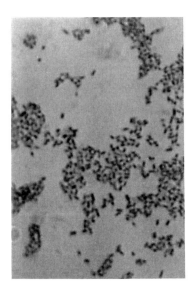

Figure 6. Sample (J), (1000x) by oil immersion, *Desulfovibrio* - short rods.

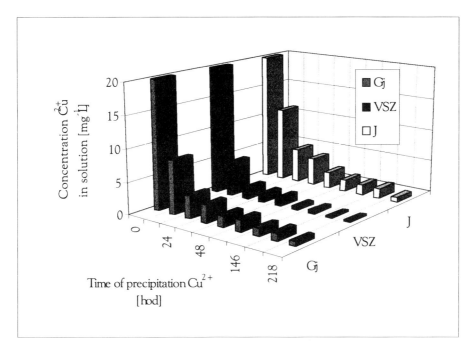

Figure 7. Precipitation Cu^{2+} with sulphate-reducing bacteria.
Gj - mixture of genera *Desulfovibrio, Desulfotomaculum* and *Desulfosarcina* from locality Gajdovka (mineral water); VSŽ - mixture of genera *Desulfovibrio* and *Desulfotomaculum* from Východoslovenské zeleziarne (works, waste water); J - genera *Desulfovibrio* from locality Jánovce (mineral water).

4. CONCLUSION

Our findings from the isolation of SRB from three Slovak water samples and testing the cultures for their ability to remove copper permit the following conclusions:
- SRB occur in sufficient numbers in sulphur mineral water from natural sources and in industrial waste water reservoirs.
- The sulphate-reducing activity can be harnessed for the purification of some industrial waste waters.
- Unmodified SRB from Slovakian sources can be effectively used to remove 98-99% of Cu^{2+} from solution in days of static incubation.

Nature possesses a great biological potential that can be exploited under certain conditions in the cleanup of environmental pollution resulting from industrial activity in the past and present.

672

REFERENCES

1. J. Paleček and J. Palatý (eds.), Toxicology, hygiene and security of work in chemistry, Czech Republic, Praha, 1985.
2. D. Kalavská and I. Holoubek (eds.), Analysis of water, Slovak Republic, Bratislava, 1987.
3. Z. Vodrážka (ed.), Biotechnolgy, Czech Republic, Praha, 1991.
4. J. Postgate (ed.), Microbes and Man, England, Middlesex, 1975.
5. H. J. Rehm and G. Reed (eds.), Biotechnology, vol.6b, Verlag Chemie GmbH, Weiheim, 1981.
6. J. M. Odom and Rivers Singleton Jr. (eds.), The sulphate-reducing bacteria: Contemporary Perspectives, Springer-Verlag, New York, 1993.
7. F. Widdel and T. A. Hansen (eds.), The dissimilatory sulphate- and sulphur- reducing bacteria, Springer-Verlag, New York, 1991.
8. J. H. Tuttle and P. R. Dugan, Appl. Microbiol., 17 (1969), 754.
9. D. J. Cork and M. A. Cusanovich, Develop. Industr. Microbiol., 20 (1979), 1024.
10. K. Imai, in R. W. Lawrence, R. M. R. Branion and H. G. Ebner (eds.), Fundamental and applied biohydrometallurgy, Elsevier, New York (1986), 383.
11. S. N. Groudev and I. I. Spasova in H. Verachtert and W. Verstraete (eds.), International Synposium „Enironmental Biotechnology‚‚ Technologisch Instituut, Oostende, 1997.
12. V. Betina and P. Nemec (eds.), The general microbiology, Slovak republic, Bratislava, 1977.

Biotransformation of oxidised anions by selected bacteria

T. Kasatkina[a], V. Podgorsky[a], S. Mikhalovsky[b], A. Tashireva[a], E.B. Lindström[c], and S. McEldowney[d]

[a]Institute for Microbiology and Virology of the National Academy of Science, Kiev, Ukraine

[b]Department of Pharmacy, University of Brighton, England

[c]Department of Microbiology, University of Umea, Sweden

[d]Department of Biosciences, University of Westminster, London, England[*]

The aim of this study was to investigate the reduction of oxidised metallic (CrO_4^{2-}, VO_3^-, MoO_4^{2-}) and non-metallic anions (ClO_3^-) by selected bacteria. Anaerobic cultures were incubated at 28°C for up to 96 hrs in a glucose minimal medium or meat broth containing 0 to 1000 mg/l oxyanion. Medium pH and Eh were monitored and oxyanion reduction followed throughout culture growth. It was found that a number of aerobic and facultatively anaerobic bacteria e.g. *Bacillus subtilis*, *Pseudomonas aeruginosa*, *Escherichia coli* and *Clostridium lituseburense*, were able to reduce VO_3^-, MoO_4^{2-} at concentrations between 100 and 1000 mg/l. Over a period of 24 hrs growth 70 - 90% of the anions were reduced to V(IV) and Mo(V). The reduction of these oxyanions correlated with bacterial activity. The metals remained in soluble form for the first 18 to 48 hrs of growth but during the next 24 hrs growth insoluble metal precipitates, probably Mo(V) oxide and V(IV) oxide, formed. The reduction of the anions was linked to a decrease in Eh from -120 to -380 mV. The pH of the cultures decreased from 6.8 to 7.5.

The ability of *Aeromonas dechromatica* KC-11 (a newly isolated species) to reduce CrO_4^{2-} and other oxyanions was also studied. The Cr(VI) (initial concentration 35 mg/l) content of the medium began to decline within 1-2 hrs of inoculation and was complete after 20 to 24 hrs. Optimum reduction occurred between pH 6.0 and 7.8 with pH remaining constant throughout the incubation. Outside this pH range Cr(VI) reduction was 1.3-2.5 times slower. The rate of Cr(VI) reduction was higher at low Eh. During culture growth Eh declined with increasing Cr(VI) reduction. *A.dechromatica* KC-11 was also capable of reducing ClO_3^-,

---

[*] This research is supported by INTAS, INTAS-UA 95-116

NO_3^- and SO_4^{2-}. Again the Eh and pH of the medium were the main factors determining the rate of reduction. During the reduction of ClO_3^- and NO_3^- the Eh decreased from +300 to -100 mV while during SO_4^{2-} reduction Eh decreased from 0 to -400 mV. *A.dechromatica* KC-11 reduced chromate, nitrate and chlorate in the first 1-2 hrs of growth. A period of adaptation to SO_4^{2-} appeared necessary since reduction only started after 10-15 hrs growth. The results suggest that the oxyanions are reduced through a dissimilative process with the anions acting as a terminal electron acceptor in anaerobic respiration. It appears that a variety of facultatively anaerobic heterotrophic bacteria are capable of reducing oxyanions such as CrO_4^{2-}, VO_3^-, MoO_4^{2-}, and ClO_3^-, with an efficiency that may make them suitable for utilization in the removal of such anions from waste water.

1 INTRODUCTION

Heavy metals have always been present in the biosphere but their concentrations have frequently been substantially raised by anthropogenic activity. Sewage and industrial plant effluents are major sources of heavy metal pollutants. Industrial effluents often contain large quantities of Cr(VI), Fe(III), Zn(II), Cu(II), Ni(II), NO_3^-, PO_4^{3-}, etc., together with a diverse range of organic compounds.

Cultures of specialised microorganisms have previously been used for the detoxification of a range of different liquid effluents. The isolation of bacteria capable of reductive detoxification of compounds such as CrO_4^{2-}, ClO_3^-, ClO_4^-, VO_3^-, MoO_4^{2-} and WO_4^-, is more likely from environments containing high concentrations of these substances [1-6]. Chromium reducing bacteria which can transform the highly toxic Cr(VI) to the less toxic Cr(III), are of special interest. These bacteria are widely used for detoxification of waste waters. Some of the characteristics of chromium reduction by bacteria has been demonstrated by Romanenko and Korenkov [7]. They isolated and characterised a chromium-reducing bacterium *Bacterium dechromaticans* sub. sp. *romanenko*.

Under anaerobic conditions Cr(VI) reduction occurs by means of membrane reductase and the cytochromes of the respiratory chain. A chromium reductase has been purified from *Pseudomonas ambigulla* [8]. The reduction of Cr(VI) by *Desulfovibrio vulgaris* occurs through the use of hydrogen as an electron donor. In this species cytochrome C_3 is the chromium reductase [9]. In addition to the anaerobic reduction of chromium, aerobic reduction has been shown to occur [10]. A chromium resistant strain of *Pseudomonas ambigulla* A-1 was shown to reduce Cr(VI) under aerobic conditions. The process involved the internal uptake and accumulation of Cr(VI) and its subsequent use as an electron acceptor [11]. Bopp and Ehrlich [12] isolated *Pseudomonas fluorescens* strain 300, which was able to reduce Cr(VI) under aerobic conditions using glucose as a carbon and energy source [12]. Internal uptake of chromium by *Pseudomonas fluorescens* 300 was found to be via the sulphate transport system [13]. The reductive transformation of toxic oxidised metal anions such as CrO_4^{2-}; VO_3^-; MoO_4^{2-}, WO_4^{2-}, and non-metallic anions e.g. ClO_3^-, and ClO_4^-, by bacteria is potentially of great importance. Such biotransformations may occur naturally at contaminated sites possibly providing an *in-situ* treatment technology. Alternatively the reductive transformations of oxidised anions may be used to remove the compounds prior to environmental release of liquid effluent. The ability of selected bacteria, either individually or in consortia, to transform these toxic anions has previously been studied [14]. The present study further examines the range of selected bacteria, including a newly isolated and identified bacterium *Aeromonas dechromatica* KC-11, capable of these biotransformations, the characteristics of the reductions, and optimum conditions for the oxyanion reduction.

2 MATERIALS AND METHODS

2.1 Organisms and growth conditions

Culture collection strains of aerobic (*Bacillus subtilis, Pseudomonas aeruginosa)*, facultative anaerobic (*Escherichia coli*) and obligatory anaerobic (*Clostridium lituseburense)* bacteria were studied. Bacteria were also isolated from samples of sewage originating from a galvanic workshop and a match factory and samples from other polluted sources (natural reservoirs, soils and sludges).

The aerobic and facultative anaerobic strains were isolated, grown and maintained on the following solid media:

- chromium- reducing bacteria were grown in KC medium consisting of (g/l): NH_4Cl - 0,3; K_2HPO_4 - 0,3; KH_2PO_4 - 0,5; $MgSO_4 \cdot 7H_2O$ - 0,1; $NaCl$ - 0,1; $CaCO_3$ - 0,03; $FeCl_3 \cdot 7H_2O$ - 0,05; K_2CrO_4 - 0,2; CH_3COONa - 0,2; and MPB-10 %(w/v).

- chlorate - reducing bacteria were grown in N1 medium consisting of (g/l): NH_4Cl - 0,3; K_2HPO_4 - 0,3; KH_2PO_4 - 0,5; $MgSO_4 \cdot 7H_2O$ - 0,1; $CaCl_2$ - 0,05; $NaCl$ - 0,1; $KClO_3$ - 2; and MPB-10%(w/v).

- sulphate-reducing bacteria were grown in Shturm medium consisting of (g/l): $(NH_4)_2SO_4$ - 4; K_2HPO_4 - 0,5; $CaSO_4 \cdot 2H_2O$ - 0,5; $MgSO_4 \cdot 7H_2O$ - 1; salt of More - 0,5; calcium lactate - 3,5.

- Giltay medium was used for the growth of denitrifying bacteria. The medium consisted of (g/l) KNO_3 - 1; aspargine – 1 in 250 ml distilled water, added to calcium citrate - 8,5; KH_2PO_4 - 1; $MgSO_4 \cdot 7H_2O$ - 1; $CaCl_2 \cdot 6H_2O$ - 0,2; $FeCl_2 \cdot 4H_2O$ – traces in 750 ml distilled water.

- the media used for vanidate and molybdate experiments were MPB-medium or a glucose medium containing mineral nitrogen, phosphorus and sulphur sources of (g/l): NH_4Cl - 1; Na_2HPO_4 - 0,5 ; Na_2SO_4 - 0,5 ; glucose - 10 ; peptone - 0,5. The anion concentrations were in both cases in the range 100-1000 mg/l.

The obligate anaerobic isolates and strains were grown and maintained on a standard nutrient medium (MPB or MPA) or the glucose medium. Anaerobic cultures were prepared using the standard anaerobic technique developed by Hungate [15]. The media were sparged with argon to ensure anaerobic conditions. All cultures were grown at 28°C and maintained at 4°C.

2.2 Determination of growth and transformation of oxyanions

Aerobic isolates and culture collection strains were grown in the appropriate liquid cultures (see above) at 28°C for up to 96 hrs. The oxyanions were added to the culture media when the bacteria were either in lag phase or the mid exponential phase of growth. Their growth and ability to transform the oxyanion was followed. Facultatively and obligately anaerobic bacteria were grown in a culture apparatus designed to allow study of the growth processes, and physiological and biochemical properties of microorganism under strictly anaerobic conditions [16]. It was possible to measure Eh, pH, H_2S, pO_2 etc., and to take sterile samples of the gas and liquid phases from the culture vessels during growth. The cultures were grown in the appropriate liquid medium (see above) at 28°C for up to 96 hrs. The transformation of CrO_4^{2-}, VO_3^-, MoO_4^{2-}, ClO_3^-, NO_3^- and SO_4^{2-} was studied.

2.3 Analytical procedures to determine bacterial growth and removal of anions

Potentiometric measurements of the redox potentials were made by ionometre with a platinum measurement electrode and silver chloride reference electrode. The concentration of anions was determined by the following analytical methods [17,18]: 1) Cr(VI), the dephenylcarbazide method; 2) SO_4^{2-}, the method for sulphate determination in sewage; 3) NO_3^- by nitratometer with an ion-selective electrode; 4) ClO_3^-, the permanganate method; 5) VO_3^- , using anthranilic acid and for MoO_4^{2-}, ammonium rhodamine. The biomass concentration was determined with a nephelometre.

2.4 Characterisation of a CrO_4^{2-} reducing isolate

An isolate found to be particularly efficient at CrO_4^{2-} reduction was identified using the following procedure. The isolate was grown in the chromium medium (see above) and incubated aerobically at 28°C. Samples were taken at different stages of expotential growth (12-36 hours) and stationary growth (36-240 hours) for cytological investigation. The morphology of cells was determined by light microscopy and the fine structure of cells was investigated by electron microscopy. Samples were prepared for electron microscopic examination by fixing and negative contrast staining using 2% (w/v) solution of phosphoric-tungsten acid (pH=6.7-7.1), 1,5% (w/v) OsO_4 and lead citrate following the standard procedure of Kellenberger-Riter [19]. The development of microcolonies from individual cells was followed over 16 hours growth on MPA. A range of biochemical and physiological tests were performed for identification of the isolate, which was carried out with reference to Bergey [20]. Nucleotydic structure of DNA was determined by the melting curves method [21].

3 RESULTS AND DISCUSSION

3.1 Biotransformation of VO_3^- and MoO_4^{2-} by selected isolates and culture collection species

A number of culture collection bacteria were able to reduce and then precipitate VO_3^- and MoO_4^{2-}. These bacteria included aerobes, and faculatively and obligately anaerobic bacteria such as *Bacillus subtilus*, *Pseudomonas aeruginosa*, *Escherichia coli*, and *Clostridium lituseburense*. These species were able to reduce VO_3^- and MoO_4^{2-} at concentrations between 100 and 1000 mg/l. Over a period of 24 hours growth 70-90 % of the anions were reduced to V(IV) and Mo(V) respectively. More reduced forms of vanadium and molybdenum were not produced. The metals remained in soluble form for the first 18 to 48 hrs of growth but during the next 24 hrs growth insoluble metal precipitates, probably Mo(V) oxide and V(IV) oxide, formed gradually. The original culture medium had an Eh of +100 mV which rose to +350 mV on addition of the metallic oxyanions. After the reductive transformation of the oxyanions by the bacteria the Eh decreased to between –120 and –350 mV. The reduction of molybdenum and vanadium correlated with metabolic activity.

Similar results were obtained with non-identified strains and microbial associations isolated from soils and river water and sediments. The same mechanism of reduction appeared to occur for culture collection species, environmental isolates and consortia. It seems that the growth related reduction of these oxyanions is wide spread among bacteria and consists of two stages:

1. The reduction of V(V) to V(IV) and Mo(VI) to Mo(V);
2. The formation of insoluble compounds followed by their precipitation.

3.2 Biotransformation of CrO_4^{2-} by *Aeromonas dechromatica* KC-11

A variety of bacterial strains isolated from CrO_4^{2-} polluted sites e.g. ships sewage, industrial effluent and sediment, were able to reduce this oxyanion. The isolates included a range of commonly encountered bacterial species including *Escherichia coli*, *Enterobacter aerogenes*, *Pseudomonas putida*. One isolate was found to be particularly efficient at reducing CrO_4^{2-}. This isolate was a previously unidentified species of the genus *Aeromonas*, and was designated *Aeromonas dechromatica* KC-11 [22]. The prevalence of chromium-reducing bacteria varied with sampling site. No chromium-reducing bacteria appear to be present in water and soil samples from sources not polluted by chromium compounds [23].

The ability of *Aeromonas dechromatica* KC-11 to reduce CrO_4^{2-} and other oxyanions was in particular studied and the mechanisms involved appear to be representative of the other CrO_4^{2-}-reducing bacteria. CrO_4^{2-} reduction by these bacteria were achieved by oxidation of organic substrates linked to dissimilative reduction of CrO_4^{2-} and was therefore growth related. *A. dechromatica* KC-11 was also able to reduce ClO_3^-, NO_3^- and SO_4^{2-}. The redox potential and pH of the medium were the main factors determining the rate of reduction. These factors controlled the phisiological state of the population in the anaerobic process. During CrO_4^{2-}, ClO_3^- and NO_3^- reduction by *A. dechromatica* KC-11 the Eh decreased from 0 to -100 mV (Figs 1-3) while during SO_4^{2-} reduction the Eh decreased from 0 to -450 mV (Fig. 4). *A. dechromatica* KC-11 used nitrate, chromate and chlorate in the first hours of growth. Sulphate, however, was reduced after 10-15 hours of growth suggesting a period of adaptation by the cells. In all cases the pH was 6.8-7.2 and changed little during the reduction. The decline of Eh during growth of the cultures was linked to increasing rates of Cr(VI) reduction (Figs 1-4).

The results suggest that the oxyanions are reduced through a dissimilative process with the anions acting as a terminal electron acceptor in anaerobic respiration. It appears that a variety of facultatively anaerobic heterotrophic bacteria are capable of reducing oxyanions such as CrO_4^{2-}, VO_3^-, MoO_4^{2-}, and ClO_3^- with an efficiency that may make them suitable for utilization in the removal of such anions from waste water. It is interesting to speculate on the role and importance of oxyanion-reducing bacteria *in-situ*. It may be the case that they contribute significantly to any biogeochemical cycling of oxyanions which occurs in polluted habitats.

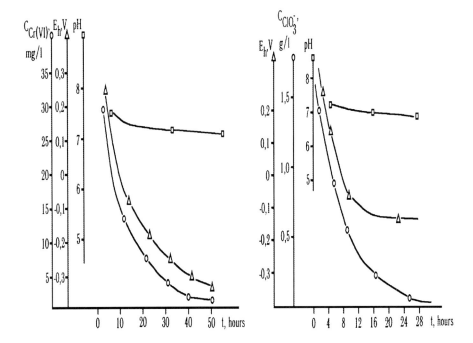

Figure 1.
Chromium, Cr(VI), reduction by
Aeromonas dechromatica KC-11.
Symbols: Δ, redox-potential (E_h);
o, Cr(VI) concentration ($C_{Cr(VI)}$); □, pH.

Figure 2.
ClO_3^--reduction by *Aeromonas dechromatica*
KC-11. *Symbols:* Δ, redox-potential (E_h);
o, ClO_3^- concentration ($C_{ClO_3^-}$); □, pH.

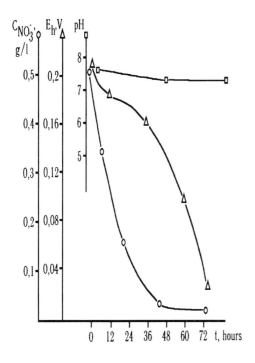

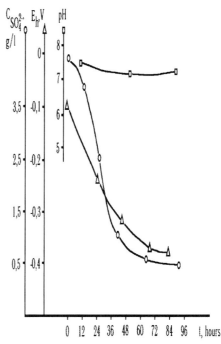

Figure 3.
NO_3^--reduction by *Aeromonas dechromatica* KC-11. *Symbols:* Δ, redox-potential (E_h); o, NO_3^--concentration ($C_{NO_3^-}$); □, pH.

Figure 4.
SO_4^{2-}-reduction by *Aeromonas dechromatica* KC-11. *Symbols:* Δ, redox-potential (E_h); o, SO_4^{2-}-concentration ($C_{SO_4^{2-}}$); □, pH.

REFERENCES

1. E.M. Chirwa and Y.T. Wang, J. Environ. Engineer., 8 (1997) 760.
2. E.I. Kvasnikov, T.M. Klyushnikova and T.P. Kasatkina, Mikrobiol. Zh., 55 (1993) 66.
3. P. Krauter, R. Martinelli, K. Williams and S. Martins, Biodegradation, 4 (1996) 277.
4. K. Fujie, T. Tsuchida, K. Urano and H. Ohtake, Water Science and Technology, 3 (1994) 235.
5. C.G. Vanginkel, C.M. Plugge and C.A. Stroo, Chemosphere, 9 (1995) 4057.
6. N.N. Lyalikova and N.A. Jurkova, in Proceedings of 9th International Symposium Biohydrometallurgy'91 (J.C. Duarte and R.W. Lawrence, eds.) (1991) 20. Forbitec editions.
7. V.I. Romanenko and V.N. Korenkov, Biologiya vnutrennikh vod. Inform. bull. (russ), 25 (1975) 8.
8. T. Suzuki and N. Miyata, J. Bacteriol., 174 (1992) 5340.
9. D.R. Lovley, E. Phillips and P. Phillips, Appl. and Environ. Microbiol, 2 (1994) 726.
10. J. Ishibashio, C. Cervantes and S. Silver, Appl. and Environ. Microbiol., 7 (1990) 2268.

11. H. Horitsu, S. Futo, I. Miyazava, S. Ogai and K. Kawai, Agric. Biol. Chem., 51 (1987) 2417.
12. L.H. Bopp and H.L. Ehrlich, Arch. Microbiol., 5 (1998) 426.
13. H. Ohtake, E. Fujii and K. Toda, Environ. Technol., 11 (1990) 663.
14. A.B. Tashirev, G.F. Smirnova and A.I. Samchuk, Mikrobiol. Zh., 59 (1997) 79.
15. R.E. Hungate, Rev. Microbiol., No15 (1985) 278.
16. D.V. Chernyshenko, Y.N. Danko, A.V. Tashirev, O.S. Radchenko, L.S. Yastremskaya. Mikrobiol. Zh., 52:6 (1990) 90
17. U.Lur'e and A. Rubnicova. Chemical analysis of waste water.- Chemistry (Moscow) (1974) 355.
18. A. Dolgorev, V. Lukachina, O. Karpova. J. Anal. chem., 4 (1974) 187.
19. F.Gerhardt (ed.), Methods of General Bacteriology (Moscow), Mir 1:2 (1983) 536.
20. Noel R. Krieg (ed.) and J.G. Holt (ed.-in-chief), Bergey's Manual of Systematic Bacterialogy, 1 (1984) 1964, Williams and Wilkins, Baltimore/London
21. I. Marmur and P.I. Doty, J. Mol. Biol., 15:1 (1962) 1109.
22. E.I. Kvasnikov, V.V. Stepanyuk, T.P. Kasatkina, Microbiology (Russian), 1 (1985) 83.
23. E.I. Kvasnikov, T.M. Klyushnikova, T.P. Kasatkina, Microbiology (Russian), 4 (1988) 680.

Experiments for the cyanide biodegradation from industrial wastewater from the processing of gold bearing ores

A. Varna, A. Michnea and A. Gavra

Mining Research and Design Institute, 62 Dr. V. Babes street, 4800 Baia Mare, Romania

Biological degradation of cyanide in a complex wastewater has been investigated. The wastewater comes from a tailing dam from the processing of gold bearing ores and contains 10-20 ppm of WAD cyanide. Detailed investigations were performed on the growth of bacteria using several wastes of the local industry (whey from a dairy and malt from a brewery) as a replacement for glucose. The tests showed that an addition of 1-5% of these carbon sources resulted in a satisfactory growth of the bacteria. The paper describes the experiments for the degradation of cyanide in the wastewater, using two bacterial strains. The tests involved the biomass growth "in situ", directly in the wastewater, in a laboratory unit running in a discontinuous mode for 96 hours. The results revealed that the degradation of the cyanide was in the range of 70-91% for the strain A and 60-78% for the strain B. This study shows that biodegradation can be an economical alternative for the destruction of cyanide in the effluent of the Bozanta dam and outlines some parameters that should be considered in a possible biological approach to wastewater treatment.

1. INTRODUCTION

Large amounts of cyanide are released in various industrial solid and aqueous wastes. They originate from the industries dealing with metal plating and mining. Gold is usually extracted from gold bearing ores or concentrates by dissolution in cyanide solutions. Some of the aqueous wastes resulting from this process contain significant quantities of cyanide as the main pollutant, along with heavy metals (metal cyanide complexes) and thiocyanates. Among the many different species, cyanide may be present in solutions as free cyanide (CN^-/HCN), 'WAD' or weak acid dissociable cyanide (such as Zn and Cu complexes) and stable cyanide complexes (ferrous and ferric complexes). This becomes a serious problem due to the high toxicity of cyanide and consequently many processes have been developed to detoxify cyanide-bearing effluents. This includes treatment by alkaline chlorination (hypoclorite), ozone oxidation, Inco's SO_2/air process, Degussa's peroxide process [1]. Current chemical treatment methods are not always efficient in regard to the final parameters of the clean water and to the required costs, related to the expensive chemicals and special equipment.

Lately, more and more scientific investigations are dealing with the biological treatment of these wastewaters, as an alternative method to the chemical procedures. Biological treatment has proven efficient when it comes to cyanide degradation at a low cost. A full scale biological treatment process is currently used to treat cyanide bearing wastewaters at

Homestake Mining Co. (Lead, SD, USA). In this plant, the cyanide compounds and thiocyanates are oxidized to ammonia and carbonate by a microbial consortium composed mainly of the bacterium *Pseudomonas paucimobilis*, immobilized in rotating biological contact reactors. Cyanides and heavy metals concentrations are reduced by 95-98% at a daily discharge of 4 million gallons of treated wastewater [2].

A number of reports related to the isolation of bacteria capable of growth on cyanide as the sole source of nitrogen and their ability to degrade the cyanide to ammonia and carbon dioxide have been published [3,4].

Cleaning of cyanide-containing wastewaters from the gold bearing ore leaching, by the alkaline treatment with hypochlorite has been a standard practice for a long time in the North Romanian mining area. In a current joint project between the Mining Research and Design Institute of Baia Mare, Romania and the Federal Institute for Geoscience and Natural Resources of Hanover, Germany, researches were performed to replace the chemical cleaning of a cyanide-bearing wastewater from the mining industry with a biological process.

The present work investigates the biological degradation of cyanide in a complex wastewater, coming from the Bozanta tailing dam, which is located near Baia Mare. Two bacterial strains that belong to the *Pseudomonas* and *Burkholderia* strains, were isolated by the microbiological research team of BGR from Hanover and used in the experimentation to confirm their ability to destroy the toxic cyanide compounds. Detailed investigations were performed on the bacteria growth. These involved several wastes from the local industry (whey from a dairy and malt from a brewery) as a replacement for glucose. Experiments were also carried out to evaluate the ability of these two strains to degrade the cyanide from the Bozanta dam wastewater.

2. MATERIALS AND METHODS

2.1. Bacterial isolates
Under selective conditions (cyanide as sole nitrogen source), a set of various bacteria was isolated from the wastewater and sediment samples from the Bozanta dam and from the oxidation ponds by specialists from BGR Hanover [5]. The best cyanide degrading bacteria were identified (DSMZ, Braunschweig). The most important of the bacteria belong to the genera *Pseudomonas* and *Burkholderia*, both common aquatic organisms. Two strains containing both bacteria were used in the experiments. They were sent to us by the specialists from BGR, and are marked in this paper with A and B to ease drafting (A and B are bacterial isolates from two different Bozanta sludge samples).

2.2. Wastewater sample
The wastewater sample taken from the Bozanta dam before the chemical treatment with hypochlorite was analyzed for heavy metal and cyanide content. The pH was 7.4. The main components are: CN^-_T (amenable to chlorination) = 13.1 ppm, SO_4^{2-} = 672 ppm, Cu = 6.2 ppm, Zn = 1.9 ppm, Fe = 0.1 ppm and Ca = 255 ppm.

2.3. Carbon source samples
The metabolic degradation of cyanides needs no special and supplementary carbon sources. However, the use of carbon sources is recommended in order to increase the bacterial activity. The best carbon source is glucose, which is rather expensive but it can be replaced by

carbohydrates contained in different wastes. We looked for some kind of waste from the local industry that would have been suitable to replace glucose. We tested two wastes with carbohydrates content:
- whey, a liquid waste resulted from the preparation of cheese out of cow milk
- malt, a liquid waste resulted from the rinsing of fermentation equipment in breweries

2.4. Experiments for bacteria growth

Bench experiments were performed in order to determine the optimum dosage of waste carbon sources that would insure the growth of both bacteria strains, A and B.

A 10 ml aliquot of the bacterial culture (O.D. = 3.0) was the inoculum for 190 ml of medium containing different amounts of wastes as carbon sources (0.5, 1, 2 and 5% v/v of malt, or whey). The medium solution contained 7 g/l Na_2HPO_4, 3 g/l K_2HPO_4, 0.5 g/l NaCl and 1 g/l $(NH_4)_2HPO_4$. The flasks were incubated at 27°C for 96 hours. Bacterial growth was monitored by O.D. determinations at 580 nm and compared with a control containing 0.4% of glucose and bacteria inoculum.

2.5. Experiments for cyanide degradation

Bench experiments were conducted in order to establish the ability of these two bacterial strains to degrade the cyanide compounds from the Bozanta wastewater. Samples of 100 ml were prepared by mixing 95 ml of Bozanta wastewater with 5 ml of bacteria inoculum (O.D.= 3.0). The samples also contained 2% of waste carbon sources (malt, or whey) and had the pH value at 7-7.5. Due to the lack of some nutritional sources in the Bozanta wastewater (e.g. phosphorous), 3 g/l of K_2HPO_4 was added. Cyanides provide the nitrogen source.

The flasks were incubated at 27°C for 96 hours. Bacterial growth was monitored by O.D. determinations at 580 nm and the cyanide concentration was determined by spectrophotometrical method using barbituric acid.

3. RESULTS AND DISCUSSION

The experiments carried out according the above procedure, concerning the choice of an appropriate waste as carbon source for replacing the glucose, resulted in the following data shown in Figures 1, 2, 3 and 4. The tests for the bacteria growth study lasted 96 hours, which is the period for the specific growing cycle for these bacterial strains. As can be concluded from these figures, the biomass grew in all conditions tested. Higher optical densities were obtained when 2 and 5% of whey was used for both strain A and B. Due to the possible occurrence of carbon sources decomposition phenomena that could decrease or stop the bacteria growth, the 5% addition of waste should be a maximum limit. According to this fact, we decided to choose 2% v/v of waste carbon source for further experiments. We consider that these carbon sources also ensure a supplementary amount of growth factors (vitamins, enzymes and other metabolites) that are favorable to the bacterial growth.

The second stage of the bench experiments dealt with the study of cyanide degradation kinetics in the Bozanta wastewater. The experiments were performed with both bacterial strains, following the above-mentioned procedure. The two waste carbon sources were added to the water at a 2% v/v dosage. The experimental results are plotted in Figures 5 and 6.

The Bozanta wastewater control sample was maintained in the same conditions as the inoculated samples. It can be noticed that a natural degradation of cyanides occurred to about

684

25% after 4 days. The degradation of the WAD cyanide was in the range of 70-91% for the bacterial strain A and 60-78% for B, thus showing a higher efficiency for strain A. A strange behavior of the biomass can be observed in these experiments. On one hand, they work better when malt is used as a carbon source, in spite of the better growth of both of them in the presence of the whey, a phenomenon that we are not able to explain. On the other hand, in the first day, the degradation process is faster in the presence of whey, but the final results are

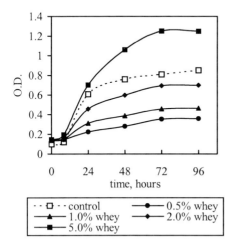

Figure 1. The influence of the whey on the growth of the A bacterial strain

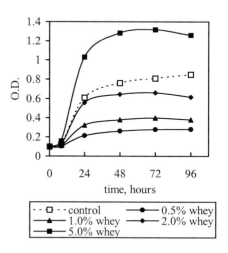

Figure 2. The influence of the whey on the growth of the B bacterial strain

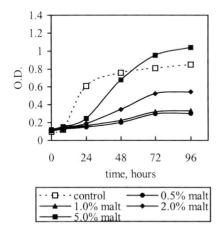

Figure 3. The influence of the malt on the growth of the A bacterial strain

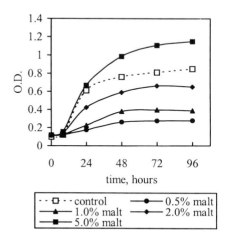

Figure 4. The influence of the malt on the growth of the B bacterial strain

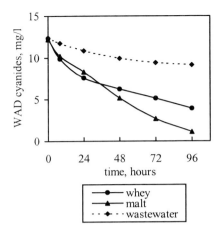

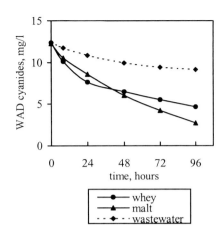

Figure 5. The cyanide degradation with the A bacterial strain

Figure 6. The cyanide degradation with the B bacterial strain

less than when malt is used. We assumed that the whey is more affected by some secondary processes that could produce compounds that induce the decrease or even the cessation of the bacterial growth. This hypothesis might also explain the shape of the growth curves, where it can be noticed that at the end of the experimental period (after 4 days) the optical densities for the bacteria grown in the presence of the whey are decreasing which indicates development of bacteria was affected.

The investigated wastewater comes from the processing of gold bearing ores with sulfide minerals matrix and it could also contain thiocyanates that are assayed by the same method as the cyanide. 90% of cyanide degradation could be a maximum limit, the remaining 10% may be thiocyanate (that the bacteria might not degrade) or undegraded cyanide.

4. CONCLUSIONS

The bench-scale experiments carried out at this stage of the research showed the following:
1. The replacement of the classical carbon source (glucose) in the culture medium of bacteria with waste carbon sources like whey and malt is possible and it ensures good biomass growth.
2. Good results for the bacteria growth were obtained in the experiments with a dosage of 2-5% v/v waste carbon sources, especially in the case of whey, for both bacteria strains.
3. The degradation of the WAD cyanides was in the range of 70-91% for the bacterial strain A and 60-78% for strain B.

We consider these results to be encouraging for suspended biomass, and we expect better results in our future research with immobilized biomass.

ACKNOWLEDGEMENTS

The authors wish to acknowledge the financial support of the Romanian Science & Technology Ministry. Various discussions with Dr. K. Bosecker and Dr. P. Blumenroth, during the joint research project development are gratefully acknowledged. A further word of thanks to the Organizing Committee of the IBS'99 for acceptance of this paper.

REFERENCES

1. M. M. Botz, J. A. Stevenson, Mining Environmental Management, June (1995) 4.
2. J. L. Whitlock, Geomicrobiology Journal, 8 (1990) 241.
3. M.M. Figueira, V.S.T. Cifninelli, V.R. Linardi, Biohydrometallurgical Processing, University of Chile, ed. by C.A. Jerez, T. Vargas, H. Toledo and J.V. Wiertz, (1995) 333.
4. H.J. Garcia, M.C. Fuerstenau, J.L. Hendrix, D.M. Lupan, Biohydrometallurgical Processing, University of Chile, ed. by C.A. Jerez, T. Vargas, H. Toledo and J.V. Wiertz, (1995) 341.
5. P.Blumenroth, K. Bosecker, A. Michnea, A. Varna, N. Sasaran, International Symposium "Technologies for mineral processing of refractory raw materials and for environmental protection in extractive industry areas", Baia Mare, Romania, 1998 (in press).

Biodegradation of some organic reagents from mineral process effluents

Namita Deo[a] . K.A. Natarajan[a]. K. Hanumantha Rao[b] and K.S.E. Forssberg[b]

[a] Department of Metallurgy, Indian Institute of Science, Bangalore 560012, India

[b] Division of Mineral Processing, Lulea University of Technology, SE-971 87 Lulea, Sweden

The utility of a soil microbe, namely *Bacillus polymyxa,* in the removal of organic reagents such as dodecylamine, ether diamine, isopropyl xanthate and sodium oleate from aqueous solutions is demonstrated. Time-bound removal of the above organic reagents from an alkaline solution was investigated under different experimental conditions during bacterial growth and in the presence of metabolites by frequent monitoring of residual concentrations as a function of time, reagent concentration and cell density. The stages and mechanisms in the biodegradation process were monitored through UV-visible and FTIR spectroscopy. Surface chemistry of the bacterial cells as well as the biosorption tendency for various organics were also established through electrokinetic and adsorption density measurements. Both the cationic amines were found to be biosorbed followed by their degradation through bacterial metabolism. The presence of the organic reagents promoted bacterial growth through effective bacterial utilization of nitrogen and carbon from the organics. Under optimal conditions, complete degradation and bioremoval of all the organics could be achieved.

1. INTRODUCTION

Froth flotation is universally used the world over for the beneficiation of sulfide and oxide ores. Alkyl xanthates are used as collector reagents for sulfide minerals while amines and fatty acids find widespread application in iron ore flotation. Although these organic reagents react preferentially with the concerned minerals in the treated ore pulp, excess and unreacted portions end up in the mill process effluents. It has been known that even smaller concentrations of these reagents in water streams are toxic to water life, besides their deleterious influence on end-stream processes during recycling.

Bioremediation has long been recognized as a cheap, and environmentally benign technique in waste water treatment. Organic flotation reagents are also amenable to biological detoxification. In this paper, the utility of a biological process for the efficient removal of some typical organic flotation collectors from aqueous solutions is illustrated. Two types of amines, namely, Deodecylammonium acetate (a primary amine) and an ether diamine, DA-16 which are extensively used in iron ore flotation, sodium isopropyl xanthate, a universal sulfide collector reagent and sodium oleate a fatty acid collector were chosen to demonstrate the potential of the bioprocess. A soil bacterium, namely, *Bacillus polymyxa,* which occurs indigenously, associated with several ore deposits was used in all the biodegradation studies.

2. MATERIALS AND METHODS

2.1. Bacterial strain and growth conditions

A pure strain of *Bacillus polymyxa* NCIM 2539 obtained from national collection of industrial microorganisms, National Chemical Laboratory, Pune, India was used in all the studies. A 10% v/v of an active inoculum was added to Bromfield medium [1] and incubated at 30^oC on a rotary shaker at 240 rpm. The bacterial growth was studied by microscopic counting using a Petroff-Hausser counter under a phase contrast microscope and also by colony counting after plating. pH changes during bacterial growth were monitored and enough cell mass (>10^9 cells/ml) was generated by continuous growth for 8 hours. The culture was filtered through Whatman No.1 filter paper to remove precipitates and centrifuged at 15000 rpm for 15 minutes. The cell pellet was washed several times and resuspended in deionised double distilled water.

2.2. Biodegradation of flotation reagents

Biodegradation of sodium oleate, sodium isopropyl xanthate, isododecyl oxypropyl aminopropyl amine (DA16) and dodecyl ammonium acetate (DAA) was studied under different conditions and concentrations. All the above collector reagents were of analytical grade and prepared in double distilled deionised water at the desired pH. Biodegradation of the above reagents was studied under different experimental conditions, namely, during bacterial growth and in the presence of cells alone and metabolites.

All flasks containing different concentrations of the collector reagents in the presence of bacterial cells or metabolites were incubated at 30^oC on a rotary shaker at 240 rpm for different periods of time. After each interval of time, the pH was checked and the residual concentrations of oleate [2], xanthate [3] DA16, and DAA [4] were measured in the cell-free supernatant by acid-titration and by UV-visible spectrophotometry.

The stages during biodegradation of the various flotation reagents, namely, oleate, xanthate, DA-16, and DAA were also monitored through UV-visible and FTIR spectra. For this purpose, an initial blank spectrum for the different reagents was obtained. The progress of degradation and disappearance from the solutions of the various collector reagents in the presence of cells, metabolite, and cells + metabolites, was then monitored through the disappearance, shifting of the initial absorption peaks and also the appearance of new peaks.

2.3. Growth of *Bacillus polymyxa* in presence of oleate, xanthate, DA-16 and DAA

To 100 ml of each previously autoclaved medium (without sucrose in 500 ml Erlenmeyer flasks) 100 mg/L of xanthate, oleate, DA-16 and DAA were added. Two of each control flasks without the above collector reagents were also taken. The pH of all the flasks was maintained at 9.0. To all these above flasks, 10% of the fully-grown culture along with the cells were added. After inoculation, the flasks were incubated on a rotary shaker at 240 rpm at 30^oC. The bacterial growth was monitored by microscopic counting using a Petroff-Hausser counter and pH changes during growth were also measured at regular intervals of time. During growth, the protein and polysaccharides produced by the bacteria were also measured frequently from the cell-free supernatant by Bradford method [5] and phenol sulphuric acid method [6] after dialysis.

2.4. Adsorption and surface chemical affinity of collector reagents on bacterial cells

Adsorption of collector reagents onto cell surfaces was estimated at different collector concentrations. A 10^{-3} M KNO_3 solution was used as the base electrolyte for the purpose. 0.5 g of each wet biomass were interacted with different concentrations (50, 100, 150, 200, 300, 400 mg/L) of xanthate, oleate, DA-16 , and DAA at pH 9.0 for different intervals of time. After each interval of time, the residual concentrations of collectors were estimated through analysis of the supernatant solution after centrifugation.

The relative surface chemical affinity of various flotation reagents towards bacterial cells was evaluated through zeta potential measurements. Cell suspensions (10^9 cells/ml) were interacted with different concentrations of oleate, xanthate, DA-16 and DAA at pH 9.0 for different periods of time. After such interaction, the cell suspension was centrifuged at 15000 rpm for 15 minutes. The cell pellet was washed several times with 10^{-3} M KNO_3 and then resuspended in 10^{-3} M KNO_3 solution and adjusted to the required pH for electrokinetic measurements. In this manner, the surface chemical changes on bacterial cell surfaces before and after interaction with various organic reagents could be established.

2.5. Surface hydrophobicity of bacterial cells

Surface hydrophobicity of bacterial cells after interaction with various collector reagents was established through microflotation. 4 g of wet biomass was interacted with 1000 mg/L of various collector reagents for 5 minutes at pH 9.0. After 5 minutes interaction, the cell suspension was centrifuged and the cell pellets were washed several times with 10^{-3} M KNO_3. Then the cells were suspended in 10^{-3} M KNO_3 solution and the flotation was carried out for five minutes under a nitrogen flow rate of 40 ml/minute using a Leaf and Knoll cell [7].

3. RESULTS AND DISCUSSION

3.1. Surface chemical affinity of collector regents towards bacterial cells

Surface affinity of the various flotation collectors towards bacterial cells was assessed initially through electrokinetic and adsorption studies. Both the amine reagents, namely, Dodecylammonium acetate (DAA) and Isododecyl oxypropyl aminopropyl amine (DA-16) exhibited enhanced surface adsorption onto cells of *Bacillus polymyxa* as attested to by a significant shift in the isoelectric point (IEP) of bacterial cells from an initial value at about pH 2 to about 3.0 and 4.3 respectively, after one hour interaction with 100 mg/L of DAA and DA-16. The IEP shifts were found to be a function of amine concentration and indicate chemical interaction between the amine and cell surface components. Also, electrostatic attraction between negatively charged cell surfaces and cationic aminium ions would facilitate biosorption. On the otherhand, interaction with 50- 100 mg/L of the anionic collectors such as sodium isopropyl xanthate and sodium oleate did not result in any significant surface chemical changes on the bacterial cells. Electrostatic repulsion between negatively charged bacterial cells and anionic collector reagents (xanthate and oleate) retard mutual interaction.

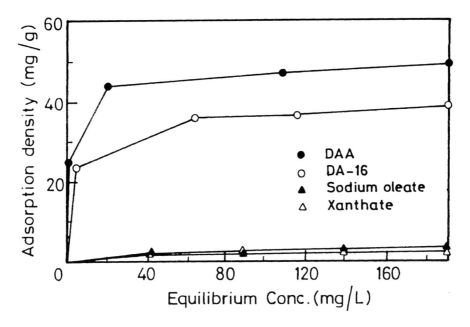

Figure 1. Adsorption isotherms for various collector reagents on bacterial cells

Adsorption behaviour of various collector reagents onto bacterial cell surfaces was then established through adsorption isotherms at a solution pH of 9 as illustrated in Figure 1. Enhanced affinity of the two amine collectors towards bacterial cell surfaces could be readily seen in comparison with the two anionic collectors. For example, the maximum adsorption density for DAA and DA-16 corresponded to about 45 and 36 mg/g, respectively, compared to only about 2.5 mg/g for xanthate and oleate. The isotherms follow a Langmuir model .

It was also observed that amine adsorption rendered the bacterial cell surfaces more hydrophobic as evident from the increased floatability of amine-interacted bacterial cells. For example, 96% and 82% of the cells after 5 minute interaction with DA-16 and DAA could be readily floated and recovered in the froth fraction from the aqueous solution at pH 9.0 On the otherhand, xanthate and oleate treated cells could not be separated through froth flotation.

3.2. Bacterial growth in the presence of collector reagents

Although only the two-amine reagents among the four flotation collectors used in this work exhibited significant biosorption tendency, all of them, promoted bacterial growth as illustrated in Figure 2. In the absence of any externally added carbon source such as sucrose, *Bacillus polymyxa* could grow effectively utilising the carbon and nitrogen associated with the flotation collector reagents. In such bacterial growth studies, the Bromfield medium contained all other constituents except sucrose, which is the essential carbon source for normal growth of *Bacillus polymyxa* in the absence of other organic reagents. Substantial increase in cell population and pH decrease due to organic acid production by bacterial metabolism could be observed.

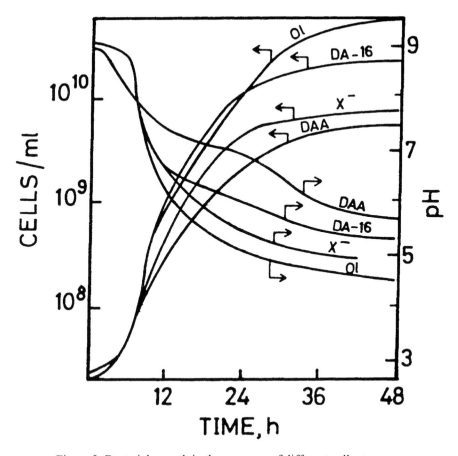

Figure 2. Bacterial growth in the presence of different collector reagents

Amounts of proteins and polysaccharides generated due to bacterial utilization of the various flotation collectors in a sucrose-free Bromfield medium are illustrated in Table 1. It becomes readily apparent that the bacteria could effectively utilize all the organic surfactants and generate polysaccharides and proteins as the major metabolic products besides organic acids. Protein production (amino acids) was found to be the highest in the presence of amines, while that of polysaccharides in the presence of xanthate and oleate. This could be expected since the amines containing nitrogen and amino groups are degraded by the cells to generate amino acid containing proteins. It has also been subsequently observed that the amino acids, polysaccharides and fatty acids are further utilised by the bacteria for metabolism and growth leading to complete dissociation of all the collector reagents. The above observation was substantiated by continuous analysis of the solution with time over a period of 4 days.

Table 1
Bacterial protein and polysaccharide in solution after bacterial growth in the presence of
flotation reagents (150 mg/L)

Added flotation collector	Protein, mg/L	Polysaccharide, mg/L
Sodium isopropyl xanthate	40.5	400.5
Sodium oleate	22.0	465.5
DA-16	195.0	250.5
Dodecyl ammonium acetate	171.5	200.0

3.3. Biodegradation of collector reagents in aqueous solutions

It becomes clearly evident from the above results that *Bacillus polymyxa* can be effectively used to degrade and remove various organic collector reagents from aqueous solutions. Biodegradation of the various collector reagents as a function of time under different conditions was then studied and the results are presented in Figures 3, a,b,c and d. The pH changes with time under different experimental conditions were also monitored. Decrease in dissolved DA-16 concentration with time in the presence of bacterial cells alone, during bacterial growth, active culture and metabolite is depicted. Higher rate of amine removal was obtained in the presence of active culture as well cells alone, while interaction with metabolite could remove only about 50% of the initial amine concentration during the period of this study. The solution pH decreased from 9 to 7 in the presence of cells alone and during growth, while in the presence of metabolite whose initial pH was 3.0, it further increased to 5.5 while in the presence of active culture whose initial pH was 4.0 it further decreased to 3.0. A similar behavior was also observed with DAA. Sodium oleate removal was found to be more efficient in the presence of bacterial cells and active culture than in the presence of bacterial metabolite alone. Thus, it could be concluded that enhanced participation of bacterial cells is involved in the biodegradation of amines as well as sodium oleate from the aqueous solution at pH 9. In contrast, to the above three cases, xanthate removal under similar conditions was observed to be the most effective in the presence of metabolite alone and active culture. 150 mg/L of DAA and DA-16 at an initial solution pH of 9 could be completely removed within about 4 hours in the presence of a fully-grown culture, whereas it took between 8-10 hours for their complete degradation in the presence of cells alone. Bacterial metabolites alone, could remove only about 50% of both the amines. In the case of sodium oleate, it took about 50 hours for complete removal in the presence of either cells alone or fully-grown bacterial culture. Bacterial metabolites alone, on the other hand, could remove only about 40% of the oleate.

Removal of isopropyl xanthate was more efficient in the presence of both fully-grown culture and the metabolites than with cells alone. For example, complete removal of all the dissolved xanthate (150 mg/L) could be achieved in about 12 hours in the presence of metabolite, while in the presence of active culture it took about 6 hrs. It took about 72 hours to attain the same level in the presence of growing cells. The above observation implies that initial high acidity of the metabolite plays an important role in xanthate degradation, while the direct influence of bacterial cells may not be that prominent in the initial stages. During

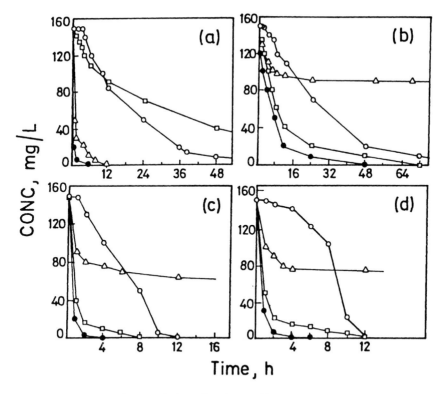

Figure 3. Biodegradation of various collector reagents (a) xanthate (b) oleate (c) DAA and (d) DA-16 o During growth ☐ cells alone △ metabolite alone • active culture

bacterial growth in the medium, organic acids are generated with time which could then influence the faster degradation of xanthate, and that could be the reason behind the prolonged interaction time required in xanthate removal in the presence of cells. On the other hand, with the metabolite, the initial pH is acidic to start with and enhanced rate of xanthate decomposition could be achieved within a short time.

Similar tests at different initial concentrations of various collector reagents resulted in the following observations:
a)   The overall rate of biodegradation of the various collector reagents was found to be influenced by their initial concentrations.
b)   There was an initial lag period for the biodegradation at higher collector concentrations
c)   The amines are biodegraded at a faster rate than oleate and xanthate

Thus the direct role of bacterial cells as well as the indirect role of their metabolites in the biodegradation of the collector reagents could be seen. Moreover, cationic collectors such as amines also exhibit enhanced bioadsorption onto cell surfaces through electrostatic and chemical forces. The surface-adsorbed amines were also subsequently utilised by the bacteria for their metabolism. Under the circumstances, higher rates of biodegradation and removal of amines could be expected. In fact, it has been observed that the cell-adsorbed amines on continuous growth of the cells in a medium were degraded as attested to by surface analysis by FTIR spectroscopy.

3.4. Probable mechanisms of biodegradation

From the above results, it is evident that *Bacillus polymyxa* efficiently utilise the various organic collector reagents for their metabolism and generate organic acids, proteins and polysaccharides in the process. The progress of the bioremoval of the various flotation collectors along with the generation of newer reaction products was monitored through UV-visible spectra. The UV-visible spectra of the bacterial metabolite alone yielded two broad peaks in the regions of 190-210 nm and 240-260 nm attributable to fatty acids, polysaccharides and protein-nucleic acids, respectively [8]. Similar spectra were also recorded during biodegradation of various collector reagents under different conditions, such as in the presence of growing cells, cells alone and metabolite alone. Time-bound removal of the various reagents could be readily seen through gradual reduction and ultimate disappearance of their characteristic absorption peaks. Bacterial reaction products were monitored through the emergence of newer absorption maxima, which also were seen to be ultimately suppressed and removed with time. Characteristic absorption maxima in the UV-visible spectra encountered in this work are illustrated in Table 2. In the presence of amines, bacterial interaction resulted in the generation of polysaccharides and organic acids (200-210 nm) as well as proteins and nucleic acids (260-280nm). In the presence of sodium oleate also, bacterial degradation resulted in the simultaneous generation of large quantities of polysaccharides and organic acids (210-220nm) with lower amounts of proteinaceous compounds. With isopropyl xanthate, the characteristic xanthate ion peaks (226 and 301nm) disappeared with time, with the generation of newer peaks corresponding to CS_2, carbohydrates and some fatty acids.

Degradation of the above flotation reagents could be brought about by growing bacterial cells as well as their metabolite products. FTIR spectroscopic studies also revealed the formation of similar reaction products and ultimate complete removal of the flotation reagents from aqueous solutions.

Table 2
Characteristic absorption maximum in the UV-Visible spectra

Species	Wave length (nm)
Amine (-NH2)	~ 190
Oleate-Oleic acid	~ 200
Fatty acids	≤ 210
Polysaccharides	210-240
Nucleic acids	260
Proteins	280
Xanthate Ions	226, 301
Carbon disulphide	205-207

The following possible reaction sequences during biodegradation of the various collector reagents can be proposed:

Sodium oleate :

Bacterial attack on the weak double bond region and formation of aldehydes and fatty acids. Bio-oxidation of the acids in the presence of bacterial cells leads to further degradation to lower aldehydes and fatty acids and ultimate decomposition to CO_2 and H_2O.

Amines :

Both carbon and nitrogen requirements for bacterial growth can be satisfied by utilization of amines. Different reaction sequences involving decarboxylation, deamination, hydrolysis and oxidation are brought about; the first essential step being degradation of an amino acid. NH_3 can be utilised by the bacteria as the nitrogen source and the CH_3COOH can enter the TCA cycle, substituting as carbon source. Formation of an aldehyde through amine oxidation is confirmed.

Xanthates :

Xanthates are unstable in acidic solutions. The production of xanthic acid and the subsequent decomposition to alcohol and CS_2 at low pH have been established [3]. The acidic metabolites generated by *Bacillus polymyxa* lead to rapid decomposition of xanthates. Moreover, direct bacterial utilization of xanthate is also evident through the formation of carbohydrates and fatty acids. As part of carbon cycle, microbes degrade organic sulfur compounds [9].

4. SUMMARY AND CONCLUSIONS

The use of a soil bacterium namely, *Bacillus polymyxa,* which is found to be indigenously associated with iron ore and bauxite deposits in the biodegradation of several organic flotation reagents has been demonstrated. The biodegradation process can be efficiently used to detoxify effluents from mineral processing plants in an environmentally acceptable fashion. The results of this study open up other possibilities such as biological stripping of adsorbed residual collector reagents from mineral surfaces, as in the case with a flotation concentrate. For example, the presence of hydrophobic coatings in an iron ore concentrate would deleteriously influence subsequent pelletisation. Residual amine or fatty acid collector reagents remaining adsorbed on iron ore flotation concentrate particles can be effectively desorbed and degraded through interaction with *Bacillus polymyxa.*

The following major conclusions could be made based on this study:

a. *Bacillus polymyxa,* a heterotrophic soil bacterium indigenously associated with several ore deposits was found to be capable of biodegradation of several organic flotation collectors such as dodecyl and di-amine, isopropyl xanthate and sodium oleate.

b. The bacteria could utilize the above organic reagents for their carbon and nitrogen requirements and grow very well, generating polysaccharides, proteins and fatty acids during metabolism.

c. Biodegradation mechanisms include both direct bacterial metabolism as well as contributions from the acidic metabolites.

d. A biological route for detoxification of mineral processing effluents containing residual organic reagents is demonstrated.

ACKNOWLEDGEMENTS

Partial financial support to this work from the STINT program of Indo-Swedish Collaborative project between Indian Institute of Science, Bangalore and Lulea University of Technology, Lulea, Sweden is gratefully acknowledged.

REFERENCES

1. A. Phalguni, J.M. Modak, K.A. Natarajan, Int. J. Miner. Process; 48 (1996) 51.
2. S.Siggia, S. Quantitative Organic Analysis via Functional Groups, John Wiley, New York, 1949.
3. J. Leja, Surface Chemistry of Froth Flotation, Plenum Press, New York, 1983.
4. F.D. Snell and C.T. Snell, Colorimetric Methods of Analysis, D. Van Nostrand Co., New York, 1954.
5. M.M. Bradford, Anal. Biochem.72, (1976). 248.
6. M. Dubois, K.A. Gilles, J.K. Hamilton, P.A. Rebers, and F. Smith, Ana. Chem. 28, (1956). 350.
7. A.F. Taggart, Hand Book of Mineral dressing, John Wiley, New York, 1967.
8. M.P. Deutscher, Methods in Enzymology (Guide to protein purification), Academic Press, New York, 1990.
9. H.L. Ehrlich, Geomicrobiology, Second Edition, Marcel Dekker, Inc., New York, 1990.

Kinetic studies on anaerobic reduction of sulphate

S. Moosa, M. Nemati and S. T. L. Harrison

Department of Chemical Engineering, University of Cape Town, Rondebosch, 7701, South Africa

The effects of sulphate concentration and its volumetric loading on the activity of anaerobic sulphate reducers and the kinetics of sulphate reduction were investigated using continuous bioreactors at a temperature of 35°C and a pH of 7.5. Media containing different concentrations of sulphate in the range of 1 to 5 kgm^{-3} were tested. For each medium the kinetics of sulphate reduction at different volumetric loading (dilution rates) were determined. With media containing 1, 2.5 and 5 kgm^{-3} of sulphate, maximum conversions of 85%, 91% and 93% were achieved respectively at a retention time of 10 days. The increasing of initial concentration of sulphate resulted in higher reduction rates, with a maximum of 0.075 kgm^{-3}h^{-1} observed with medium containing 5 kgm^{-3} of sulphate. The corresponding conversion of sulphate was 54% at a retention time of 1.5 days. Kinetics of microbial growth was also influenced by initial concentration of sulphate, with an increasing trend in both values of μ_m and K_s due to an increase in initial concentration of sulphate.

1. INTRODUCTION

Acid Mine Drainage (AMD) resulting from the uncontrolled microbial oxidation of sulphide wastes is a major problem in terrains affected by untreated acid wastes. These acidic wastes usually contain high levels of metal and sulphate. Traditionally AMD is treated by passive methods or lime neutralisation. During lime neutralisation the acidity of AMD is decreased, with subsequent precipitation of heavy metals as hydroxides.

The passive treatment is based on biological and physicochemical processes, such as oxidation, reduction, adsorption, absorption and precipitation. The passive treatment processes normally take place in large dams or reed beds and as such the process cannot be controlled[1]. Disposal of the sludge or reed mat at the end of the process requires further consideration.

Acid mine drainage and process effluents containing sulphates and heavy metals are amenable to anaerobic digestion and concomitant removal of the metal pollutants as metals sulphides. The anaerobic conversion of long chain organic substrates to acetic acid, carbon dioxide and hydrogen by acidogenic bacteria and microbial reduction of sulphate to sulphide by sulphate reducing bacteria are two major steps in the treatment of sulphate containing wastes[2]. A general description of the process is shown in Figure 1.

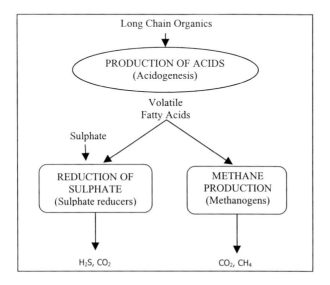

Figure 1 Schematic diagram of anaerobic treatment of acidic wastes

Anaerobic sulphate reduction has been used to treat a range of mining and process wastewaters containing sulphate at different concentrations ranging from 0.2 kgm^{-3} to 8.3 kgm^{-3} [3-7]. A review of the available literature reveals that most of the work in this area has focussed primarily on the choice of organic substrate for optimal reduction of sulphate. The organic compounds which have been tested included sugar[8], acetic acid[2], molasses[8, 12], ethanol[10], lactate[15], hydrogen and hydrogen/carbon dioxide[11], propionate and butyrate[7] ecoonomics and availability dictate the choice of the organic substrate.

Information regarding the kinetics of sulphate reduction and the effect of different parameters such as concentration of sulphate, pH, and temperature as well as inhibitory effects of sulphide and metallic ions is rather limited. From a process engineering point of view, in order to implement the biological reduction of sulphate as part of a large scale treatment process, an extensive study on the kinetics of this reaction is required.

The objective of the present work was to study the effects of sulphate concentration and its volumetric loading on the activity of sulphate reducing bacteria, as well as the kinetics of anaerobic sulphate reduction.

2. MATERIALS AND METHODS

2.1. Microorganisms

A mixed culture of anaerobic bacteria consisting of acid-producing bacteria, methane producers and sulphate reducers was obtained from the Council for Scientific and Industrial Research (CSIR) Pretoria, South Africa.

2.2. Medium and culture conditions

The medium used was a soluble complex organic mixture (Table 1). It is biodegradable and has been recommended for the growth of anaerobic microorganisms[13]. All reagents were analytical grade. Sodium sulphate was used as a source of sulphate. The stock cultures were grown in a medium containing 2.5 kgm^{-3} of sulphate. The medium was sterilised by autoclaving at 121°C and 15 psig for 20 minutes. To prevent reaction and precipitation, the organic compounds and metal salts were autoclaved separately and combined when cool.

The anaerobic culture was incubated in 1-litre bottles, each containing 500 mL of liquid medium and 50% (v/v) inoculum at a constant temperature of 35°C on a rotary shaker at 160 rpm. Prior to incubation air was stripped from the liquid using gaseous nitrogen.

Table 1

Composition of medium used for growth
and maintenance of the anaerobic mixed culture

Components	Weight (g)
Glucose	1.334
Peptone	0.400
Lab-Lemco	0.133
K_2HPO_4	0.040
$NaHCO_3$	1.250
Na_2SO_4	3.700
Deionised Water	500 ml
Trace Metals	
$CoCl_2.6H_2O$	0.0119
$FeCl_2.4H_2O$	0.0785
$MnCl_2.4H_2O$	0.0038
$NaMoO_4.2H_2O$	0.0038
$NiCl_2.6H_2O$	0.0045
Deionised Water	500 ml

2.3. Experimental procedure

Figure 2 shows a schematic diagram of the experimental set-up used for kinetic studies. The experiments were performed in three identical bioreactors with a capacity of one liter. To ensure that anaerobic conditions were maintained the glass lids of the vessels and the adapters were sealed with vacuum grease. The mixing in the bioreactor was achieved by overhead stirrers agitating at a speed of 400 rpm. Fresh medium was fed into the reactor by a multiple speed peristaltic pump. To avoid channeling, feed was introduced near the bottom of the reactor. The effluent was discharged by gravity through an overflow tube. The reaction temperature was maintained at 35°C, using a circulating waterbath and the pH was controlled at 7.5 with the addition of either concentrated HCl or a saturated NaOH solution.

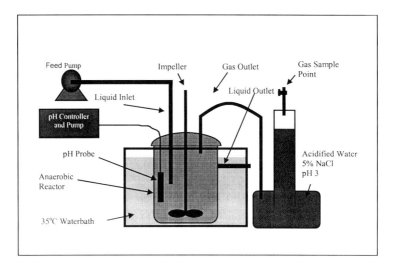

Figure 2 Schematic diagram of the experimental set-up

In order to study the effects of sulfate concentration and its volumetric loading on the kinetics of sulfate reduction and microbial growth, three independent experimental runs with media containing 1, 2.5 and 5 kgm^{-3} of sulfate were conducted. Acetate was used as the carbon source.

In the experiments with media containing 1 and 2.5 kgm^{-3} of sulphate, the stock culture was used as an inoculum (20% v/v). The bioreactors were initially operated batchwise and the glucose replaced with 2.5 kgm^{-3} of acetic acid. The acetic acid concentration was increased step wise to 17.5 kgm^{-3}. Once sufficient sulphate reduction was achieved the bioreactors were switched to continuous mode, at a retention time of 10 days. The inoculum for the experiment with 5 kgm^{-3} of sulphate experiment was taken from the effluent of the reactor reducing 2.5 kgm^{-3} of sulphate.

For each media various flow rates of the feed in the range 0.004 to 0.021 Lh^{-1} were applied. Steady state conditions were used at each flow rate to estimate the kinetics of sulphate reduction and bacterial growth. Steady state conditions were assumed to be established when both the residual sulphate concentration and bacterial concentration varied by less than 10% during a period of operation equal to three retention times. Liquid samples were taken on a daily basis and analysed for sulphate and bacterial concentrations.

2.4. Analytical procedures

The bacterial concentration was determined by measuring dry weight at 80°C. A turbidimetric method was used to measure the concentration of sulphate[14]. Sulphate forms an insoluble precipitate with barium under acidic conditions. Prior to sulphate determination, suspended solids were removed from the sample by centrifugation. After addition of 0.25 ml conditioning reagent (50 ml glycerol, 30 ml concentrated HCl, 75 g NaCl, 100 ml ethanol and 300 ml deionised water) to 5 ml of sample an excess amount of finely ground $BaCl_2$ was added and mixed for 1 minute on a vortex mixer. The absorbance of the sample was then measured at a wavelength of 420 nm. The absorbance of the sample was used to calculate the concentration of sulphate. A calibration curve for dependency of adsorption on sulphate concentration was obtained using a similar procedure.

3. RESULTS

Results of continuous anaerobic reduction of sulphate at initial concentrations of 1, 2.5 and 5 kgm^{-3} sulphate are presented in Figures 3, 4 and 5 respectively. The volumetric reduction rate, calculated in terms of the reduced sulphate multiplied by the dilution rate, was used to ascertain the kinetics of sulphate reduction.

For medium containing 1 kgm^{-3} of sulphate at dilution rates below 0.008 h^{-1}, 81% to 85% of the sulphate was reduced. This corresponds to a reduction rate of 0.005 kgm^{-3}h^{-1} to 0.007 kgm^{-3}h^{-1}. Increasing the dilution rate, while causing a decrease in the conversion of sulphate led to higher reduction rates, with a maximum of 0.007 kgm^{-3}h^{-1} at a dilution rate of 0.011 h^{-1}. Further increase in the dilution rate resulted in the continuous decrease in the reduction rate and conversion of sulphate. The maximum bacterial concentration, 0.98 kgm^{-3}, was observed at a dilution rate of 0.006 h^{-1} corresponding to a residual sulphate concentration of 0.154 kgm^{-3}. Increasing the dilution rate caused a gradual decrease in bacterial concentration and washout was observed at a dilution rate of 0.012 h^{-1}

702

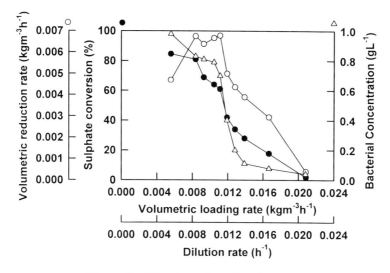

Figure 3 Kinetic results of continuous reduction of sulfate at an initial concentration of 1 kgm^{-3}

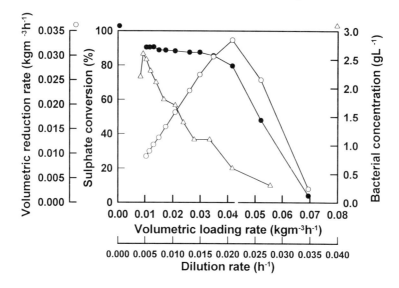

Figure 4 Kinetic results of continuous reduction of sulfate at an initial concentration of 2.5 kgm^{-3}

With a medium containing 2.5 kgm^{-3} of sulphate, the conversion was between 85% and 90.5% for dilution rates in the range 0.004 h^{-1} to 0.014 h^{-1}. The corresponding reduction rates were 0.009 kgm^{-3}h^{-1} and 0.030 kgm^{-3}hr^{-1} respectively. The maximum reduction rate of 0.033 kgm^{-3}h^{-1} was observed at a dilution rate of 0.017 h^{-1}, with a 79.6% conversion of sulphate. Bacterial concentration exhibited a gradual decrease from 2.6 kgm^{-3} to 1.1 kgm^{-3} as the dilution rate was increased from 0.005 h^{-1} to 0.017 h^{-1}. The residual concentrations of sulphate were 0.238 kgm^{-3} and 0.510 kgm^{-3} respectively. A decrease in bacterial concentration, from 1.1 kgm^{-3} to 0.6 kgm^{-3} was observed when the dilution rate was increased from 0.017 h^{-1} to 0.021 h^{-1}, while the concentration of the residual sulphate increased from 0.51 kgm^{-3} to 1.3 kgm^{-3}. In this set of experiments washout occurred at a dilution rate of 0.017 h^{-1}.

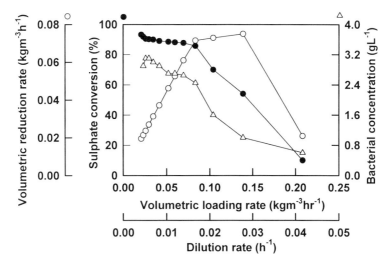

Figure 5   Kinetic results of continuous reduction of sulfate at an initial concentration of 5 kgm^{-3}

For the medium containing 5 kgm^{-3} of sulphate, increasing the dilution rate to 0.017 h^{-1} enhanced the reduction rate to 0.07 kgm^{-3}h^{-1}. Conversion of sulphate in the range 80 to 90% was observed. The highest reduction rate of 0.075 kgm^{-3}h^{-1} was observed at a dilution rate of 0.028 h^{-1} where a reduced conversion of sulphate was recorded. Washout occurred at a dilution rate was 0.017 h^{-1}.

The maximum bacterial concentration was 3.1 gl^{-1} and it showed a gradual decrease to 2.5 gl^{-1} at a dilution rate of 0.0167 h^{-1}. Thereafter the bacteria concentration decreased sharply until washout was reached.

4. DISCUSSION

The results of the present study show that the concentration of sulphate has an effect on the activity of sulphate reducers. Table 2 compares the maximum concentration of the bacteria as well as the dilution rate at which washout occurs for the various influent sulphate concentrations. As can be seen, the increase in initial sulphate concentration not only increased the maximum concentration of biomass but also resulted in washout of the cells at higher dilution rates.

Table 2
Comparison of maximum bacterial concentration and critical dilution rates observed in the presence of different concentrations of sulphate

Initial sulphate concentration (kgm^{-3})	Maximum bacterial concentration (gL^{-1})	Critical dilution rate (washout) (h^{-1})
1	0.98	0.012
2.5	2.60	0.017
5	3.10	0.017

Using the residual concentration of sulphate at each dilution rate the dependency of specific growth rate on the concentration of sulphate was correlated by the Monod equation. Table 3 summarizes the values of maximum specific growth rate (μ_m) and saturation constant (K_s).

The kinetics of anaerobic reduction of sulphate at different concentrations of 1, 2.5 and 5 $kgm^{-3}h^{-1}$ are compared in Figure 6 and Table 4. As can be seen the dependency of volumetric reduction rate and volumetric loading of sulphate were similar.

In all cases increasing volumetric loading up to a certain level (corresponding to critical dilution rate) resulted in a linear increase in the reduction rate. Further increase of volumetric loading led to a sharp decrease in the reduction rate. Higher concentrations of sulphate resulted in the sulphate reducers being more resistant to washout, and as such the maximum reduction rate in the presence of 5 $kgm^{-3}h^{-1}$ of sulphate was more than ten times faster than that with medium containing 1 $kgm^{-3}h^{-1}$.

In conclusion, it is apparent that sulphate concentration influences both the growth of the SRBs as well on the reduction capability of the system. The exact nature of this effect cannot be ascertained from the range of sulphate concentrations presented here. Further research work is being carried out to investigate the effect of higher concentrations of sulphate on the activity of sulphate reducers and the extent of sulphate reduction by these microorganisms.

Table 3

Comparison of Monod kinetic constants achieved for media containing 1, 2.5 and 5 kgm^{-3} of sulphate

Initial sulphate concentration (kgm^{-3})	μ_{max} (h^{-1})	K$_s$ (kgm^{-3})
1	0.023	0.422
2.5	0.040	1.043
5	0.077	3.912

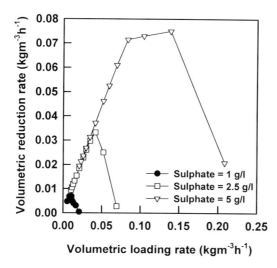

Figure 6   Performance of the continuous reactor in the presence of media containing different concentrations of sulphate

Table 4

Comparison of maximum volumetric reduction rates and its corresponding loading rates

Initial sulphate concentration (kgm^{-3})	Retention time (h)	Volumetric loading rate (kgm^{-3}h^{-1})	Maximum volumetric reduction rate (kgm^{-3}h^{-1})
1	66	0.001	0.007
2.5	60	0.042	0.033
5	36	0.140	0.075

706

REFERENCES

1. Kuyacak, N and P St-Germain, EPD Congress, 1993.
2. Middleton, A C, and A W Lawrence, Journal of the Water Pollution Control Federation, July (1977) 1659.
3. Maree, J P, A Gerber and E Hill, Water Science and Technology, 23 (1991) 1293.
4. Dvorak, D H and R W Hammack, Biotechnology and Bioengineering, 40 (1992) 609.
5. du Preez, L A and J P Maree., Environmental Technology, 13 (1992) 875.
6. Hammack, R W and H M Edenborn, Applied Microbiology and Biotechnology, 37 (1992) 674.
7. Omil, F, P Lens, A Visser, L W Hulshoff and G Lettinga, Biotechnology & Bioengineering, 57 (1998) 676.
8. Maree, J P, and W F Strydom, Water Research, 21 (1987) 141.
9. Maree, J P, A Gerber and E Hill, Journal of the Water Pollution Control Federation, 59 (1987) 1069.
10. Barnes, L J, P J M Scheeren and C J N Buisman, Transactions of the Institute of Mining and Metallurgy, 101 (1992) C183.
11. du Preez, L A and J P Maree, Proceedings of the Seventh International Symposium on Anaerobic Digestion", Cape Town, SA, January 1994.
12. Herrara, L, J Hernandez, L Bravo, L Romo and L Vera, Environmental Toxicology and Water Quality, 2 (1997.
13. Grobicki, AMW. "Hydrodynamic characteristics and performance of the anaerobic baffled reactor" PhD Theses, 1989, Imperial College of Science and Technology, UK.
14. American Public Health Association, "Standard Methods for the examination of water and wastewater", 1992, Washington DC.
15. Okabe, S and W G Characklis, Biotechnology and Bioengineering, 39 (1992) 1031.
16. Nethe-Jaenchen, R and R K Thauer, Archives of Microbiology, 137 (1984) 236.

Removal and recovery of metal-cyanides from industrial effluents

Y. B. Patil and K. M. Paknikar[*]

Division of Microbial Sciences, Agharkar Research Institute, G.G. Agarkar Road, Pune
411 004, India

Biosorption of metal-cyanide complexes viz. tetracyanocuprate(II) $\{[Cu(CN)_4]^{-2}\}$ (TCC)
and tetracyanonickelate(II) $\{[Ni(CN)_4]^{-2}\}$ (TCN) from solutions was studied using different
fungal biosorbents. Optimum sorption of TCC and TCN by the biosorbents tested occurred at
pH 4.0. Among the cultures used, *Cladosporium cladosporioides* showed maximum loading
capacity (40 μmol/g and 34 μmol/g for TCC and TCN, respectively) which was higher than
activated charcoal. Further studies on *C. cladosporioides* showed that rate of uptake of TCC
and TCN was maximum in the first 15 min (>50%), though equilibration of the system was
achieved in 30 min. Uptake of both the metal-cyanide complexes was found to increase
linearly as a function of initial metal-cyanide concentration up to 1.75 mM. The data could be
fitted to Langmuir and Freundlich models for adsorption processes. Adsorbed TCC and TCN
could be desorbed using 1 M NaOH with more than 95% efficiency. Concentrations achieved
in the eluant were 38 (TCC) and 32 (TCN) fold of the original, respectively. The traces of
metal-cyanide complexes remaining in solution after biosorption were completely biodegraded
by a bacterial consortium in 5-6 h. The treated solution fully conformed to the disposal
standards prescribed by statutory agencies.

1. INTRODUCTION

Toxicity of cyanide is due to its action as a metabolic inhibitor of the respiratory enzyme
cytochrome oxidase (1). Though toxic in nature, cyanide is one of the most indispensable
industrial chemicals used in metal extraction (mainly gold), electroplating, metal finishing,
metal hardening, steel and printed circuit board manufacturing industries. Consequently, waste
waters generated by these industries often contain cyanide in association with heavy metals
(viz. copper, nickel, zinc, cadmium, iron, silver, gold, etc.) forming complex cyanides of
variable stability and toxicity. The cyanide and metal concentrations in these effluents may
range between 5 to 600 mg/l and 30 to 60 mg/l, respectively.

The most commonly adopted method for the treatment of cyanide contaminated effluent is
alkaline-chlorination-oxidation process (2). Although this method of treatment can be very

[*] Corresponding author. Fax: +91-20-351542; E-mail: paknikar@vsnl.com
YBP is thankful to Council of Scientific and Industrial Research, New Delhi for the award of a Senior
Research Fellowship.

efficient in detoxifying free cyanide bearing wastes, it is not effective when challenged with metal-cyanide species due to slow reaction rates (3). Other physico-chemical methods, such as copper-catalyzed hydrogen peroxide oxidation, ozonation, electrolytic decomposition, etc. are highly expensive and are rarely used for treatment of metal-cyanides.

Complex cyanides of copper and nickel are commonly encountered in waste waters emanating from the electroplating and mining industries. In most countries the statutory limits for discharge of cyanide, copper and nickel in water bodies are 0.2 mg/l, 3.0 mg/l and 3.0 mg/l, respectively. Clearly, metal-cyanide containing effluents cannot be discharged in the environment without proper treatment. There are a few reports on the treatment of metal-cyanides (viz. nickel cyanide, copper cyanide, zinc cyanide, iron cyanide, etc.) using microorganisms (4-8). However, these studies mainly focus on biodegradation of metal-cyanides. Biosorption of metal cations from waste waters has been developed to a practical scale in the recent times (9). However, very few attempts have been made to adopt this technology for possible removal and recovery of anions such as cyanide (10) and metal-cyanide complexes (11). Present study explores the possibility of treating two important metal-cyanides encountered in plating industry waste waters, viz. tetracyanocuprate (TCC) and tetracyanonickelate (TCN) using a combination of biosorption and biodegradation, the former being used for recovery of metal-cyanides, and the latter for removal of unrecoverable (residual) cyanide complexes.

2. MATERIALS AND METHODS

2.1. Biosorbents
The biosorbents used in the present study comprised of fungal cultures (*Aspergillus fumigatus, Aspergillus niger, Aureobasidium pullulans, Cladosporium cladosporioides, Fusarium moniliforme, Fusarium oxysporum, Mucor hiemalis*) and mycelial wastes from fermentation industries (*Penicillium chrysogenum, Streptomyces pimprina* and *Streptoverticillium cinnamoneum*). Activated charcoal, saw dust, fly ash and pure chitin were used for comparison. Fungal cultures were grown in 400 ml Sabouraud's medium in a 1000 ml Erlenmeyer flasks at 30°C on a rotary shaker (120 rpm). Fungal biomass was harvested by filtration after five days growth, washed several times with deionized water to remove traces of medium constituents, dried at 60°C for 48 h and powdered. Mycelial wastes obtained from fermentation industry were also washed with deionized water, dried in an oven at 60°C for 48 h and powdered.

2.2. Metal-cyanide sorption/uptake studies
In order to evaluate the effect of pH on metal-cyanide biosorption, 10 g of powdered biomass and other sorbents were conditioned to desired pH (3.0, 4.0, 5.0, 6.0, 7.0, 8.0 and 9.0). Sterile 12.5 mM stock solutions of tetracyanocuprate (TCC) and tetracyanonickelate (TCN) were prepared by mixing equal volumes of 25 mM of autoclaved salts of the respective metal and 100 mM of filter-sterilized KCN solution (4,12).

Biosorbent (1 g) was contacted with 50 ml solution containing 0.5 mM of the chosen metal-cyanide (TCC and/or TCN) of desired pH in a set of 250 ml Erlenmeyer flasks. The flasks were incubated on rotary shaker (Gallenkamp, England) at 120 rpm for 30 min. Contents of the flasks were filtered and were analyzed for residual metal-cyanide and total cyanide by UV spectrophotometric method (4) and pyridine-barbituric acid method (13), respectively. Residual metal contents in the filtrates were determined by atomic absorption spectrophotometry (Unicam 929, UK). All the experiments were performed in duplicates and repeated twice along with appropriate controls.

For determination of optimum biosorbent concentration, metal-cyanide solution (0.5 mM, pH 4.0) was contacted with varying amounts of biosorbent powder, ranging from 0.5 to 4% (w/v). Rates of metal-cyanide uptake were studied by contacting the biosorbents for a period ranging between 0-100 min. Metal-cyanide loading capacity (μmol metal-cyanide bound per gram weight of the biosorbent powder) was determined by contacting 1 g biosorbent powder several times with fresh batches of 50 ml metal-cyanide solution (0.5 mM, pH 4.0) until saturation was achieved.

2.3. Adsorption isotherms

Metal-cyanide solutions (TCC and TCN) of varying concentrations (ranging from 0.125 mM to 2.0 mM, which corresponds to 13 mg/l to 208 mg/l of cyanide), were used to study the effect of initial metal-cyanide concentration on its adsorption. Biosorbent (1 g) was contacted with 50 ml of each of the solutions for 30 min on a rotary shaker at 120 rpm.

In order to obtain sorption kinetics data, the metal-cyanide uptake value (Q) was calculated using following equation:

$$Q = V(C_i - C_f)/1000m \tag{1}$$

where, Q is the metal-cyanide content (μmol/g biomass), V the volume of metal-cyanide solution (ml), C_i the initial concentration of metal-cyanide in solution (mM), C_f the final concentration of metal-cyanide in solution (mM) and m is the mass of biosorbent (g).

The 'Q' value thus obtained was used to plot an adsorption isotherm according to Freundlich and Langmuir equations:

$$\ln Q = \ln K + (1/n)\ln C_{eq} \quad \text{(Freundlich equation)} \tag{2}$$

$$C_{eq}/Q = 1/(bQ_{max}) + C_{eq}/Q_{max} \quad \text{(Langmuir equation)} \tag{3}$$

where C_{eq} is the liquid phase concentration of metal-cyanide, b is the Langmuir constant, Q the metal-cyanide uptake (μmol/g biomass) and Q_{max} the maximum metal-cyanide uptake.

2.4. Adsorption/desorption of metal-cyanides

Samples of 1 g biosorbent loaded with metal-cyanide were eluted using 2 ml 1 M sodium hydroxide solution and metal-cyanide content in eluted solutions was analyzed. Following elution of metal-cyanide, the biosorbent was washed with deionized distilled water and then conditioned to pH 4.0 prior to use in the next adsorption/desorption cycle.

2.5. Biodegradation of unrecoverable metal-cyanide complexes

Unrecoverable (residual) metal-cyanide complexes in the solutions treated by biosorption were subjected to biodegradation. Typical concentrations of TCC and TCN after biosorption were 0.14 mM and 0.16 mM (corresponding to 14.56 mg/l and 16.64 mg/l cyanide), respectively. A consortium culture comprising of one *Citrobacter* and three *Pseudomonas* strains previously isolated in our laboratory was used in the biodegradation experiments. The consortium culture was capable of utilizing metal-cyanides as a source of nitrogen and glucose was added as the source of carbon. Optimized conditions for biodegradation of metal-cyanides were found to be: pH 7.5, temperature 37°C and initial cell density of 10^9 cells/ml. During present work, residual metal-cyanide solutions (50 ml aliquots) were supplemented with 5 mM glucose and dispensed in 250 ml Erlenmeyer flasks. Biodegradation experiments were carried out under optimized conditions on a shaker at 120 rpm. Residual cyanide and metal levels in the flasks were monitored at an interval of 1 h.

3. RESULTS

3.1. Optimum pH

The results obtained showed that biosorption of metal-cyanide complexes (TCC and TCN) could not take place above pH 6.0. It was further observed that optimum sorption of both the metal-cyanide complexes was at pH 4.0. However, in case of TCN sorption by *Aureobasidium pullulans* and *Mucor hiemalis,* optimum pH values were found to be 6.0 and 5.0, respectively. Table 1 summarizes the results obtained under optimal pH conditions.

On the basis of maximum metal-cyanide uptake values obtained under optimum pH conditions, three organisms, viz. *Cladosporium cladosporioides, Aspergillus fumigatus* and *Aureobasidium pullulans* were selected for further studies. Activated charcoal was used as reference material.

Table 1
Biosorption of metal-cyanides (TCC and TCN) at optimum pH

Sorbents/Biosorbents	μmol TCC sorbed per g biomass (optimum pH)	μmol TCN sorbed per g biomass (optimum pH)
Activated charcoal	23.0 (4.0)	24.0 (4.0)
Chitin	10.0 (4.0)	08.0 (4.0)
Fly ash	06.0 (4.0)	02.0 (4.0)
Saw dust	07.0 (4.0)	02.0 (4.0)
Aspergillus niger	00.0 (4.0)	00.0 (4.0)
Aspergillus fumigatus	14.0 (4.0)	02.0 (4.0)
Aureobasidium pullulans	14.0 (4.0)	12.0 (6.0)
Cladosporium cladosporioides	18.0 (4.0)	17.0 (4.0)
Fusarium moniliforme	00.0 (4.0)	00.0 (4.0)
Fusarium oxysporum	02.0 (4.0)	00.0 (4.0)
Mucor hiemalis	07.0 (4.0)	03.0 (5.0)
Penicillium chrysogenum	10.0 (4.0)	05.0 (4.0)
Streptomyces pimprina	06.0 (4.0)	08.0 (4.0)
Streptoverticillium cinnamoneum	06.0 (4.0)	00.0 (4.0)

Table 2
Metal-cyanide loading capacity of various biosorbents at optimum pH

Biosorbents	Loading capacity (μmol/g biomass) [optimum pH]	
	TCC	TCN
Activated charcoal	30.00 [4.0]	27.50 [4.0]
Cladosporium cladosporioides	40.00 [4.0]	34.00 [4.0]
Aspergillus fumigatus	28.00 [4.0]	Not determined
Aureobasidium pullulans	26.00 [4.0]	13.00 [6.0]

3.2. Loading capacity

Table 2 shows the metal-cyanide loading capacity of biosorbents selected on the basis of maximum sorption under optimum pH, as described above. It could be seen that *C. cladosporioides* culture had maximum loading capacity (40 μmol/g for TCC and 34 μmol/g for TCN) for both the metal-cyanide complexes. Considering these results, selection of the biosorbent was further narrowed down to only *C. cladosporioides* biomass.

3.3. Biosorption of metal-cyanides by *C. cladosporioides*

Data in Figure 1 shows that in case of *C. cladosporioides*, optimum biosorbent concentration for maximum sorption (92% and 86%, for TCC and TCN, respectively) was 3.5 % (w/v). With 2% biomass concentration, uptake decreased to 70% and 65%, respectively. The curve (Fig. 2) representing the kinetics of TCC and TCN adsorption showed that rate of

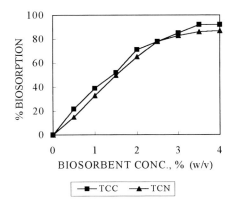

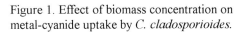

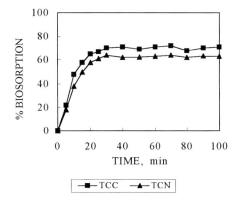

Figure 1. Effect of biomass concentration on metal-cyanide uptake by *C. cladosporioides*.

Figure 2. Rate of metal-cyanide uptake by *C. cladosporioides* biomass.

712

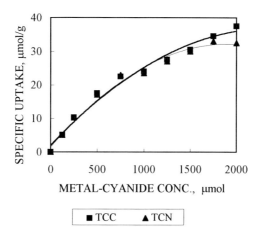

Figure 3. Effect of initial metal-cyanide concentration on uptake of *C. cladosporioides* biomass.

metal-cyanide uptake was maximum in the first 15 min, with over 50% sorption. Later, the sorption rate slowed down until it reached a plateau after 30 min, indicating equilibration of the system.

It is evident from Figure 3 that TCC and TCN uptake by *C. cladosporioides* biomass was directly proportional to the initial concentration in solution at equilibrium. Figures 4a, 4b, 5a and 5b show that metal-cyanide uptake values could be fitted to the Langmuir and Freundlich isotherm models.

3.4. Adsorption/desorption

The loaded metal-cyanide complexes could be desorbed with more than 95% efficiency using 1 M sodium hydroxide solution. Final concentrations of metal-cyanides in the eluant

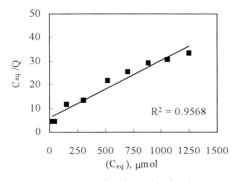

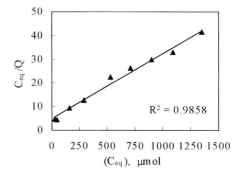

Figure 4a. Langmuir adsorption isotherm for TCC biosorption by *C. cladosporioides* biomass.

Figure 4b. Langmuir adsorption isotherm for TCN biosorption by *C. cladosporioides* biomass.

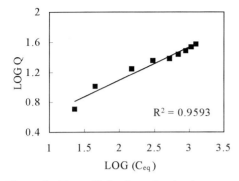

Figure 5a. Freundlich adsorption isotherm for TCC biosorption by *C. cladosporioides* biomass.

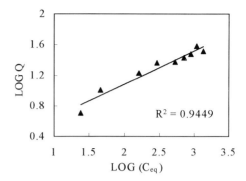

Figure 5b. Freundlich adsorption isotherm for TCN biosorption by *C. cladosporioides* biomass.

were 19 mM of TCC and 16.2 mM of TCN. These values represented an approximately 38 and 32 fold concentration of TCC and TCN, respectively, as compared to the original concentration of 0.5 mM that was used for biosorption. During the second cycle of metal-cyanide adsorption, loading capacity of the biosorbent decreased by 55%.

3.5. Biodegradation of unrecoverable (residual) TCC and TCN

When TCC and TCN solutions after biosorption were subjected to biodegradation in flasks, it was seen (Fig. 6a and 6b) that the consortium culture could degrade these metal-cyanide complexes with an efficiency exceeding 99% in 5 h (TCN) and 6 h (TCC). Total cyanide and metal concentrations in the treated solutions were <0.2 mg/l and <3.0 mg/l, respectively.

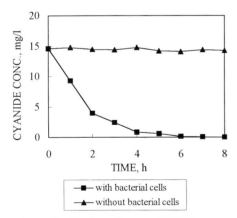

Figure 6a. Biodegradation of TCC by a mixed bacterial community.

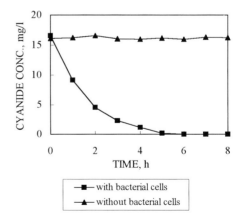

Figure 6b. Biodegradation of TCN by a mixed bacterial community.

4. DISCUSSION

Main objective of the present work was to develop an efficient method for the removal and recovery of metal-cyanide containing waste waters emanating from plating industries. Microbiological degradation of unrecoverable metal-cyanides was therefore employed as a final polishing step to meet the requisite disposal standards.

TCC and TCN dissociate in water forming anionic moieties, viz. $[Cu(CN)_4]^{-2}$ and $[Ni(CN)_4]^{-2}$ respectively. Therefore it was thought worthwhile to explore the possibility of their recovery by adsorption on microbial biomass. Various known sorbents such as activated charcoal, saw dust and fly ash were used for obtaining comparative data. The results obtained revealed that *C. cladosporioides* biomass had a higher metal-cyanide sorption capacity than activated charcoal. This observation opens up new possibilities of developing an efficient microbial technology for the recovery of metal-cyanides from waste waters.

It is well known that process of biosorption is governed by solution pH (14,15). For this reason, the first selection step during the present studies was determination of optimum solution pH. It was found that maximum biosorption of metal-cyanides could take place at pH 4.0 in most cases. No sorption was found at pH 7.0 and in alkaline conditions.

The increase in sorption of metal-cyanide complexes in acidic conditions may be due to the protonation of biosorbent surface and formation of species such as $H_2Cu(CN)_4$ and $H_2Ni(CN)_4$ on the biomass. Thus, relatively more metal-cyanide species can be accommodated on the biosorbent sites. It is worthwhile to mention that HCN gas is not formed under acidic conditions (as happens easily in case of free cyanide i.e. KCN/NaCN) because TCC and TCN are highly stable metal-cyanide complexes (16). Therefore, biosorption under acidic conditions would be a safe procedure. Metal-cyanide containing industrial waste waters generally have pH ranging from 5.5 to 12.5. Obviously, appropriate pH adjustments of the effluents will have to be carried out prior to biosorption.

It was observed that uptake of TCC and TCN by *C. cladosporioides* biomass reached a plateau when the concentration was 1.75 to 2.0 mM (Fig. 3). This might be due to the saturation of binding sites, which clearly showed that metal-cyanide uptake by *C. cladosporioides* biomass was a chemically equilibrated and saturable phenomenon. Despite the complexity of the adsorption process, which can include several mechanisms, adsorption isotherms have been used to characterize metal uptake, and they appear to be of use for projected industrial applications (17). Hence it was decided to fit the available metal-cyanide biosorption data with two most widely accepted adsorption models, viz. Freundlich and Langmuir. Linear transformation of the adsorption data using Freundlich and Langmuir models allowed computation of the metal-cyanide adsorption capacities ($r^2 = >0.94$ for both the models). Experimental data obtained in the studies were found to obey basic principles underlying these models, that is, heterogeneous surface adsorption and monolayer adsorption at constant adsorption energy respectively (18,19).

Process of biosorption is essentially a surface interaction and is characterized by rapid uptake of ions by microbial surfaces. Rapidity of the process makes it a good candidate for use in effluent treatment on a large scale. Experimental results showed that the rate of sorption of metal-cyanides by *C. cladosporioides* is maximum in the first 15 min (50% sorption) and equilibration of the system is achieved in 30 min.

In the studies on sorption/desorption with *C. cladosporioides* biomass it was found that 1 M sodium hydroxide was effective for the removal of bound metal-cyanide complexes. Recovery of TCC and TCN after biosorption was possible in a concentrated form. For example, with the biomass of *C. cladosporioides,* which showed maximum loading capacity, both the metal-cyanides could be concentrated almost 38 (TCC) and 32 (TCN) fold. This achieved the concentration of metal and cyanide in the range of 1000 to 1200 mg/l and 1600 to 2000 mg/l, respectively. Such highly concentrated solution of recovered metal-cyanide may be recycled in the plating circuit in the user industry.

When residual TCC and TCN biodegradation experiment was run under optimized conditions in batch mode it was found that the mixed bacterial community could degrade TCC and TCN in 5-6 h with very high efficiency. Resulting treated solution could pass the disposal standards prescribed by statutory agencies in India. These findings indicated that biodegradation could be used as a polishing step in the treatment of metal-cyanide containing waste waters.

REFERENCES

1. C.J. Knowles, Bacteriol. Rev., 40 (1976) 652.
2. J.J. Ganczarczyk, P.T. Takoaka and D.A. Ohashi, J. Wat. Poll. Cont. Fed., 57 (1985) 1089.
3. W.W. Eckenfelder, Industrial Water Pollution Control, McGraw-Hill, New York, USA, 1989.
4. G. Rollinson, R. Jones, M.P. Meadows, R.E. Harris and C.J. Knowles, FEMS Microbiol. Lett., 40 (1987) 199.
5. I. Finnegan, S. Toerien, L. Abbot, F. Smit and H.G. Raubenheimer, Appl. Microb. Biotechnol., 36 (1991) 142.
6. J. Silva-Avalos, M.G. Richmond, O. Nagappan and D.A. Kunz, Appl. Environ. Microbiol., 56 (1990) 3664.
7. M.M. Figueira, V.S.T. Cifninelli and V.R. Linardi, In: C.A. Jerez, T. Vargas, H. Toledo and J.V. Wiertz (eds.), Biohydrometallurgy Processing, University of Chile, Chile, 1995, 333.
8. M. Barclay, A. Hart, C.J. Knowles, J.C.L. Meeussen and V.A. Tett, Enzyme Microb. Technol., 22 (1998) 223.
9. J.A. Brierley, G.M. Goyak and C.L. Brierley, In: H. Eccles and S. Hunt (eds.), Immobilization of ions by biosorption , Horwood, Chichester, 1986.
10. M.S. Azab, H.M. El-Shora and H.A. Mohammed, Al-Azhar Bull. Sci., 6 (1995) 311.
11. S.I. Ishikawa and K. Suyama, Appl. Biochem. Biotech., 70-72 (1998) 719.
12. Y.J. Suh, J.M. Park and J.W. Yang, Enzyme Microb. Technol., 16 (1994) 529.
13. APHA, AWWA, WEF. (eds.) Standard Methods for the Examination of Water and Wastewater, Washington, DC, 1992.
14. P.R. Puranik and Paknikar, J. Biotechnol., 55 (1997) 113.
15. A.V. Pethkar and K.M. Paknikar, J. Biotechnol., 63 (1998) 121.
16. R.M.C. Dawson, D.C. Elliott, W.H. Elliott and K.M. Jones, (eds.), Data for Biochemical Research, Oxford Science Publications, Oxford, 1986.
17. M. Tsezos and B. Volesky, Biotechnol. Bioeng., 23 (1981) 583.
18. I. Langmuir, J. Am. Chem. Soc., 40 (1918) 1361.
19. H. Freundlich, Colloid and Capillary Chemistry, Methuen, London, 1926.

Reduction of soil pH using *Thiobacillus* cultures

S. K. Polumuri and K. M. Paknikar[*]

Division of Microbial Sciences, Agharkar Research Institute, G.G. Agarkar Road, Pune 411004, India

Occurrence of saline, alkaline and sulfur-deficient soils has become an important problem in the Indian subcontinent which has affected important cash crops like tea and potato. An attempt was made to reduce the soil pH by providing soils with oxidizable substrate like sulfur or pyrite alongwith inoculation of appropriate *Thiobacillus* cultures. It was seen that among the various soil amendments tested, treatment with sulfur + *T. thiooxidans* was found to be most efficient, followed by sulfur + *T. ferrooxidans*, uninoculated sulfur, and pyrite + *T. ferrooxidans*. With these treatments a decrease in soil pH by 0.5-2.5 units could be achieved. As the soil pH decreased, sulfate levels of the soils raised alongwith concomitant increase in solubilization of micronutrients such as manganese and calcium. Due to decrease in soil pH, a phytopathogenic strain of *Streptomyces scabies* was found to be completely inhibited. These results open up new possibilities of using *Thiobacillus* cultures in soil treatment for increased crop productivity and control of certain plant pathogens.

1. INTRODUCTION

Acidophilic, chemolithotrophic *Thiobacillus* sp. derive energy by oxidizing inorganic forms of ferrous iron or reduced sulfur compounds. In nature, these organisms are found to be associated with iron and sulfur containing minerals deposits. The use of thiobacilli, especially *T. ferrooxidans* in leaching of metals from sulfide ores is well documented (1).

MACS collection of Microorganisms (MCM) housed in our institute has a large number of strains of iron- and sulfur-oxidizing microorganisms belonging to *Thiobacillus* sp. that were isolated from Indian mines (2). While working on metal leaching aspects with these cultures, our attention was drawn to numerous reports on the occurrence of saline, alkaline and sulfur-deficient soils from the Indian subcontinent. Some of the areas in India reported to have such problems are: the Indo-gangetic alluvial and black soils in Gujarat and Madhya Pradesh, certain black soils in Maharashtra, tea plantation soils in Himachal Pradesh, red and laterite soils in Karnataka, Kerala and West Bengal and mixed red yellow soils in Uttar Pradesh (3).

The increase in soil pH and salinity has affected important cash crops like tea and potato in India. Tea plants are cultivated in soils of widely different geological origin and almost all

[*] Corresponding author, Fax: +91-20-351542, E-mail: paknikar@vsnl.com

physical types. Optimum soil condition for tea plantation is acidic (pH 4-6), low in calcium and generally rich in iron and manganese (4). However, due to indiscriminate use of chemical fertilizers, the soil pH is increasing making condition unfavorable for tea plantation. Moreover, increased soil pH has escalated occurrence of diseases like common scab of potato caused by *Streptomyces* sp., which are highly prevalent in alkaline soils. Scab remains one of the most important and least satisfactorily controlled of potato diseases (5).

Possible remediation measures for reclamation of saline, alkaline or sulfur-deficient soils could involve acidification of soils - a process that could be catalyzed in soils by iron- and sulfur-oxidizers. Such acidification could also increase availability of micronutrients and may control diseases caused by *Streptomyces* sp. Work reported in this paper seeks to confirm these possibilities.

2. MATERIALS AND METHODS

2.1. Microorganisms

Thiobacillus ferrooxidans (MCM B-90), *Thiobacillus thiooxidans* (MCM B-41) and *Streptomyces* sp. cultures used in this study were procured from MACS Collection of Micro-organisms (WFCC Code No. 561). *T. ferrooxidans* was maintained in 9K medium (6) while *T. thiooxidans* was maintained in 9K basal medium supplemented with 1% sulfur. Phytopathogenic strain of *Streptomyces* sp. was maintained in the starch-casein nitrate medium.

2.2. Soils

Soil samples were collected from different locations in Western Maharashtra, India. The samples were thoroughly mixed, dried at 60°C in an oven for 12 h and sieved (1.25 mm size). Typical analysis of the composite soil sample was: pH 5.9, organic carbon 2.25%, total nitrogen 0.19% and inorganic phosphorous 0.01%.

2.3 Sulfur and Pyrites

Sulfur flowers were procured from a local manufacturer (PCL Chemicals, Pune). Pyrite sample was obtained from M/s Pyrites, Phosphates and Chemical Ltd., Amjhore, Bihar, India. The proximate chemical analysis of pyrites was - iron (14%), sulfur (18%), magnesium oxide (0.3%), calcium oxide (0.2%), aluminium oxide (10%), silicon oxide (47%), carbon (3.5%), zinc (80 mg/kg), copper (80 mg/kg) and manganese (40 mg/kg).

2.4 Soil amendments

The experiments were performed in sets of polythene bags filled with 250 g soil having a moisture content 60% w/w. Soils were supplemented with 5 g of sulfur flower or 20 g of pyrites. Two sets of polythene bags supplemented with sulfur were inoculated with 10 ml cell suspension of *T. ferrooxidans* or *T. thiooxidans* (cell density 25×10^8 /ml). Another set of polythene bags that was supplemented with pyrites was inoculated with a culture of *T. ferrooxidans* of similar cell density. The bags were incubated at room temperature (ca. 28 ± 3°C) in dark. Soil samples (20 g) were removed from the bags at an interval of 7 days and were analyzed for pH, sulfates, manganese, iron, and calcium content. Appropriate uninoculated controls with and without sulfur or pyrite supplementation were also run simultaneously. All the experiments were carried out in duplicates and repeated twice.

2.5. Effect of soil amendments on phytopathogenic *Streptomyces scabies*

In this experiment sterile 250 g soils of pH 7.5 were filled in polythene bags. The soils were supplemented with 5 g of sulfur flowers that were steam sterilized. Two different sets of polythene bags were inoculated with 10 ml cell suspension of *T. ferrooxidans* or *T. thiooxidans* (cell density 25 x 10^8 /ml). Subsequently, the bags were also inoculated with *Streptomyces scabies* so as to get a final density of 3.5 x 10^4 cells/g soil. Bags were incubated at room temperature (ca. 28±3°C) in dark. Soil samples (1 g) were removed from the bags at an interval of 7 days and total viable counts of *Streptomyces scabies* were taken by plating appropriate dilutions on starch-casein-nitrate agar. Appropriate uninoculated controls with and without sulfur supplementation were also run simultaneously. All the experiments were carried out in duplicates and repeated twice.

2.6 Analyses

pH of soil in polythene bags was determined by saturating it with water (20 g of soil in 100 ml of distilled water), stirring the slurry on a shaker for 1 h and measuring the pH using a pH meter (7).

Samples of saturated soils, as described above (10 ml each) were centrifuged (Sorvall, RC 5B plus, USA) at 10,000 rpm for 15 minutes. Supernatants were then used to determine soluble sulfate levels in the soils by barium chloride turbidimetric method (7). Analysis of iron, manganese and calcium was carried out by Atomic Absorption Spectrophotometer (Unicam 929 AA spectrometer, UK).

3. RESULTS

3.1. Effect of soil amendments on soil pH and solubilization of micronutrients

It could be seen from the data given in Table 1 that there was very little decrease in pH of soils supplemented with pyrite (0.30 and 0.51 units in uninoculated and inoculated soils, respectively). However, in case of soils supplemented with sulfur, the pH was found to decrease considerably. The decrease was more pronounced when bacterial cultures were inoculated. Among the bacterial cultures tested, *T. thiooxidans* could reduce soil pH to the greatest extent, i.e. by 2.57 units in 35 days.

A similar trend could be observed when sulfate levels of soils were monitored. It was again seen that addition of pyrite with or without inoculation of bacterial culture did not contribute significantly to the increase in sulfate levels. However, with sulfur, the sulfate levels increased rapidly within 7 days. Interestingly, after 7 days the levels of sulfate were higher in soils inoculated with *T. ferrooxidans*. During further incubation, however, soils inoculated with *T. thiooxidans* recorded a fast increase and at the end of 35 days sulfate levels in these soils were to the tune of 6500 µg/g.

As regards solubilization of micronutrients, viz. manganese, calcium and iron, it was seen that none of the soil amendments improved the availability of soluble iron (speciation of iron not carried out). Manganese and calcium were solubilized by all types of soil amendments.

However, higher levels of these micronutrients could be found in soils inoculated with bacterial cultures.

In general, it was seen that as the soil pH decreased, sulfate levels raised alongwith concomitant increase in solubilization of manganese and calcium. To summarize, among the various soil amendments tested, treatment with sulfur + *T. thiooxidans* was found to be most

Table 1
Effect of soil amendments on soil pH and solubilization of micronutrients[a]

Soil Treatments	Parameters	Number of days				
		7	14	21	28	35
Pyrite	pH[b]	0	0	0	0	0.30
	Sulfate (μg/g)	57	80	247	248	256
	Mn (μg/g)	9.2	10.2	30	27.2	69.2
	Ca (μg/g)	16	17	21	36	72
	Fe (μg/g)	0	0	0	0	0
Sulfur	pH[b]	0.89	1.20	1.23	1.46	1.70
	Sulfate (μg/g)	487	860	1069	1991	2609
	Mn (μg/g)	40	99	194	261	348
	Ca (μg/g)	38	68	50	100	124
	Fe (μg/g)	0	0	0	0	0
Pyrite + *T. ferrooxidans*	pH[b]	0.17	0.19	0.11	0.22	0.51
	Sulfate (μg/g)	76	138	279	250	283
	Mn (μg/g)	21	26	28	64	95
	Ca (μg/g)	24	28	90	50	77
	Fe (μg/g)	0	0	0	0	0
Sulfur + *T. ferrooxidans*	pH[b]	1.02	1.47	2.06	1.98	2.36
	Sulfate (μg/g)	1015	1156	1254	3671	4769
	Mn (μg/g)	57.7	118.7	301.7	308.7	512.7
	Ca (μg/g)	92	122	116	135	122
	Fe (μg/g)	0	0	0	0	0
Sulfur + *T. thiooxidans*	pH[b]	1.09	1.74	2.11	2.17	2.57
	Sulfate (μg/g)	633	1279	2726	3824	6545
	Mn (μg/g)	58	133	296	415	755
	Ca (μg/g)	69	117	83	120	147
	Fe (μg/g)	0	0	0	0	0

a: values represent net increase after correction of basal levels present in soils
b: values represent decrease in pH units as compared to pH of uninoculated soils (5.9)

Table 2
Effect of soil amendments on *Streptomyces scabies*

Soil treatment	Total viable count (cfu/g soil)	
	0 days	7 days
Streptomyces scabies	3.5×10^4	6×10^4
Sulfur + *Streptomyces scabies*	3.5×10^4	3×10^4
Sulfur + *T. ferrooxidans* + *Streptomyces scabies*	3.5×10^4	nil
Sulfur + *T. thiooxidans* + *Streptomyces scabies*	3.5×10^4	nil

efficient, followed by sulfur + *T. ferrooxidans*, uninoculated sulfur and pyrite + *T. ferrooxidans*.

Considering these results, further experiments on control of phytopathogenic strain of *Streptomyces scabies* were carried out using the two most effective amendments, viz. treatment with sulfur + *T. thiooxidans* and sulfur + *T. ferrooxidans*.

3.2 Effect of soil amendments on phytopathogenic *Streptomyces scabies*

It could be clearly seen (Table 2) that phytopathogenic strain of *Streptomyces scabies* was completely inhibited by treatment of soil with sulfur and bacterial inoculants, viz. *T. thiooxidans* and *T. ferrooxidans* in a period of 7 days. The pH of the soil after these treatments was found to decrease by one pH unit.

4. DISCUSSION

The observed reduction in pH of soils supplemented with sulfur flower but without inoculation of thiobacilli could be due to oxidation of sulfur by native microflora. It is known that many types of aerobic bacteria, actinomycetes and yeast are able to oxidize sulfur in soils (8,9). This observation was corroborated by the corresponding figures obtained for enhanced sulfate levels. The fact that soils supplemented with sulfur and inoculated with *T. ferrooxidans* or *T. thiooxidans* showed significant decrease in soil pH (by 2.36 and 2.57 units, respectively) indicated that introduction of proper bioinoculants could improve sulfur oxidation. Rupela and Tauro (10) showed that alkaline soils in India possess low population of thiobacilli and suggested that rapid reclamation of these soils could be brought about by enrichment with efficient sulfur oxidizing bacteria. To test this possibility, they isolated a strain of *T. novellus* from the alkaline soils and used it in laboratory studies to inoculate the same soil amended with 1% (w/w) elemental sulfur. After inoculation, the pH of soil fell by 2.7 pH units compared to a fall of 1.9 units in the uninoculated soil. They also noted a significant decrease in soil pH when 0.5% (w/w) elemental sulfur was used. Despite these studies, the inoculation of soils with thiobacilli has not become a standard practice.

In areas where tea is cultivated in India, soil pH is in the range of 6.0 to 7.5. However, due to indiscriminate use of fertilizers pH is rising in most locations. A soil pH between 4-6 is considered most conducive for tea plantation. Therefore, if soil pH in these areas could be brought down by 1.5-2.0 units, it would help the plantation. Our studies clearly showed that soil amendments with sulfur alongwith inoculation of cultures such as *T. thiooxidans* and *T. ferrooxidans* could be used for this purpose.

Sulfate levels in soil amended with sulfur and inoculated with *T. ferrooxidans* or *T. thiooxidans* were found to increase considerably. This may be due to enhanced sulfur oxidation brought about by these cultures. Widespread deficiency of sulfur was reported from soils under groundnut and tea plantation in northern states of India. Based on soil analyses, a number of alluvial, red and laterite samples were found to be low in available sulfur (3). Our studies showed that soil amendment with sulfur/pyrite and appropriate bacterial cultures could be of help in sulfur deficient soils.

The manganese levels increased in all the soil amendments tested but the increase was more pronounced in soils treated with sulfur. Obviously, sulfuric acid generated by oxidation of sulfur reacted with soil minerals and other insoluble forms leading to nutrient mobilization. Manganese occurs in soil in the tetravalent forms and exchangeable divalent forms. However, plants are known to use assimilable divalent manganese forms. At higher pH in soil, manganese exhibits tetravalent forms which are not available to plants. Such a situation can be corrected by application of sulfur which increases the concentration of divalent forms (9).

Our observations that calcium levels increase in soils amended with pyrite and sulfur treatment is important. In soil, the native insoluble calcium carbonate reacts with sulfuric acid produced by bacterial action and gets converted into soluble calcium sulfate. If employed for reclaiming an alkaline soil high in exchangeable sodium, the transformation cycle of sulfur in soil is helpful in more than one way. Firstly, the acid produced will neutralize any sodium carbonate present in the soil, secondly, the acidic media created will lower the pH of the soil and lastly, the calcium solubilized from the soil will exchange with sodium, to convert sodium-rich infertile soil to calcium-rich fertile soil. The overall effect will be reflected in large improvement in the water infiltration and permeability characters of the soil (11).

When soluble concentration of iron was measured during our study no increase in the iron concentration of soils was noticed in any of the treatment used. It is known that factors such as high pH levels or high concentrations of phosphates retard the oxidation of pyrites (12,13). This inhibitory effect appears to be due to the formation of insoluble iron compounds like iron hydroxide and iron phosphate which block the oxidation and release of iron from pyrite surface.

Iron, manganese and zinc concentrations were found to have increased in plant tissue due to application of pyrites (14). This is because pyrite not only contains these micronutrients as impurities but also enhances the availability of native micronutrients by increasing the acidity around the pyrite particles which is of particular importance in calcareous soils that are prone to lime-induced chlorosis (15). Vlek and Lindsay (16), in their study involving addition of pyrites at the rate of 1%, found that iron levels decreased from 1.5 to 64 ppm, while manganese levels quadrupled. However, Jaggi *et al.*(17) have also reported an increase in available iron from 14.5 to 65.7 ppm with the application of pyrites.

In our study, phytopathogenic *Streptomyces scabies* was found to be completely inhibited within 7 days of incubation in the soils supplemented with sulfur and inoculated with *T. thiooxidans* and *T. ferrooxidans*. It has been reported that the total viable counts of actinomycetes were reduced in 45 days in soils supplemented with 1% sulfur and inoculated with *Thiobacillus* sp.(18). Our results are in agreement with this finding.

Scale-up experiments need to be carried out for possible practical scale utilization. The results are sufficiently interesting and open up new avenues of research on the use of *Thiobacillus* cultures in soil treatment. We believe that further research in this area would help increase crop productivity and aid in the control of certain plant pathogens.

REFERENCES

1. D.G. Lundgren and M. Silver, Ann. Rev. Microbiol., 34 (1981) 263.
2. K.M. Paknikar and A.D. Agate, MIRCEN Journal., 3 (1987) 169.
3. H.L.S. Tondon, Sulfur in Agriculture., 9 (1985) 8.
4. Handbook of Agriculture, Indian Council of Agricultural Research (ICAR), New Delhi, India, 1992.
5. G.S. Shekawat, B.D. Singh and M.D. Jeswani, Central Potato Research Institute (CPRI), Indian Council of Agriculture Research(ICAR), India, Technical Bulletin No. 41 (1993).
6. M.P. Silverman and D.G. Lundgren, J. Bacteriol., 77 (1959) 642.
7. R.K. Trivedy and R.K. Goel., Chemical and Biological Methods for Water Pollution Studies, Environmental Publication, Karad, India, 1986.
8. R.L. Starkey, Soil Sci., 70 (1950) 55.
9. M. Alexander, Introduction to Soil Microbiology, John Wiley and Sons, Inc., New York and London, 1961.
10. O.P. Rupela and P. Tauro., Soil Biol. Biochem., 5 (1973) 899.
11. Pyrite in the Reclamation of Alkali Soils, Mining for Agriculture, Pyrites, Phosphates and Chemicals Ltd., Amjhore, Bihar, India, 1981.
12. M.J. Barrow, Aust. J. Expt. Agr. Anim. Husb., 11 (1971) 217.
13. A. Quispel, G.W. Harmsen and D. Otzen., Plant Soil., 4 (1952) 43.
14. K.N. Kaul, K.L. Luthra and K.N. Tiwari., Proc. Sem. FAI-PPCL-DAUP, Lucknow (1978) 134.
15. T.N. Jaggi., Proc. Sem. FAI-PPCL-DAUP, Lucknow (1978) 37.
16. P.L.G. Vlek and W.L. Lindsay., J. Environ. Qual., 7 (1978) 111.
17. T.N. Jaggi, K.L. Luthra and K.N. Goel, Sem. on Management of Salt-affected Soils FAI (Eastern region) Calcutta, India, (1982) 139.
18. S.W. Jadhava, B.K. Konde and S.Y. Daftardar, Curr. Agric., 3 (1979) 115.

Entrapment of particles from suspensions using *Aspergillus* species

K. M. Paknikar[*], J. M. Rajwade and P. R. Puranik

Division of Microbial Sciences, Agharkar Research Institute, G.G. Agarkar Road, Pune 411 004, India

A fungal isolate belonging to *Aspergillus* sp., could entrap different particulates such as ferric oxide, granulated carbon in pellets during growth. The entrapment of suspended particulates was dependent on two factors, viz. ability of the particulates to remain in suspension and the growth morphology of the fungus. Growth morphology of the fungus was found to be governed by composition of medium, inoculum size, incubation conditions etc. It was possible to suitably modify these factors to achieve increased entrapment of the particles. In experiments with iron ore, it was found that although the weight of ore entrapped in the pellets increased from 590 mg to 640 mg with the increase in inoculum size from 1 ml to 5 ml, the iron content of entrapped ore decreased from 56.5% to 43.0%. This indicated that with increase in inoculum size and consequent decrease in the pellet size, less number of iron particles were entrapped by the fungus although overall particulate entrapment increased. These results indicated that this phenomenon could possibly be used in practice for (a) removal of suspended particulate matter from waste waters and (b) beneficiation/upgradation of ores by removing the impurities.

1. INTRODUCTION

Fungi are ubiquitous and may be even dominant in adverse environments such as metal-polluted sites. They exhibit a variety of responses towards heavy metals (1) which, in principle, could be exploited in developing biotechnological methods for the abatement of pollution caused by metals. For example, biosorption of metal ions by fungal biomass has been studied extensively (2). Fungi are known to bring about various kinds of metal transformations, viz. methylation of selenite (3), reduction of chromium (4), oxidation of manganese, solubilization of iron from iron oxide (5), etc. Fungi are well suited for use in transformation of metals as they have an intrinsic ability to tolerate high concentrations of metals and extreme pH conditions (6). Many filamentous fungi can adsorb particulates such as elemental sulfur, insoluble sulfides, charcoal, clays and magnetites (7,8). As early as 1918, Williams (9) showed that fungi could adsorb gold from colloidal solution, although this ability seems not to have been exploited industrially. During our recent studies, an *Aspergillus* sp. was found to possess an extraordinary ability of entrapping particulates from solutions inside the pellets

---

[*] Corresponding author, Fax: +91-20-351542, E-mail: paknikar@vsnl.com

during growth. When the fungus was grown in a ferric oxide containing medium, particles in suspension were entrapped in fungal pellets thereby clarifying the medium. It is well known that certain physiological and physico-chemical factors influence pellet morphology of fungi (10). Since such an entrapment property could be potentially used in clarifying effluents by removing insoluble materials, an attempt was made to investigate the phenomenon further and the results obtained are presented in this paper.

2. MATERIALS AND METHODS

2.1. Microorganisms and growth conditions

In our earlier studies on the heterotrophic leaching of iron a fungal culture was isolated from municipal sewage sample. The strain was identified as *Aspergillus* sp. and was deposited in MACS Collection of Microorganisms (World Data Center Code No. 561, Accession No. MCM F-2). The strain was maintained on Sabouraud's agar slants and stored at 5°C.

2.2. Effect of medium composition and inoculum size

During the experiments the fungus was grown on Sabouraud's agar at 30°C and the spores were harvested from 5 days old slants. They were suspended in sterile 1:1000 triton-X 100 solution. Density of the suspension was adjusted to 1×10^8 spores/ml determined by a microscopic count.

In order to study the effect of growth media components on pellet morphology, the spores were inoculated in a series of 250 ml Erlenmeyer flasks containing various combinations of glucose (1-25 g/l) and peptone (5-20 g/l). pH of the medium was adjusted to 4.0 and medium to flask volume ratio was maintained at 1:5. Flasks were incubated at 30°C on a rotary shaker (Gallenkamp, UK) at 120 rpm for 72 h. After growth, pellets were observed microscopically and their sizes were determined using a stage slide micrometer. The utilization of glucose during growth was assessed by dinitrosalicylic acid (DNSA) method. Dry weight of the fungus in the experimental set was also determined. For determination of dry weight, fungal mycelia were harvested by filtering through Whatman No. 1 filter paper, dried in an oven at 60°C and dry weights were recorded.

Effect of inoculum size on the pellet morphology was checked by inoculating a series of flasks containing 50 ml Sabouraud's broth with different volumes of spore suspension (1-5 ml).

2.3. Entrapment of particulates by *Aspergillus* sp.

The ability of the *Aspergillus* sp. MCM F-2 to entrap particulate matter was checked in Sabouraud's medium supplemented with sterilized particulates (viz. granulated charcoal, sand, ferric oxide, manganese dioxide, a combination of ferric oxide-manganese dioxide, iron ore) at 1% w/v level. All the samples had a particle size distribution in the range of 100-150 mesh (0.149-0.105 mm) according to Tyler standard screens. Flasks were inoculated with 1 ml spore suspension and incubated at 30°C for 72 h. After incubation, usual growth parameters were measured. Fungal pellets containing the entrapped particles were separated by sieving and their dry weights were recorded. In case of experiments involving iron ore, samples of the pellet-entrapped ore and the residual ore in flasks were digested in minimal amounts of concentrated HCl. These samples along with the spent media were analyzed for total iron content by using an atomic absorption spectrophotometer (Perkin Elmer, Model 2380, USA). The changes in pellet structure during growth in presence of ferric oxide was checked using an electron microscope (Stereoscan 200, Cambridge Instruments, UK). For this purpose, pellets grown in presence/absence of ferric oxide were fixed in 2% glutaraldehyde for 18 h and fixation was followed by dehydration using 5-100% ethanol. All procedures were carried out at 4°C. The samples were coated with platinum (Biorad, UK). In case of flasks containing ferric oxide scanning electron microscopic observations were made.

3. RESULTS

It was observed that by varying glucose and peptone concentrations in the media, growth morphology of the *Aspergillus* sp. MCM F-2 changed considerably (Table 1). At glucose concentration of 10 g/l, dry weight of the fungus increased in response to increasing concentrations of peptone (5-20 g/l). A similar observation was made with fixed peptone concentration (20 g/l) and varying glucose concentrations (1-10 g/l) in the media. Data in Table 1 further show that cultures containing 10 g/l glucose, and peptone concentrations in the range of 5-15 g/l had clumped, coalesced growth. With peptone concentration of 20 g/l, however, a pelleted growth was obtained. Residual sugar data demonstrated that very little glucose was utilized in cultures containing lower peptone concentrations, which indicated that growth in these cultures was limited by nitrogen levels. It was further observed that inoculum size had a marked effect on the average pellet size when pelleted growth was obtained. It

Table 1
Effect of peptone and glucose concentration on growth morphology of *Aspergillus* sp.

Medium composition		Growth characteristics of *Aspergillus* sp.				
Peptone (g/l)	Glucose (g/l)	Glucose (residual) (g/l)	Dry weight of fungus (mg/100 ml)	Growth morphology	Average pellet size	
					Inoculum	
					1 ml	5 ml
10	20	17.05	621.4	Pelleted	2.0	0.5
20	10	2.76	1021.6	Sporulated pelleted	2.5	0.5
20	5	0.32	840.8	Pelleted	2.0	0.5
20	1	0.33	703.4	Pelleted	1.0	0.5
15	10	7.12	507.6	Coalesced growth	-	-
10	10	5.26	449.0	Coalesced growth	-	-
5	10	9.66	128.0	Coalesced growth	-	-
5	25	22.37	315.0	Coalesced growth	-	-

Table 2
Growth of *Aspergillus* sp. in the presence of particulate matter

Particulates used	Growth characteristics of *Aspergillus* sp.
Activated charcoal	Uptake of activated charcoal in pellets
Sand	No uptake of sand, pelleted growth
Ferric oxide	Uptake of ferric oxide pelleted growth
Manganese dioxide	No uptake, pelleted growth
Ferric oxide + Manganese dioxide	Uptake of only ferric oxide, pelleted growth
Iron ore	Uptake of iron ore in pellets

could be seen that pellet size decreased proportionately with increase in the level of inocula.

Experiment carried out to check the ability of *Aspergillus* sp. MCM F-2 to entrap various particulates (Table 2) showed that although pellet formation was observed in all the flasks, there was no entrapment of sand and manganese dioxide. In case of manganese dioxide-ferric oxide mixture, selective entrapment of ferric oxide was observed. When an iron ore was used, uptake of the ore particles was observed.

The pellets from flasks supplemented with ferric oxide were red in color. Light microscopic observations revealed presence of amorphous ferric oxide particles entangled in the fungal filamentous structure. In order to assess whether the entrapped particulates had any influence on surface topography, pellets were observed using a scanning electron microscope. Observations revealed that diameter of mycelia decreased by almost 50% in the presence of iron oxide (Figure 1). Further, the filaments in this case were tightly arranged as compared to flasks without iron oxide.

When particulate entrapment experiment was carried out on a quantitative basis with an iron ore by varying the inoculum size, interesting observations could be made (Figure 2). It was found that although weight of ore entrapped in the pellets increased from 590 mg to 640 mg

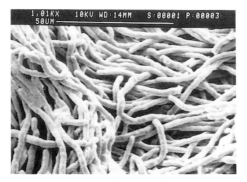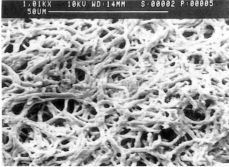

Figure 1. Scanning Electron Micrographs of *Aspergillus* pellets before (left) and after (right) entrapment of ferric oxide. Note reduction in mycelial diameter and increased compaction in ferric oxide entrapped pellets (right) (Magnification 1000X).

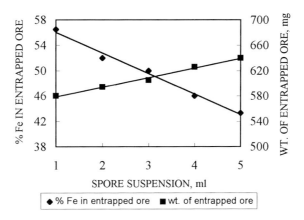

Figure 2. Effect of inoculum size on entrapment of iron by
Aspergillus sp. from iron ore.

with increase in the inoculum size from 1 ml to 5 ml, the iron content of entrapped ore
decreased from 56.5% to 43.0%. This indicated that with increase in inoculum size and
consequent decrease in the pellet size, less number of iron particles were entrapped by the
fungus although overall particulate entrapment increased.

4. DISCUSSION

Results obtained with *Aspergillus* sp. MCM F-2 showed that composition of medium
significantly affected growth morphology of the fungus. This result suggested that
pellet/filament form could be obtained by changing proportion of the media constituents.
These results were in agreement with those observed for *Rhizopus arrhizus* by Byrne and
Ward (11). The pelleted morphology is desirable in certain industrial applications such as
antibiotic fermentation, vitamin production, etc. However, filamentous morphology is
preferred in industrial production of organic acids.

Wainwright *et al.* (7,8) reported that mycelium of *Mucor flavus* adsorbed calcium silicate,
elemental sulfur, lead sulfide and zinc dust from suspension. It also adsorbed ferric hydroxide
from acid mine drainage. The adsorption increased with increase in proportion of mycelium
and addition of a carbon substrate proved inhibitory to the process. This phenomenon can be
described as "bioadsorption". However, entrapment of particulates described in this paper is
different from the above observation because particulates were not adsorbed *sensu strictu* but
were probably enmeshed in the tight mycelial structure of the pellet. Since pellet formation
was influenced by medium composition, the entrapment was also affected by these factors.

Entrapment of suspended particulates is a phenomenon dependent on two factors, viz. the
ability of the particulates to remain in suspension and growth morphology of the fungus. These
factors, in turn, are governed by various physical and physiological properties like specific
gravity of the particulates, composition of growth medium and growth conditions. Since
medium composition and the growth conditions for particulate entrapment experiments were

identical, it could be inferred that ability of the particulates to remain in suspension was a significant factor affecting entrapment. Our studies do not answer the mechanism of entrapment fully at this stage. However, they point out a new area in mineral biotechnology which could possibly be exploited for (a) separation of suspended particulate matter from waste waters and (b) beneficiation/upgradation of ores by selectively removing the impurities, when conventional methods are not feasible.

REFERENCES

1. G.M. Gadd, In: R.A. Herbert and G.A. Codd (eds.), Microbes in Extreme Environments, Academic Press, London, 1987.
2. A. Kapoor and T. Viraraghavan, Biores. Technol., 53 (1995) 195.
3. J.M. Brady, J.M. Tobin and G.M. Gadd, Mycol. Res., 100 (1996) 955.
4. K.M. Paknikar and J.V. Bhide, In: A.E. Torma, M.L. Apel and C.L. Brierley (eds.), Biohydrometallurgical Technologies Vol. II, The Minerals, Metals & Materials Society, Warrendale, PA, 1993.
5. J.C.G. Ottow and A. Von Klopotek, Appl. Microbiol., 18 (1969) 41.
6. C.L. Brierley, J.A. Brierley and M.S. Davidson, In Metal Ions and Bacteria. T.J. Beveridge & R.J. Doyle (eds.). Wiley Interscience Publication, John Wiley & Sons. Inc., New York, 1989.
7. M. Wainwright, S.J. Grayston and P. DeJong, Enzyme Microb. Technol., 8 (1986) 597.
8. M. Wainwright, I. Singleton and R.G.J. Edyvean, Biorecovery, 2 (1990) 37.
9. M. Williams, Ann. Bot., 32 (1918) 531.
10. A. Whitakar and P.A. Long, Process Biochem., 8 (1973) 27.
11. S. Byrne and O.P. Ward, Biotech. Bioeng., 33 (1989) 912.

Biological Processes for Thiocyanate and Cyanide Degradation

G.V. Rorke and R.M. Mühlbauer

Billiton Process Research, Randburg, South Africa

A common problem for bacterial oxidation plants that treat refractory gold concentrates is the low tolerance of the bacteria to thiocyanate and cyanide species. Thiocyanate forms from a reaction between cyanide and a sulphur species during cyanide leaching of the bioleached concentrate. The final plant effluent streams contain thiocyanate and cyanide, and generally report to a single tailings dam. Consequently the tailings dam return water cannot be utilised in the bacterial oxidation plant.

Environmental legislation and water balance considerations, such as dry season water shortages and wet season water surplus, require a water stream that can be recycled to the bacterial oxidation plant. Such water must be free of thiocyanate and cyanide.

A culture of micro-organisms, isolated from a tailings water containment at Fairview Gold Mine, was shown to degrade both cyanide and thiocyanate. Two different thiocyanate destruction processes have been tested on a pilot scale utilising these micro-organisms. The first process piloted was an attached growth process and the second an activated sludge process. This paper gives some results of the laboratory and pilot plant testwork.

1. INTRODUCTION

Tailings arising from the BIOX® process tend to contain relatively high levels of thiocyanate. Thiocyanate is formed through the interaction of free cyanide with sulphur species during the cyanide leaching of the bioleached material. The presence of both cyanide and thiocyanate in the tailings water prevents the option of recycling this water upstream of a BIOX® circuit, due to its toxicity to the bacterial strains. In addition, this solution cannot be used as wash water in the post BIOX® CCD, as thiocyanate is an effective gold lixiviant in an acidic environment.[1] Environmental legislation associated with the land disposal of cyanidation tailings is becoming increasingly stringent world-wide and dry season water shortages at some mine sites require recycling of solutions, thus enforcing the treatment of contaminated water streams.

The treatment of cyanide is well-established technology and various commercial processes are available, for example, the INCO SO_2 [2,3] and the Degussa hydrogen peroxide [3,4] processes. These processes are generally targeted at the destruction of cyanide and only oxidise a fraction of the thiocyanate and some of the cyanide-metals present. Older technology, such as alkaline chlorination, [3,5] destroys some of the thiocyanate and none of the ferro-cyanide complexes and also tends to be costly, and hazardous to the environment and to personnel. The biological degradation of thiocyanate and cyanide in effluents provides an alternative treatment to the more traditional processes. Several advantages are associated with biological degradation of thiocyanates; these include the production of non-toxic end- or by-products, minimal costs and operation under ambient conditions. [6,7]

The biodegradation of cyanide, thiocyanate and cyano-metals has been widely reported [8,9,10] and has been shown to occur in a large number of microorganisms which include bacteria, actinomycetes and fungi. Cyanide and thiocyanate are degraded to ammonia and carbon dioxide. Sulphate is formed as an additional end product during the breakdown of thiocyanate. [6,7,8] All forms of cyanide-metal complexes are generally removed through a combination of oxidation and sorption into the biofilm/biomass. [3,6]

A full scale biological process treating cyanidation tailings containing less than 50 mg/l Weak Acid Dissociable cyanide (WAD), has been operational at the Homestake Mine, Lead, South Dakota since 1984. [3,6] The Homestake process is, however, principally concerned with the destruction of cyanide and the subsequent conversion of ammonia to nitrite, because the feed solutions generally contain relatively low levels of thiocyanate. Since BIOX® cyanidation effluents contain fairly high concentrations of thiocyanate, in addition to cyanide and cyanide-metal complexes, an investigation was initiated at BILLITON Process Research for the efficient removal of thiocyanate and cyanide from the post-BIOX® cyanidation tailings as it has both environmental and economic advantages.

2. EXPERIMENTAL

2.1. Micro-organisms

A composite culture, comprising a mixture of various cultures isolated at the Fairview tailings dam, was used for all cyanide and thiocyanate degradation testwork as well as the attached growth and activated sludge processes.

2.2. Nutrient Medium

Laboratory scale tests were conducted using a synthetic non-sterile liquid medium containing phosphate and thiocyanate. Tailings dam water, which was supplemented with phosphate, was used as nutrient medium for the attached growth process testwork. The initial nutrient solution used for the activated sludge testwork consisted of phosphate and thiocyanate supplemented with molasses as an organic carbon source. Subsequent tests for the activated sludge process utilised tailings dam water originating from different sources.

2.3. Laboratory Scale Testwork

Laboratory scale testwork was conducted to establish the effect of relative concentrations of thiocyanate in the feed solutions. The parameters of operation have been described at a previous conference.[7]

Laboratory scale tests using synthetic solutions, containing phosphate and varying concentrations of thiocyanate, were conducted in stirred glass reactors with a working volume of 1.8 litres. Air was injected into the reactors and a dual blade flat bladed turbine provided agitation.

2.4. Attached Growth System

A pilot plant consisting of six columns of Ø0.15m x 2m was set up at Fairview Mine. Each column was packed with 16mm pall rings to increase surface area of the immobilised biomass. Aeration was provided by injecting air into the base of each of the columns. A batch system was adopted initially to establish bacterial growth in the columns whereby the overflow solution was re-circulated to the feed tank. Once bacterial growth was obtained, the columns were operated continuously in parallel. The feed was pumped from the base of a 1000 litres feed tank to the base of each column via a variable speed pump. Further testwork was conducted by changing the pilot plant configuration to a series configuration, first with three columns and then with all six, and operated continuously. The feed was pumped from the base of a 1000 litres feed tank to the base of the first column via a variable speed pump and the overflow from each column reported to the base of the next column.

2.5. Activated Sludge System

A pilot plant consisting of a 60 litre reactor with a clarifier was constructed. Aeration was provided by injecting air through a circular sparge ring at the bottom of the reactor. The feed was pumped into the reactor via a variable speed pump. The sludge from the reactor overflowed into the clarifier. The supernatant solution from the clarifier reported as the final product while the thickened sludge was returned to reactor by a variable speed peristaltic pump.

Start up of the process was achieved by half filling the reactor with activated sludge obtained from the re-aeration ponds of a local water treatment works. This sludge was then inoculated with 10 litres of solution containing biomass removed from the attached growth columns.

2.6. Process Monitoring and Chemical Analyses

All tests were monitored routinely for thiocyanate concentrations using a Hach spectrophotometer and the *Nanocolor* thiocyanate test kit. Samples were analysed for cyanide in solution where cyanide was expected to be high.

734

3. RESULTS AND DISCUSSION

The specific objective of this testwork was to produce an effluent that consistently contained less than 1 ppm of thiocyanate in order to enable the re-use of the solution in a BIOX® process. The rate of degradation was used to evaluate the performance of the test reactors. The rate equation used for the laboratory scale tests indicated an instantaneous rate, as defined in equation 1, whereas the rate used for both the activated sludge and attached growth reactors was an overall rate (Equation 2). Spatial variations in thiocyanate concentrations and degradation rate may occur within each column. However, instantaneous rates within each column were not analysed.

$$Rate = \frac{dC}{dt} \tag{1}$$

$$Rate = \frac{(C_{in} - C_{out}) * F}{V} \tag{2}$$

Where :
	C	= Thiocyanate Concentration (ppm)
	t	= Time (h)
	C_{in}	= Thiocyanate Concentration in the Feed (ppm)
	C_{out}	= Thiocyanate Concentration in Product (ppm)
	F	= Feed Rate (l/h)
	V	= Reactor or Column Volume (l)

3.1 Laboratory Scale Testwork

Several laboratory scale tests were conducted to determine the effect of thiocyanate concentration and time on the performance of the culture. The results of the experiments are depicted in Figures 1 and 2.

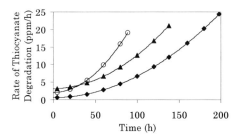
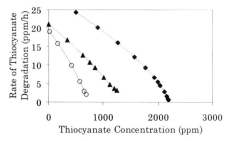

Figure 1. Effect of Time on Degradation Rate.
☐ Initial Concentration 700 ppm,
☐ Initial Concentration 1500 ppm,
◆ Initial Concentration 2000 ppm

Figure 2. Effect of Concentration on Degradation Rate
☐ Initial Concentration 700 ppm,
☐ Initial Concentration 1500 ppm,
◆ Initial Concentration 2000 ppm

The results of the three tests show that biological degradation of thiocyanate at concentrations above 2 g/l is possible. The trends in Figure 1 indicate that the kinetics are not autocatalytic and increasing the initial thiocyanate concentration increases the period of time required to reach the same rate of degradation. From Figure 2 it is evident that the same degradation rate was achievable at various thiocyanate concentrations. However, in the period required to reach the same rate of degradation, at increasing thiocyanate concentrations, a greater mass of thiocyanate was first degraded. This suggests that a larger amount of biomass is required to achieve the same rate of degradation with increasing thiocyanate concentration and thus the rate of degradation is inversely proportional to concentration. In order, therefore, to develop an efficient process (in terms of process retention), the process needs to operate at the lowest possible concentration of thiocyanate.

3.2. Attached Growth System

The attached growth testwork was conducted as single columns operating in parallel and subsequently as a train of three columns and then six columns operating in series.

Parallel Operation

The results of the parallel operation are shown in Figures 3 and 4.

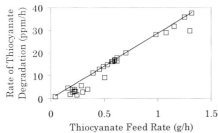

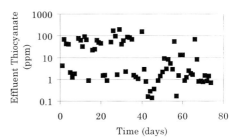

Figure 3. Feed Rate and Degradation Rates achieved by Parallel Operation of Attached Growth Columns. □ Actual, — Required for zero effluent.

Figure 4. Combined Results of Parallel Column. Effluent Thiocyanate Concentration.

A thiocyanate degradation rate of up to 37.4 ppm/h (Figure 3) was achieved but the single columns failed to achieve an effluent thiocyanate concentration of less than 1 ppm consistently (Figure 4).

Series Operation Three Columns

The results for 3 columns operated in series are shown in Figures 5 and 6.

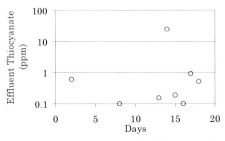

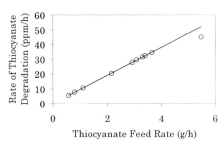

Figure 5. Results of Series Operation Effluent Thiocyanate Concentration.

Figure 6. Feed Rate and Degradation Rates Achieved by Series Operation of Attached Growth Columns. ☐ Actual, — Required for zero effluent.

The process with columns operated in series was far more stable than the parallel column operation. Consistent effluents below 1 ppm thiocyanate were achieved with the exception of day 14 (Figure 5), when the feed rate was increased to 5.5 g/h. The rates of degradation achieved are shown in Figure 6 where the solid line represents the rate required for complete thiocyanate degradation. Effluents with less than 1 ppm thiocyanate were attained up to a feed rate of 3.6 g/h, resulting in a degradation rate of 34.5 ppm/h. An increase in the feed rate to 5.5 g/h exceeded the maximum rate the process could attain while still maintaining an effluent containing less than 1 ppm of thiocyanate.

Series Operation Six Columns

The results for 6 columns operated in series are shown in Figures 7, and 8.

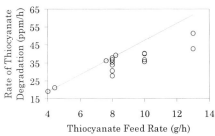

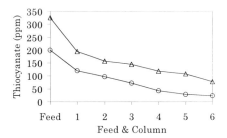

Figure 7. Feed Rate and Degradation Rates Achieved by Series Operation of Attached Growth Columns. ☐ Actual, — Required for zero effluent.

Figure 8. Average Feed and Effluent Thiocyanate Profiles of Each Column. ☐ Feed rate 13g/h. ☐ Feed Rate of 8 g/h.

The effluent for 6 columns operated in series was consistently below the 1 ppm thiocyanate limit when operating at a feed rate below 7.6 g/h, corresponding to a degradation rate of

35 ppm/h. Figure 8 shows the average profile for thiocyanate from the feed to the effluent of each column. The change in concentration from the feed to the column 1 effluent appears to contradict the laboratory tests, which indicated that the rate would be inversely proportional to the concentration. This inconsistency may be attributed to external transport limitations to the packing surface, particularly since the flow in the columns would have been well within what is considered laminar flow (Calculated Reynolds number $\approx$ 70)[11]. External transport was evaluated utilising a mass transfer correlation for non-spherical particles [12] and Colburn's mass transfer factor [13] . The calculation indicated maximum external transfer rates for this process were in the region of 1.5 g/h of thiocyanate. This is substantially lower than was achieved, probably because of inaccuracy in the correlation, and the true surface area in the columns would have been considerably larger than the packing surface due to the irregular surface of the biomass film. It does, however, confirm that mass transfer may have been a limiting factor. It is possible that oxygen may also have been a limiting factor. However, a rigorous analysis to determine the rate limiting step cannot be done in the absence of instantaneous rate data, as well as concentration profiles within the columns.

3.3 Activated Sludge Process
Artificial Feed
Initial tests were conducted with an artificial solution to determine the stability of the process. The results are shown in Figures 9 and 10.

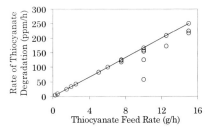
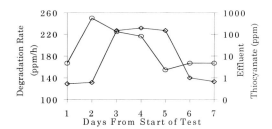

Figure 9. Feed Rates and Degradation rates achieved. □ Actual, — Required for zero effluent.

Figure 10. Effect of Exceeding Maximum Rate. □ Effluent Thiocyanate concentration, □ Degradation Rate

The process was extremely stable below a feed rate of 10 g/h and tolerated sudden increases in the feed rate, while maintaining an effluent thiocyanate concentration of less than 1 ppm. A degradation rate of up to 250 ppm/h was achieved, but stability was compromised. Significantly higher degradation rates were possible with activated sludge in comparison with the attached growth system. The kinetics established by the laboratory testwork, suggested that processes that operate at low concentrations of thiocyanate would give high overall degradation rates. A single stage activated sludge reactor operates at the same concentration as the effluent. In addition, an activated sludge system contains large amounts of biomass and is well mixed, reducing possible external transport resistance.

Figure 10 shows the effect of over-feeding the plant. The feed rate was increased from 10 g/h to 15 g/h after feed and effluent sampling were carried out on day 1. The plant appeared to tolerate the increase as a degradation rate of 250 ppm/h was achieved on day 2. By day 3, the thiocyanate in the effluent increased to 147 ppm and the rate of degradation decreased to

225 ppm/h. This trend worsened as the thiocyanate increased further, resulting in a degradation rate of 210 ppm/h. The feed was reduced to 10 g/h after feed and effluent sampling on day 4, but, the thiocyanate remained high and consequently the measured rate was lower than normally obtained at a feed of 10 g/h. The process returned to normal by day 6, achieving effluent thiocyanate levels of less than 1 ppm. This cycle seems to confirm the laboratory test data, which showed that optimum rates would be achieved at low thiocyanate concentrations. A feed rate of 15 g/h appeared to exceed the maximum rate possible, as this feed rate resulted in a rise in the thiocyanate concentration with a resultant decrease in rate. The reduced rate then elevated the concentration further with an additional decrease in degradation rate. A reduction in the feed rate did not result in an immediate decrease in the effluent thiocyanate concentration to less than 1ppm thiocyanate.

Industrial Cyanidation Effluents

Several industrial cyanidation effluents were treated. Dried samples of activated sludge were analysed for heavy metal accumulation. The results indicated that a considerable amount of gold and arsenic had accumulated on the biomass, even though these solutions only contained trace amounts of these particular metals. The effluent solutions were also tested for toxicity to BIOX® bacteria and in all cases no toxicity was evident. The details of the testwork are represented in Table 1.

Table 1
Industrial cyanidation effluents treated

	Influent		Effluent		Sludge	
	SCN⁻ (ppm)	Total CN⁻ (ppm)	SCN⁻ (ppm)	Total CN⁻ (ppm)	Au (g/t)	As (g/t)
BIOX® CIP Solution	965	n.d. [*]	0.7	n.d. [*]	n.d. [*]	n.d. [*]
BIOX® Leach Solution	184	2.0	1.4	<0.7	114	35
BIOX® Tailings Dam Return Water	405	<5 [**]	0.3	< 0.02	107	465

* Not determined
** WAD Cyanide detected but quantitative value indeterminate because of interference

The industrial solutions had no impact on the process performance.

3.4 Toxicity Effects on BIOX® Bacteria

The BIOX® process has been shown to be sensitive to several cyano-compounds and anions Testwork to evaluate the potential toxicity of components produced as a result of thiocyanate degradation was required in order to determine if the treated effluent was suitable for recycle to a bioleach plant. By-products generated during thiocyanate degradation include ammonia, nitrate if some nitrification occurs, sulphate and carbonate. In addition, residual amounts of thiocyanate and cyanide which include free, WAD and iron cyanide, may be present in the effluent stream.

Several effluent samples were taken from the different processes when effluents of less than 1 ppm thiocyanate were being achieved. These samples were tested for toxicity to BIOX® bacteria. All tests showed the solutions to be non toxic. Feed samples for the attached growth test work indicated the presence of WAD cyanide ($Cu(CN)_4^{3-}$ 21 – 0.2 ppm), free cyanide (CN^- 48 – 2 ppm) and iron cyanides ($Fe(CN)_6^{4-}$ 29 – 0.2 ppm). Feed and effluent cyanide concentrations of the activated sludge process are given in Table 1. In all cases, the lack of toxicity to the BIOX® bacteria suggests that all cyanide and thiocyanate compounds were removed.

4. CONCLUSIONS

The studies confirm that the culture isolated can be employed for the complete degradation of thiocyanate and cyanide compounds in a continuous process. Comparative data obtained from the attached growth and activated sludge mini-plants are summarised in Table 2.

Table 2
Comparison of processes using the isolated strain

	ATTACHED GROWTH		ACTIVATED
	Single column	Multiple Columns	SLUDGE
Percent Degradation	99+	99+	99+
Degradation Rate	Low	Low	High
pH Control	None	None	None
Stability	Poor	Good	Good
Feed Concentration	<1000ppm	<1000ppm	No limit*
Biomass Control	Difficult	Difficult	Easy
Design (Scale) up	Complex	Complex	Established Technology
Plant Construction	Expensive	Expensive	Simple

*No limit to feed concentration as long as feed rate is adjusted accordingly.

The operational costs of both processes are minimal, as only trace amounts of nutrients and air are required. Precious metals accumulated within the biomass may be recoverable, thereby compensating for operational costs. An added environmental benefit is the removal and concentration of toxic species such as arsenic by the biomass.

ACKNOWLEDGEMENTS

The authors wish to thank Billiton for granting permission to publish this paper.

REFERENCES

1. O. Barbosa-Filho and A.J Monhemius, Trans. Inst. Min. Metall., 103C (1994) C117.
2. E.A. Devuyst and G. Robbins, Randol Gold Forum Cairns, Australia (1991) 145.
3. A. Smith and T.I. Mudder (eds), The Chemistry and Treatment of Cyanidation Wastes, Mining Journal Books Limited, London, 1991.
4. Q. Ahlsan, E, Hang, R. Norcross, Randol Gold Forum, Squaw Valley. (1990) 311.
5. P.N. Cheremisinoff (ed.), Encyclopedia of Environmental Control Technology, Vol. 4, Gulf Publishing Company, Houston, 1990.
6. T.I. Mudder and J.L. Whitlock, Minerals Metall. Process., (1984) 161.
7. R.M. Mühlbauer and J.L. Broadhurst, Biomine, (1997) CT3.1.
8. T.I. Mudder and J.L. Whitlock, US Patent 4461834, (1984).
9. C.J. Knowles, Bacterial Review, 40 (3), (1976) 652.
10. I. Finnegan, I. Toerien, I. Abbot, F. Smit, G. Raubenheimer, Appl. Microbiol. Biotechnol. Vol. 36, (1991) 142.
11. J.R. Welty, C.E. Wicks, R.E. Wilson (eds), Fundamentals of Momentum, Heat and Mass Transfer, John Wiley & Sons, Inc. Singapore, 1984.
12. R.H. Perry and D. Green (eds), Perry's Chemical Engineers' Handbook, McGraw – Hill Book Company, Japan, 1984.
13. J.M. Smith, (ed.), Chemical Engineering Kinetics, McGraw-Hill Book Company, Singapore, 1981.

Pilot experiments to reduce environmental pollution caused by acid rock drainage

N. Cosma[a], N. Sãsãran[a], Zs. M. Kovacs[a], M. Jelea[a], M. Popa[a], A.-A. Nagy[a], A. Gavra[a], W. Sand[b], A. Schippers[b], P.-G. Jozsa[b], H. Saheli[c], and E. Gock[c]

[a] Mining Research & Design Institute, 62 Dr. Victor Babes Street, RO-4800 Baia Mare, Romania, Phone/Fax: +40/64/432208, e-mail: michneaa@icpm.ubm.ro

[b]Universität Hamburg, Institut für Allgemeine Botanik, Abteilung Mikrobiologie, Ohnhorststraße 18, D-22609 Hamburg, Germany, Phone/Fax: +49/40/82282423, e-mail: fb6a042@mikrobiologie.uni-hamburg.de

[c] Technische Universität Clausthal, Institut für Aufbereitung und Deponietechnik, Walther-Nernst-Straße 9, D-38678 Clausthal-Zellerfeld, Germany, Phone/Fax: +49/5323/722037, e-mail: gock@aufbereitung.tu-clausthal.de

Bacterial leaching experiments were performed with sulfidic mine waste from dumps located adjacent to Ilba Mine, a Romanian base metal mining operation, to test techniques for reduction of environmental pollution by acid rock drainage. Two countermeasures had an effect:
1. Inhibition of bacterial leaching by incorporating three alkaline layers in the waste heap
2. Purification of acid rock drainage by removal of metal ions through the use of a chemical barrier (buffer) at the base of the waste heap.
Experiments proceeded "in situ" in the dump in a four-chamber percolator with 65 m^3 volume per chamber. Parameters of the chemical and microbial processes have been monitored for 630 days. The application of the first measure caused an increase of the pH. As a consequence, bacterial activity was reduced, the ratio between the several *Thiobacillus* species was modified, and the metal content in the effluent was reduced, because solubilization became inhibited and metallic ions precipitated in the alkaline layers. The result of the second method was a stopped percolation due to precipitations in the chemical barrier.

1. INTRODUCTION

Underground mine development and mining as well as open pit mining result in considerable amounts of mine waste deposited in dumps. These wastes contain metal sulfides, which in the course of biooxidation cause acidic effluents containing dissolved heavy metals. For the dissolution, an autochthonous microflora is responsible [1-3].

In the Romanian district Maramures 244 waste dumps exist, belonging to the mining operations controlled by REMIN Baia Mare. These dumps incorporate 5.7 million m^3 of mine waste located on 143 ha of forest and farm land. The dumps include rocks of eruptive, sedimentary, and vulcano-sedimentary origin. Most of them contain non-ferrous metals and have a low acid buffering capacity. Generally, these dumps are located on hill and mountain slopes. Chemical and biological processes cause leaching of metals. The run off-waters contaminate the surrounding areas with acid and heavy metals [4-5].

The permanent presence of acidic, heavy metal containing waters, a typical phenomenon for all mining areas with non-ferrous sulfidic minerals, has caused severe pollution of the environment. This project shall analyze and control acid rock drainage. It is focused on an almost "in situ" implementation of countermeasures in the 4CP at the experimental dump of Ilba Mine.

2. MATERIALS AND METHODS

2.1. Experimental installation (4CP)

The 4CP is located in the Valea Ardeleană Bază dump of Mine Ilba, Maramures, Romania, and consists of 4 chambers with 65 m^3 volume each, and a central access shaft, from which solid and liquid samples may be taken from the 4 chambers.

The percolator has been built of reinforced concrete. PVC tubes of 100 mm diameter were used as drains. The concrete structure has been insulated by Lehmann&Voss&Co., Hamburg, Germany, by means of Silquest AP-134 Silane Primer mixed with Corodur 2K-V90 ISO. The 4CP is provided with run-off water tanks (basins), pumping station, and a sampling system for solid samples from the following depths: 0.5-0.7 m, 1.3-1.5 m, 2.0-2.1 m and 2.6-2.7 m.

2.2. Mine waste material

The percolator chambers were filled with mine waste originating from the Valea Ardeleană Bază dump. It had the following chemical composition: 0.05-0.07 % Cu, 0.1-0.3 % Pb, 0.05-0.23 % Zn, 11.0-12.7 % Al_2O_3, 4.3-4.6 % Fe, 0.04 % Mn, 1.9-6.2 % S, 2.8-3.1 % Ca, 0.8-1.2 % Mg, 0.9-1.0 % Na, 3.8-4.5 % K and 62-65 % SiO_2.

Chemical agents used for the alkaline layers were ground limestone with a grain size below 2 mm containing 19.5 % Ca and 15 % Mg, and powdered burned lime containing 28-30 % Ca. Quartz sand was used as filling agent.

2.3. Countermeasures

Chamber A: Control, natural leaching of metal sulfides, see [6].

Chamber B: Inhibition of bacterial leaching activity by addition of sodiumdodecylsulfate (SDS), see [6 - 8].

Chamber C: Reduced generation of heavy metal containing acidic waters by incorporating three layers of a blend of limestone and burned lime (3.5:1). The blend amounted to 7.5 kg/t of waste.

Chamber D: Purification of acid rock drainage by the use of a bottom barrier consisting of sand, limestone, and lime as a blend (1:0.75:0.35). Of this blend 30 kg/t of waste were applied.

2.4. Sampling

The paper provides the results of 630 days of experimental time. The experiments included the monitoring of the rain water passing through the chambers and the physical and chemical analysis of the collected waters. Solid samples were analyzed for the activity and the composition of the bacterial leaching biocenosis by microcalorimetry and by cell count determination using the MPN-technique or agar plating [3].

2.5. Equipment

1. PERKIN ELMER -3110 AAS for analysis of metal ions.

2. UV/V UNICAM 5126 spectrophotometer for determination of iron(II) and iron(III) ions by the ortho-phenanthroline method.

3. THERMOMETRIC 2277 microcalorimeter to determine the activity of the bacterial leaching biocenosis.

3. RESULTS AND DISCUSSION

3.1. Control experiment (chamber A)

22.5 m^3 total volume of rain water percolated through the waste pile during 630 days. A drop from pH=4.1 (at start) to pH=2.5-2.8 and an increase of the concentrations of heavy metals from 35-50 mg/l Cu, 130-250 mg/l Zn, and 45-100 mg/l Fe up to 77-122 mg/l Cu, 380-517 mg/l Zn, and 200-266 mg/l Fe was measured. In total 1025 g of Cu, 4385 g of Zn, and 2430 g of Fe were detected in the effluent (Fig. 1).

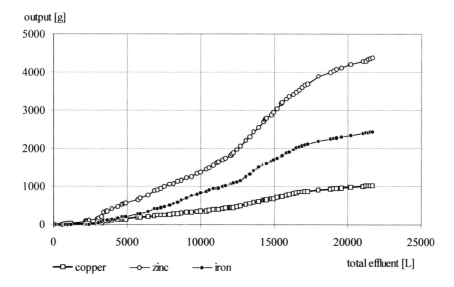

Fig. 1. Cumulative output of heavy metal ions in the control experiment A

The high amounts of metals, washed out from this chamber, as well as the acidic waters reproduce the large scale process occurring in the waste dump. During the experiment a decrease from pH 4.0 to 2.8 in the effluent was measured.

The microbiological analysis revealed the existence of the following bacteria: *Thiobacillus ferrooxidans-, T. thiooxidans-,* and *Thiomonas intermedia-*like species. During the experiment the cell numbers amounted to $2-6x10^7$ cells/g sample for *T. ferrooxidans-,* $6x10^7-8x10^8$ cells/g sample for *T. thiooxidans-* and $1x10^7-1x10^8$ cells/g sample for *Thiomonas intermedia-*like species.

The highest values of microbial activity, namely 12-34 μW/g sample were measured for samples from 1.3 m depth. Samples collected during the winter exhibited minimal microbial activity and cell counts, presumably because of low temperature and/or humidity [6, 9].

3.2. Alkaline layers (chamber C)

The pH as in the control decreased to the range of 2.8-3.0. However, the concentrations of heavy metals (20-47 mg/l Cu, 35-159 mg/l Zn, and 0.30-32 mg/l Fe) were below the concentrations detected for the control chamber.

At the end of the experiment the total amount of metals in the effluent was: 640 g Cu, 2263 g Zn, and 555 g Fe (Fig. 2). This means that, compared to the control experiment, the output of heavy metals could be reduced by 38 % for copper, by 48 % for zinc and by 77 % for iron.

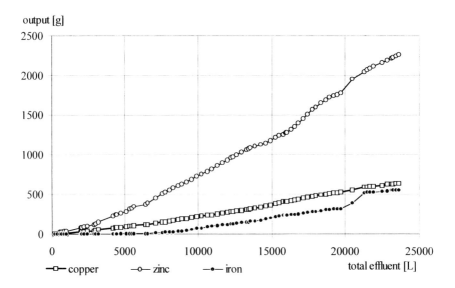

Fig. 2. Cumulative output of heavy metal ions in the experiment C

The highest values of bacterial activity (13-27 μW/g) and of cell counts ($7x10^6$ cells/g for *T. ferrooxidans-,* $2x10^6$ cells/g for *T. thiooxidans-* and $2x10^6$ cells/g for *Thiomonas intermedia-*like bacteria) were detected in the waste material located between the two upper layers. However, the bacterial activity was lower than in the control chamber.

The lowest microbial activity (0-9 μW/g) and the lowest cell numbers ($1x10^2$ cells/g for *T. ferrooxidans-,* $1x10^4$ cells/g for *T. thiooxidans-,* and $1x10^5$ cells/g for *Thiomonas intermedia-*like bacteria) at pH 4.3 were measured in the waste material located between the two bottom alkaline layers, while for the material located underneath the deepest layer the values were intermediate, respectively: 8 µW/g sample and 1×10^5 cells/g sample for *T. ferrooxidans*-, 2×10^5 cells/g sample for *T. thiooxidans*-, and 2×10^3 cells/g sample for *Thiomonas intermedia*-like bacteria at pH 4.1 [9, 10].

The results indicate that the decrease of cells counts and of microbial activity as well as precipitation of metals are caused by the pH increase induced by the alkaline layers.

The low level of metal ions in the effluent remained relatively constant during the whole experiment, indicating the efficiency of the applied method. However, the presence of heavy metal ions indicates that only incomplete precipitation occurs in the alkaline layers. The amount of alkaline material was presumably not high enough to achieve a fully satisfactory effect.

3.3. Alkaline bottom barrier (chamber D)

The pH of the effluents was in the range of 8.1-12.5 for the first 300 days. The concentration of heavy metals amounted to 0.05-0.7 mg/l Cu, 0.02-1.3 mg/l Zn, and 0.01-0.4 mg/l Fe and, thus, were far below the concentrations detected in case of the control chamber. The total amounts of heavy metals in the effluent were 3.34 g Cu, 4.31 g Zn, and 0.58 g Fe (Fig. 3).

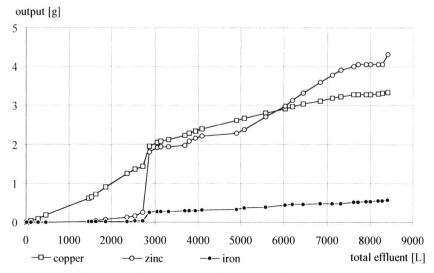

Fig. 3. Cumulative output of heavy metal ions in the experiment D

Compared to the control experiment the heavy metal content of the effluent decreased by 99.7 to 99.9 % by the application of this measure.

The highest microbial activity (26 µW/g sample) was obtained in samples from 1.4-2.1 m depth, comparable with values for the control chamber.

According to the experiment in the control chamber, the bacterial leaching produced a heavy metal containing, acidic solution within the chamber. This solution became alkaline upon contact with the bottom barrier, causing an almost complete precipitation of the heavy metals.

Due to the precipitations in the alkaline layer, an almost impermeable barrier developed, causing the percolation to stop after 300 days. As a consequence, the solution percolated through the outlet located 50 cm above the barrier.

During the following period from 330 to 630 days the pH of the percolating solution (collected from the outlet 50 cm above the bottom) ranged between 2.6-3.0, while the metal content increased up to 90-100 mg/l Cu, 420-480 mg/l Zn, and 350-450 mg/l Fe, and was, thus, similar to the effluent obtained in the control experiment.

At the end of the experiment the amount of precipitated metals in the alkaline layer was analyzed: 0.2% Cu 0.04% Pb, 0.5% Zn, 0.05% Mn, 1.5% Fe, 7.2% S (15% SO_4^{2-}), 6.0 Ca, 0.3% Mg, 1.3% Na, 0.03% K and 55.9% SiO_2, (since active lime was absent, Ca mainly occurred as gypsum).

4. CONCLUSIONS

The large-scale experiments demonstrated the efficiency of alkaline barriers or layers for a mitigation of Acid Rock Drainage, as previously shown in lab-experiments [8]. As exemplarily shown for solid samples from depth of 1.3-1.5 m (Fig. 4), usually samples from chamber C and D exhibited significantly less activity than control samples.

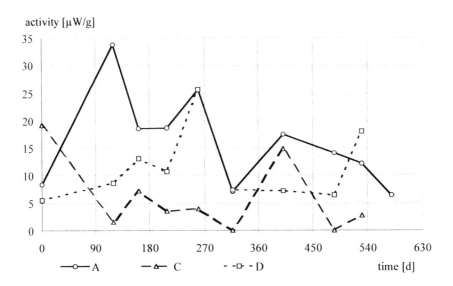

Fig. 4. Metabolic bioactivity in solid samples from a depth of 1.3-1.5 m

Extrapolation of the results of the control experiment demonstrated that during 630 days the Valea Ardeleana Baza dump of 2.1 ha and 5 m height, consisting of 231000 t of mine waste, polluted the environment with 29.8 m^3 of acidic water per day (pH 2.5-2.8) containing the following amounts of heavy metals: 2.3-3.6 kg copper, 11.3-15.4 kg zinc, and 5.9-7.9 kg iron.

The experiment with the alkaline layers resulted in a decrease of the metal content of up to 38 % for Cu, 48 % for Zn, and 77 % for Fe. A higher decrease will certainly be obtained by an application of increased amounts of alkaline material.

The experiment with the alkaline bottom barrier demonstrated a nearly complete precipitation of heavy metals by 99.7-99.9 %. This precipitation resulted in the formation of an impermeable layer (hardpan) at the bottom.

ACKNOWLEDGMENTS

The studies were founded by grants of the Romanian Ministry for Research and Technology to SC ICPM SA, Baia Mare and of the German Federal Ministry for Education, Science, Research, and Technology (BMBF) via the Federal Environmental Agency (UBA, Berlin, Mr. Wittmann) to W. S. (FKZ 1490954) and E. G. (FKZ 1490970).

REFERENCES

1. G. I. Karavaiko, G. I. Kuznetsov, and S. A. Golomzik, Roli microorganizmov v vâscelacivanii metallov iz rud, Ed. Nauka, Moskow (1972).
2. A. E. Torma and G. I. Karavaiko, Microbiological processes for the leaching of metals from ores, Center of international projects, GKNT Moskow (1985).
3. A. Schippers, R. Hallmann, S. Wentzien, and W. Sand, Appl. Environ. Microbiol., 61 (1995) 2930.
4. P. Ilie, Resurse secundare de metale neferoase, posibilitãti de valorificare, depozitare si protectia mediului, Studiu SC-ICPM.-SA Baia Mare, Romania (1990).
5. P. Ilie, Protectia mediului în zonele miniere, Curs, Universitatea Baia Mare (1991).
6. P.-G. Jozsa, A. Schippers, N. Cosma, N. Sãsãran, Zs. M. Kovacs, M. Jelea, A. M. Michnea, and W. Sand, Proceedings of the International Biohydromatallurgy Symposium, IBS 99, El Escorial, Spain, Elsevier, Amsterdam (1999) (in this volume).
7. W. Sand, Environm. Technol. Lett., 6 (1985) 439.
8. A. Schippers, P.-G. Jozsa, and W. Sand, Appl. Microbiol. Biotechnol., 49 (1998) 698.
9. Zs. M. Kovacs and M. Jelea, Lesierea naturalã a haldelor de deseuri din minerale metalifere, Studiu SC-ICPM-SA Baia Mare, Romania, 1994-1998.
10. W. Sand, H. Saheli, A. Schippers, E. Gock, and P.-G. Jozsa, Jahresbericht: Untersuchungen zur mikrobiellen Sicherung von Bergbaualtlasten, FKZ 1490954 & 1490970 (BMBF), Umweltbundesamt, Berlin, Germany (1997).

Large-scale experiments for safe-guarding mine waste and preventing acid rock drainage

P.-G. Jozsa[a], A. Schippers[a], N. Cosma[b], N. Säsäran[b], Zs. M. Kovacs[b], M. Jelea[b], A. M. Michnea[b], and W. Sand[a]

[a] Universität Hamburg, Institut für Allgemeine Botanik, Abteilung Mikrobiologie, Ohnhorststraße 18, 22609 Hamburg, Germany, Phone/Fax: +49/40/82282-423, e-mail: fb6a042@mikrobiologie.uni-hamburg.de

[b] SC ICPM SA, Str. Dr. Victor-Babes 62, 4800 Baia Mare, Romania, Phone/Fax: +40/64/432208, e-mail: michneaa@icpm.ubm.ro

In a German-Romanian joint project feasible and cheap techniques for a safe storage of mine waste to prevent acid rock drainage were tested. A percolator (4CP), consisting of four separate leaching chambers, has been installed in the main waste dump of Ilba Mine, Maramures, Romania. Each unit contains 65 m^3 of sulfidic waste material. The inhibitory effect of sodiumdodecylsulfate (SDS) on the activity of leaching bacteria and on metal mobilization was studied in a long term experiment. SDS was proven to have an inhibitory, however transient, effect on leaching bacteria. In the course of the experiment, it became for the first time possible to quantify the potential for environmental pollution resulting from microbial leaching in mine waste by the use of microcalorimetric activity measurements.

1. INTRODUCTION

Acid rock drainage (ARD) causes environmental pollution in many countries [1]. Sulfuric acid, dissolved heavy metals, and soil degradation are serious problems, calling for an inhibition of ARD in mine waste.

The construction of current and future mine waste deposits like low-grade ore dumps or tailing ponds need high safety standards, to prevent ARD generation. Several measures for a prevention of ARD and a safe storage of sulfidic mine waste are known and have thoroughly been tested in large scale experiments. Applied measures are e.g. prevention of water entry, under-water storage, reduction of oxygen permeation, pH-regulation, as well as application of inhibitory substances [1-10]. However, in none of these publications data on long term effects on the leaching bacteria were reported. Furthermore, no data are available to allow for an ascertained secured prediction of the pollution potential resulting from microbial metal dissolution (bioleaching).

Our joint work was started to acquire long-term physical, chemical, and microbiological data in experiments in real leaching conditions [11, 12], in order to improve the construction of industrial mine waste deposits and tailings ponds. Additionally, a technique based on microcalorimetry had to be developed, to allow for a rapid evaluation of the effects of countermeasures on ARD formation in leaching biotopes.

2. MATERIALS AND METHODS

2.1. Installations
A 4CP consisting of four separate, pilot-scale leaching units has been built into the main waste dump of Ilba Mine, Maramures, Romania. They allow a quantification of ARD and of the bacterial population under natural climatic conditions. The 4CP was constructed of reinforced concrete and was protected against acid attack and mechanical stress by spray-application of CORODUR 2K® (Lehmann&Voss&Co., Hamburg, Germany). Each unit contained 65 m³ of sulfidic waste material. Sampling of solid in different depth, and a separate, quantifiable collection of all effluents is ensured.

2.2. Mine waste
The mine waste originated from the main dump of Ilba Mine. It was taken in layers from the surface down to a depth of 4.5 m. The maximum grain size was limited to 400 mm and corresponded to the usual run-off-mine ore. Before excavation, the dump had been completely flooded, due to an exhaustive rain event. Thus, most water-soluble compounds, present on the surface of the waste, had probably been removed. The entire material was of a brownish colour, due to precipitated iron compounds. The rocky material showed a lot of cracks and cleavage surfaces, which were also partially brown, indicating that sulfide degradation occurred also at inner surfaces.

The material was introduced into the four chambers in the same succession, as it had been in the dump. However, the layers were reduced to 1.2 m thickness. Chemical analysis was performed for ore fractions below 30 mm. Chemical data, given in Table 1, are characteristic for a run-off-mine low-grade ore, which conventionally can only uneconomically be processed. In general, the chambers had been filled with a largely homogenous material.

Table 1
Chemical composition of the mine waste at Ilba Mine used for leaching experiments (weight %, grain size <30 mm)

Fe	Cu	Zn	Pb	Mn	S	SiO$_2$	Al$_2$O$_3$	CaO	MgO	Na$_2$O	K$_2$O
4.35	0.05	0.14	0.20	0.04	2.02	64.80	15.20	3.77	1.77	1.17	4.46

The sulfur species in the waste were approximately to 27 % in the form of sulfide and to 67 % in the form of sulfate, indicating a considerable preoxidation. Elemental sulfur was present in amounts of almost 6 %.

2.3. Analysis
For a balance of metal- and sulfate-output, all effluents were collected at least once a week and measured for pH, redox potential, concentration of Cu, Zn, Fe, Mn, Mg, Ca, Al, Si, and sulfate. Chemical analysis was performed by ICP, AAS, and IC. SDS was determined

spectrophotometrically via chloroform extraction in the presence of methylene blue (Romanian standard procedure). Microcalorimetric measurements of bacterial activity as well as microbiological analyses were performed every 6 weeks, using solid samples collected in different depth. Cell counts of thiobacilli were determined by the MPN-technique using dilution series and selective culture media. Cell numbers of acidophilic and neutrophilic chemoorganotrophic bacteria were determined by direct counting of colony forming units on selective solid agar [13].

2.4. Sodiumdodecylsulfate (SDS)

The inhibitory effect of SDS on growth and activity of leaching bacteria is based on a lysis of the cytoplasmic membranes [14] or on the inhibition of the attachment to the substratum [15]. SDS was applied by spraying SDS-solution in a concentration of 30-60 g/L on the surface of the waste body, followed several times by recirculation of the SDS-containing effluent, to achieve a thorough distribution. Two SDS-applications of 5 kg, each, were performed between day 130 and 200 and between day 330 and 365, respectively. Six additional applications of 2 kg SDS, each, were performed later on (between days 484 and 608).

3. RESULTS AND DISCUSSION

For the evaluation of the effect of SDS-application data are available for a test period of 20 months. Besides microbiological data, a comparison of the metal and sulfate output of the control and of the inhibition experiment reveals the efficacy of this method as a countermeasure.

The daily effluent volume and the cumulative effluent volume are comparable in the control and the SDS-experiment. Thus, in Fig. 1 only the data for the latter are given. Because the SDS containing effluent was recirculated several times, a total flow-through volume was calculated for Fig. 1. At the site of the 4CP it rained regularly, at least every 10 to 15 days. As an average, a daily effluent of 35 L/chamber was registered. In the cold season from November to April the total effluent amounted to about 4400 L/chamber, i.e. daily 24 L/chamber. In the warm season from May to October about 8000 L/chamber were collected, i.e. 44 L/chamber.

3.1. Control-experiment

Microbiological analyses demonstrated the ubiquitous presence of leaching bacteria (*Thiobacillus ferrooxidans*-, *Thiobacillus thiooxidans*-, and *Thiomonas intermedia*-like) with cell counts ranging from 10^3 til 10^8 cells/g sample in the course of the experiment. Acidophilic and neutrophilic chemoorganotrophic bacteria were also detected in all solid samples. The cell counts ranged from 10^3 to 10^5 cells/g sample.

Microcalorimetric activity measurements (Fig. 2) were consistent with these results, exhibiting the highest values (about 20 µW/g) at a depth of 1.3 m. The heat output is correlated with the metal sulfide oxidation rate. For calculations, standard conditions and 1500 kJ/mole of pyrite have been used [16]. Consequently, sulfate formation, iron dissolution, and acid release can be predicted by our microcalorimetric activity measurements (see below).

The formation of ARD is indicated by a decrease of the pH from 4.0 to 2.8 and an increase of the redox potential from 170 to 250 mV in the effluents within 7 months. Initially low, but increasing concentrations of the metal ions Cu, Zn, Mn, Fe, Mg, Ca, Al, Si, and of sulfate indicate the slow start of the leaching process.

752

A period of 8 months ensues with a relatively constant, increased output level. A further increase of the concentrations was noted for most of the elements 4 months later. Afterwards, the concentrations declined. An exception was observed for Ca and Si, which remained throughout at almost constant concentrations. The calcium concentration corresponded to the solubility of gypsum, irrespective of the time.

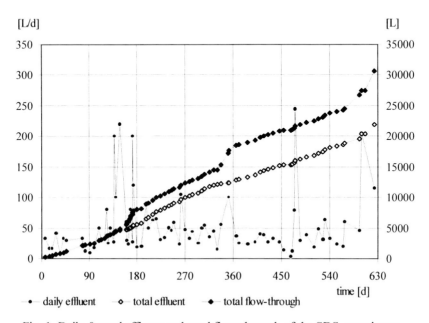

Fig. 1. Daily & total effluent and total flow-through of the SDS-experiment

Since metal (and sulfate) output occurs by a washing-out of soluble compounds, the cumulative output is presented in dependence of the total effluent volume. The data, included in Fig. 3 and 4, show graphs with extended phases of quite linear shape. For each of these domaines the cumulative metal output (Me) is linearly correlated with the total effluent volume (V). The output rate may be defined as the quantity of a certain metal, discharged from the solid waste, related to the volume of effluent, and corresponds to the slope S of the linear equation:

$$Me \ [g] = S \ [g/L] \cdot V \ [L] + N \ [g] \tag{1}$$

In this equation N is the intersection with the y-axis.

For a comparison of output rates of different experiments or of the same experiment at different times, we define acceleration factors $F_{2/1}$ from the ratio of the corresponding slopes:

$$F_{2/1} = S_2/S_1 \tag{2}$$

Acceleration factors from the first to the second year ranged from 1.6 to 2.4 for zinc, copper, manganese, iron, aluminium, magnesium, and sulfate. Therefore, we propose to designate these elements as relevant for leaching. On the contrary, the leaching rates of calcium and silicon remained unchanged. Thus, we consider them to be non-relevant for a leaching process.

The bacterial oxidation of sulfides occurs on the outer and on the inner surface of an ore, i.e. in cracks and crevices of the rocks. Mobilization and washing-out of the water-soluble salts by the effluent is dependent on the availability of water to wet all surfaces. Thus, the metal output is dependent on the volume of the effluent.

[μW/g]

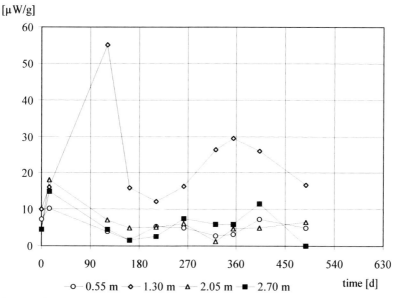

—○— 0.55 m —◇— 1.30 m —△— 2.05 m —■— 2.70 m

time [d]

Fig. 2. Microcalorimetric activity of ore from different depth of the control experiment

output [g]

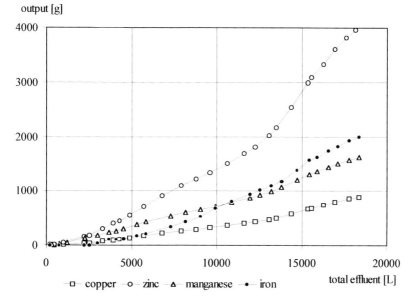

—□— copper —○— zinc —△— manganese —•— iron

total effluent [L]

Fig. 3. Cumulative output of heavy metal ions in the control experiment

The increase of the metal output, observed from the beginning, may be explained by the complete flooding of the dump and, hence, an almost complete washing-out of the soluble compounds from the waste, prior to the start of the experiments. Afterwards, the bacterial flora continued to produce sulfuric acid and metal salts, but probably only the compounds formed at the outer surfaces were washed out. Probably, after one year the pores and cracks of the inner surfaces had reached a status of saturation, thus, an increased amount of leaching products occurred in the effluent in the second year.

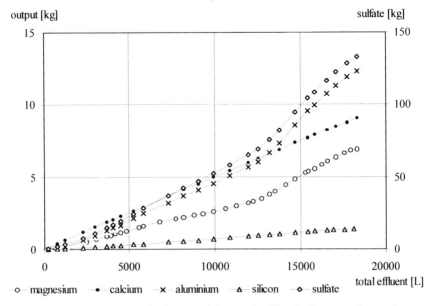

Fig. 4. Cumulative output of other metal ions and sulfate in the control experiment

Another process to be considered is the precipitation of ions. The iron output in the effluent of our control experiment amounted to only 2 kg. Instead, at least 38 kg were dissolved, as calculated from the sulfate output in the effluent (130 kg, see Fig. 4) and the stoichiometry of pyrite. Thus, most of the dissolved iron ions precipitated in the ore dump, probably as hydroxides and/or basic salts like jarosites. Based on the stoichiometry of jarosite, about 171 kg sulfate and 50 kg iron must have been mobilized by oxidation of pyrite. Thus, the effluent yielded 76 % of the sulfate and 4 % of the iron mobilized by oxidation of pyrite. This calculation does not include the products possibly kept in pores and cracks, which are not accessible to the usual wash-out. However, the values obtained may serve as a good approximation of what happens in such a mine waste heap.

3.2. SDS-experiment

Concerning the efficacy of the SDS-application, an unequivocal result was obtained. Microbiological analyses revealed that cell counts for all species of leaching bacteria were reduced compared to samples from the control (data not shown).

Microbial activity values were also lower than those of the control experiment. The differences in the microbial activity values (A) of samples from chamber A and chamber B (A_A-A_B) at different times are shown in Fig. 5.

According to these data, the SDS reduced bacterial sulfide oxidation. It became obvious, however, that the inhibition was only transient, and that the bacterial activity recovered, as soon as the SDS-concentration dropped below 1 mg/L. In Fig. 6 the data for the SDS-dosage are given.

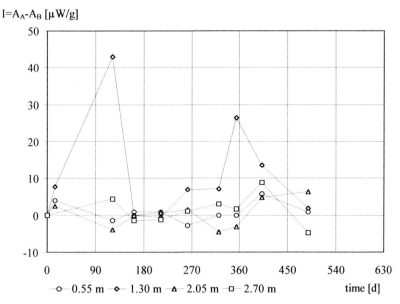

Fig. 5. Difference (I) between activity values of the control (A_A) and of the SDS-experiment (A_B)

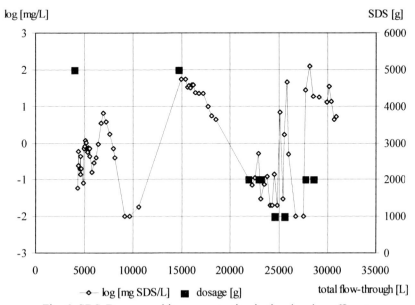

Fig. 6. SDS-Dosage and its concentration in the chamber effluent

Besides, in the SDS-chamber a proliferation of acidophilic and neutrophilic chemoorganotrophic bacteria was detected. Cell counts of these organisms increased at the time of SDS-dosage by one order of magnitude over the control samples (results not shown). Possibly, these bacteria thrived on the degradation of SDS and may also have produced some organic compounds, enhancing the mobilization of metal ions.

On the basis of the same total flow-through volume, the output of manganese, magnesium, aluminium, calcium, and, especially, sulfate is significantly higher in the control than in the SDS-chamber (Table 2). Thus, biogenic acid production and metal mobilization must have been reduced by the SDS-addition.

Table 2
Cumulative output values for metal and sulfate ions from the control and the SDS-experiments.

	control experiment					SDS-experiment				
	total effluent = rainfall volume					total flow-through=rainfall+recirculation				
[L]	3935	6795	9100	14542	16021	3920	6930	9100	14542	16042
at day	130	201	253	420	475	130	167	201	330	343
	cumulative metal and sulfate output [g]									
Zn	407	909	1259	2610	3252	634	924	1537	3184	3197
Cu	98	220	308	595	723	134	191	300	604	606
Mn	257	518	442	1183	1379	308	421	598	1037	1042
Fe	106	283	543	1342	1657	165	270	629	2146	2163
Mg	953	1882	2397	4720	5698	1056	1422	2010	3654	3678
Ca	1999	3416	4561	7290	7991	1881	2382	3248	5944	5972
Al	1412	3035	4124	8353	10152	1555	2295	3747	7333	7381
SO_4^{2-}	16407	34404	47349	92565	110695	19638	27304	42278	82693	83243

However, an inhibitory effect for zinc, copper, and iron output has not been observed. An unequivocal explanation for this finding is not yet available. It needs to be added that the reduced surface tension of the SDS-solution may cause a better wetting of the ore particles and an enhanced capillary penetration of the rocks. Thus, an increased metal output may result. In addition, by a recirculation of SDS-solutions also ferric iron ions were recirculated. These may have had a profound effect by causing an intensive chemical leaching. An answer will be available in the future only, because the control experiment will also be run with recirculation.

Whenever the SDS concentration ranged above 1 mg/L, the reduction of the metal output became detectable. As a consequence, for an application in the field the SDS-concentration has to be maintained above this level, to ensure a positive effect. Therefore, continuous applications of smaller quantities in short intervals are in progress, in order to maintain a sufficient high SDS-concentration in the effluent.

3.3. Estimation of pollution potential by microcalorimetric activity measurements

The heat output, measured under standard conditions, is correlated with the metal sulfide oxidation rate. The specific heat output, assuming total oxidation of pyrite to sulfate, is 1500 kJ/mole of pyrite [16]. The heat output of a sample depends on the following main parameters: metal sulfide content, specific reaction surface, temperature, humidity, extent of chemical oxidation, composition and activity of the bacterial population.

The microcalorimetric activity measurements allow to calculate the amount of total metal sulfide dissolution, and thus, the amount of sulfate and iron ions actually formed under standard conditions in the sample measured.

Based on the microcalorimetric activity data, the presumable sulfate formation (T) may be calculated. This value should be comparable with the actual sulfate output (M) of an ore body obtained by the effluent analysis. However, the real waste differs in its grain size and possibly in its metal sulfide concentration from the samples selected for microcalorimetric measurements and leaching conditions in the natural environment differ from standard conditions. A factor (R = T/M) summarizes the effects of all leaching parameters, without an explicit knowledge of the individual influences.

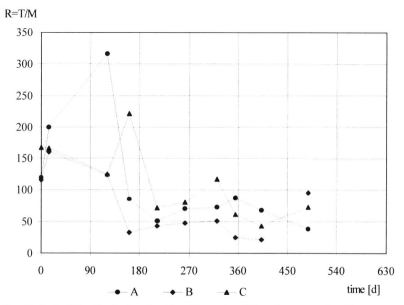

Fig. 7. Ratio (R) of calculated sulfate formation (T) and measured sulfate output (M) for the control-, SDS-, and buffer-experiment (A, B, and C, respectively)

However, the data of the control-, the SDS-, and the buffer-experiment [17] shown in Fig. 7 indicate over 20 times more dissolution under standard conditions than washing out occurred from the heaps. Obviously, the oxidation products are only partially washed out, due to precipitations of jarosites, metal sulfates, or gypsum. Thus, an exact calculation of a sulfur mass balance is not possible yet. Further research is needed to quantify the in situ precipitation and the water-soluble compounds remaining inaccessible to the wash-out processes as they usually occur in a waste heap. However, the microcalorimetric activity measurements allow to estimate the pollution potential of a certain mine waste heap. Furthermore, for active leach operations the progress of leaching may continuously be monitored by this technique.

4. CONCLUSIONS

The pilot scale installation at Ilba Mine, Maramures, Romania, the 4CP for leaching of mine waste under natural climatic conditions, proved to provide reliable, quantitative data on ARD and on the microbiological composition of leaching biotopes. By the use of the 4CP countermeasures can unequivocally be evaluated, to ensure the application of the findings and their transfer to a large scale. Valuable data have been collected in order to evaluate or predict pollution potentials of mine waste deposits. Besides SDS-testing, evaluations of a cover of organic materials and self-sealing alkaline layers in the 4CP are under way.

ACKNOWLEDGMENTS

The studies were funded by the German Federal Ministry for Education, Science, Research, and Technology (BMBF) via the Federal Environmental Agency (UBA, Berlin, Mr. Wittmann) to W. S. (1490954) and of the Romanian Ministry for Research and Technology to SC ICPM SA, Baia Mare. The experiments were performed in cooperation with Prof. E. Gock of the Technical University of Clausthal-Zellerfeld.

REFERENCES

1. V. P. B. Evangelou, Pyrite oxidation and its control, CRC Press, Boca Raton, Florida, USA, 1995.
2. P. R. Dugan and W. A. Apel, Appl. Environ. Microbiol., 46 (1983) 279.
3. S. J. Onysko, R. L. P. Kleinmann, and P. M. Erickson, Appl. Environ. Microbiol., 48 (1984) 229.
4. P. M. Erickson, R. L. P. Kleinmann, and S. J. Onysko, Control of acid mine drainage, R. C. Horton and D. P. Hodel (eds.), Bureau of Mines, U. S. Department of the Interior, IC 9027, (1985) 25.
5. C. A. Backes, I. D. Pulford, and H. J. Duncan, Mine drainage and surface mine reclamation, D. S. Brown and D. P. Hodel (eds.), Bureau of Mines, U. S. Department of the Interior, IC 9183 (1988) 91.
6. G. R. Watzlaf, Mine drainage and surface mine reclamation, D. S. Brown and D. P. Hodel (eds.), Bureau of Mines, U. S. Department of the Interior, IC 9183, (1988) 109.
7. ILPMC, Proceedings of the International Land Reclamation and Mine Drainage Conference and of the Third International Conference on the Abatement of Acidic Drainage, Pittsburgh, PA, Volumes I, II, III, and IV, 1994.
8. V. Rastogi, Annual meeting of the Society for Mining, Metallurgy, and Exploration, inc, Denver, Colorado, SME, Littleton, Colorado, USA, preprint number   95-29, 1995.
9. ICARD, Proceedings of the Fourth International Conference on Acid Rock Drainage, Vancouver, B.C., Canada, Volumes I, II, III, and IV, 1997.
10. J. Schüring, M. Kölling, H. D. Schulz, Environmental Geology, 31 (1997) 59.
11. A. Schippers, H. von Rège, and W. Sand, Minerals Engineering, 9 (1996) 1069.
12. A. Schippers, P.-G. Jozsa, and W. Sand, Appl. Microbiol. Biotechnol., 49 (1998) 698.
13. A. Schippers, R. Hallmann, S. Wentzien, and W. Sand, Appl. Environ. Microbiol., 61 (1995) 2930.
14. H. G. Schlegel, General microbiology. 7th Ed., Cambridge University Press, 1995, 223
15. T. R. Neu, Microbiological Reviews, 60 (1996) 151.
16. T. Rohwerder, A. Schippers, and W. Sand, Thermochimica Acta, 309 (1998) 79.
17. N. Cosma, N. Säsäran, Zs.-M. Kovacs, M. Jelea, M. Popa, A.-A. Nagy, A. Gavra, W. Sand, A. Schippers, P.-G. Jozsa, H. Saheli, and E. Gock, Proceedings of the International Biohydromatallurgy Symposium, IBS 99, El Escorial, Spain, Elsevier, Amsterdam (1999), (in this volume).

Sulphate reduction optimization in the presence of *Desulfotomaculum acetoxidans* and *Desulfobacter postgatei* species. Application of factorial design and factorial correspondance analysis methods

M.Crine [a], M.L. Sbai [b], J.Bouayad [b], A. Skalli [c]

[a]Université de Liège (ULG). Laboratoire de Génie chimique, Biotechnologie et Environment. Belgique.

[b]Universite Mohammed V. Départment Génie Minéral Laboratoire LAMINEMET-Bioremediation. Ecole Mohammadia d'Ingénieurs (EMI). Maroc.

[c]Universite Mohammed V. Départment Génie Industriel. Ecole Mohammadia d'Ingénieurs (EMI). Maroc.

The "green acid process" leads to the production of large quantities of phosphogypsum as solid waste. In Morocco, this waste is rejected at sea, contributing to the pollution of the environment. One of the main objectives of the present study is to develop an integrated biological process for the production of sulfur from phosphogypsum, comprising a biological sulfate reduction step.This paper presents the results of sulphate reduction rate optimization studies in the presence of *Desulfotomaculum acetoxidans* and *Desulfobacter postgatei* species.The experimental work was carried out according to the method of factorial design, and the results were analysed according to a factorial correspondence analysis technique. Four factors were studied, i.e., the initial sulphate and acetate concentrations ($[SO_4]_i$, $[Act]_i$), and the initial pH and incubation temperature (pH_i, T_i). Three response-variables were determined: the maximal average sulphate reduction and sulphide production rates and the final pH.These studies show that for *Desulfotomaculum acetoxidans* (i) optimum bacterial growth and optimum sulphate reduction are obtained for pH_i range from 6.6 to 7.6 and T_i range from 30°C to 40°C (ii) optimum initial concentrations range from 7.75 to 17.25 mM for sulphate and from 10 to 20 mM for acetate (iii) strong inhibitions occur respectively below $pH_i = 6.1$, above $pH_i = 8.1$ and above $T_i = 45°C$ (iv) Strong inhibitions are obtained below 3 mM and above 22 mM for sulfate and below 5mM and above 50 mM for acetate. For *Desulfobacter postgatei* (i) optimum bacterial growth and optimum sulphate reduction are obtained for pH_i range from 6.2 to 8.4 and T_i range from 25°C to 35°C (ii) optimum initial concentrations range from 10 to 22 mM for sulphate and from 12.5 to 27.5 mM for acetate. (iii) strong inhibitions are obtained respectively above $pH_i= 9.5$ and $T_i= 40°C$ and below $pH_i = 5.1$ and $T_i= 20°C$ (iv) strong inhibitions are obtained below 4 mM and above 28 mM for sulfate and below 5 mM and above 45 mM for acetate.

1. INTRODUCTION

The state-owned fertilizer industry (OCP) in Morocco imports large quantities of sulfur to produce sulphuric acid. The latter is used as phosphate rock leachant to produce phosphoric acid (Green Acid Process)(1,2). The leaching process also produces phosphogypsum (PG, impure $CaSO_4$; $2H_2O$) as solid waste. Having little economic value, due to various impurities often associated with it (cadmium, uranium, rare earths, organic matter) (2), this solid waste is commonly rejected at sea or stock-piled near the producing plants, which discharge contributes to environment pollution (3,4). One of the main objectives of the present research is to develop an integrated economically viable biological process for sulfur production from phosphogypsum. Phosphogypsum biodegradation in the presence of pure strains of sulphate reducing bacteria (SRB) has been studied worldwide since 1960.

While several studies have concerned the group of lactoclastic SRB (2,5), relatively little work has been undertaken on the group of acetoclastic SRB (5-9). The present research work aims to optimise the physico-chemical parameters acting on the reaction rates of sulphide production, with species of acetoclastic, chimio-organotrophic and mesophilic SRB, namely *Desulfotomaculum acetoxidans* and *Desulfobacter postgatei*.

2. MATERIALS AND METHODS

2.1. Media and culture preparation

Desulfotomaculum acetoxidans species DSM 771 and *Desulfobacter postgatei* species DSM 2034 were received in 10 ml closed containers in Postgate's medium G (5) and in Widdel and Pfennig's medium (7) respectively. Inocula of *Desulfotomaculum acetoxidans* and *Desulfobacter postgatei* species were subcultured for enrichment in Postgate (5) and in Widdel and Pfennig (5) media respectively. For both species, only B_{12} vitamin, thiamin and biotin were added in the same proportions as in Postgate (5) and by Widdel and Pfennig (5) media respectively. Inoculation was undertaken in 30 mL sterilized, hermetically closed and completely filled medicinal bottles.Upon inoculation, 0.15 mmol (Na_2SO_4)/L was added from freshly prepared, nitrogen flushed, filtered and sterilized stock solution. Inoculation was 10% in volume. Final pH of the solution was adjusted respectively to 7.1 and 7.3 for *Desulfotomaculum acetoxidans* and *Desulfobacter postgatei* species using sterilized solutions of HCl or Na_2CO_3. Bottles were finally incubated at 35°C and subcultures were prepared according to Widdel and Pfennig (8,9), and Postgate (5) for *Desulfotomaculum acetoxidans* species and according to Uphaus, Grimm, and Cork (7) for *Desulfobacter postgatei* species.

2.2. Analytical techniques

Sulphate concentration was determined by turbidimetry as described by Tabatabai (10). Sulphide concentration was determined by iodometry (11). Stock solutions of 0.1N sodium thiosulphate and 0.1N iodine were routinely standardized with potassium iodate according to the procedures described by Vogel (12).

2.3. Experimental Protocol

The effects of four factors were studied for each species i.e., initial sulphate concentration $[SO_4]_i$, initial acetate concentration (carbon source) $[Act]_i$, initial temperature T_i and initial pH_i. Three response-variables were determined: the maximal average rate of sulphate reduction V_{Sl}, the maximal average rate of sulphide production V_{Su} and the final pH (pH_f).

Rotating central composite experiment design was used (13,14). Factor levels were initialized according to the literature (6-8), as depicted in Table 1. The treatment combinations are codified by ESX (X=01, 02, 03,...., 31).

For both species, histograms presenting the number of experiments were set up for each response value and for each studied factor level (13).

The studied factors and the variable response values are classified in three modulates as depicted in Table 2. The 1st, the 2nd and the the 3rd modulates were codified respectively by (100) (010) and (001). Following this codification, a correspondence table (31x21) was processed by factorial correspondence analysis software (13) to obtain the 1st (x,y) plan projection graph as depicted in Figure 1 as example.

Carbon source inhibition was studied under the same conditions as for subculture preparation and by widening the range of carbon source (acetate) concentrations. These were 10 to 50 mM for *Desulfotomaculum acetoxidans* species and 10 to 60 mM for *Desulfobacter postgatei* species.

3. RESULTS AND DISCUSSION
3.1 *Desulfotomaculum acetoxidans* species
3.1.1. Effect of pH

Combinations (ES17, ES18) show the effect of pH on sulphate reduction rates, which are relatively low for both combinations, with V_{Sl}= 43.66 mg/l.d and V_{su}=14.64 mg/l.d for pH_1= 6.1 (ES18) and V_{Sl}= 32.50 mg/l.d and V_{su}=10.91 mg/l.d for pH_3= 8.1 (ES19).

Widdel and Pfennig (8,9) suggest that the optimal pH range for bacterial growth is within the [6.6, 7.6] pH interval. This is in accordance with the 1st order factorial design results (pH_1 = 6.6 , pH_3 = 7.6) which show that in this pH range, the sulphate reduction rate is located mainly in the 2nd modulate with $40 < V_{sl} \leq 120$ mg/l.d and $10 < V_{su} \leq 35$ mg/l.d. Outside this range, sulfate reduction is inhibited.

3.1.2. Effect of the Temperature

Published work indicates that optimal temperature range for sulphate reduction is from 20°C to 40°C (8,9). For the 1st order factorial design combinations (T_1=30°C, T_3=40°C), the reaction rates are located mainly in the 2nd modulate with $40.00 < V_{sl} \leq 120.00$ mg/l.d and $10.00 < V_{su} \leq 35.00$ mg/l.d , as the high and low levels remain within the optimal temperature range.

However, for T_3=45°C (ES20), the reaction rate is relatively low with V_{Sl}=30.00mg/l.d, and V_{su}=10.11mg/l.d whileas for T_1=25°C(ES19), which still is within the optimal range, the reaction rate is located is in the 2nd modulate with V_{sl}=51.90 mg/l.d and V_{su}=17.42 mg/l. Consequently, partial inhibition sulphate reduction occurs above T=45°C.

3.1.3. Effect of the Initial Sulphate Concentration

For the combination (ES22), the initial sulphate concentration (Sla$_3$=22mM) is above

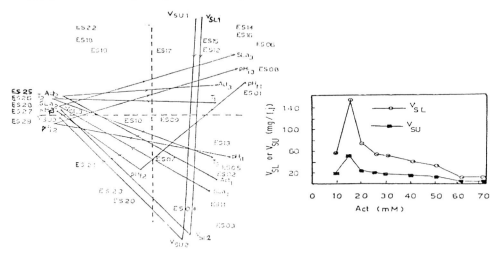

Figure 1: Projection graph
(Factoriel correspondence analysis)
Desulfobacter postgatei species

Figure 2: Effect of the initial carbon
source concentration
Desulfotomaculum acetoxidans species

the inhibition concentration, and the reaction rate is located in the 1st modulate with V_{sl}= 5.83 mg/l.d and V_{su}=2.03 mg/l.d. However for the 21st combination where the initial sulphate concentration is relatively low (Sla$_1$=3mM), the rates are located in the 2nd modulate with V_{sl}=50.00 mg/l.d and V_{su}=16.85 mg/l.d.This indicates that these species of bacteria do not tolerate high sulphate concentrations. Inhibition occurs above a sulphate concentration of 22mM, wheras in Widdel and Pfennig'work (8,9), the maximal initial sulphate concentration used did not exceed 4mM.

On the other hand, the 1st order factorial design is applied with a relatively high sulphate range (Sla$_1$=7.75mM, Sla$_3$=17.25m), which allowed satisfactory reaction rates, located mainly in the 2nd modulate (40.00<V_{sl}≤ 120.00 mg/l.d and 10<V_{su} ≤ 35.00 mg/l.d).

3.1.4. Effect of the Initial Carbon Source Concentration

Regarding the carbon source, for a relatively low acetate concentration (Act$_1$ =5mM, (ES23)), the reaction rate achieved is located in the 1st modulate (V_{sl}=38.00 mg/l.d and V_{su}=12.07 mg/l.d) and is obviously low. Thus, below this level, the carbon source becomes growth limiting. However, acetate concentrations as high as 25mM (ES24)), lead to reasonably good reaction rates such as V_{sl}=93.66 mg/l.d and V_{su}=31.16 mg/l.d, which approach those obtained in the optimal conditions corresponding to an initial carbon source concentration of 15 mM. Subsequent work has shown that strong inhibition is achieved above 40 mM as depicted for example in Figure 2.

For the treatment combinations of the 1st order factorial design, corresponding to Act_1=10mM, Act_3=20mM, the reaction rate is located mainly in the 2nd modulate with $40.00 < V_{sl} \leq 120.00$ mg/l.d and $10 < V_{su} \leq 35.00$mg/l.d. This result proves that the optimal acetate concentration range is wider than that established earlier by Widdel and Pfennig (3 to 15 mM (8,9)).

3.1.5. Optimization

Maximum reaction rates of V_{Sl}=152.70 mg/l.d and V_{Su}=51.45 mg/l.d are achieved for pH_2=7.1, T_2=35°C, Sla_2=12.5mM and Act_2=15mM.

Published works established optimal initial pH and incubation temperature to 7.1 and 36°C respectively, for initial acetate and sulfate concentrations ranging from 3 to 15 mM and from 1 to 4 mM respectively (8,9).

3.2. *Desulfobacter postgatei* Species
3.2.1. Effect of the pH

Combinations (ES17, ES18) show the effect of pH on sulphate reduction rates, which are very low for both combinations, with V_{sl} =4.00 mg/l.d, V_{su} =1.34 mg/l.d for pH_1 =5.1 (ES17) and V_{sl} =5 mg/l.d, V_{su} =1.65 mg/l.d for pH_3 =9.5.

However, the low and high pH levels for the 1st order factorial design (pH_1 =6.2 (-1), pH_3 =8.4 (+1)), indicate that sulphate reduction rate is located in the 2nd modulate with $20 \leq V_{sl} \leq 45$ mg/l.j and $10 < V_{su} \leq 20$ mg/l.d. This result is in accordance with previous work which has shown that the optimal initial pH range for bacterial growth is from 6.2 to 8.5 with an optimum at 7.3 (6,7). In the present research work, sulfate reduction is inhibited outside this range.

3.2.2 Effect of the Temperature

The combination (ES19), T_1=20°C (-2)) of the 2nd order factorial design shows the temperature effect on the reaction rate. Indeed, the rates are relatively low with V_{sl}=27.20 mg/l.d and V_{su} =9.16 mg/l.d.

Published work indicates that optimal temperature range for sulphate reduction is from 10°C to 37°C with an optimum at 30°C (6,7).

For the 1st order factorial design (T_1 =25°C (1), T_3 =35°C (+1)), the reaction rate is mainly located in the 2nd modulate with $20 \leq V_{sl} \leq 45$ mg/l.d and $10.00 < V_{su} \leq 20.00$ mg/l.d. This result shows that satisfactory reaction rate is achieved wihin this temperature range.

Table 1

Rotating central composite design treatment combinations for *Desulfotomaculum acetoxidans* (DA) and *Desulfobacter postgatei* (DP) species.

pHi		Ti °C		$[SO_4]_i$ (mM)		$[Acet.]_i$ (mM)		V_{sl} (mg/L.day)		V_{su} (mg/L.day)		pH_f	
DA	DP	DA	DP	DA	DP	DA	DP	DA	DP	DA	DP	DA	DP
6.6(-1)	6.2 (1)	30(-1)	25(-1)	7.75(-1)	10(-1)	10(-1)	12.5(1)	10.83	5.00	3.57	1.162	8.7	7.3
7.6(1)	8.4(1)	30(-1)	25(-1)	7.75(-1)	10(-1)	10(-1)	12.5(-1)	29.30	54.50	9.76	13.58	8.5	8.2
6.6(-1)	6.2(1)	40(1)	35(1)	7.75(-1)	10(-1)	10(-1)	12.5(-1)	50.00	49.63	17.95	11.83	8.5	8.3
7.6(1)	8.4(1)	40(1)	35(1)	7.75(-1)	10(-1)	10(-1)	12.5(-1)	4.00	39.26	1.35	14.66	9.1	8.3
6.6(-1)	6.2(1)	30(-1)	25(-1)	17.25(1)	22(1)	10(-1)	12.5(-1)	44.18	56.32	14.56	18.28	8.8	8.1
7.6(1)	8.4(1)	30(-1)	25(-1)	17.25(1)	22(1)	10(-1)	12.5(-1)	80.13	10.07	16.75	3.25	9.0	8.0
6.6(-1)	6.2(1)	40(1)	35(1)	17.25(1)	22(1)	10(-1)	12.5(-1)	61.50	61.56	20.89	20.59	8.8	8.3
7.6(1)	8.4(1)	40(1)	35(1)	17.25(1)	22(1)	10(-1)	12.5(-1)	69.22	13.65	23.22	4.64	8.8	8.3
6.6(-1)	6.2(1)	30(-1)	25(-1)	7.75(-1)	10(-1)	20(1)	27.5(1)	17.20	65.73	5.80	21.03	8.7	8.0
7.6(1)	8.4(1)	30(-1)	25(-1)	7.75(-1)	10(-1)	20(1)	27.5(1)	6.41	64.14	2.02	21.39	8.4	8.6
6.6(-1)	6.2(1)	40(1)	35(1)	7.75(-1)	10(-1)	20(1)	27.5(1)	9.00	58.60	3.03	19.75	9.1	8.0
7.6(1)	8.4(1)	40(1)	35(1)	7.75(-1)	10(-1)	20(1)	27.5(1)	41.70	30.00	13.92	9.90	9.2	8.1
6.6(-1)	6.2(1)	30(-1)	25(-1)	17.25(1)	22(1)	20(1)	27.5(1)	68.66	33.32	23.14	11.00	8.1	8.1
7.6(1)	8.4(1)	30(-1)	25(-1)	17.25(1)	22(1)	20(1)	27.5(1)	70.66	14.66	23.81	4.94	8.2	8.1
6.6(-1)	6.2(1)	40(1)	35(1)	17.25(1)	22(1)	20(1)	27.5(1)	61.10	30.00	20.58	9.90	8.9	8.1
7.6(1)	8.4(1)	40(1)	35(1)	17.25(1)	22(1)	20(1)	27.5(1)	58.68	25.00	11.87	8.27	8.9	8.1
6.1(-2)	5.1(2)	35(0)	30(0)	12.50(1)	16(0)	15(0)	20(0)	43.66	4.00	14.64	1.34	9.0	5.4
8.1(2)	9.5(2)	35(0)	30(0)	12.50(0)	16(0)	15(0)	20(0)	32.50	5.00	10.91	1.65	9.2	9.0
7.1(0)	7.3(0)	25(-2)	20(-2)	12.50(0)	16(0)	15(0)	20(0)	51.90	27.20	17.42	9.16	8.6	8.4
7.1(0)	7.3(0)	45(2)	40(2)	12.50(0)	16(0)	15(0)	20(0)	30.00	40.05	10.11	13.61	8.1	8.4
7.1(0)	7.3(0)	35(0)	30(0)	3(-2)	4(-2)	15(0)	20(0)	50.00	35.00	16.85	11.65	8.9	8.5
7.1(0)	7.3(0)	35(0)	30(0)	22(2)	28(2)	15(0)	20(0)	5.83	30.00	2.03	9.90	8.2	8.5
7.1(0)	7.3(0)	35(0)	30(0)	12.50(0)	16(0)	5(-2)	5(-2)	38.00	42.53	12.07	14.23	9.2	8.5
7.1(0)	7.3(0)	35(0)	30(0)	12.50(0)	16(0)	25(2)	35(2)	93.66	76.93	31.16	25.76	9.1	8.4
7.1(0)	7.3(0)	35(0)	30(0)	12.50(0)	16(0)	15(0)	20(0)	149.0	70.00	50.20	23.42	8.9	8.5
7.1(0)	7.3(0)	35(0)	30(0)	12.50(0)	16(0)	15(0)	20(0)	152.0	78.00	50.45	25.74	8.9	8.5
7.1(0)	7.3(0)	35(0)	30(0)	12.50(0)	16(0)	15(0)	20(0)	150.0	72.50	50.87	24.43	8.9	8.3
7.1(0)	7.3(0)	35(0)	30(0)	12.50(0)	16(0)	15(0)	20(0)	150.5	80.00	50.56	26.93	8.9	8.5
7.1(0)	7.3(0)	35(0)	30(0)	12.50(0)	16(0)	15(0)	20(0)	149.2	76.00	50.43	25.62	8.9	8.5
7.1(0)	7.3(0)	35(0)	30(0)	12.50(0)	16(0)	15(0)	20(0)	150.0	77.00	50.30	25.93	8.9	8.5
7.1(0)	7.3(0)	35(0)	30(0)	12.50(0)	16(0)	15(0)	20(0)	150.0	75.00	50.50	25.27	8.9	8.5

For $T_i = 40°C$ (+2) (ES20), conversly to what was predicted, the reaction rate is located in the 2nd modulate: $V_{sl} = 40.05$ mg/l.d and $V_{su} = 13.61$ mg/l.d. Thus, strong inhibition occurs for $T_i = 20°C$ (-2) whereas partial inhibition occurs for $T_i = 40°C$ (+2).

3.2.3 Effect of the Initial Sulphate Concentration

Combinations (ES21, ES22) show the effect of the initial sulphate concentration on the sulphate reduction and sulphide production reaction rates.

For the combination (ES22) where the sulphate concentration $Sla_3 = 28$mM (+2) is above the inhibition concentration, the reaction rate is located in the 1st modulate with

V_{sl} =30.00 mg/l.d and V_{su} =9.90 mg/l.d.

However, the 21st combination where the initial sulphate concentration is relatively low Sla_1=4mM(-2), the reaction rates are reasonably low: V_{sl} =35.00mg/l.d and V_{su} =11.65 mg/l.d.

This result indicates that these species of bacteria do not tolerate high and very low sulphate concentrations. Inhibition occurs above 28 mM and below 4mM

For the 1st order factorial design (Sla_1 =10.mM(-1), Sla_3=22mM (+1)), the reaction rate is mainly located in the 2nd modulate with 20.00$\leq V_{sl} \leq$45 mg/l.d and 10<$V_{su} \leq$20 mg/l.d, and sometimes even in the 3[rd] modulate with 50$\leq V_{sl} \leq$80 mg/l.d and 20.00<$V_{su} \leq$30.00 mg/l.d.

Therefore, the sulphate concentration range for the 1st order factorial design is optimal for sulphate reduction kinectics whereas in published works, the optimal sulphate concentration range is from 2 to 16 mM (6,7).

3.2.4. Effect of the Initial Carbon Source Concentration

For the1[st] order factorial design, the reaction rate is mainly located in the 2nd or even in the 3rd modulate. Consequently, the acetate concentration optimal range is Act_1=12.5 (-1), Act_3 =27.5mM (+1). In acordance to this, Widdel and Pfennig (6) suggest that optimal acetate concentration for the bacterial growth is around 20 mM.

For a relatively weak concentration (Act_1 =5mM, (ES23)), the reaction rate achieved is reasonably low as compared to the one achieved under optimal conditions i.e., V_{sl} =42.53 mg/l.d and V_{su} =14.23 mg/l.d. This partial inhibition is explained therefore by the energy source defficiency.

However, acetate concentrations as high as 35mM (ES24)), lead to rates located in the 3rd modulate V_{sl} =76.93 mg/l.d and V_{su} =25.76 mg/l.d.

For the combination (ES24) (Act_3 =35mM (+2)), conversly to what was predicted, the reaction rate is in the 3rd modulate with V_{sl} =76.93 mg/l.d and V_{su} =25.76 mg/l.d. Subsequent work has shown that strong inhibition is achieved above 45 mM.

3.2.5. Optimization

The combinations cloud for central points, pH_2=7.3 (0), T_2=30°C (0), Sla_2=16mM (0) and Act_2=20mM(0), correspond to maximum reaction rates: V_{sl}= 80.00mg/l.d and V_{su} =26.93 mg/l.d. Published works (6) suggest that the maximum sulphide production rate is obtained for T_i =32°C and for pH_i =7.3 with initial acetate and sulfate concentrations respectively of 20 mM and 16 mM..

3.3. Final pH Results Analysis and Discussion

For both species, acetate oxidation through bacterial metabolism occurs according to the following reaction (2,5):

$$17\ CH_3COO^- + 11\ H_2O \rightarrow 8\ C_4H_7O_3\ (biomass) + 2\ HCO_3^- + 15\ OH^- \tag{1}$$

Table 2. Homogenous classification in modulates of studied factors and of variable response values for *Desulfotomaculum acetoxidans* and *Desulfobacter postgatei* species.

Factors / Variable response	Range of modulates *Desulfotomaculum acetoxidans*	Range of modulates *Desulfobacter postgatei*	Notation
PH_i 1st modulate	$6.1 \leq pH_i \leq 6.6$	$5.1 \leq pH_i \leq 6.2$	pH_1
2nd modulate	$PH_i = 7.1$	$pH_i = 7.3$	pH_2
3rd modulate	$7.6 \leq pH_i \leq 8.1$	$8.4 \leq pH_i \leq 9.5$	pH_3
$T_i °C$ 1st modulate	$25 \leq T_i \leq 30$	$20 \leq T_i \leq 25$	T_1
2nd modulate	$T_i = 35$	$T_i = 30$	T_2
3rd modulate	$40 \leq T_i \leq 45$	$35 \leq T_i \leq 40$	T_3
$[SO_4]_i$ (mM) 1st modulate	$3 \leq [SO_4]_i \leq 7.75$	$4 \leq [SO_4]_i \leq 10$	Sla_1
2nd modulate	$[SO_4]_i = 12.5$	$[SO_4]_i = 16$	Sla_2
3rd modulate	$17.25 \leq [SO_4]_i \leq 22$	$22 \leq [SO_4]_i \leq 28$	Sla_3
$[Act]_i$ (mM) 1st modulate	$5 \leq [Act]_i \leq 10$	$5 \leq [Act]_i \leq 12.5$	Act_1
2nd modulate	$[Act]_i = 15$	$[Act]_i = 20$	Act_2
3rd modulate	$20 \leq [Act]_i \leq 25$	$27 \leq [Act]_i \leq 35$	Act_3
V_{Sl} (mg/l.d) 1st modulate	$0 < V_{Sl} \leq 40$	$0 \leq V_{Sl} \leq 16$	V_{Sl1}
2nd modulate	$40 < V_{Sl} \leq 120$	$20 \leq V_{Sl} \leq 45$	V_{Sl2}
3rd modulate	$120 < V_{Sl} \leq 160$	$50 \leq Vsl \leq 80$	V_{Sl3}
V_{Su} (mg/l.d) 1st modulate	$0 < V_{Su} \leq 10$	$0 \leq V_{Su} \leq 10$	V_{Su1}
2nd modulate	$10 < V_{Su} \leq 35$	$10 < V_{Su} \leq 20$	V_{Su2}
3rd modulate	$35 < V_{Su} < 55$	$20 < V_{Su} \leq 30$	V_{Su3}
pH_f 1st modulate	$8.1 < pH_f \leq 8.6$	$5.4 \leq pH_f \leq 8.1$	pH_{f1}
2nd modulate	$8.6 < pH_f \leq 8.8$	$8.2 \leq pH_f \leq 8.4$	pH_{f2}
3rd modulate	$8.8 < pH_f \leq 9.3$	$8.4 < pH_f \leq 9.0$	pH_{f3}

whereas sulphate reduction occurs according to (1) :

$$CH_3COO^- + SO_4^{2-} \rightarrow 2\ HCO_3^- + HS^- \qquad (2)$$

Bacterial metabolism seems to generate alkalinity. Indeed, The number of moles of hydroxide ions produced in the system is equal to 15/17 of acetate moles consumed. On the other hand, dissociation in basic medium of carbonic acid present in the system, produces carbonate ions accompanied by the dissociation of bisulphide ions according to the following reactions (1,2)

$$HCO_3^- + OH^- \rightarrow CO_3^- + H_2O \quad (pK_2 = 10.28\ (2)) \qquad (3)$$

$$HS^- + OH^- \rightarrow H_2O + S^{2-} \quad (pK_2 = 12.91(2)) \qquad (4)$$

The dissociation of bisulphide ions is favoured by an increase in pH above neutrality.
The present research work has shown that for *Desulfotomaculum acetoxidans* species, the final pH increases with the improvement of growth conditions and cellular metabolism. It is therefore possible, as pointed out by previous investigators (5,8,9), to link this pH increase to more active metabolism rate.

In extreme initial pH conditions ($pH_1 = 6.1$ or $pH_3 = 8.1$), total or partial sulphate reduction and sulphide production inhibition occurs following bacterial growth inhibition.

In the presence of 20 mM of sodium carbonate as co-substrate, there will be an increase in the final pH to $pH_f = 9.0$ and $pH_f = 9.2$ respectively for intial values of $pH_i = 6.1$ and $pH_3 = 8.1$.

For *Desufobacter postgatei* species, final pH increases with the improvement of growth conditions and cellular metabolism as the 3rd modulate for final pH values is obtained for optimal conditions. Regarding *Desulfotomaculum acetoxidans* species, this pH increase can be due to a more active metabolism reaction rate. In inhibition conditions, initial pH does not vary and thus for $pH_1 = 5.1$ (-2), $pH_f = 5.4$ and for $pH_3 = 9.5$ (+2), $pH_f = 9.0$. The results indicate that bacterial metabolism and sulphide production are very slow.

In fact, the final pH increase tendency is very complex. This complexity is related to the diversity of primary and secondary reactions occuring in the culture medium.

4. CONCLUSIONS

Kinetics optimization of sulphate reduction with *Desulfotomsaculum acetoxidans* and *Desulfobacter postgatei* species was studied according to the factorial design method, and the results are analysed according to the factorial correspondence analysis technique.These studies lead to the following conclusions.

For *Desulfotomaculum acetoxidans:* (i) optimum bacterial growth and optimum sulphate reduction are obtained for pH_i range from 6.6 to 7.6 and T_i range from 30°C to 40°C (ii) optimum initial concentrations range from 7.75 to 17.25 mM for sulphate and from 10 to 20 mM for acetate (iii) strong inhibitions occur respectively below $pH_i = 6.1$, above $pH_i = 8.1$ and above $T_i = 45°C$ (iv) Strong inhibitions are obtained below 3 mM and above 22 mM for sulfate and below 5mM and above 50 mM for acetate.

For *Desulfobacter postgatei* (i) optimum bacterial growth and optimum sulphate reduction are obtained for pH_i range from 6.2 to 8.4 and T_i range from 25°C to 35°C (ii) optimum initial concentrations range from 10 to 22 mM for sulphate and from 12.5 to 27.5 mM for acetate. (iii) strong inhibitions are obtained respectively above $pH_i = 9.5$ and $T_i = 40°C$ and below $pH_i = 5.1$ and $T_i = 20°C$ (iv) strong inhibitions are obtained below 4 mM and above 28 mM for sulfate and below 5 mM and above 45 mM for acetate.

For *Desulfobacter postgatei* (i) optimum bacterial growth and optimum sulphate reduction are obtained for pH_i range of 6.2 to 8.4 and T_i range of 25°C to 35°C (ii) optimum initial concentrations range from 10 to 22 mM for sulphate and from 12.5 to 27.5 mM for acetate. (iii) strong inhibitions occur respectively above $pH_i = 9.5$ and $T_i = 40°C$ and below $pH_i = 5.1$ and $T_i = 20°C$ (iv) strong inhibitions occur below 4 mM and above 28 mM for sulfate and below 5 mM and above 45 mM for acetate.

The overall optimum rates achieved so far are $V_{sl} = 152.70$ mg/l.d and $V_{su} = 51.45$ mg/l.d for *Desulfotomaculum acetoxidans* species and $V_{sl} = 80.00$ g/l.d and $V_{su} = 26.93$ mg /l.d for *Desulfobacter postgatei* species respectively.

768

ACKNOWLEDGEMENTS

This work is being developped in collaboration between the Départment de Génie Chimique, Biotechnologie et Environment, University of Liège (ULg), Belgium, and the LAMINEMET-BIOREMEDIATION Laboratory, Mohammadia School of Engineers (EMI), University Mohamed V of Rabat, Morocco. The financial support was provided by the Belgian Administration Générale de la Coopération pour le Développement AGCD.

REFERENCES

1. Dudeney, A.W.L., Sbai, M.L., Bioleaching of rare-earth-bearing posphogypsum, Biohydrometallurgical Technologies. Edited by A.E.Torma, J.E.Wey and V.L.Lakshmanan. The Minerals, Metals & Materials Society, 1993.
2. Sbai, M. L., Leaching and biodegradation of rare earth bearing phosphogypsum, Ph.D Thesis, University of London, 1989.
3. United Nations Industrial Development Organization, Vienna/ Austria, Final report: Identification of needs and development of the phosphogypsum utilization integrated program. Project No: US/UT/INT/0/02. Contract-: 89/138. Elaborated Industry Consult Berlin Gmbh.1992.
4. Oceanol. Acta, 8(4) (1985) 411.
5. Postgate, J . R ., The sulphate reducing bacteria, Cambridge University Press, 1984.
6. Widdel, F., Pfennig., Arch . Microbiol. 129 (1981) 395.
7. Uphaus, R.A., Grimm, D., Cork, J., Microbiol. 24 (1983) 435.
8. Widdel, .F., Pfenning. N., Arch. Microbiol. 112 (1977) 119.
9. Widdel. F., Pfennig. N., Arch. Microbiol. 129 (1981) 401.
10. Tabatabai, M . A., Sulfur Inst. J. 10 (1974) 11.
11. American Public Health Association, Standard methods for the examination of water and wastewater, 16th Ed. 1985, 476.
12. Vogel, A., Vogel's textbook of quantitative inorganic analysis, 4th Ed., Longman Pub., 1978.
13. Cochran . W . G., G. M . Cox ., Experimental Designs. 2nd Ed. , A Wiley International Edition,1957.
14. Espensen, J . H ., Chemical Kinetics and reaction mechanisms, Mc Graw-Hill, Inc., 1981.
15. Benzécri. J. P. et col., L'analyse des données en économie, Dunod.

Study of the conditions of forming environmentally sound insoluble arseniferrous products in the course of biohydrometallurgical processing of gold-arsenic concentrates

G.V. Sedelnikova, R.Ya. Aslanukov, E.E. Savari

Central Research Institute of Geological Prospecting for Base and Precious Metals (TSNIGRI), 113545, Varshavskoe shosse, 129 «B», Moscow, Russia

Refractory gold-arsenic concentrates of ores from six deposits, their biooxidation (BO) solid residues and residual bacterial solutions were analyzed for chemical composition and Fe/As molar ratio. The effect of the Fe/As ratio on As solubility was examined in two solid BO product types: (I) BO solid residues obtained in the course of biooxidation, and (II) precipitates that formed upon the completion of the BO procedure as a result of As precipitation from the bacterial solutions through the jarosite process. Stable arsenic-containing products can be formed provided that the Fe/As molar ratio is greater than 4. The pH governing the BO process and the iron concentration in the bacterial solutions can be varied in order to provide for a controlled increase in Fe/As. Mineragraphic, X-raying and X-ray photoelectron spectroscopic techniques were used to determine the composition arsenic-containing compounds incorporated in   insoluble BO residues $FeAsO_4 \cdot 2H_2O$ and $Fe_2(AsO_4) \cdot SO_4 \cdot OH \cdot 7H_2O$.

1. INTRODUCTION

Refractory   gold-arsenic ores are one of the main sources of precious metal commodities in Russia. They account for more than 25% of the national ore gold reserves. Till now, refractory ores have never been (and still are not) involved in exploitation on a commercial scale. During the past fifteen years, TSNIGRI keeps performing research and development in the field of biooxidation and hydrometallurgical treatment of refractory gold-arsenic concentrates [1], [2]. Samples representing more than 20 deposits from Russia and other countries have been already tested in laboratory and pilot scale. During 1997-1998, pilot tests of the biohydrometallurgical technology were run at the Olimpiadinskaya processing plant.

The possibility of environmentally safe performance of industrial procedures based on this technology critically   depends on the stability of solid arsenic-containing   products resulting from the biooxidation of sulphides. Many earlier investigations dealt with the stability of solid arsenic-containing products. A decrease in arsenic solubility was observed with increasing Fe/As ratio in soluble solids. Stable bioresidues, i.e. those with soluble arsenic less than 5 mg/l as related to the EPA limit, are obtained for Fe/As ratios greater than 4.0 [3]. Ferric iron excess is required for ferric-arsenic precipitates stable over a wide pH. Basic ferric arsenates precipitated from hydrometallurgical liquors by hydrolysis are shown to be very insoluble over a range of pHs. Precipitates with Fe/As ratio = 8 have solubilities <1 mg/l between pH 3 and 8. The As solubility did not increase during aging over a 2-3.7 yr period [4], [5], [6].

The main objectives of this study were the following:

- to gain knowledge on the composition and stability of arsenic-containing solids resulting from biooxidation of refractory gold-pyrite-arsenopyrite-pyrrhotite concentrates;
- examination of the effect of the Fe/As molar ratio, pH and aging time on the stability of the solid residues;
- determination of the conditions that are required for the formation of insoluble solid residues (1) in the course of biooxidation of concentrates, and (2) as precipitates from bacterial solutions upon the completion of the biooxidation procedure and separation of solid and liquid phases.

2. TEST TARGETS AND TECHNIQUES

2.1. Test targets

In this study, test targets were gravity-flotation concentrates obtained as a result of concentration of refractory gold-arsenic ores from four Russian deposits (No. 1-4), from one deposit in Kazakhstan (No. 5), and one deposit from Slovakia (No. 6).

The concentrates studied show different proportions of essential components

According to fire assaying data, gold content of the concentrates varies from 21.6 to 70 ppm, silver concentration from 3.8 to 121.7 ppm, and arsenic content from 1.9 to 9.6%. Phase analysis revealed the presence of oxidized arsenic species in amounts from 0.5 to 20%. In five of the six concentrate samples, the Fe/As molar ratio is greater than 4 (4.3-15.9). Only the concentrate No. 5 had Fe/As=3.4.

The samples studied differ from each other in mineral modes of major sulphides. They are represented by the following types:

- pyrite-arsenopyrite concentrates (No. 3-6);
- pyrrhotite-arsenopyrite-pyrite concentrate (No. 1);
- pyrite concentrate with a minor amount of arsenopyrite (No. 2).

All the concentrates tested contain 0.1 to 2.1% antimonite and minor amounts of other sulphides (chalcopyrite, sphalerite, galena etc.).

2.2. Biooxidation experiments

Bacterial oxidation of the concentrates No. 1-4 and No. 6 was carried out in a continuous flow mode. The pilot plant consisted of Pachuca reactors of 150 l volume.

The concentrate No. 5 was subjected to leaching under laboratory conditions, in the mode of recurrent biomass cultivation in separator funnels and in Pachuca reactors of 0.3 and 1.0 l volumes.

The pulp was stirred and aerated by air supplied from a compressor. The temperature was maintained at 28-35°C. The liquid-to-solid ratio in the contents of the reactors was 5:1. The duration of the leaching procedure was 90-120 h, and pH varied from 1.3 to 2.0.

Biooxidation was conducted using autotrophic thionic bacteria of the species Thiobacillus ferrooxidans from a culture grown in the Institute of Microbiology of the Russian Academy of Sciences which have been repeatedly used for years in experiments on biooxidation of refractory sulphide concentrates. Bacteria were grown on the 9K medium with 4 g/l Fe^2 in the presence of initial concentrates. An updated nephelometric technique [7] was used for rapid determination of bacterial concentrations in the liquid phase of pulp. The physiological activity of the biomass was estimated from the rate of Fe^{2+} oxidation and from the intensity of oxygen absorption by the pulp or by the solution [8].

2. 3. Characteristics of biooxidation products

The pulp which was produced as a result of bacterial oxidation of the concentrates No. 1-6 was then subjected to thickening and filtration. Data on the chemical compositions of resultant bacterial solutions and solid products are presented in Table. The low proportions of As (0.2-0.44%) point to a rather high (96-99%) degree of oxidation of arsenopyrite, the major gold-containing mineral in the samples. In the course of biooxidation, arsenic partially (30-40%) passes into the bacterial solution where its concentration tends to vary from 2.5 to 11.6 g/l. Pyrite and pyrrhotite undergo oxidation along with arsenopyrite. Pyrite is oxidized to a much smaller extent than other sulphides involved (65-84%), whereas pyrrhotite becomes 92% oxidized. Iron passes into the bacterial solution in amounts of 12.8-40 g/l. Following the mechanism of biooxidation of sulphides described in [7], the dissolved iron forms nearly insoluble ferric arsenate compounds according to the reaction:

$$3H_3AsO_4 + Fe_2(SO_4)_3 \rightarrow 2FeAsO_4 + H_2SO_4 \tag{1}$$

In addition, iron may be present in the solid phase of BO products in a hydroxide form:

$$Fe_2(SO_4)_3 + 6H_2O \rightarrow 2Fe(OH)_3 + 3H_2SO_4 \tag{2}$$

When monovalent cations (Na$^+$, K$^+$, NH$_4^+$) are present in the material subject to biooxidation, insoluble jarosite species are formed according to the reactions:

$$2Fe_2(SO_4)_3 + Na_2SO_4 + 12H_2O \rightarrow 2NaFe_3(SO_4)_2(OH)_6 + 6H_2SO_4 \tag{3}$$

The solid-arsenic-containing products and bacterial solutions analyzed (Table) show variable Fe/As molar ratios and pH. In solid BO products No. 1-3, the values of the Fe/As molar ratio are greater than 4, while in BO products No. 4-6 Fe/As varies from 2.7 to 3.5. A similar situation is observed for the bacterial solutions.

It is noteworthy that the Fe/As molar ratio is greater than 4 in solutions with pH=1.8 - 1.9 (No.1, 3), while Fe/As<4 (No. 4, 6) was found in bacterial solutions having pH=1.4 -1.5. The high Fe/As molar ratio in the solution No. 2 with pH=1.2 might be due to the sharp predominance of pyrite (55.2%) over arsenopyrite (5.9%) in the concentrate No. 2.

Table
Major components of BO products

Essential elements	BO solid residues, No.					
	1	2	3	4	5	6
Iron total, %	9.3	20.0	9.1	11.06	6.9	17.87
Sulphur total, %	10.1	16.4	7.1	10.71	5.13	13.4
Sulphide sulphur , %	4.0	14.2	1.6	8.2	1.33	13.3
Arsenic total, %	22.6	1.5	3.0	4.25	2.6	8.81
Sulphide arsenic , %	0.20	0.20	0.2	0.25	0.23	0.44
Fe/As molar ratio	5.5	17.4	4.0	3.5	3.5	2.7

	Bacterial solutions, No.					
	1	2	3	4	5	6
pH	1.9	1.2	1.8	1.5	1.5	1.4
Content, g/l:						
Potassium	0.010	0.014	-	-	-	0.048
Sodium	0.006	0.026	-	-	-	0.031
Iron total	44.8	25.9	26.6	12.8	15.5	34.4
Fe^{3+}	40.0	25.9	26.2	12.8	15.5	34.4
Arsenic	2.5	3.78	6.2	4.8	5.3	11.6
SO_4^{2-}	66.1	58.6	56.4	28.5	27.3	58.1
Fe/As molar ratio	24.2	9.24	5.7	3.5	3.8	3.9

3. RESULTS AND DISCUSSION

3.1. Batch tests

Solid arsenic-containing products are formed in two distinct units in the process of bacterial oxidation of gold-arsenic concentrates. Respectively, two types of arsenic products are distinguished:

- type I - solid residues formed in the course of biooxidation;
- type II - arsenic precipitates settled from bacterial solutions upon completion of the BO procedure and pulp separation by filtration into the solid (solid residues) and liquid phases (bacterial solutions).

In this study, both of these solid arsenic product types were examined for As solubility.

3.1.1. As solubility in solid BO residues

In order to asses the toxicity of solid BO residues, As solubility in tap water at pH=7-8 was tested, which was meant to simulate the conditions under which the residues are usually retained in tailings ponds.

The procedure of solubility tests consisted of the following operations. Solid BO residues, upon filtration, were water-washed to pH~7 and then kept underwater. At definite time intervals, solution samples were taken to be analyzed for arsenic.

Figure 1 shows the temporal behaviour of arsenic concentration in the water phase during aging of the solid BO residues No. 1-6. For graphic assessment of the toxicity of BO residues, a horizontal line was drawn in the figure to denote the environmentally limited concentration (ELC) of arsenic dissolved in wastewater which is specified as 0.05 mg/l under regulations set up by the environmental control survey of the Russian Federation. With respect to these constraints, the BO residues showing Fe/As≥4 (No. 1-3) could be said to be environmentally safe. Arsenic removal from this residues remained close to zero at pH~7-8 during the whole observation period which lasted 180 days (curves 1 to 3). Solid residues with Fe/As=2.7-3.5 are much less stable. A significant increase in arsenic concentration in the water solution in which these solid BO residues rested was noted as their aging time increased. The highest As solubility was shown in the BO residue No. 6, which produced more than 0.6 mg/l of dissolved arsenic in a period of 180 days.

3.1.2. Effect of BO mode on the Fe/As ratio

The unstable arsenic-containing BO residues (No. 4-6) showed respective Fe/As values equal to 3.5, 3.5 and 2.7. In an attempt to increase the stability of these solids, laboratory experiments on bacterial leaching of the concentrates No. 4 and 5 were performed, with $FeSO_4$ addition to the bacterial solutions. The portionwise addition of $FeSO_4$ controlled the Fe/As molar ratio in the bacterial solutions and, hence, in the resultant solid BO residues. An increase in Fe/As shown by solid residues was observed when the pH of the bacterial solutions was concurrently increased (Figure 2). Solid BO residues with Fe/As>4 were produced at pH>1.5-1.6.

The efficiency of the bacterial oxidation of sulphides and Fe^{2+} conversion into Fe^{3+} depends on the activity of the biomass. Examination of the biomass activity in the course of oxidation of the concentrates No. 4 and 5 within the pH range under study showed that the peak of the activity, at which the highest rate of sulphides oxidation in the concentrates No. 4 and 5 was observed, falls within the pulp pH range between 1.5-1.6 and 1.9-2.0. Further increase in pH results in a decrease in bacterial activity and in the efficiency of the BO procedure as a whole.

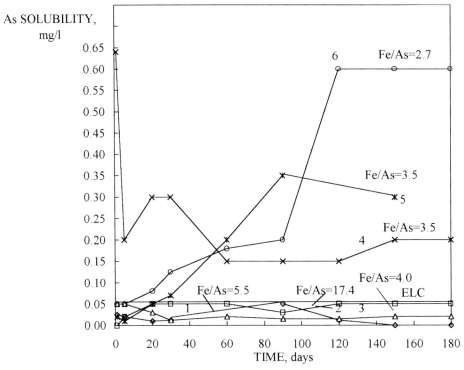

Figure 1. As solubility in BO solid residues vs. Fe/As molar ratio in solid residues.

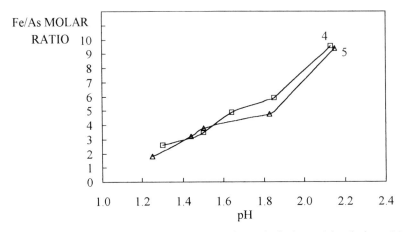

Figure 2. Effect of pH on Fe/As molar ratio in bacterial solutions (a) and biomass activity (b) in the biooxidation of the concentrates from deposits No.4 and No.5.

In summary, the control over the mode of biooxidation through varying the pH and iron concentration in the solutions made it possible to produce solid residues with Fe/As molar ratios greater than 4 from the concentrates No. 4 and 5. When the BO procedure is run in a continuous flow mode, the increase in iron concentration in the bacterial solution is accomplished by recurrent removals of definite portions of the bacterial solutions for arsenic precipitation and by returns of bacterial solutions with increased Fe/As ratios back to the process for bacterial oxidation.

3.1.3. Stability of solid BO residues after the changing the mode of the process

The solid BO residues from the concentrates No. 4 and 5, which were produced in the course of the above experiments, were then tested for As solubility [9]. The residues, upon washing to neutral pH, were treated with tap water (pH=7.4) under active stirring during 15 minutes at solid-to-liquid ratio equal to 1:125. After the pulp was allowed to settle, the solution was decanted, and the sample was subjected to arsenic analysis. The cake passed through a number of washing cycles under the same conditions.

The solid residues with Fe/As=4.8-5.9, produced under the above biooxidation mode, demonstrated a rather high stability, with As solubility varying from 0.065 to 0.01 mg/l. The solid residues with Fe/As=2.6-3.6 were partially dissolved, showing an increase in As concentration in water solutions up to 0.4-0.5 mg/l (Figure 3).

These results allow to conclude that As stability in solid BO residues from the concentrates studied could be ensured if arsenic-containing products a resulted from bacterial leaching of sulphides have the Fe/As molar ratios greater than 4. This can be achieved through the control over the pH and Fe and As concentrations in the bacterial solutions.

3.1.4. Composition of arsenic BO residues

Mineragraphy, X-raying and X-ray photoelectron spectroscopy were used to test solid BO residues in comparison with their initial concentrates. The studies showed that all the solid BO residues contained an invariant ore component made up of quartz, feldspar, sericite, hydromica and finely dispersed carbonaceous matter. Analysis of XRD patterns from BO residues shows that these contain less pyrite, antimonite and pyrrhotite as compared with the initial concentrates. Arsenopyrite is practically absent in the residues. The residual material consists essentially of precipitated, X-ray-amorphous substances which are hard to identify by means of X-ray testing techniques.

Analysis of the chemical states of the major elements based on binding energy data and XPS determinations of their relative abundances using the LAS-3000 apparatus shows that arsenic occurs in the BO residues as As^{5+}, and iron predominantly as Fe^{3+}. Both iron and arsenic form chemical bonds with oxygen.

Examination of the composition of the insoluble arsenic BO residues No. 1-3 having respective Fe/As values equal to 5.5, 4.0 and 17.4 reveals that arsenic is fixed in a scorodite phase $FeAsO_4 \cdot 2H_2O$ and in highly hydrated ferric arsenate sulphates of bukovskite type $Fe_2(AsO_4) \cdot SO_4 \cdot OH \cdot 7H_2O$. In addition, iron is bound in sulphate hydroxides. Finally, iron and calcium sulphates are also present in the residues, while calcium arsenates were not found.

The soluble solid BO residues No.4 and 6 having respective Fe/As values equal to 3.5 and 2.7 contain all the same compounds as those found in the residues No. 1-3, plus calcium arsenates. The latter are known as thermodynamically unstable compounds.

776

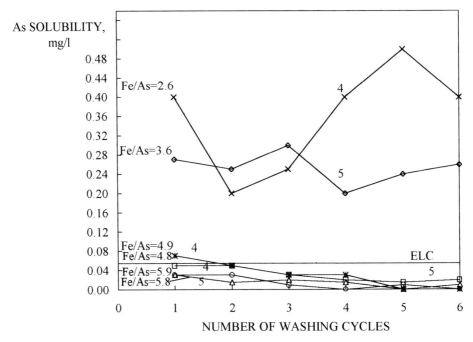

As SOLUBILITY, mg/l

Figure 3.   As solubility in BO solid residues from concentrates No.4
and No. 5 vs. Fe/As molar ratio in bacterial solutoins

4. CONTINUOUS FLOW TESTS

4.1. As precipitation using the jarosite process

L.S. Getskin and his co-authors [10] reported that ferric arsenate compounds were readily precipitated from sulphate solutions with addition of alkaline jarosite $KFe_3[SO_4]_2 \cdot (OH)_6$ as an amorphous precipitate which is chemically similar to natural scorodite. In this process, only a very minor proportion of arsenic (no more than 2%) is co-precipitated with jarosite through substitution for sulphate ions. The rate of precipitation of ferric arsenates is much higher than that of jarosite. It is noteworthy that the formation of the jarosite phase is not accompanied by co-precipitation of any base metals.

In this study, the effect of the Fe/As molar ratio in the bacterial solution No. 6 on As extraction (precipitation) using the jarosite process was examined (Figure 4, curve 1).

It has been found that 90% of dissolved arsenic is extracted as a precipitate at Fe/As=3.5. The highest As extraction (95-98%) was attained in bacterial solutions having the Fe/As molar ratio no less than 4. In this case, the resultant precipitate was easy to filtrate.

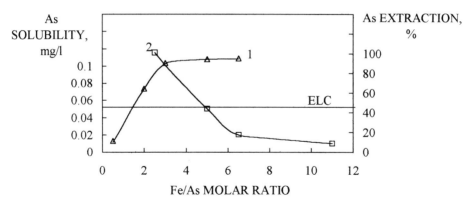

Figure 4. Effect of Fe/As molar ratio in bacterial solutions from biooxidation of concentrate No.6 on As extraction (1) and As solubility (2).

4.2. Stability of arsenic-containing precipitates produced using the jarosite process

Stability tests of arsenic precipitates produced in the jarosite process were carried out according to the procedure reported in [9] which was described in Section 3.1.3. Curve 2 in Figure 4 illustrates the behavior of As solubility from the precipitates obtained from bacterial solutions with different Fe/As ratios. The As solubility curve crosses the ELC line (0.05 mg/l) in the point corresponding to Fe/As=4.8. The stability of arsenic-containing precipitates tends to increase as the Fe/As molar ratio is increased up to 11.

5. CONCLUSIONS

Examination of the solubility of solids produced in the course of biohydrometallurgical processing of gold-arsenic concentrates of ores from six different deposits revealed that:

- environmentally safe stable arsenic-containing BO products exhibiting As solubility lower than 0.05 mg/l are produced at Fe/As molar ratios greater than 4;
- the Fe/As molar ratio in solid BO residues depends on the composition of initial concentrates and on the Fe/As molar ratios in these concentrates. Insoluble solid residues with Fe/As equal to 5.5, 17.4 and 4.0 were produced from concentrates with respective Fe/As molar ratios of 7.8, 15.9 and 4.7. Unstable solid BO residues with Fe/As = 3.5, 3.5 and 2.7 were produced from concentrates with respective Fe/As values equal to 4.3, 3.4 and 4.3. There is reason to conclude that the optimum Fe/As molar ratio in an initial concentrate would be close to 5, which could ensure the stability of solid BO residues provided that proper parameters of the bioleaching process are complied with;

- in order to produce insoluble solid BO residues, the Fe/As molar ratio should be controlled by varying the pH within the range between 1.5-1.6 and 1.9-2.0 and Fe^{3+} concentration. The removals of some portions of bacterial solutions for As precipitation and their returns to the process for biooxidation help to solve the problem of increasing Fe/As»4 when dealing with concentrates in which this ratio is «unfavourable»;

- mineragraphic, XRD and XPS studies revealed that arsenic occured here as As^{5+}, and iron as Fe^{3+}. Arsenic compounds incorporated in the insoluble solid BO residues were found to be $FeAsO_4 \cdot 2H_2O$ and $Fe_2(AsO_4) \cdot SO_4 \cdot OH \cdot 7H_2O$. In addition, thermodynamically unstable calcium arsenates were found in the unstable solid BO residues.

REFERENCES

1. G.I. Karavaiko, G.V. Sedelnikova, I.D. Fridman et al. Symposium on Biotechnology for Progress in Economy. Szeged, Hungary, June (1985).
2. R.Ya. Aslanukov, G.I. Karavaiko, G.V. Sedelnikova et al. Base Metals, No. 4 (1992) 27.
3. K. Adam, F. Battaglia, J.N. Mau, D. Morin, N. Papassiopi, N.E. Tidy and T. Pooley. Proceedings of the XX International Mineral Processing Congress. Aachen, Germany, September (1997) 525.
4. E. Krause and V.A. Ettel. Proceedings of the 15 th Annual Hydrometallurgical Meeting of CJM, Vancouver, Canada (1985) 5/1-5/20.
5. E. Krause and V.A. Ettel. Crystallization and Precipitation, G.L. Stratdess, M.O. Klein and L.A. Melis (eds.), Pergamon Press (1987) 195.
6. N. Papassiopi, M. Stefanakis and A. Kontopoulos. Metallurgy - Fundamentals and Applications, R.G. Reddy, J.L. Hendrix and P.B. Queneau (eds.), TMS of AEME (1988) 321.
7. G.I. Karavaiko. Biotechnology of Metals, Centre for International Projects, SCSE, Moscow (1989) 51 (in Russian).
8. T.I. Kovalenko, R.Ya. Aslanukov. Methods of bacterial biomass determination in the course of metal leaching from ores and concentrates. Moscow, Proc. TsNIGRI (1985) 49 (in Russian).
9. Unified techniques for water analysis. Yu.Yu. Lurie (ad.), Moscow, Khimia (1973) 14 (in Russian).
10. L.S. Getskin, V.A. Grebenyuk, A.S. Yaroslavtsev. Base Metals, No. 2 (1976) 17.

Transformation of arsenic and tellurium in solution by fungi

P.M .Solozhenkin[a], V. P. Nebera[b] and N. N. Medvedeva-Lyalikova[c]

[a]Institute of the Problems of Complex Utilization of Mineral Resources RAS,
Kryukovsky tupik, 4, Moscow, E-20, Russia ,111020

[b]Moscow State Geological Prospecting Academy,
Miclucho-Maklaya, 23, Moscow, Russia, 117873

[c]Institute of Microbiology RAS,
Prospect 60-letiya Oktyabrya, 7, korp. 2, Moscow, Russia ,117811

The fungus *Scopulariopsis brevicaulis* was studied for treatment of arsenic-bearing solutions. The fungus changed dissolved arsenic AsO_4^{3-} into the gases, arsine and trimethylarsine. From 93 to 99% of the arsenic were recovered under optimal conditions. The gases were transformed into high quality arsenic by thermal treatment. The fungus *Penicillium chrysogenum* was studied for reduction of tellurium to elemental tellurium and the gas dimethyltelluride. The fungus was used to extract tellurium from electrolytic slimes of complex composition in copper refining. From 89 to 98% of tellurium were extracted.

1. INTRODUCTION

Bacterial leaching is adventageous in the xtraction of arsenic from different materials. Optimal conditions for dissolution of arsenic have been determined [1-4]. The recovery of arsenic from solution remains a difficult problem. Existing recovery methods, which envolve arsenic immobilization, can be divided into three groups [5]: 1) sorption using activated coals, aluminates, clays, titanium-based materials, or metallic sorbents (Fe, Zn); 2) precipitation reactions, including thermal precipitation of As, coprecipitation, precipitate flotation, sorptive colloidal flotation, sulfide precipitation of arsenic, or formation of calcium magnesium arsenate Fe(III) and Al(III); 3) other technologies, including electrolysis, solvent extraction, diffusion membrane screaning, ion flotation, chemical redox processes, and biological processes.

Regardless of which of the above treatment are used for recovering arsenic, certain environmental conditions may cause some re-solution of immobilized arsenic. In none of the technologies cited above is the immobilization of arsenic absolutely irreversible. New processes need to be developed for arsenic extraction so that immobilized aresenic in the elemental state or as arsenic(V) sulfide, can not subsequently polute the environment. We consider less common microbiological methods of arsenic immobilization preliminary.

The fungus *Scopulariopsis brevicaulis* id known to possess an ability to reduce arsenic and transform it into trimethylarsine, a gas [6,7]. This characteristic of fungus is exploited for removing arsenic from liquids [8,9].

Choice of ways for cleaning of industrial sewers and sewages from harmful admixtures of tellurium depends on their qualitative and quantitative compositions, as well as from required degrees of cleaning and possibility of using an extracted metal. Full recovering of tellurium from solutions may be achieved by coprecipitation with aluminium hydroxide. When tellurium is cemented, and then reduced, 93% of it can be recovered [10]. For purification of tellurium a number of researchers favor thermochemical process (destillation) because less material is needed, labour costs are lower, and recovery is faster [11].

A process for removal of tellurium from industrial wastes exists and is needed. However, there is a problem in removing tellurium from sewers microbiologically, which resides in initial atage of the process. Insufficient attention ahs been paid to micelial fungi, though they are known to grow rapidly and have a great potential for forming different useful metabolic products, and their biomass has great sorption capacity [12,13]. Lower fungi are able to reduce tellurium and selenium to their elemental state [14]. This fungal traits have been used by the authors in the extraction of tellurium from solutions and from various ore treatment wastes. Of all fungi tested, the most tellurium-tolerant was a culture of the fungus *P. Chrysogenum*. Its most useful traits were its ability to reduce oxidized tellurium compounds and to methylate tellurium [14,15].

2. MATERIALS AND METHODS

2.1. Organisms and enriched cultures

The fungi, *Scopulariopsis brevicaulis* and *Pennicillium chrysogenum*, used in this study, were obtained from the National culture collection in Moscow. They were routinely grown in slightly modified liquid Chapek's medium. To make cultivation inexpensive, complex media with all the nutrients required by the fungi were used [16]. To stimulate growth, microbial phospholipides (FMF), cotton seeds extract (ECS), an aqueous extract of weathered brown coal (CE) were added to Chapek's medium. These suplements decreased tyhe consumption of sucrose (50%) in the medium. CE substitudes all mineral salts. CE contained 50-70% organic materials and 30-50% mineral salts. The organic component of CE contained (in %): C 24-29, H 4.0-4.5, and (N+0) 66-72. The chief mineral components of CE were Si, Al, Mg, and Mn.

These medium modification enhanced biomass accumulation and speeded up arsenic transformation. The stimulatory effect of CE is attributed to the presence of physiologically active compounds in it, such as ferric-, magnesium-, calcium- as well as silicon-organic compounds [13]. They stimulated the early developmental stages of the fungi by halving the length of the lag phase. The additives accelerate the remediation of arsenic-containing solutions and thereby reduce the primary treatment cost.

2.2. As and Te samples

Test were performed with various solutions with different concentrations of arsenate from *thiobacillus* bacterial leaching, production effluents from Alawerdsky mining-metallurgical complex, containing arsenic with different degree of oxidation. Were also used electrolytic slimes from copper refining, containing, %: Te 0.6-1.0; Sb 1-3; Pb 0.5-1; Fe 0.5-1; Cu 2-7; CaO 0.7; MgO 0.3-1.

3. RESULTS AND DISCUSSION

3.1. Inoculum preparation, incubation, metal ions tolerance of the fungi

3.1.1. *Scopulariopsis brevicaulis*

Non-adapted microorganisms, in particular when working with the ore raw material, are influenced upon their vital activity by different cations and anions [10]. We therefore studied adaptation of *Sc. br.* to cations and anions in various arsenic-containing sewage samples. Adaptation of the fungus *Sc. br.* to mercury, stibium, copper, lead, cadmium, and cobalt was realized by the resowing on Chapek's agarized medium with gradual increasing the concentrations of cations. The accumulation of biomass and radial growth rate of fungi was followed at the different ion concentration. Adaptation of the fungus *Sc. br* to cations extracted by the water from the arsenic containing ore, and after leaching arsenic by thio bacteria from the gold- stibium ore were also studied. The ore was wet-grinded to 80 % of class - 74 μm and filterred. The filtrate was sterilized before addition to the growth medium, with following addition of the cations (Table. 1).

The sodium arsenate solutions with various concentrations As or sewers of Alawerdsky mining metallurgical combine, containing arsenic with different degree of oxidation, were used.

From data of Table 1 follows, that presence of heavy metals cations oppressed fungi growth. Ore extract improved vital activity of fungi. This difference may have to do with the , presence of biologically active materials in the extract.

Table 1
Influence of metal cation concentrations on radial growth of fungi (Arsenic contents - 1.5 g/l, duration of growth - 5 days)

Cations concentration, mg/l	Average diameter of fungal colony, mm					
	Hg	Pb	Cu	Sb	Co	Cd
0	54.0	54.0	54.0	54.0	55.0	54.0
25	16.5	18.0	14.5	16.0	17.0	10.5
50	12.7	12.4	7.6	10.4	10.0	6.0
100	4.0	8.5	4.5	0	0	0
200	0	3.0	4.2	0	0	0
600	0	0	0	0	0	0
Extract from arsenic-containing ore						
0	55.0	54.0	54.0	54.0	56.0	57.0
12	30.4	42.4	40.5	34.5	28.5	24.0
25	25.0	36.0	24.0	26.0	21.5	20.0
50	12.5	16.5	19.5	21.0	10.4	0
100	5.0	11.0	14.0	16.0	0	0

Table 1 shows that as the fungal mass increases, arsine + trimethylarsine (gas) production increases. When the initial arsenic concentration is too high, gas production by *Sc. Br.* is depressed. It was possible to increase the arsenic tolerance of the fungus to increased concentrations of arsenic to 10 g/l. The adapted to high concentrations of arsenic fungus *Sc. Br.* produced significant quantity of biomass, however gas production was three to four times slower that at low As concentration.

3.1.2. *Penicillium chrysogenum*

To make the reduction of dissolved tellurium by *P.chrysogenum* inexpensive, 2-6 % of waste-liquor from acetyl-butyl fermentations (ABF) from Grozny were added to Chapek's medium. Organic compounds in the ABF replaced expensive sucrose in the medium. Table 2 shows how biomass yield depends on concentration of various medium supplements in arsenic removal.

Table 2
The role of medium supplements in biomass yield and arsenic removal (initial concentration of arsenate – 5 g/l)

Length of fungal growth, days	Yield of fungal biomass: dry mass, % to control	Recovery As in gas, %	Comments
	Control optimal conditions		Lower fungus was
3	100	99.5	grown on Chapek's medium
1	60.4	84.4	For inactivation of
3	81.1	80.9	fungal growth
6	92.2	87.7	corrosive sublimate
9	98.8	96.8	was used (0.2 mg/l)
1	140.0	98.6	For activation of
3	186.4	99.4	fungus growth extract
6	121.4	98.1	of brown coal was
9	111.4	96.5	added (0.2 mg/l)

The fungus was adapted to increasing concentrations of tellurium by consequent resowing on agarized Chapek's medium with gradual increasing concentrations of tellurium over a range of 0.5-7.5 g/dm^3. The fungus was also adapted to arsenic-containing ore by cultivation in ore extract produced by soaking 1 part of ore, ground to a particle size 90% -0.074 mm in 3 parts of water during 24 hours, with following filtrations. In addition, the fungus was adapted to increased contents of Hg, Sb, Pb, Cu, Co, Cd by consequent resewing on agarized Chapek's medium, prepared on the ore extract with gradual increasing of respective cation concentrations (Table. 1). Table 3 shows, that presence of heavy metals cations inhibits development of fungus, however, in the ore extract, growth of fungus was intensified, that bound, much more likely, with presence of biologically active materials required for fungal growth.

3.2 Production of fungi mutants

3.2.1. *Ultra-violet irradiation of Scopulariopsis brevicaulis*
Fungus cultures were irradiated with bactericidal lamp (DB-30).

Mutants of *Sc. Br*. With enhanced arsine + trimetilarsine (gas) producing capacity were induced by ultra-violet irradiation [16]. Survival of irradiated fungus were defined by method of macro colonies. Under repeated UV-irradiating of the fuingus colonies vere selected mutants, capable to translate arsenic in gaseous form quicker than inicial species. As can be seen from the figure 1, the fungus survived under significant length of irradiation.

Yield of fungi biomass growing on the medium, containing 40-50 % standard Chapek's medium and 60-50% dissolve, containing 4% ABF, increased by 6-8 times in contrast with control. Introduction to the media a departure of shuger production - molassa (3 % of sugar), sodium nitrate (2 g/dm^3) and mono-substituted potassium phosphate (1 g/dm^3) have allowed to enlarge yield of biomass more then 3 times.

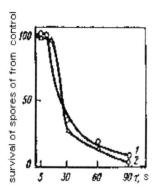

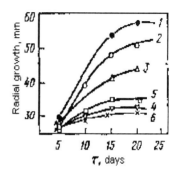

Fig 1. Survival *S. brevicalis* under different duration of UV-irradiation: 1-Chapek's medium; 2- Chapek's medium with arsenic-containing solution.

Fig.2. The radial growth of fungus on Chapek's medium with arsenic at different duration of UV irradiation. Mutants: 1-N8; 2-N7; 3-N4; 5-N6; 6-N5.

Under short UV-irradiating (<30s) presence of the arsenic in the medium renders stimulating action. This may have been due to production of secondary metabolites that help to protect the fungus against arsenic toxicity. Asenic toxicity was assessed by measuring the rate of radial growth. The growth rates and biomass production of different mutants differed (Figure 2.).

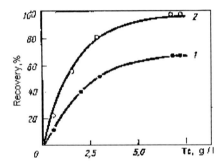

Figure 3. Tellurium recovery from solutions depending of growth
method: 1 - inoculation; 2 - eddition of pregrown fungal biomass.

3.2.2. *Ultra-violet irradiation of Penicillium chrysogenum*

Fungus cultures were irradiated with bactericidal lamp (DB-30). The mutants selected after irradiation produced twice as much biomass and exhibited a shorter lag phase, as compared to initial culture.

3.3. Arsenic removal by *Scopulariopsis brevicaulis*

Arsenic removal from waste waters and sediment from Alaverdsky plant, containing in mg/l: 3.8 As; 710 Fe; 0.91 Mo; 1.18 Zn; 1.04 Pb; 341.67 C; and having pH 3.4 was tested. To optimize fungal activity, the pH of waste water was brought to neutrality and the effluents were then filtered. The filtrate and the sediment contributed a culture of fungus for the following tests. The intensifying additives - phosphatides of microbial fat were added (0,2 mg/l). In 6 days, the fungus converted 99.5% of the dissolved arsenic and 97.7 % of arsenic from the sediment to arsine + trimethylarsine.

The observed arsenic removal became the basis for making an installation [8,9].The reactor was a round-bottomed glas flask, in a thermostatic water bath. The reactor was filled with arsenic-containing solution and nutrient medium and inoculated with the fungus. The system had hydrogen gas passed trough it. After 5 days of fungus growth the arsine + trimethylarsine produced by the fungus was blown off and collected in a trap. Carrier gas was than used to transfer the arsines to a furnace for combustion at 560°C. On the way to furnace gas was cleaned from the hydrogen sulfide with the lead acetate and dried by passing trough calcium chloride. Thermal decomposition of the arsines led to formation of elemental arsenic of high-quality with commercial value. The reaction in the furnace can be presented as following:

$$2(CH_3)_3As = 2As + 6C + 9H_2 \tag{1}$$

The residual gas is passed into a receptacle with potassium permanganate for final cleaning.

3.4. Purification of industrial effluents from arsenic using sulfate-reducing bacteria (SRB)

Many examples of removing heavy metal ions from industrial effluents with hydrogen sulfide generating by SRB can be found in publications [18-20]. Thus, Cu, Hg, Pb, Co, and Cd can be removed from waste water. In the case of As removal, As(III) is first oxidized to As(V) being less toxic, than As(III). The As(V) is then precipitated as arsenic sulfide with hidrogen sulfide generated by SRB.

From the dilute solutions, containing As less that $1g/dm^3$, last can be removed by biosorption using spent biomass from antibiotic production. Such bioremediation techniques are useful ecologically by immobilizing toxic metals temporarily or permanently.

3.5. Recovery of Te with *Penicillium.chrysogenum*

The water extract of weathered brown coal (CE) was used as intensifying additive for increasing the yield of fungal biomass. The growth of fungus *P. chrys.*, adapted to 5 g/dm^3 Te was stimulated two-fold by the addition of CE. At a concentration of CE 0.1 g/dm^3 the time required for removal of tellurium from solution decreased from 16 to 9 days and tellurium recovery increased from 89.8 to 98.5%.

Two methods of tellurium immobilization were studied: 1) direct dissolution with a spore suspension of the fungus; 2) dissolution by pregrown biomass. The second method was more efficient than the first. (Figure 3). The following factors may influence the results: lack of grown inhibition of fungi activity by the presence of cation that increased a possibility to sorptions of tellurium on the biomass of fungus and realization of fermentation reactions.

Reduction of tellurium requires living biomass. Telurium recovery from solution requires its reduction to elemental Te and methylated tellurium, gaseous compound (15-20 %). The methylated tellurium is, probably, dimethyltelluride, which can be sorbed on the natural zeolite with particles size 0.5-0.1 mm (Dzegvi deposit).

Table 3
Reduction of dissolved Te by *P. Chrysogenum* (initial Te contents – 0.5g/l; duration of experiment – 9days)

Contence of biomass, g/l	The amount of reduced Te			Notes
	Sum.	gas	metal	
3.63	89.2	8.1	81.1	Pregrown
4.52	93.83	9.63	84.2	biomass added to
5.38	97.76	11.76	86.0	solution
6.23	98.8	15.7	83.1	
7.60	98.4	22.4	77.0	
3.71	68.80	6.4	62.4	Inoculation with
4.68	71.73	7.23	64.5	fungal spores
6.01	78.38	8.08	70.3	
6.64	80.41	8.21	72.2	
7.84	86.51	10.11	76.4	

Received elemental tellurium was analyzed by physicochemical methods. Comparison of characteristic peaks of X-ray spectra at 0,223, 0,234 and 0,323 nm for pure tellurium and spectral characteristic of recovered product (admixtures Al, Ca, Mn, Fe, Sn in the amount $1-10^{-5}$ % and admixtures Mg, Pb, Sb in the amount $1-10^{-3}$ %), showed that recovered tellurium was technicaly clean product.

P. chr. only reduced tellurium from tellurium-containing materials, such as electrolytic slime in particular. It did not act on other accompanying cations. Fungus strains that were adapted to increased concentrations of some components of electrolytic slimes, containing, %: Te 0.6-1.0; Sb 1-3; Pb 0,5-l; Fe 0.5 -1; C 2-7; CaO 0.7; MgO 0.3-1 exhibited enhanced activity (Table 3).

4. CONCLUSION

It is possible to remove arsenic microbiologically from effluents of mining and metallurgical complexes. Microbiological removal of tellurium from industrial waste streams can substitute labor-intensive storage of electrolytic slimes produced in copper and lead-zinc production, and result in the production of high-quality tellurium. Fungi can be successfully used in mineral processing, hydrometallurgy, and treatment of industrial waste streams. The biohydrometallurgy of arsenic and tellurium and the characteristics of the microbiological processes involved are important for assessing their techno-economic application in the future. These recovery processes may lower the costs of collection, storage and recycling of such materials and the treatment of vast quantities of industrial waste waters.

REFERENCES

1. G. I. Karavayko, S. I. Kuznetsov, A. I. Golomzik, The role of microorganisms in leaching of metals. Moscow, Science, 1972.
2. S. I. Polkin, E. V. Adamov, V. V. Panin, The technology of bacterial leaching of non-ferrous and rare metals. Moscow, Nedra, 1982.
3. G. G. Kulebakin, Bacterial leaching of sulfide minerals. Novosibirsk, Science, 1978.
4. A. N. Ilyaletdinov, Microbiological conversion of metals. Alma-Ata, Science, 1984.
5. V. Nenov, A. I .Zouboulis, N. Dimitrova, I. Dobresky, Environ. Pollut. 83 (1994).
6. Z. E. Becker, Physiology of fungi and their practical use. Moscow, MSU (1963).
7. D. Foster, Chemical activity of fungi. Moscow, Inostr. Literature, 1950.
8. P. M. Solozhenkin, L. L. Lyubavina, S. A Sherepova., N. N. Lyalikova, Recent Progress in Biohydrometallurgy, G. Rossi & A. E. Torma (eds.) Associazione Mineraria Sarda, Iglesias, Italy (1983).
9. P. □. Solozhenkin, L. L. Lyubavina, N. N. Buyanova, Tsvetniye Metally, 6 (1987) 24.
10. A. A. Kudryavtsev, Chemistry and technology of selenium and tellurium. Moscow, Metallurgy, 1968, 153.
11. L. A. Niselson, A. A. Titov, The reception and analysis of materials of high purity. Moscow, Science, 1978, 98.
12. A. N. Ilyaletdinov, Trans. Internat. Seminar, Sofia (1982) 349.
13. W. J. Nickerson, W. A Taber, Y Falcone, Can. J. Microbiolog., 2 (1956) 575.
14. Y. Falcone and W. J. Nickerson, Enzymatic reduction of selenite. Bacteriol. Proc. (1960) 152.
15. P. M. Solozhenkin, L. L. Lyubavina, N. N. Buyanova and I. V. Kirshenina, Tsvetnie Metally, 11 (1992) 71.
16. Yu. T. Lyakov, Science (1980) 45.
17. Z. A. Avakyan, Itogy Nauky i Techniky, Microbiology, 2 (1973) 96.
18. F. G. Vafina, Z. A. Rumyantseva, Z. I. Pevzner, Gumine fertilizers and their using, Theory and practice, v. IV, Dnepropetrovsk, Agrycultural Inst., 1980, 34.

19. A. M. Ilyaletdinov, Microbiological transformations, Alma-Ata, Nauka, 1984.
20. P. M. Solozhenkin, V. P. Nebera, I. G. Abdulmanov, Innovations in Mineral and Coal Processing, Eds. Atak, Onal & Celik, Balkema, Rotterdam, Netherlands, 1998, 495.

Bioprecipitation of copper from a leaching solution by a moderately thermophilic iron-oxidizing bacterium

T. Sugio[a], K. Matsumoto[a], M. Takai[b], S. Wakasa[b], T. Sogawa[a], and K. Kamimura[a]

[a]Division of Biological Function and Genetic Resources Science, Faculty of Agriculture, Okayama University, 1-1-1 Tsushima Naka 700-8530, Japan

[b]Miura Institute of Research and Development, 7 Horie-cho, Matsuyama-shi, Ehime 799-2696, Japan

A moderately thermophilic iron-oxidizing bacterium strain TI-1 produced 74 μmol H_2S extracellularly when grown at 45°C for 15 days in 20 mL of ferrous sulfate-5AAG medium (pH 4.0) containing 1.0 % elemental sulfur and 0.31 % L-glutamic acid. Precipitation of copper from a leaching solution by the H_2S produced by strain TI-1 was investigated. A leaching solution containing Cu^{2+} was prepared by the cultivation of *Thiobacillus ferrooxidans* AP19-3 on a copper concentrate (5.0 %)-medium (pH 2.5) for 15 days. When strain TI-1 was grown for 8 days in a ferrous sulfate-5AAG medium (pH 4.0) containing 1.0 % elemental sulfur, 0.31 % L-glutamic acid and 5 mM Cu^{2+} from a leaching solution, the Cu^{2+} in the medium was completely precipitated by the H_2S produced by strain TI-1. Growth of strain TI-1 in ferrous sulfate-5AAG medium was not inhibited by 10 mM Cu^{2+}. A moderately thermophilic sulfate-reducing bacterium *Desulfotomaculum nigrificans* IFO 136987 produced 45 μmol H_2S extracellularly when grown at 45°C for 15 days in 20 ml of lactate (0.35 %)-medium. The maximum productivity of H_2S by *D. nigrificans* was observed in the medium with an initial pH of 7.2. The bacterium did not produce H_2S in the lactate-medium with a pH less than 6.0. Growth of *D. nigrificans* in the lactate-medium was completely inhibited by 0.56 mM Cu^{2+}. These results suggest that strain TI-1 is superior to *D. nigrificans* in the ability to precipitate copper from the acidic leaching solution containing Cu^{2+}.

1. INTRODUCTION

The iron-oxidizing bacteria *Thiobacillus ferrooxidans* and *Leptospirillum ferrooxidans* are acidophilic chemolithoautotrophs. They play a crucial role in the solubilization of metals from sulfide ores (1-5). In addition to these mesophilic iron-oxidizing bacteria, the existence of moderately thermophilic iron-oxidizing bacteria is known (6-14). These bacteria derive their energy for growth from the oxidation of Fe^{2+} or pyrite and have an optimum growth temperature of around 50 °C. The moderately thermophilic iron-oxidizing bacteria are considered to be useful for metal extraction from metal sulfides in or near the hot zones of leach dumps. The recovery of metals from a bioleaching solution is one of the most important processes in bacterial leaching of sulfide ores. To reduce the large amount of electrical energy required for this process, the development of a low-priced method for the precipitation of metals is absolutely needed. It has been known that hydrogen sulfide (H_2S) reacts with many kinds of metal ions to give insoluble metal sulfides. Thus, bioprecipitation of metals with the H_2S produced by microorganisms is interesting and promising. Sulfate-reducing bacteria are well-known as hydrogen sulfide producers. However, the bacteria seem to be unsuitable for direct inoculation to acidic leaching solutions to precipitate Cu^{2+} because they can not grow and produce H_2S under the acidic conditions. Recently, we showed that a moderately thermophilic iron-oxidizing bacterium strain TI-1 has two kinds of enzymes involved in H_2S production, namely, thiosulfate reductase and sulfur reductase (15,16). When strain TI-1 is grown in ferrous sulfate-5AAG medium (pH 1.8) containing elemental sulfur and L-glutamic acid, the bacterium produced H_2S extracellularly (17, 18). The optimal culture medium for H_2S production by the bacterium has been developed (19). When directly applied to an acidic leaching solution to precipitate copper, strain TI-1 seems to be superior to sulfate-reducing bacteria because the former has an optimum pH for growth at 1.8. In this report we compare the ability of a moderately thermophilic iron-oxidizing bacterium strain TI-1 to precipitate copper from an acidic leaching solution with that of a moderately thermophilic sulfate-reducing bacterium *Desulfotomaculum nigrificans*.

2. MATERIALS AND METHODS

The moderately thermophilic iron-oxidizing bacterium strain TI-1 (13) and a moderately thermophilic sulfur-reducing bacterium *Desulfotomaculum nigrificans* IFO 136987 were used throughout this study. The ferrous sulfate-5AAG medium (pH 4.0) was used for the cultivation of strain TI-1 (19). The medium contained $FeSO_4 \cdot 7H_2O$ (1.7 %), $(NH_4)_2SO_4$ (0.2 %) , L-glutamic acid (0.31 %), and 0.01 % each of K_2HPO_4, $MgSO_4 \cdot 7H_2O$, KCl, L-aspartic acid, L-serine, L-arginine and L-histidine. Strain TI-1 was grown at 45 ℃ in 20 mL of a ferrous sulfate-5AAG medium. *D. nigrificans* IFO 136987 was grown at 45 ℃ in 20 mL of a lactate-medium (pH 7.2) containing sodium lactate (0.35 %), yeast extract (0.1 %), NH$_4$Cl (0.05 %), K$_2$HPO$_4$ (0.1 %), MgSO$_4$ · 2H$_2$O (0.2 %), CaSO$_4$ · 7H$_2$O (0.1 %), and Fe$_2$(SO$_4$)$_3$ · (NH$_4$)$_2$SO$_4$ · 24H$_2$O (0.05 %). Strain TI-1 cells were separated from the particles of ferric hydroxide by filtration of cultures through a Toyo No. 5C paper filter. The number of cells in the filtrates was counted with a hemacytometer (Kayagaki Irika Kogyo Co., Ltd., Tokyo) after dilution with 0.1 N sulfuric acid when necessary.

The production of H$_2$S by strain TI-1 was measured as follows. A moderately thermophilic iron-oxidizing bacterium strain TI-1 was grown for 4 days at 45℃ in 20 mL of a ferrous sulfate-5AAG medium (pH 4.0) under aerobic conditions to give a cell yield of approximately 4.0 × 108 cells/ mL. After the 4d cultivation, a small test tube containing 2.5 mL of 2N NaOH, a trapping reagent for H$_2$S, was aseptically inserted into the 50 mL-culture flask (18). In addition to the NaOH solution, elemental sulfur was also added to the medium at a concentration of 1.0 % and the culture was further incubated under anaerobic conditions at 45 ℃. The H$_2$S produced by the cells during the anaerobic incubation was trapped into the 2.5 ml of 2N NaOH solution in the small test tube and analyzed by the methylene blue method. The H$_2$S produced in a lactate-medium (pH 7.2) by *D. nigrificans* IFO 136987 was also trapped into the 2.5 mL of 2N NaOH solution in the small test tube and analyzed by the methylene blue method. Copper concentration was measured by atomic absorption spectroscopy with a Shimadzu AA-625-01 spectrophotometer, using an air-acetylene flame.

3. RESULTS

3.1 Production of H$_2$S by a moderately thermophilic iron-oxidizing bacterium strain TI-1

When strain TI-1 was grown in ferrous sulfate-5AAG medium containing elemental sulfur and L-gulutamic acid, the bacterium produced H$_2$S extracellularly (17, 18). The optimal culture conditions for the H$_2$S production by strain TI-1 are shown in Table 1 (19). Strain TI-1 produced 74 μmol H$_2$S extracellularly when cultured in the optimal culture medium for H$_2$S production. The effect of cupric ion on the growth of strain TI-1 was studied with ferrous sulfate-5AAG medium supplemented with cupric chloride. 10 mM of cupric chloride did not inhibit cell growth (Fig. 1). However, 50 mM of cupric chloride completely inhibited the growth of strain TI-1. Growth of strain TI-1 in a ferrous sulfate-5AAG medium (pH 4.0) was not inhibited by 7 mM lead chloride, 10 mM zinc chloride, or 5 mM nickel chloride, respectively. Mercuric chloride completely inhibited the growth of strain TI-1 at 0.05 μM.

Table 1

Optimal culture conditions for H_2S production by a moderately thermophilic iron-oxidizing bacterium strain TI-1 (19).

Electron donor	0.31	%	L-gulutamic acid
Electron acceptor	1.0	%	elemental sulfur
Concentration of Fe^{2+}	1.7	%	ferrous sulfate
Temperature	55	°C	
pH	4.0		

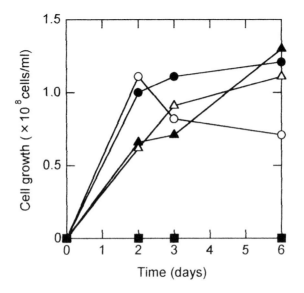

Figure 1. Effect of cupric ion on the growth of strain TI-1 in a ferrous sulfate-5AAG medium (pH 4.0) supplemented with 1.0 (○) , 5.0 (▲) , 10 (△) , and 50 mM Cu^{2+} (■) or without Cu^{2+} (●) .

The concentration of Cu^{2+} in a ferrous sulfate-5AAG medium (pH 4.0) with Cu^{2+} was measured. Strain TI-1 was grown at 45 °C in a ferrous sulfate-5AAG medium with 1.0 mM Cu^{2+} for 4 days under aerobic conditions to give a cell yield of approximately 4.0×10^8 cells/mL. The culture was further incubated under anaerobic conditions or aerobic conditions at 45 °C. After 6, 7, 8, and 10 days of cultivation, each of the media was centrifuged at

12,000 × g for 15 min to obtain a clear supernatant and the concentration of copper in the supernatant was measured with atomic absorption spectroscopy. In the medium obtained after 3 days anaerobic incubation, Cu^{2+} was not detected in the supernatant, suggesting that the Cu^{2+} supplemented with the medium made a complex with the H_2S produced by strain TI-1 and was precipitated at the bottom of the flask (Fig. 2). Cupric ion in the medium was not

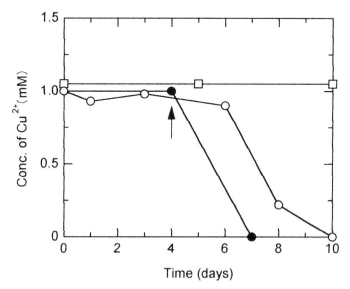

Figure 2. The concentration of copper in a ferrous sulfate-5AAG medium with 1.0 mM Cu^{2+}. After 4 days cultivation air in the head space of the medium was exchanged for nitrogen gas (●) . Strain TI-1 was grown under aerobic conditions (○) . Copper concentration in the medium without ferrous iron was measured (□) .

precipitated when strain TI-1 was grown in 5AAG medium without ferrous iron. Copper precipitation by strain TI-1 was also observed under aerobic conditions. These results suggest that the Cu^{2+} added to the medium was precipitated by the H_2S produced by strain TI-1 not only under anaerobic but also aerobic conditions.

3.2 Precipitation of copper from a leaching solution by a moderately thermophilic iron-oxidizing bacterium strain TI-1

A leaching solution containing Cu^{2+} was prepared by the cultivation of *Thiobacillus ferrooxidans* AP19-3 on a copper concentrate (5 %)-medium (pH 2.5). The copper concentrate used in this study was black ore (425 mesh) and composed of Cu (20.48 %), Fe

(31.61 %), S (38.22 %), Pb (3.84 %), and Zn (4.22 %). After the cultivation of *T. ferrooxidans* AP19-3 for 15 days in a modified 9K medium (pH 2.5), the medium was centrifuged at 12,000 × g for 15 min to obtain supernatant. The supernatant (leaching solution) thus obtained (pH 2.0) contained iron (16.1 mM), copper (125 mM), zinc (16.9 mM), and lead (0.019 mM), respectively. A diluted leaching solution was added to a ferrous sulfate-5AAG medium (pH 4.0), in which copper concentrations were adjusted to 1.25, 2.5, or 5.0 mM, respectively. Strain TI-1 was grown in the ferrous sulfate-5AAG medium (pH 4.0) supplemented with a leaching solution under aerobic conditions for 4 days to give a cell yield of approximately 4.0×10^8 cells/ mL, and further incubated under anaerobic or aerobic conditions for 4 days. The Cu^{2+} in the medium reacted with the H_2S produced by strain TI-1 and was precipitated at the bottom of the flask (Fig. 3). The 5 mM Cu^{2+} in the medium was completely precipitated by the cultivation of strain TI-1 in a ferrous sulfate-5AAG medium

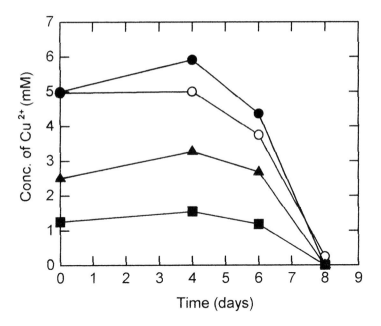

Figure 3. Precipitation of copper from a leaching solution by strain TI-1. Strain TI-1 was grown in a ferrous sulfate-5AAG medium (pH 4.0) with a copper leaching solution, in which the concentration of copper was adjusted to 1.25 (■) , 2.5 (▲) , and 5.0 mM (●) . Strain TI-1 was also grown in a ferrous sulfate-5AAG medium (pH 1.8) with a copper leaching solution (○)

with pH 1.8 and 4.0. After 8 days cultivation, the precipitate formed at the bottom of the culture flask was dissolved with hydrochloric acid and the amount of copper in the precipitate were measured. Recoveries of copper from the media containing 1.25, 2.5 and 5.0 mM Cu^{2+} were 48, 64, and 48 %, respectively.

3.3 Precipitation of copper from a lactate-medium containing Cu^{2+} by a moderately thermophilic sulfate-reducing bacterium

Sulfate-reducing bacteria of genera *Desulfovibrio* and *Desulfotomaculum* are know to produce H_2S by reducing sulfate ion with an organic electron donor under neutral pH conditions. The effect of pH on the production of H_2S by a moderately thermophilic sulfate-reducing bacterium *Desulfotomaculum nigrificans* IFO 136987 was studied in the medium with sodium lactate as an electron donor and sulfate ion as an electron acceptor. The maximum productivity of H_2S was observed when *D. nigrificans* was grown at 45 °C under anaerobic conditions in the medium with an initial pH of 7.2 (Fig. 4). Interestingly, *D. nigrificans* could not produce H_2S in the media with a pH lower than 6.8 and the maximum H_2S productivity of a moderately thermophilic iron-oxidizing bacterium strain TI-1 was approximately 2 fold higher than that of *D. nigrificans*.

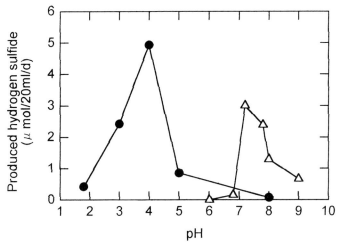

Figure 4. Effect of pH on the production of H_2S by a moderately thermophilic iron-oxidizing bacterium strain TI-1 (●) and a moderately thermophilic sulfur-reducing bacterium *Desulfotomaculum nigrificans* IFO 136987 (△)

The effect of cupric ion on the growth of *D. nigrificans* IFO 136987 was studied with a lactate (0.35 %)-medium (pH 7.2). Concentration of Cu^{2+} and cell number in 20 mL of lactate-medium were measured after 7 days cultivation. *D. nigrificans* did not grow and produce H_2S in a lactate-media supplemented with 0.56 and 1.27 mM Cu^{2+}, suggesting that the bacterium is more sensitive to Cu^{2+} than strain TI-1 (Table 2). When grown in a lactate-medium with 0.1 mM Cu^{2+}, *D. nigrificans* completely precipitated the copper ion.

Table 2
Effect of cupric ion on the growth of *D. nigrificans* IFO 136987 in a lactate-medium

Concentration of Cu^{2+}		H_2S production	Cell growth
Culture time (days)		(μmol/20 mL/d)	($\times 10^6$ cells/mL)
0	7		
0.00	0.00	3.78	36
0.11	0.00	3.52	30
0.56	0.30	0.00	0
1.27	1.48	0.00	0

4. DISCUSSION

In this report we first show that a moderately thermophilic iron-oxidizing bacterium strain TI-1 completely precipitates copper from an acidic leaching solution containing 5 mM Cu^{2+}. Moreover, we show that a bioprecipitation of copper with a moderately thermophilic iron-oxidizing bacterium strain TI-1 is superior to a moderately thermophilic sulfate-reducing bacterium *Desulfotomaculum nigrificans* because the former is much more resistant to Cu^{2+} than the latter. The formation of insoluble metal sulfides is usually the result of an interaction of iron compounds with the H_2S that originated from bacterial reduction of sulfate, and sulfate-reducing bacteria which produce H_2S from sulfate ion at neutral pH play an important role in the formation of certain sulfide minerals (20). Thus, the findings that a moderately thermophilic iron-oxidizing bacterium strain TI-1 produces H_2S extracellularly under acidic conditions are important because they suggest an involvement of this bacterium in the formation of metal sulfides under acidic conditions.

The abilities shown below of a moderately thermophilic iron-oxidizing bacterium strain TI-1 suggest that the bacterium can be used in a bioprecipitation process of an acidic leaching solution containing copper. (1) Strain TI-1 produces H_2S at a pH between 1.8 and 4.0 under aerobic and anaerobic conditions. (2) The level of H_2S produced by strain TI-1 was high and comparable to that of a moderately thermophilic sulfur-reducing bacterium *D. Nigrificans*. (3) Strain TI-1 was much more resistant to Cu^{2+} than *D. nigrificans* and the growth of strain TI-1 in a ferrous sulfate-5AAG medium was not inhibited by 10 mM Cu^{2+}. (4) The H_2S produced by strain TI-1 completely precipitated 5 mM Cu^{2+} in a ferrous sulfate-5AAG medium supplemented with a leaching solution. (5) Strain TI-1 used a low-priced L-glutamate as an electron donor for the reduction of elemental sulfur and gave H_2S.

REFERENCES

1.   A.T. Torma, Adv.   Biochem. Eng., 6 (1977) 1.
2.   D.G. Lundgren and M. Silver, Ann. Rev. Microbiol., 34 (1980) 263.
3.   S.R. Hutchins, M.S. Davidson, J.A. Brierley and C.L. Brierley, Ann. Rev. Microbiol., 40 (1986) 311.
4.   C.L. Brierley, Sci. Am., 247 (1982) 42.
5.   A.A. Nicolaidis, J. Chem. Tech. Biotechnol., 38 (1987) 167.
6   N.W. Le Roux, D.S. Wakerley and S.D. Hunt, J. Gen. Microbiol, 100 (1977) 197.
7.   J. A. Brierley and S. J. Lockweed, FEMS Microbiol., 2 (1977) 163.
8.   J.A. Brierley, P.R. Norris and N.W. Le Roux, Eur. J. Appl. Microbiol. Biotech., 5 (1978)   291.
9.   J. A. Brierley, Appl. Environ. Microbiol., 36 (1978) 523.
10.   P.R. Norris, J.A. Brierley and D.P. Kelly, FEMS Microbiol Lett., 7 (1980) 119.
11.   A.P. Wood and D.P. Kelly, FEMS Microbiol Lett., 17 (1983) 107.
12.   R.M. Marsh and P.R. Norris, FEMS Microbiol Lett., 17 (1983) 311.
13.   T. Sugio, M. Takai and T. Tano, Biosci. Biotech. Biochem., 57 (1993) 1660.
14.   P.R. Norris, D.A. Clark, J.P. Owen and S. Waterhouse, Microbiol., 142 (1996) 775.
15.   T. Sugio, K. Kishimoto and K. Oda, Biosci. Biotech.   Biochem., 61 (1997) 470.
16.   T. Sugio, K. Oda, K. Matsumoto, M. Takai, S. Wakasa and K. Kamimura, Biosci. Biotech. Biochem., 62 (1998) 705.
17.   T. Sugio, K. Kishimoto, M. Takai, K. Oda, and T. Tano, J. Ferment. Bioeng., 79 (1995) 290.
18.   T. Sugio, K. Oda and K. Kishimoto, Biosci. Biotech. Biochem., 60 (1996) 1356.
19.   K. Oda, K. Matsumoto, M. Takai, A. Wakasa, K. Kamimura and T. Sugio, Ferment. Bioeng., 84 (1997) 372.
20.   H.L. Ehrlich, Geomicrobiology, J, Marcel Bekker, Inc., New York, 1996, 578.

Recycling CIP process water to bacterial oxidation circuits via a thiocyanate degrading bioreactor

H. R. Watling[a], L. Quan[a], T. L. Williams[b], M. B. Stott[a,c], B. J. Clark[a], P. C. Miller[b], M. R. Houchin[a] and P. D. Franzmann[c]

[a] A.J. Parker Cooperative Research Centre for Hydrometallurgy, CSIRO Minerals, PO Box 90, Bentley, Western Australia 6982.

[b] BacTech (Australia) Pty. Ltd., 5 Belmont Avenue, Belmont, Western Australia 6104.

[c] Centre for Groundwater Studies, CSIRO Land & Water, Underwood Avenue, Floreat Park, Western Australia 6014.

Significant concentrations of thiocyanate are generated in effluent from the CIP/CIL gold extraction process. As thiocyanate is toxic to iron- and sulphide-oxidising bacteria, the effluent cannot be returned directly to a bacterial oxidation plant. To overcome this problem, a rotating cage laboratory-scale bioreactor was inoculated with two strains of known thiocyanate-degrading bacteria. The reactor, which had a 20 L solution volume and surface area of 20 m^2 available for biomass support, degraded thiocyanate from a concentration of 2000 to <1 mg L^{-1} (flow rate 30 mL min^{-1}; residence time 11 h).

The efficiency of the thiocyanate-degrading bioreactor to detoxify process water was evaluated in bacterial oxidation tests using Fe^{2+} and arsenopyrite as energy sources. The experimental results indicated that water detoxified using the thiocyanate-degrading bioreactor could be recycled to a biological oxidation plant.

1. INTRODUCTION

Gold can be present in both "free-milling" and "refractory" ores. Free-milling ores can be leached by cyanide, without pre-treatment, to recover the gold values. Refractory sulphide ores must be oxidized prior to cyanide leaching to release the gold from the sulphide matrix. Bacterial oxidation offers an environmentally acceptable alternative to roasting, as a pre-treatment.

However, significant concentrations (up to 250 mg L^{-1}) of thiocyanate are generated in the CIP/CIL gold extraction effluent. Williams [1] reported that thiocyanate is toxic to bacterial cultures in a ferrous sulphate medium at concentrations as low as 2 mg L^{-1} but that, in an arsenopyrite slurry, bacteria can tolerate up to 5 mg L^{-1} thiocyanate. These results are consistent with those of Morin [2], who stated that bacteria used in the leaching of refractory gold ores could not tolerate thiocyanate concentrations above 5 mg L^{-1}. Weston *et al.* [3] also reported that thiocyanate was toxic to sulphide-oxidising bacteria at concentrations below detection limits (the method of analysis was not described). Thus process effluent and tailings dam water are not suitable for direct recycle through a bacterial oxidation plant.

Chemical and physical methods of cyanide (and possibly thiocyanate) removal are costly. Consequently, for mines in sparsely inhabited and arid regions such as Western Australia, operators rely mainly upon attenuation by natural degradation to lower the concentrations of these toxins. Volatilization, chemical oxidation, hydrolysis, photodegradation and bacterial oxidation all contribute to natural degradation in a tailings dam [4]. In particular, the removal of the more stable thiocyanate depends upon a combination of photodegradation and bacterial oxidation [5,6].

Natural degradation only achieves the desired remediation where toxin levels are low. For example, weak acid-dissociable cyanide levels need to be <50 mg L^{-1} for effective degradation. In addition, the process must be optimized by careful design of tailings dams. However, the cost of changing the design of tailings dams to achieve only minor improvements in degradation is not always economically feasible for operational mines.

Gold-mine tailings dams support a heterogeneous bacterial community which have acclimatized to the contaminants in the wastewater. These bacteria may adsorb and/or utilize and degrade toxins, particularly cyanide [7-9]. This characteristic has been utilised successfully to degrade both cyanide and thiocyanate in wastewater and has been implemented on an industrial scale at Homestake mine, South Dakota [4].

Two strains of thiocyanate-degrading bacteria, isolated from the gold-processing effluent and tailings dam at Youanmi, Western Australia (Figure 1), were enriched and inoculated into a laboratory-scale rotating biological contactor (RBC). The biofilm was established and stabilized with synthetic nutrient medium of composition similar to that of Youanmi wastewater.

Figure 1. Map of Western Australia

Flow rates and thiocyanate concentrations were varied to meet the needs of the project and bioreactor performance was assessed by measuring thiocyanate concentrations in bioreactor feed and effluent [10].

The aim of this study was to determine whether gold-process water containing thiocyanate could be detoxified by passage through a thiocyanate-degrading RBC to the extent that it could be recycled to a biological oxidation plant utilizing moderately thermophilic, iron- and sulphide-oxidising bacteria. A matrix of tests was devised in which bacterial oxidation rates in several synthetic media were compared, including the effluent from the RBC. Bacteria were provided with one of two energy sources, ferrous ion or Youanmi concentrate (arsenopyrite and pyrite).

2. EXPERIMENTAL

2.1. Media

Synthetic process water was prepared by dissolving 9.8 g L^{-1} of synthetic sea salts in deionized water to give a conductivity of 17.8 mS cm^{-1}, equivalent to the salinity and conductivity of Youanmi effluent and tailings dam water.

Synthetic process water containing 250 mg L^{-1} SCN^- in addition to the above constituents was also prepared. A portion of this was passed through the thiocyanate-degrading bioreactor and the effluent from the reactor collected for use in the bacterial oxidation tests. This effluent contained <1 mg L^{-1} thiocyanate. The method for the measurement of thiocyanate concentration is given in (5).

Makeup water for media for comparing the rates of Fe^{2+} and arsenopyrite oxidation was one of four types; deionized water, synthetic process water, synthetic process water to which thiocyanate (250 mg L^{-1}) was added and thiocyanate-rich process water after passage through the RBC. Aliquots of nutrient solutions were added to each of the test make-up waters to give initial concentrations of 0.4 g L^{-1} $(NH_4)_2SO_4$, 0.8 g L^{-1} $MgSO_4.7H_2O$ and either 0.2 g L^{-1} KH_2PO_4 (for Fe^{2+} tests) or 0.4 g L^{-1} KH_2PO_4 (for arsenopyrite tests to counter the precipitation of phosphate compounds). Solution pH was adjusted to 1.6-1.8 with concentrated H_2SO_4.

For tests in which ferrous ion was the bacterial energy source, $FeSO_4.7H_2O$ was added to the flasks to give an initial concentration of 9000 mg L^{-1} Fe^{2+}. For tests in which arsenopyrite was the energy source, Youanmi concentrate (Table 1) was added to each flask to give 5% (w/v) solids content.

The inoculum was comprised of a mixed culture of stationary phase, moderately thermophilic, iron- and/or sulphide-oxidising bacteria [BacTech (Australia) Pty Ltd.] that had been grown on Youanmi arsenopyrite concentrate. It contained 14 mg L^{-1} Fe^{2+}, 3899 mg L^{-1} $Fe_{(total)}$ and 486 mg L^{-1} As.

2.2. Experimental matrix and methods

Duplicate flasks were prepared according to the experimental matrix (Table 2, Figure 2).

Table 1
Composition of Youanmi concentrate*

Element	%	Element	%	Element	%
S	24.9	Ti	0.6	Cr	0.012
Sulphide[#]	19.5	Na	0.24	V	0.012
Fe	20.0	Mn	0.05	Ni	0.012
Si	14.2	Zn	0.03	Co	0.011
Al	4.5	Cu	<0.03	Pb	0.002
As	0.99				

*Dissolution by borate fusion and elemental analyses by ICP-AES;
[#] Sherritt/Leco method (Ammtec Ltd.)

Table 2
Experimental matrix for bacterial oxidation tests

| | Energy source | |
Nutrient solution prepared with...	ferrous ion	arsenopyrite
Synthetic process water with 250 mg L^{-1} SCN⁻	T1	T5
Effluent from SCN⁻-degrading bioreactor	T2	T6
Synthetic process water without added SCN⁻	T3	T7
Deionized water (no added salts or SCN⁻)	T4	T8
Deionized water (no additives, no inoculum)	C1	C2

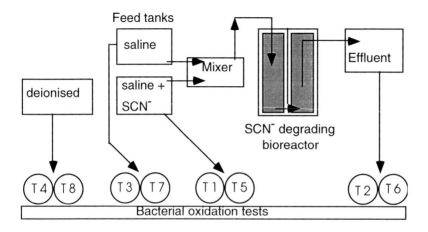

Figure 2. Schematic illustrating the relationship between the thiocyanate-degrading bioreactor and the bacterial oxidation tests.

Once the solutions/slurries had been prepared and the pH adjusted to 1.6-1.8, the flasks were incubated in a New Brunswick environmental incubator/shaker (160 rpm 37 °C). The pH was re-measured, adjusted if necessary, and 25 mL of inoculum added to all but the controls. After a further 15 min in the incubator, the solution pH was again measured and adjusted if necessary. The total volume of solution and mass of each flask were recorded, and the flasks returned to the incubator.

Solutions/slurries were sampled at selected intervals after correction for any evaporation (assessed by mass loss and compensated by the addition of deionized water). A 1.00 mL sample was withdrawn, acidified with 0.5 mL concentrated HCl and allowed to stand for 20 min. The solution was centrifuged and a 1.0 mL aliquot collected for total iron and arsenic analysis. A second sample (160 μL) was separated for Fe^{2+} analysis. Solution pH and the mass of each flask were recorded before it was returned to the incubator.

2.3. Analyses

Ferrous iron concentrations were determined colorimetrically using 2,2-dipyridyl in acetate buffer to form an iron-dipyridyl complex with an absorption maxima at 525 nm. The colour formation was immediate and stable for more than 30 min. Excess ferric ions in solution contributed a positive interference and were compensated with a blank determination. The method is a modification of that described by Wilson [11], and can be used in the range 0-200 mg L^{-1} ferrous ion with a detection limit of <0.1 mg L^{-1}.

Total iron and arsenic in solution were determined by ICP-AES after appropriate dilution.

Errors were estimated taking into account sampling, dilution and instrumental inaccuracies. The relative error was about 8.0%.

Sulphide contents of residues were determined by the Sherritt/Leco method (Ammtec Ltd.).

3. RESULTS AND DISCUSSION

3.1. Fe^{2+} as energy source

Bacterial oxidation was arrested in the presence of thiocyanate (T1) and the Fe^{2+} concentration over time was comparable to its concentration in the un-inoculated control medium. Bio-oxidation proceeded rapidly in the thiocyanate-free solutions, including the effluent from the bioreactor, as indicated by the decrease in ferrous ion concentrations as a function of time in the ferrous sulphate-containing medium (Figure 3). *Hence, the thiocyanate-degrading bioreactor has clearly reduced the thiocyanate concentration to a level whereby it no longer inhibits the bacterial oxidation of ferrous ion to ferric ion.*

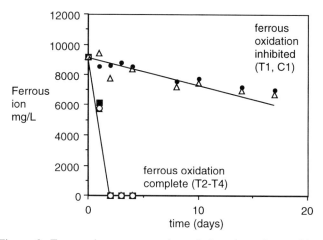

Figure 3. Ferrous ion concentrations during the culture of iron-oxidizing bacteria in media containing ferrous sulphate dissolved in media prepared with ● T1 synthetic process water with SCN$^-$; ▲ T2 effluent from bioreactor; ■ T3 synthetic process water with no added SCN$^-$; ○ T4 deionized water; △ C1 control (deionized water, no inoculum).

3.2   Arsenopyrite as energy source

Bacterial oxidation of the Youanmi concentrate (arsenopyrite and pyrite) was monitored using solution iron and arsenic analyses (Figures 4A & B). [Note that the results have been adjusted for the contributions of iron (988 mg L^{-1}) and arsenic (123 mg L^{-1}) that were carried over with the inoculum.] Oxidation proceeded more slowly than was observed in the ferrous ion tests, and only occurred in thiocyanate-free solutions, including synthetic process water in which the thiocyanate had been degraded by passage through the bioreactor. On the basis of iron concentrations, bacterial oxidation was inhibited in saline solution as compared with deionized water, but the difference was less marked with respect to arsenic release. Oxidation was arrested in the presence of thiocyanate (synthetic process water).

Oxidation of the controls (deionized water with no added thiocyanate and no inoculum) was observed after about 20 days, suggesting that either the natural bacterial population of the concentrate had increased to a significant level, or that this flask had been cross-contaminated during sampling. *In particular, thiocyanate levels in the effluent from the bioreactor have been reduced to the extent such that bacterial oxidation of Youanmi concentrate (arsenopyrite and pyrite) was not inhibited.*

Sulphide concentrations were determined for all residues at the end of the experiment (25 days) and the amount of sulphide oxidised (converted to other sulphur-containing species) estimated (Table 3). This estimation is formulated on the basis that the withdrawal of each 1 mL sample from the flasks does not carry with it a significant amount of concentrate.

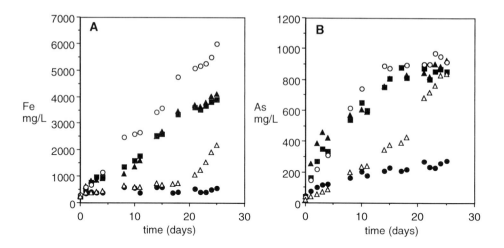

Figure 4. Iron and arsenic concentrations in solution during culture of iron- and sulphide-oxidizing bacteria prepared in media containing Youanmi concentrate suspended in make-up water of different types. **A**: Total iron and **B**: arsenic concentrations as a function of time. Media prepared with ● T5 synthetic process water with SCN⁻; ▲ T6 effluent from bioreactor; ■ T7 synthetic process water with no added SCN⁻; ○ T8 deionized water; △ C2 control (deionized water, no inoculum).

The conversion (%) of sulphide relates to the overall sulphide content of the Youanmi concentrate but does not provide information on the contributions from arsenopyrite and pyrite oxidation. Calculation of the amounts of arsenic and iron solubilized in the same period, as a percentage of the total ore content, is not possible because

- an indeterminate amount of the solubilized iron is re-precipitated, and contains adsorbed arsenic; and
- the partial contributions of arsenic from arsenopyrite and from pyrite in the ore are not known.

Nevertheless, concentrations of iron solubilized by bacterial oxidation correlate well with the extent of sulphide conversion for the different experimental conditions (Figure 5).

Table 3
Sulphide contents of experimental residues after 25 days (in duplicate) and estimated conversion (%) to other sulphur containing species.

Make-up water	Flask	$Fe^{\#}$ mg L^{-1}	S^{2-} in residues mg g^{-1}	Sulphide Conversion* %
Synthetic process water without added SCN⁻	T7	3386	16.5	40.5
	T7A	4400	13.2	55.2
Synthetic process water with 250 mgL⁻¹ SCN⁻	T5	574	20.5	11.1
	T5A	530	20.8	10.4
Effluent from SCN⁻-degrading bioreactor	T6	3985	14.7	47.7
	T6A	4225	13.3	54.3
Deionised water (no added sea salts or SCN⁻)	T8	6014	9.4	72.5
	T8A	5991	11.5	63.1
Control (no added sea salts, SCN⁻, or inoculum)	C2	2193	19.4	25.1

Iron concentrations in solution as a result of bacterial oxidation are included for comparison.

* Calculation of sulphide conversion to other sulphur-containing species as
% sulphide conversion = ((feed g x sulphide %) - (residue g x sulphide %)) x 100
(feed g x sulphide %)
For this experiment the initial mass of Youanmi concentrate (feed) was 7.5 g with a sulphur as sulphide content of 19.5%.

806

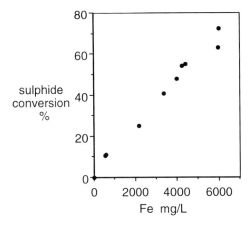

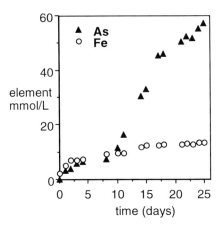

Figure 5. Correlation between iron in solution and conversion (%) of sulphide in the Youanmi concentrate (arsenopyrite/pyrite) after 25 days.

Figure 6. Iron and arsenic release during bacterial oxidation of Younami concentrate in the presence of bioreactor effluent.

3.3. Bacterial oxidation and mineral energy source

Youanmi concentrate contains about 2.5% arsenopyrite (which contains almost all of the ~40 mg kg^{-1} gold) and 31% pyrite; alpha quartz (~35%), muscovite (~27%) and chlorite (~4%) constitute the remainder. Sulphide-oxidising bacteria can acquire their energy from either the arsenopyrite or the pyrite in the concentrate.

Taking the results for bacterial oxidation in the bioreactor effluent medium as representative of the results obtained for the other thiocyanate-free media, iron release to solution as a function of time followed a typical bacterial oxidation curve for arsenopyrite (Figure 6). When the Youanmi concentrate was introduced into solution there was an immediate release of iron to solution (~5 mmol L^{-1} or 300 mg L^{-1}). The source of this soluble iron could have been surface oxidation of iron minerals during processing and handling of the ores. Following inoculation, there was a steady increase in iron concentration, a short-duration plateau followed by a further increase in iron concentration.

When a correlation factor was applied to the solution data with respect to the "soluble" iron, it was found that iron released to solution as a result of bacterial oxidation increased in concentration at a rate slightly greater than that predicted from the 1:1 Fe:As stoichiometry of arsenopyrite. Clearly arsenopyrite rather than pyrite was the preferred energy source for sulphide-oxidising bacteria under these experimental conditions, accounting for approximately 70% of the total iron released in the first four days. However, pyrite also contributed some of the iron during the early part of the experiment.

After about ten days most of the arsenopyrite had been oxidised and the arsenic concentration reached a maximum. It was at this point that bacterial oxidation of the pyrite proceeded more strongly. This indicated that there was a small secondary lag followed by exponential bacterial growth.

4. CONCLUSIONS

Two strains of thiocyanate-degrading bacteria, previously isolated from Youanmi tailings waste water, were enriched and inoculated into a laboratory-scale rotating biological contactor. The biofilm which was established in thiocyanate-rich synthetic process water was capable of degrading thiocyanate to carbon dioxide, sulphate and ammonium ions [10], all of which are relatively harmless to the environment when compared with thiocyanate.

Effluent from the bioreactor was used as the basal medium for testing bacterial oxidation of ferrous sulphate or Youanmi concentrate (containing arsenopyrite and pyrite). The rates of bacterial oxidation of ferrous ion and of arsenopyrite and pyrite were similar to those obtained with synthetic process water. This indicated that effluent from the thiocyanate-degrading bioreactor did not inhibit bacterial oxidation of sulphide ores.

Arsenopyrite was the preferred energy source for sulphide-oxidising bacteria under these experimental conditions. Significant pyrite oxidation only occurred after arsenopyrite was almost completely oxidised.

The by-products of thiocyanate degradation proved to be non-inhibitory to iron-and sulphide-oxidising bacteria. Ammonium ion, which was produced in the degradation of thiocyanate, is an additive for the bacterial-oxidation reaction and thus could potentially save money on reagents.

The successful bacterial oxidation experiment using bioreactor effluent in the experimental medium proved that process water can be recycled to bacterial oxidation plants, with the proviso that the bacteria (both thiocyanate degrading and sulphide oxidising) can adapt to the environmental conditions at a particular mine site. Selection of a mixed population of thiocyanate-degrading bacteria from the waste stream targeted for treatment is recommended for application in the field. Recycling of process water will significantly reduce the cost of the quality water supply needed for a bacterial oxidation plant. In addition, it may be feasible to recycle tailings sediment in order to reprocess and extract gold from non-leached ore. Previously this was prevented by the high concentration of thiocyanate in the tailings water and sediments.

ACKNOWLEDGMENTS

We thank the Western Australian Government for financial support of this research through its Western Australian Innovation Support Scheme (WAISS). This study forms part of the research program of the A. J. Parker Cooperative Research Centre for Hydrometallurgy. The financial support of the Australian Federal Government to the A. J. Parker CRC is gratefully acknowledged.

REFERENCES

1. T. Williams, Thiocyanate inhibition levels on bacteria operating on Youanmi material. BacTech Australia Ltd., Perth, WA (1998) unpublished report 34 p.
2. D. Morin, Bacterial oxidation of refractory gold sulphide ores. *In* Bioextraction and biodeterioration of metals edited by C. C. Gaylorde and H. A. Videla, Cambridge University Press, Cambridge, UK, (1995) 25.

3. T. Weston, J. Perkins, I. C. Ritchie, and H. Marais, Concentrate biological oxidation in a hypersaline environment for the Kanowna Belle project. In Biomine '94, Australian Mineral Foundation, Glenside, SA, (1994) 9.1.
4. A. Smith, and T. Mudder (eds.), The Chemistry and Treatment of Cyanidation Wastes, Mining Journal Books, London, (1991).
5. L. Clesceri, A. Grennberg, and R. Trussell, *Standard methods for the examination of water and wastewater.* 17th edition. American Public Health Association, Washington, DC, (1989) 4.42.
6. J. L. Whitlock, and G. R. Smith, Operation of Homestake's cyanide biodegradation wastewater system based on multi-variable trend analysis. *In* R. Lawrence, R. Branion, and H. Ebner (eds), International Symposium on Biohydrometallurgy, Jackson, WY (1989).
7. J. L. Whitlock, Biological detoxification of precious metal processing wastewaters. *Geomicrobiology J.*, 8 (1990) 241.
8. S. K. Dubey, and D. S. Holmes, Biological cyanide destruction mediated by microorganisms. *World Journal of Microbiology and Biotechnology*, 11 (1995) 257-265.
9. D. E. Rawlings, and S. Silver, Mining with microbes. *Bio/Technology*, 13 (1995) 775.
10. M. B. Stott, L. R. Zappia, P. D. Franzmann, P. C. Miller, H. R. Watling, and M. R. Houchin, Bacterial degradation of thiocyanate in saline process water. (this conference).
11. A. D. Wilson, The micro determination of ferrous iron in silicate minerals by a volumetric and colorimetric method. *Analyst (London)*, 85 (1960) 823.

Bacterial degradation of thiocyanate in saline process water

M. B. Stott[a,b], L. R. Zappia[b], P. D. Franzmann[b], P. C. Miller[c], H. R. Watling[a] and M. R. Houchin[a].

[a] A. J. Parker Cooperative Research Centre for Hydrometallurgy, CSIRO Minerals, PO Box 90, Bentley WA 6102, Australia.

[b] Centre for Groundwater Studies, CSIRO Land and Water, Underwood Ave., Floreat Park WA 6014, Australia.

[c] BacTech (Australia) Pty. Ltd., 5 Belmont Ave., Belmont WA 6104, Australia.

The lack of quality water supplies in the Western Australian gold fields is a limiting factor in the use of biological oxidation to treat refractory gold ores prior to the extraction of gold by cyanidation. Clearly, the recycling of water from the CIP process back to a biological oxidation plant is desirable in regions where supplies of fresh water are a limiting resource. However, thiocyanate, a by-product of the cyanidation process and a major contaminant in tailings dams, must be removed if water is to be recycled because it is toxic to sulphide-oxidising bacteria.

Two strains of bacteria that were capable of degrading thiocyanate were isolated from the Youanmi mine in Western Australia. Both strains demonstrated the ability to utilise thiocyanate as their sole energy and nitrogen source. Physiological characterization of the strains indicated that both species of bacteria had the potential to tolerate the variable conditions encountered in Youanmi tails water. Phylogenetic analysis showed that one strain was a member of the genus *Thiobacillus* and the other was a member of the genus *Halomonas* but neither could be accommodated within any described species. Phosphate was the only additional nutrient required for thiocyanate degradation.

Both strains were inoculated into a laboratory-scale Rotating Biological Contactor (RBC). The success and extent of colonization was confirmed by recovery of signature lipids of both strains from the fixed-film biomass. The bacteria degraded thiocyanate to ammonium, sulphate and carbon dioxide. The biomass supported on the reactor surface area ($20 \ m^2$) was capable of degrading approximately 2800 mg L^{-1} to less than 1 mg L^{-1} thiocyanate at a flow rate of 30 mL min^{-1}. The process operated in saline, low nutrient water.

1. INTRODUCTION

More than 70% of Australian gold production occurs in Western Australia. However, the gold present in arsenopyrite deposits, estimated to be 15-30% of the total gold resource, is not readily available to extraction without some prior treatment. Bacterial oxidation of arsenopyrite-rich gold ores is a favoured alternative to roasting because it eliminates the release of toxic sulphur and arsenic oxides (products of roasting) to the environment.

Until recently, two gold mines in Western Australia utilised bacterial oxidation as a pretreatment for refractory gold ores prior to the extraction of gold by cyanidation. One of the reasons that other mines have not adopted bacterial oxidation as a pretreatment is the need for a plentiful supply of good quality water which is often a limiting factor in Western Australia. The cost of importing water for a bacterial oxidation plant would increase the overall costs of a mining operation substantially. However, the ability to recycle plant wastewater may assist in overcoming water limitations.

Wastewater from CIP circuits contains thiocyanate, a product of the reaction between cyanide and sulphur from sulphide-based ore. Concentrations in tailings dams are generally in the order of 200 mg L^{-1}. Thiocyanate is a persistent environmental contaminant because it is non-hydrolysable and is non-volatile (1). It is also toxic to sulphide-oxidising bacteria, inhibiting their growth at concentrations as low as 5 mg L^{-1} (2).

Traditionally thiocyanate destruction is accomplished using physico-chemical methods of cyanide degradation; all have inherent disadvantages for thiocyanate removal prior to water reuse for bacterial oxidation plants (3). Bacteria capable of degrading thiocyanate have been described from tailings dams and other environments exposed continuously to thiocyanate and cyanide (3,4).

The mechanisms of thiocyanate degradation by bacteria are complex and have been the subject of several studies (5). The reactions may be represented as the hydrolysis of thiocyanate to cyanate and sulphide (Eq. 1), the hydrolysis of cyanate to carbon dioxide and ammonia (Eq. 2) and oxidation of sulphide to sulphate (Eq. 3). Bacteria obtain energy from the oxidation reaction, and the nitrogen in thiocyanate is utilised as nutrient .

$$SCN^- + H_2O \rightarrow CNO^- + S^{2-} + 2H^+ \qquad (1)$$

$$CNO^- + 2\,H_2O \rightarrow CO_2 + NH_3 + OH^- \qquad (2)$$

$$S^{2-} + 2O_2 \rightarrow SO_4^{2-} \qquad (3)$$

The removal of up to 99% of available thiocyanate from mining wastewater using thiocyanate-degrading bacteria in fixed film in full-scale and laboratory-scale bioreactors has been demonstrated (4,6,7). It is a cost-effective alternative to physical and chemical treatments and the discharge is compatible with strict environmental standards.

The method has been applied to tailings dam water from Youanmi. Bacteria have been enriched and isolated, a bioreactor constructed and the efficiency of thiocyanate degradation assessed in a series of experiments. The purpose of the study was to determine whether bacteria from Western Australian tailings dams could be used to detoxify thiocyanate-rich wastewater.

2. EXPERIMENTAL

2.1. Bacterial isolation and enrichment

Tailings dam water samples from three mine sites were inoculated with 3.5 g L^{-1} of thiocyanate and incubated at 30 °C for 10 days. The resulting enrichment was then sub-cultured into a thiocyanate medium containing synthetic sea salts at a concentration to give a final conductivity of 17.8 mS cm^{-1}, thiocyanate and phosphate. The pH was set at 7.85 and the incubation temperature at 30 °C. The pH and salinity levels were modeled on the CIP wastewater chemistry at the Youanmi mine site.

Pure strains were obtained through a series of rapid transfers in the enrichment medium followed by serial dilutions and then multiple transfers of single colonies on solid media.

2.2. Microbiology

Strains of thiocyanate-degrading bacteria were characterized according to cellular morphology, biochemical characteristics, thiocyanate degradation rates and phospholipid fatty acid (PLFAMEs) profiles. The temperature and pH profiles, thiocyanate concentration growth range, salinity levels and growth in the presence of arsenic were examined. Bacterial growth on alternative carbon and energy sources was also tested.

Detailed methods for bacterial isolation, bacterial enrichment and phylogenetic studies are given in (5).

2.3. Bioreactor

A Rotating Biological Contactor (RBC) was constructed using polyvinyl chloride. The biological support consisted of polypropylene mesh and commercially available "bioballs" with a high surface to volume ratio. No metallic parts were in contact with the thiocyanate medium or bacterial biofilm (Figure 1). Aquarium immersion heaters were used to maintain the solution temperature at 30 °C.

The cylindrical biological supports were rotated continuously at 1.5 rpm so that the biofilm was alternately immersed in thiocyanate solution then aerated. The solution flow-rate to the bioreactor was controlled by a peristaltic pump.

The bioreactor had a surface area of 20 m^2 and contained 20 L solution in a two-chamber arrangement which would permit batch, continuous or parallel-flow, as required.

2.4. Colonisation of bioreactor

Initially, the bioreactor was filled with 20 L of thiocyanate medium (2500 mg L^{-1} SCN⁻, 0.01% (w/v) phosphate) and inoculated with a 200 mL volume of turbid culture of each bacterial strain. The bioreactor was operated in batch mode until the thiocyanate concentration had been reduced to <500 mg L^{-1}.

Sucrose to yield a final concentration of 1.0% (w/v) was added to the thiocyanate-rich solution to promote polysaccharide formation and bacterial attachment.

After 100 hours the bioreactor was reinoculated with the two strains of thiocyanate-degrading bacteria, the flow stabilized at 500 mg L^{-1} thiocyanate medium with a feed rate of 250 mL h^{-1}. The bioreactor influent and effluent concentrations were measured daily and changes in bioreactor conditions were noted.

Figure 1. Thiocyanate-degrading bioreactor shown in continuous-flow configuration. The arrows represent the direction of solution flow through the bioreactor. The bioballs [inset - bioball without biofilm] filled each cage and supported the biofilm.

3. RESULTS AND DISCUSSION

3.1. Selection and characterization of bacteria

Bacterial enrichments from mine wastewater contained a multitude of bacterial types that exhibited a diverse range of morphotypes. This bacterial growth occurred at thiocyanate concentrations 17 times greater than those found in Youanmi tailings water. The increase in bacterial number in the enrichments indicated that thiocyanate-degrading bacteria populated the wastewater samples.

As the Youanmi mine was the proposed site for a thiocyanate-degrading bioreactor, the pH and salinity of the thiocyanate medium was adjusted to simulate wastewater from the tails influent at Youanmi. It was therefore expected that, of the three mine wastewater samples used as inoculum, the bacterial enrichment inoculated with Youanmi samples would exhibit the greatest rate of thiocyanate degradation (Figure 2). Indigenous bacteria probably had the advantage over those from other mines because of their pre-existing acclimatization to both the salinity and pH levels, as well as to other potentially toxic contaminants in the tails influent at Youanmi. Thus, the selection and isolation of thiocyanate degrading bacteria was undertaken using wastewater from the Youanmi mine.

Characteristics of both strains are given in Table 1. Physiological characterization of the bacteria indicated that both strains had adapted well to the wastewater composition at Youanmi. Variations in the environmental conditions of temperature, salinity and pH and concentrations of arsenic and thiocyanate usually encountered in Youanmi wastewater were tested and produced no inhibition of growth (Table 1). As both bacteria were able to grow chemolithotrophically, the only nutrient addition needed to produce growth and thiocyanate degradation, was phosphate.

Both strains (Figures 3a and 3b) were slow growing and, thus, unable to maintain a self-sustaining population density in a continuous stirred-tank bioreactor at the high wastewater flow rates of a typical mine. In addition, in stirred tanks, each bacterium is more likely to be subjected to the full effect of any contaminants in solution. Thus, fluctuations in toxin concentrations could be detrimental to the population. Both factors were overcome by choosing a fixed-biofilm reactor. Washout was avoided (3) and adsorption of contaminants by surface cells minimized the detrimental effects of toxins to the multiple cell layers contained within the biofilm (8). Fixed film bioreactors have the added advantage of being able to degrade low concentrations of contaminants in high flow environments, such as occur at the Homestake mine. The bioreactor design (Figure 1) was based on the full-scale bioreactors operating at Homestake (3,9) and rotating cage bioreactors used in other bioremediation facilities (10). What differed in this case were the types of bacteria involved in thiocyanate oxidation which probably reflects the site-specific nature of contaminant-degrading bacteria.

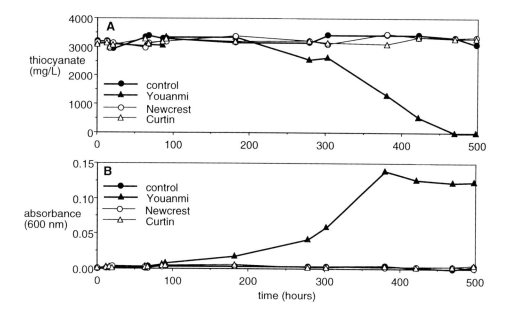

Figure 2. Thiocyanate loss (A) and bacterial turbidity (B) in media containing 3.5 g L^{-1} of thiocyanate and inoculated with mine tailings water collected from three separate mines and incubated for 500 hours.

Table 1
Bacterial characterization.

Parameter	strain MS01b	strain MS02
cell morphology	Gram –ve; pleomorphic rod (1-2 x 0.5-0.7 µm); up to 6 lateral flagella (~7 µm).	Gram –ve; pleomorphic rod (1-2 x 0.4-1.0 µm); single polar flagellum.
respiration	aerobic	aerobic
major PLFAMEs	hexadecanoate-Me (16:0) : 17.7 % cis-9-hexadecanoate-Me (16:1 c9) : 9.2 % cis-11-octadecanoate-Me (18:1 c11) : 55.2 % cis-9,10-methyleneoctadecanoate-Me (19:0 cy) : 14.9 %	hexadecanoate-Me (16:0) : 25.2 % octadecanoate-Me (18:0) : 25.5 % cis-11-octadecanoate-Me (18:1 c11) : 41.9 %
closest relative based on 16S rRNA sequence comparisons	*Halomonas meridiana*	*Thiobacillus hydrothermalis*
growth range: pH temperature SCN⁻ salinity arsenate	$6 - 9.3$ $20 - 37°C$ $0.01 - 12$ g L^{-1} $1 - 100$ g L^{-1} $0 - 2$ mM	$6 - 9.3$ $20 - 37°C$ $0.01 - 12$ g L^{-1} $1 - 100$ g L^{-1} $0 - 2$ mM
energy sources	thiocyanate thiosulphate metabisulphate glucose acetate sucrose	acetate
carbon sources	carbon dioxide acetate glucose sucrose	carbon dioxide acetate

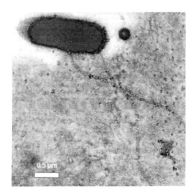

Figure 3a. Transmission electron micrograph of strain MS01b.

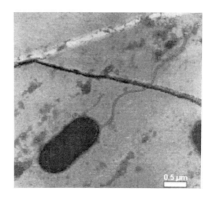

Figure 3b. Transmission electron micrograph of strain MS02.

3.2. Thiocyanate degradation in an RBC

The bioreactor was stabilized with a 500 mg L^{-1} thiocyanate feed and 250 mL h^{-1} flow rate once the biofilm had been established. Influent and effluent thiocyanate concentrations were monitored daily for two months, during which time a number of changes were made (Figure 4).

Thiocyanate concentrations in the effluent remained high for 7 days (Figure 4, point A), after which bacterial degradation to very low concentrations was achieved (Figure 4, point B). Thiocyanate concentrations in the feed to the bioreactor were reduced to ~400 mg L^{-1} with a flow of 250 mL h^{-1}. The thiocyanate concentration in the effluent diminished to and remained at 0 mg L^{-1} for a period of 10 days (Figure 4, point C).

At this time (Figure 4, point C), both the flow rates and the thiocyanate concentrations in the bioreactor feed were increased. At the highest flow rate, the increased levels of thiocyanate exceeded the degrading-capacity of the available biofilm (Figure 4, point D), but with time, there was an increase in biomass accompanied by more efficient thiocyanate degradation (Figure 4, point E). The response time was about seven days, reflecting the slow growth rates of these bacterial species. Only the bioballs in the first chamber showed significant biofilm formation (compare Figures 5A and B).

Based on the results obtained with this laboratory scale bioreactor, it was calculated that a combined surface area of about 35,000 m^2 would be required to treat 600 m^3 d^{-1} process water at ~250 mg L^{-1} thiocyanate. Indeed, this could represent a substantial overestimate of the required surface area as the laboratory-scale bioreactor was not fully colonized (see Figures 5A and 5B).

The thiocyanate degradation products were not determined in the pure cultures of MS01b and MS02. The thiocyanate degradation products within the bioreactor in which strains MS01b and MS02 had colonised the biofilm, were ammonium and sulphate ions and carbon dioxide. The product concentrations balanced stoichiometrically with the amount of thiocyanate degraded. During operation the oxygen concentration in the reactor solution was measured at 4 mg L^{-1}, so oxygen was not limiting to the reactor performance. As the bioreactor was not a closed system, some part of the observed thiocyanate degradation might

816

come from microorganisms other than the isolated strains. Analysis of the PLFAMEs from the bioreactor indicated that a diverse population of microorganisms inhabited the bioreactor. It was evident, however, that the thiocyanate degrading biomass and the PLFAME profile was consistent with a mixture of strains MS01b and MS02 contained within the biomass.

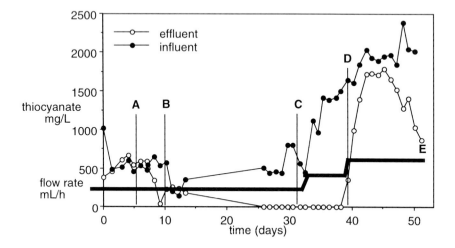

Figure 4. Capability of the thiocyanate-degrading bioreactor using thiocyanate medium as feed.

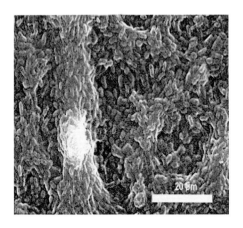

Figure 5A. Scanning electron micrograph of a section of bioball from the first chamber of the bioreactor.

Figure 5B. A scanning electron micrograph of a section of bioball from the second chamber of the bioreactor. Note the limited bacterial colonization.

4. SUMMARY

Two strains of thiocyanate-degrading bacteria were isolated from Youanmi tailings waste water. The bacterial cultures were enriched and inoculated into a laboratory-scale rotating biological contactor. The biofilm which was established in thiocyanate-rich synthetic process water was capable of degrading thiocyanate to carbon dioxide, sulphate and ammonium ions, all of which are relatively harmless to the environment when compared with thiocyanate.

While removing thiocyanate from mine wastewater has been demonstrated in a number of studies (4,7,9) the chemistry of Youanmi tailings water differed from those of the other mines and different types of bacteria were responsible for thiocyanate degradation. In previous studies, the treated wastewater contained a sufficient concentration of organic nutrients to allow heterotrophic growth in conjunction with thiocyanate degradation. Water in the gold fields of Western Australia contains only low amounts of dissolved organic compounds and, invariably, has a high salinity, producing a restricted environment for bacterial growth.

ACKNOWLEDGEMENTS

We thank the Western Australian government for financial support of this research through its Western Australian Innovation Support Scheme (WAISS). This study forms part of the research program of the A. J. Parker Cooperative Research Centre for Hydrometallurgy. The financial support of the Australian Federal Government to the A. J. Parker Cooperative Research Centre is gratefully acknowledged.

REFERENCES

1. Clesceri, L., Grennberg, A., and Trussell, R., Standard methods for the examination of water and wastewater. American Public Health Association, Washington, D.C., (1989) 4.42.
2. Morin, D., Bacterial oxidation of refractory gold sulphide ores, in Gaylorde, C. C. and Videla, H. A. (eds.), Bioextraction and Biodegradation of Metals. Cambridge University Press, Cambridge, Great Britain, (1995) 25.
3. Smith, A. and Mudder, T., The chemistry and treatment of cyanidation wastes. Mining Journal Books Ltd., London, England, (1991) 1.
4. Boucabeille, C., Bories, A., Ollivier, P., and Michel, G., Microbial degradation of metal complexed cyanides and thiocyanate from mining wastewaters. Environmental Pollution, 84 (1994) 59.
5. Stott, M. B., The development of a thiocyanate-degrading bioreactor for detoxifying mining wastewater (Honours Thesis). Microbiology Department, University of Western Australia, Perth, Australia, (1997).
6. Whitlock, J. L., Biological detoxification of precious metal processing wastewater, Geomicrobiology Journal, 8 (1990) 241.
7. Mühlbauber, R. M. and Broadhurst, J. L., Biodegradation of thiocyanate and cyanide contained in biooxidation cyanidation tailings, in Ritchie, A. I. M. (ed.), Biomine '97, Australian Mineral Foundation, Glenside, South Australia, (1997) CT3.1.

818

8.   Brown, M. L. and Gauthier, J. J. (1993). Cell density and growth phase as factors in the resistance of a biofilm of *Pseudomonas aeroginosa* (ATCC 2785) to iodine. Applied and Environmental Microbiology, 59 (1993) 2320.

9.   Mudder, T. I. and Whitlock, J. L., Biological treatment of cyanidation wastewaters, *in* Bell, J. M. (ed.), Proceedings of the 38th Industrial Waste Conference. Boston, Butterworth, (1984) 279.

10. Warner, R. H., Litchfield, C. D., Martin, E. L., and Elliot, E., Packed-cage RBC with combined cultivation of suspended and fixed-film biomass. Water Science and Technology, 22 (1990) 101.

Degradation of thiocyanate by immobilized cells of mixed and pure cultures

L. H. Rosa[a], E. M. Souza-Fagundes[a], M. H. Santos[a], J. C. T Dias[b], P. F. Pimentel[a], N. C. M. Gomes [a,c]*

[a]Department of Biotechnology and Chemical Technology – CETEC, Belo Horizonte, MG, Brazil

[b]Graduate Program in Microbiology, Federal University of Minas Gerais, Brazil

[c]Institute of Microbiology Prof. Paulo de Góes, Federal University of Rio de Janeiro, Brazil

Thiocyanate is frequently found in industrial wastewater such as those from coal, gold and silver mining. Although the microbial degradation of thiocyanate has been well documented, scarce information is available on the heterotrophic degradation of thiocyanate. A pure bacteria strain (BMV8) and a mixed culture of bacteria and fungi were isolated by an enrichment method from a fixed bed bioreactor for cyanide and thiocyanate removal and they were able to metabolize thiocyanate heterotrophically as a nitrogen source. The efficiency of thiocyanate biodegradation decreased when the cells were immobilized in citric pectin. The immobilized mixed culture was able to degrade thiocyanate more efficiently than the BMV8 strain.

1. INTRODUCTION

Cyanide, a well-known metabolic inhibitor, is frequently used in industrial processes, including synthetic production and gold and silver extraction [1, 2]. Thiocyanate is commonly found in mining wastewater as a result of the interaction of free cyanide and sulfur.

Unlike cyanide, thiocyanate presents low toxicity at low concentrations, but its toxicity increases at high concentration [1]. Chronic absorption of thiocyanate can cause dizziness, skin eruption, running nose, vomiting and nausea [3], so it must be removed before disposal procedures.

Traditionally, cyanide and related compounds were detoxified by alkaline chlorination, hydrogen peroxide and SO_2/air methods. These conventional techniques are characterized by their universal applicability because of their insensitivity towards the nature of the waste material and by their relatively low costs. However, their main handicaps are need of hazardous reagents like chlorine and their potential for creating toxic residues requiring a post

* CX Postal. 2306, 31.170-000, Belo Horizonte, Brasil. Fax (031) 489 2200. e-mail: gomesncm@tulipa.cetec.br

treatment [4]. After Homestake Mine (Lead, SD) began operating a biological treatment plant in 1984, a lot of interest was generated in biodegradation of cyanide and thiocyanate. However, biological treatment as an alternative to those chemical methods is still in its infancy [5]. Although microbial treatment has been used for removing thiocyanate, the efficiencies of the process are variable. This is particularly observed when the treatment also involves mixtures of pollutants such as phenol and cyanide. Thiocyanate is known as an aggravating ion for the low efficiency in these process. Therefore, a better understanding of the biochemistry and microorganism interactions involved in thiocyanate degradation in bioprocess would be of fundamental importance to create consistent systems with efficient designs for the biotreatment of thiocyanate containing wastes [6].

The application of immobilized cells in the treatment of wastewaters offers the possibility of degrading higher concentrations of toxic pollutants that can be achieved with free cells [7]. Immobilized cells have advantages over free cells because they do not suffer cell loss during washing and also provide higher population densities at any flow [8].

This paper reports the results obtained on thiocyanate degradation by free and citric pectin immobilized cells of a mixed culture and a bacterial strain isolated from a fixed bed bioreactor for cyanide and thiocyanate removal.

2. MATERIAL AND METHODS

2.1. Culture enrichment and isolation

Liquid samples were collected from a laboratory scale fixed bed bioreactor used for mining effluent treatment studies, and inoculated into Erlenmeyer flasks containing 25 ml of modified minimal salts medium M9 as described by Miller [9]. The M9 medium composition in (g/l) was Na_2HPO_4 0.6; KH_2PO_4 0.3 and NaCl 0.05, pH adjusted to 7.0. To this medium were added 5 mM of thiocyanate and 10 mM of glucose as a nitrogen and carbon source, respectively. The M9 medium was supplemented with 100 μM of calcium chloride, 1.0 mM of magnesium sulfate and 0,1% (v/v) of trace metal solution [10] containing (g/l): MgO, 10.75; $CaCO_3$, 2.0; $FeSO_4.7H_2O$, 4.5; $ZnSO_4.7H_2O$, 1.44; $MnSO_4.5H_2O$, 1.12; $CuSO_4.5H_2O$, 0.28; $CoSO_4.7H_2O$, 0.28; H_3BO_3, 0.06; 3 ml HCl 5 M.

The thiocyanate degrading microorganisms were isolated by spreading the growth onto Petri dishes containing agar M9 medium after incubation at 30°C for 3 to 5 days. The colonies were replica-plated onto the same medium and three successive replicates were performed. Isolated bacteria were characterized by microscopic examination and Gram staining.

2.2. Cells immobilization (pectin-microorganism)

Citric pectin (4 % w/w) (Citrus Colloids - low methoxy - amide gel) was dissolved in distilled deionized water and mixed with a magnetic stirrer for 1 hour at room temperature. To this solution was added the mixed bacterial culture or BMV8 strain, containing 10% (w/v) dry weight. This final solution (pectin-microorganism) was dropped into a 0.2 M solution of $BaCl_2$. A spontaneous cross-linking reaction occurred resulting in spherical beads with an average diameter of about 3 mm. These beads were stored at 4°C for 24 hours. After that period the beads were washed with distilled deionized water to remove the excess of $BaCl_2$.

2.3. Kinetic study

The thiocyanate degradation profile was studied by culturing the mixed culture and the BMV8 strain in M9 medium supplemented with 5 mM of thiocyanate. Erlenmeyer flasks containing 25 ml of M9 medium were inoculated with 1% (v/v) inoculum of a 48 - hour culture grown in the same medium. At 12 hours intervals, the cultures were harvested by centrifugation at 27 200 x g for 10 minutes. The supernatants were retained for thiocyanate analysis and the cell pellets were dried at 60°C until constant weight, for dry weight determination.

The mixed culture and BMV8 strain immobilized in pectin citric (20 g wet weight), were incubated in Erlenmeyer flasks containing 25 ml of M9 medium supplemented with 5 mM of thiocyanate and 10 mM of glucose. At 12 hours intervals, the supernatant were harvested by centrifugation at 27 200 x g for 10 minutes and retained for thiocyanate, glucose, sulfate, nitrite and nitrate analysis.

In all steps, non-inoculated controls containing the M9 medium with thiocyanate and glucose were run as control. A negative control was accomplished (M9 medium with citric pectin).

Thiocyanate was determined by the method of Stafford & Callely [11]. Nitrite, nitrate and sulfate were determined by the method of APHA [12].

3. RESULTS AND DISCUSSION

Continued growth and complete thiocyanate removal were observed following several serial transfers to fresh medium, from which the mixed culture was obtained. Three bacterial strains were isolated from these mixed culture and named BMV7, BMV8 and BMV9. These strains are in taxonomic classification tests and they were initially characterized as Gram-negative, rod shaped bacteria, and were catalase positive and oxidase negative. In the non-inoculated control M9 medium, thiocyanate levels did not decrease during the incubation period. This eliminated the possibility of thiocyanate removal due to a chemical reaction with other medium components. The mixed culture and the isolated strains were able to degrade thiocyanate heterotrophically using this compound as a nitrogen source. Most of the thiocyanate utilizing microbes reported in the literature are autotrophs, and only few heterotrophic with ability to degrade thiocyanate have been described [13-16]. Although differences in the thiocyanate removal by each strain isolated were observed, the strains were physiologically very similar and probably they belong to the same species, therefore, only the isolated BMV8 was examined in detail due to its greatest ability to remove thiocyanate from growing culture (data not shown).

The kinetics of growth of the BMV8 strain and the mixed culture in presence of 5 mM of thiocyanate showed that both cultures were able to grow and remove thiocyanate from the medium in 36 hours of incubation (Figure 1 a, b). The thiocyanate utilization occurred mainly during the exponential phase of growth. Similar results were previously reported for *Arthrobacter* species by Betts et al. [17] and for an isolated strain by Stratford et al. [16]. In BMV8 strain cultures, unlike the mixed culture, was observed formation of 1,46 mM of ammonia after 36 hours of incubation. This amount decreased to 1,23 mM in 48 hours. Youatt [13] and Katayama et al. [18] suggested ammonia as a possible product from thiocyanate consumption. In a similar way Happold et al. [19] also observed that ammonia is originated by thiocyanate degradation. The oxidative pathway of thiocyanate has been known to proceed as

822

follows. First thiocyanate is hydrolyzed to cyanate and sulfide, then the cyanate is hydrolyzed to ammonia and bicarbonate, whereas sulfide is oxidized to sulfate. However some authors did not found ammonia in degrading bacterial cultures and it can be attributed to a rapid utilization of ammonia by the bacterial cells for further metabolism during growth [6]. Nitrite, nitrate and sulfate were not identified as end products of thiocyanate degradation.

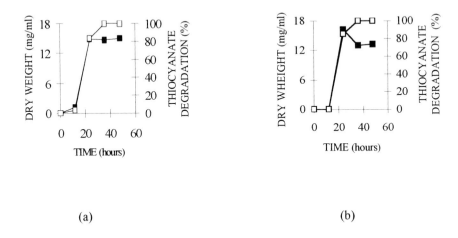

(a)   (b)

Figure 1: Kinetic of growth (—■—) and thiocyanate degradation (—□—) of BMV8 (a) and the mixed culture (b) on 5 mM thiocyanate as a nitrogen source and 10 mM glucose as a carbon and energy source.

The kinetics of thiocyanate degradation by citrate immobilized cells of BMV8 strain and the mixed culture in presence of 5 mM of thiocyanate showed that both cultures kept the capacity of removing this compound from the medium (figure 2). However, there were a decrease of 12% (mixed culture) and 45% (BMV8) in comparison to the end of the incubation period. Numerous authors have reported differences in specific metabolic activity between free and immobilized microorganisms. The cause of these changes of cellular activity is still unclear and could be attributed to a modification of the physical and chemical environment of immobilized cells, or actual change of cellular physiology induced by immobilization [20]. So, more studies should be done in order to understand the mechanism of thiocyanate degradation in the immobilized system and optimization of the parameters as well for determining its efficiency as internal and external mass transfer, cell density, effects on metabolic state and osmose pressure. Some authors reported an increase in the efficiency of cyanides removal by immobilized cells suggesting that this technology can be employed in the environment for the remediation of inorganic cyanides [21].

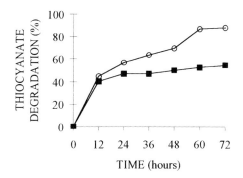

Figure 2 - Kinetics of thiocyanate degradation by immobilized BMV8 strain (——■——) and mixed culture (——□——) on 5 mM thiocyanate as a nitrogen source.

4. CONCLUSIONS

From results presented in this work, it can be seen that a mixed culture of microorganisms and the pure culture strain (BMV8) were able to metabolize thiocyanate heterotrophically as a nitrogen source. Despite of similar kinetics of thiocyanate degradation in the growing cultures, the mixed culture presented better efficiency when immobilized than BMV8 strain. This advantage could be useful in bioprocesses for the treatment of thiocyanate containing effluents and provides further information about the physiological conditions and biochemical mechanism of thiocyanate degradation.

ACKNOWLEDGEMENTS

We thank the "Mineração Morro Velho" for the laboratory and field facilities. This work was supported by Fundação de Amparo à Pesquisa de Minas Gerais - FAPEMIG.

REFERENCES

1. C. Boucabeille; A. Bories; P. Olliver, Biotechnol. Lett., 16 (1994) 425.
2. K. D. Chapatwala,; G.R.V Babu, E. R. Armstead, E. M. White, J. H. Wolfram, Appl. Biochem. Biotechnol., 51/52 (1995) 717.
3. Y. L. Paruchuri, N. Shivaraman, P. Kumaran, Environ. Pollu., 68 (1990) 15.
4. S. Basheer, O. M. Kut, J. E. Prenosil and J. R. Bourne, Biotecnol. Bioeng. 39 (1992) 629.
5. J. B. Mosher, L. Figueroa, Minerals Engineering, 9 (1996) 573.
6. E. X. Oliveira Dias, Thiocyanate degradation by a novel isolate. Ph.D. Thesis. Faculty of Natural Sciences. University of Kent, 1993.

7. D. A. Kunz, O. Nagappan, J. Silva-Avalos and G. T. Delong, Appl. Environ. Sc. Technol. 58 (1992) 2022.
8. S. K. Dubey and D. S. Holmes, W. J. Microbiol. Biotechnol., 11 (1995) 257.
9. J. H. Miller, Experiments in molecular genetics. Cold Spring Harbor Laboratory. Cold Spring Harbor. New York, (1972) 431.
10. T. Bauchop & S. R. Elsden, J. Gen Microbiol., 23 (1960) 457.
11. D. A. Stafford and A. G. Callely, J Gen Microbiol., 55 (1969) 285.
12. APHA. Methods for the examination of water and wastewater. Standard Methods, 19 ed. Washington, 1995.
13. J. B. Youatt, J.Gen.Microbiol., 11 (1954) 139.
14. T. I. Mudder and J. L. Whitlock, United State Patent 4,461,834 (1983).
15. V. Andreoni, A. Ferrari, A. Pagani, C. Sorlini, V. Tandoi, and V. Treccani, Anal. Microbiol., 38 (1988) 193.
16. J. Stratford, A.E.X.O. Dias, and C. J. Knowles, Microbiol. 140 (1994) 2657.
17. P. M. Betts, D. F. Rinder, and J. R. Fleeker, Can. J. Microbiol., 25 (1979) 1277.
18. Y. Katayama, Y. Narahara, Y. Ioue, F. Amano, T. Kanagawa,. and H. Kuraishi, J. Biol .Chem., 267 (1992) 9170.
19. F. C. Happold, K. I. Johnstone, H. J. Rogers and J. B Youatt, J. Gen. Microbiol., 10 (1954) 261.
20. S. Norton and T. Dte Amore, Enzyme Microb. Technol., 16 (1994) 365.
21. K. D. Chapatwala, G. R. V. Babu and J. H. Wolfram, J. Ind. Microbiol., 11 (1993) 69.

Biohydrometallurgy and the environment toward the mining of the 21st century

PART B

MOLECULAR BIOLOGY, BIOSORPTION, BIOREMEDIATION

AUTHOR INDEX

Biohydrometallurgy and the environment toward the mining of the 21st century

PART A

BIOLEACHING, MICROBIOLOGY

AUTHOR INDEX

Biohydrometallurgy and the environment toward the mining of the 21st century

PART B

MOLECULAR BIOLOGY, BIOSORPTION, BIOREMEDIATION

SUBJECT INDEX

Biohydrometallurgy and the environment
toward the mining of the 21st century

PART A

BIOLEACHING, MICROBIOLOGY

SUBJECT INDEX